ASTRODYNAMICS 1989
Part I

Thanks are due Diane M. Massey, Univelt, Incorporated, for final preparation of the manuscript for publication.

Front Cover Illustration:

The *Galileo* spacecraft was launched from the Space Shuttle on October 19, 1989 into a six-year flight path to Jupiter. The mission will be the first to make direct measurements from an instrumented probe within Jupiter's atmosphere, and the first to conduct long-term observations of the planet and its magnetosphere and satellites from orbit around Jupiter. It will be the first orbiter and atmospheric probe for any of the outer planets. *Galileo* is expected to arrive at Jupiter in December, 1995.

AN AMERICAN ASTRONAUTICAL SOCIETY PUBLICATION

ASTRODYNAMICS 1989

Volume 71
Part I
ADVANCES IN THE ASTRONAUTICAL SCIENCES

Edited by
Catherine L. Thornton
Ronald J. Proulx
John E. Prussing
Felix R. Hoots

Proceedings of the AAS/AIAA Astrodynamics Conference held August 7-10, 1989, Stowe, Vermont

Published for the American Astronautical Society by
Univelt, Incorporated, P.O. Box 28130, San Diego, California 92128

AAS Publications Office
P.O. Box 28130
San Diego, California 92128

Affiliated with the American Association for the Advancement of Science
Member of the International Astronautical Federation

First Printing 1990

Library of Congress Card No. 57-43769

ISSN 0065-3438

ISBN 0-87703-317-X (Hard Cover)
ISBN 0-87703-318-8 (Soft Cover)

Published for the American Astronautical Society
by Univelt, Incorporated, P.O. Box 28130, San Diego, California 92128

Printed and Bound in the U.S.A.

[redacted]

FOREWORD

These astrodynamics conferences have been held annually since the mid-1960s, managed alternately by the American Astronautical Society and the American Institute of Aeronautics and Astronautics. Every second year the American Astronautical Society publishes the proceedings. The proceedings usually consist of a hard-copy volume or set of volumes plus a microfiche supplement. Proceedings volumes for earlier conferences are still available either in hard copy or in microfiche form. The following page and the appendix at the end of Part II of this volume list astrodynamics proceedings available through the American Astronautical Society.

Astrodynamics 1989 consists of two parts plus a microfiche supplement. All available papers appear in full in either the volume or its supplement. In all cases, at least an abstract appears in the volume itself with reference to the supplement. The decision to publish papers in the microfiche supplement rather than in the hard-copy volume was not based upon technical content but upon the following criteria: Papers that are too long or not in AAS format or not of good reproducible quality appear in full in the supplement. A chronological index and an author index are appended to the second part of the volume.

In our proceedings volumes the technical accuracy and editorial quality are essentially the responsibility of the authors. The session chairmen and our editors do not review all papers in detail; however, format and layout are improved when necessary by our editors. Papers by guests from non-English speaking countries are reviewed for editorial quality.

The current proceedings are valuable to keep specialists abreast of the state of the art; however, even older volumes contain some articles that have become classics and all volumes have archival value.

Horace Jacobs
Series Editor

AAS/AIAA ASTRODYNAMICS CONFERENCES

An astrodynamics conference is held every year under the joint sponsorship of the American Astronautical Society and the American Institute of Aeronautics and Astronautics. The two societies alternate in publishing proceedings of these conferences.

Astrodynamics 1989 appears as Volume 71, *Advances in the Astronautical Sciences* along with a microfiche supplement, Volume 59, *AAS Microfiche Series*. The two publications present the complete proceedings of the AAS/AIAA Astrodynamics Conference 1989.

Earlier astrodynamics proceedings available through the American Astronautical Society's publishers are:

Astrodynamics 1987	Volume 65 I & II, *Advances in the Astronautical Sciences* Volume 55, *AAS Microfiche Series*
Astrodynamics 1985	Volume 58 I & II, *Advances in the Astronautical Sciences* Volume 51, *AAS Microfiche Series*
Astrodynamics 1983	Volume 54 I & II, *Advances in the Astronautical Sciences* Volume 45, *AAS Microfiche Series*
Astrodynamics 1983	Volume 46 I & II, *Advances in the Astronautical Sciences* Volume 37, *AAS Microfiche Series*
Astrodynamics 1979	Volume 40 I & II, *Advances in the Astronautical Sciences* Volume 32, *AAS Microfiche Series*
Astrodynamics 1977	Volume 27, *AAS Microfiche Series*
Astrodynamics 1975	Volume 26, *AAS Microfiche Series*
Astrodynamics 1973	Volume 21, *AAS Microfiche Series*
Astrodynamics 1971	Volume 20, *AAS Microfiche Series*
Astrodynamics 1968	Volume 7, *AAS Microfiche Series*
Astrodynamics 1966	Volume 11, *Science and Technology* Volume 2, *AAS Microfiche Series*
Astrodynamics 1965	Volume 9, *Science and Technology*

All of these proceedings are available from **Univelt, Inc., P.O. Box 28130, San Diego, California 92128**, Publishers for the American Astronautical Society.

PREFACE

The 1989 Astrodynamics Conference was hosted by the American Astronautical Society and co-sponsored by the American Institute of Aeronautics and Astronautics. The conference was held August 7-10, 1989 at the Stoweflake Lodge in Stowe, Vermont. Technical presentations focused on analytical and numerical aspects of astrodynamics and celestial mechanics with emphasis on the following areas: planetary mission design, trajectory design and analysis, orbit determination, attitude determination and control, guidance analysis, and dynamics and control of large, flexible space structures. Special sessions highlighted papers on the neutral upper atmosphere density and orbiting debris.

Registration totaled 221 with attendees representing over 7 countries. There were 113 papers presented in 21 sessions. Professor Victor Szebehely of the Department of Aerospace Engineering and Engineering Mechanics at the University of Texas at Austin gave an entertaining lecture titled "Adventures in Celestial Mechanics".

We would like to thank the Session Chairmen for their critical role in making this conference a success. A special word of thanks is also extended to those who assisted in organizational planning, registration and numerous other functions required for a successful conference.

Ronald J. Proulx
AAS General Chairman

Catherine L. Thornton
AAS Technical Chairman

Felix R. Hoots
AIAA General Chairman

John E. Prussing
AIAA Technical Chairman

ADVENTURES IN CELESTIAL MECHANICS

Victor Szebehely*

INTRODUCTION

It is a great honor to be invited to deliver this lecture at the Astrodynamics Conference and it is my pleasure to express my thanks for the invitation to the organizers of this meeting especially to Dr. Catherine L. Thornton, Dr. Ronald J. Proulx, Felix R. Hoots and Professor John E. Prussing. It was the instruction I received with the invitations that I should leave all my equations at home. Consequently, this will be a difficult lecture, because equations usually make the delivery of a lecture rather easy. Concerning the title of this lecture I would like to mention that this is the title of a recent book of mine. The title of the book was first jokingly suggested by a colleague of mine, Professor Raynor L. Duncombe at the University of Texas. When I mentioned this possible title to the publisher of the book, he immediately adapted it and was not even willing to discuss any other possible titles.

In order to put the audience at ease, may I quote a short poem which I think all speakers should memorize, never forget and always apply.

The audience cheered the speaker,
The applause was beyond belief,
It was not for what he said, though,
But for being so very brief.

In my introduction I would like to mention a short story which is absolutely not true and certainly not applicable to this lecture. According to this story, there was this well known professor who was invited quite frequently to deliver lectures and he usually took a graduate student with him. Since the invitations came rather frequently, the professor did not have a chance to write completely new lectures every time but had to repeat some of his lectures. Consequently, the graduate student heard the same lecture several times and learned how to deliver it. When the next invitation came the professor told him that he should deliver the lecture and the professor would just sit in the audience. So they walked into the lecture room and the graduate student did a beautiful job delivering the professor's lecture and everything went wonderfully until the end of the lecture someone from the

* Professor and R.B. Curran Centennial Chair, Department of Aerospace Engineering, University of Texas, Austin, Texas 78712.

audience asked a rather difficult question and the graduate student did not have the faintest idea how to answer it. But he was a quick thinker and his answer was: "Sir, this question is so simple that even my graduate assistant can answer it." And with this he pointed to the professor in the audience, who of course answered the question. I wish to mention that this story has no application whatsoever to the present lecture.

With the invitation also came a suggestion concerning the subject of this lecture which was going to be the past, the present and the future of orbit mechanics and celestial mechanics. I found this an interesting idea but when I started preparing the lecture I found that I better put emphasis on the past rather than on the present. We have many famous names and heroes in our field and I will speak about them but one must be very careful with our present heroes because if one forgets to mention just one name, one creates a new enemy. So I will emphasize the past, will delete the present which is anyway well represented by the present meeting, and I shall say a few words about the future which I find also very difficult since in general, to predict the future accurately is impossible. In fact all of us who participate in this conference are getting paid to predict and if anybody would come up with a method to accurately predict a trajectory for an arbitrary length of time, we would lose our jobs. Consequently, to predict the future of our field is very difficult but I will try to do it. I will conclude this introduction with a few definitions since it might not hurt to know what we are talking about.

Celestial mechanics is often considered a branch of astronomy dealing with the dynamics and the motion of natural bodies, mostly in the solar system. I heard once a quotation that this "field is one of the greatest achievements of the human intellect." This is certainly not a very modest and humble introduction but I hope you will like it and I will be forgiven. *Astrodynamics* is a modern version of celestial mechanics and it deals mostly with artificial bodies. If we want to include artificial and natural bodies, motion of stars and of galaxies, we can speak about *dynamical astronomy*. The concept or the definition of celestial mechanics is not very well known. It happened to me once that the dean of one the prestigious Eastern Universities gave a small party to which I was invited. His wife asked me what field I was working in and when I said, celestial mechanics, she asked me what was that. I said, wc are studying the motion of heavenly bodies. She was a little bit shocked but she was, of course, used to strange things at the university and so she asked me: "And do we pay you for doing this?" I, being very respectful and young at that time said: "Yes, thank you." Today, I would say: "Yes, but not very well, thank you."

THE PAST OF ORBIT MECHANICS AND CELESTIAL MECHANICS

We shall discuss very shortly the private lives and the major contributions of some of the great men in our field, such as Copernicus, Kepler, Galilei, Euler, Lagrange, Laplace, Gauss and Poincare. As a special subject, we shall discuss Newton's activities and a short history of his life. We start with **Nicolaus Copernicus** (1473-1543) who was one of the first to propose scientifically the heliocentric theory of the Solar system. It is very interesting to note that he and his students treated and defended the heliocentric view not as a natural fact but they claimed that this model of the Solar system allowed easier computations of the positions of the planets than the geocentric model. We also should remember that Copernicus considered that all planets were moving on circular orbits around the Sun, which is a very good approximation.

The second name I would like to mention is **Johannes Kepler** (1571-1630) about whom I would like to tell you a short story which is very much applicable to all of us attending this meeting. Kepler was invited by the Emperor Rudolf II to become his astronomer, which was of course a wonderful invitation. Kepler thought that he would just do his research and observations and could dedicate his life to science. Everything went well for a while until the Emperor asked him for his report. We all have this problem of course, that the sponsor every once in a while wants a report and Kepler should not have been surprised. The Emperor told him that the report was to be his horoscope. Kepler, of course, did not believe in astrology but he considered it as a source of income so he prepared the requested horoscope and in this way he became probably the first principal investigator who satisfied the request of the sponsor against his better judgment.

Galileo Galilei (1564-1642) introduced once again the concept of the heliocentric Solar system but not only as a way to compute planetary orbits but as a physical fact. This way he ran into trouble with the church but the worse thing that actually happened to him was that the last eight years of his life he had to spend in house arrest as the penalty for his conflict with the church. Now, think of a professor who is put in house arrest, whereby his food, his apartment, his computer and his secretary are taken care of. He does not have to attend any committee meetings, he does not have to deliver any lectures and he can just sit at home and do his science. So do not feel bad about the church, or do not think that Galileo really had a very hard time.

We may wish to mention **Leohard Euler** (1707-1783) who was one of the many famous Swiss scientists. He was invited by the Czarina to visit Petersburg (Leningrad) as a visiting scientist, which is a delightful position today and also was, in the past. The subject of his contract or grant was to establish a method to compute the orbit of the Moon. He liked this assignment but I do not think he realized what the reason was for the invitation. The Russian Navy needed some assistance in navigation, which during the night was made much simpler by knowing the location of the Moon. I do not have to tell you that ever since that time the military makes much use of orbital mechanics and celestial mechanics.

Euler died at 77 and the way people described his death was that "He ceased to live and to calculate." In spite of his blindness during the last 17 years of his life he still kept on working and if you ever tried to handle equations with your eyes closed you would appreciate his difficulties. But graduate students usually find a solution to any problem. In his case, his assistants were taking dictations and putting equations on the blackboard. They all worked with the great man Euler and established new results.

He had 13 children and two wives. One of his interesting statements we might wish to remember expressed his strong feelings about mathematics. "Pure mathematics is for young people, while applied mathematics for the old." He was indeed the competent and qualified man to make this statement because he has contributed to pure as well as to applied mathematics.

Our next hero is **Joseph Louis Lagrange** (1736-1813) who was born in Italy, his name was French and he spent many years of his life in Germany. But you should not be surprised at this, since we are used to the international character of our field. One of his important results was the discovery of the so-called triangular libration points of the restricted problem of three bodies. His bodies of interest were the Sun, Jupiter and the asteroids located around the fourth and fifth triangular libration points. Lagrange predicted the existence of these stable positions and urged astronomers to look in these directions and search for asteroids. Actually, 134 years after Lagrange's announcement, astronomers located the so-called Trojan asteroids at these libration points. Lagrange made the point that "astronomers are strange, they will not believe in a theory unless it agrees with their observations."

Pierre Simon Laplace's (1749-1827) name you probably connect with the introduction of the concept of potential but let me tell you something else about him. Many students of the literature of our field claim that he introduced in our field the expression: "il est aisé à voir," meaning "it is easy to see," in connection with derivations. If we have a lengthy derivation, sometimes we would like to cut out some of it and we make the statement "from this equation it follows" or "it is easy to see." Laplace gets the questionable credit for this method.

Henri Poincare (1854-1912) is credited with establishing the concept of non-integrability. Several people in our field claim that Poincare has done more disservice to our field than anybody else by establishing the non-integrability principle. I like to look at it as an opportunity to practice our humbleness and humility and accept the limitations of our field.

Sir Isaac Newton was probably the most important contributor to our field. He was born in 1642 on Christmas day. His birth was premature and at the time he was born his father was already dead. He was very weak physically and mentally which was also true about Einstein. Albert Einstein according to a story, was very late to start speaking and it happened during a dinner when eating his soup he suddenly put down the spoon and uttered his first sentence: "The soup is too hot." At which point his mother jumped up and said: "But this is wonderful Albert! You can speak. Why didn't you speak before?" To which Albert answered: "Well, up to now everything was all right." There is a delightful book, drawing a parallel between Newton and Einstein by Aaron Lerner, published in 1973.

Sir Isaac Newton received his first degree in Cambridge when he was 23 and became a professor when he was 27. His professor was Isaac Barrow who resigned so that Newton could have his position. (I am always concerned mentioning this and I am sure all the professors at this meeting prefer not to discuss the subject in front of their students). The famous *Principia* was published in 1687 when Newton was 45 years old. We all know about the nervous fatigue which slowed down Newton when he was around fifty. He became the Warden and later the Master of the Royal Mint and he became the President of the Royal Society at 60. Newton died on the 20th of March 1727 and he is one of the very few scientists who was buried in Westminster Abbey with national honours. The unverified apple incident probably occurred in 1665 when Newton left Cambridge because of the plague and went back to his birthplace, the city of Woolsthorpe. There were no students, no committee meetings, no lectures and no pressure so he could think and discover the analogy between the fall of an apple and the motion of the Moon, both phenomena associated with gravity. The story goes, and I am sure this is not a true story, that he applied for a research grant to study gravity and mentioned the motion of the Moon and the falling of the apples. He was denied the grant for two reasons. Number one was that the grantors at that time, did not find any interest in the motion of the Moon, and the second reason was that the grantors wanted him to study the fall of other fruits besides apples.

Sir Isaac was never married but he got close to it twice in his life. One of the stories about this is that when he had dinner with a young lady and was holding her hand getting ready to propose, his pipe stopped working and he used her finger to fix it. She screamed and ran away and according to the story Sir Isaac decided at that point that marriage was not for him.

Mewton's scientific importance was described best by Alexander Pope who said:

> "*Nature and Nature's laws lay hid in night*
> *God said: 'Let Newton be' and all was light.*"

Two quotations by Newton will give an idea of his humbleness and humility. The first is: "If I have seen further, it is by standing upon the shoulders of giants." I often mention this to my students and try to remind them that after Newton there were many more giants, consequently, they, meaning my students, should be able to look much much further.

The second quotation by Newton is: "I do not know what I may appear to the world but to myself I seem to have been like a boy playing on the seashore and I am diverting myself in now and then finding a smoother pebble or a prettier shell than ordinary while the great ocean of truth lay all undiscovered before me."

I am sure you will all agree with me that this is one of the best ways to express humbleness and humility. Think of Sir Isaac who established the general law of gravitation and he considered himself a little boy looking for pretty shells and pebbles.

Newton was a human being and had to deal with the conflict between his humility and his extreme sensitivity. As you know there were significant conflicts with Leibniz who made many negative remarks about Newton's approach to science and the priority concerning the discovery of calculus.

In order to demonstrate and emphasize the humbleness and humility which are deeply embedded in our field, I offer just a few numbers to show what we are dealing with in our field. Consider the size of the Earth and the size of a human being. One way to look at this is that if the Earth is the size of six football fields, then the size of a man is represented by one grain of sand. Note that on a one-inch line there are 254 grains of sand. Another way to describe us, consider the following: If the Earth is the size of a basketball, then the Moon going around the Earth could be represented by an orange which is located 30 feet away from the Earth. In this system the Sun is represented by a 100 foot balloon and the Earth going around this balloon is three miles away. So we have the 100 foot balloon, the basketball, and the orange. Of course in this system we would be much much smaller than a grain of sand. I might mention two more short examples. One is that if we go around the Earth on the Equator ten times, we reach the Moon. On the other hand, driving around the Earth once, takes about one month (following the speed limits). Driving with the same speed to the Moon takes ten months. A spaceship to the Moon usually takes about three days, so it goes a hundred times faster than a car. (Light takes only one and one third seconds to go th the Moon). There is just no end to playing these number games. But I will conclude these examples of teaching humbleness and humility by mentioning that according to some recent estimates there are about 1000 planets with life in our galaxy. This, depending of course on the size of the planets, roughly would mean that we have 4000 billion brothers and sisters in the galaxy. Now take a grain of sand for all of these 4000 billion people and put these grains of sand next to each other on a line. This line would reach from the Earth to the Moon and you would be one of the grains of sand. I do believe that this suggests the dimensions of our natural environment and how we fit into it.

FUTURE

One of the major problems concerning our field is the establishment of the limits of meaningful predictions with a required accuracy. This is a very serious problem if the required time of predictions is long and if high accuracy is needed. The input to our calculations, such as the initial conditions and the physical parameters of the system are not known exactly. These initial conditions and the physical parameters are known exactly only in textbooks and in classroom exercises. We tell our students to compute the speed of a rolling sphere on a given slope if the initial velocity is ten ft/sec, if the angle of the incline is 30 degrees and if the sphere has certain given properties. They are supposed to compute the arrival velocity at the bottom of the slope. We do not usually tell them that we do not know the angle exactly, the initial velocity exactly, the constant of gravity exactly, not to mention the frictions involved, we do not know the properties of the sphere exactly and so on. So as long as we hide the reality from the students they can compute exact results.

These problems are often referred to in our field as the direct method, whereby we compute the motion of a dynamical system. The inverse method is when we make trajectory observations and from the observations we try to establish the force field. Here, one of the major problems is that the solution is not unique, meaning that if we obtain very precise observations of a satellite in order to compute higher order gravitational coefficients of the Earth, it is not inconceivable that another satellite on another orbit might give slightly different numbers for these coefficients. One way to put this is to ask the question: "Does the Earth's gravitational field depend on the satellite's orbit?" Clearly the answer is negative but there are some practical and fundamental problems here still to be solved.

Probably the most important reason for our inability to predict the future of our field is that it is saturated with non-linear, non-integrable problems. If in our dynamical system the number of participating bodies is more than two, Poincare tells us that it becomes a so-called non-integrable system, meaning that the general solution for any given set of initial conditions is not known. I believe we can stop at this point because we have established the fact that it is very difficult to predict the future in general and consequently, it is very difficult to predict the future of our field. Unquestionably, the future is strongly influenced by the stability of the system and we have difficulty in making predictions for unstable systems. We still do not have precise information concerning the stability of the Solar System and our prediction regarding the stability of the lunar orbit depends on the model we are using to compute the orbit of the Moon. We should consider stability when we wish to predict the future of our field since it depends on the stability of the organizations which sometimes use our methods and techniques and sometimes are responsible to support and sponsor our field. If these organizations show instability our future becomes definitely unpredictable.

CONTENTS

Page

Part II

Page

TRAJECTORY COMPUTATION TECHNIQUES

SESSION 1

Chairman: Thomas J. Eller
Kaman Sciences Corporation

AAS 89-385

LAMBERT PROBLEM IN TRUE-ANOMALY REGULARIZED VARIABLES

Paul W. Schumacher, Jr.*

True–anomaly regularization of the two–body problem permits a natural and efficient formulation of the Lambert problem once the time equation has been cast in a universal form. Using a new universal direct relation between time and true anomaly, the Lambert problem can be solved entirely without reference to Lambert's theorem. The iterative solution involves only one transcendental function evaluation per cycle. The parameters which appear naturally in the true–anomaly regularization permit easy interpretation and exploitation of a well known geometric property of the Lambert problem (linear locus of transfer–orbit Laplace vectors), and also avoid certain spurious singularities found in some other methods.

INTRODUCTION

The two–point boundary–value problem for Keplerian satellite motion has inspired a vast amount of analysis. The classical results by Lambert, Lagrange and Gauss continue to be developed in ways that the founders of the subject would have considered most intriguing[1-4]. For instance, it is the rule nowadays to use universal formulae to solve the Lambert problem since some spacecraft must traverse both elliptic and hyperbolic arcs. Moreover, even if only elliptic orbits are to be considered, the unknown parameter in the time equation may, during the iterative solution, take on a range of values corresponding to elliptic, parabolic and hyperbolic arcs before converging to its final value.

Completely general universal formulations of Keplerian motion have been available for many years. One especially convenient way to obtain such formulations is by a regularization of the differential equations of motion. For example, a regularization based on eccentric anomaly as the independent variable not only removes the collision singularity but also results in rigorously linear equations of motion when Cartesian or KS coordinates are used. These equations are easily solved in a form which is valid for all orbits, including rectilinear ones. The time transformation can be integrated to give a form of Kepler's equation which likewise is valid for all types of orbits. Because the relation between eccentric anomaly and transfer angle (true anomaly) is orbit–dependent, considerable insight

---

* Analysis and Software Department, U.S. Naval Space Surveillance Center, Dahlgren, Virginia 22448.

is needed in order to formulate a successful solution method for the boundary–value problem. Nevertheless, many such analyses have been made[5-9].

There is another regularization which has only recently been used to formulate the boundary–value problem. Burdet[10] uses true anomaly as the independent variable and represents position by the reciprocal radius and the unit radial vector. The governing equations turn out to be rigorously linear and are easily solved in a form which is valid for all orbits. The integration of the time transformation is more complicated than in the eccentric–anomaly case, and when the time of flight is expressed in terms of elementary functions then different equations are required for elliptical, parabolic or hyperbolic orbits. Heretofore, the lack of a universal true–anomaly time equation has prevented a solution of the boundary–value problem in Burdet–type variables. Yet there are at least two reasons why such a formulation is desirable. First, the Keplerian solution of Burdet's regularized equations is the most natural setting of the boundary–value problem since the independent variable is simply the transfer angle itself. Second, the several true–anomaly time equations are readily put in forms in which at most one transcendental function, an arctangent, appears in each equation. Formulae of this type have recently been published by Jezewski[11], and offer the possibility of notable computational efficiency. Also recently, the missing link in the analysis was forged, namely, the reduction of the different forms of the true–anomaly time equation to a common universal form suitable for practical computation[12]. The development of the universal true–anomaly time equation and its application in solution of the initial–value problem are addressed in another paper[13]. Its application to the two–position–and–time boundary–value problem is addressed in this paper. Only the mathematical developments are given here; numerical results in comparsion with other methods are reserved for a future publication.

ANALYSIS

Burdet's regularization can be summarized as follows. The equation of motion

$$d^2\underline{r}/dt^2 + \mu r^{-3}\underline{r} = \underline{0} \qquad (1)$$

is subjected to a differential time transformation

$$dt = Cr^2 ds \qquad (2)$$

where C is either strictly constant or a function of the orbital elements. New coordinates are introduced as

$$\underline{r} = u^{-1}\underline{\xi} \qquad (3)$$

where $\underline{\xi} = r^{-1}\underline{r}$ is the unit radial vector. Then the transformed equations of motion become

$$\underline{\xi}'' + C^2h^2\underline{\xi} = \underline{0} \qquad (4)$$

$$u'' + C^2h^2u = C^2\mu \qquad (5)$$

$$t' = Cu^{-2} \tag{6}$$

where h is the angular momentum magnitude given by

$$h^2 = \underline{h} \cdot \underline{h}, \quad \underline{h} = \underline{r} \times d\underline{r}/dt \ , \tag{7}$$

and $(\cdot)'$ denotes differentiation with respect to new independent variable:

$$(\cdot)' = d(\cdot)/ds = Cu^{-2}d(\cdot)/dt = Cr^2 d(\cdot)/dt. \tag{8}$$

Thus $\underline{\xi}$ and u are governed by simple harmonic oscillator equations, and time t is given by a quadrature. There are several useful choices for the factor C. Burdet used C=1 in his original analysis[10], a choice which permits continuous transition from rectilinear (h=0) orbits to nonrectilinear orbits as the parameter h is varied. Here the value $C=h^{-1}$ will be used. This choice excludes rectilinear orbits from consideration, since the time transformation is singular, but in most satellite problems the singular condition is not actually approached because of the finite size of the attracting body. The governing equations become

$$\underline{\xi}'' + \underline{\xi} = \underline{0} \tag{9}$$

$$u'' + u = \mu h^{-2} \tag{10}$$

$$t' = h^{-1}u^{-2} \tag{11}$$

Indicating initial values by subscript 0, final values by subscript 1, and denoting the independent variable by η, the solutions of the oscillator equations have the simple form

$$\underline{\xi}_1 = \underline{\xi}_0 \cos \eta + \underline{\xi}'_0 \sin \eta \tag{12}$$

$$\underline{\xi}'_1 = -\underline{\xi}_0 \sin \eta + \underline{\xi}'_0 \cos \eta \tag{13}$$

$$u_1 = \mu h^{-2}(1 + \alpha \cos \eta + \beta \sin \eta) \tag{14}$$

$$u'_1 = \mu h^{-2}(-\alpha \sin \eta + \beta \cos \eta) \tag{15}$$

The dimensionless parameters α and β are

$$\alpha = h^2 u_0/\mu - 1 \text{ and } \beta = h^2 u'_0/\mu \tag{16}$$

Realizing that u is the reciprocal radius, it is easy to verify that η is the transfer angle (which is the reason for the choice of C) and that α and β are proportional to eccentricity. In fact, $\alpha^2 + \beta^2$ is just the square of the eccentricity. Values for $\underline{\xi}_0$ and u_0 in terms of $\underline{r}_0$ can be obtained directly from the coordinate transformation. Values for $\underline{\xi}'_0$ and u'_0 in terms of $\underline{r}_0$ and $(d\underline{r}/dt)_0 = \underline{v}_0$ are obtained by considering the transformation of derivatives.

$$\underline{v} = d\underline{r}/dt = C^{-1}u^2 d(u^{-1}\underline{\xi})/ds = C^{-1}(u\underline{\xi}' - u'\underline{\xi}) \tag{17}$$

Similarly,

$$u' = du/ds = Cr^2 d(r^{-1})/dt = -Cr^{-1}\,\underline{r}\cdot\underline{v} \tag{18}$$

These relations, which hold at every instant and for a general value of C, allow ξ' and u' to be computed in terms of $\underline{r}$ and $\underline{v}$. Of course, $s=\eta$ when $C=h^{-1}$. In that case, these relations can be used to show that ξ' is actually a unit vector transverse to ξ in the plane of motion, pointing in the same sense as the motion. For any other choice of C, ξ' will still indicate the plane and sense of the motion but will not be a unit vector. Once initial values of ξ_0, u_0, ξ'_0 and u'_0 have been assigned, the corresponding final values are computed, for some value of η, from the oscillator solutions above. Then final values of position and velocity, $\underline{r}_1$ and $\underline{v}_1$, are computed from the coordinate transformation and velocity relation.

The manipulations just described are suitable for solving the inital–value problem: given $\underline{r}_0$, $\underline{v}_0$ and time of flight $\Delta t = t_1 - t_0$, find $\underline{r}_1$ and $\underline{v}_1$. In a companion paper[13] and in earlier work[12], the time equation has been obtained in a universal form

$$\Delta t = \frac{h^3}{\mu^2} K(\eta; \alpha, \beta) \tag{19}$$

Given values for Δt, h, α and β, this equation determines η. Newton's method, or another iterative procedure, is usually appropriate for approximating the value of η. (Slightly out of context, it is worth noting that Ioakimidis and Papadakis[14,15] have proposed noniterative quadrature methods for finding the root of any nonlinear equation. To date, however, these direct methods require more computation to achieve a given accuracy than does a well–designed iteration.) The boundary–value problem is more complicated: given $\underline{r}_1$, $\underline{r}_2$ and Δt, find $\underline{v}_0$. Values for ξ_0, u_0, ξ_1, and u_1 are known but u'_0 and u'_1 are unknown. The transverse unit vector ξ'_0 is determined if $\underline{r}_0$ and $\underline{r}_1$ are linearly independent:

$$\xi'_0 = (\xi_0 \times \xi_1)/|\xi_0 \times \xi_1| \times \xi_0 \tag{20}$$

(For the "long way" transfer, this vector should be reversed.) In the time equation, η is known, being the angle between $\underline{r}_0$ and $\underline{r}_1$, but h, α, and β are unknown. It might be reasoned naively that the boundary–value problem must involve three unknowns since $\underline{v}_0$ has three independent components. But, in fact, the situation is as follows. Substituting the oscillator solutions into the velocity relation produces the succinct result

$$\underline{v}_1 = \mu h^{-1}[\xi'_0(\alpha + \cos\eta) - \xi_0(\beta + \sin\eta)] \tag{21}$$

$$\underline{v}_1 = \underline{v}_0 - \mu h^{-1}[\xi'_0(1 - \cos\eta) + \xi_0 \sin\eta] \tag{22}$$

where

$$\underline{v}_0 = \mu h^{-1}[\xi'_0(1 + \alpha) - \beta\xi_0] \tag{23}$$

To solve the boundary–value problem, it is necessary to specify all the quantities on the right–hand side of this equation. Angular momentum magnitude h can be expressed in terms of α according to the definition of the latter:

$$h = \sqrt{\mu(1+\alpha)/u_0} \tag{24}$$

Then

$$\underline{v}_0 = \sqrt{\mu u_0/(1+\alpha)} \quad [(1+\alpha)\xi'_0 - \beta\, \xi_0] \tag{25}$$

The presence of two unknowns, α and β, in this equation merely reflects the fact that there are only two unknown components of initial velocity when that vector is resolved in the known orbital plane. If the transfer angle is π (or a multiple) then the orbit plane is not uniquely defined and must be determined from other data besides that given in the bare boundary–value problem. In most practical situations the extra information is available. But even if the orbital plane is not specified, the magnitudes of the radial and transverse components of $\underline{v}_0$ can still be evaluated.

An important feature of the boundary–value problem is that a constraint exists between α and β. Substituting for h in terms of α in the oscillator solution for u, one obtains

$$u_1(1+\alpha) = u_0(1+\alpha \cos \eta + \beta \sin \eta) \tag{26}$$

Hence, one can solve for either α or β in terms of the other so that the components of $\underline{v}_0$ are determined completely by a single scalar parameter. The fact that α and β are related linearly is also significant. The Laplace vector for the transfer orbit may be written as

$$\mu\underline{e} = \underline{r}_1(\underline{v}_1 \cdot \underline{v}_1) - \underline{v}_1(\underline{r}_1 \cdot \underline{v}_1) - \mu\underline{r}_1/r_1 \tag{27}$$

Expressing the right–hand side in Burdet–type variables, one finds that

$$\mu\underline{e} = \mu(\alpha\xi_0 + \beta\, \xi'_0) \tag{28}$$

Hence, because α and β are linearly related, the locus of transfer–orbit Laplace vectors is a straight line in space. Each point on the line corresponds to a different value of the unknown parameter and also, according to the time equation, to a different value of the transfer time. Several other demonstrations of this property of the boundary–value problem are known[2,4], but here the property is being exploited in a completely natural way.

It remains only to use the time equation to determine the value of the unknown parameter which corresponds to the given transfer time. When h has been replaced in terms of α, that equation takes the form

$$(1 + \alpha)^{3/2}\, K(\eta;\alpha,\beta) - \Delta t \sqrt{\mu u_0^3} = 0 \tag{29}$$

Now suppose α is selected as the unknown parameter. The linear constraint is solved for β in terms of α as

$$\beta = \Big[[(u_1/u_0) - \cos \eta]\, \alpha + (u_1/u_0) - 1\Big] / \sin \eta \tag{30}$$

and this expression is to replace β in the time equation. Unfortunately, the $\beta(\alpha)$ relation cannot be used if the transfer angle is π. If β is chosen as the unknown parameter, the linear constraint gives

$$\alpha = \frac{1 - (u_1/u_0) + \beta \sin \eta}{(u_1/u_0) - \cos \eta} \tag{31}$$

The $\alpha(\beta)$ relation cannot be used if $r_0 = r_1 \cos \eta$. This geometrical circumstance is entirely unexceptional, so it would be desirable to avoid introducing a singularity here unnecessarily. The remedy for these difficulties is to make use of a second linear relation between α and β. Replacing h in terms of α in the oscillator solution for u' produces

$$u'_1(1+\alpha) = u_0(-\alpha \sin \eta + \beta \cos \eta) \tag{32}$$

The parameter u'_1 is the sole remaining unknown in the boundary value problem. Letting nondimensional parameters be

$$\rho = u_1/u_0 = r_0/r_1 \text{ and } q = u'_1/u_0, \tag{33}$$

the two linear relations can be solved for α and β in terms of q:

$$\alpha = \frac{(\rho - 1) \cos \eta - q \sin \eta}{1 - \rho \cos \eta + q \sin \eta} \tag{34}$$

$$\beta = \frac{(\rho - 1) \sin \eta - q(1 - \cos \eta)}{1 - \rho \cos \eta + q \sin \eta} \tag{35}$$

Then q must be chosen to satisty the time equation. Notice that the time equation itself is not modified; only these two additional relations for α and β must be included. As a sidenote, by expressing u' in terms of $\underline{r}$ and $\underline{v}$, one can show that

$$q = \rho \tan \gamma_1 \tag{36}$$

where γ_1 is the flight path angle at $\underline{r}_1$.

Whatever numerical method one uses to solve the time equation, care must be taken to choose q always such that $\alpha > -1$. The value of the other parameter β is unrestricted, but from the definition of α it is apparent that a value $\alpha \leq -1$ is not physically possible. In fact, values of α larger than 1 in magnitude correspond to hyperbolic arcs, and $\alpha \leq -1$ would mean that the initial position lies outside the asymptotic limits of the motion. The implied restriction on q is obtained from

$$1 + \alpha = \frac{1 - \cos \eta}{1 - \rho \cos \eta + q \sin \eta} > 0 \tag{37}$$

The sign of this expression is controlled by the denominator. Then, provided $\sin \eta \neq 0$, we must have

$$q > (\rho \cos \eta - 1)/\sin \eta \text{ if } \sin \eta > 0 \tag{38}$$

$$q < (\rho \cos\eta - 1)/\sin \eta \text{ if } \sin \eta < 0. \tag{39}$$

The critical value $q = (\rho \cos \eta - 1)/\sin \eta$ therefore separates "long way" trajectories ($\eta > \pi$) from "short way" trajectories ($\eta < \pi$). For the special values $\eta = 0, \pi, 2\pi, \ldots$ the value of α is independent of q. No difficulties are encountered in the important special case $\eta = \pi$, but the case $\eta = 0$ cannot be handled by the present method. For unequal terminal radii, a rectilinear transfer is implied, a situation which was excluded from the governing equations of motion and which is signalled here by the value $\alpha = -1$. Hence, numerical difficulties can be expected in the present method for small transfer angles. Using Burdet's original regularization (C=1 in the time transformation), it is possible to formulate the boundary–value problem in terms that are valid for all arcs, including rectilinear ones[12]. The resulting formulae show no difficulty or inaccuracy for small transfer angles.

The time equation to be used in this analysis is a new direct relation between time and true anomaly[12,13], the derivation of which is lengthy and will not be repeated here. It is valid for all nonrectilinear orbits and for transfer angles up to (but not including) $\eta = \pm 2\pi$. In the form given here, it has several characteristics in common with time equations used by Battin in the boundary–value problem[1-3] and by Shepperd[16] in the initial–value problem. It contains one transcendental function (essentially an arctangent) which can be evaluated efficiently either by a power series or by a continued fraction. Simple transformations, described below, can be used to improve dramatically the convergence of either type of evaluation. The continued fraction is offered here since it converges better than the series when no auxiliary transformations are used. The explicit form is

$$K(\eta;\alpha;,\beta) = \frac{4zN}{E(D+\sqrt{E})^2(1+C_1)} \tag{40}$$

where

$$N = [8z^2 + (1-z^2)(D + \sqrt{E})](D + \sqrt{E}) - 4z^2\left[D + \frac{1}{(3+C_2)}[D-(1-z^2)(1-\alpha^2-\beta^2)]\right] \tag{41}$$

$$z = \tan (\eta/4) \tag{42}$$

$$D = (1+\alpha)(1-z^2) + 2\beta z \tag{43}$$

$$E = D^2 + 4z^2(1-\alpha^2 - \beta^2) \tag{44}$$

$$C_1 = y/(3 + C_2) \tag{45}$$

$$C_2 = 2^2 y/(5+C_3) \tag{46}$$

$$C_3 = \cfrac{3^2 y}{7 + \cfrac{4^2 y}{9 + \cfrac{5^2 y}{11 + \ldots}}} \tag{47}$$

$$y = 4z^2(1-\alpha^2-\beta^2)/(D + \sqrt{E})^2 \tag{48}$$

The continued fraction is derived from an expansion of the arctangent function discussed by Wall[17]. The preferred method of evaluating any contined fraction is a recursion develped by Gautschi[18] based on another theorem given in Wall's book. For a continued fraction of the form

$$f = \cfrac{a_1}{b_1 + \cfrac{a_2}{b_2 + \cfrac{a_3}{b_3 + \ldots}}} \tag{49}$$

the algorithm proceeds in "top–down" fashion as

$$u_{k+1} = \left[1 + (a_{k+1}/b_k b_{k+1}) u_k\right]^{-1} \tag{50}$$

$$v_{k+1} = v_k(u_{k+1} - 1) \tag{51}$$

$$w_{k+1} = w_k + v_{k+1} \tag{52}$$

where $k = 1,2,3,\ldots$, $u_1 = 1$ and $v_1 = w_1 = a_1/b_1$. Provided the continued fraction is convergent, $w_k \to f$ as $k \to \infty$.

It can be shown that

$$y = \tan^2(\theta/4) = -\tanh^2(\phi/4) \tag{53}$$

where θ is the transfer eccentric anomaly and ϕ is the transfer hyperbolic anomaly[12,13]. Thus the interval of interest is $-1 < y < \infty$. The continued fraction happens to converge over this entire interval. However, it is found that the convergence is slow when y takes on large values. With some auxiliary computations, one can compensate for this slow convergence, or obtain improved convergence even when y is nc⁺ large. The details are explained elsewhere[12,13], but the basic idea is as follows. The half–angle identity

$$\tan(\delta/2) = \frac{\tan\delta}{1 + \sqrt{1 + \tan^2\delta}} \tag{54}$$

shows that

$$\tan^2(\theta/8) \equiv x = \frac{y}{(1 + \sqrt{1 + y}\,)^2} . \tag{55}$$

Because $|x| < |y|$, $C_3(x)$ converges faster than $C_3(y)$. This in itself would be of little use, but because the continued fraction derives from an arctangent function it is possible to relate $C_3(x)$ to $C_3(y)$:

$$C_3(y) = \frac{8(1+\sqrt{1+y}\,)^2[5+C_3(x)] + 8x}{3(1+\sqrt{1+y}\,)[5+C_3(x)] + 4(1+\sqrt{1+y}\,)-6} - 5 \tag{56}$$

If enough recursions are saved through reducing the magnitude of the function argument in this way, then the net effect will be a reduction in the computational load. In principle, the transformation x(y) could be applied repeatedly before C_3 is evaluated, producing drastic reductions in the number of recursions needed. Then, of course, the relation between $C_3(y)$ and $C_3(x)$ would have to be invoked the same number of times. In practice, one obtains diminishing returns from each successive transformation. Mainly because of the extra square roots involved, the auxiliary calculations soon become as burdensome as the extra recursions would have been. Exerience has shown that making one or two transformations of the function argument is almost always worthwhile, however.

The function

$$F(q) = (1 + \alpha(q))^{3/2}\, K(\eta;\alpha(q), \beta(q)) - \Delta t\sqrt{\mu u_0{}^3} = 0 \tag{57}$$

has enough algebraic complexity that a root–finding method should be chosen judiciously. For smoothly varying functions such as this one, Newton's method usually requires fewer iterations than a derivative–free method. In this case, the derivative turns out to be an extensive expression itself, so that it is reasonable to accept several extra iterations in a simpler scheme. The secant method

$$q_{k+1} = q_k - F_k\left[\frac{q_k - q_{k-1}}{F_k - F_{k-1}}\right] \tag{58}$$

has been used successfully[12]. In a group of 37 test cases, this method produced the root to at least 9 decimal places in an average of 5 iterations. No special precautions were taken to begin near the root or to ensure that the root lay between successive estimates. This behavior bodes well for the application of Newton's method, though to date the latter has not been tested.

The derivative needed for Newton's method is

$$\frac{dF}{dq} = (3/2)\sqrt{1+\alpha}\, K \frac{d\alpha}{dq} + (1 + \alpha)^{3/2}\left[\frac{\partial K}{\partial \alpha}\frac{d\alpha}{dq} + \frac{\partial K}{\partial \beta}\frac{d\beta}{dq}\right] \tag{59}$$

Derivatives of α and β with respect to q are readily obtained.

$$\frac{d\alpha}{dq} = \frac{-\sin \eta \, (1- \cos \eta)}{(1- \rho \cos \eta + q \sin \eta)^2} \tag{60}$$

$$\frac{d\beta}{dq} = \frac{-(\rho - \cos \eta)(1- \cos \eta)}{(1 - \rho \cos \eta + q \sin \eta)^2} \tag{61}$$

The partial derivatives of K with respect to α and β are more extensive. Inspection shows that eventually the derivative dC_2/dy will be needed and that this is really the main complication in obtaining the partial derivatives. It is not feasible to differentiate the continued fraction "term by term"; some indirect means must be used. At a similar juncture in his analyses, Battin[1,3] uses the fact that the continued fraction can be expressed in terms of a hypergeometric function. A rich variety of identities are available for hypergeometric functions and their derivatives, so it proves possible for him to obtain efficient expressions for the derivative. A similar approach will be used here, though all the manipulations will involve only elementary functions. The arctangent function can be represented by [17]

$$\tan^{-1}\sqrt{y} = \frac{\sqrt{y}}{1+C_1(y)} = \frac{\sqrt{y}}{1 + \dfrac{y}{3 + C_2(y)}} \tag{62}$$

so that

$$3 + C_2 = \frac{y \, (\tan^{-1}\sqrt{y}\,)/\sqrt{y}}{1-(\tan^{-1}\sqrt{y}\,)/\sqrt{y}} \equiv \frac{yT}{1-T} \tag{63}$$

For completeness, let it be noted that

$$(\tan^{-1}\sqrt{y}\,)/\sqrt{y} = F(\tfrac{1}{2}, 1; \tfrac{3}{2}; -y) \tag{64}$$

$$(\tanh^{-1}\sqrt{y})/\sqrt{y} = F(\tfrac{1}{2}, 1; \tfrac{3}{2}; +y) \tag{65}$$

where F is a hypergeometric function in the usual notation (not to be confused with the notation F(q) used for the time equation). This equation for C_2 is real–valued only for elliptic orbits ($y>0$), and is indeterminate for parabolic orbits ($y=0$), but it does provide a convenient vehicle for calculating the derivative.

$$\frac{dC_2}{dy} = \frac{T}{1-T} + \frac{y}{1-T}\frac{dT}{dy} + \frac{yT}{(1-T)^2}\frac{dT}{dy} \tag{66}$$

where

$$\frac{dT}{dy} = -\frac{T}{2y} + \frac{1}{2y(1+y)} \tag{67}$$

Then

$$\frac{dC_2}{dy} = \frac{4y^2 - 3yC_2 + y^2C_2 - yC_2^2}{2y^2(1+y)} \tag{68}$$

Calculating C_2 in terms of C_3 allows cancellation of y^2 from the denominator:

$$\frac{dC_2}{dy} = \frac{1}{2(1+y)} \left[(4+C_2) - \frac{4}{(5+C_3)}(3+C_2)\right] \tag{69}$$

This expression is valid for all non–rectilinear orbits ($-1< y < \infty$). Now the chain rule applied to K gives

$$\frac{\partial K}{\partial \alpha} = \frac{4z}{E(D+\sqrt{E})^2(1+C_1)} \frac{\partial N}{\partial \alpha}$$

$$- K\left[\frac{1}{E}\frac{\partial E}{\partial \alpha} + \frac{2}{D+\sqrt{E}}\left[\frac{\partial D}{\partial \alpha} + \frac{1}{2\sqrt{E}}\frac{\partial E}{\partial \alpha}\right] + \frac{1}{(1+C_1)}\frac{\partial C_1}{\partial \alpha}\right] \tag{70}$$

$$\frac{\partial K}{\partial \beta} = \frac{4z}{E(D+\sqrt{E})^2(1+C_1)} \frac{\partial N}{\partial \beta}$$

$$- K\left[\frac{1}{E}\frac{\partial E}{\partial \beta} + \frac{2}{D+\sqrt{E}}\left[\frac{\partial D}{\partial \beta} + \frac{1}{2\sqrt{E}}\frac{\partial E}{\partial \beta}\right] + \frac{1}{(1+C_1)}\frac{\partial C_1}{\partial \beta}\right] \tag{71}$$

The remaining derivatives are calculated straightforwardly as

$$\frac{\partial N}{\partial \alpha} = [8z^2 + 2(1-z^2)(D + \sqrt{E})]\left[\frac{\partial D}{\partial \alpha} + \frac{1}{2\sqrt{E}}\frac{\partial E}{\partial \alpha}\right]$$

$$-4z^2\left[\frac{\partial D}{\partial \alpha} - \left[\frac{D-(1-z^2)(1-\alpha^2-\beta^2)}{(3+C_2)^2}\right]\frac{\partial C_2}{\partial \alpha} + \frac{1}{(3+C_2)}\left[\frac{\partial D}{\partial \alpha} + 2\alpha(1-z^2)\right]\right] \tag{72}$$

$$\frac{\partial N}{\partial \beta} = [8z^2 + 2(1-z^2)(D + \sqrt{E})\left[\frac{\partial D}{\partial \beta} + \frac{1}{2\sqrt{E}}\frac{\partial E}{\partial \beta}\right]$$

$$-4z^2\left[\frac{\partial D}{\partial \alpha} - \left[\frac{D-(1-z^2)(1-\alpha^2-\beta^2)}{(3+C_2)^2}\right]\frac{\partial C_2}{\partial \beta} + \frac{1}{(3+C_2)}\left[\frac{\partial D}{\partial \beta} + 2\beta(1-z^2)\right]\right] \tag{73}$$

$$\frac{\partial D}{\partial \alpha} = 1 - z^2 \text{ and } \frac{\partial D}{\partial \beta} = 2z\,. \tag{74}$$

$$\frac{\partial E}{\partial \alpha} = 2D\frac{\partial D}{\partial \alpha} - 8\alpha z^2 \text{ and } \frac{\partial E}{\partial \beta} = 2D\frac{\partial D}{\partial \beta} - 8\beta z^2 \tag{75}$$

$$\frac{\partial C_1}{\partial \alpha} = \frac{1}{(3+C_2)}\frac{\partial y}{\partial \alpha} - \frac{y}{(3+C_2)^2}\frac{dC_2}{dy}\frac{\partial y}{\partial \alpha} \tag{76}$$

$$\frac{\partial C_1}{\partial \beta} = \frac{1}{(3+C_2)}\frac{\partial y}{\partial \beta} - \frac{y}{(3+C_2)^2}\frac{dC_2}{dy}\frac{\partial y}{\partial \beta} \tag{77}$$

$$\frac{\partial C_2}{\partial \alpha} = \frac{dC_2}{dy}\frac{\partial y}{\partial \alpha} \text{ and } \frac{\partial C_2}{\partial \beta} = \frac{dC_2}{dy}\frac{\partial y}{\partial \beta} \tag{78}$$

$$\frac{\partial y}{\partial \alpha} = \frac{-8\,\alpha z^2}{(D+\sqrt{E})^2} - \frac{8z^2(1-\alpha^2-\beta^2)}{(D+\sqrt{E})^3}\left[\frac{\partial D}{\partial \alpha} + \frac{1}{2\sqrt{E}}\frac{\partial E}{\partial \alpha}\right] \tag{79}$$

$$\frac{\partial y}{\partial \beta} = \frac{-8\,\beta z^2}{(D+\sqrt{E})^2} - \frac{8z^2(1-\alpha^2-\beta^2)}{(D+\sqrt{E})^3}\left[\frac{\partial D}{\partial \beta} + \frac{1}{2\sqrt{E}}\frac{\partial E}{\partial \beta}\right] \tag{80}$$

No attempt has been made to simplify these expressions; it will be obvious how to reduce them to more efficient forms for computation. The important feature to note is that C_3 is the only transcendental function needed to evaluate both the time equation and its derivative. Two square roots, $\sqrt{E}$ and $\sqrt{1+\alpha}$, are also required. Besides this, only rational algebraic operations are needed. The second square root could be avoided by treating the square of the time equation

$$G(q) = (1+\alpha(q))^3\, K^2(\eta;\alpha(q),\beta(q)) - \Delta t^2 \mu u_0^3 = 0 \tag{81}$$

but this is not necessarily recommended since G(q) cannot be monotonic.

CONCLUSION

True–anomaly regularization of the two–body problem provides a natural setting in which to formulate and solve the orbital boundary–value problem. Knowledge of Lambert's theorem is not required. However, the essence of Lambert's theorem, which is that the unknown velocity vector is determined completely by given data and a single scalar parameter, appears here in a different guise. The velocity has two unknown components when resolved in the orbital plane, and these components are determined by two eccentricity parameters. The equation of the orbit provides a linear constraint between the two eccentricity parameters, so the velocity is effectively a function of only one unknown parameter. The geometric interpretation of the constraint is well known: the locus of transfer–orbit Laplace vectors is a straight line in space, with each point on the line corresponding to a different transfer time. Then, of course, some numerical root–finding technique must be used to determine the value of the unknown parameter which corresponds to the given time of flight. The time equation used here is a new universal formula in terms of true anomaly. Like some other universal time equations in current use, it contains only a single transcendental function and this function can be evaluated efficiently in terms of a continued fraction. In the version offered here, it is valid for transfer angles up to (but not including) 2π. Hence, no special logic is required to evaluate the time equation for either "long–way" or "short–way" trajectories. The time equation is complicated enough algebraically that one should consider using some derivative–free root–finding method. The derivative required for Newton's method turns out to be an extensive expression, though no extra transcendental function or extra square roots are required to evaluate it.

REFERENCES

1. Richard H. Battin, "Lambert's Problem Revisted", AIAA Journal, May 1977, vol.15. no. 5, pp. 707–713.

2. Richard H. Battin, Thomas J. Fill and Stanley W. Shepperd; "A New Transformation Invariant in the Orbital Boundary–Value Problem", Journal of Guidance and Control, Jan.–Feb. 1978, vol. 1, no. 1, pp. 50–55.

3. R. H. Battin and R. M. Vaughn, "An Elegant Lambert Algorithm", Journal of Guidance, Control and Dynamics, Nov.–Dec. 1984, vol. 7, no.6, pp. 662–670.

4. Richard H. Battin, "Elegant Propositions of the Boundary Value Problem", Acta Astronautica, 1987, vol. 15, no. 12, pp. 965–971.

5. Edward T. Pitkin, "A General Solution of the Lambert Problem", Journal of the Astronautical Sciences, 1968, vol. 15, pp. 270–271.

6. Donald J. Jezewski, "KS Two–Point Boundary–Value Problems", Celes. Mech., 1976, vol. 14, pp.105–111.

7. J. Kriz, "A Uniform Solution of the Lambert Problem", Celes. Mech., 1976, vol. 14, pp. 509–513.

8. R. C. Engels and J. L. Junkins, "The Gravity–Perturbed Lambert Problem: a KS Variation of Parameters Approach", Celes. Mech., 1981, vol. 24, pp. 3–21.

9. George L. Hand, "The Guidance Problem – Solution for All Types of Trajectories with Oblate Gravity", Charles Stark Draper Laboratory Technical Report No. CSDL–C–5560, Cambridge, MA, Oct. 1982.

10. Claude A. Burdet, "Le Mouvement Keplerien et les Oscillateurs Harmoniques", Journal für die Reine und Angewandte Mathematik, 1969, band 238, pp. 71–84.

11. D. Jezewski, "An Ideal Reference Frame for Perturbed Orbital Motion", AIAA Paper 88–4218, AIAA/AAS Astrodynamics Conference Proceedings, Aug. 1988.

12. Paul W. Schumacher, Jr., "Results of True–Anomaly Regularization in Orbital Mechanics", Ph.D. dissertation, Virginia Polytechnic Institute and State University, Blacksburg, VA, Nov. 1987.

13. Paul W. Schumacher, Jr., "Universal True–Anomaly Time Equations", AAS Paper 89–448, AAS/AIAA Astrodynamics Conference, Aug. 1989.

14. N. I. Ioakimidis and K. E. Papadakis, "A New Simple Method for the

Analytical Solution of Kepler's Equation", Celes. Mech., 1985, vol. 35, pp. 305–316.

15. N. I. Ioakimidis and K. E. Papadakis, "A New Class of Quite Elementary Closed–Form Integral Formulae for Roots of Nonlinear Equations", Applied Mathematics and Computation, 1989, vol. 29, pp. 185–196.

16. Stanley W. Shepperd, "Universal Keplerian State Transition Matrix", Celes. Mech., 1985, vol. 35, pp. 129–144.

17. Hubert Stanley Wall, *Analytic Theory of Continued Fractions*, Van Nostrand Col., Inc., New York, 1948.

18. Walter Gautschi, "Computational Aspects of Three–Term Recurrence Relations", SIAM Review, Jan. 1967, vol. 9, no. 2, pp. 24–82.

AAS 89-387

A NUMERICAL INTEGRATION PROCEDURE FOR PERIODICALLY-STIFF, SECOND-ORDER SYSTEMS

David L. Richardson*

A multistep numerical procedure similar to the conventional Störmer/Cowell method has been developed to integrate efficiently the special second-order, initial-value system $d^2y_i/dt^2 = f_i(y_1, \ldots, y_n; t)$ whenever the solutions to this system are known to be oscillatory. Standard methods require that the stepsize h be a small percentage of the time scale $T_1 = 2\pi/\omega_1$ where ω_1 is the fundamental system frequency with the largest modulus. If at least two of the system frequencies are widely separated, i.e., $\omega_1 >> \omega_j$, then conventional methods of integration can be inefficient, and the system is termed "periodically stiff." The method presented here allows each separate dependent variable y_i of the system to be integrated with its own, separate stepsize h_i determined in accordance with criteria selected by the analyst.

INTRODUCTION

Quite often the conventional k-step Störmer/Cowell predictor-corrector procedure is used to obtain solutions to initial value problems described by the special second-order system

$$\frac{d^2y_i}{dt^2} = f_i(y_1, \ldots, y_n; t), \quad i = 1, \ldots, n. \tag{1}$$

However, it is not desirable to apply this method whenever the following conditions prevail:

1. The solution is oscillatory under a broad range of initial conditions. That is, the set of eigenvalues $\{\lambda\}$ of the $n \times n$ Jacobian matrix $\dfrac{\partial f_i}{\partial y_j}$ are purely imaginary with zero real parts.

2. There are at least two of these eigenvalues whose moduli are widely separated: $\|\lambda_i\| >> \|\lambda_j\|$.

* Professor, Department of Aerospace Engineering, University of Cincinnati, 797 Rhodes Hall (#70), Cincinnati, Ohio 45221-0070.

Systems which conform to these conditions are defined (in the paper) as periodically-stiff (P-stiff).

Accurate integration of P-stiff systems using standard linear multistep methods requires that the stepsize h be a small percentage of the time scale T_1,

$$T_1 \triangleq 2\pi/\omega_1, \tag{2}$$

where ω_1 is the fundamental frequency corresponding to the eigenvalue λ_1 with the largest modulus in $\{\lambda\}$. This situation leads to inefficient procedures. In particular, if the span of the integration is to extend over many thousands of cycles in T_1, the round-off error buildup in the solution could be considerable and unacceptable.

To try to address these difficulties, a multistep procedure with characteristics similar to Störmer/Cowell has been developed which can in effect integrate each dependent variable y_i of the system in accordance with a separate time scale selected by the analyst. That is, the method allows multiple stepsizes to be active independently as the numerical solution progresses. The grid spacing corresponding to these stepsizes is that used in the correction process. The method is made practicable by incorporating into the prediction procedure various quantities calculated at off-grid points. Use of values at off-grid points has been the subject of many papers in the open literature. Of particular interest to the present work was the paper of Forrington [1] who detailed a multi-stepsize, fourth-order method for first-order systems.

The P-stiff procedure has smaller computational overhead than that for conventional multistep methods of like stepnumber. Depending upon the selected stepsizes, the method can execute up to 30% faster. Its general accuracy and ability to suppress global error growth is now being tested on numerical integrations of the rectangular, heliocentric equations of motion for the planets. One idea which is now under investigation defines the various integration steps as a percentage of the system's intrinsic time scales $2\pi/\omega_i$. These stepsizes h_i are defined by the n relations

$$h_i = \frac{2\pi}{M_i\omega_i}, \tag{3}$$

where M_i is a nonzero real number chosen to effectively minimize the local truncation error. Preliminary results of this approach and that using alternate grid spacings are discussed in the sequel.

GENERAL PROCEDURE

The development of the integration procedure generally follows the construction elaborated by Henrici [2] for classical second-order methods. For the special second-order system of Eq. (1), it is assumed that each of the n functions $f_i(y_1, \ldots, y_n; t)$ is real-valued and of class C^2 for all x in the interval $[0, b]$, $b > 0$. For notational purposes, prime ($'$) signifies the operation of differentiation d/dx, and y and f (no subscripts) denote any y_i and f_i, respectively. Let $h = h_i$ of Eq. (3), and consider the construction of an expression for $y(x + ah)$ for $0 < a \leq 1$. By manipulation of the identities

$$\int_{x1}^{x2} y''(x)\,dx = y'(x_2) - y'(x_1),$$

$$\int_{x1}^{x2} y'(x)\,dx = x_2 y'(x_2) - x_1 y'(x_1) - \int_{x1}^{x2} x y''(x)\,dx,$$

we obtain

$$y(x_2) = y(x_1) + (x_2 - x_1)y'(x_1) + \int_{x1}^{x2} (x_2 - z)y''(z)\, dz.$$

The fundamental equation is derived from the above expression by eliminating the derivative y' term using two equations obtained from the following sets of limits:

$$\begin{aligned} x_1 &= x, & \qquad x_1 &= x, \\ x_2 &= x + ah. & \qquad x_2 &= x - h. \end{aligned}$$

The result is

$$y(x+ah) = (1+a)y(x) - ay(x-h) + \int_x^{x+ah} (x+ah-z)y''(z)\, dz + a\int_x^{x-h} (x-h-z)y''(z)\, dz,$$

which, by appropriately selecting the new dummy variable of integration w, is rewritten as

$$\begin{aligned} y(x+ah) &= (1+a)y(x) - ay(x-h) \\ &\quad + ah^2 \int_0^1 (1-w)\,[af(x+ahw) + f(x-hw)]\, dw. \end{aligned} \tag{4}$$

This equation provides $y(x+ah)$ given information (explicit or implicit) about y and f in the domain $[x-h, x+ah]$. To perform the indicated integrations, the functions $f(x+ahw)$ and $f(x-hw)$ in the integrands are replaced by backward difference schemes whose backpoints are equally spaced linear combinations of the stepsize: x, $x-h$, $x-2h$, ...

This replacement process begins with the definition of the backward difference operator ∇ defined via the relation

$$\nabla f(x) \triangleq f(x) - f(x-h).$$

For arbitrary β, the Taylor-series expansion for $f(x \pm \beta h)$ is the expression

$$f(x \pm \beta h) = e^{\pm \beta h D} f(x),$$

where D is the differential operator d/dx. Combining these two equations provides the standard relationship between ∇ and the exponential operator:

$$\nabla = 1 - e^{-hD}.$$

Consequently, we have the following binomial expansions

$$\begin{aligned} f(x+ahw) &= (1-\nabla)^{-aw} f(x), \\ &= \sum_{k\geq 0} \binom{-aw}{k} (-1)^k \nabla^k f(x), \\ f(x-hw) &= (1-\nabla)^{w} f(x), \\ &= \sum_{k\geq 0} \binom{w}{k} (-1)^k \nabla^k f(x), \end{aligned}$$

which when substituted in Eq. (4) gives the basic result

$$y(x+ah) = (1+a)y(x) - ay(x-h) + ah^2 \sum_{k\geq 0} \sigma_k \nabla^k f(x), \tag{5}$$

where

$$\sigma_k = (-1)^k \int_0^1 (1-w)\left[a\binom{-aw}{k} + \binom{w}{k}\right] dw. \tag{6}$$

Eq. (5) defines an arbitrary-order *predictor* for the special second-order system given in Eq. (1). Substituting $a = 1$ into this expression will give the explicit method of Störmer. Using a similar approach, a corresponding "Cowell-like" corrector could be constructed. However, if corrections are deemed unnecessary at every step, then they can be often times deferred sufficiently so that the "classical" Cowell corrector can be applied. This deferred-corrections approach has its own built-in efficiency and is demonstrated in the sequel.

Recursion Relation for the σ_k

Evaluation of the coefficients σ_k is most easily accomplished using the notion of a generating function $G(t)$. To this end, define

$$\begin{aligned} G(t) &= \sum_{k\geq 0} \sigma_k t^k, \\ &= \sum_{k\geq 0} (-t)^k \int_0^1 (1-w)\left[a\binom{-aw}{k} + \binom{w}{k}\right] dw. \end{aligned}$$

Noting that

$$(1-t)^{\pm s} = \sum_{k\geq 0} (-1)^k \binom{\pm s}{k} t^k,$$

allows $G(t)$ to be written as

$$G(t) = \int_0^1 (1-w)\left[a(1-t)^{-aw} + (1-t)^w\right] dw,$$

which upon integrating gives

$$\begin{aligned} G(t) &= \frac{1}{a \ln^2(1-t)}\left[(1-t)^{-a} - at - 1\right], \\ &= \sum_{k\geq 0} \sigma_k t^k. \end{aligned}$$

Consequently,

$$a \ln^2(1-t) \sum_{k\geq 0} \sigma_k t^k = (1-t)^{-a} - at - 1.$$

The recursion for the σ_k is found by equating the coefficients of like powers of t in this expression after substitution of the following expansions:

$$\begin{aligned} (1-t)^{-a} &= \sum_{j\geq 0} (-1)^j \binom{-a}{j} t^j, \\ \ln^2(1-t) &= 2 \sum_{j\geq 0} \frac{\alpha_{j+1}}{j+2} t^{j+2}, \end{aligned}$$

where in the latter equation, α_m is the mth partial sum of the harmonic series:

$$\alpha_m = 1 + \frac{1}{2} + \frac{1}{3} + \ldots + \frac{1}{m}.$$

Ultimately, this process produces,

$$\sum_{k=0}^{m} \frac{\alpha_{m-k+1}}{m-k+2}\sigma_k = \frac{1}{2(m+2)!}\prod_{k=1}^{m+1}(a+k),$$

which after a minor rearrangement gives

$$\sigma_m = \frac{1}{(m+2)!}\prod_{k=1}^{m+1}(a+k) - 2\sum_{k=0}^{m-1}\frac{\alpha_{m-k+1}}{m-k+2}\sigma_k. \tag{7}$$

The recursion begins with the initial coefficient

$$\sigma_0 = \frac{1}{2} + \frac{1}{2}a. \tag{8}$$

The above equation is easily evaluated for arbitrary index m and scale parameter a. Expressions for σ_m, $m = 0, 1, \ldots, 10$ have been constructed using our interactive symbolic processor PARSEC [3]. The results are displayed below in terms of the *scaled* coefficients σ_k^*:

$$\begin{aligned}
\sigma_0^* &= \frac{1}{2} + \frac{1}{2}a, \\
\sigma_1^* &= -\frac{1}{12} + \frac{1}{12}a^2, \\
\sigma_2^* &= -\frac{1}{48} + \frac{1}{24}a^2 + \frac{1}{48}a^3, \\
\sigma_3^* &= -\frac{1}{60} + \frac{1}{24}a^2 + \frac{1}{32}a^3 + \frac{1}{160}a^4, \\
\sigma_4^* &= -\frac{7}{320} + \frac{1}{16}a^2 + \frac{11}{192}a^3 + \frac{3}{160}a^4 + \frac{1}{480}a^5, \\
\sigma_5^* &= -\frac{107}{2688} + \frac{1}{8}a^2 + \frac{25}{192}a^3 + \frac{7}{128}a^4 + \frac{1}{96}a^5 + \frac{1}{1344}a^6, \\
\sigma_6^* &= -\frac{995}{10752} + \frac{5}{16}a^2 + \frac{137}{384}a^3 + \frac{45}{256}a^4 + \frac{17}{384}a^5 + \frac{5}{896}a^6 + \frac{1}{3584}a^7, \\
\sigma_7^* &= -\frac{6031}{23040} + \frac{15}{16}a^2 + \frac{147}{128}a^3 + \frac{203}{320}a^4 + \frac{49}{256}a^5 + \frac{25}{768}a^6 + \frac{3}{1024}a^7 + \frac{1}{9216}a^8, \\
\sigma_8^* &= -\frac{40187}{46080} + \frac{105}{32}a^2 + \frac{1089}{256}a^3 + \frac{3283}{1280}a^4 + \frac{6769}{7680}a^5 + \frac{35}{192}a^6 + \frac{23}{1024}a^7 + \frac{7}{4608}a^8 \\
&\quad + \frac{1}{23040}a^9, \\
\sigma_9^* &= -\frac{1129981}{337920} + \frac{105}{8}a^2 + \frac{2283}{128}a^3 + \frac{29531}{2560}a^4 + \frac{5607}{1280}a^5 + \frac{1069}{1024}a^6 + \frac{81}{512}a^7 \\
&\quad + \frac{91}{6144}a^8 + \frac{1}{1280}a^9 + \frac{1}{56320}a^{10}, \\
\sigma_{10}^* &= -\frac{1306707}{90112} + \frac{945}{16}a^2 + \frac{21387}{256}a^3 + \frac{58635}{1024}a^4 + \frac{4523}{192}a^5 + \frac{12825}{2048}a^6 + \frac{9039}{8192}a^7 \\
&\quad + \frac{525}{4096}a^8 + \frac{29}{3072}a^9 + \frac{9}{22528}a^{10} + \frac{1}{135168}a^{11}.
\end{aligned}$$

The σ_k are then found from the relationship

$$\sigma_k = \frac{2^k}{k!}\sigma_k^*. \tag{9}$$

The Algorithm

Eq. (5) is manipulated into a form which is appropriate for computational purposes. We first apply the formal definition of the ∇ operator k times to obtain

$$\nabla^k f(x) = \sum_{j=0}^{k}(-1)^j \binom{k}{j} f(x - jh).$$

This expansion applied to Eq. (5) gives

$$y(x + ah) = (1 + a)y(x) - ay(x - h) + ah^2 \sum_{k\geq 0}\sum_{j=0}^{k}(-1)^j \binom{k}{j} \sigma_k f(x - jh),$$

which is rewritten as the $m + 1$-step method

$$\boxed{y(x + ah) = (1 + a)y(x) - ay(x - h) + ah^2 \sum_{k=0}^{m} \tau_k f(x - kh),} \tag{10}$$

where

$$\begin{aligned}\tau_k &= (-1)^k \sum_{n=k}^{m} \binom{n}{k} \sigma_n, \\ &= (-1)^k \sum_{n=k}^{m} \frac{2^n}{k!(n-k)!}\sigma_n^*. \end{aligned} \tag{11}$$

The algebra indicated above was submitted to PARSEC, and the expansions for τ_k were obtained for all k, $k \leq 10$. These results are too extensive to be listed here. Instead, we list the 10 coefficients corresponding to a maximum $k = 9$. These coefficients in conjunction with Eq. (10) constitute the 10-step prediction procedure which was used in comparison trails on the planetary equations described in the next section:

$$\begin{aligned}\tau_0 = {} & \frac{13119671}{59875200} + \frac{1}{2}a + \frac{7129}{15120}a^2 + \frac{6515}{24192}a^3 + \frac{4523}{45360}a^4 + \frac{19}{768}a^5 + \frac{3013}{725760}a^6 \\ & + \frac{5}{10752}a^7 + \frac{29}{870912}a^8 + \frac{1}{725760}a^9 + \frac{1}{39916800}a^{10}, \\ \tau_1 = {} & \frac{3250433}{5322240} - \frac{3}{2}a^2 - \frac{4609}{3360}a^3 - \frac{14139}{22400}a^4 - \frac{7667}{43200}a^5 - \frac{7807}{241920}a^6 - \frac{11}{2880}a^7 \\ & - \frac{59}{207360}a^8 - \frac{11}{907200}a^9 - \frac{1}{4435200}a^{10}, \\ \tau_2 = {} & -\frac{17760409}{19958400} + 3a^2 + \frac{5869}{1680}a^3 + \frac{20837}{11200}a^4 + \frac{24901}{43200}a^5 + \frac{6787}{60480}a^6 + \frac{563}{40320}a^7 \\ & + \frac{7}{6480}a^8 + \frac{43}{907200}a^9 + \frac{1}{1108800}a^{10},\end{aligned}$$

Table 1

τ_k Coefficients for a Conventional 10-Step Störmer Method

τ_0	1153247/725760
τ_1	−1408913/453600
τ_2	7409783/907200
τ_3	−12642403/907200
τ_4	29850337/1814400
τ_5	−2460113/181440
τ_6	6973151/907200
τ_7	−2599333/907200
τ_8	328541/518400
τ_9	−8183/129600

$$\tau_3 = \frac{5150479}{3991680} - \frac{14}{3}a^2 - \frac{6289}{1080}a^3 - \frac{72569}{21600}a^4 - \frac{4013}{3600}a^5 - \frac{13873}{60480}a^6 - \frac{401}{13440}a^7 - \frac{31}{12960}a^8 - \frac{7}{64800}a^9 - \frac{1}{475200}a^{10},$$

$$\tau_4 = -\frac{3117517}{2217600} + \frac{21}{4}a^2 + \frac{6499}{960}a^3 + \frac{6519}{1600}a^4 + \frac{122249}{86400}a^5 + \frac{36769}{120960}a^6 + \frac{3313}{80640}a^7 + \frac{353}{103680}a^8 + \frac{41}{259200}a^9 + \frac{1}{316800}a^{10},$$

$$\tau_5 = \frac{44073661}{39916800} - \frac{21}{5}a^2 - \frac{265}{48}a^3 - \frac{1091}{320}a^4 - \frac{5273}{4320}a^5 - \frac{32773}{120960}a^6 - \frac{305}{8064}a^7 - \frac{67}{20736}a^8 - \frac{1}{6480}a^9 - \frac{1}{316800}a^{10},$$

$$\tau_6 = -\frac{89609}{147840} + \frac{7}{3}a^2 + \frac{6709}{2160}a^3 + \frac{84307}{43200}a^4 + \frac{10279}{14400}a^5 + \frac{9823}{60480}a^6 + \frac{313}{13440}a^7 + \frac{53}{25920}a^8 + \frac{13}{129600}a^9 + \frac{1}{475200}a^{10},$$

$$\tau_7 = \frac{1468913}{6652800} - \frac{6}{7}a^2 - \frac{967}{840}a^3 - \frac{4101}{5600}a^4 - \frac{2939}{10800}a^5 - \frac{3817}{60480}a^6 - \frac{373}{40320}a^7 - \frac{151}{181440}a^8 - \frac{19}{453600}a^9 - \frac{1}{1108800}a^{10},$$

$$\tau_8 = -\frac{2395}{49896} + \frac{3}{16}a^2 + \frac{3407}{13440}a^3 + \frac{1823}{11200}a^4 + \frac{10579}{172800}a^5 + \frac{3487}{241920}a^6 + \frac{347}{161280}a^7 + \frac{41}{207360}a^8 + \frac{37}{3628800}a^9 + \frac{1}{4435200}a^{10},$$

$$\tau_9 = \frac{1129981}{239500800} - \frac{1}{54}a^2 - \frac{761}{30240}a^3 - \frac{29531}{1814400}a^4 - \frac{89}{14400}a^5 - \frac{1069}{725760}a^6 - \frac{1}{4480}a^7 - \frac{13}{622080}a^8 - \frac{1}{907200}a^9 - \frac{1}{39916800}a^{10}.$$

As a partial test of the accuracy of the expansions, we verify that by setting the scaled parameter $a = 1$, the Störmer coefficients of Table 1 are obtained.

Table 2

β_k Coefficients for Conventional 10-Step and 16-Step Cowell Methods

	10-step	16-step
β_{-1}	3250433/53222400	111956703448001/2134124568576
β_0	3124027/3193344	11204909309477/10260214272
β_1	−57128921/159667200	−568165679865403/533531142144
β_2	16745741/19958400	191448070609867/533531142144
β_3	−88645069/79833600	−229681583662231/266765571072
β_4	42375577/39916800	217522529315683/133382785536
β_5	−2342533/3193344	−487426618987613/19760412672
β_6	7139837/19958400	2665769427531911/88921857024
β_7	−18674153/159667200	−2087227264749581/711374856192
β_8	1838819/79833600	615627443742221/266765571072
β_9	−330157/159667200	−7728394515471857/533531142144
β_{10}	...	34733145712199/48502831104
β_{11}	...	−727783498508069/266765571072
β_{12}	...	257826699983/33345696384
β_{13}	...	−655449026249/4268249137152
β_{14}	...	1696568778353/88921857024
β_{15}	...	−264713507083/237124952064

The correction process was based on conventional Cowell correctors written in the form

$$\boxed{y(x+h) = 2y(x) - y(x-h) + h^2 \sum_{k=-1}^{m} \beta_k f(x-kh).} \tag{12}$$

For the 10-step predictor, most of our integration trials were conducted using the limit $m = 9$ which required 11 coefficients β_k. These coefficients are listed in Table 2. Application of this corrector generally improved values by one or two orders of magnitude. However, alternate and separate k-step correctors (k arbitrary) can be applied to each y_i of the system. The use of distinct correctors on each dependent coordinate has not been investigated. However, we did utilize the 16-step corrector (Table 2) to improve the positions of the outer planets in some of our investigations with the planetary equations. The increased computational overhead was not significant, but the coding logic (bookkeeping) was more involved (and consequently harder to check and debug).

Because the method is not self-starting, a one-step method must be used to generate the starting backpoints. We elected to use the 10th-order Nyström method developed by E. Hairer [4] because its truncation error is nearly the same as that of the conventional 10-step Störmer procedure.

Table 3

Intra-Grid Divisions and Step Data for the 9 Planets

	anomalistic period (days)	divisions per step N	actual stepsize h corresponding to N (days)
Mercury	87.97	1	.2500
Venus	224.7	2	.5000
Earth[a]	365.2	4	1.000
Mars	686.9	8	2.000
Jupiter	4333.	10	2.500
Saturn	10740.	20	5.000
Uranus	30910.	25	6.250
Neptune	60360.	40	10.00
Pluto	91520.	50	12.25

[a] Earth + Moon

From the structure of Eq. (12), it is seen that the correction process is based solely on backpoints whose spacing in the time domain is determined by the individual stepsizes $h = h_i$. The stepsizes need not be related in any way to each other, however, it is convenient to select the h_i as multiples of *the* fundamental step h_1:

$$h_i = N_i h_1,$$

where from Eqs. (2) and (3), we define

$$h_1 = \frac{2\pi}{M_1 \omega_1}.$$

Here, N_i is a positive integer called the ***intra-grid division number***, and M_1 is an appropriately-selected real number.

Eqs. (10) and (12) constitute the P-stiff procedure. A variety of implementations as mentioned above is possible. The "best" strategy for a given system can be determined through numerical experiments. However, the production of codes needed to investigate alternatives is a laborious task that is highly error prone. A general purpose code encompassing alternative strategies has not been developed. Because of its assumed complexity, it would certainly contain significant overhead (which is one thing we strive to avoid).

APPLICATION TO THE PLANETARY EQUATIONS

The efficient and accurate integration of the rectangular, heliocentric equations of motion of the nine-body solar system was the principal motivation for development of the P-stiff procedure. Multimillion year simulations using standard multistep methods require

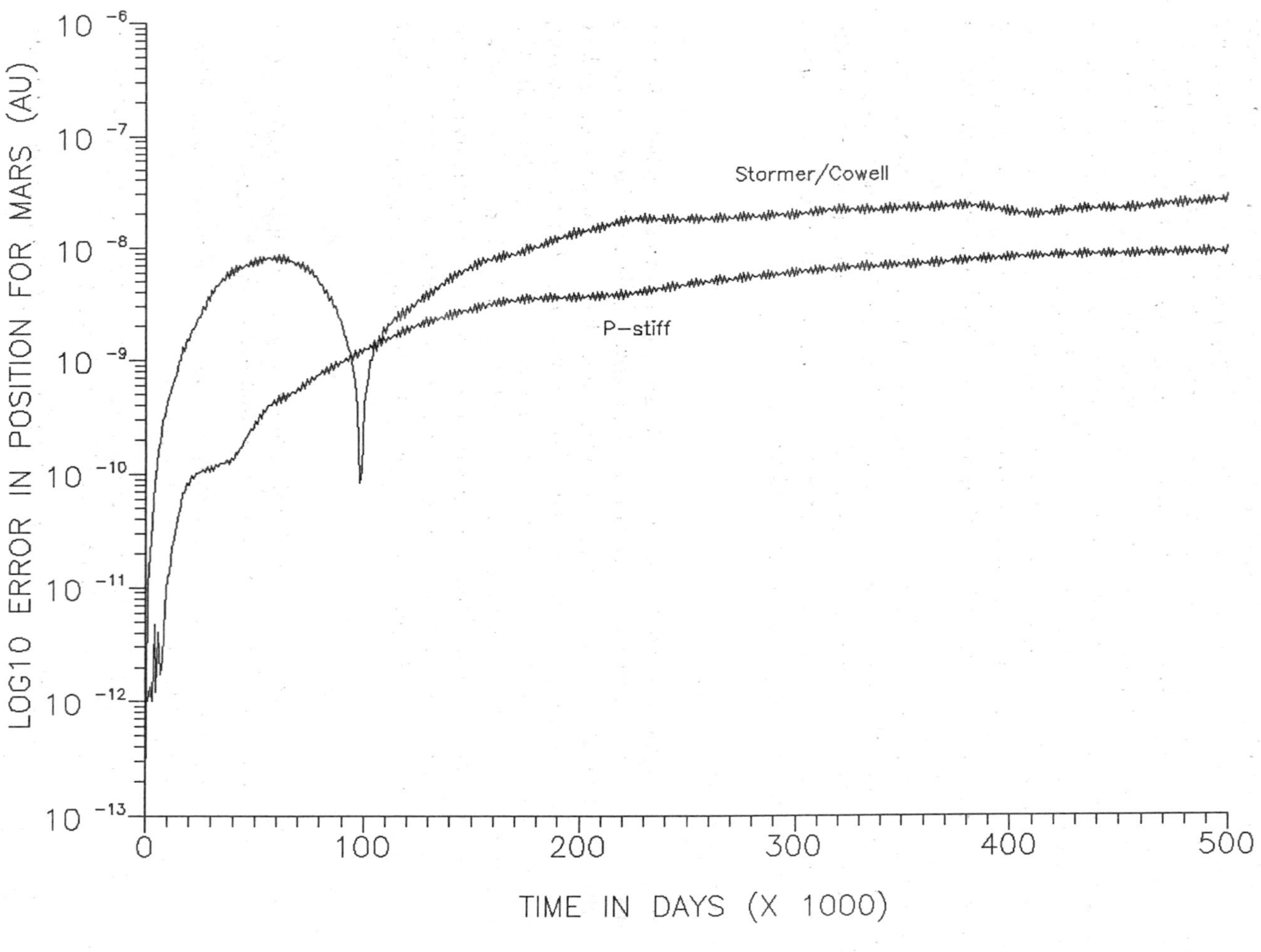

Fig. 1 Errors in the Position of Mars for 500,000-Day Simulations.

enormous computation times. Accompanying such long execution timings is the unacceptable build up of roundoff error in the planetary positions. We hoped to address both these difficulties with the P-stiff method.

The dual problems of efficiency and accuracy were assessed in comparison integrations between a conventional 10-step Störmer/Cowell and a 10-step P-stiff procedure. Positional data for 500,000 days were compared against data generated by an extended-precision integration using a 16-step Störmer/Cowell method. Initial data were taken from DE200 [5]. All "conventional" integrations used 1/4-day stepsizes. The fundamental step h_1 for the P-stiff procedure was also 1/4 day. Parameters pertinent to the P-stiff integration are displayed in Table 3. Positional error data were generated for all planets. The error in position for Mars is typical of these data for the inner planets and is shown in Fig. 1. It is seen that the P-stiff procedure produced errors that were approximately one order of magnitude smaller than those coming from Störmer/Cowell. The P-stiff execution time was approximately 26% less.

The numerical experiment described above was subsequently modified to use the 16-step Cowell corrector (Table 2) applied to the outer planets. The 10-step corrector was used on the inner planets. The resulting errors in position for the inner planets remained virtually the same, but the positional errors for the outer planets decreased by two to three orders of magnitude. P-stiff execution time increased about 1%.

CONCLUDING REMARKS

Preliminary numerical investigations of the P-stiff method show that it is both efficient and accurate. However, theoretical investigations of its error characteristics have not yet been conducted. It would be ideal to base the intra-nodal divisions for each dependent variable in accordance with a fixed percentage of its fundamental time scale. However, if the frequencies are very widely separated, then serious errors are introduced for the variables having the longest fundamental time scales. This condition is partially remedied by using Cowell correctors having greater step numbers. However, the practical limit of this approach is quickly reached. The problem of intra-nodal spacing is certainly related to frequency content and sample rates as dictated by the Sampling Theorem of spectral analysis.

REFERENCES

1. C.V. Forrington, "Extensions of the Predictor-Corrector Method for the Solution of Systems of Ordinary Differential Equations," *Computer J.*, **4**, 1961, pp. 80-84.

2. P. Henrici, *Discrete Variable Methods in Ordinary Differential Equations*, John Wiley, New York, 1962, pp. 289-343.

3. D.L. Richardson, "PARSEC: An Interactive Poisson Series Processor for Personal Computing Systems," *Celesial Mech.*, **45**, 1989, pp. 267-274.

4. E. Hairer, "A One-Step Method of Order 10 for $y'' = f(x, y)$," *IMA J. Num. Analysis*, **2**, 1982, pp. 83-94.

5. Standish, E.M., and Williams, J.G., *Development Ephemeris DE200-LE200*, Jet Propulsion Laboratory, Pasadena, 1982.

AAS 89-388

REGULARIZED LAGRANGE'S PLANETARY EQUATIONS FOR NUMERICAL USE

Daniel E. Snow*

Lagrange's Planetary Equations are developed for numerical integration for two different dynamical systems (Oblate Central Body and Circular Restricted Three Body Problems) which have conserved quantities (e.g. energy, Jacobian, etc.). Time transformations are used to remove singularities due to close approaches between the satellite and the attracting bodies and the existence of small divisors. The "regularized" systems are then integrated numerically for a highly inclined, highly eccentric orbit. The integration of these equations is compared to those of the cartesian position and velocity of the satellite on the basis of computer run time and the respective degradations of the conserved quantities. It is seen that using Lagrange's Planetary Equations improves the accuracy of the data generated while reducing the computer run time, and that these results are further improved by using the regularization. The performance of the various time transformations is then analyzed from an analytic viewpoint. Finally, sets of equations for the rate of change of the orbital elements using candidate time transformations are developed for the motion of the satellite as it experiences Solar, Lunar , and J_2 perturbations. These sets of equations are integrated and, as before, compared to those of the cartesian state vector. The results are shown to be in agreement with the analytical interpretations developed earlier. This study is intended to assist in the selection of time transformations to improve the accuracy of numerical integrations of systems of equations.

INTRODUCTION

Almost every research project will yield results which apply outside the original scope of the research project. Such is the case with this work. While the original goal of this research was to find a better way to generate data for a specific dynamic system with specific initial conditions, the end result was an improved understanding of the numerical integration of systems of equations and the differences between analytical and numerical integration. So while the method described here was developed for a specific system, it also has general applications.

---

* Aerospace Engineer, Space Division, Headquarters, U.S. Air Force Space Command/DOA, Peterson Air Force Base, Colorado 80914.

The specific system that is the object of this research is a satellite in a highly eccentric, highly inclined orbit around the Earth. Initially, the value of the semi-major axis of the spacecraft's orbit is approximately the mean Earth-Moon distance. The spacecraft experiences both lunar and solar perturbations, as well as perturbations due to the oblateness of the Earth. This dynamic model is referred to as the Solar-Lunar-J_2 (SLJ_2) model.

In an attempt to improve the numerical integration of this system, various analytic methods are used to modify the equations to be integrated. The spacecraft state is integrated using Lagrange's Planetary Equations for the orbital elements a,e,I,E, ω and Ω. Time transformations are also used to remove nearly singular terms in the equations. This "numerical regularization" is not used to regularize the equations in a purely analytic sense, but is used to remove terms which may cause difficulties during the numerical integration of the equations for the time rates of change of the orbital elements.

When one speaks of improving numerical integrations, both computer run time and accuracy of the data generated must be considered. While evaluation of computer time used is easy, a criterion needs to be established as to how to measure accuracy. To this end, constants of integration (for systems which have them) are used. Two somewhat simpler dynamical systems (the Circular Restricted Three Body Problem and the Oblate Central Body Problem) which have constants of integration are initially studied. These two systems are chosen not only because of the existence of constants of integration but also because during certain portions of this satellite's trajectory the dynamics is similar to one or the other of these systems. The results from these studies are used to understand and make decisions concerning the integration of the SLJ_2 model.

The Circular Restricted Three Body Problem

The Circular Restricted Three Body Problem is used in this study for several reasons. Preliminary studies showed the Lunar perturbation to be the most troublesome disturbance to the satellite's trajectory [1]. The Circular Restricted Three Body Problem is also useful because of the existence of a well known constant, the Jacobian Constant. For further information about the Circular Restricted Three Body Problem the reader is referred to Szebehely [2].

Lagrange's Planetary Equations for the set of orbital elements a, e, I, E, ω and Ω can be written in the general form

$$\dot{a} = -\frac{a}{r}(a, M_0)\frac{\partial R}{\partial E}$$

$$\dot{e} = -\frac{a}{r}(e, M_0)\frac{\partial R}{\partial E} - (e, \omega)\frac{\partial R}{\partial \omega}$$

$$\dot{I} = -(I, \omega)\frac{\partial R}{\partial \omega} - (I, \Omega)\frac{\partial R}{\partial \Omega}$$

$$\dot{E} = \frac{a}{r}\left(n - (M_0\,, a)\,\frac{\partial R}{\partial a} - (M_0\,,e)\,\frac{\partial R}{\partial e} \right)$$

$$\dot{\omega} = -\,(\omega,e)\,\frac{\partial R}{\partial e} - (\omega,I)\,\frac{\partial R}{\partial I} \qquad (1)$$

$$\dot{\Omega} = -\,(\Omega,I)\,\frac{\partial R}{\partial I}$$

where

$$(a, M_0) = \frac{-2}{na}$$

$$(e, M_0) = \frac{-(1 - e^2)}{n\,a^2\,e}$$

$$(e, \omega) = \frac{(1 - e^2)^{1/2}}{n\,a^2\,e}$$

$$(I, \omega) = \frac{-\cot I}{na^2\,(1 - e^2)^{1/2}}$$

$$(I, \Omega) = \frac{1}{na^2\,(1 - e^2)^{1/2}\,\sin I}$$

and by definition

$$(\alpha, \beta) = -\,(\beta, \alpha)$$

Note that the equations are formulated using the partial derivatives of the disturbing function (R) with respect to the various orbital elements. Furthermore, the disturbing function must be written with respect to a non-rotating coordinate system with the origin located at the body about which the initial two-body motion is assumed. In this coordinate system, the disturbing function is

$$R = k^2\,m'\,[\,(1/r_2) - (xx' + yy' + zz')\,/r'] \qquad (2)$$

where

m' = mass of the perturbing body
x', y', z', r' = position quantities of the perturbing body
x, y, z = position quantities of the satellite
r2 = distance from the perturbing body to the satellite

Since the disturbing function depends only on position, it is useful to express the partial derivative of R with respect to an orbital element α as

$$\frac{\partial R}{\partial \alpha} = \frac{\partial R}{\partial x}\frac{\partial x}{\partial \alpha} + \frac{\partial R}{\partial y}\frac{\partial y}{\partial \alpha} + \frac{\partial R}{\partial z}\frac{\partial z}{\partial \alpha} \qquad (3)$$

The partial derivatives of R with respect to the position components are simply the components of the perturbing forces in the respective directions. By defining the following quantities

$$B = \cos E - e$$

$$C = \sqrt{1 - e^2}\,\sin E$$

$$\vec{P} = P_x\hat{i} + P_y\hat{j} + P_z\hat{k}$$

$$\vec{Q} = Q_x\hat{i} + Q_y\hat{j} + Q_z\hat{k}$$

$$\vec{R} = R_x\hat{i} + R_y\hat{j} + R_z\hat{k}$$

$$P_x = \cos\omega\cos\Omega - \sin\omega\cos I\sin\Omega$$

$$P_y = \cos\omega\sin\Omega + \sin\omega\cos I\sin\Omega$$

$$P_z = \sin\omega\sin I$$

(4)

$$Q_x = -\sin\omega\cos\Omega - \cos\omega\cos I\sin\Omega$$

$$Q_y = -\sin\omega\sin\Omega - \cos\omega\cos I\cos\Omega$$

$$Q_z = \cos\omega\sin I$$

$$R_x = \sin I\sin\Omega$$

$$R_y = -\sin I\cos\Omega$$

$$R_z = \cos I$$

The position vector may be written as

$$\vec{r} = a(B\vec{P} + C\vec{Q}) \tag{5}$$

and the partial derivatives of the position components with respect to the orbital elements are

$$\frac{\partial x}{\partial a} = BP_x + CQ_x$$

$$\frac{\partial y}{\partial a} = BP_y + CQ_y$$

$$\frac{\partial z}{\partial a} = BP_z + CQ_z$$

$$\frac{\partial x}{\partial e} = -a\left[P_x + \frac{e}{(1-e^2)^{1/2}}\sin E\, Q_x\right]$$

$$\frac{\partial y}{\partial e} = -a\left[P_y + \frac{e}{(1-e^2)^{1/2}}\sin E\, Q_y\right]$$

$$\frac{\partial z}{\partial e} = -a\left[P_z + \frac{e}{(1-e^2)^{1/2}}\sin E\, Q_z\right]$$

$$\frac{\partial x}{\partial l} = a(B\sin\omega + C\cos\omega)R_x \tag{6}$$

$$\frac{\partial y}{\partial l} = a(B\sin\omega + C\cos\omega)R_y$$

$$\frac{\partial z}{\partial l} = a(B\sin\omega + C\cos\omega)R_z$$

$$\frac{\partial x}{\partial E} = a(-\sin E\, P_x + \sqrt{1-e^2}\,\cos E\, Q_x)$$

$$\frac{\partial y}{\partial E} = a(-\sin E\, P_y + \sqrt{1-e^2}\,\cos E\, Q_y)$$

$$\frac{\partial z}{\partial E} = a(-\sin E\, P_z + \sqrt{1-e^2}\,\cos E\, Q_z)$$

$$\frac{\partial x}{\partial \omega} = a(B\,Q_x - C\,P_x)$$

$$\frac{\partial y}{\partial \omega} = a(B\,Q_y - C\,P_y)$$

$$\frac{\partial z}{\partial \omega} = a(B\,Q_z - C\,P_z)$$

$$\frac{\partial x}{\partial \Omega} = a(BP_y + CQ_y)$$

$$\frac{\partial y}{\partial \Omega} = -a(BP_x + CQ_x)$$

$$\frac{\partial z}{\partial \Omega} = 0$$

Defining

$$\vec{F} \equiv \frac{\partial R}{\partial x}\hat{i} + \frac{\partial R}{\partial y}\hat{j} + \frac{\partial R}{\partial z}\hat{k} \qquad (7)$$

allows the dot product to be used to write the equations for the time rate of change of the orbital elements in a somewhat concise manner as follows:

$$\dot{a} = \frac{2}{n(1 - e\cos E)}\vec{F} \bullet \left\{-\sin E\,\vec{P} + \sqrt{1 - e^2}\,\cos E\,\vec{Q}\right\}$$

$$\dot{e} = \frac{(1 - e^2)^{1/2}}{na(1 - e\cos E)}\vec{F} \bullet \left\{-C\cos E\,\vec{P} + [\,1 + (B - e)\cos E]\,\vec{Q}\right\}$$

$$\dot{I} = \frac{1}{na(1 - e^2)^{1/2}\sin I}\vec{F} \bullet \left\{-C\cos I\,(B\,\vec{Q} - C\,\vec{P}) + (BP_y + CQ_y)\hat{i} - (BP_x\,CQ_x)\hat{j}\right\}$$

$$\dot{E} = \frac{1}{1 - e\cos E}\left\{n - \frac{1}{nae}\vec{F} \bullet [(2eB - (1 - e^2) - C^2)\,\vec{P} + C\cos E\,\vec{Q}]\right\} \qquad (8)$$

$$\dot{\omega} = \frac{1}{nae(1 - e^2)^{1/2}\sin I\,(1 - e\cos E)}\vec{F} \bullet \Big\{\sin I\,[(-C^2 - (1 - e^2)(1 - e\cos E))\,\vec{P} + BC\vec{Q}]$$

$$- e(1 - e\cos E)\cos I\,(B\sin\omega + C\cos\omega)\vec{R}\Big\}$$

$$\dot{\Omega} = \frac{1}{nae\,(1 - e^2)\sin I}\,\vec{F} \bullet (B \sin \omega + C \cos \omega)\vec{R}$$

While these equations could be numerically integrated in their current form, there exist terms which could make the time rate of change of the orbital elements become large. When this occurs, the data generated quickly loses accuracy. These "singularities" arise from two sources; the Lagrange brackets and the perturbing forces. The singularities due to these two sources shall be referred to as geometric and dynamic singularities, respectively. Geometric singularities tend to be functions of the satellite' s orbital elements such as sin I,

$\sqrt{1 - e^2}$, etc., whereas dynamic singularities tend to be distances between the satellite and the perturbing bodies. While most of the terms which are a cause of concern are terms appearing in the denominator of a fraction, every term appearing in a denominator is not a cause for concern. For example, since in the circular restricted three body problem the distance between the two primaries (r') is constant, the fraction 1/r' is not a singularity but is in fact simply 1.0.

To eliminate the anticipated problems, a new independent variable s is introduced through a differential relation between the real time t and the transformed time s. This differential relation is chosen in such a way that the terms which we anticipate will cause problems during the integration are eliminated. Note that since the system of equations we wish to regularize is of first order, changing independent variables is a much simpler matter than when the system of equations is of second order. The problem of regularizing our system of equations can be summarized as "find $F(a, e, I, E, r_2)$ such that the relation

$$dt = F(a, e, I, E, r_2)\,ds \qquad (9)$$

removes the numerical singularities from the system of equations of interest." If the time transformation

$$dt = a^{-3/2}\, e \sin I\, (1 - e^2)^{1/2}\, (1 - e \cos E)\, r_2^2\, ds \qquad (10)$$

is used, our new system of equations is

$$a' = \frac{2e(1 - e^2)^{1/2} \sin I}{\mu^{1/2}}\,\tilde{F} \bullet \left\{ -\sin E\, \vec{P} + \sqrt{1 - e^2}\, \cos E\, \vec{Q} \right\}$$

$$e' = \frac{e \sin I}{\mu^{1/2} a} \tilde{F} \bullet \left\{ -C \cos E \vec{P} + [1 + (B - e) \cos E] \vec{Q} \right\}$$

$$I' = \frac{e(1 - e \cos E)}{\mu^{1/2} a} \tilde{F} \bullet \left\{ \cos I (B \vec{Q} - C \vec{P}) + (BP_y + CQ_y)\hat{i} - (BP_x + CQ_x)\hat{j} \right\}$$

$$E' = \sqrt{1 - e^2} \sin I \left\{ \frac{\mu^{1/2} e}{a^3} - \frac{1}{\mu^{1/2} a} \tilde{F} \bullet [(2eB - (1 - e^2) - C^2)\vec{P} + C \cos E \vec{Q}] \right\}$$

$$\omega' = \frac{1}{\mu^{1/2}} \tilde{F} \bullet \left\{ \sin I [(-C^2 - (1 - e^2)(1 - e \cos E)) \vec{P} + B C \vec{Q}] - e(1 - e \cos E) \cos I \right.$$

$$\left. (B \sin \omega + C \cos \omega) \vec{R} \right\}$$

$$\Omega' = \frac{e(1 - e \cos E)}{\mu^{1/2} a} \tilde{F} \bullet \left\{ (B \sin \omega + C \cos \omega) \vec{R} \right\} \qquad (11)$$

$$t' = a^{3/2} e(1 - e \cos E) \sqrt{1 - e^2} \sin I \, r_2^2$$

where

$$\tilde{F} = -k^2 m_2 \left[\hat{r}_2 - \left(\frac{r_2}{r'} \right)^2 \hat{r}' \right]$$

$$\hat{r}_2 = \frac{(x - x')}{r_2} \hat{i} + \frac{(y - y')}{r_2} \hat{j} + \frac{z}{r_2} \hat{k}$$

$$\hat{r}' = \frac{x'}{r'} \hat{i} + \frac{y'}{r'} \hat{j}$$

Careful inspection of this new system of equations shows that the equations for the rate of change of the seven dependent variables in the fictitious time domain do not contain any of the numerical singularities which are of concern.

The numerical integrations were performed using the Cyber 175/750 system at the University of Texas. Test cases were run using different initial conditions, error tolerances, length of integration and integration methods. These differences produced little change in the relative results between the integrations of the two systems of equations.

Instead of discussing the results from every numerical integration individually, a typical case shall be examined to illustrate the merits of the method described earlier. The initial conditions of the test case are as follows:

(1) Orbital Elements of the Probe

semi-major axis, a	368,520,292.8 meters
eccentricity, e	0.94873885
inclination, I	78.15920325 degrees
longitude of node, Ω	131.63502778 degrees
argument of perigee, ω	334.83129531 degrees
mean anomaly at t_0, M_0	0.0 degrees

(2) Orbial Elements of the Moon

Semi-major axis, a	384,400,000.0 meters
eccentricity, e	0.0
inclination, I	5.14540 degrees
longitude of node, Ω	259.18328 degrees
argument of perigee, ω	244.76372 degrees
mean anomaly at t_0, M_0	0.0 degrees

To evaluate the results of the numerical integrations of the regularized Planetary Equations, the components of the Cartesian position and velocity vectors of the probe were also integrated for the same time span (three years). The CPU time used for this run was called 1.0 for comparison purposes. During this time, the Jacobian Constant (C) remained unchanged to six significant figures. When the regularized Planetary Equations were integrated, the CPU time used was approximately 0.70 and the Jacobian Constant remained unchanged to ten significant figures. These results are typical of all test cases run (210 test cases).

The Oblate Central Body Problem

For this study, we consider only the perturbations due to the oblateness of the Earth (no third body perturbations). Due to previous work [1], it has been found that for a satellite in this type of orbit the geopotential can be truncated after the J_2 term. Because of this truncation, there exist two constants of integration for this dynamical system; the specific energy and the z component of angular momentum.

The disturbing function for this system is

$$R = -\frac{\mu}{r}\frac{J_2}{2}\left[\frac{Re}{r}\right]^2\left[3\frac{z^2}{r^2} - 1\right] \tag{12}$$

and the components of the perturbing forces are

$$F_x = \frac{\mu x}{r^3} J_2 \left(\frac{3}{2}\right)\left(\frac{R_e}{r}\right)^2 \left(5\frac{z^2}{r^2} - 1\right)$$

$$F_y = \frac{\mu y}{r^3} J_2 \left(\frac{3}{2}\right)\left(\frac{R_e}{r}\right)^2 \left(5\frac{z^2}{r^2} - 1\right) \tag{13}$$

$$F_z = \frac{\mu z}{r^3} J_2 \left(\frac{3}{2}\right)\left(\frac{R_e}{r}\right)^2 \left(5\frac{z^2}{r^2} - 3\right)$$

By redefining the vector $\vec{F}$ for this section as

$$\vec{F} = F_x\hat{i} + F_y\hat{j} + F_z\hat{k} \tag{14}$$

with F_x, F_y, and F_z as stated in this section, equations (8) are valid for this dynamic system. The ability to make this simple substitution means that the geometric singularities are the same for both this system and the Circular Restricted Three Body Problem. The dynamic singularity in this case is $1/r$ as $r \Rightarrow 0$.

By defining the unit vector $\hat{r}$ as

$$\hat{r} = \frac{x}{r}\hat{i} + \frac{y}{r}\hat{j} + \frac{z}{r}\hat{k} = \tilde{x}\hat{i} + \tilde{y}\hat{j} + \tilde{z}\hat{k} \tag{15}$$

and using the time transformation

$$dt = a^{-5}\, e \sin I\, (1 - e^2)^{1/2}\, r^5\, ds \tag{16}$$

the regularized equations are

$$a' = \frac{2e(1 - e^2)^{1/2} \sin I}{\mu^{1/2} a^{7/2}} \tilde{F} \bullet \left\{-\sin E\, \vec{P} + \sqrt{1 - e^2}\, \cos E\, \vec{Q}\right\}$$

$$e' = \frac{e(1 - e^2) \sin I}{\mu^{1/2} a^{9/2}} \tilde{F} \bullet \left\{- C \cos E\, \vec{P} + [\, 1 + (B - e) \cos E]\, \vec{Q}\right\}$$

$$I' = \frac{e(1 - e^2) \sin I}{\mu^{1/2} a^{9/2}} \tilde{F} \bullet \left\{\cos I\, [B\, \vec{Q} - C\, \vec{P}] + (BPy + CQy)\, \hat{i} - (BPx - CQx)\, \hat{j}\right\}$$

$$E' = \sqrt{1 - e^2}\, \sin I \left\{ a^{3/2} e(1 - e \cos E) - \frac{1}{\mu^{1/2} a^{9/2}} \tilde{F} \bullet [(2eB - (1 - e^2) - C^2)\, \vec{P} + C \cos E\, \vec{Q}\,] \right\} \tag{17}$$

$$\omega' = \frac{1}{\mu^{1/2} a^{9/2}} \tilde{F} \bullet \left\{\sin I\, [(-C^2 - (1 - e^2)(1 - e \cos E))\, \vec{P} + B\, C\, \vec{Q}] - e(1 - e \cos E) \cos I\, (B \sin \omega + C \cos \omega) \vec{R} \right\}$$

$$\Omega' = \frac{e(1 - e^2) \sin I}{\mu^{1/2} a^{9/2}} \tilde{F} \bullet \{B \sin \omega + C \cos \omega\}\, \vec{R}$$

$$t' = e(1 - e \cos E)^5\, (1 - e^2)^{1/2} \sin I$$

where

$$\tilde{F} = \mu J_2 R_e^2 \left(\frac{3}{2}\right) [(5\tilde{z} - 1)\tilde{x}\hat{i} + (5\tilde{z} - 1)\tilde{y}\hat{j} + (5\tilde{z} - 3)\tilde{z}\hat{k}]$$

The numerical work for this section used satellites with the same initial conditions as in the Circular Restricted Three Body Problem. Once again, the Cartesian position and velocity vectors were integrated for three years, and the CPU time required to generate the data was chosen as the unit of time for comparison of the numerical integrations. During the integrations of the Cartesian state vector, the energy stayed constant to eight significant digits and the z component of momentum stayed constant to twelve digits. This difference has been explained by Brouwer [3]. When the time transformation involving $(1 - e \cos E)^5$ was used, the accuracy of the data increased slightly but the CPU time used increased by forty percent. To try to understand why the computer time used increased, integer powers of (1 - e cos E) from negative two to positive six, inclusive, were used in time transformations to derive new sets of equations. These equations were also integrated to see if performance was improved. It was found that by using the time transformation involving (1 - e cos E) to the first power the accuracy was the same as the first set of regularized equations mentioned, but the CPU time used was 0.5.

Analytic Explanation

Before proceeding further, some understanding of the results stated previously is required. This understanding will prove useful in choosing time transformations for the SLJ_2 model. We begin by examining integrations of the two-body problem.

Suppose we choose to integrate the equations for the time rate of change of the variables a, e, I, E, ω, Ω, and t with time as the independent variable. Most of these equations would be trivial to handle, since a, e, I, ω, and Ω are constants and (obviously) time varies linearly with time. The only variable left to consider is the eccentric anomaly. Since

$$\frac{dE}{dt} = \frac{na}{r} \qquad (18)$$

this equation can be rewritten as a Fourier series in time

$$\frac{dE}{dt} = n\left\{1 + 2\sum_{k=1}^{\infty} J_k(Ke)\cos\left\{K\left[n(t - t_0) + M_0\right]\right\}\right\} \qquad (19)$$

where J_k is a Bessel function.

If we numerically integrated this expression, the step size will be determined by the highest frequency term which has a non-negligible coefficient. There are two reasons for this. First, since computers have a finite word length, terms with extremely small coefficients will not contribute to dE/dt. Also, during the integration there will occur a differencing process to determine if error tolerances are met. Due to D'Alembert's principle, a small change in t will produce changes in the low frequency terms of dE/dt which are larger than the amplitudes of the high frequency terms. Therefore, the difference obtained will tend to be dominated by the low frequency terms. Because of these two facts, there will exist terms whose amplitudes are large enough and whose frequencies are high enough that they will consistently make significant contributions to the differences used in determining if error tolerances are met Therefore, these terms will drive the selection of the optimal step size.

Next, consider integrating the equations for the rate of change of the variables a, e, I, E, ω, Ω and t with eccentric anomaly as the independent variable. In this case, all of the equations are trivial, with the most difficult equation being

$$\frac{dt}{dE} = \frac{1}{n}(1 - e\cos E) \tag{20}$$

an equation having only one frequency. We would expect the second example to be integrated faster and more accurately, and indeed this is the case.

We now consider the perturbed two-body system, in particular J_2 and lunar perturbations for a satellite with the initial conditions used earlier. For the case where the satellite experiences J_2 perturbations, all of the orbital elements will now have terms with varying frequencies. However, these terms all have a 10^{-3} factor included in the amplitude and so these terms will have a limited effect upon the step size selection. Essentially, this means that the two-body effects are much larger than the J_2 effects throughout the range of the satellite's motion. For the case where the satellite experiences lunar perturbations, the Earth's two-body effects upon the satellite do not dominate the motion throughout the region. Since the lunar "perturbations" can be the same order of magnitude (and often are) as the Earth's two-body force, the Moon-satellite distance needs to be included in the time transformation. This method of considering only the zeroth order terms in the regularization is similar to what has been done in previous work [4, 5] and, more recently, has been used by Dr. Heggie in his numerical integrations of the many-body problem [6].

The Solar - Lunar - J2 (SLJ2) Model

Previous analysis has shown that the motion of this satellite could be accurately modeled by including solar and lunar perturbations along with perturbations due to the J_2 term of the Earth's geopotential. A further simplification is to assume that the Sun and Moon move in circular orbits with respect to the Earth. This simplification saves a great deal of computer time and does not affect the accuracy of the data generated to an appreciable level for the length of time of the integrations (three years). This is the model to be studied in this section (hereafter referred to as the SLJ2 model).

Due to the assumptions made, no constants of integration exist for this dynamical system. This is the reason for examining the two previous systems instead of beginning with the SLJ2 model. The Circular Restricted Three Body Problem shall also be further utilized during the analysis of the SLJ2 model.

The coordinate system used is the Earth centered Ecliptic system, with the X-axis pointing in the direction of the vernal equinox, the Z-axis pointing to the celestial north pole, and the Y-axis chosen such that $\hat{i} \times \hat{j} = \hat{k}$. In this system the partial derivatives with respect to the coordinate are

$$\frac{\partial R}{\partial x} = -k^2\left(m_2\left[\frac{x - x_2}{r_2^3} + \frac{x_2}{L_2^3}\right] + m_3\left[\frac{x - x_3}{r_2^3} + \frac{x_3}{L_3}\right]\right) + \frac{\mu x}{r^3} J_2\left(\frac{3}{2}\right)\left(\frac{R_e}{r}\right)^2 \left(5\frac{(y \sin\varepsilon + z\cos\varepsilon)^2}{r^2} - 1\right) \tag{21}$$

$$\frac{\partial R}{\partial y} = -k^2\left(m_2\left[\frac{y - y_2}{r_2^3} + \frac{y_2}{L_2^3}\right] + m_3\left[\frac{y - y_3}{r_2^3} + \frac{y_3}{L_3}\right]\right) + \frac{\mu}{r^3} J_2\left(\frac{3}{2}\right)\left(\frac{R_e}{r}\right)^2 \left\{(y\cos\varepsilon - z\sin\varepsilon)\left(5\frac{(y\cos\varepsilon - z\sin\varepsilon)^2}{r^2} - 1\right)\cos\varepsilon - (y\sin\varepsilon + z\cos\varepsilon)\left(5\frac{(y\sin\varepsilon + z\cos\varepsilon)^2}{r^2} - 3\right)\sin\varepsilon\right\}$$

$$\frac{\partial R}{\partial z} = -k^2\left(m_2\left[\frac{z-z_2}{r_2^3}+\frac{z_2}{L_2^3}\right]+m_3\left[\frac{z-z_3}{r_2^3}+\frac{z_3}{L_2^3}\right]\right)+\frac{\mu}{r^3}J_2\left(\frac{3}{2}\right)\left(\frac{R_e}{r}\right)^2$$

$$\left\{(y\cos\varepsilon - z\sin\varepsilon)\left(5\frac{(y\cos\varepsilon - z\sin\varepsilon)^2}{r^2}-1\right)\sin\varepsilon + (y\sin\varepsilon + z\cos\varepsilon)\right.$$

$$\left.\left(5\frac{(y\sin\varepsilon + z\cos\varepsilon)^2}{r^2}-3\right)\cos\varepsilon\right\}$$

By defining

$$\vec{F} = \frac{\partial R}{\partial x}\hat{i} + \frac{\partial R}{\partial y}\hat{j} + \frac{\partial R}{\partial z}\hat{k} \tag{22}$$

equations (8) are valid for this system.

The singularities present in the equations are the geometric singularities discussed earlier and the following dynamical singularities:

i) $r \Rightarrow 0$

ii) $r_2 \Rightarrow 0$

iii) $r_3 \Rightarrow 0$.

The r_3 singularity has the physical meaning of the satellite approaching the Sun. Since the problem has lost all practical significance when the satellite has left Earth orbit (for the purposes of this study), r_3 going to zero will not be considered a singularity. (In fact, r_3 is essentially constant throughout the integrations.) The two other dynamical singularities can indeed cause problems.

To study the effect of various time transformations on the numerical integration of this dynamical system, the Planetary Equations were developed with time transformations in which the integer power of the Moon-probe distance involved varied from negative one to positive three, inclusive. This was done for both the SLJ_2 and the circular restricted three body problem dynamic models. The reason for this is that the optimal time transformation varies with the model studied. Since the equations are first order, the Planetary Equations (after the application of the time transformation) can in all the above mentioned cases be written as

$$a' = \frac{2e(1 - e^2)^{1/2} \sin I}{\mu^{1/2}} \tilde{F}_n \bullet \left\{- \sin E \vec{P} + \sqrt{1 - e^2} \cos E \vec{Q}\right\}$$

$$e' = \frac{e \sin I}{\mu^{1/2} a} \tilde{F}_n \bullet \left\{-C \cos E \vec{P} + [1 + (B - e) \cos E] \vec{Q}\right\} \qquad (23)$$

$$I' = \frac{e(1 - e \cos E)}{\mu^{1/2}} \tilde{F}_n \bullet \left\{\cos I (B\vec{Q} - C\vec{P}) + (BP_y + CQ_y)\hat{i} - (BP_x + CQ_x)\hat{j}\right\}$$

$$E' = \sqrt{1 - e^2} - \sin I \left\{\frac{\mu^{1/2}}{a^3} - \frac{1}{a\mu^{1/2}} \tilde{F}_n \bullet [(2eB - (1 - e^2) - C^2)\vec{P} + C \cos E \vec{Q}]\right\}$$

$$\omega' = \frac{1}{a\mu^{1/2}} \tilde{F}_n \bullet \left\{\sin I [(-C^2 - (1 - e^2)(1 - e \cos E))\vec{P} + BC\vec{Q}] - e(1 - e \cos E) \cos I (B \sin \omega + C \cos \omega)\vec{R}\right\}$$

$$\Omega' = \frac{e(1 - e \cos E)}{a\mu^{1/2}} \tilde{F}_n \bullet (B \sin \omega + C \cos \omega)\vec{R}$$

$$t' = a^{-3/2} e(1 - e \cos E) \sqrt{1 - e^2} \sin I \, r_2^n \equiv G_n$$

where

$$\tilde{F}_n = r_2^n \left[\frac{\partial R}{\partial x}\hat{i} + \frac{\partial R}{\partial y}\hat{j} + \frac{\partial R}{\partial z}\hat{k}\right] \qquad (24)$$

and, as before, the prime stands for rate of change with respect to the new independent variable. Note that since the geometric singularities are the same as in the circular restricted three body problem, the portion of the time transformation involving these singularities is the same.

For the numerical testing, two schemes were devised; one involving local error, the other global. For the local error test, all the computer problems involving the equations for the SLJ2 model were run with the same value of error tolerance and the run times were compared. The values varied from 1.0 x 10^{-09} to 1.0 x 10^{-13}, and fifteen different sets of initial conditions for the satellite were used at each value of the error tolerance. Normalizing by the run time of the problem which integrates cartesian position and velocity, the results are summarized in Table 5-1.

As can be seen from the table, the time transformation involving the Moon-probe distance to the first power (G_1) produced the best results, followed by the time transformation which did not involve this distance. The relative results did not vary significantly between the various values of the error tolerance.

The global error test was developed because Shampine and Gordon state that ODE's local error tolerance does not directly control the growth of global error (ODE was the numerical integrator used for this portion of the testing). Since in most applications the user is more interested in global error, it was felt that this test was necessary and relevant. The first step in the test was to choose a value for the error tolerance, run the circular restricted three body problem model which integrates cartesian position and velocity, and note the change in the Jacobian. This change in the Jacobian is a measure of global error. Then for each time transformation, a value of the error tolerance was found which produced approximately the same change in the Jacobian when the transformed Planetary Equations were integrated. These values were used to integrate the various sets of equations for the SLJ_2 model. This model assumed that since the global error accumulated in integrating the various sets of equations in the circular restricted three body problem was approximately the same, the global error accumulated in integrating the various sets of equations for the SLJ_2 model was also approximately the same. While this assumption could not be proven, it does not seem unreasonable.

During this testing, several facts were learned about using time transformations and Lagrange's Planetary Equations for the integration of the circular restricted three body problem. These are as follows:

1. The "best" time transformation for this system was the G_2 transformation. This was the one used earlier. This time transformation produced the shortest run times and the least amount of global error for a given value of the error tolerance.

2. The G_{-1} and G_3 time transformations were the worst, performing poorer than even the cartesian integrations in all aspects.

3. In the integrations of the Planetary Equations which used the G_0, G_1 or G_2 time transformations, the value of the error tolerance needed to match the global error seen in the integration of the cartesian coordinates was typically ten to one hundred times larger than that used in the integration of the cartesian coordinates.

4. As a consequence of 3, integrations of the Planetary Equations were capable of far smaller values of global error, by two or three orders of magnitude, than integrations of the cartesian coordinates.

The values of the error tolerance used in the integrations of the cartesian coordinates varied from 1.0×10^{-9} to 1.0×10^{-13} and fifteen different sets of initial conditions for the satellite were tested at each value of the error tolerance. The results (normalized by the run time of the corresponding cartesian position and velocity integration) are summarized in Table 5-2.

Table 5-1

Local Error Test Results For SLJ_2 Model

Model	Mean Run Time	Range of Run Time
Cartesian	1.0	1.0-1.0
Planetary, G_{-1}	1.4	1.34-1.46
Planetary, G_0	.90	.87-.93
Planetary, G_1	.85	.83-.87
Planetary, G_2	1.08	1.04-1.11
Planetary, G_3	1.57	1.52-1.62

Table 5-2

Global Error Test Results for SLJ_2 Model

Model	Mean Run Time	Range of Run Time
Cartesian	1.0	1.0-1.0
Planetary, G_0	.76	.69-.87
Planetary, G_1	.69	.65-.80
Planetary, G_2	.86	.79-.98

NOTE: Study of G_{-1} and G_3 time transformations were discontinued due to poor initial results.

In both the local and global error tests for SLJ_2, the time transformation involving the Moon-satellite distance to the first power produced the best results, followed by the time transformation which did not involve this distance. This result differs from that of the study of the Circular Restricted Three Body Problem, the reason being that the addition of Solar perturbations considerably reduces the space in which the Lunar perturbations have a significant effect upon the choice of stepsize used in the integration of the satellite's trajectory. Therefore it is advantageous to use a smaller power of the Moon-satellite distance in the time transformation, since for much of the integration this force acts as a perturbation and not as a "zeroth order term".

Conclusions

The Regularized Lagrange's Planetary Equations discussed are useful for special perturbation studies of satellites in highly inclined, highly eccentric orbits with large semi-major axes. The techniques used to develop the equations used within may prove useful in developing equations to study other classes of satellite orbits. In particular, the "regularization" of zeroth order terms (where such terms are well defined) should prove to be useful in later studies of different classes of satellite orbits.

Acknowledgements

The author acknowledges Air Force Space Command for its support of this project. Also, the author thanks his colleagues in AFSPACECOM/DOA, in particular Dr. Joseph J.F. Liu, for their constructive critiques of this work. Finally, the author thanks Mrs. Chris Nutting for her editorial assistance.

References

1. Segrest, J.F., and Liu, J.J.F., "Preliminary Numerical Analysis of a 27-Day Orbit," Analysis Memo 86-2, Directorate of Astrodynamics, HQ AFSPACECOM, PAFB CO.

2. Szebehely, V.: Theory of Orbits: The Restricted Problem of Three Bodies. Academic Press, Orlando, Florida, 1967

3. Brouwer, D., 1937, "On the Accumulation of Errors in Numerical Integration," Astron, J., 46, 149-153

4. Stiefel, E. and Scheifele, G.: Linear and Regular Celestial Mechanics. Springer - Verlag, New York, 1971.

5. Saari, D., "Regularization and the Artificial Earth Satellite Problem," Celes. Mech. 9, 55.

6. Heggie, D.: "The N-Body Problem in Stellar Dynamics," Long-Term Dynamical Behaviour of Natural and Artificial N-Body Sytems (A. Roy, ed.), Kluwer, Dordrecht, 1988.

AAS 89-389

ON SOLVING KEPLER'S EQUATION*

L. G. Taff† and T. A. Brennan‡

Intrigued by the recent advances in research on solving Kepler's equation, we have attacked the problem too. Our contributions emphasize the unified derivation of all known bounds and several starting values, a proof of the optimality of these bounds, a very thorough numerical exploration of a large variety of starting values and solution techniques in both mean anomaly/eccentricity space and eccentric anomaly/eccentricity space, and finally the best and simplest starting value/solution algorithm; $M + e$ and Wegstein's secant modification of the method of successive substitutions. The very close second is Broucke's bounds coupled with Newton's second-order scheme.

INTRODUCTION

There has been a recent burst of activity on optimal methods of solving Kepler's equation. We have three items to add to this endeavor. First, we rederived Smith's (Ref. 1) and Broucke's (Ref. 2) starting values and bounds from Serafin's (Ref. 3) perspective [an earlier version of this work was completed in 1985; Taff and Sayer (Ref. 4)]. One can generalize and show that all known bounds and many starting values follow from this method. This unification is easily seen to be optimal too. Thus, apparently no better bounds nor closer starting values can be found. Serafin (Ref. 3) does not stress this. Second, we not only explore starting values and solution techniques as a function of mean anomaly and eccentricity, but as a function of eccentric anomaly and eccentricity too. Our numerical trials are also much more

---

* This work, performed at the Massachusetts Institute of Technology's Lincoln Laboratory, was sponsored by the Department of the Air Force under Contract F19628-85-C-0002. The views are those of the authors and do not reflect the official policy or position of the U.S. Government.

† Dr. Taff is now the Instrument Scientist for the Fine Guidance Sensors on the Hubble Space Telescope. He is at the Space Telescope Science Institute, 3700 San Martin Drive, Baltimore, Maryland 21218.

‡ Theresa Brennan has returned to school at the Massachusetts Institute of Technology.

extensive than others have reported. Finally, we report the simplest, most efficient method yet for solving Kepler's equation; Wegstein's (Ref. 5) iteration coupled with a starting guess of $M + e$ for E.

THE PROBLEM

Kepler's equation is a transcendental equation for eccentric anomaly E as a function of mean anomaly M with parameter equal to the eccentricity e, viz.,

$$f(E) = E - e \sin E - M = 0 \tag{1}$$

Because of the properties of the sine function (for real arguments), only the ranges

$$M \in [0, \pi] \quad \text{and} \quad e \in [0, 1)$$

need be considered; $e = 0$ is trivial and $e = 1$ is a different problem (*i.e.*, Barker's equation—a cubic in the tangent of $E/2$). Note too that for $M = 0$ or π, $E = 0$ or π also. We remind the reader that

$$dE/dM = 1/(1 - e \cos E) > 0$$

This enhances the stability of many iterative solution techniques.

STARTING VALUES AND BOUNDS

Broucke (Ref. 2) provides an explanation of the tangent-chord method apparently used by Smith (Ref. 1);

$$\begin{aligned} (\pi/2)M/(\pi/2 - e) < E < M + e, &\quad \text{for } M \in [0, \pi/2 - e] \\ (M/2 + e)\pi/(\pi/2 + e) < E < M + e, &\quad \text{for } M \in [\pi/2 - e, \pi] \end{aligned} \tag{2}$$

Before we read Broucke's paper, and with Smith providing no hints, we set out to derive these results. As Serafin (Ref. 3) has since pointed out, they follow from the inequalities

$$1 \geq \frac{\sin z}{z} \geq 2/\pi, \quad z \in [0, \pi/2] \tag{3}$$

and the form of Kepler's equation (1).

To see this set $u = \pi M/(\pi - 2e)$. Then, from the range of M given by the top line of (2) it follows that $u \in [0, \pi/2]$. As

$$f(u) = e(2u/\pi - \sin u) \leq 0$$

the veracity of the lower bound for this range of M immediately follows from the right-hand inequality in (3), the fact that $f(M+e) = e[1-\sin(M+e)] \geq 0$, and the continuity of f. Similarly, if $v = \pi(\pi - M)/(\pi + 2e)$ for $M \in [\pi/2 - e, \pi]$, then $v \in [0, \pi/2]$ and

$$f(v) = e(2v/\pi - \sin v) \leq 0$$

Using an identical chain of reasoning completes the verification of the bottom line of (2). Note that the only other fact used are the inequalities

$$1 \geq \sin x \geq -1 \qquad \text{for all real } x \tag{4}$$

A generalization of this derivation will also yield Broucke's bounds. Finally, we note that the upper bound of $M/(1-e)$ for E also follows from (1) and (3). Similarly, all other upper and lower bounds known to us are consequences of inequalities (3) and (4). Whence, all non-trivial bounds and starting values may be deduced from a unified theoretical perspective.

Inequalities (3) and (4) are *optimal* in the sense that both the greatest lower and least upper bounds are actually attained in the interval of interest. Hence, the possibility of the equal sign is real and no better (*i.e.*, tighter) bounds can be found. Since Kepler's equation (1) does not inspire other functional forms to investigate, it appears that this is the best that can be done.

Another consequence of the inequality (3) is the starting guess

$$E = M/(1 - 2e/\pi) \tag{5}$$

This is a lower bound to the solution when $M \in [0, \pi/2 - e]$ and an upper bound when $M \in [\pi/2 - e, \pi]$. When it is a lower bound it is also the greatest lower bound amongst 0, M, $M/(1+e)$, and itself. Of the upper bounds of π, $M+e$, and $M/(1-e)$ it always exceeds π and $M+e$ while it is always less than $M/(1-e)$.

The starting value (5) has three other interesting properties. Since,

$$\langle \sin E \rangle = (1/\pi)\int_0^\pi \sin E \, dE = (1/\pi)\int_0^\pi \sin E \, dM = 2/\pi$$

it is the average "best" starting value. In the same context it is the minimum mean square value for

$$\Sigma^2(k) = (1/\pi)\int_0^\pi (\sin E - k)^2 dE = (1/\pi)\int_0^\pi (\sin E - k)^2 \, dM$$

is a minimum for $k = 2/\pi$. The last point is how poor it is in practice (see Tables 1, 2, and 3)!

Two other types of starting bounds are found in the literature. One is a consequence of an exercise in Battin's (Ref. 6) book,

$$E = M + \alpha(1 - \alpha^2/2), \quad \alpha = e \sin M/(1 - e \cos M) \tag{6}$$

This has been advocated because it matches the Fourier series for $E(M)$ through all terms in e^3. The formula (6) is only one of infinitely many analytic approximations to the $E(M)$ Fourier series through cubic terms in the eccentricity. It also suffers from undesirable behavior as $e \to 1^-$ and $M \to 0^+$. Thus, we have rejected it as a realistic starting value because it is neither unique nor a priori good. Similarly, the Hermite interpolation formulas for an $E(M)$ starting guess all suffer from a singularity at $e = 1$, $M = 0$.

We made an effort to find a very good approximation of the form

$$E = M + eg(e, M)$$

A glance at Fig. 1 shows that the eccentricity dependence of g is muted, g is asymmetrical with respect to $M = \pi/2$, and g is nearly parabolic in M. This suggests

$$g(e, M) = \frac{(2\pi/3)^{3/2} M(\pi - M)^{3/2}}{2} \tag{7}$$

where $g(\pi/3) = 1$ has been enforced. This formula fares poorly compared to $g = 1$ (*i.e.*, $E = M + e$)!

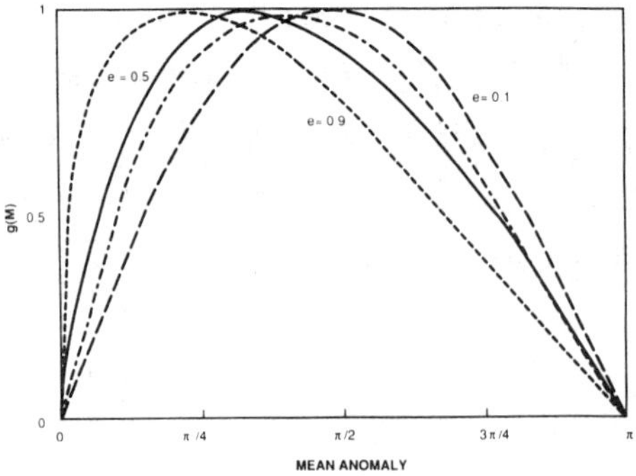

Fig. 1. Curves of $g(e, M)$ for eccentricity 0.1 (long dashes), $e = 0.5$ (dot-dash), $e = 0.9$ (short dashes), and the real curve (solid).

MATRIX OF NUMERICAL SOLUTIONS

We have used a uniform spacing of $\Delta M = \pi/151$ rad and $\Delta e = 1/151$ to span M, e space. We have also used a uniform spacing of $\Delta E = \pi/151$ rad and $\Delta e = 1/151$ to cover E, e space. Our electronic computer has a precision of 11 decimal digits.

Seven different methods of solution have been investigated. They were (1) Newton's first-order method

$$E_n = E_o - f(E_o)/f'(E_o)$$

wherein E_o is the old value of the eccentric anomaly and E_n is the new one, (2) Newton's second-order method (in the correct form)

$$E_n = E_o - f(E_o)/[f'(E_o) - f(E_o)f''(E_o)/2f'(E_o)] \tag{8}$$

(3) the method of successive substitutions, (4) Steffensen's method as suggested by Danby and Burkhardt (Ref. 7) using Isaacson and Keller's (Ref. 8) formulation to speed the convergence of the method of successive substitutions, (5) Wegstein's iteration method (Ref. 5), (6) successive linear interpolation between upper and lower bounds, and (7) Mueller's successive bisection, inverse parabolic interpolation between upper and lower bounds (see Ref. 9).

Thirteen different starting values have been evaluated with the one-point solution methods. Four were based on the zero-, first-, second-, and third-order Taylor series for $E(M)$. The next quadruplet were simple functions of M; $E = \pi, E = M + e, E = M/(1 - 2e/\pi)$, and Eq. (7). Two more were based on Smith's 1979 paper (Ref. 1); namely, the lower bound of M, the upper bound of $M + e$, and linear interpolation to the root [*i.e.*, his Eq. (5)] and his more refined limits [his Eqs. (9) and (11)] coupled with linear interpolation to the root. Another one utilizes Broucke's (Ref. 2) generalization [*i.e.*, his Eqs. (5)] of Smith's bounds followed by linear interpolation. Finally, we tried the $(6M)^{1/3}$ suggestion of Ng (Ref. 10) and the values in Serafin's (Ref. 3) Table I plus linear interpolation.

For the solution techniques of inverse linear or quadratic interpolation utilizing upper and lower bounds to bracket the solution, a quintet of bound pairs was investigated. These were 0 and π, M and $M + e$, the couplet of bounds given by Smith depending upon the values of the mean anomaly and eccentricity, the quartet of bounds given by Broucke depending upon the values of M and e, and the greatest lower bound and the least upper bound from the standard set [*e.g.*, 0, M, $M/(1+e)$, $M/(1 - 2e/\pi)$, $M + e$, $M/(1 - e)$, and π].

Each starting value/solution method was executed for the 22,500 E and e pairs and the 22,500 M and e pairs to a level of convergence of 10^{-9} (or 0.2 milli-arc seconds) in E. After convergence in eccentric anomaly was obtained the vanishing of $f(E)$ was separately tested; the absolute value of $f(E)$ had to be less than 10^{-7}.

GENERAL SUMMARY OF THE TRIALS

Although we admired the visual impact of the figures in Danby and Burkardt (Ref. 7), lack of space in this Journal led us to keep tabular statistics on the efficiencies of our candidate starting value/solution method pairs. In particular, for a fixed eccentricity we computed the average and the standard deviation about the average over all mean (or eccentric) anomalies of the total number of trigonometric functions required to reach convergence.

Fig. 2 shows an example of the detailed eccentricity dependence of the convergence statistics. We show the mean $\pm$ one standard deviation of the number of trigonometric functions necessary to reach convergence. The eccentric anomaly version is shown in Fig. 3. Note the divergence as the eccentricity approaches unity. From the statistics (for each starting value and solution method) we then calculated the average over all eccentricity values. These doubly averaged results are in Tables 1 and 2. Finally, using our detailed mean anomaly/eccentricity statistics, we drew "iso-trigonometric function evaluation" level surfaces for selected starting value/solution pairs. An example is shown in Fig. 4 with its eccentric anomaly counterpart in Fig. 5.

The "iso-evaluation" contour maps were constructed by smoothing our 150×150 convergence statistics array. We broke it up into local areas of 5×5, averaged the more detailed numbers, plotted the locally averaged convergence statistics, and then drew the contour map. Figs. 4 and 5 show the well-known problems as $e \to 1^-$ and $M \to 0^+$ but relatively little structure in M, e space compared to that in the E, e exploration.

Tables 1 and 2 have two types of entry. A value such as 7.5 or 5.8 is the doubly averaged value (*i.e.*, over the appropriate anomaly variation and eccentricity range) of the total number of trigonometirc functions necessary for convergence. All integers flagged by an astrerisk (*e.g.*, 1* or 12,345*) denote the number of times that starting value/solution method failed to correctly converge after 20 iterations out of the 22,500 trials.

For the method of successive substitutions there always were failures because of an excessive number of iterations. Similarly, the linear interpolation method never succeeded all the time because of slowed convergence. In this instance the convexity of f is the culprit; the approach to the root was always one-sided. Conjecturing that a quadratic interpolation would produce better results we chose Mueller's bisection/parabolic interpolation technique. The convergence characteristics are better but still not competitive with Newton's second-order method; at best 7.6 in M and 7.7 in E. We do not show the detailed results herein nor did we try Brent's (Ref. 11) improvement.

Not surprisingly Broucke's starting values tend to work the best because they divide the range of E most finely [but Serafin's Table 2 should have been even better]. Newton's second-order scheme and Wegstein's secant modification (see the Appendix) are the best solution methods. The robustness and quickness of the $M + e$ starting value is a mystery.

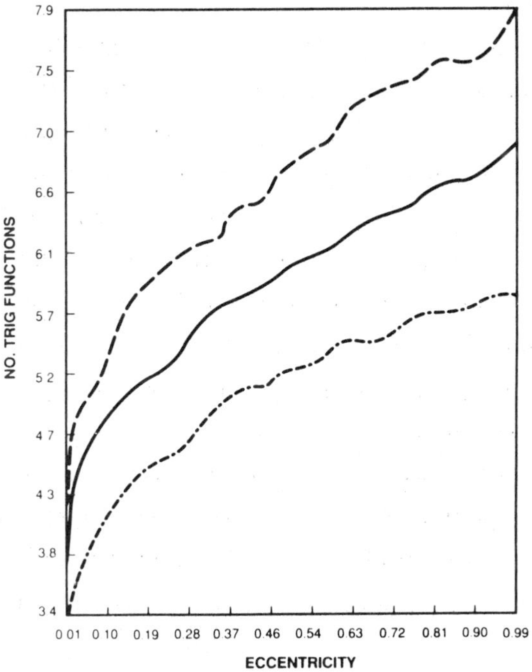

Fig. 2. The average number of trigonometric function evaluations needed to reach convergence for Wegstein's method and an $M+e$ starting value (solid curve). Averaging is over $M \in [0, 2\pi)$. Dashed and dot-dashed curves are a standard deviation above and below the mean.

THE REGION $e \to 1^-$, $M \to 0^+$

Since we counted the number of trigonometric function evaluations necessary to achieve convergence, we can ask for that eccentricity/mean anomaly pair that required the most. Not suprisingly this occurred for large eccentricity and small mean anomaly (see Figs. 2 and 3 too). Hence, we decided to more thoroughly explore this region of mean anomaly/eccentricity space.

We probed the region $e \in [0.85, 1)$, $M \in (0, 0.15\pi]$ with a mesh of 150×150 equally spaced points. An almost full suite of starting value/solution algorithms were tested. The exceptions were the method of successive substitutions, linear interpolation, and parabolic interpolation. Results for the remaining 44 possibilities are given in Table 3 (same format as Table 1). Fig. 6 shows the detailed eccentricity dependence curve for the best combination (*i.e.*, Newton's second-order with $M + e$) while Fig. 7 shows the corresponding contour map.

THE METHOD OF STEEPEST DESCENT

Conway (Ref. 12) used an unconventional solution method by regarding Kepler's equation as a polynomial. We tried a less aggressive exploration—instead of looking for a zero of f we look for a minimum of f^2. Then we can employ the method of steepest descent to find the location of the extrema (*i.e.*, root; see Ref. 13). Basically, one looks for a minimum in $z = F(y) \geq 0$ by going down the gradient from some

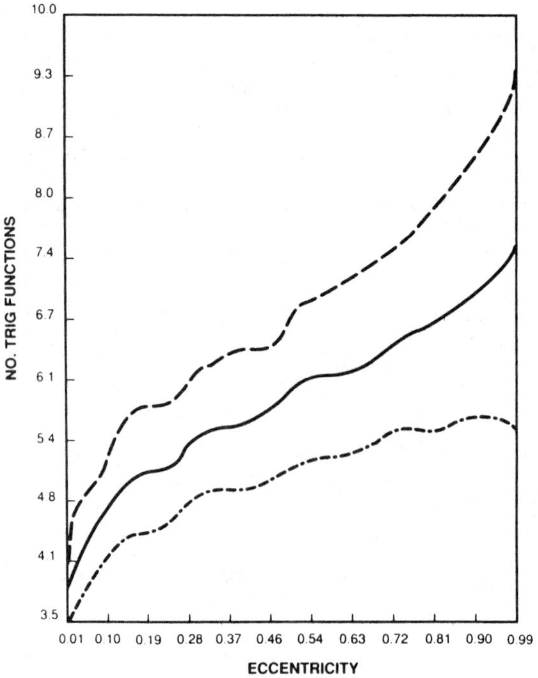

Fig. 3. Same as Fig. 2 but averaged over $E \in [0, 2\pi)$.

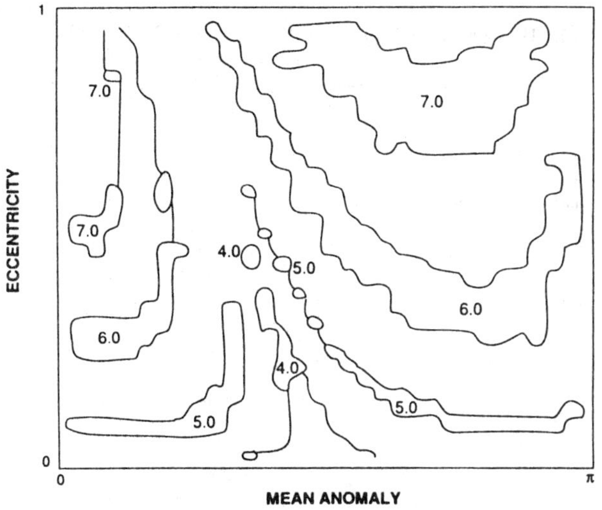

Fig. 4. Iso-number trigonometric function contours needed to reach convergence for Wegstein's method and an $M+e$ starting value.

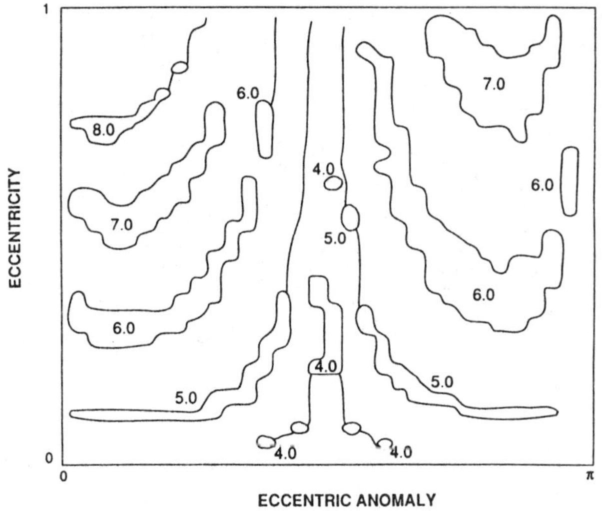

Fig. 5. Same as Fig. 4 but for eccentric anomaly testing.

hypothetical value for the minimum. The amplitude of the step is adjusted utilizing the second-order Taylor series approximation to F. Thus, if we write

$$z = F(y_o + \lambda Y) = F(y_o) + F'(y_o)(\lambda Y) + (1/2)F''(y_o)(\lambda Y)^2$$

we take $Y = -F'(y_o) = -F'_o$ and then $\lambda = 1/F''_o$, $F''_o = d^2F/dy^2$ at y_o. The new guess for y is given by

$$y_n = y_o + \lambda Y = y_o - F'_o/F''_o$$

Table 1

STATISTICS OF TRIGONOMETRIC EVALUATION IN M, e SPACE

Starting Value	Solution Method: Newton First-Order	Newton Second-Order	Successive Substitutions	Steffensen	Wegstein's
M	42*	6.8	8443*	7.3	6.5
$M + e \sin M$	8.4	6.9	8006*	7.6	6.8
$M + e \sin M + (e^2/2) \sin 2M$	8.8	7.5	7668*	8.1	7.4
$M + e \sin M + (e^2/2) \sin 2M + (e^3/8)(3 \sin 3M - \sin M)$	8.4	7.2	7278*	7.8	7.1
π	9.4	7.9	8921*	8.3	7.3
$M + e$	7.5	6.2	8307*	6.7	5.9
$M/(1 - 2e/\pi)$	1604*	273*	8715*	7.5	6.5
Eq. (7)	8.2	6.7	8322*	7.3	6.4
Smith Eq. (5)	7.9	6.9	6859*	7.3	6.8
Smith/Interpol.	7.3	6.3	6214*	6.8	6.4
Broucke	6.5	5.8	4442*	6.2	5.9
$(6M)^{1/3}$	8.3	6.5	8277*	7.2	6.2
Serafin	6.8	6.1	4716*	6.4	6.1

Table 2

STATISTICS OF TRIGONOMETRIC EVALUATION in E, e SPACE

Starting Value	Solution Method: Newton First-Order	Newton Second-Order	Successive Substitutions	Steffensen	Wegstein's
M	75*	6.9	8467*	7.5	6.6
$M + e \sin M$	8.7	7.0	8010*	2*	6.9
$M + e \sin M + (e^2/2) \sin 2M$	9.1	7.6	7649*	8.3	7.5
$M + e \sin M + (e^2/2) \sin 2M + (e^3/8)(3 \sin 3M - \sin M)$	8.6	7.3	7225*	7.9	7.2
π	10.3	8.4	8940*	8.6	7.5
$M + e$	7.7	6.2	8229*	6.9	5.9
$M/(1 - 2e/\pi)$	869*	128*	8532*	1*	6.4
Eq. (7)	8.3	6.6	8222*	7.3	6.3
Smith Eq. (5)	8.1	7.0	6889*	7.5	6.9
Smith/Interpol.	7.6	6.5	6340*	7.1	6.5
Broucke	6.8	5.9	4779*	6.4	6.0
$(6M)^{1/3}$	8.6	6.7	8290*	7.3	6.3
Serafin	7.1	6.3	5038*	6.7	6.2

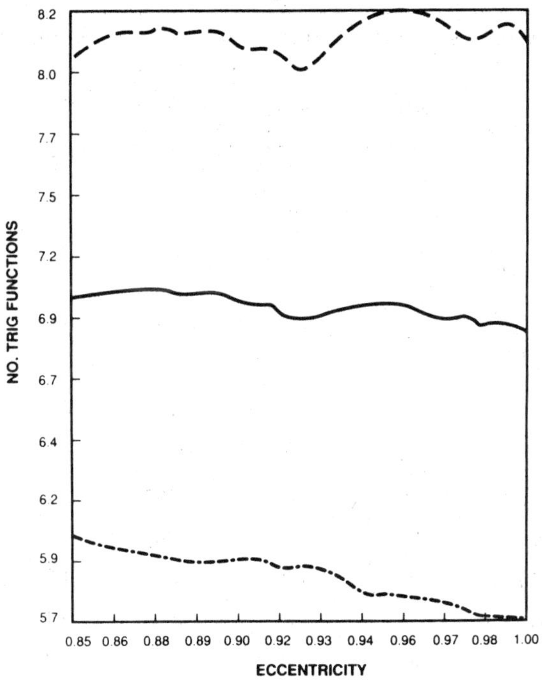

Fig. 6. Same as Fig. 2 but for $M \in [0, 0.15]$ and $e \in [0.85, 1)$. The curves are for Newton's second-order method and the $M + e$ starting value.

We need to choose $F(f)$. We can deduce $F(f)$ by requiring that the "miss distance" $y_n - Y$, where $F[f(Y)] = 0$, be as small as possible. Regarding

$$Y - y_o = -F_o'/F_o''$$

as a differential equation for $F(y_o)$, subject to $F(Y) = 0$, the solution is

$$F(y) = A(y - Y)^2$$

Our iteration would be

$$E_n = E_o - f(E_o)/[f'(E_o) + f(E_o)f''(E_o)/f'(E_o)]$$

(A drops out and is irrelevant). This is a sign-reversed Newton second-order scheme (within an unimportant factor of 2)! It only always converged for starting values of π [as it must because $f'(\pi) = 1 + e > 0$ and $f''(\pi) = 0$] and $(6\text{M})^{1/3}$. Even then it required 10.2 or 9.4 trigonometric function evaluations (on the average). This is true for the E, e exploration also (11.4 and 9.7).

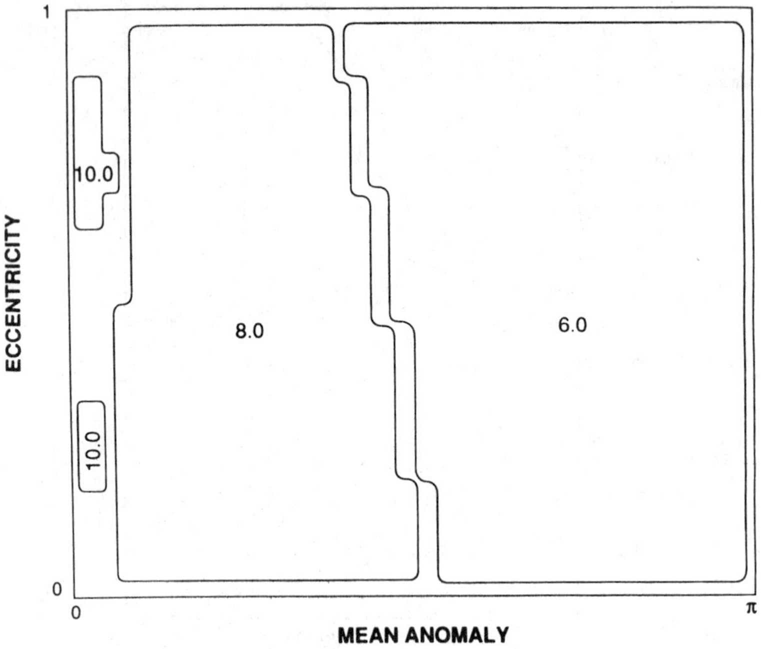

Fig. 7. Same as Fig. 4 but for the Fig. 6 case.

Table 3

STATISTICS OF TRIGONOMETRIC EVALUATION
IN THE REGION OF $e \to 1^-$, $M \to 0^+$

	Solution Method			
Starting Value	Newton First-Order	Newton Second-Order	Steffensen	Wegstein's
M	2067*	9.1	5*	16*
$M + e \sin M$	75*	9.4	4*	9*
$M + e \sin M +$ $(e^2/2) \sin 2M$	4*	10.0	11.7	3*
$M + e \sin M +$ $(e^2/2) \sin 2M +$ $(e^3/8)(3 \sin 3M$ $- \sin M)$	11.9	9.1	10.8	2*
π	13.8	10.5	1*	25*
$M + e$	9.6	7.0	8.4	7.5
$M/(1 - 2e/\pi)$	14*	8.1	7.4	5*
Eq. (7)	9.8	7.1	8.6	1*
Smith Eq. (5)	11.2	8.6	10.1	9.3
Smith/Interpol.	10.5	8.3	9.6	8.8
Broucke	10.2	8.2	9.3	8.6
$(6M)^{1/3}$	8.8	6.6	7.9	7.0
Serafin	12.2	9.3	10.8	9.5

CONCLUSION

We have explored a large number of starting value/solution technique combinations for Kepler's equation. In addition, we report extensive numerical trials in both mean anomaly/eccentricity and eccentric anomaly/eccentricity space. After all, the eccentric anomaly is a regularizing variable for the two-body problem. Hence, a uniform sampling in E/e space can weight the cumulative statistics of the efficiency of a starting value/solution method differently then a comparable uniform sampling in M/e space.

Our trials show no numerical instabilities. This is because of our use of a reasonable starting value, the demand that there be convergence in both the eccentric anomaly iteration as well as the value of $f(E)$, and because we have not misused higher-order schemes. As the steepest descent failures show, the assumption necessary to simplify Newton's second-order method scheme is dangerous.

Finally, we have found a clear winner in the starting value/solution method sweepstakes. It is $M + e$ and Wegstein's secant modification of the method of successive substitutions.

REFERENCES

1. G. R. Smith, "A Simple, Efficient Starting Value for the Iterative Solution of Kepler's Equation," *Cel. Mech.* Vol. 19, 1979, pp. 163–166
2. R. Broucke, "On Kepler's Equation and Strange Attractors," *J. Astronaut. Sci.* Vol. 28, No. 3, July–Sept. 1980, pp. 255–265
3. R. A. Serafin, "Bounds on the Solution to Kepler's Equation," *Cel. Mech.* Vol. 38, 1986, pp. 111–121
4. L. G. Taff and S. Sayer, *On Solving Kepler's Equation*, Lincoln Laboratory Technical Memorandum 94L-0120, December 1985
5. J. H. Wegstein, "Accelerating Convergence of Iterative Processes," *Comm. Assoc. Comput. Mach.* Vol. 1, No. 6, June 1958, pp. 9–13
6. R. H. Battin, *Astronautical Guidance*, McGraw-Hill Book Co., New York, 1964, p. 22
7. J. M. A. Danby and T. M. Burkardt, *Cel. Mech.* Vol. 31, 1983, pp. 95–107
8. E. Isaacson and H. B. Keller, *Analysis of Numerical Methods*, Wiley, New York, 1966, p. 102
9. G. K. Kristiansen, "Zero of an Arbitrary Function," *B.I.T.* Vol. 3, 1963, pp. 205–207
10. E. W. Ng, "A General Algorithm for the Solution of Kepler's Equation for Elliptic Orbits," *Cel. Mech.* Vol. 20, 1979, pp. 243–249
11. R. P. Brent, "An Algorithm With Guaranteed Convergence," *The Computer J.* Vol. 14, 1971, pp. 422–425
12. B. A. Conway, "An Improved Algorithm Due to Laguerre for the Solution of Kepler's Equation," *Cel. Mech.* Vol. 39, 1980, pp. 199–211
13. T. L. Saaty and J. Bram, *Nonlinear Mathematics*, McGraw-Hill Book Co., New York, 1964, p. 75
14. T. A. Jeeves, "Secant Modification of Newton's Method," *Comm. Assoc. Comput. Mach.* Vol. 1, No. 8, 1958, pp. 9–10

APPENDIX

WEGSTEIN'S ITERATION METHOD

Wegstein's iteration scheme (Ref. 5) is used to accelerate convergence from the initial guess x_o of a root to the general nonlinear equation $x = f(x)$. The function $f(x)$ must have a continuous first derivative in the range in which the iteration moves.

Following Fig. 8, set $x_1 = y_o = f(x_o)$ and $y_1 = f(x_1)$. Refinement of x_1 is done by determination of the intersection of $y = x$ and the secant through the points (x_o, y_o) and (x_1, y_1), viz.

$$x_2 = x_1 + \frac{x_1 - x_o}{\frac{x_o - y_o}{x_1 - y_1} - 1}$$

and then

$$y_2 = f(x_2)$$

The next step is performed by starting at (x_2, y_2) and defining (x_3, y_3) by

$$x_3 = x_2 + \frac{x_2 - x_1}{\frac{x_1 - y_1}{x_2 - y_2} - 1}$$

and then

$$y_3 = f(x_3)$$

This represents the intersection between $y = x$ and the secant through the points (x_1, y_1) and (x_2, y_2). Therefore, Wegstein's iteration scheme is often referred to as a secant modification of $x_{i+1} = f(x_i)$.

The final result is the iteration scheme

$$x_{i+1} = x_i + \frac{x_i - x_{i-1}}{\frac{x_{i-1} - y_{i-1}}{x_i - y_i} - 1} - 1 \quad i = 1, 2, \ldots$$

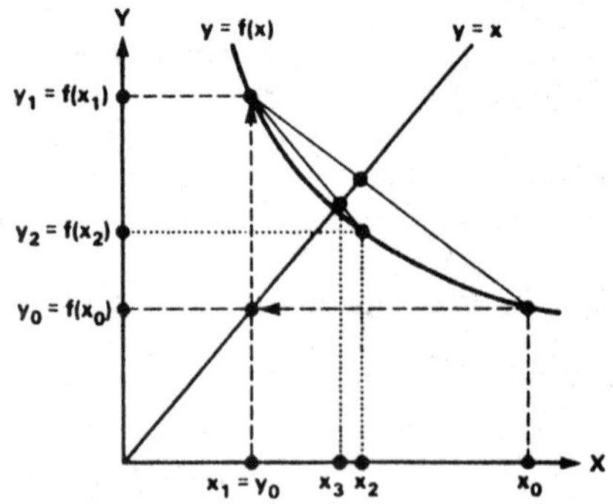

Fig. 8. Diagram showing definition of starting values for Wegstein's secant iteration method. Curve is $y = f(x)$ and the first secant passes through the initial guess (x_o) and its refinement [$x_1 = f(y_o)$]. Then the intersection with $y = x$ completes the iteration cycle for $x = f(x)$.

$$y_{i+1} = f(x_{i+1})$$

See Jeeves (Ref. 14) for a fuller discussion.

AAS 89-386

THE USE OF NONLINEAR PROGRAMMING TO CORRECT THE NUMERICAL INTEGRATION OF THE N-BODY PROBLEM*

James R. Downey† and Bruce A. Conway‡

A nonlinear programming technique is described for improving the numerical integration of the equations of motion for the n-body problem. If any integrals of the motion exist, this technique can be used to correct the trajectories of the n bodies. The corrections are made in an optimal fashion using a least-squares approach. The constants of the motion serve as constraints in the optimization. Results are presented for two-, five-, and 25- body systems which represent the range of n-body systems likely to be of interest. Depending on the complexity of the system this correction can reduce the error in the integration by several orders of magnitude.

* This paper appears in full in the microfiche supplement to this volume, **AAS Microfiche Series**, Volume 59.

† Graduate Research Assistant, University of Illinois at Urbana-Champaign, 104 S. Mathews Ave., Urbana, Illinois 61801.

‡ Associate Professor, Department of Aeronautical and Astronautical Engineering, University of Illinois at Urbana-Champaign, 104 S. Mathews Ave., Urbana, Illinois 61801.

A NEW METHOD OF DIFFERENTIAL CORRECTION*

L. G. Taff†

The differential correction of orbital element sets is a standard statistical adjustment procedure in celestial mechanics. Up until now the rules of estimation have been based on manipulation of the errors of observation, for example the method of least squares or the maximum likelihood method. This is a very unphysical point of view, emphasizes the unknown and the unknowable, is not coordinate-free in formulation, and has no experimental basis. In the simplest gravitational two-body problem we know quite a lot; the energy, the angular momentum vector, and the Laplace (-Runge-Lenz) vector are constants of the motion. These quantities have straightforward physical interpretations, are known to be fixed in value, are coordiate-free (being scalars and vectors), and can be measured. Thus, a natural question to ponder is whether or not a new maxim of statistical estimation can be invented which rests on physical grounds rather than arbitrary ones. The answer is in the affirmative, the new principle has wide applicability, leads to verifiable predictions, and is illustrated herein in the differential correction context. Moreover, as the physical problem is enlarged to include dissipative or symmetry-breaking forces the natural extension of the principle is exhibited via adiabatic invariants (i.e., Delaunay variables). Reference to allied work is included too.

* This paper appears in full in the microfiche supplement to this volume, **AAS Microfiche Series**, Volume 59. This work was sponsored by the Department of the Air Force under Contract F19628-85-C-002 at the Massachusetts Institute of Technology's Lincoln Laboratory. The views are those of the author and do not reflect the official policy or position of the U.S. Government.

† Instrument Scientist for the Fine Guidance Sensors of the Hubble Space Telescope. Dr. Taff is currently at the Space Telescope Science Institute, 3700 San Martin Drive, Baltimore, Maryland 21218.

GEOCENTRIC SATELLITE ORBIT DETERMINATION

SESSION 2

Chairman: Thomas P. Yunck
Jet Propulsion Laboratory

AAS 89-356

RECOVERY OF A GEOCENTRIC REFERENCE FRAME USING THE PRESENT-DAY GPS SYSTEM

Rajendra P. Malla* and Sien-Chong Wu†

A geocentric reference frame adopts the center of mass of the earth as the origin of the coordinate axes. The center of mass of the earth is the natural and unambiguous origin of a geocentric satellite dynamical system. But in practice a kinematically obtained terrestrial reference frame may assume an origin other than the geocenter. The establishment of a geocentric reference frame, to which all relevant observations and results can be referred and in which geodynamic theories or models for the dynamic behavior of earth can be formulated, requires the ability to accurately recover a given coordinate frame origin offset from the geocenter. GPS measurements, because of their abundance and broad distribution, provide a powerful tool to obtain this origin offset in a short period of time. Two effective strategies have been devised. Data from the First Central And South America (Casa Uno) geodynamics experiment has been studied, in order to demonstrate the ability of recovering the geocenter location with present day GPS satellites and receivers.

INTRODUCTION

Reference frames are established in order to represent positions and motions of objects with respect to the earth (terrestrial frames) or celestial bodies, including the earth, in space (celestial inertial frame). One of the geophysical requirements of a reference frame is that other geophysical measurements can be related to it; for example, the reference frame used for expressing the earth's gravity field as spherical harmonic expansion adopts the center of mass of the earth as origin. This frame must

* Sterling Software, Pasadena, California; now Jet Propulsion Laboratory, California Institute of Technology, Mail Stop 238-640, 4800 Oak Grove Drive, Pasadena, California 91109.

† Jet Propulsion Laboratory, California Institute of Technology, 4800 Oak Grove Drive, Pasadena, California 91109.

be related to the adopted terrestrial frame as well as to the inertial frame in which the satellite orbits are calculated. The geocenter is at a focus of the orbital ellipse of a geocentric satellite dynamical system. It is therefore directly accessible through the dynamical methods. But the origin assumed by a kinematically obtained terrestrial reference frame, in practice, can be some other point than the center of mass of the earth. Any time-dependent offset of a geocentric terrestrial frame origin from the geocenter can be misinterpreted as plate motions. In order to avoid such confusion it is important to determine the translational offset of the adopted terrestrial reference frame origin from the geocenter as accurately as possible.

The eighteen (or possibly twenty four) satellites of the fully operational Global Positioning System (GPS) distributed in six orbital planes will allow a user, anywhere on the earth or in a low earth orbit, to view at least five satellites most of the time. The basic GPS observables, P-code pseudorange and carrier phase at two L-band frequencies provide the opportunity to produce geodetic measurements accurate to the centimeter level (Ref. 1) and orbit determination of low earth orbiters to the subdecimeter level (Ref. 2). The broadcast ephemerides of the GPS satellites are based upon the World Geodetic System (WGS 84) (Refs. 3,4) and their accuracies are of the order of ten meters (Ref. 5). For high precision applications it is essential to adjusted the GPS orbits to a much higher precision along with all other parameters in the network. GPS satellites can be observed simultaneously from several sites in a geodetic network, which includes a few fiducial tracking sites (Ref. 6). The relative positions between these fiducial sites are known to a high level of precision, typically a few centimeters, as a result of repeated measurements of the baselines using VLBI (Very Long Baseline Interferometry) (Ref. 7). Filter strategies can be designed to adjust the satellite orbits to enhance their accuracy to better than one meter (Ref. 8). The ephemerides thus adjusted refer to the same coordinate frame in which the fiducial baselines are known. It is generally believed that the best VLBI coordinate system origin approximates the geocenter to about one meter. The Satellite Laser Ranging (SLR) technique is capable of recovering the geocenter offset to better than 10 cm, but this generally requires a long period of observations.

Two effective strategies were devised to determine the geocenter offset from the reference frame origin using only GPS observations (Ref. 10). A covariance analysis was performed estimating the accuracy with which the geocenter position can be determined with these two strategies. The analyses indicate that the geocenter position can be determined to an accuracy of a few centimeters with just one day of precision pseudorange and carrier phase data. Such precise knowledge of absolute position of the coordinate system origin is essential for tracking deep space vehicles and for accurate geodetic positioning, and plays an important role in precise orbit determination of earth observing satellites, such as NASA's Ocean Topography Experiment (Topex). Topex is seeking an altitude accuracy of 13 cm or better.

The results of geocenter estimation using GPS data collected during the Casa Uno Geodynamics Experiment are presented here. The purpose is to demonstrate the ability to estimate the geocenter with the present day GPS constellation, data quality and a semi global network of tracking sites. A multi-day arc is used in order to increase the data strength.

STRATEGIES TO DETERMINE THE ORIGIN OFFSET FROM THE GEOCENTER

In the past several years the fundamental concept behind accurate GPS orbital adjustment has been that of the fiducial network (Ref. 9). A fiducial network consists of three or more tracking stations whose (relative) positions have been determined in an earth fixed coordinate frame to a very high accuracy, usually by VLBI. Several receivers at other, less accurately known, stations observe the GPS satellites along with the fiducial network. The data are then brought together to simultaneously adjust GPS orbits and the positions of the non-fiducial sites. Thus the fiducial stations established by VLBI provide a self-consistent earth fixed coordinate system with respect to which the improved GPS satellite orbits and the non-fiducial stations can be expressed to a greater accuracy. At the same time the coordinate frame origin offset from the geocenter can also be estimated using the same set of data. Past experience in this area has indicated that an over-constrained network, where more baselines or sites than necessary are fixed, can in fact produce a degraded solution. This is because in an over-constrained network the a priori uncertainty on the fixed parameters, which are more than necessary, will result in a suboptimal filter weighting. The solution will then be highly influenced by the mismodeling of these parameters.

In the first strategy proposed, one or two fiducial baselines are fixed or constrained by their a priori weighting (Ref. 10). The baselines define the orientation of the adopted coordinate frame. The absolute scaling can be fixed either by the length of these baselines or by the Earth's gravitational constant, GM. Both are known to an accuracy of about one part in 10^8. The baseline length is used to define the absolute scaling so that the resulting coordinate frame will be consistent with the VLBI frame defined by the fiducial baselines. The filter process is so designed that the baselines between the reference site and all other non-fiducial sites are adjusted along with the Earth Orientation Parameters (EOP), namely polar motion (x_p, y_p) and (UT1–UTC) rate, GPS orbits and the absolute coordinates of the reference site, which in turn infers the adjustment of the geocenter position coordinates. The Earth's GM is also adjusted, although the data strength may not be great enough to improve the value of GM appreciably.

In the second strategy, only the longitude of a reference site is held fixed; all other site coordinates are adjusted simultaneously with GPS orbits. Here, the GM of earth provides the absolute scaling. The geocentric radius at a station can be derived from the adjusted periods of GPS orbits and pseudorange measurements. The coordinate system thus defined will be an earth centered, earth fixed (ECEF) coordinate frame. The time signature of the measurements defines the latitude. Figure 1 schematically demonstrates the time signature of the measurements for two hypothetical cases. The first graph shows the periodic signature generated by the pseudorange (ρ) measurements to an orbiting GPS satellite from a stationary receiver. The period is equal to the GPS orbit period which is nearly 12 hours; the amplitude is proportional to the geocentric position vector of the receiver projected onto the orbital plane. The second graph shows the case when a stationary GPS satellite is above the equator of a spinning earth. The period is now 24 hours; the amplitude is proportional to the cosine of the receiver latitude. The variation of the signature with respect to the receiver latitude is depicted in the sketch. Because of the difference in period, the effects due to a rotating receiver can be separated from the GPS orbit signature and the latitude can be solved unambiguously.

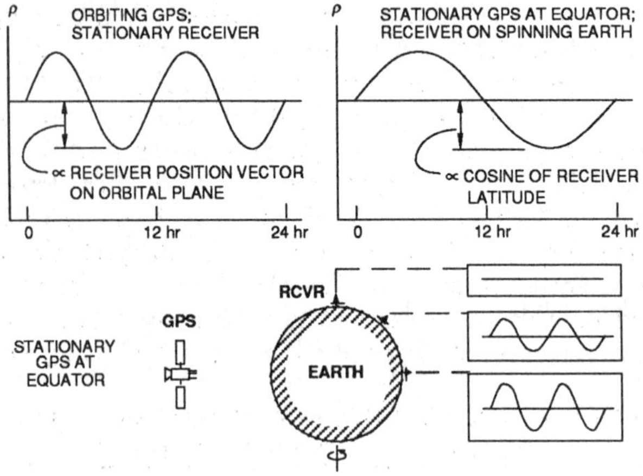

Fig.1 Time Signature of GPS Measurements

COVARIANCE ANALYSIS AND RESULTS

A covariance analysis was carried out to assess the accuracy with which the geocenter offset from the origin of the adopted coordinate frame can be determined with each of the approaches proposed in previous section. A constellation of 18 GPS satellites distributed in six orbital planes was assumed. A data arc spanning over 34 hours from a network of six globally distributed tracking stations was also assumed (Ref. 10). The relative positions of the fiducial sites have been measured repeatedly by VLBI over many years and are known to an accuracy about 3 cm. P-code pseudorange and carrier phase data noise were assumed to be 5 cm and 0.5 cm respectively when integrated over 30 minutes and corrected for ionospheric effects by dual frequency combination. Carrier phase biases were adjusted with a large a priori uncertainty. Table 1 lists the error sources assumed for the first strategy. The abundance and broad distribution of the GPS measurements allow all the GPS and station clocks to be treated as white-noise processes and adjusted to remove their effects on the solutions. Also adjusted are the zenith tropospheric delays at all ground sites, which were treated as random-walk parameters to model the temporal variation. Such models have proved to be effective in removing tropospheric errors without heavily depleting the data strength (Refs. 8,11). The second strategy assumes the same network of six tracking sites. The estimated quantities are the coordinates of all six sites except the longitude of the reference site, together with the GPS satellite states, white-noise clocks, random-walk troposphere parameters and carrier phase biases. The Orbit Analysis and Simulation Software system, OASIS (Ref. 12), developed at JPL, was used to carry out the study.

Data arcs of various lengths were used to study the solution convergence. In Strategy 1 the value of GM was adjusted although it was found that the data strength of the GPS measurements is not great enough to improve on its a priori value. It should be noted that adjusting earth's GM makes GPS satellite states consistent with the absolute scaling as implied by the baselines. Figure 2 shows the total error of the origin offset as the length of the data span increases from 6 hours to 34 hours when two baselines are fixed. At the end of 34 hours the origin offset error is 4.0 cm (RMS of all three components). The geocenter can be determined only up to the a priori accuracy of the fiducial baselines. Therefore, with this strategy, any improvement on the baseline accuracy can improve the accuracy of the origin offset from the geocenter (Ref. 10). For instance, it is customary to find baselines reported with a higher accuracy in length than in the other two components. When a smaller error of 1 cm is assumed for the fiducial baseline length, along with 3 cm for the transverse and vertical components, the RMS error on the origin offset from the geocenter reduces to 3.5 cm with a 34-hour arc of GPS measurements. The analysis here has indicated that a constant bias for polar motion and a (UT1–UTC) rate can be included in the filter as additional adjusted parameters without significantly degrading the performance. However, GPS measurements are insensitive to any constant (UT1–UTC) bias error.

TABLE 1. ERROR SOURCES AND OTHER ASSUMPTIONS FOR STRATEGY 1 (*FIXING BASELINES*)

Reference Site:	Goldstone
Other Fiducial Sites:	Canberra, Madrid
Non-fiducial Sites:	Brazil, Japan, South Africa
GPS Constellation:	18 Satellites distributed in 6 Orbital Planes
Cut-off Elevation:	10 degrees
Data Type:	P-code pseudorange; carrier phase
Data Span:	6 – 34 hours
Data Interval:	30 minutes
Data Noise:	5 cm — pseudorange; 0.5 cm—carrier phase
Carrier Phase Bias:	10 km (adjusted)
Clock Bias:	3 μsec—white noise (adjusted)
GPS Epoch State:	10 m; 1 mm/sec (adjusted)
Geocenter Position:	10 m each comp. (adjusted)
Baseline Coordinates:	3 cm each comp. fiducial; 10 cm each comp. non fiducial; (adjusted)
Zenith Troposphere:	random-walk parm. (adjusted); 20 cm bias; 1.3cm batch to batch
Earth's GM:	one part in 10^8
Solar Pressure:	10%
(UT1–UTC) rate:	10 m/day (adjusted)
Polar Motion (x_p, y_p):	10 m (adjusted)

In the second strategy no tracking site coordinates, except the longitude of the reference site, were held fixed. As before, simultaneous adjustment of all GPS satellite states, tracking site coordinates, carrier phase biases and zenith tropospheric corrections were carried out for various arc lengths ranging between 6 and 34 hours. Fig. 3 plots the variation of the RMS error of the origin offset from the geocenter as a function of data arc length. The errors affecting the origin offset from the geocenter in this strategy are the data noise and GM of Earth, which defines absolute scaling. At the end of 6 hours the RMS error of the origin offset is 143.7 cm; it reduces to 8 cm at the end of 12 hour arc. This indicates that the control on the absolute scaling and the orientation in latitude is greatly improved after all the GPS satellites have been tracked by the globally distributed sites for a complete orbit cycle. At the end of 34 hours the RMS error reduces to 2.1 cm. The graph shows a strong trend of decreasing RMS error as the arc length increases. This indicates that the origin offset determination is limited only by the data noise. The Earth's GM is known accurately enough so that its effect is of the order of 0.2 cm after 12 hours and is 0.1 cm at the end of 34 hours.

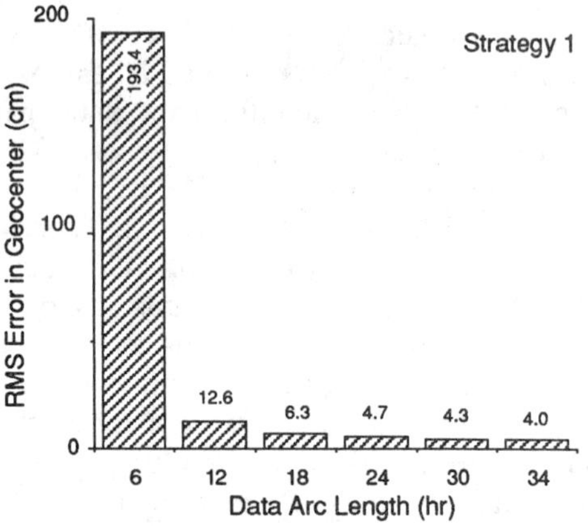

Fig. 2 Geocenter offset with two Baselines fixed

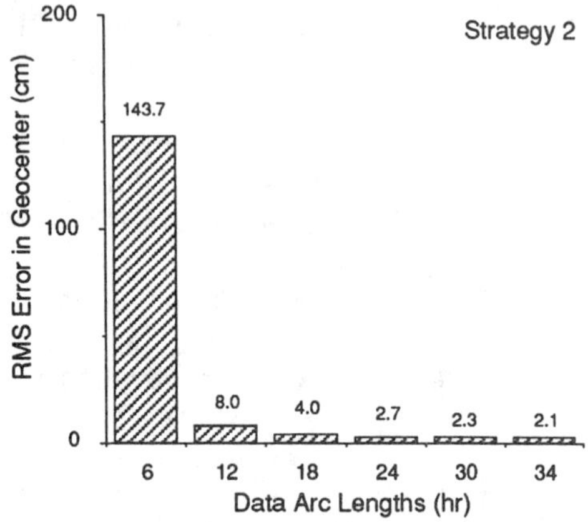

Fig. 3 Geocenter offset with one longitude fixed

GEOCENTER OFFSET STUDY USING CASA UNO DATA

The first Central And South America (Casa Uno) geodynamics experiment was carried out from January 18 to February 5, 1988. This was the first civilian effort at implementing an extended international (15 nations) GPS satellite tracking network (Ref. 13). For the geocenter study 12 globally distributed sites were selected (Fig. 4). There were seven GPS satellites in the constellation distributed in two orbital planes with a separation in the right ascension of the ascending node of approximately 120°. Strategy 1 is more appropriate because the data set is weakened by a lack of both a sufficient number of satellites and a well distributed set of global tracking sites. The object of this exercise is to determine the achievable accuracy on the estimation of the geocenter position, although the situation is far weaker than the one presented in covariance analysis. The baselines known to higher accuracies are available only within the continental US, which are relatively short compared to the extent the tracking network. One baseline, OVRO to Haystack, was held fixed with OVRO as the reference point. Figure 4 shows the distribution of all tracking sites used in the experiment. Seven satellites distributed in only two orbital planes do not provide a good global coverage, and data quality was very poor.

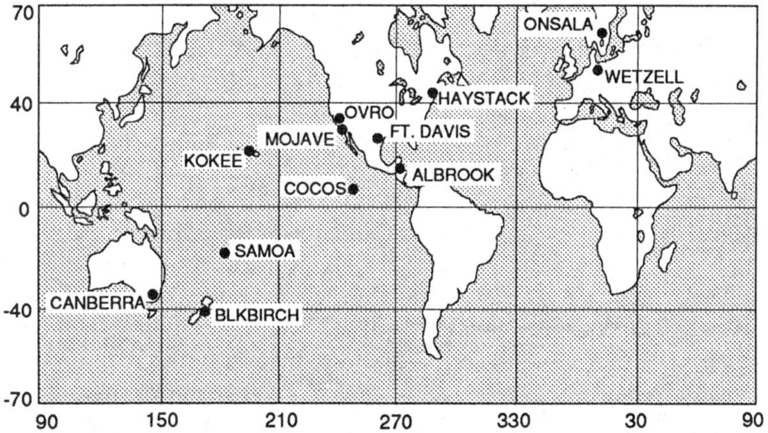

Fig. 4 The Casa Uno Tracking Sites for Geocenter Studies

Although the tracking site network was designed to collect a maximum amount of data with the available constellation, it does not provide a well balanced global coverage. In order to compensate for these suboptimal circumstances, the data strength was increased by using a 5-day data arc along with a network geometry controlled by constraining the well known sites to their a priori weights. Table 2 lists the error budget differences from Table 1 used for this experiment. In the covariance analysis presented earlier the GPS Block I ROCK 4 model for solar radiation pressure was used, where the three parameters G_x, G_y and G_z were considered with 10% error. G_y is the constant acceleration in the y axis, called the y-bias parameter (Ref. 14). In the Casa Uno multi-day analysis a new approach was adopted in which two constant solar pressure parameters G_y and G_{xz} were estimated along with two tightly constrained process noise parameters G_x and G_z (Ref. 11). The parameter G_{xz} is a combined effect of $(G_x + G_z)$.

Figure 5 shows the results in terms of RMS error in geocenter. The effect due to data noise dominates the error, showing the poor quality of data; however, the error due to the fixed baseline is also significant. This is because the ground tracking sites are not uniformly distributed in this experiment and the fixed baseline is not long enough to provide a global control. The tracking sites used in the covariance study, however, were distributed quite evenly around the globe and the eighteen satellite constellation has given a much stronger data than the seven satellite constellation available here. The estimated geocenter offset along X, Y and Z axes are found to be –142, –33 and –43 cm respectively. The error associated with the geocenter offset estimation due to data noise is 31 cm and that due to baselines is 7.1 cm, resulting in a total estimated RMS error in the geocenter offset solution of 31.8 cm. Despite these unfavorable circumstances, the results obtained are encouraging. The well known Deep Space Network (DSN) sites are not yet included in the present Casa Uno network. With better quality of data collection and the inclusion of DSN sites, the results will continue to improve. Strategy 2 was not examined because the data strength is not great enough with the present data set. In order to implement this strategy a better global distribution of tracking sites along with a longer data arc is necessary.

TABLE 4. ERROR BUDGET FOR GEOCENTER STUDY ON CASA UNO DATA USING (*ONE BASELINE FIXED*)

Reference Site:	OVRO
Other Fiducial Sites:	Haystack
Non-fiducial Sites:	Blkbirch, Canberra, Kokee, Samoa, Cocos, Albrook, Ft. Davis, Mojave, Wetzell and Onsala
GPS Constellation:	GPS 11, GPS 3, GPS 4, GPS 6, GPS 8, GPS 10, and GPS 9; distributed in 2 Orbital Planes
Cut-off Elevation:	15 degrees
Data Span:	up to 5 days
Data Interval:	6 minutes
Data Noise:	175 cm - pseudorange; 1 cm - carrier phase
GPS Epoch State:	20 km and 20 m/sec, (adjusted)
Baseline Coordinates: *(relative to OVRO, each comp. adjusted)*	*Fiducial:* Haystack 4 cm; *Non-fiducial:* Ft. Davis and Mojave - 4 cm; Wetzell and Onsala - 10 cm; *Others:* 1 km each comp.;
Solar Pressure:	G_Y and G_{XZ} (adjusted)

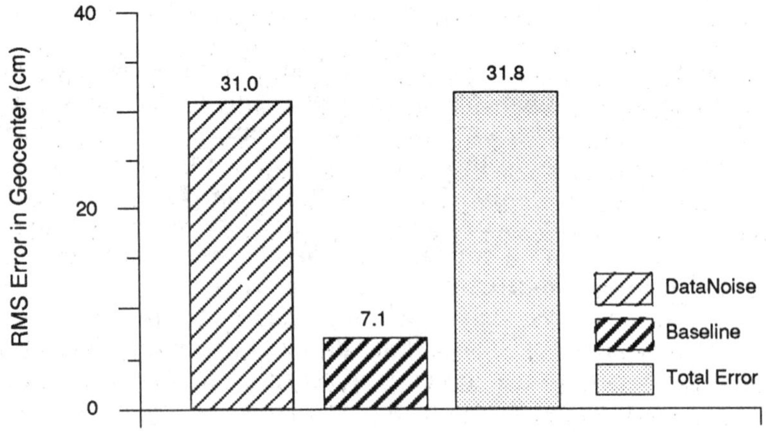

Fig. 5 Geocenter Error Using Casa Uno 5-day arc Data

SUMMARY AND CONCLUSION

A geocentric coordinate frame provides a practical global reference system with a physically meaningful and unambiguous definition of the coordinate origin. Two basic strategies for establishing a geocentric coordinate frame for GPS measurements have been investigated. The first strategy makes use of the precise relative positions which have been predetermined by VLBI to fix the frame orientation and the absolute scaling, while the offset from the geocenter is determined from GPS measurements. The reference frame thus adopted is consistent with the VLBI coordinate system. The second strategy establishes a reference frame by holding only the longitude of one of the tracking sites fixed. The absolute scaling is inferred by the adopted gravitational constant (GM) of earth; the orientation in latitude is inferred from the time signature of earth rotation in the GPS measurements. The coordinate system thus defined is a geocentric earth fixed coordinate system. Covariance analysis has shown that geocentric positioning to an accuracy of a few centimeters can be achieved with just a one-day arc of precise GPS pseudorange and carrier phase data.

Despite the poor data quality, uneven global coverage and limited satellite constellation the results of the Casa Uno 5-day data arc indicate that the geocenter position can be recovered to an estimated accuracy of 32 cms. The results will continue to improve with superior data quality, an evenly distributed global network including longer fiducial baselines, and an increased number of satellites in more orbital planes. Inclusion of the DSN sites will provide the global baselines which are lacking in the Casa Uno network. Longer baseline ties to the well known sites in North America from Europe and the Southern Hemisphere will provide a much stronger network geometry for geocenter studies.

Each of the two strategies has its advantages in different applications. The first strategy should be adopted in applications requiring a coordinate frame consistent with the VLBI reference frame. Among these applications are the monitoring of crustal motions in areas which have been investigated by VLBI observations and the determination of the earth rotation parameters, namely, polar motion and variation of UT1–UTC. The second strategy, which holds the longitude at a reference site fixed, strictly limits itself in an ECEF frame established by the adopted values for the fixed longitude and the GM of earth, and by GPS measurements. This method provides a superior result as long as the precise applications are within the same ECEF frame. Applications in which such an ECEF coordinate frame can be effective include datum definition and network densification in an area where ECEF coordinates are appropriate. Various topographic and oceanographic surveys, and prospecting surveys can benefit from its simplicity. In precise orbit determination of the Topex satellite, this method can also be convenient if a ECEF frame such as WGS 84 is adopted.

ACKNOWLEDGMENT

The work described in this paper was carried out by the Jet Propulsion Laboratory, California Institute of Technology, under contract with the National Aeronautics and Space Administration.

REFERENCES

1. T. P. Yunck, W. G. Melbourne and C. L. Thornton, Jul. 1985, "GPS-Based Satellite Tracking System for Precise Positioning," IEEE Trans. Geosci. & Remote Sensing, Vol. GE-23, No. 4.
2. T. P. Yunck, S. C. Wu and J. T. Wu, Nov. 1986, "Strategies for Sub-Decimeter Satellite Tracking with GPS," Proc. 1986 IEEE Position Location and Navigation Symp., Las Vegas, NV.
3. Department of Defense World Geodetic System 1984, Sept. 1987, DMA Tech. Rep. 8350.2, The Defense Mapping Agency.
4. B. Louis Decker, Apr. 1986, " World Geodetic System 1984," Proc. Fourth Int. Geodetic Symp on Satellite Positioning, Austin, Vol 1, pp 69–92.
5. E. R. Swift, Apr. 1985, "NSWC's GPS Orbit/Clock Determination System," Proc. First Int. Symp. on Precise Positioning with GPS, Rockville, MD, pp. 51–62.
6. J. M. Davidson, et al., Apr. 1985, "The March 1985 Demonstration of the Fiducial Concept for GPS Geodesy: A Preliminary Report," Proc. First Int. Symp. on Precise Positioning with GPS, Rockville, MD, pp. 603–611.
7. O. J. Sovers, et al., Sept. 1984, "Radio Interferometric Determination of Intercontinental Baselines and Earth Orientation Utilizing Deep Space Network Antennas: 1971 to 1980," J. Geophys. Res., Vol 89, No. B9,pp. 7597–7607.
8. S. M. Lichten and J. S. Border, Nov. 1987, "Strategies for High-Precision Global Positioning System Orbit Determination," J. Geophys. Res., Vol. 92, No. B12, pp. 12751–12762.
9. C. Thornton, J. Fanselow and N. Renezetti, 1986, "GPS-Based Geodetic Measurement Systems, in Space Geodesy and Geodynamics," A. Anderson and A. Cazenave, eds, Academic Press.
10. R. P. Malla and S-C Wu, Nov. 1988, "Deriving a Unique Reference Frame for GPS Measurements," Proc. IEEE PLANS '88 Position Location and Navigation Symposium, Orlando, Florida.
11. W. Bertiger, S. M. Lichten and E. C. Katsigris, Nov. 1988, "A Demonstration of Submeter GPS Orbit Determination and High Precision User Positioning," Proc. IEEE PLANS '88 Position Location and Navigation Symposium, Orlando, Florida.
12. S. C. Wu and C. L. Thornton, Apr. 1985, "OASIS — A New GPS Covariance and Simulation Analysis Software System," Proc. First Int. Symp. on Precise Positioning with GPS, Rockville, MD, pp. 337–346.
13. R. E. Neilan, T. H. Dixon, T. K. Meehan, W. G. Melbourne, J. H. Scheid, J. N. Kellogg and J. L. Stowell, April 1989, "Operational Aspects of Casa Uno '88 - The First Large Scale International GPS Geodetic Network," IEEE Trans. of Instr. and Meas., Vol.38, No. 2
14. H. F. Fliegel, et al., Apr.1985, "The GPS Radiation Force Model," Proc. First Int. Symp. on Precise Positioning with GPS, Rockville, MD, pp. 113–120.

AAS 89-357

ENSEMBLING CLOCKS OF THE GLOBAL POSITIONING SYSTEM (GPS)

Arthur L. Satin* and Cornelius T. Leondes†

The GPS is a satellite navigation system capable of providing 15 meter position accuracy. Its system time reference is currently one of the monitor station clocks. This paper analytically shows, using a simple two-clock example, that improved reference time stability and overall state estimation accuracy can be achieved by constructing GPS time as an ensemble of all system clocks and that the problem of covariance divergence can be handled by the introduction of pseudomeasurement processing.

INTRODUCTION

The Global Positioning System (GPS) is a Department of Defense space-based navigation and time dissemination system currently in the production phase.[1] It will consist of a 21 space vehicle (SV) constellation, a host of user receivers, and a control segment that tracks the SVs with its monitor stations (MSs), determines SV ephemeris and clock states along with tracking station clock states, and periodically uploads predicted SV states that the user requires to navigate his position.

Space vehicle range measurements depend on the geometric range and the phase offset between the SV clock and the MS clock. The control segment, by solving for system clock states from the measurements, effectively synchronizes the clocks to GPS time. In the current system, the GPS time reference is taken to be one of the MS clocks. States for that MS (i.e., the "master" MS) are defined to be zero, meaning that all other clock states, computed by the control segment

* Project Engineer, GPS Program Office, The Aerospace Corporation, Mail Station M5/685, P.O. Box 92957, Los Angeles, California 90009-2957.

† Professor of Engineering, University of California, Los Angeles, California 90024.

using a Kalman filter, are offsets relative to the master. When there is no master clock, i.e., when the filter is allowed to estimate all clock states, the configuration is called a composite clock system. This paper predicts the characteristic differences in these two reference time concepts by analyzing a simple two-clock system.

COMPOSITE CLOCK IMPLEMENTATION

Composite Clock Versus Master Clock

Filter dynamics, with a composite clock implementation instead of a master clock, are best illustrated by considering a simple two-clock example (Fig. 1). In this system X_1 is the phase state of Clock No.1, X_2 the phase state of Clock No. 2, and Z a measurement of the phase offset between the two.

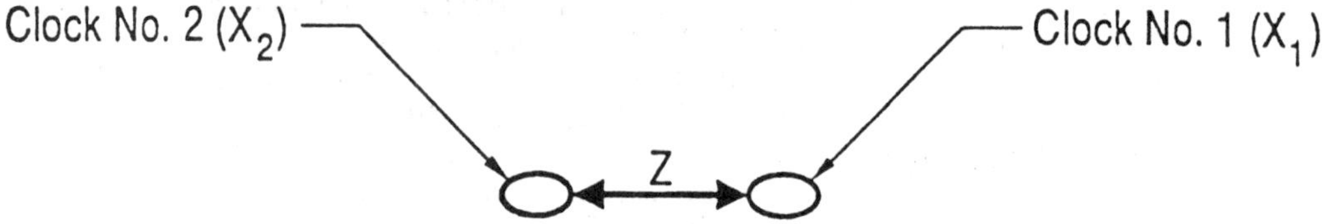

Fig. 1 A Simple Two-Clock Example

The phase error of both clocks is driven by zero mean, white noise on frequency (ω), which integrates into random walk on phase.[2] Eqs. (1), (2), and (3) describe the actual state and measurement processes.

$$\dot{X}_1 = \omega_1 \tag{1}$$

$$\dot{X}_2 = \omega_2 \tag{2}$$

$$Z = X_1 - X_2 + \nu \tag{3}$$

where ν also is a zero mean, white measurement noise. The various stochastic processes have covariance matrices given by

$$E[\omega_1(t)\omega_1(\tau)^T] = Q\delta(t-\tau) \tag{4}$$

$$E[\omega_2(t)\omega_2(\tau)^T] = Q\delta(t-\tau) \tag{5}$$

$$E[\nu(t)\nu(\tau)^T] = R\delta(t-\tau) \tag{6}$$

In this section and further explained in the paper, three different filter models of the simple two-clock system will be analyzed. These are numbered 1 through 3, and they provide insight into some of the unique properties of the composite clock.

Consider first the master clock model (filter model No. 1). In this case the filter estimates state X_1 (relative to state X_2) using measurements Z. The model is therefore

$$\dot{X} = \dot{X}_1 - \dot{X}_2 = \omega_1 - \omega_2 \tag{7}$$

$$Z = X + \nu \tag{8}$$

where

$$X = X_1 - X_2 \qquad \text{and} \qquad E[(\omega_1-\omega_2)(\omega_1-\omega_2)^T] = 2Q.$$

The processing compensates for Clock No. 2 random walk by doubling the Clock No. 1 process noise covariance matrix Q. According to the general linear system model description[3]

$$\dot{X} = Fx + G\omega \tag{9}$$

$$Z = Hx + \nu \tag{10}$$

The master clock model has

$$H = 1,\ F = 0 \text{ and } G = 1$$

The matrix differential equation for the covariance matrix of estimation error in the state X simplifies to:

$$\dot{P} = -\ PR^{-1}P + 2Q \tag{11}$$

with all variables having dimension 1. Since $P = \sqrt{2RQ}$ is a particular solution to Eq. (11), the equation can be linearized by transforming to the variable

$$Y = \frac{1}{P - \sqrt{2RQ}}$$

This leads to the following solution for P:

$$P = \sqrt{2RQ} + 2\sqrt{RQ}\ \left\{e^{2\sqrt{2QR}\,(t\ +\ c)}{}_{-1}\right\}^{-1} \tag{12}$$

where c is a constant of integration. Taking the limit as $t \to \infty$ in Eq. (12), it is seen that the steady-state value for P is:

$$P \to \sqrt{2RQ} \tag{13}$$

Also, the steady-state filter gain, given by

$$K = PH^TR^{-1} \text{ is}$$

$$K = \frac{1}{R}P - \sqrt{2Q/R} \tag{14}$$

So, the filter gain and covariance reach constant steady-state values. If the observability matrix $OB = [H^T, F^TH^T, \ldots (F^T)^{n-1}H^T]$ is computed one finds that it is of rank 1, indicating that according to this filter model the system is completely observable. In reality, however, the master clock is not observable, because it has no states and random walk in its phase propagates directly into error in X.

The filter model for the system [Eqs. (7) and (8)] can be improved upon by replacing the master clock model for GPS reference time with the composite clock model (filter model No. 2). In this model the filter estimates X_1 and X_2. In this way, some of the master clock error is estimated, and therefore removed from X_1, thereby improving overall accuracy. The filter model for the composite clock is the same as the actual system model described in Eqs. (1), (2), and (3). When written more conveniently as:

$$\begin{bmatrix} \dot{X}_1 \\ \dot{X}_2 \end{bmatrix} = \begin{bmatrix} 1 & 0 \\ 0 & 1 \end{bmatrix} \begin{bmatrix} \omega_1 \\ \omega_1 \end{bmatrix}$$

$$Z = [1 \ -1] \begin{bmatrix} X_1 \\ X_2 \end{bmatrix} + \nu$$

it is apparent that this model has

$$F = 0, \ G = \begin{bmatrix} 1 & 0 \\ 0 & 1 \end{bmatrix} \text{ and } H = [1 \ -1].$$

The model is now two-dimensional with 2 x 2 covariance matrix P. Since P is symmetric and both clocks have the same Q, it follows that $P_{12} = P_{21}$ and $P_{11} = P_{22}$, where these are the matrix elements of P. The matrix differential equation for P, i.e.,

$$\dot{P} = -PH^TR^{-1}HP + G \begin{pmatrix} Q & 0 \\ 0 & Q \end{pmatrix} G^T \tag{15}$$

leads to the following nonlinear differential equations for P_{11} and P_{12}:

$$\dot{P}_{11} = -\frac{1}{R}(P_{11} - P_{12})^2 + Q \tag{16}$$

$$\dot{P}_{12} = \frac{1}{R}(P_{11} - P_{12})^2 \tag{17}$$

Subtracting Eq. (17) from (16) yields a differential equation similar to the one that was solved in the master clock case [Eq. (11)]. This equation is

$$\dot{P}_{11} - \dot{P}_{12} = -\frac{2}{R}(P_{11} - P_{12})^2 + Q$$

Solving this equation and then taking the limit as $t \to \infty$ yields the steady-state value for $P_{11} - P_{12}$ namely,

$$P_{11} - P_{12} \to \sqrt{RQ/2} \tag{18}$$

Substituting the value of $P_{11} - P_{12}$, from Eq. (18) in Eqs. (16) and (17) yields the steady-state values for $\dot{P}_{11}$ and $\dot{P}_{12}$:

$$\dot{P}_{11} \to Q/2 \tag{19}$$

$$\dot{P}_{12} \to Q/2 \tag{20}$$

This means that the covariances P_{11} and P_{12} will grow linearly with time once steady state is reached. Even though the covariances grow, the Kalman gains reach constant values at steady state so that system state estimates are well behaved. This 2 x 1 gain vector is computed as:

$$K = R^{-1}\left[P_{11} - P_{12},\ P_{12} - P_{11}\right]^T = \left[\sqrt{Q/2R},\ -\sqrt{Q/2R}\right]^T$$

The observability matrix for this system is rank deficient (rank = 1) implying that the system is not completely observable. The difference between this system and the master clock system (that was found to be completely observable) is that now, if each phase state is increased by the same amount the measurement Z does not change; whereas it did in the master clock case. The covariance matrix of state error then grows with time to reflect the absolute indeterminability of phase states for each clock in the system. Another way of looking at this, is that with the composite clock, at any time t, there is always one more state than there are independent measurements. The fact that the covariance matrix elements grow with time and become very nearly equal means that, potentially, numerical problems can arise with time. The next section discusses a way (filter model No. 3) of avoiding the problem of covariance growth.

Covariance Matrix Reduction Using Pseudomeasurements

The measurement equation for the pseudomeasurement,

$$0 = X_1 + X_2 + \nu'$$

or

$$0 = [1 \;\; 1] \begin{bmatrix} X_1 \\ X_2 \end{bmatrix} + \nu' \tag{21}$$

may be looked upon as an additional linear dependence relationship between the states that remove the single degree of freedom degeneracy that makes the covariance matrices grow with time. Note in Eq. (21) that the measurement residual is defined to be zero (hence, "pseudomeasurement").

It is now possible to re-solve the original composite clock system [Eqs. (1), (2), and (3)] with the pseudomeasurement of Eq. (21). The only changes are that the H matrix becomes 2 x 2, i.e.,

$$H = \begin{bmatrix} 1 & -1 \\ 1 & 1 \end{bmatrix}$$

and

$$E[\nu'(\tau)\nu'(\tau)^T] = R'$$

where R' may be arbitrarily specified. Again solving the matrix differential equation for P yields:

$$\dot{P}_{11} + \dot{P}_{12} = - 2/R'[P_{11} + P_{12}]^2 + Q \tag{22}$$

and

$$\dot{P}_{11} - \dot{P}_{12} = - 2/R\,[P_{11} - P_{12}]^2 + Q \tag{23}$$

The steady-state solution of this system is:

$$P_{11} - P_{12} \to \sqrt{RQ/2} \tag{24}$$

$$P_{11} + P_{12} \to \sqrt{R'Q/2} \text{ so that} \tag{25}$$

$$P_{11} \to 1/2\left(\sqrt{RQ/2} + \sqrt{R'Q/2}\right) \tag{26}$$

$$P_{12} \to 1/2\left(\sqrt{R'Q/2} - \sqrt{RQ/2}\right) \tag{27}$$

Since the pseudomeasurement residual is zero, it is only necessary to compute a 2 x 1 gain which is:

$$K = \left[\sqrt{Q/2R} \quad - \sqrt{RQ/2(R')^2}\right]^T$$

If R' = R these gains are identical to those computed for the composite clock case considered earlier and the states will be identical. The covariance elements, however, will now reach steady-state values according to: $P_{11} \rightarrow \sqrt{RQ/2}$, $P_{12} \rightarrow 0$. Note that this P_{11} is 1/2 the P_{11} computed for the master clock case.

SUMMARY AND CONCLUSIONS

The Global Positioning System (GPS) reference for time is currently the monitor station (MS) clock designated as the "master." All SV and MS clock phase and frequency offsets are determined relative to this master MS. The master clock, by definition, has no phase or frequency states computed by the MCS Kalman filter. Master MS instabilities map directly into other SV and MS states and frequently problems with the master MS require that operators switch the master to another MS (i.e., redesignate another MS as the master MS). To avoid these detriments to a stable system time reference and to gain more inherent reference stability, the operational control segment is transitioning to a composite clock reference for GPS time. This entails treating the master clock as just another clock in the system by letting the filter estimate its states as it does all others. A series of examples, based on a two-clock system, shows analytically that the composite clock configuration produces more accurate estimates for all clock states, and that the clock state estimation error covariances grow linearly with time since there is an indeterminability for absolute clock phase and frequency. This growth is proportional to clock process noise variance (Q). States are well behaved asymptotically, but diverging covariances will cause numerical problems eventually. To preclude this, a covariance reduction scheme using pseudomeasurements will be implemented.

REFERENCES

1. Specifications for the Navstar Global Positioning System, SS-GPS-300C, USAF Space Division, 10 June 1985.

2. J. S. Meditch, Clock Error Models for Simulation and Estimation, TOR-0076(6474-01)-2, The Aerospace Corporation, 7 July 1975.

3. J. S. Meditch, Stochastic Optimal Linear Estimation and Control, McGraw Hill, 1969.

AAS 89-358

THE EFFECTS OF GENERAL RELATIVITY ON NEAR-EARTH SATELLITES

J. C. Ries*, C. Huang†, M. M. Watkins* and B. D. Tapley‡

Whether one uses a solar system barycentric frame or a geocentric frame when including the general theory of relativity in orbit determination for near-Earth satellites, the results should be equivalent to some limiting accuracy. The purpose of this paper is to clarify the effects of relativity in each frame and to demonstrate their equivalence through the analysis of three years of laser tracking data taken on the Lageos satellite. It is demonstrated that the simpler formulation in the geocentric frame is adequate for the purpose of near-Earth satellite orbit determination. A correction to the conventional barycentric equations of motion is shown to be required.

INTRODUCTION

Since the advent of laser ranging to near-Earth satellites during the late 1960's, the observation precision has steadily improved to the point where modern lasers routinely obtain centimeter precision. This requires that the analysis of the data has comparable accuracy. One of the factors to be considered is the inclusion of general relativity in the data reduction procedure. The relativistic treatment of the near-Earth satellite orbit determination problem involves relativistic equations of motion, corrections to the measurement model, and time transformations between frames. The problem can be formulated in either a solar system barycentric reference frame[1] or a geocentric reference frame[2].

As a result of the generalized principle of equivalence, the main relativistic effects on a near-Earth satellite should be due only to the Schwarzschild field of the Earth itself[3]. The

* Research Associate, Center for Space Research, The University of Texas, Austin, Texas 78712-1085.

† Visiting Scholar, Center for Space Research, The University of Texas, Austin, Texas 78712 (Permanent Address: Shanghai Observatory, Academia Sinica, Shanghai, China).

‡ AIAA Fellow; Professor and Director for Center for Space Research, The University of Texas, Austin, Texas 78712.

largest dynamical effect of general relativity on a satellite orbit is the well-known precession of perigee. This effect is illustrated in Figure 1, where the size of the relativity perturbation is compared to the precision with which the argument of perigee for Lageos is determined. The small variations in the perigee residuals are caused by errors remaining in the modeling of the Newtonian forces on Lageos while the large slope is the effect of general relativity. Relativity theory also predicts several small effects, including a change in the mean motion of the satellite and a precession of the longitude of the node due to the angular momentum of the rotating Earth, the Lense-Thirring[4] effect. In addition, there is a precession of the orbit plane due to the effect of geodesic (or de Sitter[5]) precession. This is the only effect not due to the Earth's mass, but rather to the motion of the Earth through the Sun's gravitational field. It is the result of the choice of a geocentric reference frame which is non-rotating with respect to the barycentric frame instead of a truly inertial geocentric reference frame, since the latter is difficult to realize in practice.

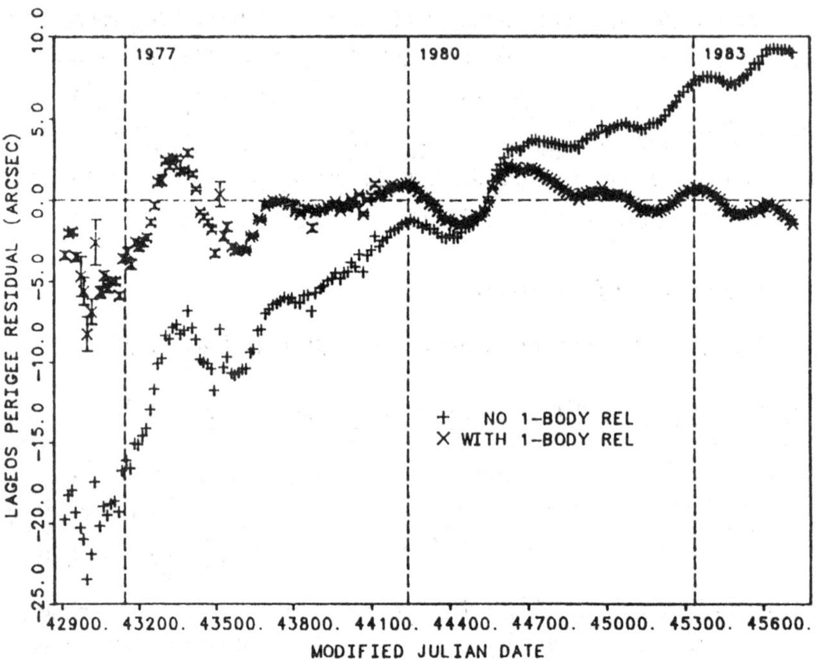

Fig. 1 Comparison of Lageos Perigee Residuals Showing Relativistic Effect

SCALE EFFECTS AND COORDINATE TIME

One of the results of the inclusion of relativity is the introduction of a scale change between these two frames caused by the definition of Barycentric Dynamical Time (TDB) and Terrestrial Dynamical Time (TDT). Standard clocks on the surface of the Earth, whose rates are defined by the atomic second in the International System of Units (SI), would run slow in comparison with coordinate time in the barycentric frame due mainly to the gravitational potential of the Sun and the Earth's orbital motion. The long term average of the ratio between the rates has been determined to be $(1-L)$ where L is approximately

1.5505×10^{-8}. However, if the secular term is removed, as is required by the International Astronomical Union (IAU) definition of barycentric time[6], then a renormalization has taken place in the time coordinate associated with the barycentric frame. Assuming that the speed of light, c, is to retain the same numerical value in every reference frame, the renormalization that has occurred in the barycentric time coordinate must carry over into the space coordinates as well, since the choice of unit for length is not independent of the unit of time. Scaling the barycentric space coordinates by $(1-L)$ to define the TDB coordinates results in no changes to the form of the equations of motion for massive bodies or photons. It is only when relating TDB units of length to SI units of length that L appears. Noting that the quantity GM/c^2 has units of length, the value for the gravitational coefficient GM of a body in TDB units must be related to its value in SI units according to $GM_{TDB} = (1-L)\, GM_{SI}$[7,8].

A similar scale difference exists in the geocentric frame[9]. The scale difference K in the geocentric frame is due to the contribution of the geoid, where K has approximately a value[10] of 0.0697×10^{-8}. The value for the GM of a body in TDT units must then be related to its value in SI units according to $GM_{TDT} = (1-K)\, GM_{SI}$. It can be seen that the scale difference between the geocentric (TDT) frame and the barycentric (TDB) frame is then $\widetilde{L} = (L-K)$, or 1.4808×10^{-8}, and that the numerical value for the mass of a body is related according to $GM_{TDB} = (1-\widetilde{L})\, GM_{TDT}$.

GEOCENTRIC RELATIVITY MODEL

The relativistic correction to the acceleration of an artificial Earth satellite is:

$$\ddot{\mathbf{r}} = \frac{GM_E}{c^2r^3}\left[\,[\,(2\beta+2\gamma)\,\frac{GM_E}{r} - \gamma(\,\dot{\mathbf{r}}\bullet\dot{\mathbf{r}})\,]\,\mathbf{r} + (2+2\gamma)\,(\,\mathbf{r}\bullet\dot{\mathbf{r}})\,\dot{\mathbf{r}}\,\right]$$

Schwarzschild solution

$$+\,2\,(\,\Omega\times\dot{\mathbf{r}}\,)$$

Geodesic precession

$$+\;(1+\gamma)\frac{GM_E}{c^2r^3}\left\{\frac{3}{r^2}\left[\,\mathbf{r}\times\dot{\mathbf{r}}\,\right]\left[\,\mathbf{r}\bullet\mathbf{J}\,\right]+\left[\,\dot{\mathbf{r}}\times\mathbf{J}\,\right]\right\}$$

Lense-Thirring precession

where

$$\Omega \cong (\tfrac{1}{2}+\gamma)\left(\dot{\mathbf{R}}\times\left[\frac{-GM_S\,\mathbf{R}}{c^2R^3}\right]\right),$$

and $\mathbf{J}$ is the Earth's angular momentum per unit mass ($|\mathbf{J}| \cong 9.8\times10^{8}$ m^2/sec). The vector $\mathbf{r}$ is the geocentric position of the satellite, $\mathbf{R}$ is the position of the Earth with respect to the Sun, and γ and β are the only parameterized post-Newtonian parameters which are non-zero in Einstein's theory. The quantities GM_E and GM_S are the gravitational coefficients of the Earth and Sun, respectively. The value of GM_E should be the geocentric value, as described above. The form of the Lense-Thirring acceleration is from Weinberg[11], and the acceleration which generates the geodesic precession is derived in Huang and Ries[12].

Because space is curved in the vicinity of a massive body, a delay in the measured travel-time of a photon is observed. The maximum effect on a one-way Lageos light-time, in terms of range, is approximately 1 cm. The form of the correction to be added to the computed (geometric) range ρ is given by [13]:

$$\Delta\rho_{rel} = (1+\gamma)\frac{GM_E}{c^2}\ln\left[\frac{r_1 + r_2 + \rho}{r_1 + r_2 - \rho}\right]$$

The quantity r_1 is the distance between the station and the geocenter, and r_2 is the distance between the satellite and the geocenter.

The value of unity for γ and β in Einstein's theory has been verified to better than 1% [14,15], and limits have been placed on the other parameters such that most alternate theories are either invalid or virtually indistinguishable from Einstein's theory for solar-system applications. The main effect of the Schwarzschild solution is the well-known precession of perigee, which for Lageos is about 9 mas/day. The geodesic precession has been verified to at least the 2% level [16], and it amounts to about 17.6 mas/yr in the Lageos orbit node. There is also a relativistic effect due to the oblateness of the Earth which has been partly investigated by Soffel *et al.*[17], and some analysis still remains to be done in this area. Its effect, however, is estimated to be about 0.5 mas/yr on the Lageos node.

The Lense-Thirring precession is a gravitomagnetic effect in which the rotating mass of the Earth tends to 'drag' the orbit plane with it, causing, for example, a 31 mas/yr precession in the node of the Lageos orbit. One proposed method of directly measuring this effect is the GP-B relativity experiment, in which extremely precise gyroscopes will be flown on a drag-free satellite in a polar orbit, and the precession of the gyroscopes due to geodesic and Lense-Thirring precession will be measured with respect to the distant stars[18]. Another experiment has been proposed in which the 'gyroscope' would consist of two Lageos-type satellites in orbits with supplementary inclinations, so that the major part of the Newtonian precession of the orbit nodes, due to the even zonal harmonics of the Earth's gravity field, will cancel[19]

BARYCENTRIC RELATIVITY MODEL

The equations of motion in the barycentric frame have most recently been derived for the case of finite rotating bodies in Dallas[20], but the only body whose rotation is of any consequence for near-Earth satellites is the Earth, and centimeter accuracy is obtained by simply adding the Lense-Thirring perturbation to the well-known point mass equations of motion[21]. The masses for the Sun and planets have generally been determined in the barycentric reference frame and thus are appropriate for use in the equations of motion without scaling, but a geocentric value for the mass of the Earth must be scaled to TDB units. (The error has been found to be negligible when using TDB quantities for computing the Newtonian n-body perturbations in the geocentric frame.) It is shown later that there is an additional correction required to the equations of motion in order to properly account for the nonspherical geopotential in the barycentric frame.

Martin *et al.*[1] describe the full barycentric measurement model for the light-time computation, including an approximate formula for computation of the light time without a full iterative solution which accounts for the motion of the station and satellite during the photon travel time to first order. Because the interval of time recorded by a station clock is different from the interval of time computed in TDB, the measured time intervals must be transformed into the barycentric frame using formulas such as those given by Moyer[22] or Fairhead *et al.*[23]. This transformation will include a location-dependent correction[22]. It is also necessary to transform geocentric station coordinates into the barycentric frame as described in Martin *et al.*[1].

EQUIVALENCE OF GEOCENTRIC AND BARYCENTRIC FRAMES

Considering only the point mass effects from the Earth and Sun, it has been shown analytically that the barycentric and geocentric frames are nearly equivalent for near-Earth satellites[24,25]. This was verified to at least the few centimeter level by comparison of 560 day ephemerides for the Lageos satellite generated by numerical integration of the Newtonian and relativistic equations of motion for point masses in each frame. The comparison was performed by transforming the barycentric position into the geocentric frame at the instant which would be measured by a barycentric frame clock coincident with a geocentric frame clock, and which would correspond to the event of the observation of the satellite position in the geocentric frame. The Earth's *GM* and the initial position and velocity of the satellite were adjusted when one ephemeris was fit to the other in order to account for the scale difference between the frames.

When the oblateness of the Earth, represented predominantly by the geopotential coefficient C_{20}, was included in the comparison, the resulting position differences exhibited a distinct signal with 100 m amplitude in the along-track component of the orbit with a period of 280 days (half the nodal period of the LAGEOS orbit with respect to the Sun). Comparison of the two cases indicated that evaluating the conventional formula for the effect of Earth oblateness using the barycentric position can introduce considerable error. The oblateness perturbation must be evaluated in a manner consistent with the barycentric formulation. A relativistic geopotential correction was developed which involved transforming the barycentric position into the geocentric frame, evaluating the acceleration due to the nonspherical geopotential in that frame, and then transforming the resulting acceleration back into the barycentric frame[26].

An experiment was then performed with the purpose of 1) verifying the near equivalence of the two frames with actual data and a complete force model, and 2) illustrating the effects of the scale difference on the Earth's *GM*, geopotential coefficients and tracking station coordinates. Aside from the relativistic perturbations appropriate to each frame, the force and measurement models were identical and included the nonspherical geopotential perturbations (GEM-L2[27] to degree and order 20) and the Newtonian n-body perturbations due to the Sun, Moon, and planets with positions and masses given by the JPL planetary ephemeris DE-200[28]. The effects of the solid Earth and ocean tides, solar radiation pressure, Earth radiation pressure, and empirical along-track acceleration (or "drag") were also modeled. The measurement model for the experiment included a full light time solution with corrections due to atmospheric delay, station motion due to plate tectonics, satellite center-of-mass offset, and relativistic light time correction. The data set consisted of LAGEOS laser ranges from 19 stations compressed into three minute normal points of 3-5 cm accuracy (and better than 1 cm precision) spanning the period from 1 January 1984 to 31 December 1986.

The experiment consisted of two least-squares fits to the laser ranges simultaneously adjusting initial position and velocity for LAGEOS at the beginning of the three year integration, Earth GM, C_{20}, a solar radiation reflectivity parameter, selected ocean tides, mean station coordinates, and an empirical along-track acceleration every 15 days. In addition, to clearly show the effect of the relativistic geopotential correction, two other fits were performed with a different set of adjusted parameters. For these cases, only GM and the initial conditions were adjusted, with all other parameters held fixed to values described below.

A summary of the experimental results is given in Table 1. The geocentric formulation was used in Case 1, and all the parameters described above were adjusted. The range residual (observed minus computed) rms (root mean square) was 9 cm, indicating an excellent fit for an integration of this length. The recovered value of GM in geocentric units was 398600.440 km^3/sec^2. The parameters in Case 2 were fixed to the Case 1 values (except the initial conditions and GM), but the barycentric formulation for the relativistic perturbations without the relativistic geopotential correction was used. If the relativistic geopotential correction is negligible, and the two frames are equivalent, the predicted result would be a fit to the data comparable to Case 1 and a value of GM differing from the Case 1 value only by the predicted scale difference of 0.006. Instead, the residual rms increased to over 30 meters, and the value for GM was seriously in error, indicating that the two formulations were not consistent. These anomalous values were caused by neglecting the relativistic geopotential correction, which was applied in Case 3. The results of Case 3 indicate that much better agreement with the geocentric formulation of Case 1 was achieved. The residual rms was 10 cm, and the GM value is equal to the predicted 398600.434 km^3/sec^2. Finally, in Case 4, the same parameters as in Case 1 were adjusted with the relativistic geopotential correction applied in the barycentric frame. Thus Case 4 was the barycentric analogue of Case 1. The residual rms was 9 cm, identical to that for Case 1, and the estimated value of GM was as predicted by the scale difference between the frames. (The precise difference was 0.0059 km^3/sec^2, which agrees very well with the scale difference of 1.4808×10^{-8}.) The estimated station coordinates should agree with those in the geocentric frame since the relativistic station coordinate correction[1] had been applied, and indeed the changes were at the millimeter level (a part in 10^{10}). The relativistic geopotential correction allows the geopotential coefficients to remain geocentric quantities, and the recovered values for C_{20} agreed to better than a part in 10^9.

Table 1
LAGEOS LASER RANGE ANALYSIS IN THE GEOCENTRIC AND BARYCENTRIC FRAMES

	Case 1 *	Case 2 †	Case 3 †	Case 4 *
Residual rms (cm)	9	3030	10	9
Earth *GM* (km^3/sec^2)	398600.440	398600.614	398600.434	398600.434
C_{20} (normalized) x 10^3	−0.484168298	--------	--------	−0.484168299

* all parameters adjusted † only initial conditions and GM adjusted

CONCLUSION

When three years of laser ranging data to LAGEOS were processed in the barycentric and geocentric frames, the rms of the residuals from fits to the range data were identical, and the estimates for the Earth's *GM* differed by the scale factor that exists due to the IAU convention for coordinate time. Furthermore, the station coordinates and C_{20} estimated in each case agreed to better than 1 part in 10^9, indicating no significant difference between the two frames for the determination of geodetic quantities from laser range data. Considering current measurement and force model accuracies, the geocentric reference frame is an adequate approximation for modeling the motion of artificial satellites of the Earth. The advantages are that, with the exception of geodesic precession, the modifications to include relativity must only account for the effects of the Earth's mass, and no transformation between frames is required when comparing computed geodetic quantities with measured ones (for example, sea surface height from satellite altimeter measurements). The barycentric formulation is more complete but also more complicated, and equivalent results are obtained only if the oblateness of the Earth is properly taken into account.

ACKNOWLEDGMENTS

This research was supported in part by NASA grants NAS5-28192 and NAGW-1330. Additional computing resources for this investigation were provided by The University of Texas Center for High Performance Computing.

REFERENCES

1. Martin, C. F., M. H. Torrence and C. W. Misner, "Relativistic Effects on an Earth-Orbiting Satellite in the Barycenter Coordinate System," *J. Geophys. Res.*, **90**, B11, 9403, 1985.

2. Fukushima, T., "Terrestrial Coordinate System in General Relativity," in *Proceedings of International Association of Geodesy (IAG) Symposia*, (Bureau Central de l'A.I.G., Paris), 4 13, 1987.

3. Ashby, N. and B. Bertotti, "Relativistic Perturbations of an Earth Satellite," *Phys. Rev. Let.*, **52**, 485, 1984.

4. Lense, J., H. Thirring, "Über die Einfluss der Eigenrotation der Zentralkörper auf die Bewegung der Planeten und Monde nach der Einsteinschen Gravitationstheorie," *Phys. Zeitschr.*, **19**, 156, 1918; English translation by B. Mashhoon *et al.*, *Gen. Relativ. Gravit.*, **16**, 711, 1984.

5. de Sitter, W., "On Einstein's Theory of Gravitation and its Astronomical Consequences," *Mon. Not. Roy. Astron. Soc.*, **77**, 155, 1916.

6. Kaplan, G. H., *The IAU Resolutions on Astronomical Constants, Time Scale and the Fundamental Reference Frame*, U.S. Naval Observatory, Circular No.163, Washington, D.C., 1981..

7. Misner, C. W., "Scale Factors for Relativistic Ephemeris Coordinates," EG&G Washington Analytical Services Center, NASA contract NAS5-25885 report, 1982.

8. Hellings, R. W., "Relativistic Effects in Astronomical Timing Measurements," *Astron. J.*, **91** (3), 650–659, 1986.

9. Guinot, B. and Seidelmann, P. K., "Time Scales: Their History, Definition and Interpretation," *Astron. Astrophys.*, **194**, 304, 1988.

10. Fukushima, T., "Time Systems in General Relativity," *Reference Frames*, J. Kovalevsky *et al.* (eds.), Kluwer Academic Publishers, Dordrecht, 417, 1989.

11. Weinberg, S., *Gravitation and Cosmology: Principles and Application of the General Theory of Relativity*, John Wiley and Sons, New York, 230-241, 1972.

12. Huang, C. and J. C. Ries, *The Effect of Geodesic Precession in the Non-inertial Geocentric Frame*, Center for Space Research TM 87-4, The University of Texas at Austin, 1987.

13. Holdridge, D. B., "An Alternate Expression for Light Time Using General Relativity," Space Programs Summary 37-48, Vol. III, Jet Propulsion Laboratory, 2–4, 1967.

14. Shapiro, I. I., C. C. Counselman, III, and R. W. King, "Verification of the Principle of Equivalence for Massive Bodies", *Phys. Rev. Let.*, **36**, 555, 1976.

15. Hellings, R. W., "Testing Relativity with Solar-System Dynamics," in: *General Relativity and Gravitation*, B. Bertotti, ed., D. Reidel, Dordrecht, 365, 1984.

16. Shapiro, I. I., R. D. Reasenberg, J. F. Chandler, and R. W. Babcock, "Measurement of the de Sitter Precession of the Moon: A Relativistic Three-Body Effect", *Phys. Rev. Let.*, **61**, 2643, 1988.

17. Soffel, M., R. Wirrer, J. Schastox, H. Ruder, and M. Scheider, "Relativistic Effects in the Motion of Artificial Satellites: The Oblateness of the Central Body I", *Celest. Mech.*, **42** (81), 1988.

18. Everitt, C. W. F., W. M. Fairbank, and L. I Schiff, "Theoretical Background and Present Status of the Stanford Relativity-Gyroscope Experiment", ESRO SP-52, proceedings of a Colloquium by the European Space Research Organization, Interlaken, Switzerland, September 4, 1969.

19. Ciufolini, I., "Measurement of the Lense-Thirring Drag on High-Altitude Laser-Ranged Artificial Satellites," *Phys. Rev. Let.,* **56**, 278,1986.

20. Dallas, S. S., "Equations of Motion for Rotating Finite Bodies in the Extended PPN Formalism,"*Celest. Mech.*, **15**, 111, 1977.

21. Moyer, T. D., *Mathematical Formulation of the Double-precision Orbit Determination Program*, JPL Technical Report 32-1527, 1971.

22. Moyer, T. D., "Transformation from Proper Time on Earth to Coordinate Time in Solar System Barycentric Space-Time Frame of Reference: Part 1," *Celest. Mech.*, **23**, 33, 1981

23. Fairhead, L., P. Bretagnon, and J. F. Lestrade, "The Time Transformation TDB-TDT: An Analytical Formula and Related Problem of Convention," *The Earth's Rotation and Reference Frames for Geodesy and Geodynamics*, A. K. Babcock and G. A. Wilkins (eds.), Kluwer Academic Publishers, Dordrecht, 419, 1988..

24. Ashby, N. and B. Bertotti, "Relativistic Effects in Local Inertial Frames," *Phys. Rev. D*, **34**, 2246, 1986.

25. Zhu, S. Y., E. Groten, R. S. Pan, H. J. Yan, Z. Y. Cheng, W. Y. Zhu, C. Huang, and M. Yao, "Motion of Satellites - The Choice of Reference Frames", *Proceedings of International Astronomical Union Colloquim No. 96: "The Few Body Problem"*, Turku, Finland, 1987.

26. Ries, J. C., C. Huang, and M. M. Watkins, "Effect of General Relativity on a Near-Earth Satellite in the Geocentric and Barycentric Reference Frames", *Phys. Rev. Let.,* **61**, 903, 1988.

27. Lerch, F. J., S. M. Klosko, and G. B. Patel, C. A. Wagner, "Gravity Model for Crustal Dynamics (GEM-L2)," *J. Geophys. Res.*, **90** (B11), 9301, 1985

28. Newhall, X X, E. M. Standish, and J. G. Williams, "DE 102, a Numerically Integrated Ephemeris of the Moon and Planets Spanning Forty-four Centuries", *Astron. Astrophys.*, **125**, 150, 1983.

TOWARD DECIMETER TOPEX ORBIT DETERMINATION USING GPS

Sien-Chong Wu*, Thomas P. Yunck† and George A. Hajj‡

Several practical aspects of precision GPS-based Topex orbit determination are investigated. Multipath signals contaminating Topex pseudorange data are greatly reduced by placing the GPS antenna on a conducting backplate consisting of concentric choke rings to attenuate signals coming in from the Topex horizon and below, and by elevating it on a boom to keep it well above all reflecting surfaces. A proper GPS antenna cutoff view angle is chosen so that a sufficient number of GPS satellites with good geometry are in view while reception of reflected signals is minimized. The geometrical strength of the tracking data is optimized by properly selecting GPS satellites to be observed so as to provide data with moderate continuity, low PDOP, and common visibility with ground tracking sites. The tracking performance is greatly enhanced when three complementary sites are added to the minimum ground tracking network consisting of the three NASA DSN sites.

INTRODUCTION

The US/French Ocean Topography Experiment satellite known as Topex/ Poseidon (Ref. 1) will be launched in mid 1992. For its primary scientific mission, it will carry a microwave altimeter that will precisely measure ocean topography. To fully exploit the few-centimeter accuracy of the altimetry measurements, the radial component of the satellite orbit will have to be independently determined to the same level of accuracy. A nominal goal of 13 cm has been set, but Topex would benefit from an altitude accuracy well below a decimeter. Although the baseline tracking system will be ground-based laser ranging, an experimental GPS receiver will also be

* Technical Group Leader, Tracking Systems and Applications Section, Jet Propulsion Laboratory, California Institute of Technology, Mail Stop 238-624, 4800 Oak Grove Drive, Pasadena, California 91109.

† Technical Group Supervisor, Tracking Systems and Applications Section, Jet Propulsion Laboratory, California Institute of Technology, 4800 Oak Grove Drive, Pasadena, California 91109.

‡ Member of the Technical Staff, Tracking Systems and Applications Section, Jet Propulsion Laboratory, California Institute of Technology, 4800 Oak Grove Drive, Pasadena, California 91109.

carried onboard to demonstrate the potential of recently proposed precise GPS tracking techniques.

Differential GPS techniques have been demonstrated to be capable of producing precise relative positioning for ground sites (Refs. 2–4). An accuracy of a few centimeters for baselines over continental distances has been reported (Refs. 3,4). The techniques can be extended to the orbit determination of low earth satellite, provided that the satellite is equipped with a high performance GPS receiver. Extensive analysis has been carried out at JPL for precise Topex orbit determination using differential GPS (Refs. 2, 5, 6). The analysis shows that various forms of differential GPS are capable of meeting the 13 cm Topex accuracy goal. Of these, the most promising form makes use of the reduced-dynamic tracking technique (Refs. 5,6). With this technique, the satellite dynamics are partly inferred from the precise GPS carrier phase measurements so that the dependence on an accurate dynamic model is reduced, thus lowering the damaging effects of mismodeling the dynamics such as the earth's gravity. The published analysis assumed idealized data acquisition conditions. In this paper, some practical aspects are investigated. These include the onboard GPS satellite selection algorithm, minimum GPS antenna view angle, ground tracking network configuration, and means to reduce multipath effects.

BACKGROUND AND ASSUMPTIONS

GPS-based precision orbit determination for Topex is based on a technique called differential GPS. Each GPS satellite transmits a signal at two L-band frequencies (1227.6 and 1575.42 MHz) with a P-code square-wave modulation (Ref. 7). Two data types can be derived from the GPS signal acquired with a properly designed receiver, the P-code pseudorange and carrier phase. For the Topex GPS demonstration, Topex and a network of uniformly distributed ground receivers will be equipped with high performance GPS receivers capable of making simultaneous precise pseudorange and carrier phase measurements to multiple GPS satellites. The ground network will include tracking sites at NASA's three Deep Space Network (DSN) stations and several complementary sites. Fig. 1 shows the hypothetical 6-site network and the ground track of the Topex orbit that have been used in error analyses in the past.

Two major error sources contaminating the GPS pseudorange and carrier phase measurements are the ionospheric delays and the transmitter and receiver clock errors. These two error sources can be eliminated by proper combination of the GPS pseudorange and carrier phase measurements. The two data types are combined in two steps. First, measurements at two L-band frequencies are linearly combined to remove the ionospheric delay effects; then the combined measurements from different receivers observing different GPS satellites are doubly differenced to eliminate any GPS and receiver clock (and instrumental) errors. Double differencing also helps to reduce GPS ephemeris error. The double differencing operation can be carried out explicitly in a pre-processing stage or, more efficiently, in the filter by solving for clock biases as white-noise processes (Refs. 2–6). The Topex GPS demonstration will adopt the reduced-dynamic technique, in which both pseudorange and carrier phase measurements are used in the filtering process. A 3-D force on Topex, treated as process noise, is estimated together with other parameters to desensitize the orbit

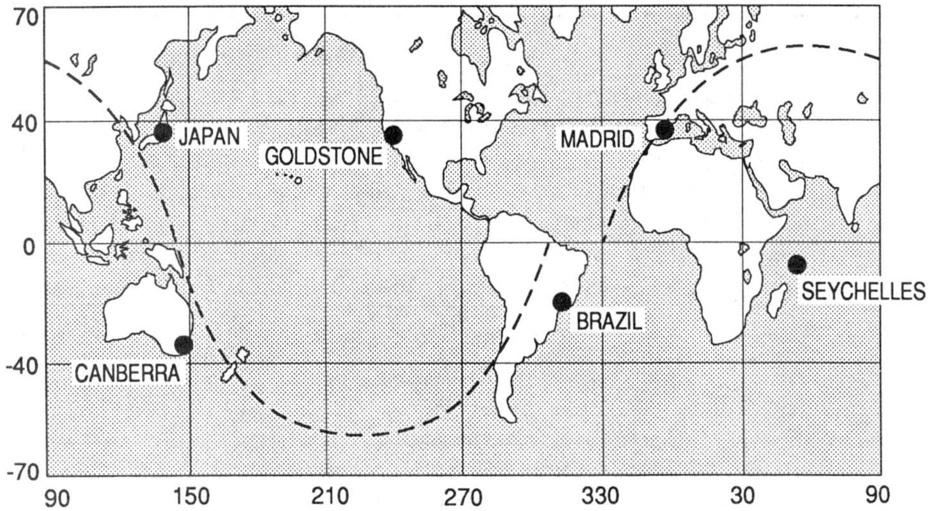

Fig. 1. A hypothetical ground tracking network and a Topex ground track

solution to the dynamic model errors. The degree of desensitization is determined by the parameters characterizing the process-noise force. These are the a priori uncertainty σ_0, the steady-state uncertainty σ_p, and the correlation time τ. In the following analysis, the reduced-dynamic technique with proper values for these parameters will be used for a near-optimum Topex orbit solution.

The ground tracking network and the nominal orbits of Fig. 1 will be adopted in the current analysis. Table 1 includes the error model and other assumptions used, unless otherwise stated. Here, we assume a constellation of 18 GPS satellites uniformly placed in six orbit planes. The Topex receiver is capable of simultaneously observing six GPS satellites while the ground receivers are capable of observing all visible GPS satellites above 10° in elevation.

ON-BOARD GPS SATELLITE SELECTION ALGORITHM

With the 18-satellite GPS constellation, up to nine are visible above the Topex horizon (90° from the zenith). However, the GPS receiver onboard Topex will be equipped with only six independent dual-frequency channels. In other words, signals from only six GPS satellites can be received at any one time. To optimize the tracking data strength, proper selection of six GPS satellites out of the available set is required. Should one or more channels fail, careful satellite selection becomes even more important. There are three criteria that need to be considered. First, switching from one GPS satellite to another should be minimized to maintain data continuity, an important factor for carrier phase data. Secondly, any selected GPS satellite should have common view with at least another selected GPS satellite from both Topex and at least one ground tracking site so that instantaneous elimination of clock error by double differencing is possible. Thirdly, the selected set of GPS satellites should have a low

TABLE 1
ERROR MODEL AND OTHER ASSUMPTIONS USED IN COVARIANCE ANALYSIS

User Satellite:	TOPEX (1,334 km in altitude)
Number of Stations:	6 (cf. Fig. 1)
Number of GPS Satellites:	18
Cut-off Elevation:	10° at stations
	0° at TOPEX
Data Type:	P-code pseudorange
	carrier phase
Data Span:	2 hours
Data Interval:	5 minutes
Data Noise:	5 cm (pseudorange)
	0.5 cm (carrier phase)
Carrier Phase Bias:	10 km (adjusted)
Clock Bias:	3 μsec (adjusted as white process noise)
TOPEX Epoch State:	2 km; 2 m/sec (adjusted)
GPS Epoch States:	2 m; 0.2 mm/sec (adjusted)
Station Location:	5 cm each component
Zenith Troposphere:	1 cm
Earth's GM:	1 part in 10^8
Gravity:	scaled GEM10 – GEML2 (20×20 lumped)
Solar Pressure:	10%

PDOP (Position Dilution of Precision), a measure of performance degradation due to observing geometry. A practical requirement for the GPS satellite selection algorithm for Topex is that it should be sufficiently compact to fit into 2.8 kbytes of the onboard computer memory. The following selection algorithm is developed to meet these criteria.

There are three levels of selection, with increasing complexity. A lower level may be used when all six channels of the Topex receiver are working; higher levels are needed when one or more channels fail.

Level 1: The selection will start with those GPS satellites having the longest remaining view periods. Thereafter any setting GPS satellite will be replaced by the one having the longest remaining view period.

Level 2: In addition, each new GPS satellite selected should have common view with at least one ground station which in turn has at least one other selected GPS satellite in common view with Topex so that doubly differenced data can be formed. This criterion is ignored if the number of eligible GPS satellites is smaller than the number of receiver channels available.

Level 3: In addition, the new set should have a low PDOP value.

A flow chart for the selection algorithm is shown in Fig. 2. The algorithm is made simple by combining the three levels into the calculation of a penalty function. The selection of the proper set of GPS satellites becomes the selection of the lowest value for the penalty function. For Level 3, the penalty function is the square of the weighted PDOP for the selected set of GPS satellites.

$$PDOPSQ = [\text{ TRACE of last } 3 \times 3 \text{ of } (\mathrm{A}^{\mathrm{T}}\,\mathrm{W}\,\mathrm{A})^{-1}\,]\,\Delta T$$

where

$$\mathrm{A} = \begin{pmatrix} 1 & \cos\alpha_1\cos\gamma_1 & \sin\alpha_1\cos\gamma_1 & \sin\gamma_1 \\ 1 & \cos\alpha_2\cos\gamma_2 & \sin\alpha_2\cos\gamma_2 & \sin\gamma_2 \\ \vdots & \vdots & \vdots & \vdots \\ 1 & \cos\alpha_{N_s}\cos\gamma_{N_s} & \sin\alpha_{N_s}\cos\gamma_{N_s} & \sin\gamma_{N_s} \end{pmatrix},$$

with α_i and γ_i being the azimuth and elevation angles of the i-th GPS satellite observed, and N_s the number of GPS satellites selected; W is a diagonal matrix with its diagonal elements equal to the remaining view periods and ΔT is the minimum channel switch interval (nominally 5 minutes). For level 1 and 2,

$$PDOPSQ = 20\,\Delta T \,/\, \Sigma\ (\text{remaining view period})$$

where the summation is over N_s. An additional penalty, $P_1 = 10$, is added to $PDOPSQ$ when a switch of channel is made. Another penalty, $P_2 = 200$, is added when a GPS satellite selected cannot form doubly differenced data (not used for Level 1). After all possible combinations of available GPS satellites are tested, the one with the lowest value of $PDOPSQ$ is selected as the proper set to be observed by the onboard receiver.

Because the set of GPS satellites thus selected is the result of a tradeoff between low PDOP and other considerations, it does not in general yield the lowest PDOP. However, under most circumstances the PDOP is only slightly higher than the lowest possible, as shown in Fig. 3.

GPS ANTENNA VIEW ANGLE

A key factor affecting the performance of GPS-based Topex orbit determination is the GPS antenna view angle. Raising the minimum view angle well above the Topex horizon will not only reduce the number of visible GPS satellites but also weaken the observing geometry. On the other hand, GPS signals coming from too low a view angle (say, near or below the Topex horizon) will suffer from large multipath effects, or even total blockage. A simulation over a 2-day period indicates that a cutoff angle of 85° from the zenith (5° elevation) will allow six or more satellites to be visible 55% of the time, five or more 95% of the time, and at least four at all times, as shown in Fig. 4. If a full hemispherical field of view (0° cutoff elevation) is available, there will be at least five GPS visible at all times and not fewer than six 87% of the time. On the

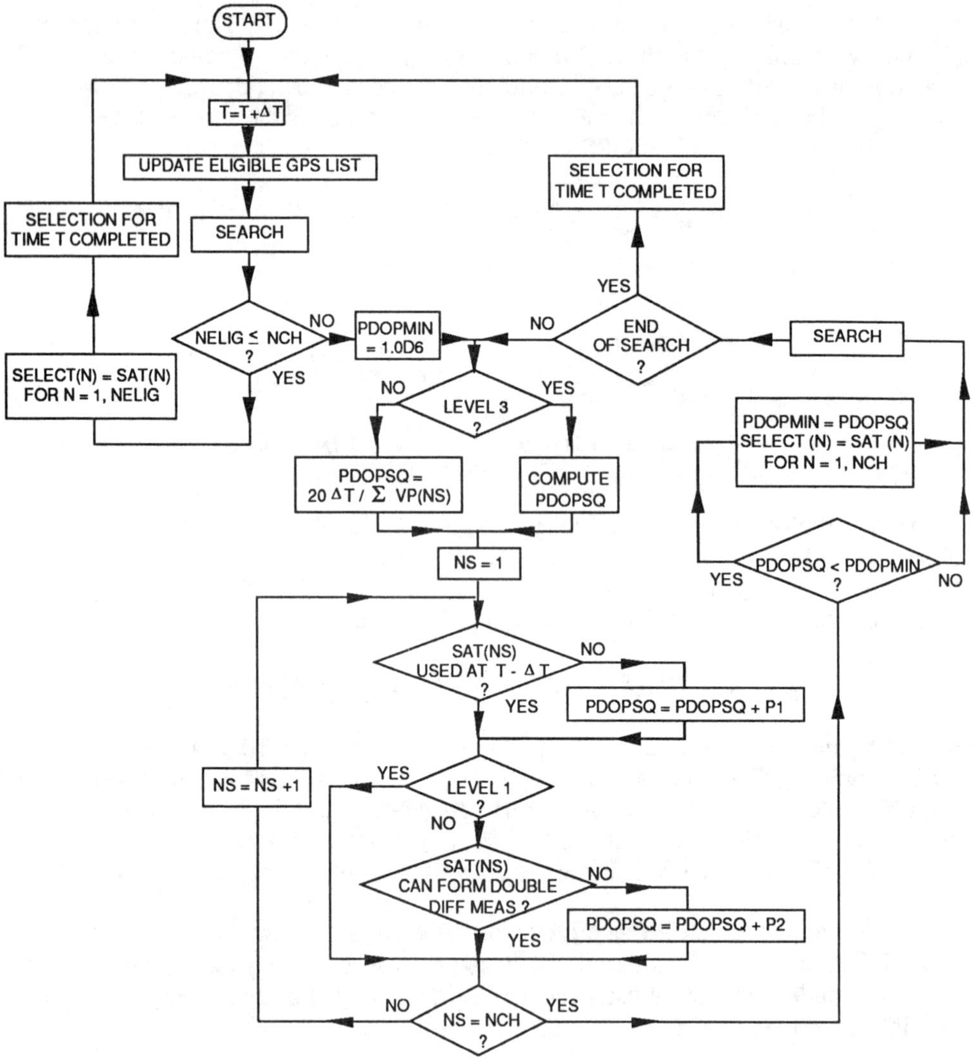

T: CURRENT TIME	
ΔT: SWITCH INTERVAL	SAT(NS): SEARCHED GPS SET
NELIG: NUMBER OF ELIGIBLE (VISIBLE) GPS	SELECT(NS): SELECTED GPS SET
NCH: NUMBER OF CHANNELS	VP(NS): REMAINING VIEW PERIOD
NS: CHANNEL NUMBER (1 — NCH)	P1, P2: PENALTIES (P1 = 10, P2 = 200)

SEARCH: SUBROUTINE FOR SEARCING ALL POSSIBLE COMBINATIONS OF GPS

Fig. 2. GPS selection algorithm for Topex receiver

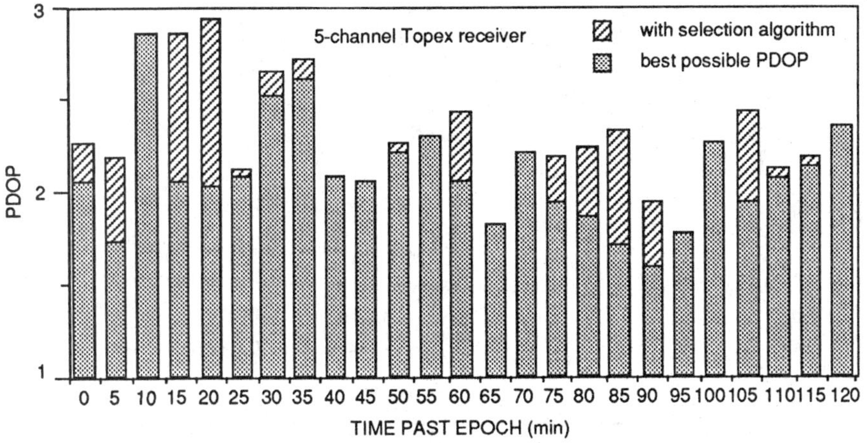

Fig. 3. Increase of PDOP with GPS selection algorithm

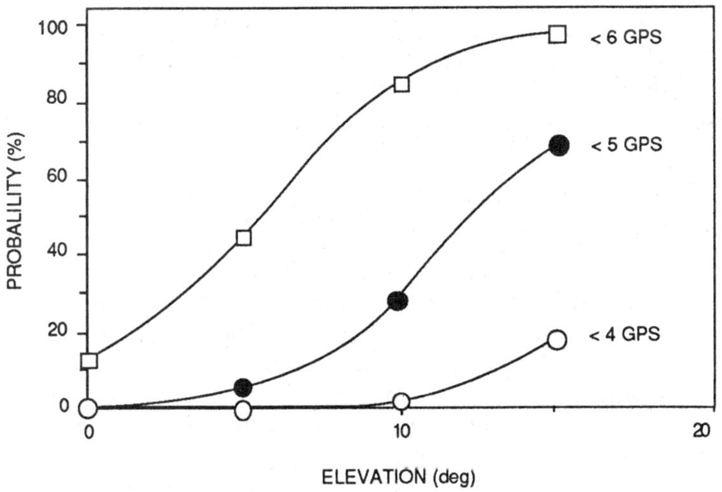

Fig. 4. Probability of fewer than 4, 5, and 6 GPS satellites visible from Topex with different cutoff angle

other hand, a cutoff angle of 80° from zenith will allow fewer than 5 GPS satellites to be visible about 30% of the time.

The performance of Topex orbit determination with different cutoff view angles, as predicted by a covariance analysis, is shown in Fig. 5 for a Topex receiver with 5-channel capability. The multipath effects are ignored in this analysis. When the minimum elevation angle is raised from 0° to 5° the error in Topex altitude increases

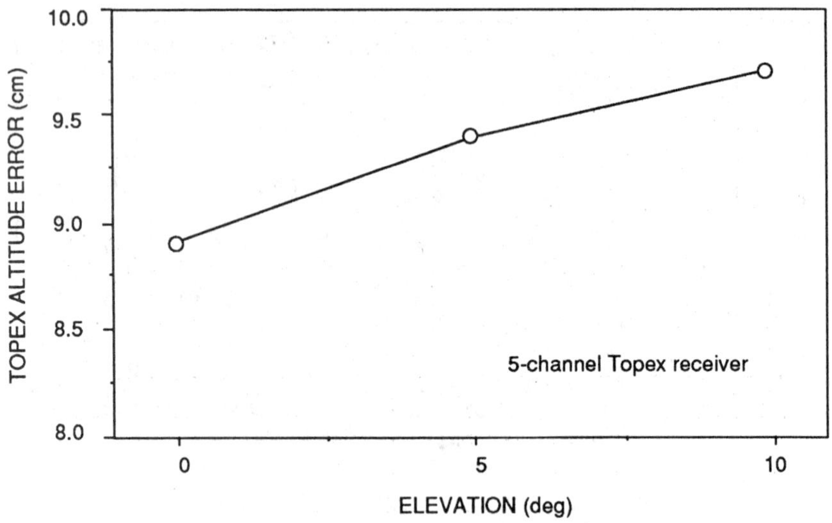

Fig. 5. Comparison of performance with different view angle limits

only marginally (from 8.9 to 9.4 cm). Further increasing the minimum elevation angle to 10° again increases the error by an insignificant amount despite there being fewer than 5 GPS satellites in view about 30% of the time. This is because the reduced-dynamic technique tends to rely more heavily on the dynamic information whenever the geometry is poor, thus smoothing overall error.

GROUND TRACKING NETWORK CONFIGURATION

In addition to establishing an earth-fixed coordinate system, ground tracking sites help to reduce GPS clock, instrumental and ephemeris errors as mention above. The three DSN sites will be equipped with the newly developed high-performance Rogue GPS receivers (Ref. 8) and form the minimum ground tracking network. Clearly, a larger network will mean larger maintenance and data processing cost; however, complementary tracking sites at other parts of the world are needed to broaden the geographical coverage and hence increase the data strength. Earlier analyses have focused on using a six-site ground network. Here, an analysis with the three DSN sites alone is compared with a six-site solution to illustrate the value of the added sites.

Although up to 9 GPS satellites are visible from Topex and each of the ground receivers, the number of GPS satellites visible that can form double differences between Topex and the ground measurements can be as few as zero depending on the number of ground sites used. Fig. 6 compares, for 3-site and 6-site networks, the points along a 2-day Topex ground track where 4 or more GPS satellites can form double differences. Note that the 6-site network provides nearly full global coverage while the 3-site network leaves a large area uncovered. The histogram in Fig. 7 shows that, with the 6-site network, 6 or more GPS satellites can form double differences about 84% of the time, and never will there be fewer than 4; with the 3-site network there are no more than 5 GPS satellites involved for about 84% of the time, and at times no double difference can be formed from any of the visible GPS satellites.

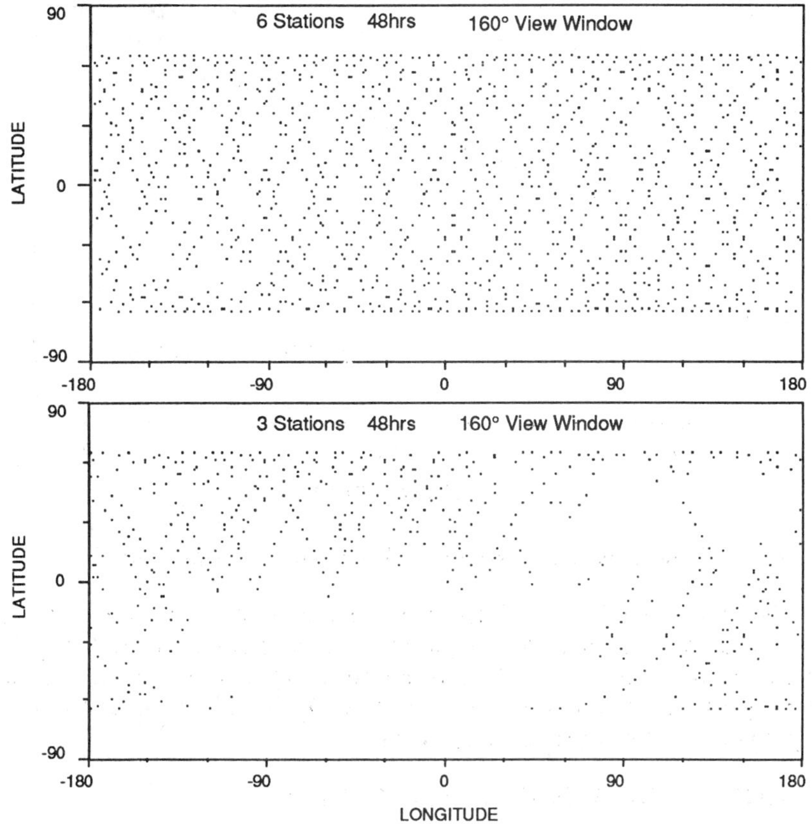

Fig. 6. Two-day coverage of Topex ground track where 4 or more GPS satellites observed can form double difference

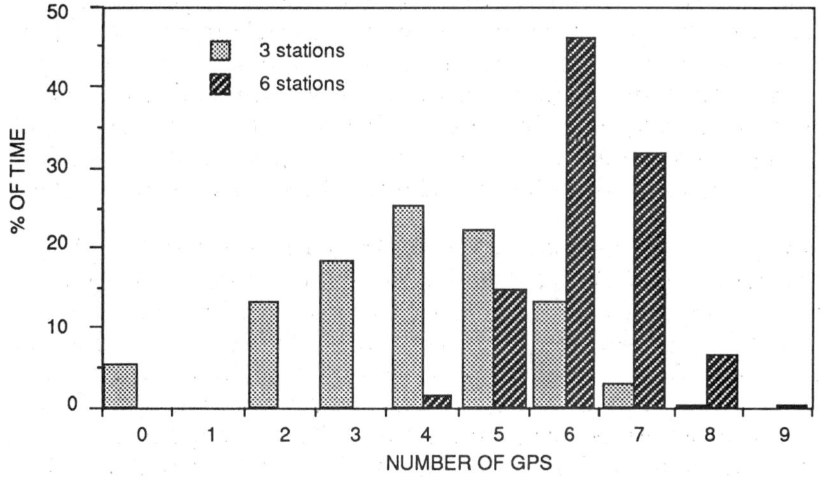

Fig. 7. A histogram of the number of GPS satellites that can form double difference measurements between Topex and ground tracking sites

Fig. 8. compares the performance between the two networks based on the results of a covariance analysis. As expected, the RMS (over a 2-hr period) Topex altitude error with the reduced network is far poorer (by more than a factor of 2) than with the full six-site network. The poor regional coverage results in occasional high error in Topex altitude determination, as can be clearly seen in Fig. 9 for a 5-channel Topex receiver.

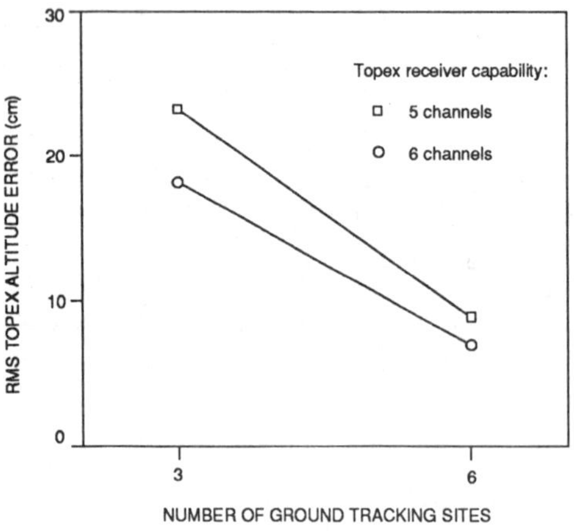

Fig. 8. Comparison of Topex altitude determination performance between a 3-site and a 6-site networks

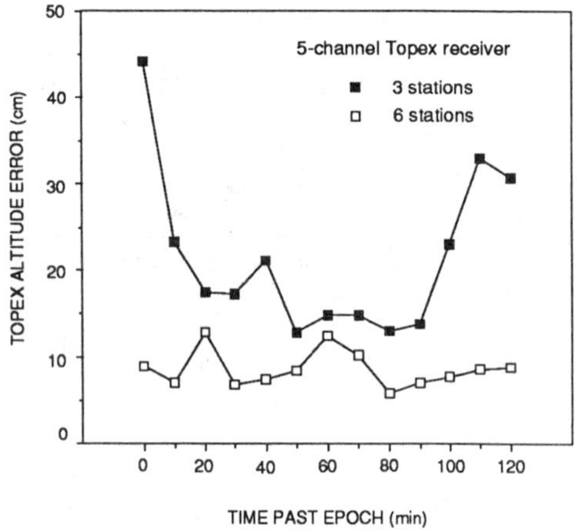

Fig. 9. Topex altitude determination errors using a 3-site and a 6-site networks over a 2-hr period

REDUCTION OF MULTIPATH EFFECTS

Ground-based GPS demonstrations carried out over the past few years have indicated that multipath is a major limitation to accurate pseudorange data. Currently, the thermal noise on the pseudorange measurement of a well-designed GPS receiver is one or two centimeters after a few minutes of averaging (Ref. 8). However, error due to multipath can be as high as several meters. Means to reduce the multipath effects have been investigated. One approach which has proved effective in reducing multipath around ground receivers is placing a GPS antenna in a conducting backplate made up of concentric choke rings (Ref. 9).

The major source of multipath for the Topex flight receiver is the reflected signals from the TDRSS (Tracking and Data Relaying Satellite System) high-gain antenna, the solar panels, and the spacecraft body, as shown in Fig. 10. The effects of multipath on pseudorange measurements will depend on the gain pattern of the GPS antenna and where it is placed with respect to these surfaces. A simple approach to reducing Topex multipath is to elevate the GPS antenna on a boom to keep it well above all reflecting surfaces, while employing a backplate to minimize gain in the direction of the reflected signals.

To assess the multipath effects on GPS pseudorange measurements at Topex a software simulator is being developed. With this new analysis tool, the GPS antenna can be placed at arbitrary heights above the Topex instrument module, and signals reflected from the solar panels, the instrument module, the high-gain antenna, and the attitude control module, as well as the direct GPS-to-Topex signals, can be simulated. The Topex attitude and the orientation of the solar array and of the high-gain antenna are all modeled. The measured radiation pattern of a cross-dipole antenna with a choke-ringed backplate is used. Preliminary results of the analysis indicate that when the antenna is raised to a height of 4.3 m above the top face of the instrument module (1.5 m above the top of the TDRSS antenna) the multipath signals can be greatly reduced, as shown in Fig. 11, where the strength of multipath signals from different reflecting objects is compared with that of the direct signals for three different GPS antenna heights, *h*. These results clearly show that the multipath effects on Topex GPS pseudorange measurements can indeed be reduced with proper placement of the onboard GPS antenna.

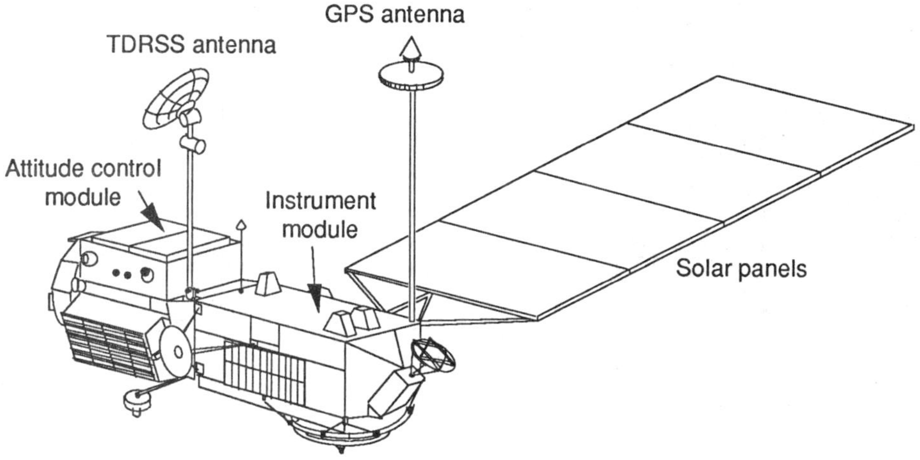

Fig. 10. Topex/Poseidon spacecraft

CONCLUDING REMARKS

Precision Topex altitude determination using differential GPS techniques requires careful design and placement of the Topex onboard antenna, proper siting of the ground tracking network, intelligent satellite selection and an effective estimation scheme. While earlier analysis has indicated that the reduced-dynamic technique is the ideal estimation technique, this paper has focused on other important considerations in achieving decimeter orbit determination. A GPS satellite selection algorithm has been developed. The algorithm is compact enough to fit in the limited onboard computer memory while still selecting a near-optimal set of GPS satellites to observe. The selected set will have reasonably long passes for continuous carrier phase measurements, common visibility with ground tracking sites for double differencing, and low PDOP for good geometrical strength.

The current analysis has shown that a ground tracking network comprising the three NASA DSN stations alone does not provide good global coverage. With this three-site network Topex altitude can be determined only to about 20 cm. Adding

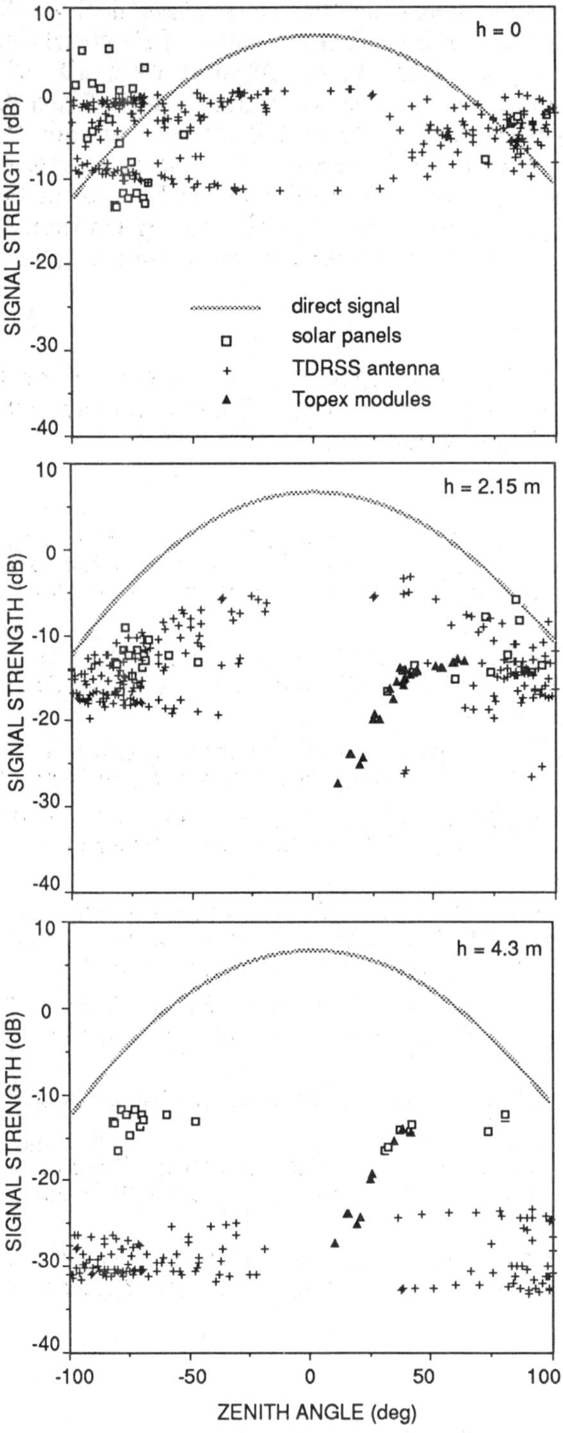

Fig. 11. Reduction of Topex multipath signals by raising the GPS antenna

three complementary sites improves performance dramatically. With the six-site network Topex altitude error is reduced below 10 cm. GPS signals incident from an angle too close to the Topex horizon (0° elevation) or lower will suffer from strong multipath effects. Raising the elevation cutoff angle from 0° to 10° to suppress multipathing has only a slight effect on overall solution strength. Hence, the view window may be narrowed to 160° to reduce the multipath effects on the Topex received signals. Multipath error can be further reduced by raising the GPS antenna above the reflecting objects using a boom. Further analysis and simulation are needed to better understand the characteristics and reduction of multipath effects.

ACKNOWLEDGMENT

The work described in this paper was carried out by the Jet Propulsion Laboratory, California Institute of Technology, under contract with the National Aeronautics and Space Administration.

REFERENCES

1. Born, G. H., Stewart, R. H. and Yamarone, C. A., "TOPEX—A Spaceborne Ocean Observing System," in *Monitoring Earth's Ocean, Land, and Atmosphere from Space—Sensors, Systems, and Applications,* A. Schnapf (ed.), AIAA, Inc., New York, NY, 1985, pp. 464–479.
2. Yunck, T. P., Melbourne, W. G. and Thornton, C. L., "GPS-Based Satellite Tracking System for Precise Positioning," *IEEE Trans. Geosci. & Remote Sensing,* Vol. GE-23, No. 4, Jul. 1985, pp. 450–457.
3. Blewitt, G., Yunck, T.P., Lichten, S. M., W. I. Bertiger, and Wu., S. C., "GPS Geodesy: A Status Report," *High Precision Navigation*, Springer-Verlag, Berlin, 1989, pp. 74–85.
4. Lichten, S. M. and W. I. Bertiger, "Demonstration of Sub-Meter GPS Orbit Determination and 1.5 Parts in 108 Three-Dimensional Baseline Accuracy." to appear in *Bulletin Geodesique.*
5. Wu, S. C., Yunck, T. P. and Thornton, C. L., "Reduced-Dynamic Technique for Precise Orbit Determination of Low Earth Satellites," *Astrodynamics*, Vol 65, Advances in the Astronautical Sciences, AAS, San Diego, CA, 1987, pp. 101–113.
6. Yunck, T. P., Wu, S. C. and Wu, J. T., "Strategies for Sub-Decimeter Satellite Tracking with GPS," *Proc. 1986 IEEE Position Location and Navigation Symp.*, Las Vegas, NV, Nov. 1986.
7. Spilker, J. J., "GPS Signal Structure and Performance Characteristics," *Navigation*, Vol. 25, Summer 1978, pp. 121–146.
8. Meehan, T., *et al.*, "Rogue: A New High Accuracy Digital GPS Receiver," International Union of Geodesy and Geophysics, XIX General Assembly, Vancouver, BC, Canada, Aug. 1987.
9. Meehan, T., *et al.*, "Baseline Results of the ROGUE Digital GPS Receiver," presented at the 1988 American Geophysical Union spring Meeting, Baltimore, MD, May 1988.

AAS 89-360

HOW WELL CAN GRAVITY BE RECOVERED USING TOPEX AND GPS DATA?

Willy Bertiger*, Sien-Chong Wu*, J. T. Wu* and Lisa Skrumeda*

When Topex is launched in mid-1992 it will carry a high quality GPS receiver which will operate in concert with a worldwide network of precision GPS ground receivers. The data from these receivers can be used to recover new information about the earth's gravity field at longer wavelengths. We have developed software and algorithms which will allow this gravity field information to be recovered with much greater efficiency than with traditional techniques. The basis for these algorithms is the gravity bin formulation and related filtering techniques that exploit the repeat orbit of Topex and the sparse matrix structure of the problem. We have used this new software to evaluate the expected improvement in the gravity field using multiple ten-day arcs of GPS data from Topex.

INTRODUCTION

The 1992 launch of NASA's Ocean Topography Experiment satellite, Topex/Poseidon, with its experimental Global Positioning System (GPS) receiver, and a succession of similar GPS-equipped missions thereafter, will enable a marked improvement of current earth gravity models as a result of the comprehensive, multi-dimensional global tracking coverage provided by GPS. For the first time, high precision tracking data will be acquired continuously over the entire globe, including the vast ocean basins and extensive land masses in eastern Europe and Asia that are now inaccessible to western ground based tracking systems. To quantify the information contained in the Topex-GPS data set we have developed new processing techniques which we call the "gravity bin algorithm" (Refs. 1,2). Here we extend previously reported results (Refs. 1,2) to include multiple 10-day arcs coinciding with the Topex ground track repeat cycle.

Topex will carry a dual frequency microwave altimeter accurate to a few centimeters for measuring ocean topography. The scientific potential of this precise altimetry can be fully realized only if Topex geocentric altitude can be independently determined to a few centimeters (Ref. 3). With conventional dynamic tracking techniques, this requires a highly accurate gravity model. Under the current mission plan, the gravity model for Topex will be refined with laser ranging data, which will

* Jet Propulsion Laboratory, California Institute of Technology, Mail Stop 238-640, 4800 Oak Grove Drive, Pasadena, California 91109.

be the baseline tracking data type, collected in the early months of Topex flight. Topex will also carry an experimental GPS flight receiver, capable of observing six GPS satellites simultaneously, to demonstrate the capability of GPS carrier phase and P-code pseudorange for precise orbit determination. In previous studies we have determined that sub-decimeter orbit accuracy can be achieved with just two hours of GPS tracking data, provided that simultaneous measurements are also made at six well distributed ground sites (Refs. 3,4). The precision GPS data from Topex can also be valuable for refining the earth gravity model.

The common procedure for gravity field recovery is to solve simultaneously for a large number of coefficients (possibly thousands) of a spherical harmonic expansion (Refs. 4-6). This procedure is computationally demanding and is usually performed on powerful supercomputers. The gravity bin algorithm reduces the computational demand by reducing the effective number of simultaneously estimated parameters to a few hundred satellite positional adjustments, or gravity bins, relating to the gravity field along the ground track of a single orbit. Because the earth's gravity field is fixed and the satellite ground track repeats, the perturbations due to the gravity field experienced by the satellite repeat nearly perfectly for each repeat of a given track. Data from many repeats of the same track can thus be efficiently combined to recover a small set of gravity perturbations along that track; the perturbations along distinct ground tracks are recovered separately, then combined in a final fast transformation that yields a conventional global harmonic solution that is essentially identical to that produced with the conventional simultaneous estimation approach. In short, by exploiting the repeat ground track of the satellite we are able to partition the global solution geographically, achieving small batches of simultaneously estimated parameters which can then be efficiently combined.

Once the gravity bins are determined, finite differencing of neighboring bins is performed to compute local accelerations. With the data from all distinct tracks we can form a grid of accelerations covering the globe (or at least that portion of the globe overflown by the satellite). These accelerations give a measure of the gravity field information contained in the GPS measurements of Topex.

GRAVITY BIN ALGORITHM

The basic algorithm is described in Refs. 1 and 2. Gravity bins are 3-D positional deviations of the low earth orbiter at each measurement time point over a period of a few hours. The epoch state of the orbiter and other pertinent parameters are also simultaneously adjusted. Assuming no other dynamic errors, the gravity bins are related to current state by the linearized equation,

$$\mathbf{r}(.) = \frac{\partial \mathbf{r}}{\partial \mathbf{r}_0}\mathbf{r}_0 + \frac{\partial \mathbf{r}}{\partial \mathbf{v}_0}\mathbf{v}_0 + \delta(t) \tag{1}$$

where $\mathbf{r}(t)$ is the adjustment to the current position vector of the spacecraft, $\mathbf{r}_0$ and $\mathbf{v}_0$ are the adjustments to the position and velocity vectors at epoch, and $\delta(t)$ is the vector of 3-D gravity bin parameters at time t. In the estimation process, these local gravity bin parameters replace the spherical harmonic coefficients. The local gravity correction $\Delta \mathbf{g}$ can be computed from the solutions of gravity bin parameters δ by

$$\Delta \mathbf{g}(t) = \ddot{\delta}(t) - \frac{\partial \mathbf{g}}{\partial \mathbf{r}} \delta(t) \qquad (2)$$

where $\mathbf{g}$ is the acceleration due to the nominal gravity field. Since the spherical harmonics are orthogonal, the *global* gravity model expressed in terms of spherical harmonics can be constructed from the *local* gravity correction by a process similar to the Fourier transformation. Using the accelerations, $\Delta \mathbf{g}$, and the orthogonality properties ignores some information and will not give exactly the same results as estimating the spherical harmonic functions directly. To get the same result, the bins (δ's) may be used as measurements on the spherical harmonic coefficients. A priori information about the gravity field is combined with the information from the bins after the conversion to spherical harmonics. Converting bins to spherical harmonics is inexpensive compared to measurement processing.

The number of gravity bin parameters required for one Topex orbit is about 150, which compares with a thousand or so spherical harmonic terms to which Topex is sensitive and which might be adjusted in a full gravity solution with an extended data arc. Thus, for each orbit we have a large reduction in the number of parameters treated simultaneously. The price we pay for this reduction is that we must account for the correlations between bins over multiple orbits for a long arc, say, ten days of data. (The Topex ground track repeats every ten days.) The correlation between bin parameters arises from parameters such as the Topex state and station locations which must be treated as common parameters over a long arc. Our algorithms account for the correlations, with very little extra cost compared to the data processing of each single rev of low earth orbiter data.

To detail the algorithm that accounts for the correlation of the bin parameters we assume 90 days of Topex tracking data as an example. Since Topex ground tracks repeat every ten days, we first break the data into 9 contiguous ten day arcs. Since each ten-day arc consists of 127 revs, we further break each ten day arc into 127 individual orbits (about 2 hrs of data). Now we classify parameters into several categories:

1. Nuisance parameters — parameters whose values are independent from 2-hour arc to 2-hour arc and whose values are not needed, e.g. clocks, carrier phase biases.
2. Common parameters — parameters that are common for the entire ten day arc, e.g. Topex epoch state. A new Topex epoch state is solved for on each 10 day arc.
3. Repeating parameters — These are the gravity bins. They are called repeating since the same names are used on each 2-hour arc. They are distinct parameters on each 2-hour arc, but then are common on multiple ten day arcs since the effects of gravity repeat every 10 days.

All individual 2-hour arcs are first processed using standard Square Root Information Filtering (SRIF) techniques (Ref. 7). These are then combined in an optimal way into one grand solution for the 90 days. The CPU time for filtering the 2-hour arcs entirely dominates the computing time (Ref. 2). We detail the steps below.

Removing Nuisance Parameters

The first step toward efficient combining of multiple filter solutions into one grand solution is to permute all the nuisance parameters to the left-hand corner of the 2-hour SRIF array R and remove them from R. This is demonstrated in the following.

Write **x** for the parameter estimates and **z** for the transformed residuals. Then the least-squares estimate of **x** is given by

$$\mathbf{x} = R^{-1}\,\mathbf{z} \tag{3}$$

with a covariance

$$P = R^{-1}\,R^{-t}. \tag{4}$$

Let $\mathbf{x} = (\,\mathbf{x}_n^t, \mathbf{x}_r^t\,)^t$ where $\mathbf{x}_n$ includes all nuisance parameters and $\mathbf{x}_r$ other parameters of interest. Thus R can be written as

$$R = \begin{pmatrix} A & B \\ 0 & C \end{pmatrix} \tag{5}$$

with A associated with the nuisance parameters and C other parameters of interest, and B their correlation. The inverse of R is

$$R^{-1} = \begin{pmatrix} A^{-1} & -A^{-1}BC^{-1} \\ 0 & C^{-1} \end{pmatrix} \tag{6}$$

and the product $R^{-1}\,R^{-t}$ is

$$R^{-1}\,R^{-t} = \begin{pmatrix} A^{-1}A^{-t} + D\,D^t & D\,C^{-t} \\ C^{-1}D^t & C^{-1}C^{-t} \end{pmatrix} \tag{7}$$

where $D = -A^{-1}\,B\,C^{-1}$. Hence, the least-squares estimate of $\mathbf{x}_r$ is

$$\mathbf{x}_r = [\,0,\;C^{-1}\,]\,\mathbf{z} \tag{8}$$

with a covariance

$$P_r = C^{-1}\,C^{-t} \tag{9}$$

In other words, removing parameters which are associated with the left-hand corner of an upper triangular SRIF matrix does not affect the solutions and covariances of the remaining parameters.

For orbit determination of a low earth satellite using GPS measurements, a large number of the estimated parameters are nuisance parameters. Hence the removal of these nuisance parameters substantially reduces the size of the problem involved in the subsequent combining process.

Combining Multiple One-Rev Solutions into a Ten-Day Solution

With the nuisance parameters removed, the remaining parameters are either repeating or common parameters. All common parameters are arranged to reside to the right of all repeating parameters in the SRIF matrix. When combining multiple one-rev solutions over a complete 10-day repeat cycle, one needs to combine only the lower right corner, which now contains information associated with common parameters only. That is, a Householder transformation need only be applied to a string of *small* matrices, as shown in Fig. 1. The upper part of the SRIF matrix remains unchanged by the combining process. Since there are only a handful of common parameters, the combining process is fast. Therefore, a full 10-day repeat cycle of a filtering process containing tens of thousands of parameters (mostly gravity bins) can be carried out at a computational cost essentially the same as processing each one-rev arc with only a few hundred parameters.

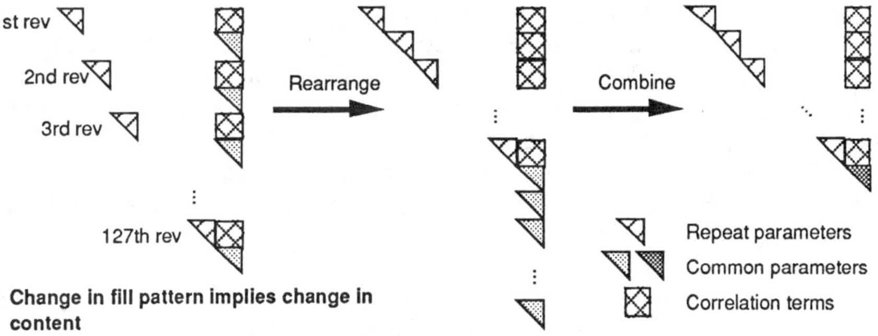

Fig. 1. Combining information of multiple one-rev arcs into a single 10-day arc

The rearrangement requires no computation and the combine step is accomplished with a sparse Householder transformation that takes advantage of the triangle structure of the common parameters. This transformation has sufficient vector length to take good advantage of vectorization on Cray type computers.

Combining Multiple Ten Day Solutions

In this second combining process, the gravity bin parameters are common to all 10-day arcs since they now represent the gravity effects on the Topex orbit over the same ground tracks. Topex initial epoch states are to be treated as (10-day) arc-dependent parameters to account for deviation from nominal orbits over different periods of time. Two combining steps are involved in this process, as shown in Fig. 2.

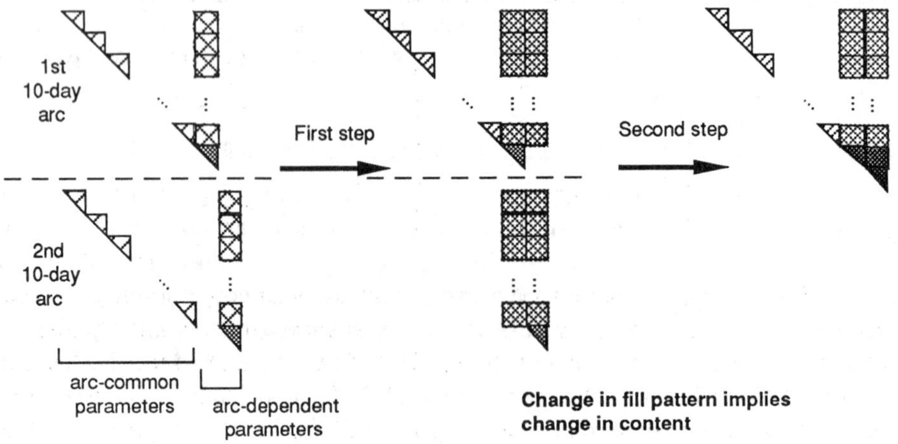

Fig. 2. Combining information of two 10-day arcs

The first step combines the information associated with the arc-common parameters (gravity bins) within each one-rev span from the first 10-day solution with those within the corresponding one-rev span from the second 10-day solution. Note that due to the block structure the rows containing *other* one-rev spans are not affected by this combining process. Hence the process is applied to a pair of one-rev spans at a time; it involves the zeroing of a triangular matrix corresponding to the bin parameters in the second 10-day solution and the transformation of the triangle corresponding to the bins in the first 10-day solution. The two rectangular matrices (which contain the correlation between the arc-common parameters and the arc-dependent parameters) are also transformed.

The second step combines all the transformed rectangular matrices in the second 10-day solution with the two (as yet unchanged) triangular matrices which contain the arc-dependent parameters, to form a single triangular matrix double in size. Since there are only a few arc-dependent parameters involved in this step, the combining process is very fast. To combine more than two 10-day arcs we simply repeat the above process by adding one new 10-day arc at a time to the combined multiple 10-day solutions. The first step will have the same number of operations while the second step will involve an increasing number of arc-dependent parameters; however, this will still be less than 250 parameters for 90 days of data.

Inversion of Combined SRIF Matrix

The result of the above combining process is an upper triangular SRIF matrix. The quantities of interest are the least-squares estimates $\hat{\mathbf{x}} = \mathrm{R}^{-1}\,\mathbf{z}$ and its covariance P $= \mathrm{R}^{-1}\,\mathrm{R}^{-t}$ where R, $\hat{\mathbf{x}}$ and $\mathbf{z}$ now refer to all parameters involved in the combined solution. To extract these quantities, the SRIF matrix must be inverted and postmultiplied by its transpose. Directly inverting a matrix with a dimension of the order of 10^4 is prohibitively time consuming. This of course is not actually necessary, as a dominant portion (containing the gravity bins) of the combined SRIF matrix R has a block diagonal structure; each block of R can be separately inverted. The inverted matrix preserves the same block diagonal structure:

$$R^{-1} = \begin{pmatrix} B_1 & & & & D_1 \\ & B_2 & & & D_2 \\ & & \ddots & & \vdots \\ & & & B_N & D_N \\ & & & & C \end{pmatrix}^{-1} = \begin{pmatrix} B_1^{-1} & & & & -B_1^{-1} D_1 C^{-1} \\ & B_2^{-1} & & & -B_2^{-1} D_2 C^{-1} \\ & & \ddots & & \vdots \\ & & & B_N^{-1} & -B_N^{-1} D_N C^{-1} \\ & & & & C^{-1} \end{pmatrix}$$

where $B_1, B_2, \ldots, B_N$ are triangular matrices, each associated with the gravity bins in a one-rev span; C is also a triangular matrix associated with the arc-dependent parameters and the D's are rectangular matrices. Blank spaces denote null matrices. Therefore, the inversion of the huge combined SRIF matrix is reduced into inversions of small triangular matrices $B_1, B_2, \ldots, B_N$ and C, and the problem is greatly simplified.

COVARIANCE ANALYSIS

A covariance analysis was performed for the problem of determining gravity bins using GPS measurements made onboard Topex and at six globally located ground tracking sites, as shown in Fig. 3. These include the three NASA Deep Space Network (DSN) tracking sites at Goldstone, California; Madrid, Spain; and Canberra, Australia; and three other sites in Japan, Brazil and South Africa. A six-orbit-plane, 18-satellite constellation was assumed for GPS. Carrier phase and P-code pseudorange data were simulated at a rate of 50 samples per Topex revolution. At this sampling rate the gravity bins resolution is comparable to that of a 36x36 spherical harmonic expansion of the global field. Data over 10 days were used. Table 1 lists all the parameters estimated in the filter process, together with their a priori uncertainties. Although tight a priori uncertainties were assumed for station location and GPS orbits, which could be attained with ground tracking of GPS (Ref. 8), the results are in fact insensitive to these values. The a priori information for nuisance and repeating parameters is applied in *all* one-rev filter runs but to only one of these runs for common parameters. The Orbit Analysis and SImulation Software (OASIS) (Ref. 9) was used for the filtering process of the 127 one-rev data arcs. The nuisance parameters were then removed after permuting them to the top of the list. Only the repeating parameters (the gravity bins) and the common parameters were saved and arranged in that order. The 127 one-rev solutions were then combined, adding one at a time, to give a 10-day solution. Multiple 10-day arc solutions were formed by combining the first 10-day arc solution with itself. This approximation introduces only a small error since GPS has such uniform coverage and the Topex ground track repeats.

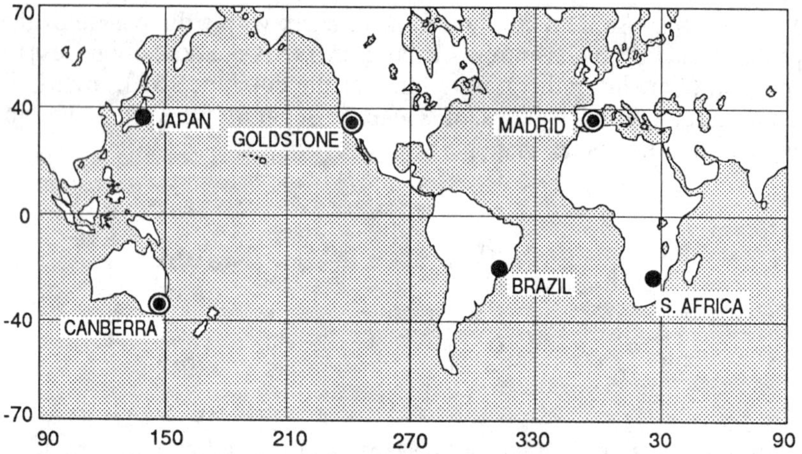

Fig. 3. A global tracking network

TABLE 1. ESTIMATED PARAMETERS AND A PRIORI UNCERTAINTIES

	Parameter Name	a priori σ	# of Parameters
Data:	P-code pseudorange	5 cm	
	carrier phase	0.5 cm	
	Data Rate	50 samples/rev	
	Cut-off Elevation	10 degrees	
Nuisance Parameters:	Carrier Phase Bias	10 km	144
	Clock Bias (white noise)	3 μsec	24
	GPS Epoch State	1 m; 0.1 mm/sec	108
	Zenith Troposphere (random walk)	20 cm bias; 1.3 cm batch to batch	6
Repeat Parameters:	Gravity Bins	1 km	150/rev
Common Parameters:	Station Location	5 cm each component	18
	Topex Epoch State	1 km; 1 m/sec	6

Results

In Ref. 2 it was shown that the sigma for acceleration due to gravity (Δg, Eq. 2) determined by the gravity bins was essentially independent of arc length up to 10 days. This can be explained as follows. Although longer data arcs will improve the estimation of the common parameters (in this case Topex epoch states and station location) their effects on gravity bins are highly correlated from one time to the next.

Since the gravity adjustment Δg is calculated from the second time difference of bin parameters, it is insensitive to such common parameters. Hence, improving these parameters by combining longer data arcs does not help improve gravity recovery accuracy. However, combining multiple 10-day repeat cycles will improve the accuracy. Fig. 4 shows the RMS error for the three acceleration components for arc 63 after combining 10, 20, and 40 days of data. The results scale as expected by approximately $\sqrt{n}$, where n is the number of ten day arcs. Extrapolating the result to 9 ten day arcs yields an average RMS error of .04 mgals (1 mgal = 1×10^{-3} cm/sec^2).

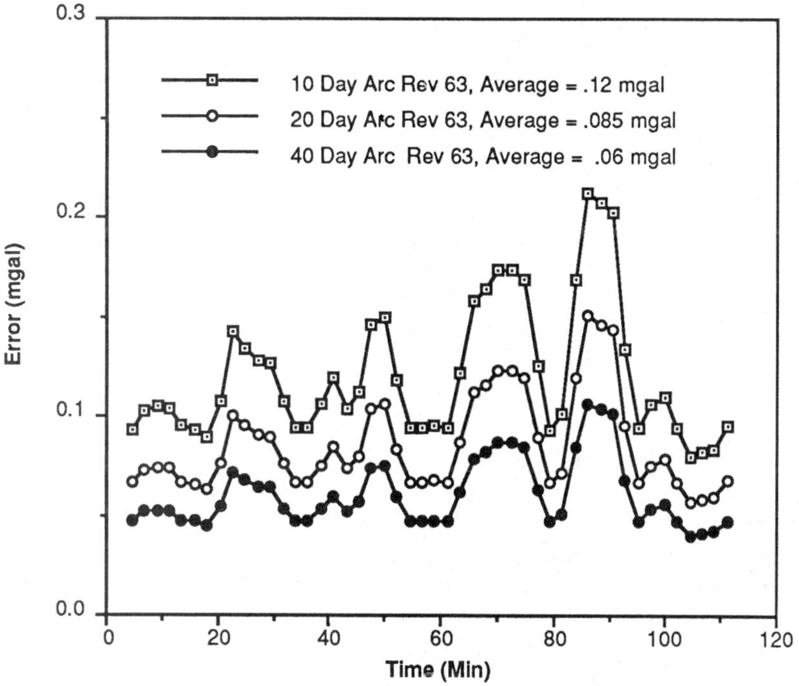

Fig. 4 Predicted RMS Error Arc 63

Arc 63 was chosen as typical; gravity sigmas vary slightly due to geometry from arc to arc. Fig. 5 shows the geometric variation in predicted error for arc 1 and arc 63. Note that although GPS coverage is very uniform there are only six ground stations. We believe it is the non-uniform distribution in ground coverage that causes the geometric variation.

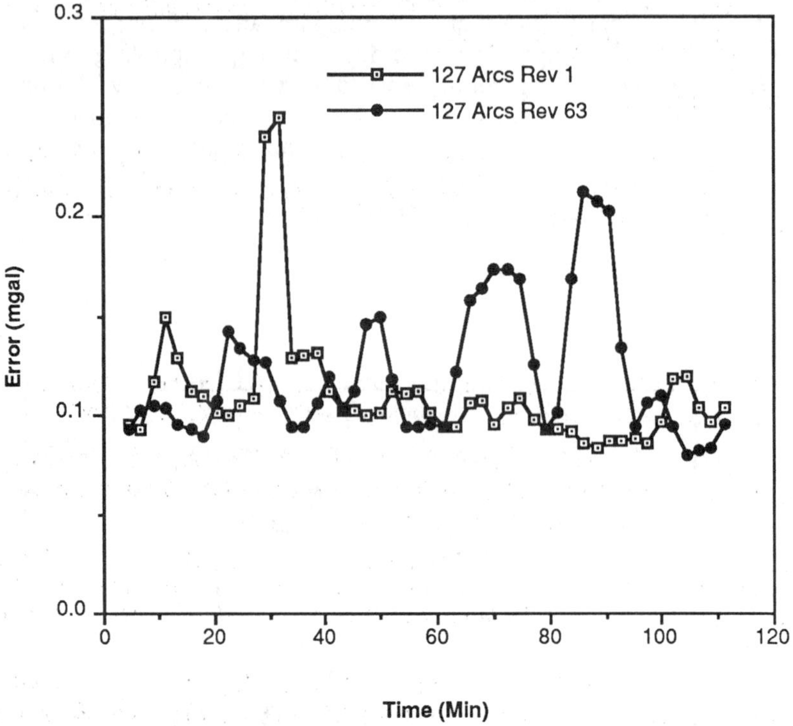

Fig. 5 Predicted RMS Error Arc 1 and 63

It would be interesting to compare the error predicted for the gravity bin technique with the error estimates contained in the covariance matrices of the current high precision gravity fields such as Goddard's GEM-T2 (Ref. 10). Software which will read the GEM-T2 covariance and compute the error in the acceleration at any point in a satellite orbit is currently being completed and results of the comparison will be given in a later report. The study by Marsh, et al. (Ref. 10) reports a predicted contribution to Topex altitude error of 12 cm from the GEM-T2 gravity model used in a conventional dynamic orbit solution. With a large ensemble of repeat arcs, the bin algorithm is projected to reduce Topex altitude error to well under 10 cm (Refs. 1,2) and shows promise for efficiently refining the gravity field for the Topex orbit.

Note that the process of averaging errors on the bins has limitations. Eventually the errors are limited by such things as non-gravitational force model errors on Topex, Earth Orientation errors, and station location errors. Future work will consider some of these error sources.

CONCLUSIONS

We have shown that the accelerations due to gravity may be recovered with GPS data from Topex by means of a highly efficient new filtering techniques. The recovered accelerations are estimated to be accurate to an average RMS value of .04

mgals after processing nine 10-day arcs. The errors are essentially uniform over the portion of the globe overflown by Topex.

ACKNOWLEDGMENT

The work described in this paper was carried out by the Jet Propulsion Laboratory, California Institute of Technology, under contract with the National Aeronautics and Space Administration.

REFERENCES

1. Wu, J. T., "TOPEX Orbit Determination by Solving Gravity Parameters with Multiple Arcs," AAS/AIAA Astrodynamics Specialists Conf., Vail, CO, Aug. 1985.

2. Wu, S. C., Wu, J. T., and Bertiger, W. I., "An Efficient Technique for Gravity Recovery Using a Low Earth Satellite," *Fith International Geodetic Symposium on Satellite Positioning*, Las Cruces, New Mexico, March 13-17, 1989.

3. Born, G. H., Stewart, R. H. and Yamarone, C. A., "TOPEX—A Spaceborne Ocean Observing System," in *Monitoring Earth's Ocean, Land, and Atmosphere from Space—Sensors, Systems, and Applications*, A. Schnapf (ed.), AIAA, Inc., New York, NY, 1985, pp. 464–479.

4. Lerch, F. J., Klosko, S. M., Laubscher, R. E. and Wagner, C. A., "Gravity Model Improvement Using Geos 3 (GEM 9 and 10)," *J. Geophys. Res.*, Vol. 84, No. B8, Jul. 1979, pp. 3897–3916.

5. Lerch, F. J., Klosko, S. M., Patel, G. B. and Wagner, C. A., "A Gravity Model for Crustal Dynamics (GEM-L2)," *J. Geophys. Res.*, Vol. 90, No. B11, Sept. 1985, pp. 9301–9311.

6. Tapley, B. D., Shum, C. K., Yuan, D. N., Ries, J. C., and Schutz, B. E., "An Improved Model for the Earth's Gravity Field," Chapman Conference on the Progress in the Determination of the Earth's Gravity Field, Sept. 13-16, 1988, Fort Lauderdale, FL, pp. 8-11.

7. Bierman, G. J., *Factorization Methods for Discrete Sequential Estimation*, Academic Press, New York, NY, 1977.

8. Yunck, T. P., Melbourne, W. G. and Thornton, C. L., "GPS-Based Satellite Tracking System for Precise Positioning," *IEEE Trans. Geosci. & Remote Sensing*, Vol. GE-23, No. 4, Jul. 1985, pp. 450–457.

9. Wu, S. C. and Thornton, C. L., "OASIS—A New GPS Covariance and Simulation Software System," *Proc. First Int. Symp. on Precise Positioning with GPS*, Rockville, MD, May 1985, pp. 337–345.

10. Marsh, J. G., Lerch, F. J., Putney, B. H., Felsentreger, T. L., Sanchez, B. V., Smith, D. E., Klosko, S. M., Pavlis, E. C., Robbins, J.W., Williamson, R. G., Colombo, O. L., Rowlands, D. D., Eddy, W. F., Chandler, N. L., Rachlin, K. E., Patel, G. B., Bhati, S., and Chinn, D. S., "Gravitational Model Improvement at the Goddard Space Flight Center," Chapman Conference on the Progress in the Determination of the Earth's Gravity Field, Sept. 13-16, 1988, Fort Lauderdale, FL, pp. 3-7.

AAS 89-361

AN AUTONOMOUS ORBIT DETERMINATION SYSTEM FOR EARTH SATELLITES*

Kerry D. Hicks† and William E. Wiesel, Jr.‡

A new system for autonomous satellite navigation is developed and investigated. Unlike many previous studies, however, this system is not limited to the determination of an Earth satellite's current position. By using a two-step technique combined with a general perturbations model, this method allows the air-drag effect on the orbit to be estimated and applied to a future position prediction. Simulations using existing hardware have demonstrated that the algorithm presented is capable of sub-kilometer accuracy.

* This paper appears in full in the microfiche supplement to this volume, **AAS Microfiche Series**, Volume 59.

† Department of Aeronautics and Astronautics, Air Force Institute of Technology, Wright-Patterson AFB, Ohio 45433-5000.

‡ Professor of Astronautical Engineering, Department of Aeronautics and Astronautics, Air Force Institute of Technology, Wright-Patterson AFB, Ohio 45433-5000.

ATTITUDE DETERMINATION AND CONTROL

SESSION 3

Chairman:

Lester L. Sackett
Control and Dynamics Division
C. S. Draper Laboratory

SESSION 8

Chairman:

Robert R. Dasenbrock
U. S. Naval Research Laboratory

AAS 89-364

ATTITUDE DETERMINATION AND PARAMETER ESTIMATION USING VECTOR OBSERVATIONS: APPLICATION

F. Landis Markley*

This paper presents tests of a new method for the simultaneous estimation of spacecraft attitude and sensor biases, based on a quaternion estimation algorithm minimizing Wahba's loss function. The new method is compared with a conventional batch least-squares differential correction algorithm. The estimates are based on data from strapdown gyros and star trackers, simulated with varying levels of Gaussian noise for both inertially-fixed and Earth-pointing attitudes. Both algorithms solve for the spacecraft attitude and the gyro drift rate biases. In the majority of tests performed, the two methods converge to the same estimates in the same number of iterations, but the new algorithm requires about 60 percent more computational effort. Some cases were found in which the new method converges in fewer iterations than the differential correction, and some for which the differential correction requires fewer iterations.

INTRODUCTION

When determining the three-axis attitude of a spacecraft, it is often necessary to simultaneously estimate sensor biases and misalignments. An extended Kalman filter or a batch least-squares differential correction procedure is generally used for this process.[1] These methods, collectively referred to as state estimation methods, require that the nonlinear estimation problem be linearized about *a priori* estimates of the attitude, biases, and misalignments. The purpose of this paper is to compare the standard batch least-squares differential correction procedure with a new algorithm[2] based on the q-method for minimizing Wahba's least-squares loss function.[3] The new algorithm computes the parameters iteratively, but does not linearize about an *a priori* attitude estimate, so it is expected to be more robust than the usual state estimation methods if the problem is highly nonlinear or if initial estimates are poor.

The development of the new algorithm is presented in detail in reference 2, along with some historical background, so it will not be repeated here. The following iterative

* Head, Control System Software and Simulation Section, Code 712.1, NASA Goddard Space Flight Center, Greenbelt, Maryland 20771.

procedure estimates the attitude at time t and the vector $\mathbf{x}$ comprising the m non-attitude parameters. An *a priori* estimate $\mathbf{x}^0$ of $\mathbf{x}$ is used to compute the matrix

$$B(t, \mathbf{x}) \equiv \sum_{i=1}^{n} a_i \, \Phi(t, t_i; \mathbf{x}) \mathbf{b}_i(\mathbf{x}) \mathbf{r}_i^T(\mathbf{x}), \tag{1}$$

where the unit vectors $\mathbf{r}_i$ are representations in an inertial reference frame of the directions to some observed objects, the $\mathbf{b}_i$ are the unit vector representations of the corresponding observations in the spacecraft body frame, the a_i are positive weights, and n is the number of observations. The 3×3 attitude propagation matrix $\Phi(t, t_0; \mathbf{x})$ is the solution of the differential equation

$$\mathrm{d}\Phi(t, t_0; \mathbf{x})/\mathrm{d}t = -[\omega(t, \mathbf{x})\times] \, \Phi(t, t_0; \mathbf{x}) \tag{2}$$

with initial value

$$\Phi(t_0, t_0; \mathbf{x}) = I \equiv \text{the } 3\times 3 \text{ identity matrix}, \tag{3}$$

where the column vector $\omega(t, \mathbf{x})$ contains the components in the body frame of the spacecraft angular velocity relative to inertial space. The matrix $[\mathbf{v}\times]$ is defined for an arbitrary three-vector $\mathbf{v}$ by

$$[\mathbf{v}\times] \equiv \begin{bmatrix} 0 & -v_3 & v_2 \\ v_3 & 0 & -v_1 \\ -v_2 & v_1 & 0 \end{bmatrix} \tag{4}$$

The parameters in $\mathbf{x}$ may enter the matrix $B(t, \mathbf{x})$ through the kinematics expressed by $\Phi(t, t_i; \mathbf{x})$, the observation modeling in $\mathbf{b}_i(\mathbf{x})$, or the reference vector models in $\mathbf{r}_i(\mathbf{x})$. The m matrices $\partial B/\partial x_j$, and the $m(m+1)/2$ independent matrices $\partial^2 B/\partial x_j \partial x_k$ expressing the derivatives of $B(t, \mathbf{x})$ with respect to the parameters must also be computed.

Standard methods[4] are next used to compute the largest eigenvalue $\lambda_{max}(\mathbf{x})$ and corresponding normalized eigenvector $\mathbf{q}_{opt}(t, \mathbf{x})$ of the symmetric 4×4 matrix

$$K(t, \mathbf{x}) \equiv \begin{bmatrix} B(t, \mathbf{x}) + B^T(t, \mathbf{x}) - I \,\mathrm{tr}B(t, \mathbf{x}) & \mathbf{p}(t, \mathbf{x}) \\ \mathbf{p}^T(t, \mathbf{x}) & \mathrm{tr}B(t, \mathbf{x}) \end{bmatrix} \tag{5}$$

with the three-component column vector $\mathbf{p}(t, \mathbf{x})$ defined by

$$[\mathbf{p}(t, \mathbf{x})\times] = B^T(t, \mathbf{x}) - B(t, \mathbf{x}). \tag{6}$$

Then the optimal attitude matrix for parameter vector $\mathbf{x}$, $A_{opt}(t, \mathbf{x})$, is computed from $\mathbf{q}_{opt}(t, \mathbf{x})$ by

$$A_{opt}(t, \mathbf{x}) = (q^2 - \mathbf{Q}^T\mathbf{Q}) \, I + 2\mathbf{Q}\mathbf{Q}^T - 2q[\mathbf{Q}\times], \tag{7}$$

where the three-vector part $\mathbf{Q}$ and scalar part q of the quaternion $\mathbf{q}_{opt}(t, \mathbf{x})$ are given by

$$\mathbf{q}_{opt}^T(t, \mathbf{x}) = [\mathbf{Q}^T, q]. \tag{8}$$

The parameter vector update is given by

$$\delta\mathbf{x}_{opt} = W^{-1}(\mathbf{x})[\mathbf{h}(\mathbf{x}) - W^0(\mathbf{x} - \mathbf{x}^0)], \tag{9}$$

where, for $j, k = 1, \ldots, m$,

$$W_{jk}(\mathbf{x}) = [W^0 - N^T(t, \mathbf{x})M^{-1}(t, \mathbf{x})N(t, \mathbf{x})]_{jk} - \mathrm{tr}[A_{opt}{}^T(t, \mathbf{x})\partial^2 B(t, \mathbf{x})/\partial x_j \partial x_k], \tag{10}$$

$$h_j(\mathbf{x}) = \mathrm{tr}\{[\partial B(t, \mathbf{x})/\partial x_j]A_{opt}{}^T(t, \mathbf{x})\}, \tag{11}$$

$$N_{1j}(t, \mathbf{x}) = \{[\partial B(t, \mathbf{x})/\partial x_j]A_{opt}{}^T(t, \mathbf{x})\}_{23} - \{[\partial B(t, \mathbf{x})/\partial x_j]A_{opt}{}^T(t, \mathbf{x})\}_{32}, \tag{12a}$$

$$N_{2j}(t, \mathbf{x}) = \{[\partial B(t, \mathbf{x})/\partial x_j]A_{opt}{}^T(t, \mathbf{x})\}_{31} - \{[\partial B(t, \mathbf{x})/\partial x_j]A_{opt}{}^T(t, \mathbf{x})\}_{13}, \tag{12b}$$

$$N_{3j}(t, \mathbf{x}) = \{[\partial B(t, \mathbf{x})/\partial x_j]A_{opt}{}^T(t, \mathbf{x})\}_{12} - \{[\partial B(t, \mathbf{x})/\partial x_j]A_{opt}{}^T(t, \mathbf{x})\}_{21}, \tag{12c}$$

and

$$M(t, \mathbf{x}) = \lambda_{max}(\mathbf{x})\, I - B(t, \mathbf{x})A_{opt}{}^T(t, \mathbf{x}). \tag{13}$$

In these equations $\mathbf{x}^0$ is the *a priori* estimate of $\mathbf{x}$ and W^0 is a symmetric positive-semidefinite matrix of weights assigned to this estimate; it is permissible to assign zero weights to the *a priori* estimate by setting $W^0 = 0$. The update $\delta\mathbf{x}_{opt}$ is added to $\mathbf{x}$ to get the new parameter estimate. If the update is small enough, the procedure is complete; otherwise the computations are repeated from Eq. (1) until convergence is achieved.

The attitude covariance $P_{\theta\theta}$, the parameter covariance $P_{\mathbf{xx}}$, and the cross-covariance $P_{\theta\mathbf{x}}$ of the converged estimate can be computed as follows:

$$P_{\theta\theta}(t) = \sigma_{tot}{}^2[M^{-1}(t, \mathbf{x}) + M^{-1}(t, \mathbf{x})N(t, \mathbf{x})W^{-1}(\mathbf{x})N^T(t, \mathbf{x})M^{-1}(t, \mathbf{x})], \tag{14}$$

$$P_{\mathbf{xx}} = \sigma_{tot}{}^2\, W^{-1}(\mathbf{x}), \tag{15}$$

and

$$P_{\theta\mathbf{x}}(t) = P_{\mathbf{x}\theta}{}^T(t) = \sigma_{tot}{}^2 M^{-1}(t, \mathbf{x})N(t, \mathbf{x})W^{-1}(\mathbf{x}), \tag{16}$$

where

$$\sigma_{tot}{}^2 \equiv [\sum_{i=1}^{n} \sigma_i{}^{-2}]^{-1} \tag{17}$$

with σ_i^2 equal to the variance of the ith vector measurement. The covariance computation assumes the weights to be

$$a_i = \sigma_{tot}{}^2/\sigma_i{}^2 \tag{18}$$

and

$$W^0 = \sigma_{tot}{}^2(P^0)^{-1}, \tag{19}$$

where P^0 is the covariance of the *a priori* parameter vector estimate. An analytic expression for $M^{-1}(t, \mathbf{x})$ is derived in the Appendix, eliminating the need for numerical inversion of this matrix.

APPLICATION TO GYRO DRIFT DETERMINATION

For the application to be treated in this paper, we assume that the kinematic information for attitude propagation is obtained from three gyros aligned with the spacecraft body axes. In this case the estimation algorithm assumes that the body rate vector $\omega(t, \mathbf{x})$ is

$$\omega(t, \mathbf{x}) = \omega_g(t) - \mathbf{x}, \tag{20}$$

where $\omega_g(t)$ is the column vector of gyro outputs and $\mathbf{x}$, a three-component vector of gyro drifts, is the vector of parameters to be estimated. These parameters enter $B(t, \mathbf{x})$ through the attitude propagation matrices $\Phi(t, t_i; \mathbf{x})$. The first and second partial derivatives of $\Phi(t, t_i; \mathbf{x})$ with respect to the components of $\mathbf{x}$ are needed to evaluate the corresponding partial derivatives of $B(t, \mathbf{x})$. The partial derivative of Eq. (2) with respect to x_j is, using Eq. (20),

$$d[\partial\Phi(t, t_0; \mathbf{x})/\partial x_j]/dt = -[\omega(t, \mathbf{x})\times][\partial\Phi(t, t_0; \mathbf{x})/\partial x_j] + [\mathbf{e}_j\times]\,\Phi(t, t_0; \mathbf{x}), \tag{21}$$

where $\mathbf{e}_j$ is the unit vector along the jth spacecraft axis. The solution of this differential equation consistent with Eqs. (2) and (3) is

$$\partial\Phi(t, t_0; \mathbf{x})/\partial x_j = \int_{t_0}^{t} \Phi(t, t'; \mathbf{x})[\mathbf{e}_j\times]\,\Phi(t', t_0; \mathbf{x})\,dt'. \tag{22}$$

Using the group property and orthogonality of the attitude propagation matrix,

$$\Phi(t', t_0; \mathbf{x}) = \Phi(t', t; \mathbf{x})\Phi(t, t_0; \mathbf{x}) = \Phi^T(t, t'; \mathbf{x})\Phi(t, t_0; \mathbf{x}), \tag{23}$$

and the relation

$$\Phi[\mathbf{v}\times]\,\Phi^T = [(\Phi\mathbf{v})\times], \tag{24}$$

which holds for any proper orthogonal matrix Φ and three-vector $\mathbf{v}$, gives

$$\partial\Phi(t, t_0; \mathbf{x})/\partial x_j = -[\psi_j(t, t_0; \mathbf{x})\times]\Phi(t, t_0; \mathbf{x}), \tag{25}$$

where $\psi_j(t, t_0; \mathbf{x})$ is the jth column of the matrix

$$\Psi(t, t_0; \mathbf{x}) \equiv -\int_{t_0}^{t} \Phi(t, t'; \mathbf{x})\,dt' = [\psi_1(t, t_0; \mathbf{x})\;\; \psi_2(t, t_0; \mathbf{x})\;\; \psi_3(t, t_0; \mathbf{x})]. \tag{26}$$

The matrices $\Phi(t, t_0; \mathbf{x})$ and $\Psi(t, t_0; \mathbf{x})$ are also used in standard state estimation methods.[5,6]

The matrix of second partial derivatives can be found by differentiating Eq. (25) with respect to x_k, giving

$$\partial^2\Phi(t, t_0; \mathbf{x})/\partial x_j\,\partial x_k = -[\partial\psi_j(t, t_0; \mathbf{x})/\partial x_k\times]\Phi(t, t_0; \mathbf{x}) - [\psi_j(t, t_0; \mathbf{x})\times]\partial\Phi(t, t_0; \mathbf{x})/\partial x_k,$$

$$= \{[\psi_j(t, t_0; \mathbf{x})\times][\psi_k(t, t_0; \mathbf{x})\times] - [\partial\psi_j(t, t_0; \mathbf{x})/\partial x_k\times]\}\Phi(t, t_0; \mathbf{x}). \tag{27}$$

Now the identities

$$[\mathbf{u}\times][\mathbf{v}\times] = \mathbf{v}\mathbf{u}^T - (\mathbf{u}^T\mathbf{v})I = \tfrac{1}{2}(\mathbf{v}\mathbf{u}^T + \mathbf{u}\mathbf{v}^T) + \tfrac{1}{2}(\mathbf{v}\mathbf{u}^T - \mathbf{u}\mathbf{v}^T) - (\mathbf{u}^T\mathbf{v})I$$

$$= \tfrac{1}{2}(\mathbf{v}\mathbf{u}^T + \mathbf{u}\mathbf{v}^T) - (\mathbf{u}^T\mathbf{v})I + \tfrac{1}{2}[([\mathbf{u}\times]\mathbf{v})\times] \tag{28}$$

give

$$\partial^2\Phi(t, t_0; \mathbf{x})/\partial x_j\, \partial x_k = \{\tfrac{1}{2}[\psi_j(t, t_0; \mathbf{x})\psi_k{}^T(t, t_0; \mathbf{x}) + \psi_k(t, t_0; \mathbf{x})\psi_j{}^T(t, t_0; \mathbf{x})]$$

$$- [\psi_j{}^T(t, t_0; \mathbf{x})\psi_k(t, t_0; \mathbf{x})]\, I + [\mathbf{f}_{jk}(t, t_0; \mathbf{x})\times]\}\Phi(t, t_0; \mathbf{x}), \tag{29}$$

where

$$\mathbf{f}_{jk}(t, t_0; \mathbf{x}) \equiv \tfrac{1}{2}[\psi_j(t, t_0; \mathbf{x})\times]\psi_k(t, t_0; \mathbf{x}) - \partial\psi_j(t, t_0; \mathbf{x})/\partial x_k. \tag{30}$$

The matrices $\Phi(t, t_0; \mathbf{x})$ and $\Psi(t, t_0; \mathbf{x})$ and the three-vectors $\mathbf{f}_{jk}(t, t_0; \mathbf{x})$ are computed recursively by adding up contributions over time intervals $t_{i+1} - t_i$, which are chosen to be short enough that variations in ω over the interval can be neglected. Thus

$$\Phi(t_{i+1}, t_0; \mathbf{x}) = \Phi(t_{i+1}, t_i\,; \mathbf{x})\Phi(t_i, t_0; \mathbf{x}), \tag{31}$$

where

$$\Phi(t_{i+1}, t_i\,; \mathbf{x}) = I - [\mathbf{u}\times]\sin\phi + [\mathbf{u}\times]^2(1 - \cos\phi), \tag{32}$$

u is the rotation axis unit vector

$$\mathbf{u} \equiv \omega/|\omega|, \tag{33}$$

and ϕ is the rotation angle

$$\phi \equiv |\omega|(t_{i+1} - t_i\,). \tag{34}$$

Similarly,

$$\Psi(t_{i+1}, t_0; \mathbf{x}) = \Psi(t_{i+1}, t_i\,; \mathbf{x}) + \Phi(t_{i+1}, t_i\,; \mathbf{x})\Psi(t_i, t_0; \mathbf{x}), \tag{35}$$

with

$$\Psi(t_{i+1}, t_i\,; \mathbf{x}) = -(t_{i+1} - t_i\,)\{I - [\mathbf{u}\times]\phi^{-1}(1 - \cos\phi) + [\mathbf{u}\times]^2(1 - \phi^{-1}\sin\phi)\}; \tag{36}$$

and

$$\mathbf{f}_{jk}(t_{i+1}, t_0; \mathbf{x}) - \mathbf{f}_{jk}(t_{i+1}, t_i\,; \mathbf{x}) + \Phi(t_{i+1}, t_i\,; \mathbf{x})\mathbf{f}_{jk}(t_i, t_0; \mathbf{x})$$

$$+ \tfrac{1}{2}\{[\psi_j(t_{i+1}, t_i\,; \mathbf{x})\times]\psi_k(t_{i+1}, t_0; \mathbf{x}) + [\psi_k(t_{i+1}, t_i\,; \mathbf{x})\times]\psi_j(t_{i+1}, t_0; \mathbf{x})\}, \tag{37}$$

with

$$\mathbf{f}_{jk}(t_{i+1}, t_i; \mathbf{x}) = \tfrac{1}{2}(t_{i+1} - t_i\,)^2\phi^{-2}\{[2\sin\phi - \phi\,(1 + \cos\phi)](u_j\mathbf{e}_k + u_k\mathbf{e}_j - 2u_ju_k\mathbf{u})$$

$$- 2(\phi - \sin\phi\,)(I - \mathbf{u}\mathbf{u}^T)_{jk}\mathbf{u} - [2(1 - \cos\phi) - \phi\sin\phi][\mathbf{u}\times](u_j\mathbf{e}_k + u_k\mathbf{e}_j). \tag{38}$$

Starting the recursive computations of Eqs. (31), (35), and (37) requires initial values for the matrices: the identity for $\Phi(t_0\,, t_0; \mathbf{x})$ from Eq. (3), and zero for $\Psi(t_0, t_0; \mathbf{x})$ and $\mathbf{f}_{jk}(t_0, t_0; \mathbf{x})$. Note from Eqs. (37) and (38) that

$$\mathbf{f}_{jk}(t, t_0; \mathbf{x}) = \mathbf{f}_{kj}(t, t_0; \mathbf{x}). \tag{39}$$

OBSERVATION MODELING

Fixed-head star tracker (FHST) data are used to estimate the spacecraft attitude and gyro drifts. Each FHST measurement is a two-component vector $\mathbf{y}_i$ giving the location of the star image in the focal plane of the sensor. For attitude estimation with the new method we need to compute the star unit vector in the spacecraft body frame $\mathbf{b}_i$ in terms of the measurement data $\mathbf{y}_i$. The star unit vector in the sensor frame is

$$\mathbf{s}_i = (1 + |\mathbf{y}_i|^2)^{-1/2}[(y_i)_1, (y_i)_2, 1]^T, \tag{40}$$

and then $\mathbf{b}_i$ is given by

$$\mathbf{b}_i = C_i^T \mathbf{s}_i, \tag{41}$$

where C_i is the proper orthogonal 3×3 matrix defining the orientation in the body frame of the star tracker making this observation.

DATA SIMULATION

Simulated gyro data and FHST data are used to test the algorithm. The simulation assumes a constant angular velocity vector ω_{true}. The gyro data are simulated by adding varying levels of Gaussian noise to the components of ω_{true}. A true attitude matrix is computed by integrating the angular rates;

$$dA_{true}(t)/dt = -[\omega_{true}\times]\, A_{true}(t), \tag{42}$$

with some specified initial attitude matrix $A_{true}(t_0)$.

A star is initially simulated for each star tracker by randomly generating a measurement vector $\mathbf{y}_i$ within the star tracker field of view. Eqs. (40) and (41) then give the star unit vector $\mathbf{b}_i$ in the body frame, and the star unit vector in the inertial reference frame is given by

$$\mathbf{r}_i = A_{true}{}^T(t_i)\mathbf{b}_i, \tag{43}$$

where t_i is the simulation time. For successive simulation times t_i, the reference vector $\mathbf{r}_i$ is held fixed and the vector in the body frame is computed as

$$\mathbf{b}_i = A_{true}(t_i)\mathbf{r}_i. \tag{44}$$

Then the corresponding vector in the star tracker reference frame is given by the inverse of Eq. (41),

$$\mathbf{s}_i = C_i\mathbf{b}_i; \tag{45}$$

the measurement vector $\mathbf{y}_i$ by the inverse of Eq. (40),

$$\mathbf{y}_i = (s_i)_3{}^{-1}[(s_i)_1, (s_i)_2]^T, \tag{46}$$

and Gaussian noise is added to the two components of $\mathbf{y}_i$. This process is continued until the star has been tracked for some fixed number of observations or until it leaves the field of view, at which time a new star is randomly placed in the field of view. Earth and Sun interference are neglected in these simulations.

COMPARISON ALGORITHM

The algorithm chosen for comparison is a batch least-squares differential correction (DC) algorithm similar to that employed in the attitude ground support system of the Upper Atmosphere Research Satellite (UARS).[6] The algorithm provides a least-squares estimate of a six-component state vector

$$\delta\mathbf{X}^T \equiv [\delta\boldsymbol{\theta}^T \; \delta\mathbf{x}^T], \tag{47}$$

where $\delta\boldsymbol{\theta}$ is the attitude error vector at epoch, and $\delta\mathbf{x}$ is the error in the gyro drift estimates. This is updated iteratively as follows. At the start of each iteration, an estimate $\mathbf{x}$ of the gyro drifts and $A_{est}(t_0)$ of the epoch attitude are available. For each measurement $\mathbf{y}_i$, a predicted value $\mathbf{g}_i$ is computed from the known reference vector $\mathbf{r}_i$ by equations identical to (44) – (46), but with the unknown attitude matrix $A_{true}(t_i)$ replaced by

$$A_{est}(t_i) = \Phi(t_i, t_0; \mathbf{x}) A_{est}(t_0). \tag{48}$$

The computed value $\mathbf{g}_i$ is seen to depend on both $\mathbf{x}$ and $A_{est}(t_0)$. The optimal state update is the solution of

$$F\,\delta\mathbf{X} = \sum_{i=1}^{n} \sigma_i^{-2} G_i^T(\mathbf{y}_i - \mathbf{g}_i) - \begin{bmatrix} \mathbf{0} \\ (P^0)^{-1}(\mathbf{x} - \mathbf{x}^0) \end{bmatrix}, \tag{49}$$

where

$$F \equiv \sum_{i=1}^{n} \sigma_i^{-2} G_i^T G_i + \begin{bmatrix} 0 & 0 \\ 0 & (P^0)^{-1} \end{bmatrix}, \tag{50}$$

with σ_i^{-2}, P^0, and $\mathbf{x}^0$ as defined previously, and with $\mathbf{0}$ a three-vector of zeros and 0 a 3×3 matrix of zeros. The 2×6 matrix G_i of partial derivatives of the errors of the ith measurement with respect to $\delta\mathbf{X}$ is given by

$$G_i = \begin{bmatrix} (y_i)_1 (y_i)_2 & -1 - (y_i)_1^2 & (y_i)_2 \\ 1 + (y_i)_2^2 & -(y_i)_1 (y_i)_2 & -(y_i)_1 \end{bmatrix} C_i \; [\Phi(t_i, t_0\,; \mathbf{x}) \;\; \Psi(t_i, t_0; \mathbf{x})]. \tag{51}$$

This state update gives new estimates of the gyro drifts and attitude:

$$\mathbf{x}_{new} = \mathbf{x} + \delta\mathbf{x}, \tag{52}$$

and

$$A_{new}(t_0) = \{I - |\delta\theta|^{-1}[\delta\theta\times]\sin|\delta\theta| + |\delta\theta|^{-2}[\delta\theta\times]^2(1 - \cos|\delta\theta|)\} A_{est}(t_0). \tag{53}$$

This iterative procedure is repeated until convergence is achieved. An estimate of the covariance matrix is provided by

$$\begin{bmatrix} P_{\theta\theta} & P_{\theta x} \\ P_{x\theta} & P_{xx} \end{bmatrix} = F^{-1}. \tag{54}$$

The initial attitude to begin the first iteration is provided by the q-method, as embodied in Eqs. (1) - (7).

NUMERICAL EXAMPLES: NOMINAL CASES

Tests were performed for both inertially-fixed and Earth-pointing spacecraft attitudes, with FHST orientations and other parameters corresponding to the Gamma Ray Observatory (GRO)[7] and UARS[6] spacecraft, respectively. Two star trackers were modeled with 8 degree by 8 degree fields of view and with an angle of approximately 73 degrees between their boresights. The time interval between FHST measurements was 32.768 seconds, the interval used by the UARS onboard computer. The tests with Earth-pointing attitude used a constant pitch rate of -1.083073×10^{-3} rad/sec, which corresponds to an orbit period of 96.7 minutes. For these tests, a simulated star was tracked until it left the FHST field of view. For tests with inertially-fixed attitude, each simulated star was observed only five times. Simulated data spans ranged from ten time intervals, about five minutes, up to 175 time intervals, approximately one full orbit of data. In some tests, the estimate used data from one FHST only, so some test cases had very high observability (one full orbit of data, two FHSTs), and some had very low observability (five minutes of data, one FHST, only two stars).

The data were simulated with no gyro bias, and the estimations were performed with non-zero initial bias estimates, so the bias estimate is the same as the bias error for these tests. The initial bias error was 10^{-4} radians/sec along either the spacecraft roll or pitch axis; the converged estimates of the attitude and gyro drift error should not depend on the initial error. The first "iteration" of the DC is not really a DC iteration; it is an initial attitude estimation using the q-method, as explained above. Thus the gyro bias error after one DC "iteration" is the *a priori* error.

Some tests were performed with perfect star tracker measurements and some with nominal Gaussian noise on each FHST output with standard deviation of 8 arc seconds, or 3.88×10^{-5} radians. The tests with perfect FHST measurements also incorporated perfect gyro mesasurements, but the cases with nominal FHST noise were performed with both perfect gyro measurements and with nominal Gaussian noise on each gyro output of one degree/hour, or 4.848×10^{-6} radians/sec. At least ten iterations were performed for both the new method and the DC in each case, and the computation was considered to have converged when the estimates of the three attitude error angles and the three components of the gyro drift vector remained constant to three significant digits.

The epoch time t_0 for the estimation was taken to be the time of the first observation. In all but four tests, the true attitude matrix at epoch was set equal to the identity matrix. The tests for one simulation case were repeated with four different true attitude matrices at epoch:

$$A_{true}(t_0) = \begin{bmatrix} 0.352 & 0.864 & 0.360 \\ -0.864 & 0.152 & 0.480 \\ 0.360 & -0.480 & 0.800 \end{bmatrix}, \tag{55a}$$

$$A_{true}(t_0) = \mathrm{diag}[1, -1, -1], \tag{55b}$$

$$A_{true}(t_0) = \mathrm{diag}[-1, 1, -1], \tag{55c}$$

and

$$A_{true}(t_0) = \mathrm{diag}[-1, -1, 1], \tag{55d}$$

where diag[...] denotes a 3×3 matrix with the given elements on the main diagonal and zeros elsewhere. The attitude and gyro bias errors for these different initial attitudes were identical to those for $A_{true}(t_0) = I$ within the precision of the output, as they should be. The covariance P^0 of the *a priori* bias estimates was taken to be infinite for all tests.

The most important aspect of the tests, as concerns this paper, is the comparison of the results of the new method to those of the DC. Both algorithms were implemented in double-precision Fortran and executed on a DEC VAX 8810. The gyro bias and attitude errors were not the same at each iteration, but both the general rate of convergence and the final converged errors are almost identical. The cases using perfect simulated FHST and gyro data did not converge in the sense of unchanging estimates to three significant digits, since these computations are dominated by computer algebra errors that prevent convergence to the exact values. The attitude errors for the new method were reduced to 10^{-15} radians and the gyro drift errors to 10^{-17} radians/sec, however, which is a check on the validity of the algorithms and of their computer implementation. The corresponding errors for the DC were 10^{-17} radians for the attitude error and 10^{-19} radians/sec for the gyro drift error. This indicates that the DC would be preferable to the new method in the rare cases where computational errors are more important than measurement errors.

Twenty cases were run with nominal FHST noise and with various choices of the other parameters within their nominal ranges. In all these cases, the DC converged in four to six iterations, with convergence to the same solution for initial gyro drift errors along the roll and pitch axes. The new method converged to the same solution as the DC in all cases to within one percent, requiring the same number of iterations in fifteen test cases, one fewer iteration in four cases, and two more iterations in one case. The largest errors were obtained in tests with data from only one star tracker and with nominal gyro noise, as would be expected. The worst-case attitude estimation error at epoch and gyro drift error were 0.0185 radians (1.1 degrees) and 9.28×10^{-5} radians/sec (19 degrees/hour) for the lowest observability case with ten observations, and 0.0140 radians (0.8 degrees) and 4.72×10^{-6} radians/sec (1.0 degrees/hour) for a span of 175 observations. All these errors were rotations about axes within 12 degrees of the boresight of the single FHST employed. The bias in naturally less well determined in the shorter run, but the epoch attitude error is about the same in the two cases. The worst estimation errors when both star trackers were employed were about one order of magnitude less than the single-FHST errors. In the cases with FHST noise but without gyro noise, the covariance matrix is a good indicator of the estimation errors. When gyro noise is included, the actual errors exceed the covariance estimates; this is not surprising since the covariance computation does not take gyro errors into account, nor does any other part of the estimation process.

The computational effort required by the two algorithms was also measured. The CPU times consist of a part that is independent of the number of iterations performed and a part proportional to the number of iterations. The iteration-independent time is the same for both methods and is equal to

$$t_{CPU,\,both} = 564 + 3.0\,n \ \text{msec}, \tag{56}$$

where n is the number of observations. This includes system overhead and the time required to create the vectors $\mathbf{b}_i$ by Eqs. (40) and (41). The time per iteration for the iteration-dependent part is

$$t_{CPU,\,new} = 3.4 + 1.8\,n \ \text{msec} \tag{57a}$$

for the new method and

$$t_{CPU,\,DC} = 3.1 + 1.1\,n \ \text{msec} \tag{57b}$$

for the DC. The coefficient of n in the iteration-dependent times is the time required to process a measurement, including propagation of the attitude transition matrix, partial dervative matrices, and so forth. The n-independent term represents the end-of-iteration computations, including matrix inversions and computation of updates to the bias vector and attitude matrix. Thus Eq. (57) shows that both the end-of-iteration computations and the measurement processing are more expensive for the new method, with the difference in the measurement processing being greater. The exact CPU times will vary from case to case, but the new method appears to require about 60 percent more computational effort per iteration than the DC for the numbers of measurements typically processed.

NUMERICAL EXAMPLES: NON-NOMINAL CASES

Since the nominal test cases showed no significant differences between the new method and the DC, some single-FHST cases were tried with large sensor errors. For some tests, the gyro noise level was increased by two orders of magnitude to 100 degree/hour, and in other tests the FHST noise was increased by two orders of magnitude to 800 arcsec; all other parameters remained within their nominal ranges. These unrealistically large errors, sometimes caused the nominally positive-definite matrix W defined by Eq. (10) to develop a negative element on its main diagonal. This resulted in a large jump in the estimate, which would then converge to a local minimum of the loss function far from the global minimum. For some of these false solutions, the computed covariance matrix at convergence was not positive definite. The DC did not become singular, since the matrix F of Eq. (50), unlike W, is manifestly positive-semidefinite; but some cases converged extremely slowly.

In order to prevent these large jumps in the estimate, the parameter update Eq. (9) was modified in a fashion similar to Marquardt's technique,[1,8] which interpolates between a gradient search and a Newton-Raphson method, depending on the estimated rate of convergence of the computation. At each iteration the maximum diagonal element of the W matrix

$$W_{max}(\mathbf{x}) \equiv \max_i \; |W_{ii}(\mathbf{x})| \tag{58}$$

and the scaled minimum element of W

$$\alpha(\mathbf{x}) \equiv [\min_i W_{ii}(\mathbf{x})] \,/\, W_{max}(\mathbf{x}) \tag{59}$$

are computed. A minimum allowed value α_{min} for $\alpha(\mathbf{x})$ is also computed as follows. An initial value $\alpha_{min} = 1$ is established for the first iteration. At each iteration after the first, the loss function $1 - \lambda_{max}$ is tested to see if the attitude estimate is improving.[4] The value of α_{min} is doubled if the attitude estimate degrades (the loss function increases by more than 0.1 percent), and it is halved if the attitude estimate improves (the loss function does not increase by more than 0.1 percent), except that α_{min} is not halved in an iteration immediately following one in which it was doubled. If $\alpha(\mathbf{x}) \geq \alpha_{min}$, Eq. (9) is used for the update; but if $\alpha(\mathbf{x}) < \alpha_{min}$, Eq. (9) is replaced by

$$\delta\mathbf{x}_{opt} = \{W(\mathbf{x}) + [\alpha_{min} - \alpha(\mathbf{x})]W_{max}(\mathbf{x})\,I\,\}^{-1}[\mathbf{h}(\mathbf{x}) - W^0(\mathbf{x} - \mathbf{x}^0)]. \tag{60}$$

If the attitude estimate improved, the $\mathbf{x}$ from the previous iteration is used on the right side of Eq. (9) or (60), but if the attitude estimate degraded, the last value of $\mathbf{x}$ that led to an improved attitude estimate is used.

With these modifications, the new method always converged to an estimate close to the DC estimate. The results for 800 arcsec FHST noise variances are given in Table 1, and Table 2 contains the results for 100 degree/hour gyro noise. The results in Table 1 are very reasonable. The converged estimates for the two methods are not identical because the minimized loss functions are not exactly the same. In all cases the attitude error and the product of gyro drift error and the time span of the data are less than 11 degrees. Whenever either of these errors exceeds 0.6 degrees, it is a rotation about an axis within 12 degrees of the boresight of the single FHST employed.

Table 1
Results for 800 arcsec FHST Noise

Attitude Data points	Method	Iterations from roll, pitch error	Epoch attitude error (rad)	Gyro drift error (rad/sec)	Loss function $1-\lambda_{max}$
inertial 175	new DC	7, 6 6, 6	1.07×10^{-2} 1.07×10^{-2}	5.76×10^{-6} 5.76×10^{-6}	1.4×10^{-5} —
inertial 10	new DC	6, 6 16, 18	3.49×10^{-2} 3.70×10^{-2}	9.42×10^{-5} 1.08×10^{-4}	1.2×10^{-5} —
Earth-pointing 175	new DC	8, 7 6, 6	5.19×10^{-3} 5.20×10^{-3}	1.88×10^{-6} 1.88×10^{-6}	1.5×10^{-5} —
Earth-pointing 10	new DC	6, 6 6, 6	0.103 0.103	5.62×10^{-4} 5.60×10^{-4}	1.2×10^{-5} —

Table 2
Results for 100 degrees/hour Gyro Noise

Attitude Data points	Method	Iterations from roll, pitch error	Epoch attitude error (rad)	Gyro drift error (rad/sec)	Loss function $1-\lambda_{max}$
inertial 175	new DC	6, 6 70, 75	0.285 0.285	1.38×10^{-4} 1.37×10^{-4}	4.25×10^{-3} —
inertial 10	new DC	10, 9 55, 56	1.56 1.57	6.38×10^{-4} 6.34×10^{-4}	6.45×10^{-5} —
Earth-pointing 175	new DC	11, 21 10, 9	1.30 1.30	5.97×10^{-4} 5.97×10^{-4}	4.23×10^{-3} —
Earth-pointing 10	new DC	8, 8 8, 8	1.40 1.40	8.72×10^{-3} 8.72×10^{-3}	3.31×10^{-5} —

The results in Table 2, on the other hand, show that neither the new method nor the DC perform well in the presence of large amounts of gyro noise. In particular, attitude estimation errors at epoch of 80 and 90 degrees are found in the two ten-observation cases, and the products of the gyro drift estimation error and the data time span are 196 and 164 degrees for the two cases with Earth-pointing attitude. For the inertial attitude case with ten observations, a much better solution exists, with epoch attitude and gyro drift estimation errors of 3.73×10^{-3} radians and 2.04×10^{-5} radians/sec, respectively, and with a loss function of 1.75×10^{-8}; but neither the DC nor the new method converged to this solution from the prescribed initial values. This failure is not surprising, since neither estimation method takes process noise into account. For cases like these, either an extended Kalman filter[1,5] or a sequential version of the method presented in this paper similar to Shuster's "filter QUEST"[9] would be preferable.

CONCLUSIONS

These tests establish the validity of a new method for the simultaneous estimation of spacecraft attitude and sensor biases, based on a quaternion estimation algorithm minimizing Wahba's loss function. The new algorithm performs as well as a batch least-squares differential correction in most tests of spacecraft attitude and gyro drift estimation, in the sense of converging to equally accurate estimates in the same number of iterations. The computational burden of the new method is about 60 percent greater in typical cases, though. In cases with large amounts of dynamic noise, the new method converges much more rapidly than the differential correction; but the results of neither method are reliable in these cases, since neither algorithm takes dynamic noise explicitly into account. A sequential version of the new algorithm, incorporating process noise, would be expected to give better results when dynamic noise is significant.

REFERENCES

1. WERTZ, JAMES R. (editor) *Spacecraft Attitude Determination and Control*, D. Reidel, Dordrecht, the Netherlands, 1978
2. MARKLEY, F. LANDIS "Attitude Determination and Parameter Estimation Using Vector Observations: Theory," *Journal of the Astronautical Sciences*, Vol. 37, No. 1, January-March, 1989, pp. 41-58
3. WAHBA, GRACE "A Least Squares Estimate of Spacecraft Attitude," *SIAM Review*, Vol. 7, No. 3, July 1965, p. 409.
4. SHUSTER, M. D. and OH, S. D. "Three-Axis Attitude Determination from Vector Observations," *Journal of Guidance and Control*, Vol. 4, No. 1, January-February 1981, pp. 70-77
5. LEFFERTS, E. J., MARKLEY, F. L., and SHUSTER, M. D. "Kalman Filtering for Spacecraft Attitude Estimation," *Journal of Guidance, Control, and Dynamics*, Vol. 5, No. 5, September-October 1982, pp. 417-429
6. KRACK, K. et al., *Upper Atmosphere Research Satellite (UARS) / Extreme Ultraviolet Explorer (EUVE) Flight Dynamics Support System (FDSS) Functional Specifications*, Computer Sciences Corporation, CSC/TR-88/6032, November 1988
7. BLEJER, D. et al., *Gamma Ray Observatory (GRO) Flight Dynamics Support System (FDSS) Specifications: II Functional Specifications, Update 1*, Computer Sciences Corporation CSC/TR-85/6005UD1, February 1987
8. MARQUARDT, D. W.,"An Algorithm for Least-Squares Estimation of Non-Linear Parameters," *SIAM Journal*, Vol. 11, 1963, p. 141
9. SHUSTER, MALCOLM D., "A Simple Kalman Filter and Smoother for Spacecraft Attitude," *Journal of the Astronautical Sciences*, Vol. 37, No. 1, January-March 1989, pp. 89-106
10. MARKLEY, F. LANDIS "Attitude Determination Using Vector Observations and the Singular Value Decomposition," *Journal of the Astronautical Sciences*, Vol. 36, No.3, July-September 1988, pp. 245-258

APPENDIX

The matrix $B(t, \mathbf{x})$ has the singular value decomposition[10]

$$B = U_+ S' V_+^T, \tag{A1}$$

where U_+ and V_+ are proper orthogonal matrices, and

$$S' = \text{diag}[S_1, S_2, S_3], \tag{A2}$$

a 3×3 matrix with S_1, S_2, and S_3 on the main diagonal and zeros elsewhere. The arguments t and $\mathbf{x}$ are omitted from this and all subsequent equations for notational simplicity. The optimal attitude estimate is given in terms of these matrices by[10]

$$A_{opt} = U_+ V_+^T. \tag{A3}$$

The maximum eigenvalue λ_{max} of the matrix K defined by Eq. (5) is related to the optimal attitude by[2]

$$\lambda_{max} = \text{tr}(A_{opt} B^T), \tag{A4}$$

where tr denotes the trace. Eqs. (A1) – (A4) and (13) show that

$$\det B = S_1 S_2 S_3, \tag{A5}$$

$$\lambda_{max} = S_1 + S_2 + S_3, \tag{A6}$$

$$M = U_+ \,\text{diag}[S_2 + S_3, S_3 + S_1, S_1 + S_2] U_+^T. \tag{A7}$$

and

$$\det M = (S_2 + S_3)(S_3 + S_1)(S_1 + S_2). \tag{A8}$$

We now define the scalar

$$\kappa \equiv \tfrac{1}{2}\,[\lambda_{max}^2 - \text{tr}(BB^T)]. \tag{A9}$$

A little algebra shows that

$$\kappa = S_2 S_3 + S_3 S_1 + S_1 S_2, \tag{A10}$$

$$\kappa \lambda_{max} - \det B = \det M, \tag{A11}$$

and

$$\kappa I + BB^T = U_+ \,\text{diag}[(S_3 + S_1)(S_1 + S_2), (S_1 + S_2)(S_2 + S_3), (S_2 + S_3)(S_3 + S_1)] U_+^T$$
$$= \text{adj}\, M, \tag{A12}$$

where adj denotes the adjoint matrix. Eqs. (A11) and (A12) give the desired result

$$M^{-1} = (\kappa \lambda_{max} - \det B\,)^{-1} (\kappa I + BB^T). \tag{A13}$$

The evaluation of M^{-1} by means of Eqs. (A9) and (A13) does not require the singular value decomposition of B.

ATTITUDE DETERMINATION FOR THE STAR TRACKER MISSION

H. L. Fisher*, M. D. Shuster† and T. E. Strikwerda‡

An efficient Kalman filter for spacecraft attitude has been developed for the Star Tracker mission which makes special use of the presence of a CCD star camera. This device is capable in its current configuration of measuring the directions of five stars simultaneously. For such a device the prior-free maximum-likelihood estimate of the attitude based on these simultaneous measurements provides a sufficient statistic which greatly reduces the computational burden of the filter. On the basis of the outputs of this sensor alone, the on-board attitude determination system will determine both three-axis attitude and angular velocity using a dynamical model in which the spacecraft torque is modeled as a Gaussian white noise. Initial simulations of the attitude determination system performance show good performance of the filter.

INTRODUCTION

The Star Tracker experiment will demonstrate near real-time, autonomous satellite attitude determination, accurate to one arc second, using a state-of-the-art charge-coupled-device (CCD) star camera. The experiment, sponsored by the U.S. Army Engineer Topographic Laboratories, will fly on a NASA spacecraft which will be launched for a forty hour mission from the space shuttle and later retrieved for return to Earth. The new camera and accompanying attitude software represent a significant step toward spacecraft autonomy. Extensive simulation software and hardware have been developed for testing the experiment in different configurations during the project[1].

Because the Star Tracker spacecraft carries only a single precise attitude sensor, the CCD star camera, from which both attitude and angular velocity must be determined, it is evident that a Kalman filter[2] must be implemented. However, because the star camera furnishes simultaneous sightings of as many as five different stars in its 7.2 deg by 9. deg field of view, it is also possible to determine three-axis attitude at any sample time without the filter. Thus, it seemed advantageous to design a Kalman filter which would take as its input the attitude computed from one frame of data. The single-frame attitude thus becomes a sufficient statistic for the attitude and angular

* Associate Engineer, Guidance and Control Analysis Section, Space Department, The Johns Hopkins University, Applied Physics Laboratory, Johns Hopkins Road, Laurel, Maryland 20707-6099.

† Senior Professional Staff, Guidance and Control Group, Space Department, The Johns Hopkins University, Applied Physics Laboratory, Laurel, Maryland 20707-6099.

‡ Principal Professional Staff and Supervisor, Guidance and Control Analysis Section, Space Department, The Johns Hopkins University, Applied Physics Laboratory, Laurel, Maryland 20707-6099.

velocity which is used by the Kalman filter as an effective measurement. This approach has several advantages. Firstly, very efficient algorithms exist for computing this single-frame attitude[3]; and, secondly, the fact that attitude can be reliably calculated outside the filter provides a check on the filter performance. In simulations the filter has performed very well.

We begin this report by presenting the state model and the prediction algorithms of the filter. Since the spacecraft dynamics are not presented explicitly, the spacecraft torque is modeled simply as a Gaussian white noise process. The results here are straightforward and follow closely the methods developed in Ref. 2. Next we develop an effective measurement model based on the outputs of the QUEST attitude determination algorithm[3]. Since the QUEST algorithm is a maximum likelihood estimator and the Kalman filter is simply a sequential mechanization of the maximum likelihood estimate when the measurement and process noise are Gaussian, it follows that using the QUEST algorithm as a front end to the filter entails no approximation. A second innovation of the present work is the manner in which the filter is initialized. Generally, when there is no *a priori* information, the Kalman filter is initialized by using a very large initial covariance and any convenient initial condition, it being hoped that the filter will eventually converge to the statistically correct behavior. This brutal treatment is usually unnecessary since it is easy enough to determine a batch maximum likelihood estimate and estimate error covariance from the first two frames of data. Thus, the filter begins with the correct statistical parameters and can provide useful results immediately. The steps needed for the initialization of the filter are described in detail.

A number of simulations have been carried out to demonstrate the filter performance.

DYNAMICAL MODEL AND THE FILTER PREDICTION

The State Equation and State Prediction

The dynamical model for the Star Tracker attitude determination system is simply

$$\frac{d}{dt}\bar{q}(t) = \frac{1}{2}\Omega(\boldsymbol{\omega}(t))\,\bar{q}(t) \quad , \tag{1}$$

$$\frac{d}{dt}\boldsymbol{\omega}(t) = \boldsymbol{\eta}(t) \quad . \tag{2}$$

Here $\bar{q}(t)$ is the attitude quaternion and $\boldsymbol{\omega}(t)$ is the body-referenced angular velocity. $\Omega(\boldsymbol{\omega})$ is given by

$$\Omega(\boldsymbol{\omega}) \equiv \begin{bmatrix} 0 & \omega_3 & -\omega_2 & \omega_1 \\ -\omega_3 & 0 & \omega_1 & \omega_2 \\ \omega_2 & -\omega_1 & 0 & \omega_3 \\ -\omega_1 & -\omega_2 & -\omega_3 & 0 \end{bmatrix} \quad , \tag{3}$$

and $\boldsymbol{\eta}(t)$ is white Gaussian noise with power spectral density $Q(t)$. The spacecraft state vector is simply

$$\mathbf{x}(t) = \begin{bmatrix} \bar{q}(t) \\ \boldsymbol{\omega}(t) \end{bmatrix} \quad . \tag{4}$$

Following Ref. 2 the attitude error state is defined by*

$$\delta\mathbf{x}(t) = \begin{bmatrix} \Delta\boldsymbol{\theta}(t) \\ \Delta\boldsymbol{\omega}(t) \end{bmatrix} \tag{5}$$

where

$$\bar{q}(t) \equiv \frac{1}{\sqrt{1+|\Delta\boldsymbol{\theta}(t)|^2/4}} \begin{bmatrix} \Delta\boldsymbol{\theta}(t)/2 \\ 1 \end{bmatrix} \otimes \bar{q}^*(t) \quad , \tag{6}$$

*Note that we use asterisks to denote estimates and carets to denote unit vectors. Often, for example, when the subscript is $k|k$ or $k|k-1$, the asterisk will be left off when no confusion can occur.

$$\boldsymbol{\omega}(t) = \boldsymbol{\omega}^*(t) + \Delta\boldsymbol{\omega}(t) \quad , \tag{7}$$

and the predicted state vector, $\mathbf{x}^*(t)$, satisfies

$$\frac{d}{dt}\bar{q}^*(t) = \frac{1}{2}\Omega(\boldsymbol{\omega}^*(t))\,\bar{q}^*(t) \tag{8}$$

$$\frac{d}{dt}\boldsymbol{\omega}^*(t) = \mathbf{0} \tag{9}$$

The operation indicated in equation (6) is quaternion composition. Note that $\delta\mathbf{x}(t)$ is six-dimensional while $\mathbf{x}(t)$ is seven-dimensional.

The state-error vector satisfies

$$\frac{d}{dt}\Delta\boldsymbol{\theta}(t) = [[\boldsymbol{\omega}^*(t)]]\,\Delta\boldsymbol{\theta}(t) + \Delta\boldsymbol{\omega}(t) \quad , \tag{10}$$

$$\frac{d}{dt}\Delta\boldsymbol{\omega}(t) = \boldsymbol{\eta}(t) \quad , \tag{11}$$

where

$$[[\,\mathbf{v}\,]] \equiv \begin{bmatrix} 0 & v_3 & -v_2 \\ -v_3 & 0 & v_1 \\ v_2 & -v_1 & 0 \end{bmatrix} \quad . \tag{12}$$

On the basis of these equations, the state prediction equations become in standard Kalman filter notation

$$\bar{q}_{k|k-1} = M_{k-1}\,\bar{q}_{k-1|k-1} \quad , \tag{13}$$

$$\boldsymbol{\omega}_{k|k-1} = \boldsymbol{\omega}_{k-1|k-1} \quad , \tag{14}$$

where

$$M_{k-1} = \cos(\phi_{k-1}/2)\,I_{4\times 4} + \sin(\phi_{k-1}/2)\,\Omega(\,\hat{\mathbf{n}}_{k-1}) \quad , \tag{15}$$

with the angle of rotation given by

$$\phi_{k-1} \equiv |\boldsymbol{\omega}^*_{k-1|k-1}|(t_k - t_{k-1}) \quad , \tag{16}$$

and the axis of rotation by

$$\hat{\mathbf{n}}_{k-1} \equiv \boldsymbol{\omega}^*_{k-1|k-1}/|\boldsymbol{\omega}^*_{k-1|k-1}| \quad . \tag{17}$$

Note that the estimated angular velocity is constant between filter updates as a result of the state equation and not a further approximation.

Covariance Prediction

The state error covariance matrix is defined as

$$P(t) = E\{\delta\mathbf{x}(t)\,\delta\mathbf{x}^T(t)\} \quad , \tag{18}$$

and the covariance prediction equation takes on the usual form

$$P_{k|k-1} = \Phi_{k-1}\,P_{k-1|k-1}\,\Phi^T_{k-1} + Q_{k-1} \quad . \tag{19}$$

To construct the state-error transition matrix we note that the state error vector satisfies

$$\frac{d}{dt}(\delta\mathbf{x}(t)) = F(t)\,\delta\mathbf{x}(t) + G(t)\,\boldsymbol{\eta}(t) \quad , \tag{20}$$

where the state-error sensitivity matrices are

$$F(t) = \begin{bmatrix} [[\boldsymbol{\omega}^*(t)]] & I_{3\times3} \\ 0_{3\times3} & 0_{3\times3} \end{bmatrix} \quad , \tag{21}$$

and

$$G(t) = \begin{bmatrix} 0_{3\times3} \\ I_{3\times3} \end{bmatrix} \quad . \tag{22}$$

Then the state-error transition matrix satisfies

$$\frac{d}{dt}\,\Phi(t,\, t') = F(t)\,\Phi(t,\, t') \quad , \tag{23}$$

with

$$\Phi(t',\, t') = I_{6\times6} \quad . \tag{24}$$

From Ref. 2 it follows that

$$\Phi_{k-1} \equiv \Phi(t_k,\, t_{k-1}) = \begin{bmatrix} \Theta_{k-1} & \Psi_{k-1} \\ 0_{3\times3} & I_{3\times3} \end{bmatrix} \quad , \tag{25}$$

with

$$\Theta_{k-1} = A(t_k)\,A^T(t_{k-1}) \quad , \tag{26}$$

$$\Psi_{k-1} = \int_{t_{k-1}}^{t_k} A(t_k)\,A^T(t')\,dt' \quad , \tag{27}$$

and $A(t)$ is the attitude matrix. Since $\boldsymbol{\omega}^*(t)$ is constant over the interval $(t_{k-1},\, t_k)$, it follows that

$$\Theta_{k-1} = I_{3\times3} + \sin(\phi_{k-1})\,[[\,\hat{\mathbf{n}}_{k-1}\,]] + (1 - \cos(\phi_{k-1}))\,[[\,\hat{\mathbf{n}}_{k-1}\,]]^2 \quad , \tag{28}$$

and

$$\Psi_{k-1} = a_{k-1}\,I_{3\times3} + b_{k-1}\,[[\,\hat{\mathbf{n}}_{k-1}\,]] + c_{k-1}\,[[\,\hat{\mathbf{n}}_{k-1}\,]]^2 \quad , \tag{29}$$

with

$$a_{k-1} = t_k - t_{k-1} \quad , \tag{30a}$$

$$b_{k-1} = \frac{1 - \cos(\phi_{k-1})}{\phi_{k-1}}\,(t_k - t_{k-1}) \quad , \tag{30b}$$

$$c_{k-1} = \left(1 - \frac{\sin(\phi_{k-1})}{\phi_{k-1}}\right)(t_k - t_{k-1}) \quad . \tag{30b}$$

The process noise covariance is given by

$$Q_{k-1} = \int_{t_{k-1}}^{t_k} \Phi(t_k,\, t')\,G(t')\,Q(t)\,G^T(t')\,\Phi^T(t_k,\, t')\,dt' \tag{31}$$

$$= \int_{t_{k-1}}^{t_k} \begin{bmatrix} L_{k-1}(t')\,Q(t')\,L_{k-1}^T(t') & L_{k-1}(t')\,Q(t') \\ Q(t')\,L_{k-1}^T(t') & Q(t') \end{bmatrix} dt' \quad , \tag{32}$$

where

$$L_{k-1} = a_{k-1}(t)\, I_{3\times3} + b_{k-1}(t)\, [[\hat{\mathbf{n}}_{k-1}]] + c_{k-1}(t)\, [[\hat{\mathbf{n}}_{k-1}]]^2 \quad , \tag{33}$$

with $a_{k-1}(t)$, $b_{k-1}(t)$, and $c_{k-1}(t)$ defined analogously to a_{k-1}, b_{k-1}, and c_{k-1} but with $t_k - t_{k-1}$ replaced by $t_k - t$. (Note that this change must be made in ϕ_{k-1} as well.)

THE MEASUREMENT MODEL AND THE FILTER UPDATE

The vector measurement model we consider is that of the QUEST model[3], which is

$$\hat{\mathbf{W}}_{i,k} = A_k\, \hat{\mathbf{V}}_{i,k} + \Delta\hat{\mathbf{W}}_{i,k} \quad , \tag{34}$$

where A_k is the attitude matrix at time t_k, $\hat{\mathbf{W}}_{i,k}$ is the observed star direction, $\hat{\mathbf{V}}_{i,k}$ is the reference star direction, and $\Delta\hat{\mathbf{W}}_{i,k}$ is the measurement noise, assumed to be Gaussian and white with covariance matrix, $R_{\hat{\mathbf{W}}_{i,k}}$, given by

$$R_{\hat{\mathbf{W}}_{i,k}} = \sigma^2_{\hat{\mathbf{W}}_{i,k}} \left[I - (A_k \hat{\mathbf{V}}_{i,k})(A_k \hat{\mathbf{V}}_{i,k})^T \right] \quad . \tag{35}$$

Here, k is the temporal index while i labels the sensor. The model of equations (34) and (35) has received frequent use.

An efficient algorithm, namely QUEST[2], exists for computing the maximum-likelihood estimate (MLE) of the attitude given the set of measurements $\{\hat{\mathbf{W}}_{i,k}, i = 1, \ldots, n_k\}$. QUEST furnishes an MLE quaternion, $\bar{p}^*_k$ and an estimate error covariance (defined in terms of $\Delta\boldsymbol{\theta}_k$, as above), which we may write as

$$\bar{p}^*_k = \delta\bar{p}_k \otimes \bar{q}_k \quad , \tag{36}$$

with

$$\delta\bar{p}_k = \frac{1}{\sqrt{1 + |\mathbf{v}_k|^2/4}} \begin{bmatrix} \mathbf{v}_k/2 \\ 1 \end{bmatrix} \quad . \tag{37}$$

Assuming that the individual star measurements are uncorrelated, $\mathbf{v}_k$ is a discrete white Gaussian process with covariance matrix R_k given by the QUEST algorithm as

$$R_k^{-1} = \sum_{i=1}^{n_k} \frac{1}{\sigma^2_{i,k}} \left[I - (A_k \hat{\mathbf{V}}_{i,k})(A_k \hat{\mathbf{V}}_{i,k})^T \right] \quad . \tag{38}$$

During the update step we define the state error $\delta\bar{q}_k$ according to

$$\bar{q}_k = \delta\bar{q}_k \otimes \bar{q}_{k|k-1} \quad . \tag{39}$$

If we define the derived measurement as

$$\bar{z}_k \equiv 2\, \bar{p}^*_k \otimes (\bar{q}_{k|k-1})^{-1} \quad , \tag{40}$$

then the vector components of this quantity satisfy simply

$$\mathbf{z}_k = \Delta\boldsymbol{\theta}_k + \mathbf{v}_k \quad , \tag{41}$$

$$= H_k\, \delta\mathbf{x}_k + \mathbf{v}_k \quad , \tag{42}$$

where

$$H_k = [I_{3\times3} \;\vdots\; 0_{3\times3}] \quad , \tag{43}$$

and

$$\delta\bar{q}_k \equiv \frac{1}{\sqrt{1+|\Delta\boldsymbol{\theta}_k|^2/4}}\begin{bmatrix}\Delta\boldsymbol{\theta}_k/2\\ 1\end{bmatrix} \quad , \tag{44}$$

which may now be implemented directly into the update step of the Kalman filter.

The Kalman filter update equations[4] are

$$B_k = H_k\, P_{k|k-1}\, H_k^T + R_k \tag{45}$$
$$= (P_{\theta\theta})_{k|k-1} + R_k \quad , \tag{46}$$

$$K_k = P_{k|k-1}\, H_k^T\, B_k^{-1} \quad , \tag{47}$$

$$\boldsymbol{\nu}_k = \mathbf{z}_k - H_k\, \delta\mathbf{x}_{k|k-1} \quad ,$$
$$= \mathbf{z}_k \quad , \tag{48}$$

$$\delta\mathbf{x}_{k|k} = \delta\mathbf{x}_{k|k-1} + K_k\, \boldsymbol{\nu}_k$$
$$= K_k\, \mathbf{z}_k \quad , \tag{49}$$

$$P_{k|k} = (I - K_k H_k)\, P_{k|k-1} \tag{50}$$
$$= (I - K_k H_k)\, P_{k|k-1}\, (I - K_k H_k)^T + K_k\, R_k\, K_k^T \quad . \tag{51}$$

Here, $\boldsymbol{\nu}_k$ is the innovation, B_k is the innovation covariance, and K_k is the Kalman gain matrix. The matrix $P_{\theta\theta}$ is the submatrix of the state-error covariance matrix between angle components. R_k is given by equation (38). Note that by definition the *a priori* state increment in equation (49) vanishes because it is measured from the predicted estimate. Once $\delta\mathbf{x}_{k|k}$ is computed $\bar{q}_{k|k}$ and $\boldsymbol{\omega}_{k|k}$ are computed using

$$\bar{q}_{k|k} = \delta\bar{q}_{k|k} \otimes \bar{q}_{k|k-1} \quad , \tag{52}$$

and

$$\boldsymbol{\omega}_{k|k} = \boldsymbol{\omega}_{k|k-1} + \Delta\boldsymbol{\omega}_{k|k} \quad . \tag{53}$$

INITIALIZATION

In general, no *a priori* initialization scheme is available for the filter since we do not know the statistical parameters of the distribution from which the initial condition is sampled. Thus, the initialization of the filter must be prior-free. Common practice in these cases has been to select an arbitrary initial condition with monstrously large initial covariance. This brutal methodology is dangerous and can result in poor convergence and large numerical errors if the system is poorly conditioned. Poor convergence properties would be particularly problematic in the Star Tracker mission, in which the attitude system will be reinitialized frequently. We wish to avoid the implementation of a square-root information filter, which would be a reliable, if more costly, alternative to the covariance filter presented here. Our approach instead is to compute the correct covariance given the filter model.

The QUEST algorithm provides complete attitudes at every time. Thus, it is sufficient to know these estimates at times t_1 and t_2 to estimate $\mathbf{x}_1$ and $\mathbf{x}_2$ unambiguously. Let us represent the two QUEST attitudes by the attitude matrices, C_1^* and C_2^*, rather than by the quaternion, $\bar{p}_1^*$ and $\bar{p}_2^*$. This will simplify the derivation because we are more accustomed to matrix multiplication than quaternion composition. Then,

$$C_1^* = e^{[[\mathbf{v}_1]]}\, A_1 \quad , \tag{54}$$
$$C_2^* = e^{[[\mathbf{v}_2]]}\, A_2 \quad , \tag{55}$$

where the $\mathbf{v}_k$ are the same as defined earlier. These measurements have realizations $C_1^{*\prime}$ and $C_2^{*\prime}$, the specific values given by QUEST*. The dynamical equations provide *uncertain* constraints on A_1, $A_2, \boldsymbol{\omega}_1$, and $\boldsymbol{\omega}_2$, namely

$$A_2 = e^{[[\,\Delta\boldsymbol{\theta}_1\,]]}\, e^{[[\boldsymbol{\omega}_1\,(t_2-t_1)]]}\, A_1 \quad , \tag{56}$$

$$\boldsymbol{\omega}_2 = \boldsymbol{\omega}_1 + \Delta\boldsymbol{\omega}_1 \quad , \tag{57}$$

and

$$\begin{bmatrix} \Delta\boldsymbol{\theta}_1 \\ \Delta\boldsymbol{\omega}_1 \end{bmatrix} \sim \mathcal{N}(\mathbf{0}, \mathcal{Q}_1) \quad . \tag{58}$$

(Note that the calculation of $\mathcal{Q}_1$ requires that we know $\boldsymbol{\omega}_{1|1}$, which must itself come from the initialization. This turns out not to be a problem since the calculation of $\boldsymbol{\omega}_{1|1}$ (actually of the smoothed estimate $\boldsymbol{\omega}_{1|2}$) in the initialization step can be carried out before that of $\mathcal{Q}_1$.)

If we solve equations (56) and (57) for A_1 and $\boldsymbol{\omega}_1$ and substitute these in equations (54) and (55), we will obtain two very non-linear stochastic equations for A_2 and $\boldsymbol{\omega}_2$, namely,

$$C_1^* = e^{[[\,\mathbf{v}_1\,]]}\, e^{-[[\boldsymbol{\omega}_2-\Delta\boldsymbol{\omega}_1\,]]\,\Delta t}\, e^{-[[\,\Delta\boldsymbol{\theta}_1\,]]}\, A_2 \quad , \tag{59}$$

$$C_2^* = e^{[[\,\mathbf{v}_2\,]]}\, A_2 \quad . \tag{60}$$

Where we have written Δt for $t_2 - t_1$. Substituting equation (60) into equation (59) yields

$$C_1^*\, C_2^{*\,T} = e^{[[\,\mathbf{v}_1\,]]}\, e^{-[[\boldsymbol{\omega}_2-\Delta\boldsymbol{\omega}_1\,]]\,\Delta t}\, e^{-[[\,\Delta\boldsymbol{\theta}_1\,]]}\, e^{-[[\,\mathbf{v}_2\,]]} \quad , \tag{61}$$

which depends only on $\boldsymbol{\omega}_2$, while equation (60) depends only on A_2. We now define $\boldsymbol{\beta}$, which has units of angular velocity, by

$$C_2^{*\prime}\, C_1^{*\prime\,T} \equiv e^{[[\boldsymbol{\beta}]]\Delta t} \quad , \tag{62}$$

If we reparameterize the state now in terms of $\Delta\boldsymbol{\varphi}_2$ and $\Delta\boldsymbol{\tau}_2$ defined by

$$A_2 = e^{[[\,\Delta\boldsymbol{\varphi}_2\,]]}\, C_2^{*\prime} \quad , \tag{63}$$

$$\boldsymbol{\omega}_2 = \boldsymbol{\beta} + \Delta\boldsymbol{\tau}_2 \quad , \tag{64}$$

then we obtain in terms of new measurement matrices, Z_1 and Z_2 ,

$$\begin{aligned} Z_1 &\equiv C_2^*\, C_2^{*\prime\,T} \\ &= e^{[[\,\mathbf{v}_2\,]]}\, e^{[[\,\Delta\boldsymbol{\varphi}_2\,]]} \quad , \end{aligned} \tag{65}$$

$$\begin{aligned} Z_2 &\equiv C_2^{*\prime}\, C_1^{*\prime\,T}\, C_1^*\, C_2^{*\,T} \\ &= e^{[[\boldsymbol{\beta}]]\Delta t}\, e^{[[\,\mathbf{v}_1\,]]}\, e^{-[[\boldsymbol{\beta}+\Delta\boldsymbol{\tau}_2-\Delta\boldsymbol{\omega}_1\,]]\Delta t}\, e^{-[[\,\Delta\boldsymbol{\theta}_1\,]]}\, e^{-[[\,\mathbf{v}_2\,]]} \quad . \end{aligned} \tag{66}$$

Note that we have defined Z_1 and Z_2 so that each has realization (Z_1' and Z_2') $I_{3\times3}$.

We wish to linearize these two effective measurements by combining the exponents, taking the matrix logarithms, and extracting the arguments of the antisymmetric matrices. This operation cannot take place yet because $\boldsymbol{\beta}$ is not infinitesimal. However, it is true that

$$e^{[[\boldsymbol{\beta}]]\Delta t}\, e^{[[\,\mathbf{v}_1\,]]}\, e^{-[[\boldsymbol{\beta}]]\Delta t} = e^{[[\,T_{\beta}\,\mathbf{v}_1\,]]} \quad , \tag{67}$$

where

$$T_{\beta} = e^{[[\boldsymbol{\beta}]]\Delta t} \quad , \tag{68}$$

---

*In the remainder of this report, primed quantities will indicate realizations and unprimed quantities the random variables.

which follows immediately from the transformation properties of orthogonal matrices, and

$$e^{[[\boldsymbol{\beta}+\Delta\boldsymbol{\tau}_2-\Delta\boldsymbol{\omega}_1]]\Delta t} = e^{[[\Delta\boldsymbol{\beta}]]\Delta t}\, e^{[[\boldsymbol{\beta}]]\Delta t} \quad , \tag{69}$$

where

$$\Delta\boldsymbol{\beta} = D^{-1}(\boldsymbol{\beta}\Delta t)\,(\Delta\boldsymbol{\tau}_2 - \Delta\omega_1) \quad , \tag{70}$$

$$D(\boldsymbol{\beta}\Delta t) = I - \frac{1}{2}\,[[\boldsymbol{\beta}\Delta t]] + \frac{2-\beta\Delta t\,\cot((\beta\Delta t)/2)}{2\,(\beta\Delta t)^2}\,[[\boldsymbol{\beta}\Delta t]]^2 \quad , \tag{71}$$

and β is the magnitude of $\boldsymbol{\beta}$. (This formula is derived in the Appendix.) Using these expressions in equation (66), all of the exponents become infinitesimal with the equivalent vectorial linearized result

$$\boldsymbol{\zeta}_1 = \Delta\boldsymbol{\varphi}_2 + \mathbf{v}_2 \quad , \tag{72}$$

$$\boldsymbol{\zeta}_2 = -D^{-1}(\boldsymbol{\beta}\Delta t)\,(\Delta\boldsymbol{\tau}_2 - \Delta\boldsymbol{\omega}_1)\Delta t + T_\beta\,\mathbf{v}_1 - \mathbf{v}_2 - \Delta\boldsymbol{\theta}_1 \quad . \tag{73}$$

Since $\boldsymbol{\zeta}_1$ and $\boldsymbol{\zeta}_2$ have realization $\mathbf{0}$, it follows immediately that the maximum likelihood estimates of $\Delta\boldsymbol{\varphi}_2$ and $\Delta\boldsymbol{\tau}_2$ vanish. Hence, the initialization of the state vector in the Kalman filter at t_2 is

$$A_{2|2} = C_2^* \quad , \tag{74}$$

$$\boldsymbol{\omega}_{2|2} = \boldsymbol{\beta} \quad , \tag{75}$$

The covariance of these estimates is $P_{2|2}$, which is readily calculated as the covariance of

$$\delta\tilde{\mathbf{x}}_{2|2} \equiv \delta\mathbf{x}_2 - \delta\mathbf{x}_{2|2} = \begin{bmatrix} \mathbf{v}_2 \\ -\Delta\boldsymbol{\omega}_1 - (\Delta t)^{-1}\,D(\boldsymbol{\beta}\Delta t)\,(T_\beta\,\mathbf{v}_1 - \mathbf{v}_2 - \Delta\boldsymbol{\theta}_1) \end{bmatrix} \tag{76}$$

using equations (38) and (58). Smoothed values of the state variables, $A_{1|2}$ and $\boldsymbol{\omega}_{1|2}$, can be obtained by predicting the estimates at t_2 backwards in time. However, it is easily seen in the present case that

$$A_{1|2} = C_1^* \quad , \tag{77}$$

$$\boldsymbol{\omega}_{1|2} = \boldsymbol{\omega}_{2|2} = \boldsymbol{\beta} \quad . \tag{78}$$

Strickly speaking, the computation of Q_1 requires that we know $\boldsymbol{\omega}_{1|1}$. However, the estimate $\boldsymbol{\omega}_{2|2}$, will be a sufficiently close approximation for calculating Q_1.

As an approximation we note that neglecting the terms in the initial covariance in Q_1, which should be smaller than those in R_1 and R_2 if any benefit is to be derived from the Kalman filter, and the change in the attitude from t_1 to t_2, the updated covariance at t_2 becomes approximately

$$P_{2|2} \simeq \begin{bmatrix} R_2 & (1/\Delta t)\,R_2 \\ (1/\Delta t)\,R_2 & (1/(\Delta t)^2)\,(R_1 + R_2) \end{bmatrix} \quad , \tag{79}$$

whose implementation in much simpler.

SIMULATION RESULTS

The filter has been tested using a dynamical model in which the attitude is given by the Gibbs vector[5]

$$\mathbf{g}(t) = \mathbf{g}_2(t) \circ \mathbf{g}_1(t) \quad , \tag{80}$$

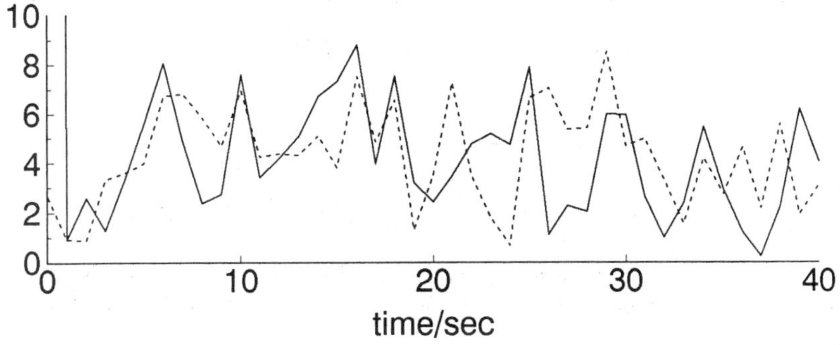

Fig. 1 Star Tracker Pointing Errors

The solid line gives the filter result. The dashed line is the single-frame estimate using QUEST. The angles are given in arc sec.

where "∘" denotes Gibbs-vector composition and

$$\mathbf{g}_1(t) = \hat{\mathbf{z}}\ \tan(\omega_o t/2) \quad , \tag{81}$$

with $\omega_o = .001\,\text{sec}^{-1}$ is just one rotation per orbit motion of the spacecraft and

$$\mathbf{g}_2(t) = \hat{\mathbf{x}}\ \tan(a_1 \sin\lambda_1 t) + \hat{\mathbf{y}}\ \tan(a_2\cos\lambda_2 t) + \hat{\mathbf{z}}\ \tan(a_3 \sin\lambda_3 t) \quad , \tag{82}$$

with

$$a_1 = 2.\,\text{deg} \quad , \quad a_2 = 1.\,\text{deg} \quad , \quad a_3 = 3.\,\text{deg} \quad , \tag{83a}$$

$$\lambda_1 = .02\,\text{sec}^{-1} \quad , \quad \lambda_2 = .025\,\text{sec}^{-1} \quad ,\text{and} \quad \lambda_3 = .015\,\text{sec}^{-1} \quad , \tag{83b}$$

an arbitrary high-frequency component of the attitude motion which is roughly at the limits of angular velocity and acceleration expected for the normal operation of the Star Tracker mission. We assume conservatively that the star tracker has an accuracy of 10 arc sec per axis and that the instrument is sampled at one-second intervals for 40 seconds. The model power spectral density was chosen to be a multiple of the identity matrix with the constant of proportionality adjusted so that

$$\text{trace}\, Q\Delta t = \text{trace}\left\langle \delta\boldsymbol{\omega}(t)\delta\boldsymbol{\omega}^T(t)\right\rangle \quad , \tag{84}$$

where $< \cdot >$ denotes a time average and

$$\delta\boldsymbol{\omega}(t) \equiv \boldsymbol{\omega}(t+\Delta t) - \boldsymbol{\omega}(t) \simeq 2\frac{d^2\mathbf{g}_2(t)}{dt^2}\Delta t \quad . \tag{85}$$

Figures 1 and 2 show plots of the attitude estimate errors as a function of time. The pointing error is defined as the angle (defined to be positive) between the true and the estimated bore-sight directions. The roll error is the magnitude of the error in the angle of rotation about the bore sight. In these curves the solid line gives the filter result and the dashed line gives the single-frame result from QUEST. Figure 3 shows the estimate errors for the three components of the angular velocity. There is a large improvement in the roll error using the filter since the process noise covariance is small compared with that of the single frame roll errors. The improvement in the pointing error is less noticeable because the process noise covariance is large compared with that of the pointing error.

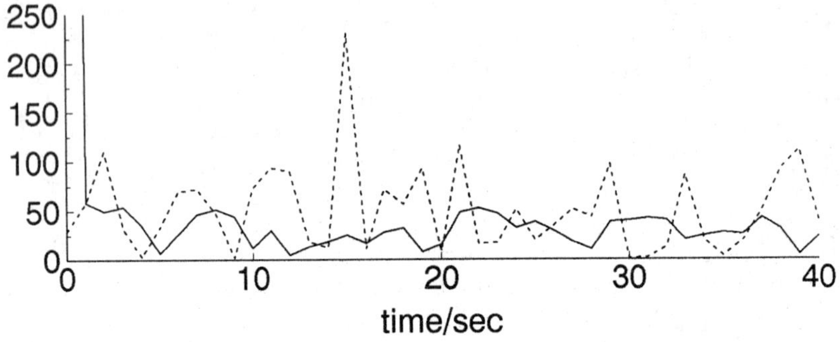

Fig. 2 Star Tracker Roll Errors

The solid line gives the filter result. The dashed line is the single-frame estimate using QUEST. The angles are given in arc sec.

We have also compared the three initialization methods. These are: (i) the statistically correct initialization scheme of Equations (74)–(76); (ii) the statistically and dynamically approximate scheme of Equations (74), (75), and (79); and (iii) the more common approximate scheme in which $P_{1|0}$ is set to

$$P_{1|0} = \begin{bmatrix} \sigma_{\theta 0}^2 \, I_{3\times3} & 0_{3\times3} \\ 0_{3\times3} & \sigma_{\omega 0}^2 \, I_{3\times3} \end{bmatrix} , \tag{86}$$

where $\sigma_{\theta 0}$ and $\sigma_{\omega 0}$ are large numbers. Here we have chosen $\sigma_{\theta 0} \simeq 1000.$ deg, and $\sigma_{\omega 0} \simeq 1000.$ deg/sec. Beyond the first frame of data the three methods turn out to be indistinguishable.

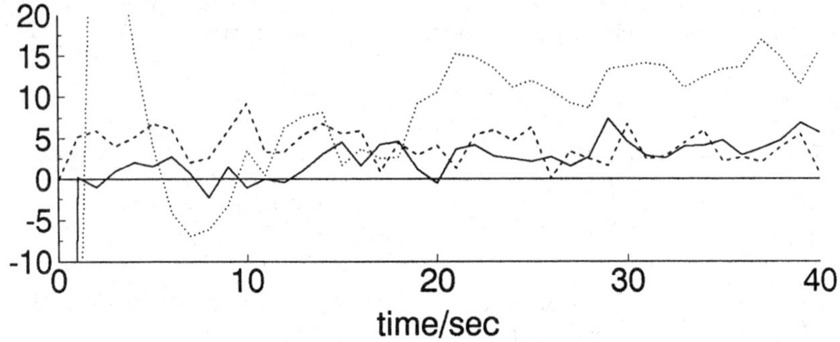

Fig. 3 Angular Velocity Estimate Errors

The solid line gives the x-component, the dashed line the y-component, and the dotted line the z-component. The rates are given in arc sec/sec.

DISCUSSION AND CONCLUSIONS

An efficient Kalman filter has been developed for a spacecraft for which attitude information is supplied solely by a true star camera, i.e., capable of sighting many stars simultaneously. The Kalman filter uses the QUEST algorithm to compute a sufficient statistic which provides a means of data compression within the filter. In addition, an algorithm is developed for initializing the filter at the correct prior-free estimate and covariance, rather than relying on brute-force initialization strategies and hoping that the filter will converge quickly enough. In the present example, however, it turns out that the brute-force initialization strategy works very well. A great advantage of the present method is that the sufficient statistic, being an attitude, can be compared with the filtered attitude as a check on consistency. This methodology, then, provides a safety net for mission analysts who have learned to be wary of Kalman filters in attitude applications.

The lack of an accurate dynamic model or attitude-rate sensors poses important restrictions on the utility of the filter. If the uncertainty in the attitude due to the unmodeled dynamics is greater than $(Q\Delta t)^{1/2}$, and in this case the unmodeled dynamics is the entire dynamics of the system, then the Kalman filter will simply provide an attitude estimate which is close in accuracy to the single frame result. Thus, the filter proposed here will have limited usefulness in situations where the spacecraft is very agile. It is expected for the Star Tracker mission, however, that this limitation will not be significant during normal operation.

The idea of using a least-squares estimator as a preprocessor for a Kalman filter is not new and originated in earlier research on autonomous attitude determination systems[6], which, in fact, considered also the use of CCD star trackers. An implementation for a gravity-gradient stabilized spacecraft of a two-step filter with a QUEST preprocessor and a dynamical model for attitude prediction has been considered by Varotto[7], whose treatment is similar in spirit to the present work but very different in execution. It may be noted that the use of sufficient statistics as data compressors for Kalman filters, of which the present work is an example, is not new. Extensive references may be found in Ref. 8.

APPENDIX – ROTATION VECTOR ALGEBRA

The rotation vector is defined as

$$\boldsymbol{\theta} = \theta\,\hat{\mathbf{n}} \quad , \tag{A1}$$

where θ is the angle of rotation and $\hat{\mathbf{n}}$ is the axis of rotation. The rotation vector is related to the attitude matrix by

$$A(\boldsymbol{\theta}) = \exp\{[[\boldsymbol{\theta}]]\} \quad , \tag{A2}$$

where exp denotes matrix exponentiation and is defined by the Taylor series for the exponential function. Evaluating the Taylor series leads to the usual expression

$$A(\boldsymbol{\theta}) = I_{3\times3} + \frac{\sin\theta}{\theta}\,[[\boldsymbol{\theta}]] + \frac{1-\cos\theta}{\theta^2}\,[[\boldsymbol{\theta}]]^2 \quad , \tag{A3}$$

where $\theta = |\boldsymbol{\theta}|$. The inverse relation is most easily obtained by solving for θ using

$$\operatorname{trace} A = 1 + 2\cos\theta \quad , \tag{A4}$$

and

$$[[\boldsymbol{\theta}]] = \left(\frac{\theta}{2\sin\theta}\right)\left(A - A^T\right) \quad . \tag{A5}$$

The kinematic equation for the rotation vector is

$$\frac{d}{dt}\boldsymbol{\theta}(t) = D(\boldsymbol{\theta})\boldsymbol{\omega}(t) \quad , \tag{A6}$$

where

$$D(\boldsymbol{\theta}) \equiv I_{3\times3} - \frac{1}{2}[[\boldsymbol{\theta}]] + \frac{2-\theta\cot(\theta/2)}{2\theta^2}[[\boldsymbol{\theta}]]^2 \quad , \tag{A7}$$

Thus, if $\Delta\boldsymbol{\theta}$ is the rotation vector for an infinitesimal rotation, we may define an addition operation for the rotation vector when the second vector is infinitesimal, which is

$$\boldsymbol{\theta} \oplus \Delta\boldsymbol{\theta} \equiv \boldsymbol{\theta} + D(\boldsymbol{\theta})\Delta\boldsymbol{\theta} \quad . \tag{A8}$$

The operation $\oplus$ is not commutative. In a similar fashion we may define the difference of two rotation vectors $\boldsymbol{\theta}_1$ and $\boldsymbol{\theta}_2$ when the difference is infinitesimal by insisting that

$$\boldsymbol{\theta}_2 = \boldsymbol{\theta}_1 \oplus (\boldsymbol{\theta}_2 \ominus \boldsymbol{\theta}_1) \quad , \tag{A9}$$

which leads to

$$\boldsymbol{\theta}_2 \ominus \boldsymbol{\theta}_1 = D^{-1}(\boldsymbol{\theta}_1)(\boldsymbol{\theta}_2 - \boldsymbol{\theta}_1) \quad . \tag{A10}$$

As an example we may consider

$$\boldsymbol{\theta}_1 = \boldsymbol{\beta}\Delta t \quad , \quad \boldsymbol{\theta}_2 = (\boldsymbol{\beta} + \Delta\boldsymbol{\tau}_2 - \Delta\boldsymbol{\omega}_1)\Delta t \quad , \tag{A11}$$

which leads to equation (70) above.

ACKNOWLEDGEMENTS

This project is supported under contract by the U. S. Army Engineer Topographic Laboratories, Fort Belvoir, Virginia. The encouragement and support of Connie Gray is gratefully acknowledged.

REFERENCES

1. Strikwerda, T. E., and Fisher, H. L., "A CCD Star Camera Used for Satellite Attitude Determination," *Proceedings, Summer Computer Simulation Conference,* Seattle, Washington, 1988.

2. Lefferts, E. J., Markley, F. L., and Shuster, M. D., "Kalman Filtering for Spacecraft Attitude Estimation," *Journal of Guidance, Control and Dynamics,* Vol. 5, No. 5, Sept.–Oct. 1982, pp. 417–429.

3. Shuster, M. D., and Oh, S. D., "Three-Axis Attitude Determination from Vector Observations," *Journal of Guidance, Control and Dynamics,* Vol. 4, No. 1, Jan.–Feb. 1981 pp. 70–77.

4. Shuster, M. D., "Kalman Filtering of Spacecraft Attitude and the QUEST Model," submitted to the *Journal of the Astronautical Sciences.*

5. Wertz, James R., (ed.), *Spacecraft Attitude Determination and Control,* D. Reidel Publishing Company, Dordrecht, the Netherlands, 1978.

6. Strikwerda, T. E. and Junkins, J. L., "Star Pattern Recognition and Spacecraft Attitude Determination," Engineering Topographic Laboratores Report ETS–0260, May 1981.

7. Varotto, S. E. C., "Determinação da Atitude de Satélites Artificiais através da Aplicação Conjunta de Técnicas de Estimação Estática e Dinámica," Instituto de Pesquisas Espaciais, São José dos Compos (SP), Brazil, Publicação N^{O}. INPE-4415-TDL/306, October 1987.

8. Medan, Y. and Bar-Itzhack, I. Y., "Batch Recursive Data Compression Schemes for INS Error Estimation," *IEEE Transactions on Aerospace and Electronics Systems,* Vol. AES-21, No. 5, September 1985, pp. 688–697.

AAS 89-366

ANALYSIS OF THE FLIGHT DYNAMICS OF THE SOLAR MAXIMUM MISSION (SMM) OFF-SUN SCIENTIFIC POINTING*

D. S. Pitone†, J. R. Klein† and B. J. Twambly†

This paper presents the algorithms created and implemented by the Goddard Space Flight Center's (GSFC's) Solar Maximum Mission (SMM) attitude operations team to support large-angle spacecraft pointing at scientific objectives. The mission objective of the post-repair SMM satellite was to study solar phenomena. However, because the scientific instruments, such as the Coronagraph/ Polarimeter (CP) and the Hard X-ray Burst Spectrometer (HXRBS), were able to view objects other than the Sun, attitude operations support for attitude pointing at large angles from the nominal solar-pointing attitudes has been required. Subsequently, attitude support for SMM has been provided for scientific objectives such as Comet Halley, Supernova 1987A, Cygnus X-1, the Crab Nebula, and the Pleiades Cluster. In addition, the analysis has been extended to include the reverse problem, computing the right ascension and declination of a body given the off-Sun angles. This analysis has led to the computation of the orbits of nine new solar comets seen in the field-of-view (FOV) of the CP. The activities necessary to meet these large-angle attitude-pointing sequences, such as slew sequence planning, viewing-period prediction, and tracking-bias computation, are described. Analysis is presented for the computation of manuevers and pointing parameters relative to the SMM-unique, Sun-centered reference frame. Finally, science data and independent attitude solutions are used to evaluate the large-angle pointing performance.

INTRODUCTION

This paper presents

- Algorithms used in support of large-angle attitude pointing maneuvers by the Solar Maximum Mission (SMM) spacecraft to nonsolar scientific objectives
- A brief history of the mission and its primary scientific objectives
- Background of the problem of pointing at celestial objects, especially as pertains to the SMM

---

* This work was supported by the National Aeronautics and Space Administration (NASA)/Goddard Space Flight Center (GSFC), Greenbelt, Maryland, under Contract NAS 5-31500.

† Computer Sciences Corporation (CSC), 6521 Arlington Blvd., Falls Church, Virginia 22046.

The algorithms used to compute the necessary parameters for the observations are derived and results are presented that show the algorithms worked correctly, with accuracy significantly better than that specified by the SMM project. This work was performed by the Flight Dynamics Division (FDD) attitude determination and control analysts in the Flight Dynamics Facility (FDF) at Goddard Space Flight Center (GSFC).

Mission History

The SMM spacecraft was launched in February 1980 from the Eastern Test Range into an approximately circular low-Earth orbit, with an inclination of nearly 28 degrees (deg) (Reference 1). The spacecraft functioned normally until November 1980 when the standard reaction wheel (SRW) package that controls the spacecraft attitude failed. To preserve the mission, the spacecraft was put into a spin, approximately 1 deg per second (sec), about the minor principal axis. While the spacecraft was in this spin mode, very little scientific work was accomplished. In April 1984, the spacecraft was repaired in-orbit by astronauts aboard the Space Transportation System (STS) orbiter. The entire modular attitude control system (MACS) was replaced, and the spacecraft returned to normal operations.

Some time before the repair mission, scientists at the GSFC Laboratory for Astronomy and Solar Physics proposed that the SMM be used to observe Comet Halley for approximately 1 month before and after perihelion of the comet's orbit (February 9, 1986). During this time, the comet, as seen from the Earth, would be between 7 and 45 deg from the Sun, and very little useful ground-based observation was anticipated.

SMM Scientific Objectives and Payload

The original scientific objective of the SMM was to study solar phenomena, especially the solar maximum of 1981. The mission was planned to be three-axis stabilized to keep the scientific payload pointed at the Sun. However, because of the SRW failure, most of the scientific instruments missed the solar maximum. Since the repair of the spacecraft, a tremendous amount of solar radiation data at a variety of wavelengths has been amassed. In addition, beginning with the Comet Halley observations, several interesting nonsolar targets have been studied.

A Coronagraph/Polarimeter (CP), designed to study the solar corona in the visible spectrum, was used to observe Comet Halley and the Moon. In addition, nine solar grazing comets have been discovered during routine operation of the CP. A Hard X-Ray Burst Spectrometer (HXRBS) was pointed at Cygnus X-1 (a suspected black hole) and the Crab Nebula. The Gamma Ray Spectrometer (GRS) was used to study gamma ray output from Supernova 1987A, and the Ultra-Violet Spectrometer (UVSP) observed a transit of the Sun by Mercury. All of these observations required attitude excursions from the primary solar objective and provided valuable data for the scientific community on a wide range of celestial objectives.

Maneuver Requirements

The goal of this analysis is to provide algorithms for attitude control of the SMM so that the scientific payload points at the desired celestial body and tracks it, even if it is moving relatively fast, to keep the body in the instrument field-of-view (FOV). In addition, the periods during which the celestial body is visible to the spacecraft (i.e., not occulted by the Earth) must be determined.

Requirements for attitude support of the large-angle attitude pointing maneuvers were not defined until several years after the spacecraft had been launched. For this reason, mathematical algorithms were developed for each nonsolar objective, as needed. The algorithms were later combined to form one software system that could efficiently plan the large-angle off-solar pointing maneuvers for any celestial body. The software was designed to plan the maneuver sequence, compute the viewing periods available, and determine the tracking parameters. Once computed, these data are uplinked to the spacecraft.

The constraints imposed by the SMM experimenters were few and were concerned with getting the best observations possible. To accomplish normal pointing, an accuracy of a tenth of a degree was desired. Another consideration was the determination accuracy in the viewing periods. Because of uncertainties in the spacecraft ephemeris, pointing accuracy, and spacecraft gyro drift, accuracy in computing the start and stop times of the viewing periods was estimated to be 10 sec (3σ).

Consideration for spacecraft safety placed other constraints on the maneuvers. The most important constraint was that the spacecraft had to remain in a power-positive mode to ensure survival of the spacecraft following any observation period contingency. This effectively meant that the body-fixed Solar Panel Array would always need to be absorbing at least as much energy as the spacecraft was using. During nominal operation, the solar panels are approximately perpendicular to the spacecraft-to-Sun vector. As a result of the analysis by the power engineers, the maximum off-Sun pointing angle was set at 65 deg.

PROBLEM ANALYSIS

The SMM Reference Frames

The SMM spacecraft attitude is represented in a solar reference frame, referred to as the SUN frame. The X-axis of the SUN frame is defined by the apparent spacecraft-to-Sun unit vector, $\underline{V}_{ss}$. The Y-axis is the unitized cross-product of the Sun's North Pole vector, $\underline{S}_{np}$, and $\underline{V}_{ss}$. The Z-axis completes the right-handed coordinate system. The equations for the unit vectors along the Sun frame axes are (Reference 2)

$$\underline{S} = \underline{X} = \underline{V}_{ss} \qquad (1)$$

$$\underline{U} = \underline{Y} = a\,(\underline{S}_{np} \times \underline{X}) \qquad (2)$$

$$\underline{N} = \underline{Z} = a\,(\underline{S}_{np} - (\underline{S}_{np} \cdot \underline{X})\,\underline{X}) \qquad (3)$$

where

$$a = [1 - (\underline{S}_{np} \cdot \underline{X})^2]^{-1/2} \tag{4}$$

Figure 1 illustrates the SUN coordinate system.

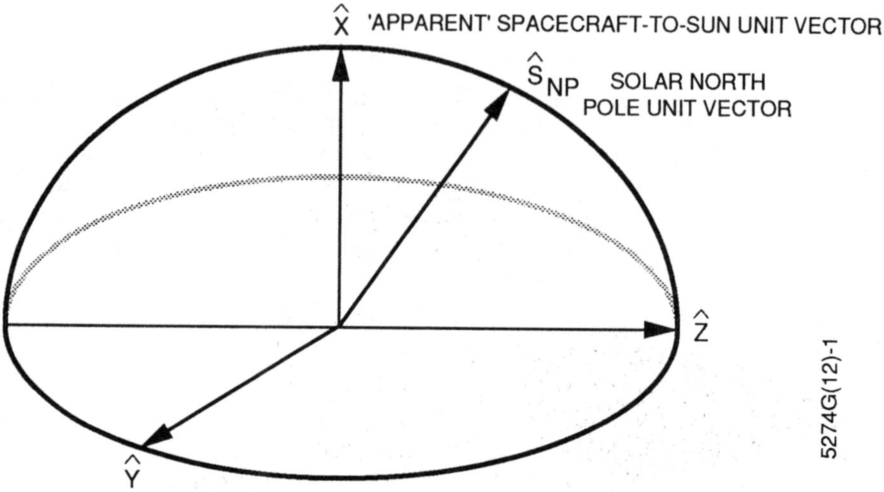

Fig. 1 Solar Coordinate System

The transform, A, from geocentric inertial (GCI) coordinates to the SUN frame, is computed from the actual spacecraft and Sun positions with consideration for the Earth and spacecraft velocities and the speed of light to correct for aberration. Thus, for attitude determination, reference vectors must be transformed using A to get the reference vectors in the SUN frame. Due to the spacecraft's orbit about the Earth and the Earth's orbit about the Sun, the SUN frame is constantly changing. Thus, each time the attitude of the spacecraft is computed by the spacecraft's onboard computer (OBC) or the ground system, the SUN frame has to be recomputed.

The SMM spacecraft attitude is represented as the transformation from the SUN frame to the SMM body frame. As shown in Figure 2, the fine-pointing Sun sensor (FPSS) and the scientific instruments are at one end of the spacecraft. During normal operations, the attitude control system is designed to point this end of the spacecraft at small areas of the Sun. The X-axis (roll) of the spacecraft body frame is defined to coincide with the direction of the FPSS boresight. Two orthogonal axes perpendicular to the FPSS boresight and corresponding to the FPSS outputs are defined as the Y- (pitch) and Z- (yaw) axes of the body frame. This alignment is advantageous because the spacecraft pitch and yaw rotations are read directly from the output of the FPSS when the Sun is in the FOV of the FPSS, which is 4-square deg. Thus, during normal operations, the spacecraft is observing the Sun, and the attitude accuracy in pitch and yaw is 5 arc-sec (3σ). The roll is determined by the fixed-head star trackers (FHSTs) to an accuracy of approximately 30 arc-sec (3σ). During periods of solar occultation or spacecraft pitch or yaw offpoint, the Sun is not in the FPSS FOV, and the attitude of all three axes is determined by the FHSTs to an accuracy of approximately 30 arc-sec (3σ). Most of the scientific instrument boresights are parallel to the boresight of the FPSS; thus, the output from the FPSS can be used directly to determine where the instruments are pointing to within 5 arc-sec when the Sun is being observed.

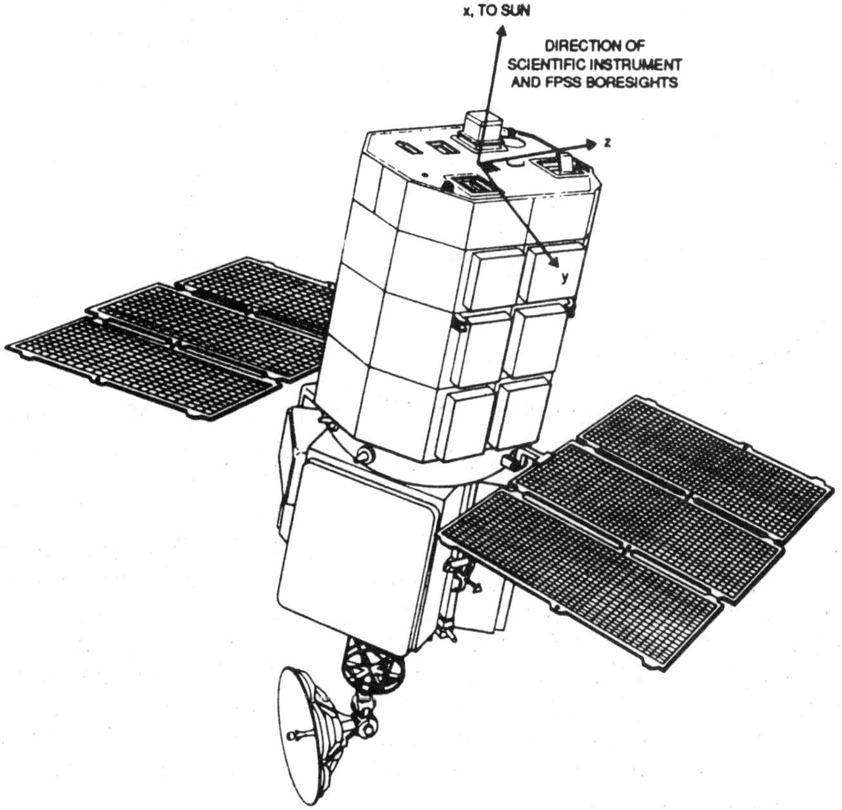

Fig. 2 Scientific and FPSS Boresight Directions

The SMM attitude matrix, B, is the matrix that transforms a vector from the SUN frame to the SMM body frame. If the roll angle is ϕ, the pitch angle θ, and the yaw angle β, the matrix B is (Reference 2)

$$B = T_3(\beta)\, T_2(\theta)\, T_1(\phi) \tag{5}$$

where T_i are the individual matrices for each Euler rotation on the intermediate axes. Thus,

$$B = \begin{bmatrix} \cos\beta\cos\theta & \cos\beta\sin\theta\sin\phi + \sin\beta\cos\phi & -\cos\beta\sin\theta\cos\phi + \sin\beta\sin\phi \\ -\sin\beta\cos\theta & -\sin\beta\sin\theta\sin\phi + \cos\beta\cos\phi & \sin\beta\sin\theta\cos\phi + \cos\beta\sin\phi \\ \sin\theta & -\cos\theta\sin\phi & \cos\theta\cos\phi \end{bmatrix} \tag{6}$$

Calculation of Pointing Maneuvers

Pointing at a celestial object involves rotating the SMM to align its X-axis with the spacecraft-to-target vector. In general, pointing from an initial target to any other target on the celestial sphere can be accomplished by consecutive rotations about any two orthogonal axes (Reference 3). Thus, any pair of rotations of roll, pitch, or yaw will suffice. For convenience, the first maneuver is chosen to be a roll. After this, the four choices for the second maneuver are positive or negative pitch or yaw. The magnitude of any of these second maneuvers is simply the angular distance from the Sun to the target objective. For this algorithm, the roll corresponding to each choice of second rotations is computed, and the smallest roll is chosen.

To compute the maneuvers, the roll-yaw combination is computed first. The attitude matrix, B, for this case is $T_3(\beta)\, T_1(\phi)$ or

$$B = \begin{bmatrix} \cos\beta & \sin\beta\cos\phi & \sin\beta\sin\phi \\ -\sin\beta & \cos\beta\cos\phi & \cos\beta\sin\phi \\ 0 & -\sin\phi & \cos\phi \end{bmatrix} \tag{7}$$

This transforms vectors from the SUN frame to the body frame after the maneuvers have been completed.

Since the scientific instrument boresights are parallel to the X-axis, the representation of the celestial target vector in the body frame after the maneuver is desired to be

$$\underline{V}_{scb} = \begin{bmatrix} 1 \\ 0 \\ 0 \end{bmatrix} \tag{8}$$

The GCI coordinate representation of the celestial target vector, unchanged by the maneuver is

$$\underline{V}_{gci} = \begin{bmatrix} V_{gci1} \\ V_{gci2} \\ V_{gci3} \end{bmatrix} \tag{9}$$

To represent the celestial target vector in the SUN frame, the matrix A is calculated for the time of the maneuvers. The representation of the celestial target vector, $\underline{V}_s$, in the Sun frame is

$$\underline{V}_s = A\, \underline{V}_{gci} \tag{10}$$

while $\underline{V}_{scb}$ is related to $\underline{V}_s$ as follows:

$$\underline{V}_s = B^T\, \underline{V}_{scb} \tag{11}$$

or

$$\begin{bmatrix} V_{s1} \\ V_{s2} \\ V_{s3} \end{bmatrix} = \left[\begin{array}{c|c|c} \cos\beta & -\sin\beta & 0 \\ \sin\beta\cos\phi & \cos\beta\cos\phi & -\sin\phi \\ \sin\beta\sin\phi & \cos\beta\sin\phi & \cos\phi \end{array}\right] \begin{bmatrix} 1 \\ 0 \\ 0 \end{bmatrix} \tag{12}$$

Thus, $\underline{V}_s$ can be computed from Equation (10) and substituted into the left-hand side of Equation (12). The unknowns in Equation (12) are then the maneuver angles β and ϕ. Further reduction of this equation leads to

$$V_{s1} = \cos\beta \tag{13}$$

$$V_{s2} = \sin\beta \cos\phi \tag{14}$$

$$V_{s3} = \sin\beta \sin\phi \tag{15}$$

Dividing Equation (15) by Equation (14)

$$V_{s3}/V_{s2} = \tan\phi \tag{16}$$

Thus, the equations for the roll-yaw maneuvers are

$$\text{Yaw} = \beta = \cos^{-1}(V_{s1}) \tag{17}$$

$$\text{Roll} = \phi = \tan^{-1}(V_{s3}/V_{s2}) \tag{18}$$

The sign of the roll is resolved by using the function DATAN2 to compute the inverse tangent. The roll-pitch sequences are calculated from the above roll-yaw results by adding and subtracting 90 deg from the roll and selecting the sign of the pitch accordingly. As stated earlier, the maneuver sequence requiring the smallest roll rotation is the favored sequence. For targets that are further than 65 deg from the Sun, the maneuver to the attitude closest to the celestial body is computed. In this case the angular distance from the target to the planned closest attitude is also computed and delivered to the experimenters.

Comet Position Determination

To date, nine Sun-grazing comets have been discovered in the CP FOV during normal observations of the solar corona. The CP uses an occulting disk on its sensor FOV to block out the main part of the Sun, enabling it to observe the solar corona. These comets could not have been observed without this special instrumentation of the CP. Several observations of each comet in CP instrument coordinates have been reported. In support of these observations, it is desired to transform the CP comet coordinate positions to GCI coordinates to determine the orbits of the comets.

The determination of the comet positions in GCI coordinates is the inverse of the maneuver planning. It is assumed that the CP boresight is parallel to the X-axis of the spacecraft. The comet position in CP coordinates is transformed to an equivalent spacecraft roll-pitch maneuver sequence that would point the spacecraft directly at the comet. The spacecraft, however, is never actually maneuvered; the maneuver is only computed so that the maneuver planning procedure can be reversed, and the comet position computed. The position vector in the Sun frame is then determined from the roll-pitch angles. The comet position vector in GCI coordinates is computed by multiplying the comet position vector in the Sun frame by the transpose of A at the time of the observation.

The CP uses a two-dimensional polar coordinate system. The first coordinate is the counterclockwise rotation about the X-axis (roll) from the positive Z-axis (yaw). Thus, this is the negative roll necessary to align the positive Z-axis (yaw) with the Sun-to-comet vector. The second coordinate is the angular separation between the center of the Sun and the comet, along the positive Z-axis (yaw). This corresponds to a negative pitch angle. The result is the roll-pitch angle pair that would be required to maneuver the spacecraft and align the boresight with the spacecraft-to-comet vector.

If the matrix, B', for a roll (ϕ)-pitch (θ) maneuver from a null attitude in the Sun frame to the comet is

$$B' = \begin{bmatrix} \cos\theta & \sin\theta\sin\phi & -\sin\theta\cos\phi \\ 0 & \cos\phi & \sin\phi \\ \sin\theta & -\cos\theta\sin\phi & \cos\theta\cos\phi \end{bmatrix} \tag{19}$$

then the position vector, $\underline{V}_c$, of the comet in the Sun frame is simply

$$\underline{V}_c = B'^T \begin{bmatrix} 1 \\ 0 \\ 0 \end{bmatrix} \tag{20}$$

or

$$\underline{V}_c = \begin{bmatrix} \cos\theta & 0 & \sin\theta \\ \sin\theta\sin\phi & \cos\phi & -\cos\theta\sin\phi \\ -\sin\theta\cos\phi & \sin\phi & \cos\theta\cos\phi \end{bmatrix} \begin{bmatrix} 1 \\ 0 \\ 0 \end{bmatrix} \tag{21}$$

The comet position in GCI coordinates, $\underline{V}_{cgci}$, is then

$$\underline{V}_{cgci} = A^T \underline{V}_c \tag{22}$$

The vector, $\underline{V}_{cgci}$, is then reduced to right ascension and declination (mean of 1950) and delivered to the International Astronomical Union (IAU), where it is used for orbit computations.

Calculation of Specific Orientation Pointing Maneuvers

As discussed in the section titled "Calculation of Pointing Maneuvers," two consecutive rotations about orthogonal axes will align the SMM X-axis with a target; however, if the target is required to appear with a specific orientation in the scientific instrument FOV, rotations about three axes are required. This situation occurred during the planning of the SMM observations of the Pleiades star cluster. The three rotations may be computed by calculating the rotation matrix between the representation of two stars in the cluster in the body frame before and after the maneuvers. This method is based on the TRIAD algorithm from Reference 4.

To compute the rotation matrix, the positions of two stars in the cluster are specified within the FOV of the scientific instrument. These positions are then transformed into the body frame and become the representations of the two stars in the body frame after the maneuvers are completed, known as $\underline{V}_b$. The corresponding target vectors in GCI coordinates are known and are transformed into SUN coordinates by the **A** matrix, which is calculated for the time of the maneuvers. Since the starting attitude for the maneuvers is null (i.e., the SUN frame and the body frame coincide before the maneuvers are started), the target vectors in the SUN frame are the positions of the stars in the body frame before the maneuvers and are called $\underline{V}_S$. The rotation matrix, **B**, is then the matrix that transforms the star positions from $\underline{V}_S$ to $\underline{V}_b$. The necessary roll, pitch, and yaw maneuvers may then be extracted from **B**.

The SMM CP was the scientific instrument used for the Pleiades observations. The desired position of each star within the CP FOV after the pointing maneuvers was given as a clockwise rotation from the solar north pole, called θ, and a radial distance from the Sun center, called r. Each vector in the body frame is thus

$$\underline{V}_{b1} = \begin{bmatrix} \cos r_1 \\ \sin r_1 * \sin \theta_1 \\ \sin r_1 * \cos \theta_1 \end{bmatrix} \qquad \underline{V}_{b2} = \begin{bmatrix} \cos r_2 \\ \sin r_2 * \sin \theta_2 \\ \sin r_2 * \cos \theta_2 \end{bmatrix} \tag{23}$$

From these vectors an orthogonal coordinate system may be computed:

$$M_b = [\underline{R}_1 \mid \underline{R}_2 \mid \underline{R}_3]$$

where

$$\underline{R}_1 = \underline{V}_{b1}, \quad \underline{R}_2 = \frac{\underline{V}_b \times \underline{V}_{b2}}{|\underline{V}_{b1} \times \underline{V}_{b2}|}, \quad \frac{\underline{R}_1 \times \underline{R}_2}{|\underline{R}_1 \times \underline{R}_2|} \tag{24}$$

Each target vector in GCI is first transformed into the SUN frame by the matrix **A**, which is calculated for the time of the maneuvers. These vectors are then

$$V_{S1} = A\,V_{GCI1} \text{ and } V_{S2} = A\,V_{GCI2} \tag{25}$$

from which an orthogonal coordinate system is also constructed as

$$M_S = [\underline{S}_1 \mid \underline{S}_2 \mid \underline{S}_3]$$

where

$$\underline{S}_1 = \underline{V}_{S1}, \quad \underline{S}_2 = \frac{\underline{V}_{S1} \times \underline{V}_{S2}}{|\underline{V}_{S1} \times \underline{V}_{S2}|}, \quad \underline{S}_3 = \frac{\underline{S}_1 \times \underline{S}_2}{|\underline{S}_1 \times \underline{S}_2|} \tag{26}$$

The rotation matrix, **B**, is then

$$B = M_b\,M_S^T \tag{27}$$

and, using Equations (6) and (22), the required maneuvers are

$$\text{Roll} = \tan^{-1}\left(-\frac{B_{32}}{B_{33}}\right) \tag{28}$$

$$\text{Pitch} = \sin^{-1}(B_{31}) \tag{29}$$

$$\text{Yaw} = \tan^{-1}\left(-\frac{B_{21}}{B_{11}}\right) \tag{30}$$

Celestial Body Viewing Period Prediction

Because of the geometry of the SMM orbit, every point in the celestial sphere of the spacecraft is occulted by the Earth at some time in one revolution of the ascending node of the spacecraft's orbit. Thus, a list of periods when the celestial target can be viewed by the spacecraft is necessary for all times that observations of nonsolar targets are being planned. The computation of these periods is performed by comparing the relevant angles in the geometry of the spacecraft, Earth, and the target objective.

The unit vector, $\underline{W}_{sce}$, from the spacecraft to the center of the Earth, is constantly changing as the spacecraft orbits the Earth and is known from the spacecraft ephemeris. The vector from the Earth to the target body, $\underline{W}_{eb}$, is normally computed from the right ascension and declination of the target body in the Earth's celestial sphere. To compute the viewing periods, the spacecraft-to-target vector, $\underline{W}_{sb}$ is needed. Since the distance to the viewing target in most cases has not been relevant, the Earth-to-target vector is represented as a unit vector. Thus, vector addition of the Earth-to-target vector and the Earth-to-spacecraft vector does not give the spacecraft-to-target vector. Instead, it is assumed that the parallax of the target body's position due to the spacecraft's orbit around the Earth is negligible, and, therefore,

$$\underline{W}_{eb} \approx \underline{W}_{sb} \tag{31}$$

For objects out of the solar system and the outer planets, this approximation will yield results that are well within the constraints. For the inner planets and comets, this approximation leads to an error in the vector of between 1 arc-sec and 20 arc-sec. This will translate to an error of less than 1 sec in the occultation times, which can be considered negligible compared to the 10-sec constraint on the accuracy. If observations of the Moon are planned, the position error caused by the above assumption can reach 1 deg. This will translate to approximately a 15-sec error in the occultation times, which is greater than the maximum error allowed. To resolve this, the Earth-to-Moon vector obtained from lunar ephemerides is used to calculate $\underline{W}_{sb}$ directly by vector addition.

Once $\underline{W}_{sb}$ and $\underline{W}_{sce}$ are known, the angle, Φ, between the vectors is computed by taking the inverse cosine of the dot product of the two vectors as

$$\Phi = \cos^{-1} (\underline{W}_{sb} \cdot \underline{W}_{sce}) \tag{32}$$

This geometry is illustrated in Figure 3.

To compute whether the celestial target is blocked by the Earth, Φ is compared to the subtended Earth half angle, Ω. The computation of Ω assumes a perfectly circular Earth, with a radius equal to the equatorial radius. This translates to a maximum error of 8 sec for celestial objects with declinations near positive or negative 90 deg. However, nearly all of the celestial targets observed by the SMM have had declinations close to the Sun's. Thus, the typical error induced by the Earth's oblateness is typically near 3 or 4 sec.

The half angle subtended by the Earth is given by

$$\Omega = \sin^{-1} (R/D) \tag{33}$$

where R is the equatorial radius of the Earth and D is the magnitude of the Earth-to-spacecraft vector obtained from the spacecraft ephemeris file. The relationship

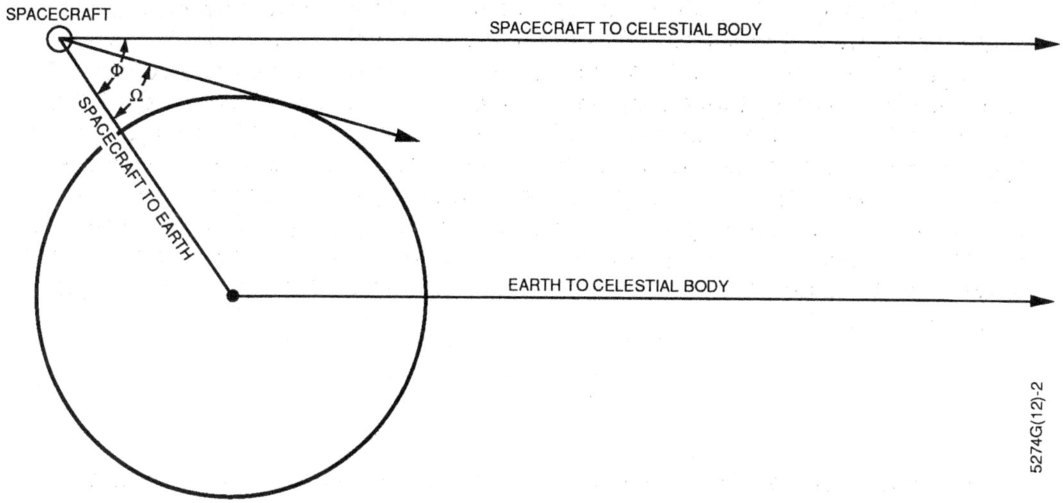

Fig. 3 Occultation Geometry

of the computed angles is also shown in Figure 3. The only variable on the right-hand side of Equation (33) is D. This varies with the orbital eccentricity and more slowly with the rate of decay of the spacecraft's altitude above the Earth. Thus, for SMM, Ω changes very slowly and is approximately 69 deg. The celestial object is considered to be occulted by the Earth from the spacecraft if Ω is larger than Φ. Conversely, if Φ is larger than Ω, the celestial object is considered in view of the spacecraft. This makes sense intuitively, because if the spacecraft is between the Earth and the celestial object, Φ will be obtuse, while if the Earth is between the spacecraft and the celestial object, Φ will be acute.

Earth occultation computations are performed every 4 minutes (min) for the period of interest. When a change occurs, a binary search is used to narrow the change time to the nearest second, which will typically have a maximum error (3σ) of less than 10 sec (Reference 5).

Tracking Parameter Computation

Targets not in the solar system, such as distant stars, do not move significantly relative to the SMM due to the spacecraft's orbit about the Earth or the Earth's motion about the Sun. Thus, during observations of very distant objects, provisions need not be made to track motion of the celestial target due to parallax and proper motion. However, bodies in heliocentric (planets or comets) or geocentric (the Moon) orbits will move through an angle δ, which can be represented as a rotation about a unit vector, $\underline{e}$, in the spacecraft inertial reference unit (IRU) coordinate frame as illustrated in Figure 4. This motion is due to the spacecraft's motion about the Earth, the Earth's motion about the Sun, and the target objects motion in its orbit about the Sun or Earth. During an observation period at a constant attitude, this motion will cause the celestial target to move through, and eventually out of, the scientific sensor FOV. To compensate, the spacecraft is rotated about the axis $\underline{e}$ by the angle δ from the beginning to the end of the observation period. This motion is called tracking.

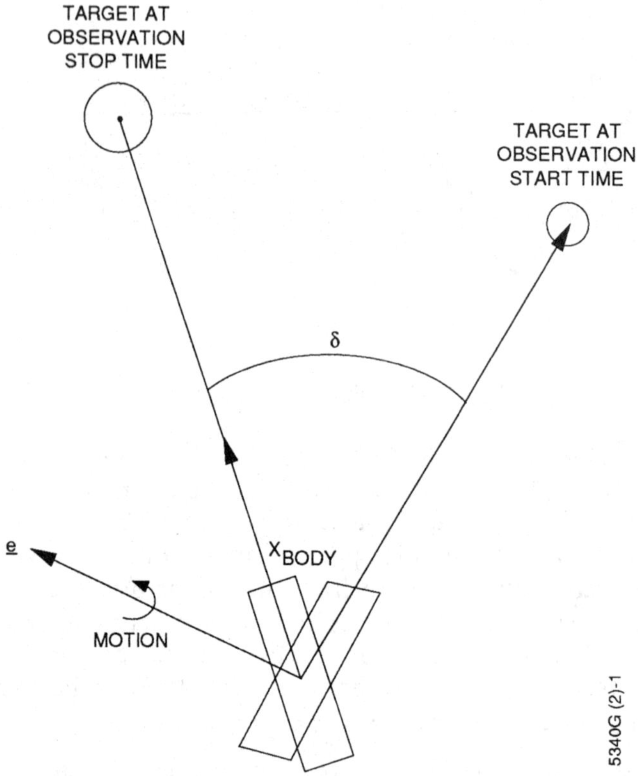

Fig. 4 Tracking Geometry

Tracking by the spacecraft is accomplished by computing the angle, δ, and axis of rotation, $\underline{e}$, of the target body in the IRU sensor frame of reference and combining δ and $\underline{e}$ with the observation time Δt to form a command gyro rate bias vector (historically called BGDT) which is uplinked to the spacecraft. These BGDTs cause the gyros to show motion opposite to the way the target is moving. This apparent (in that the spacecraft is not actually moving) motion causes the SRWs to counteract the apparent motion and move the spacecraft in the direction of the celestial objective. Thus, tracking is accomplished. To compute the BGDTs, the transformation matrix, H, between the SMM IRU-frame attitude at the start of the observation period and the attitude at the end of the observation period is computed and used to determine average rates about the three IRU axes.

To compute H, the maneuvers to point the spacecraft at the target are computed by the methods described earlier at the observation start and stop times. The maneuvers are assumed to be relative to an attitude of roll, pitch, and yaw equal to zero. Thus, using Equation (6), the SUN-frame-to-body-frame matrices, B_1 and B_2, at the start and stop times are computed. However, the coordinate frame that B is referenced to at all times on the SMM mission is the SUN frame, which, as stated above, is not inertial. The attitude motion for target tracking must be referenced to an inertial frame. For this, the GCI-to-SUN frame matrix, A, is used. In addition, because the BGDTs are for the IRUs, they must be represented in the IRU sensor frame. Defining G_1 and G_2 to be the GCI-to-IRU transformation matrices at the start and stop times, respectively, and M to be the IRU-to-body transformation matrix, then

$$G_1 = M^T B_1 A_1 \tag{34}$$

$$G_2 = M^T B_2 A_2 \tag{35}$$

and

$$G_2 = H G_1 \tag{36}$$

Solving for H yields

$$\begin{aligned} H &= G_2 G_1^T = M^T B_2 A_2 (M^T B_1 A_1)^T \\ &= M^T B_2 A_2 A_1^T B_1^T M \end{aligned} \tag{37}$$

If the Sun frame were an inertial frame, A_1 would be equal to A_2, and Equation (37) would be simplified. Now that H has been found, δ and $\underline{e}$ can be computed from H, and the BGDTs can be computed.

To compute δ and $\underline{e}$, H is written as an Euler rotation with unit vector $\underline{e}$ in the IRU frame and rotation angle δ. Thus, using the Euler convention for a general rotation, H is (Reference 5)

$$H = \cos(\delta)\, I + (1 - \cos(\delta))\, \underline{e}\underline{e}^T - \sin(\delta)\, [[\underline{e}]] \tag{38}$$

where $[[\underline{e}]]$ is the skew symmetric matrix representation of a vector. The general form for a three-component vector is

$$[[\underline{x}]] = \begin{bmatrix} 0 & -x_3 & x_2 \\ x_3 & 0 & -x_1 \\ -x_2 & x_1 & 0 \end{bmatrix} \tag{39}$$

Thus, it is desired to manipulate Equation (38) to solve for $\underline{e}$ and δ. Defining the parameter k^2 as

$$k^2 = (H_{23} - H_{32})^2 + (H_{31} - H_{13})^2 + (H_{12} - H_{21})^2 \tag{40}$$

substitution from Equation (38) leads to

$$\begin{aligned} k^2 = {} & [e_2\, e_3\, (1 - \cos\delta) + e_1 \sin\delta - e_2\, e_3\, (1 - \cos\delta) + e_1 \sin\delta]^2 \\ & + [e_1\, e_3\, (1 - \cos\delta) + e_2 \sin\delta - e_1\, e_3\, (1 - \cos\delta) + e_2 \sin\delta]^2 \\ & + [e_1\, e_2\, (1 - \cos\delta) + e_3 \sin\delta - e_1\, e_2\, (1 - \cos\delta) + e_3 \sin\delta]^2 \end{aligned} \tag{41}$$

Some simple algebra leads to

$$\begin{aligned} k^2 &= 4e_1^2 \sin^2\delta + 4e_2^2 \sin^2\delta + 4e_3^2 \sin^2\delta \\ &= 4 * \sin^2\delta * (e_1^2 + e_2^2 + e_3^2) \end{aligned} \tag{42}$$

Since $\underline{e}$ is a unit vector, the last quantity on the right-hand side of Equation (42) is equal to 1. Therefore,

$$k^2 = 4 * \sin^2\delta \tag{43}$$

or taking the square root of both sides and solving for δ

$$\delta = \sin^{-1} (k/2) \tag{44}$$

where k is defined by Equation (40).

To solve for $\underline{e}$, the off-diagonal terms in Equation (38) are used. To demonstrate this using e_1,

$$H_{23} - H_{32} = e_2\, e_3\, (1 - \cos \delta) + e_1 \sin \delta - e_2\, e_3\, (1 - \cos \delta) + e_1 \sin \delta \tag{45}$$

$$= 2 * e_1 * \sin \delta$$

Solving for e_1

$$e_1 = (H_{23} - H_{32})/(2 * \sin \delta) \tag{46}$$

Similarily,

$$e_2 = (H_{31} - H_{13})/(2 * \sin \delta) \tag{47}$$

$$e_3 = (H_{12} - H_{21})/(2 * \sin \delta) \tag{48}$$

To compute the BGDTs, the angle of rotation in the IRU coordinate frame is divided by the observation time. The rate about each of the IRU axes is then

$$x' = e_1 * (\delta/\Delta t) \tag{49}$$

$$y' = e_2 * (\delta/\Delta t) \tag{50}$$

$$z' = e_3 * (\delta/\Delta t) \tag{51}$$

where Δt is the observation time. These rates are converted to the units of the SMM IRU, radians per 0.512 sec, and uplinked to the spacecraft for use in the control loop.

RESULTS AND CONCLUSIONS

Results

To date, these algorithms have been used to support all the SMM scientific off-Sun maneuvers. Since the computed parameters are designed to support pointing to celestial objects other than the Sun, the performance of the algorithms is ultimately measured by the quality of these observations. Thus far, the science observations have been completed without incident and have yielded excellent results each time. Images obtained from the SMM scientific instruments have always had the target object in the sensor FOV and at the correct times. In addition, for the heliocentric and geocentric objectives, the tracking system has continuously kept the object in view.

As an independent check, attitude determination solutions during the observations have indicated the spacecraft pointing accuracy is within 180 arc-sec. The accuracy of the solutions was approximately 30 arc-sec (3σ). The bulk of the pointing error was due to uncertainty in the IRU bias. During the observations, the spacecraft was inertially stabilized by the IRU package. However, spacecraft attitude drift occurred because of uncertainty in the knowledge of the IRU bias. This bias was calibrated periodically, and the error in the pointing did not exceed 120 arc-sec due to the error in the bias.

Most of the remaining error in the pointing accuracy was due to uncontrollable timing problems with the maneuvers. Since the geometry of the spacecraft, Earth, Sun, and target-body system was constantly changing, the required maneuvers were constantly changing. The algorithm derived above assumes the maneuvers were instantaneous. However, the SMM maneuvered at approximately 5 deg per min; consequently, pointing errors were incurred. In addition, the start time of the maneuvers was calculated at 5-sec intervals. Thus, if the maneuvers did not start at these times, greater error was incurred. The pointing errors due to the timing errors did not exceed 50 arc-sec. Other potential error sources include uncertainties in the position of the target body, the initial spacecraft attitude from which the maneuvers were performed, and the IRU scale factors that controlled the magnitude of a commanded maneuver.

Conclusions

Algorithms to support off-Sun maneuvers of the SMM have been presented. These algorithms compute parameters for maneuver planning, viewing period start and stop times, and target tracking rate. All of these items are necessary to maneuver the spacecraft to observe and track the celestial target objective. Because the SUN reference frame is non-inertial, the equations differ from ones normally seen. This work can be adapted to spacecraft that are controlled relative to the inertial frame; however, the primary application is to missions with non-inertial reference frames, such as Earth-pointing (1 revolution per orbit) missions.

ACKNOWLEDGMENTS

The authors wish to acknowledge the generous support of the following GSFC personnel: Mr. Frank Vaughn, Assistant Technical Representitive, SMM Analytical Task; Mr J. B. Joyce, Head, Mission and Network Support Section; and Mr. Gary Meyers, Head, Flight Dynamics Support Branch.

The authors also wish to thank Dr. Milton Phenneger of Computer Sciences Corporation (CSC) for his help in preparing this work for publication.

Acknowledgement for contributions to this paper are also extended to Mr. Daniel R. Kulp (CSC), Mr. Arnold Eudell (CSC), Ms. Deborah Roberts (CSC), and Mr. Robert Feiertag (CSC).

REFERENCES

1. Computer Sciences Corporation, CSC/TM-84/6131UD1, *Solar Maximum Mission Attitude Operations Handbook, Revision 4,* A. Eudell et al., July 1988

2. --, CSC/SD-78/6082, *Solar Maximum Mission (SMM) Attitude System Functional Specifications and Requirements*, R. Byrne et al., September 1978

3. Wertz, J. R., editor, *Spacecraft Attitude Determination and Control*, CSC/PR-76/6001, Dordrecht - Holland: D. Reidel

4. Shuster, M. D., and Oh, S. D., "Three Axis Attitude Determination From Vector Observations," *Journal of Guidance and Control*, Vol. 4, No. 1, pp. 70-77, 1981

5. Computer Sciences Corporation, CSC/SD-88/6138, *Solar Maximum Mission (SMM) Celestial Body Pointing and Occultation Prediction Program (CBPOPP) User's Guide*, D. S. Pitone, November 1988

AAS 89-367

DESIGN OF AN ON-BOARD ANTENNA DRIVE CONTROL SYSTEM WITH HIGH POINTING ACCURACY

Hiroshi Hojo*

The high accuracy requirement of antenna pointing makes it necessary to make the antenna drive control system (ADCS) independent of the conventional satellite attitude control system. This paper describes the design method of the ADCS control law which achieves high pointing accuracy, and also presents the result designed for a 2-ton class satellite.

The factors which mainly disturb the antenna pointing are satellite attitude changes and antenna reflector vibrations. The parameter defining the attitude change is the maximum attitude rate, which determines the lower limit of the ADCS bandwidth. Sub-reflector resonant vibration is avoided by ensuring that the eigen frequency is higher than the ADCS bandwidth. The difference between the eigen frequency and the ADCS bandwidth is determined by calculating the gain change of the vibration transfer function. The influence of the main-reflector vibration reduces the gain and phase margins, and it will become a serious problem if the size of the main-reflector becomes larger.

By using the parameters of a 2-ton class satellite, it was proven that the ADCS bandwidth and sub-reflector eigen frequency must be higher than 2 and 12 Hz, respectively. As a result of computer simulation, it is also confirmed that the pointing error can be controlled within the control accuracy requirement of 0.002°.

INTRODUCTION

Accurate antenna pointing control is one of the key technologies in developing a multibeam satellite communication system because of the narrow beamwidth involved. The antenna pointing accuracy possible with a conventional satellite attitude control is generally limited to about 0.05°. In order to achieve higher antenna pointing accuracy, new on-board control systems are being studied. In the NASA ACTS multibeam antenna system[1], for example, a new monopulse tracking system with gimbal drive mechanism is being developed to achieve a pointing accuracy of 0.01°. TV-SAT[2] utilizes antenna fine control loops in which two antenna reflectors are controlled with an RF sensor.

In Japan, NTT plans to carry out an orbit experiment for multibeam satellite communication systems in conjunction with the sixth Japanese Engineering Test Satellite (ETS-VI). The beamwidth of ETS-VI is about 0.3°

* Research Engineer, NTT Radio Communication Systems Laboratories, 1-2356 Take, Yokosuka 238-03, Japan.

to cover the main Japanese islands with 13 beams. In order to maintain high antenna gain and to decrease interference from other beams, the pointing accuracy of ETS-VI must be 0.015°, which makes it necessary to control the direction of the antenna beams independently of the satellite attitude control system. The antenna system configuration of ETS-VI is shown in Fig.1. The antenna pointing control system on ETS-VI is now being designed[3].

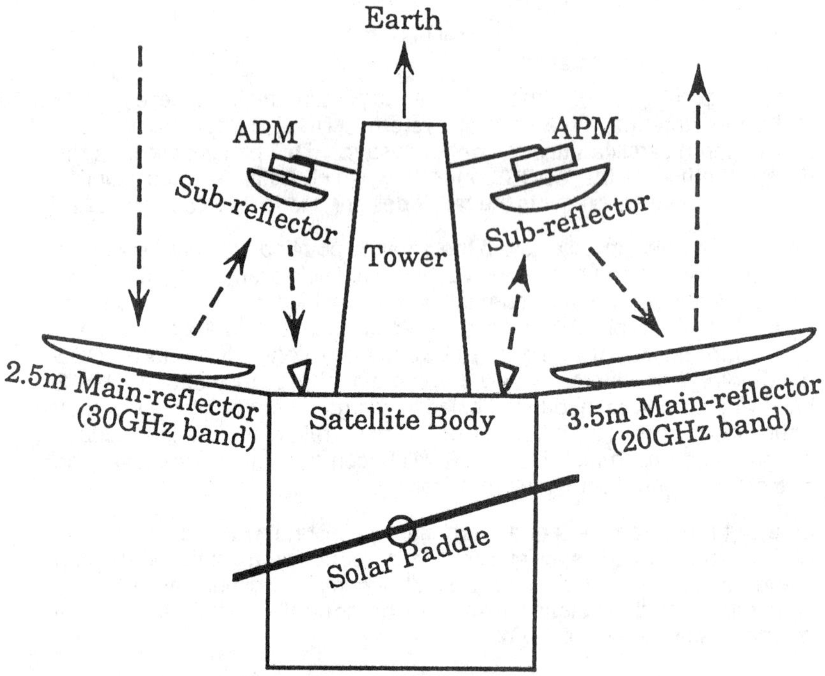

Fig.1 Antenna System Configuration

To design the antenna drive control system of 2-ton class 3-axis stabilized satellites, two important problems must be overcome. One is the coupling between two control loops (the satellite attitude control and antenna drive control) and the other is the disturbance caused by the antenna reflector vibrations. The coupling, however, is negligible if the moment-of-inertia of the driven antenna reflector is much smaller than that of the satellite[4]. Therefore, we need estimate only attitude changes which directly change the antenna beam directions.

This paper presents the design method of the ADCS control law which achieves high pointing accuracy. First, the factors which mainly disturb the antenna pointing are selected by the equations of motion. Then the effects of the disturbances are estimated in the frequency domain. Finally the parameters necessary to design the ADCS control law are chosen by using the transfer functions of the control systems, and the relations between the parameters and the ADCS control law are obtained. In addition, computer simulations were also performed to confirm the results.

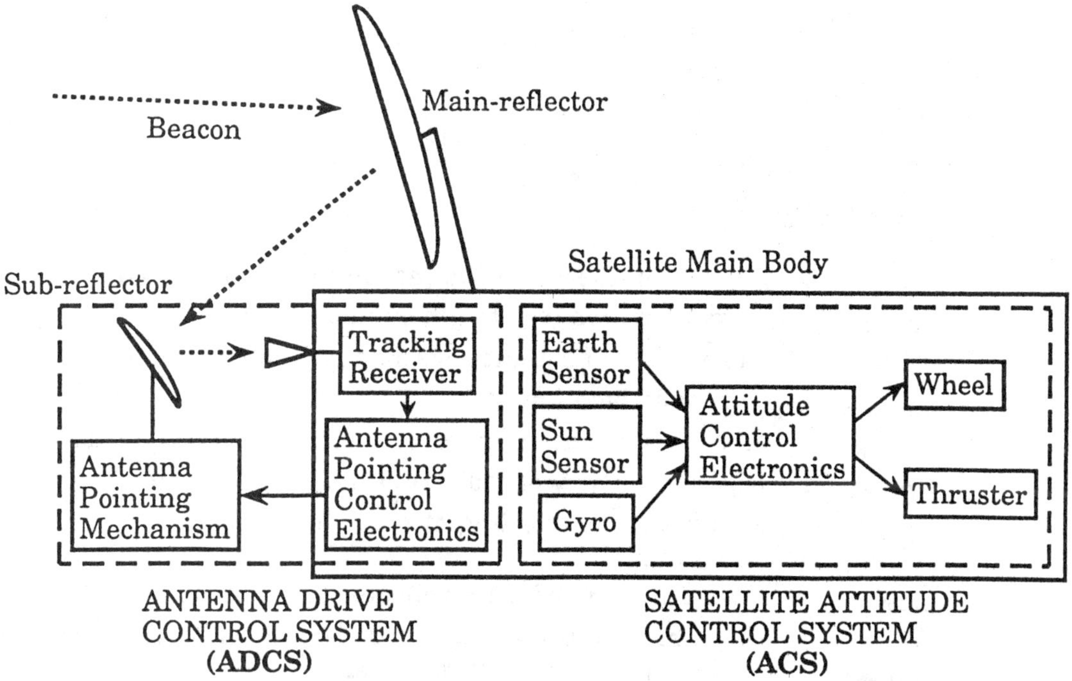

Fig. 2 Antenna Pointing Control System

ANTENNA DRIVE CONTROL SYSTEM

Configuration

The typical configuration of the antenna pointing control system (APCS) is shown in Fig.2. This control system consists of two control systems, a conventional attitude control system (ACS) and an antenna drive control system (ADCS). The ADCS consists of an RF sensor, antenna pointing control electronics (APE) and an antenna pointing mechanism (APM). The antenna beam direction is controlled by rotating the sub-reflector with APM according to the error signals detected by the RF sensor which receives a beacon transmitted from an earth station. The function of the ADCS is to compensate for the antenna pointing errors due to the satellite attitude errors, antenna misalignments and thermal deformation of the antenna structure.

Equations of Motion

The equations of motion of the satellite, antenna and vibrations are represented in the following form[4].

$$I_s\ddot{\theta}_s = T_s - \Omega_m\ddot{\eta}_m - \Omega_s\ddot{\eta}_s - I_a\ddot{\theta}_a \quad (1)$$

$$I_a\ddot{\theta}_a = T_a - \Omega_{as}\ddot{\eta}_s - I_a\ddot{\theta}_s \quad (2)$$

$$\ddot{\eta}_m + 2\zeta_m\omega_m\dot{\eta}_m + \omega_m^2\eta_m = -\Omega_m\ddot{\theta}_s \quad (3)$$

$$\ddot{\eta}_s + 2\zeta_s\omega_s\dot{\eta}_s + \omega_s^2\eta_s = -\Omega_s\ddot{\theta}_s - \Omega_{as}\ddot{\theta}_a \quad (4)$$

$$\theta_b = \theta_s + K_b\theta_a + K_m\eta_m + K_s\eta_s \quad (5)$$

I_s, I_a	:	Moment-of-inertia of the satellite and sub-reflector
T_s, T_a	:	Torque applied to satellite and sub-reflector
θ_s, θ_a	:	Error angle of satellite attitude and sub-reflector
θ_b	:	Antennas beam angle
η_m, η_s	:	Modal coordinate of main-reflector and sub-reflector
ω_m, ω_s	:	Eigen frequency of main-reflector and sub-reflector
Ω_m, Ω_s	:	Flexible mode coupling of main-reflector and sub-reflector
ζ_m, ζ_s	:	Damping ratio of main-reflector and sub-reflector
K_b	:	Angle ratio of beam angle to sub-reflector drive angle
K_m, K_s	:	Ratio of beam angle to main-reflector and sub-reflector modal coordinate

Eqs. (1) and (2) represent the motions of the satellite attitude (pitch axis) and sub-reflector angle (E-W axis), and Eqs.(3) and (4) represent the vibrations of the main-reflector and sub-reflector, respectively. Eq. (5) represents the relation between the beam angle and other angles. The vibrations are modelled by using the cantilever method[6], and solar paddle vibration is neglected because the vibration does not directly influence the antenna pointing control system. Figure 3 is a block diagram of the APCS. The factors which disturb the antenna beam angle θ_b, are not only sub-reflector drive angle θ_a but also the satellite attitude changes θ_s and antenna reflector vibrations η_m, η_s.

Model Parameters

We estimated the values of these disturbances by using the tentative parameters shown in Table 1 for a 2-ton class 3-axis stabilized satellite. The antenna pointing control accuracy, which is the most important parameter in the design of the ADCS, must be less than 0.002° according to the pointing error budget[4].

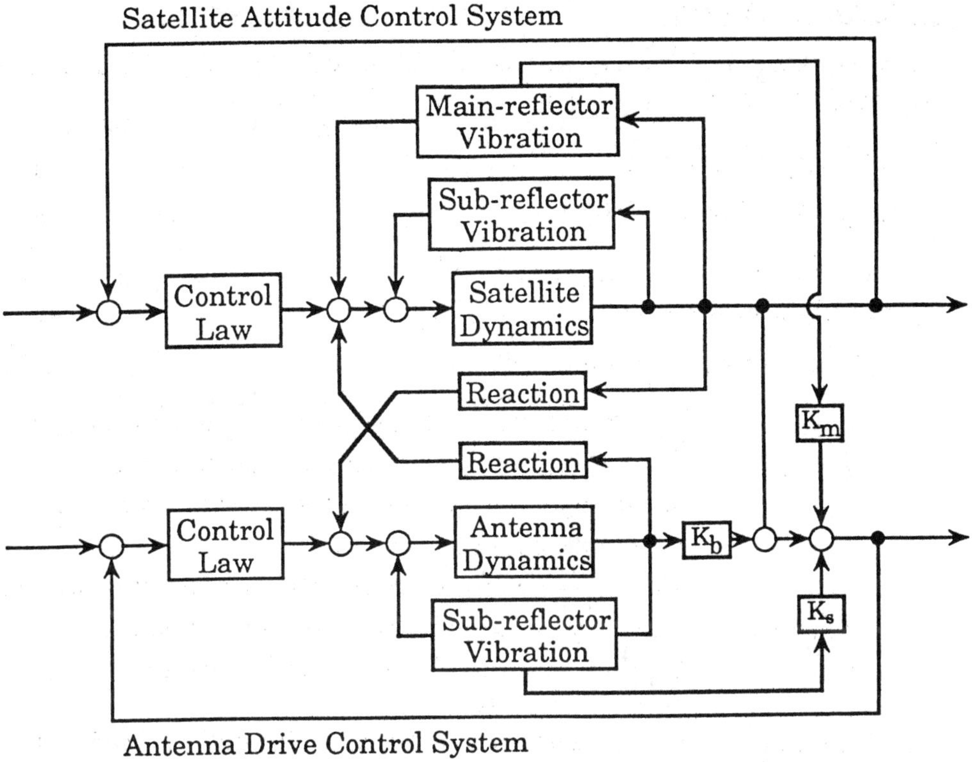

Fig.3 Block Diagram of the Antenna Pointing Control System

Table 1

TENTATIVE PARAMETERS OF A 2-TON CLASS SATELLITE

Antenna Pointing Accuracy	0.015°
Antenna Pointing Control Accuracy	0.002°
Maximum Satellite Attitude Rate	0.02 °/sec.
Main-reflector Diameter	3.5 m
Sub-reflector Diameter	1.0 m

DSTURBANCE FOR ANTENNA DRIVE CONTROL SYSTEM

Satellite Attitude Changes

Two factors must be considered in order to evaluate the direct coupling between the ADCS and ACS. One is the reaction torque of the sub-reflector drive. The other is the antenna beam shift due to attitude changes. For the model shown in Table 1, the torque created by the sub-reflector drive is so small that the ADCS control law can be designed independently of the ACS control law[4]. Therefore, the ADCS control law is designed considering attitude changes only.

The worst case for the ADCS is that the attitude angle varies at the highest frequency in the ACS bandwidth and the amplitude equals the maximum attitude errors. Therefore, the attitude change θ_s can be written in the following sine wave form.

$$\theta_s = \theta_{sm} \sin(\omega_{sb} t + \alpha) \quad (6)$$

ω_{sb} : Control frequency of the ACS
θ_{sm} : Maximum attitude error
α : Initial phase
t : Time

The maximum attitude rate ω_{sm} can be derived from differentiating Eq.(6) in the following form

$$\omega_{sm} \leq \omega_{sb}\, \theta_{sm} \quad (7)$$

To compensate the attitude change of Eq.(6) and keep the pointing error within the required control accuracy θ_{am}, the ADCS open loop gain at the ACS control frequency ω_{sb} must satisfy the following relation.

$$G(\omega_{sb}) \geq \theta_{sm} / \theta_{am} \quad (8)$$

Since the gradient of ADCS gain against frequency is -20dB/decade or less, $G(\omega_{sb})$ can be written as the following approximate relation

$$G(\omega_{sb}) \approx \omega_{ab} / \omega_{sb} \; . \quad (9)$$

ω_{ab} : Control frequency of the ADCS

Combining of Eqs.(8) and (9) yields

$$\omega_{ab} \geq (\theta_{sm}/\theta_{am})\, \omega_{sb} \quad (10)$$

Substituting Eq.(7) into Eq.(8), we finally obtain the relation of the ADCS control frequency ω_{ab} as follows.

$$\omega_{ab} \geq \omega_{sm}/\theta_{am} \quad (11)$$

This relation between the ADCS bandwidth and antenna pointing accuracy is depicted in Fig.4. Assuming that θ_{am}=0.002° and ω_{sm}=0.02°/s for example, the bandwidth must be more than about 2 Hz.

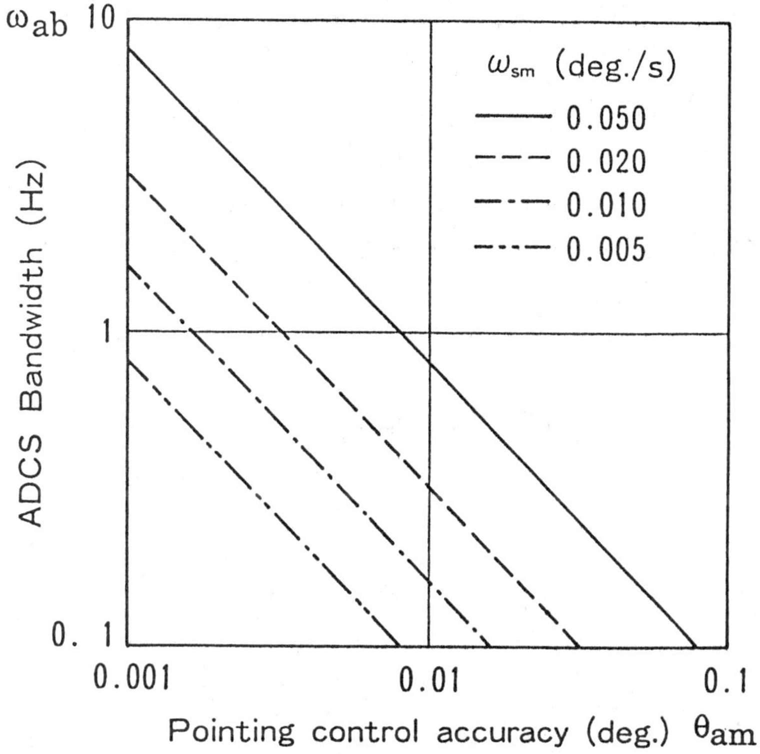

Fig.4 Relation between the ADCS Bandwidth and the Pointing Control Accuracy

Sub-reflector Vibration

The sub-reflector is so rigid that its vibration frequencies are expected to be higher than the ADCS bandwidth. Therefore, in order to control these vibrations, it is appropriate to gain-stabilize the sub-reflector vibrations. This condition imposes the lower limit on the sub-reflector eigen frequency.

Since the ACS bandwidth is much lower than the eigen frequency of the sub-reflector, the influence of the attitude change on the ADCS at the eigen frequency is neglected. Therefore, from Eqs. (2) and (4), the transfer function T_a/θ_b takes the following form[5]

$$\frac{T_a}{\theta_a} = \frac{s^2+2\zeta_s\omega_s s+\omega_s^2}{(I_a-\Omega_{as})s^4+2I_a\zeta_s\omega_s s^3+I_a\omega_s^2 s^2} \qquad (12)$$

It is evident from this equation that the control loop has both resonance and anti-resonance frequencies and the loop gain changes occur at these frequencies. To stabilize the control loop, the open loop gain must be less than 0 dB at the resonance frequency. Assuming that the open loop gain decreases by -40dB/decade, gain as a function of frequency can be approximately represented as

$$G(\omega_{sr}) \approx (\omega_{ab}/\omega_{sr})^2 G_v \qquad (13)$$

ω_{sr} : Resonance frequency calculated from Eq.(12)

where G_v is the gain change caused by the sub-reflector vibration, which is given by Eq. (12). Therefore, the resonance frequency must satisfy the following equation

$$\omega_{ab} \leq \frac{\omega_{sr}}{\sqrt{G_v G_m}} \qquad (14)$$

where G_m is gain margin for the vibration.

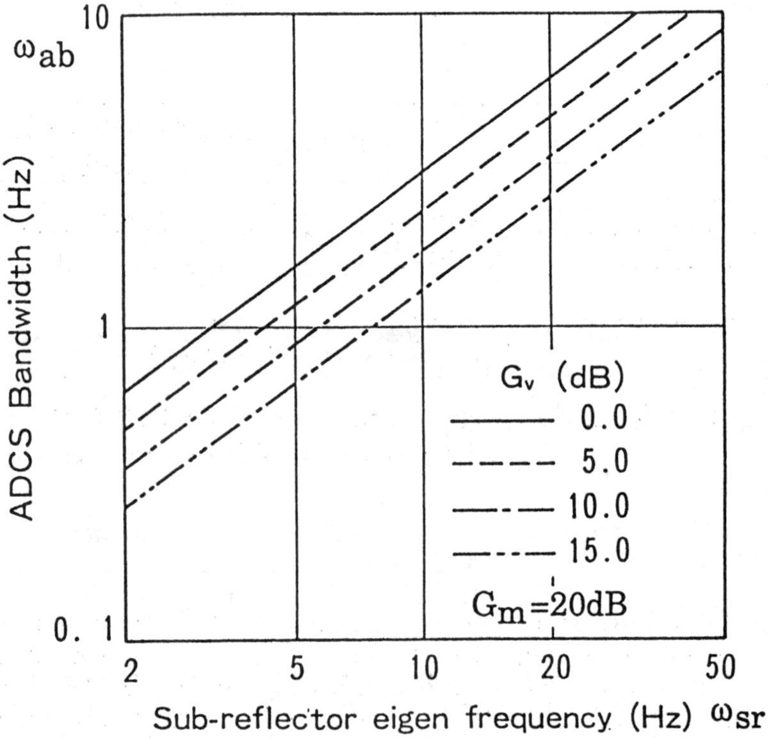

Fig.5 Relation between the ADCS Bandwidth and the Sub-reflector Eigen Frequency

Figure 5 shows the relation between ω_{ab} and ω_{sr} when G_m=20dB. It was found that the eigen frequency must be about 6 times higher than the ADCS bandwidth if G_v=10dB.

Main-reflector Vibration

The vibration of the main-reflector alters the antenna beam direction directly and the ADCS reaction torque is likely to induce this vibration. Therefore, if the main-reflector eigen frequency falls within the ADCS bandwidth, the ADCS may become unstable due to resonance.

Since the ACS bandwidth is much lower than the frequency of the main-reflector, the gain of the ACS control law is nearly equal to zero at the main-reflector eigen frequency. The sub-reflector vibration is also neglected because its eigen frequency is much higher than that of main-reflector. As a result, Figure 3 can be modified to a block diagram as shown in Fig.6. If K_m is not equal to zero, the resonance term is added to the beam direction θ_b. Therefore, changes in open loop gain and phase occur around the resonance frequency. In order to keep the control loop stable, extra margins in gain and phase are necessary.

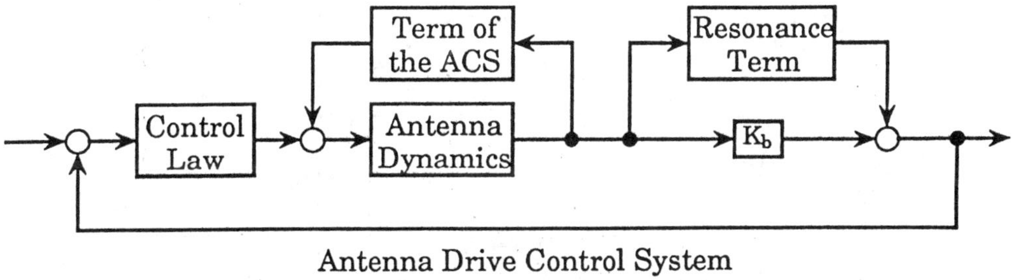

Fig.6 Block Diagram Equivalent to Fig.3 at the Main-reflector Eigen Frequency

The Bode chart of the transfer function from T_a to θ_b is shown in Fig.7 calculated from the parameters listed in Table 2. The broken line shows the resonance term, which is much less than the whole transfer function (solid line) for the 2-ton class satellite. Although the gain and phase changes seem to be very small, it must be noticed that the resonance frequency is lower than the anti-resonance frequency, which causes the phase delay around the resonance frequency. Since the resonance term becomes larger with the sub-reflector moment-of-inertia, the phase delay can't be neglected for large sub-reflectors.

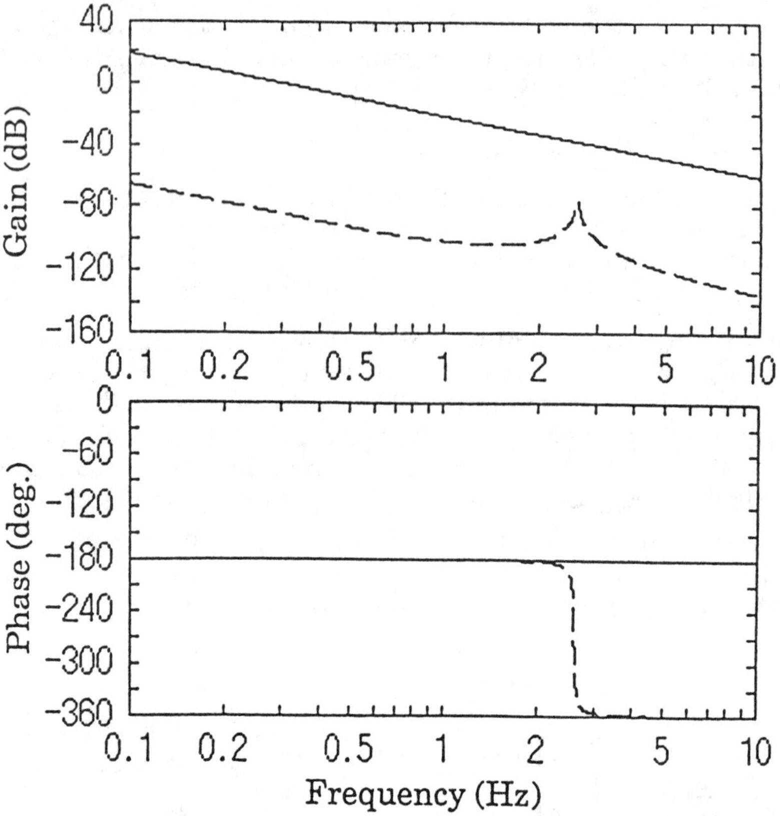

Fig.7 Bode Chart of the System Shown in Fig.6

Table 2

MODEL PARAMETERS OF A 2-TON CLASS SATELLITE

Moment-of-inertia of Sub-reflector	I_a	0.1 kgm^2
Eigen Frequency of Sub-reflector	ω	2.5 Hz
Flexible Mode Coupling of Sub-reflector Vibration	Ω	20.0 $\sqrt{kg}$m
Damping Ratio of Sub-reflector Vibration	ζ	0.01
Ratio of Beam Angle to Main-reflector Modal Coordinate	K_v	0.2 rad/$\sqrt{kg}$m
Ratio of Beam Angle to Sub-reflector Drive Angle	K_b	0.35
Moment-of-inertia of Satellite	I_s	5000 kgm^2

DESIGN OF THE ANTENNA DRIVE CONTROL SYSTEM

Antenna Drive Control Law

The ADCS must be designed to overcome the problems caused by the disturbances mentioned above. Bandwidth is the most important parameter in the design of the ADCS. The conditions which determine the ADCS bandwidth are to satisfy both Eqs.(11) and (14). Combining Eqs.(11) and (14), we obtain

$$\frac{\omega_{sm}}{\theta_{am}} \leq \omega_{ab} \leq \frac{\omega_{sr}}{\sqrt{G_v G_m}} \qquad (15)$$

It is practically convenient to make ω_{ab} and ω_{sr} as small as possible. Therefore, ω_{ab} and ω_{sr} are determined as follows

$$\omega_{ab} = \omega_{sm}/\theta_{am} \qquad (16)$$

$$\omega_{sr} = \omega_{ab}\sqrt{G_v G_m} \qquad (17)$$

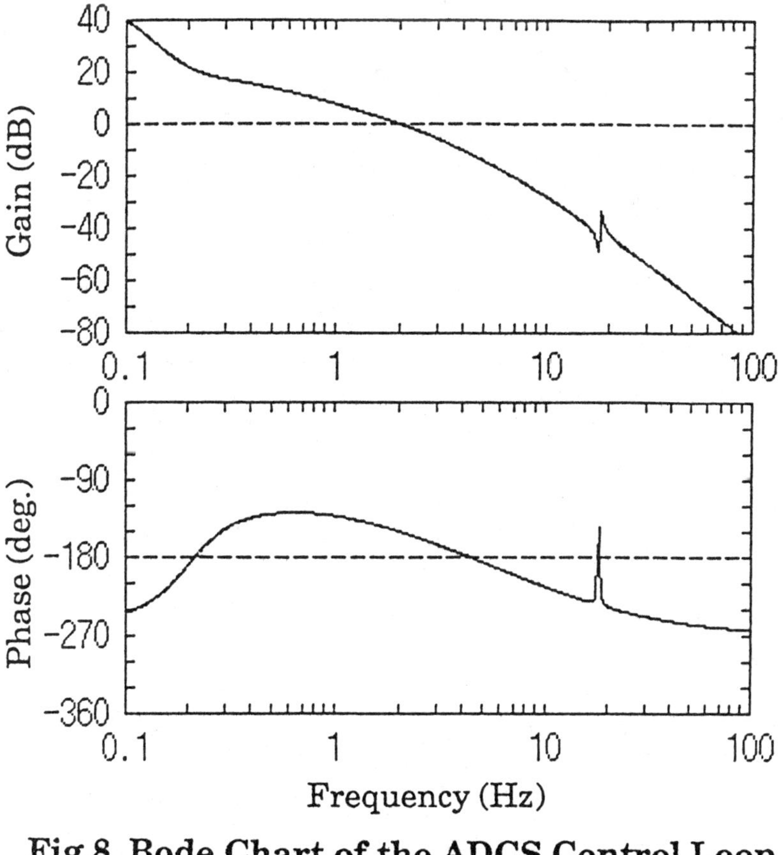

Fig.8 Bode Chart of the ADCS Control Loop

Calculated from the parameters for the 2-ton class satellite, ω_{ab} and ω_{sr} are about 2 Hz and 12 Hz, respectively.

The control law is easily designed by using the filters of the classical control methods such as PID and lead-lag filters. Figure 8 shows the designed Bode chart of ADCS control loop satisfying the condition of Eq.(15). It is clearly evident from Fig.8 that the system is stabilized at the resonance frequencies.

Simulation

A computer simulation was performed to estimate the control performance of the designed ADCS. Figure 9 shows the attitude error and antenna pointing error for the orbit correction mode. The parameters used in the simulation are shown in Table 3. Solar paddle vibration was also considered in the simulation. The antenna pointing error is kept within the required control accuracy of 0.002°.

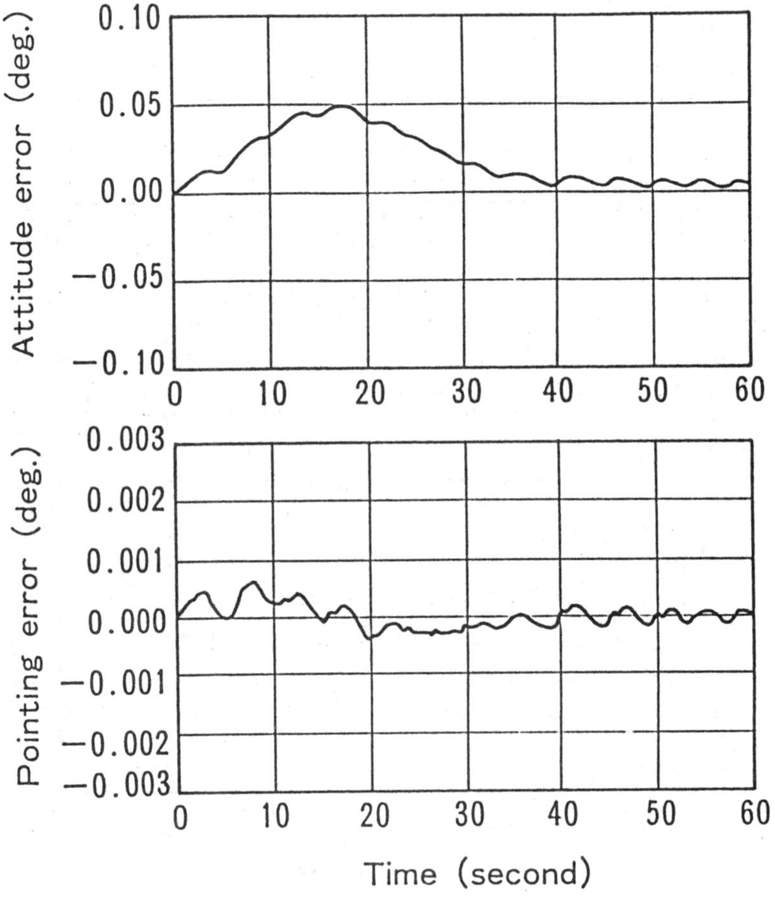

Fig.9 Computer Simulation Results

Table 3

PARAMETERS USED IN THE SIMULATION

Simulation Time	60 second
External Torque	0.5 Nm
ACS Bandwidth	0.3 Hz
ADCS Bandwidth	2.0 Hz
Main-reflector Eigen Frequency	2.5 Hz
Sub-reflector Eigen Frequency	18.0 Hz
Paddle Eigen Frequency	0.2 Hz

CONCLUSION

The design basis of the ADCS for a 2-ton class satellite has been explained. The major conclusions about the disturbances which affect ADCS control performance are as follows.

- Since the satellite attitude change directly tilts antenna direction, the ADCS must reduce the pointing error to within the required accuracy. The parameter defining the attitude change is the maximum attitude rate, which determines the lower limit of the ADCS bandwidth.
- The sub-reflector is so rigid that its eigen frequency is assumed to be higher than the ADCS bandwidth. The difference between the eigen frequency and the ADCS bandwidth is determined by calculating the gain change of the vibration transfer function. In case of the 2-ton class satellite, the eigen frequency must be more than 6 times higher than the ADCS bandwidth.
- Main-reflector vibration reduces the gain and phase margins. Although the reductions are very small for the 2-ton class satellite, it will become a serious problem if the main-reflector is large and the eigen frequency falls in the ADCS bandwidth.

The ADCS control law is easily designed by considering the disturbances mentioned above. By applying the parameters of a 2-ton class satellite, it was proven that the ADCS bandwidth must be about 2 Hz and sub-reflector vibration frequency must be more than 12 Hz. From computer simulations, we confirmed that the pointing error is controlled within the control accuracy requirement of 0.002°. The proposed design method is very useful for designing an ADCS suitable for a satellite with similar size antenna reflectors.

ACKNOWLEDGMENTS

The author would like to thank Dr. Heiichi Yamamoto, Executive Manager of Radio Communication systems laboratories, Dr. Kozo Morita, Executive Manager, Isao Ohtomo, Research Group Leader and Yoichi Kawakami, Senior Research Engineer, Supervisor, both of Communication Satellite Technology Laboratory for their valuable guidance and suggestions during discussions pertaining to this study.

REFERENCES

1. Y. H. Choung, W. H. Stiles, J. Wu, W. C. Wong, C. H. Chen and K. Oye, "NASA ACTS Multibeam Antenna(MBA) System", *EASCON' 86*, 1986, pp.191-196.

2. H. Bittner, A. Brauch, E. Brüderle, C. Roche, A. Scheit, J. Starke and Surauer, "The Attitude and Orbit Control Subsystem of the TV-SAT/TDF1 Spacecraft", Proc. of the Ninth IFAC/ESA Symposium Automatic Control in Space, July 1982, pp.83-102.

3. Y. Kawakami, H. Kumazawa, I. Ohtomo, M. Minomo and Y. Yamada, "On-board Ka-band Multibeam Antenna System with High Pointing Accuracy for ETS-VI", *GLOBECOM'87*, November 1987, pp.952-956.

4. Y. Kawakami, H. Hojo and M. Ueba, "Design of an On-board Antenna Pointing Control System for Communication Satellites", *AIAA/AAS Astrodynamics Conference*, AIAA-88-4306-CP, August 1988, pp.689-694.

5. Y. Kawakami, H. Hojo and M. Ueba, "Control Design of an Antenna Pointing Control System with Large On-board Reflectors", *XIth IFAC Symposium on Automatic Control in Aerospace*, July 1989.

6. P. C. Hughes, "Modal Identities for Elastic Bodies with Application to Vehicle Dynamics and Control", *J. of Applied Mechanics*, Vol.47, No.1, 1980, pp.177-184.

AAS 89-390

THE EFFECTS OF SEASONAL AND LATITUDINAL EARTH INFRARED RADIANCE VARIATIONS ON ERBS ATTITUDE CONTROL*

M. Phenneger, J. Dehen, D. Foch, E. Harvie and M. Virdy†

Analysis performed in the Flight Dynamics Facility by the Earth Radiation Budget Satellite (ERBS) Attitude Determination Support team illustrates the pitch attitude control motion and roll attitude errors induced by Earth infrared (IR) horizon radiance variations. IR scanner and inertial reference unit (IRU) pitch and roll flight data spanning 4 years of the ERBS mission are analyzed to illustrate the changes in the magnitude of the errors on time scales of the orbital period, days, months, and seasons.

The analysis represents a unique opportunity to compare prelaunch estimates of radiance-induced attitude errors with flight measurements.

INTRODUCTION

This paper presents analysis performed in the Goddard Space Flight Center (GSFC) Flight Dynamics Division (FDD) by the Earth Radiation Budget Satellite (ERBS) Attitude Determination Support team. The analysis was performed to measure the ERBS infrared (IR) horizon scanner sensing errors induced by seasonal and latitudinal variations in the Earth's IR horizons. The ERBS mission attitude data offer a unique opportunity to compare prelaunch and early postlaunch estimates of radiance-induced attitude errors with flight measurements of these errors. In addition, this analysis attempts to corroborate the conclusions about the FDD Earth Horizon Radiance Data Base (HRDB) from earlier analysis of data from the Nimbus-7 Limb Infrared Monitor of the Stratosphere (LIMS) experiment. The effects on estimates of these errors, due to adjustments to the radiance data base derived from the earlier LIMS analysis, are evaluated. Radiance errors are derived by a difference between pitch and roll angles obtained from processed IR scanner telemetry and pitch and roll propagations from batch least squares estimates of pitch and roll reference attitudes using the inertial reference unit (IRU). Averages of these differences are used to illustrate the changes in the magnitude of

* This work was supported by the National Aeronautics and Space Administration (NASA)/Goddard Space Flight Center (GSFC), under Contract NAS 5-31500.

† Computer Sciences Corporation (CSC), 6565 Arlington Blvd., Falls Church, Virginia 22046.

the horizon radiance-induced pitch and roll errors on time scales of the orbital period, a day, 1 month, and 1 year.

The remaining sections of this paper present an overview that relates this analysis to earlier analysis of IR radiance errors, a description of the ERBS orbit characteristics and attitude system, and a brief description of the ERBS IR scanner sensing geometry and Earth pulse processing. A description of how the errors are caused by seasonal radiance variations, the concept and procedures applied here to extract the errors using the ERBS flight system telemetry data, and a description of the flight data errors is presented. A brief explanation of the IR sensor modeling software system and the result of attempts to model the flight data with this system using two schemes for rescaling the original ERBS radiance profiles are presented. Finally, the results and future applications of this analysis are summarized.

OVERVIEW

Analysis performed between 1977 and 1984 evaluated the methods of applying an Earth IR horizon radiance model to correct IR scanner flight data (Ref. 1). Flight data from 12 spacecraft, including early postlaunch data from the ERBS, were used to compare the actual IR scanner response to the modeled response using the Horizon Radiance Modeling Utility (HRMU) (Ref. 2). Differences in the actual Earth horizon radiance pitch and roll errors relative to the model were found to occur due to limitations in the Earth IR model and IR scanner sensitivity to short duration cold cloud effects.

Data from the Nimbus-7 LIMS experiment, which included horizon radiance measurements in two IR spectral passbands similar to those used for IR horizon scanners, were compared with a model of the LIMS data using a data base of Earth IR spectra referred to here as the HRDB (Ref. 3). The HRDB was developed using the LOWTRAN computer program (Ref. 4) and a data base of worldwide radiosonde temperature profiles (Ref. 5). The LIMS comparison indicated that the modeled IR horizon intensities for the polar latitudes were underestimated for the summer season and overestimated for the winter season.

An overview of the ERBS attitude system and orbital characteristics is provided in Table 1.

ERBS IR Scanner Description

The ITHACO IR Scanwheels employ a rotating prism lens and a single-flake thermistor bolometer to sense the Earth with a 1x2-degree (deg) field of view (FOV), which sweeps along a 45-deg scan cone at 2000 revolutions per minute (rpm). The IR passband is between 14 and 16.1 microns. The scanner cone axes are on opposite sides of the spacecraft in the pitch-yaw plane and are canted 10 deg down from the pitch axis. Figure 1 illustrates the inflight geometry of the scanner optics for nominal attitude (Ref. 6).

The IR scanner uses normalized threshold locator logic. For this, the Earth IR pulse is averaged in a 5-deg window centered at 22.5 deg and 17.5 deg from the

Table 1. ERBS Orbit and Attitude Characteristics

Orbit:

Semimajor axis: 6891 km

Inclination: 57 deg

Eccentricity: 0.0014 (near-frozen orbit)

Attitude Parameters:

Angular momentum biased, Earth oriented, 1 revolution per orbit

Nominal geodetic pitch and roll = 0.0 deg

Nominal yaw = 0.0 or 180.0 deg for solar array illumination

Attitude Sensors:

Two Adcole fine Sun sensors 64x64 deg 0.004 deg (l.s.b.)

Two ITHACO Scanwheel IR sensors 0.025 deg (l.s.b.)

One Schoenstedt three-axis fluxgate magnetometer 4.68 mg (l.s.b.)

Two IRUs with three Northrop rate gyros 0.001 deg/sec (l.s.b.)

One gyrocompass onboard analog processor 0.03125 deg (l.s.b.)

Attitude Actuators:

One pitch momentum wheel

Two ITHACO Scanwheels

Four orbit adjust and pitch/roll hydrazine thrusters

Two pairs of yaw turn hydrazine thrusters

One roll axis and one yaw axis 50-ampere turn meter squared (ATm^2) magnetic dipole torque rod for pitch momentum control

Two pitch axis 50 ATm^2 dipole torque rods for roll control

inward and outward horizons, respectively, to determine the horizon triggering threshold voltage. A typical Earth pulse is shown in Figure 2. A magnetic pickoff mounted on the sensor body provides a reference pulse from which the acquisition of signal (AOS) and loss of signal (LOS) to index angles are computed (Ω_{in} and Ω_{out} in Figure 2). The pitch angle is computed as

$$P = \pm K_p \left[(\Omega_{los}^R - \Omega_{aos}^R) + (\Omega_{los}^L - \Omega_{aos}^L)\right] \tag{1}$$

where R and L designate the right and left side scanner angles, respectively, and where K_p is a geometry-dependent constant. For ERBS $K_p \approx 0.2462$.

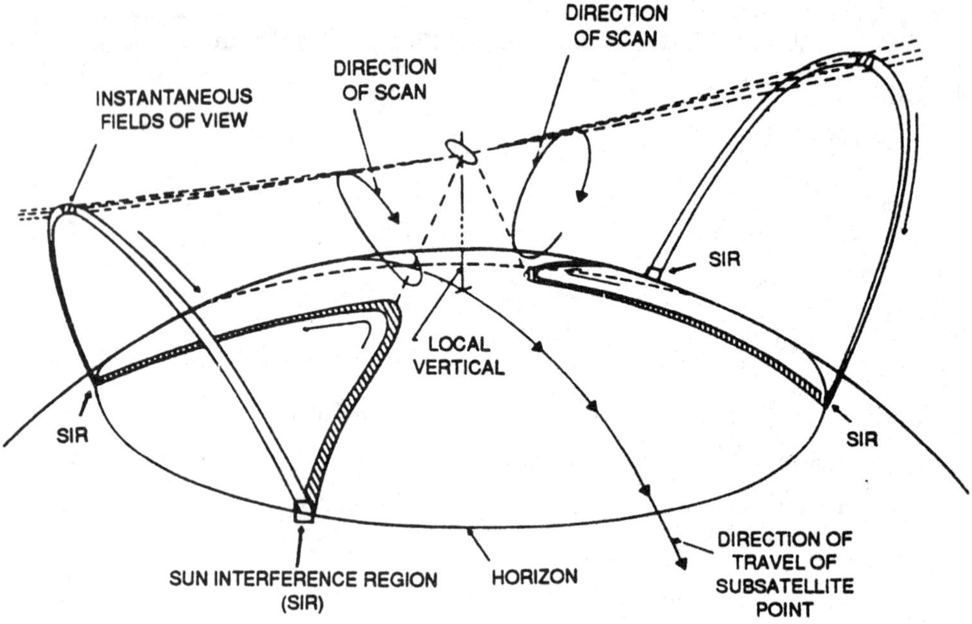

Fig. 1 In-Flight Geometry of the ERBS IR Horizon Sensing System for Pitch = 0, Roll = 0, Yaw = 0

Roll is computed as

$$R = K_r \left(\Omega^R - \Omega^L\right)$$

and (2)

$$\Omega^{R,L} = \left(\Omega^{R,L}_{aos} + \Omega^{R,L}_{los}\right)$$

where $K_r \approx 0.247$.

Figure 3 shows the scanner ground traces at 5-minute intervals for both an equatorial and a polar view of the Earth. It can be seen that the AOS and LOS threshold computation regions (indicated by hashmarks in the figure) are separated by a wide range of latitudes in the equatorial regions, and that the left and right scanners are widely separated at the northern and southern extremes of the orbit.

HORIZON RADIANCE ERRORS FROM FLIGHT DATA

The horizon radiance-induced pitch and roll errors in the IR scanner input to the ERBS magnetic control system (MCS) control loop are caused by radiance

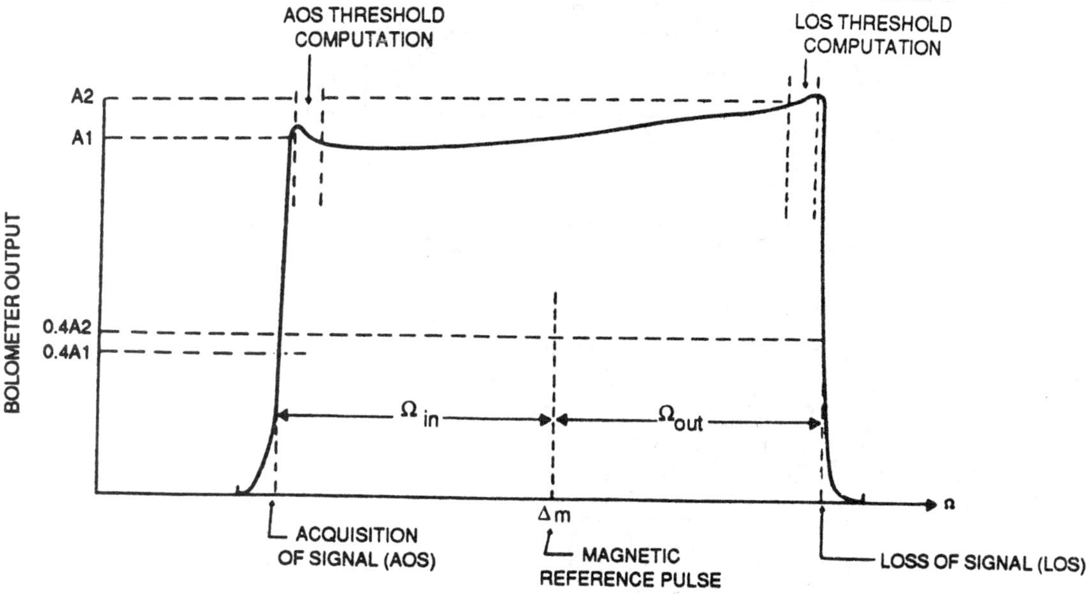

Fig. 2 The Horizon Locator Logic and the IR Scanner Earth Pulse

gradients along the scan ground traces. The gradients are most severe in the winter and summer seasons between the polar latitudes and the temperate latitudes. The gradient causes the threshold normalization region intensity and the rising edge of the Earth pulse intensity to vary relatively. A brightening at the horizon causes an increased Earth width for a given threshold voltage. Likewise, a diminished radiance at the horizon will decrease the sensed Earth width. When the ERBS is on the Equator, a minimal north-south gradient occurs for any month and pitch errors are near zero. Roll errors at this location are dependent on east-west gradients that on the average will be zero. At the midlatitudes, near 40 deg, the AOS and LOS horizons for either scanner are at maximum latitudinal separation and include the latitude regions where the stratosphere experiences the greatest seasonal radiance variation. These are latitudes between 40 deg and the poles. Thus, the pitch errors will be maximum. At the highest and lowest latitudes, the left and right scanner traces are at north and south extremes, where differences between the 80-deg and 40-deg radiance intensities will determine the peak roll errors. At these points the latitudes of the AOS and LOS horizon points are the same for each scanner, and pitch errors are expected to be near zero.

Error Computation Concepts

The flight system horizon radiance-induced errors are computed by subtracting IRU-based pitch and roll determined by the ERBS Attitude Ground Support System (AGSS) (Ref. 7) Fine Attitude Determination Subsystem (FADS) from the IR scanner pitch and roll. The IR scanner pitch error is the error input to the ERBS pitch control loop. Horizon radiance-induced pitch errors from the IR scanner thus cause pitch motion to null the IR scanner error signal that is also received in the

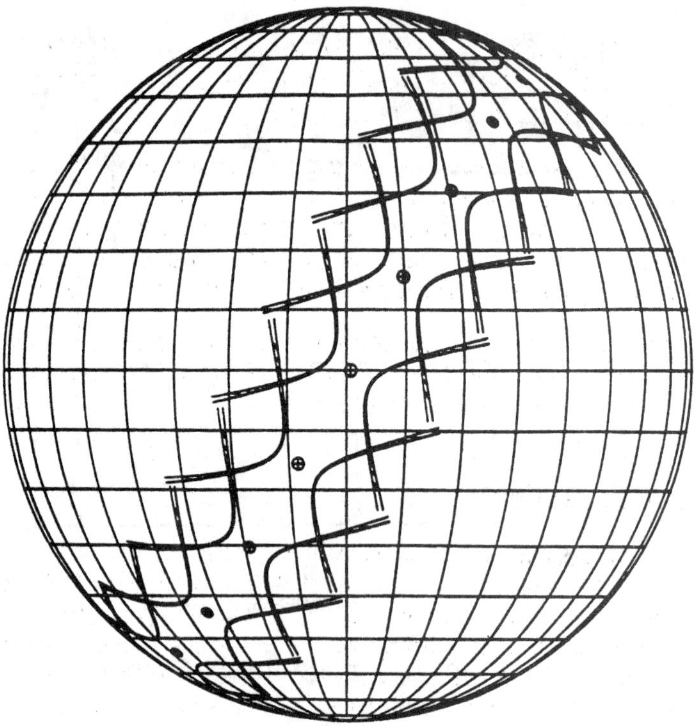

EQUITORIAL VIEW

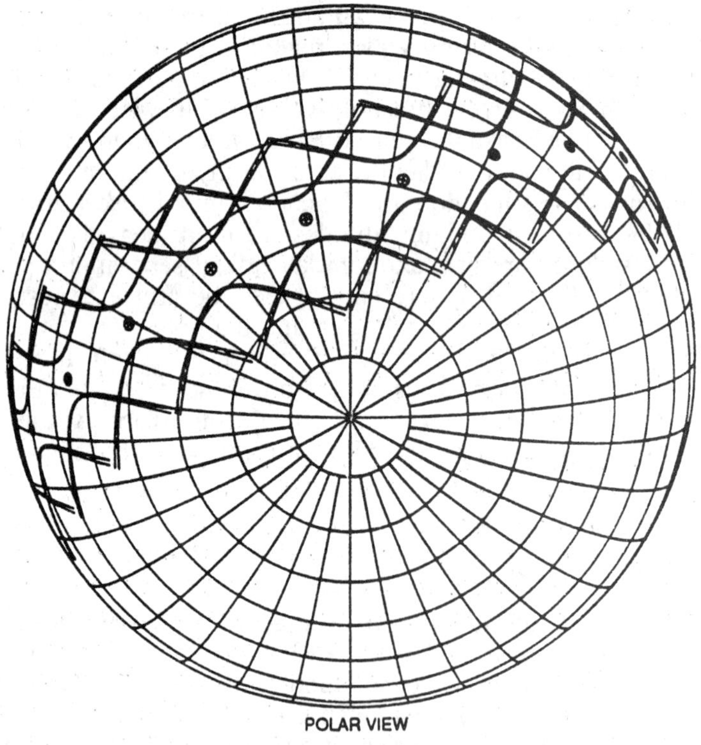

POLAR VIEW

Fig. 3 ERBS IR Horizon Scanner Ground Tracks on the Earth at 5-Minute Intervals

downlink telemetry. However, this motion is sensed by the pitch IRU. The difference, IR scanner pitch minus FADS-IRU pitch, is approximately equal to the radiance-induced pitch error in IR scanner output.

The IR scanner roll error signal is not used for continuous roll control but for intermittent activation of the magnetic dipoles for nutation and precession control. These torques cause the spacecraft pitch axis to precess along the 33-deg latitude line at the ascending node rate of 4-deg per day. IR scanner roll minus FADS-IRU roll is approximately equal to the radiance-induced roll error in IR scanner output.

Since true spacecraft motion caused by control system and environmental torques is registered by both the IR scanners and the IRUs, this motion will not contribute to the difference in the measurement.

Error Computation Procedure

Processing is performed on selected orbits of ERBS attitude telemetry from archived data spanning 4 years of mission operations, between 1984 and 1988. The IR scanner pitch and roll angles are subtracted from the IR scanner pitch and roll angles to produce the IR scanner horizon radiance-induced errors. Errors are averaged over 1-deg latitude bins on the northward and southward sides of each orbit to statistically improve the accuracy and reduce the data volume. To remove the effect of IR scanner and AGSS processing biases, the monthly averaged error representations are shifted to be zero at 0-deg latitude. The bias values were determined by the averaged pitch and roll errors between –5- ,and +5-deg latitude. The three to five one-orbit representations of the latitude averaged errors were again averaged over the 4 years to form an overall average for each month.

Flight Data Analysis Results

The pitch and roll errors obtained from flight data analysis for each month are plotted versus subsatellite latitude in Figures 4a and 4b. The standard deviations from the mean and the northward (N) and southward (S) direction of flight are indicated by the size and type of the plot symbols. The number of orbits averaged is noted to the right of each plot. The tick marks on the ordinate at 0-deg latitude are at intervals of 0.1 deg.

The following characteristics should be noted. For most of the months the pattern of the IR radiance-induced pitch and roll errors is as expected. December and February pitch errors are unusual when compared with adjacent months. The magnitude of the pitch and roll errors in the winter and summer seasons is two to five times higher than the prelaunch analysis predicted. The prelaunch analysis pitch errors are illustrated in Figure 5 for comparison. The Southern Hemisphere summer and winter errors are significantly different from their Northern Hemisphere counterparts. These 4-year averaged data zshow effects of the gradual change in the radiance with season and also indicate a high level of annual similarity in the stratosphere. Finally, for June and July, the months when the errors are the largest, the standard deviations do not exceed the error amplitude.

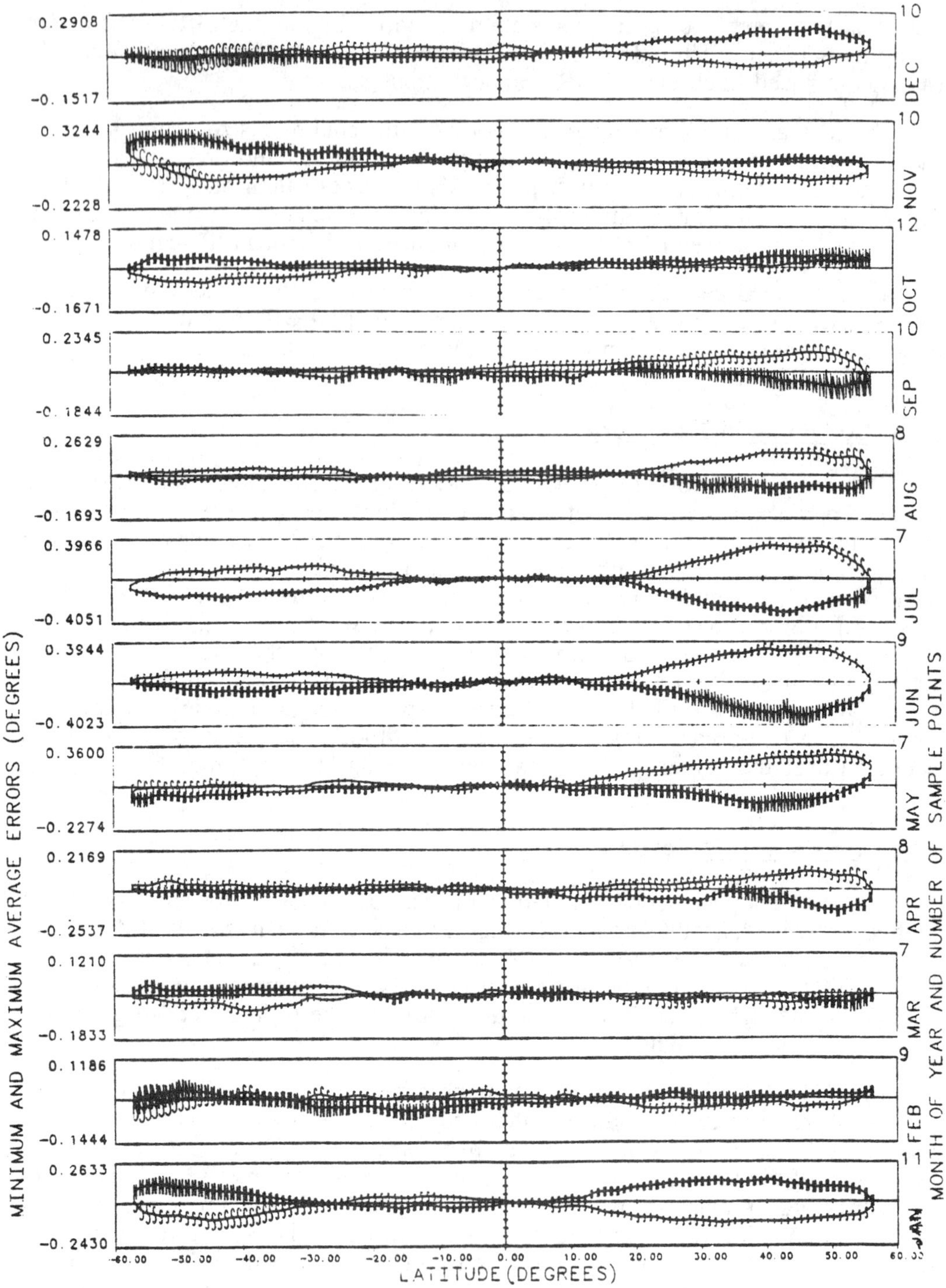

Fig. 4a Flight Data Pitch Errors Versus Latitude and Month (With Standard Deviations)

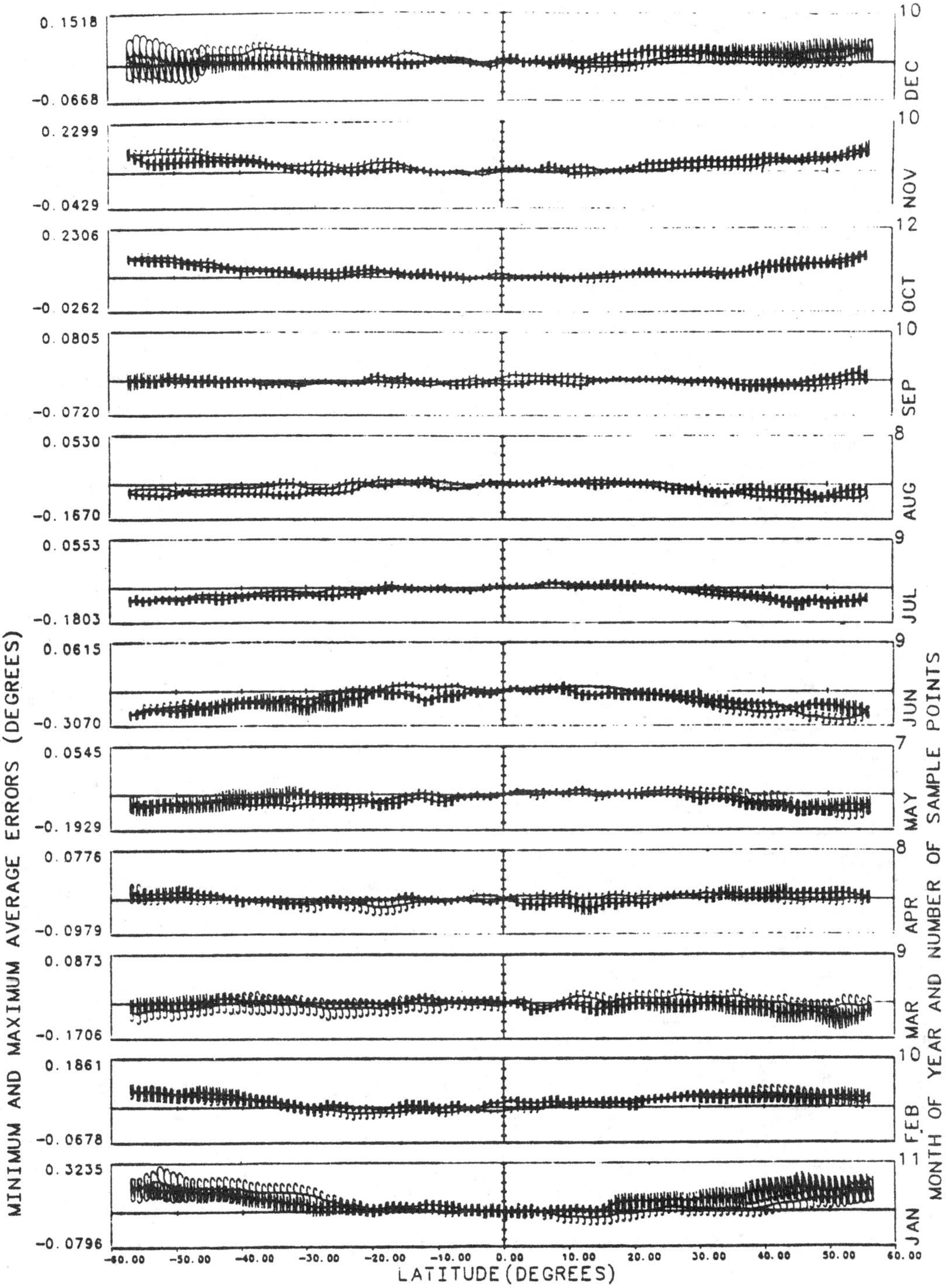

Fig. 4b Flight Data Roll Errors Versus Latitude and Month (With Standard Deviations)

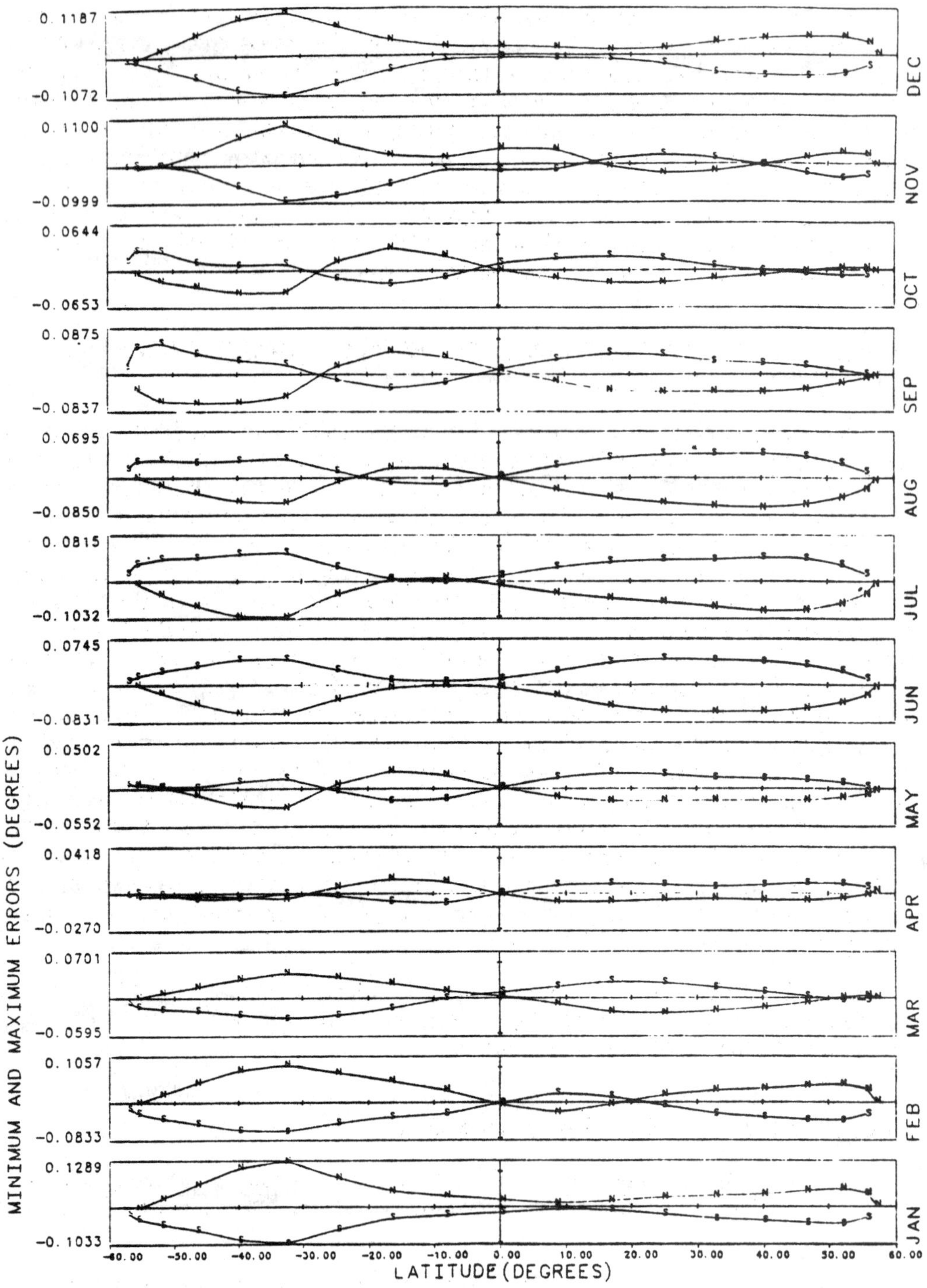

Fig. 5 Pitch Errors Versus Latitude and Month (Original Model)

The effects of longitudinal variations in stratospheric IR radiance can be seen by comparing errors from 15 consecutive orbits. The pitch errors from July 24, 1987, are illustrated in Figure 6 in a stacked format. The unfolded plot runs from the south latitude extreme for one orbit. Error for each orbit is displaced by one deg on the ordinate from the previous starting at the bottom of the figure. The "figure 8" pattern of previous figures is still evident in this format and the pitch errors at 30 deg south vary from -0.5 deg to near zero on the 14th and 15th orbit. The triangles on this figure indicate terminator crossing times, and the X's are times when the Sun crosses the IR scanner FOV scan cone. No evidence for errors due to these phenomena is apparent in this figure.

MODELING THE FLIGHT DATA WITH THE HORIZON RADIANCE MODELING UTILITY

The horizon radiance errors modeled by the HRMU are derived from a detailed computation using an Earth IR model and an IR scanner model composed of the orbital geometry, the IR scanner optics, and the signal processing electronics. The input characteristics of the IR scanner electronics include threshold normalization parameters and time constants, scanner mounting and FOV angles, and orbital radius and inclination. HRMU input is the IR horizon profile of brightness versus FOV tangent height and angle of incidence to the Earth. These data are in a set of nine profiles for each month in 20-deg latitude bins between -90 and +90 deg. This horizon profile data set is derived from the HRDB by integrating the HRDB Earth IR spectra over the optical IR passband for each of the 51 viewing angles represented in the profile. The HRMU computes sensor response to the Earth radiance by integrating the IR radiance, from a latitude interpolated function of the profiles, incident on the scanner optics. As the scanner FOV sweeps across the Earth, a model of the bolometer energy pulse is computed. The scanner step response function is convolved with the input radiance pulse to compute the electronics output Earth pulse. The horizon crossing angle is determined from this output pulse as onboard the spacecraft. The pitch and roll error signals expected over one orbit for a specific month of the year are computed using the computed horizon crossing angles in Equations (1) and (2).

The HRMU was run on the radiance profile data adjusted using two schemes. The first uses LIMS analysis results; the second adjusts the profiles until a reasonable match to the flight data is obtained.

Horizon Radiance Analysis Using LIMS Data

To investigate the cause of the difference between the prelaunch predictions and the flight measurements of the errors, modifications were made to the HRMU input radiance profiles.

Profile adjustment scale factors were determined from plots of 0-kilometer tangent height radiance intensities from LIMS and HRDB models of LIMS profiles, illustrated in Figure 7 from Reference 3. For the months in which LIMS data were not available, the radiance profiles in the Northern and Southern Hemispheres

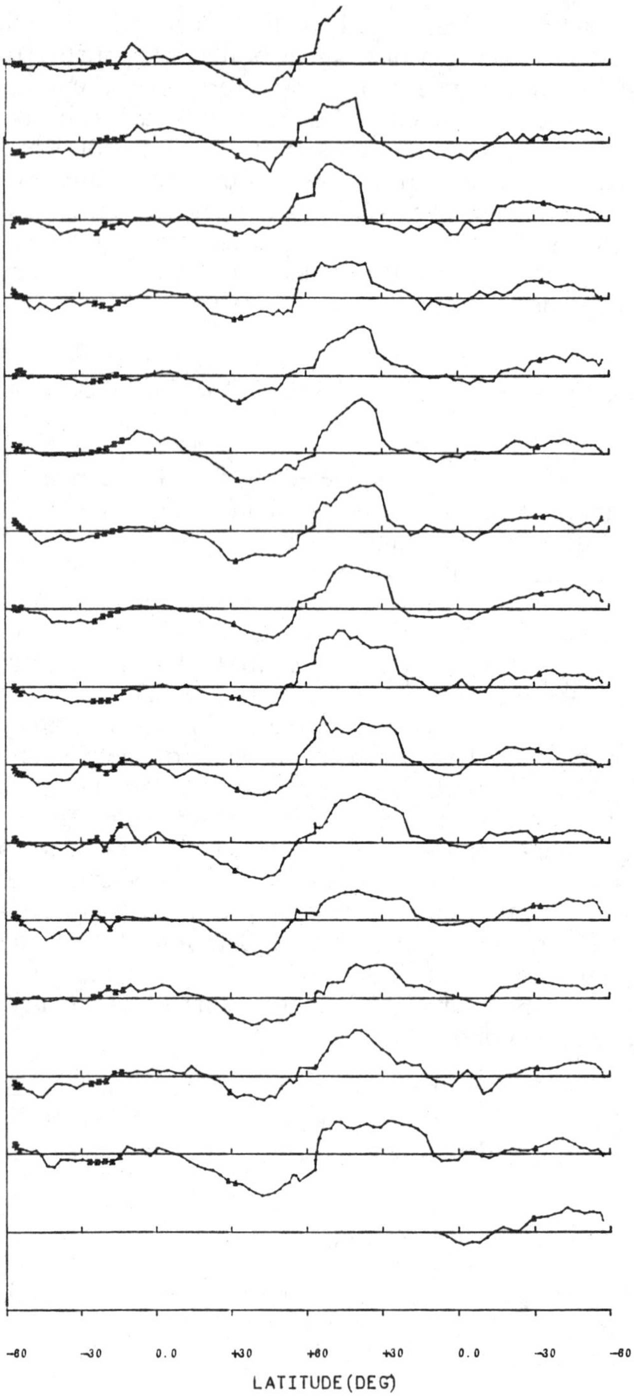

Fig. 6 Pitch IR Errors, July 24, 1987

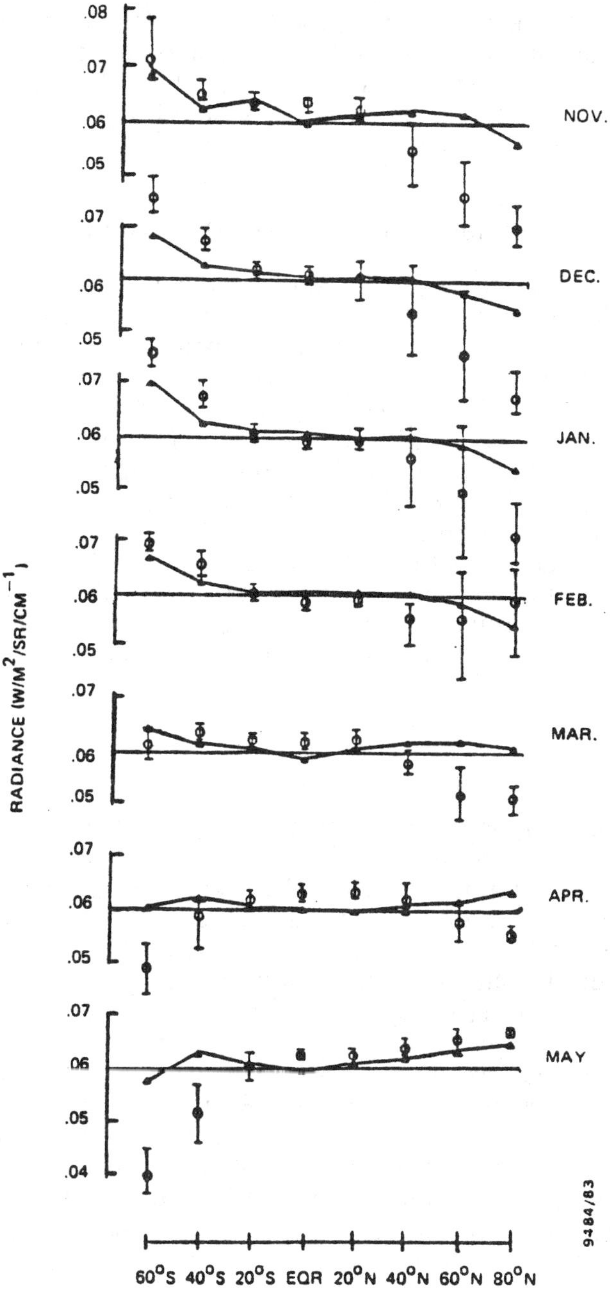

Fig. 7 CO_2N Radiance Intensities at a 0-Kilometer Tangent Height Observed by LIMS (O) and Predicted by the HRDB (▲) (Bars Indicate Maximum and Minimum LIMS Values)

were assumed to be seasonally symmetric. The pitch and roll errors resulting from this rescaled profile data are illustrated in Figures 8a and 8b. Improved agreement between predicted and measured pitch and roll errors for ERBS is demonstrated; however, the differences in the error magnitudes of 0.2 deg still remained. Because of this, the second adjustment scheme was tried.

Adjusting the Radiance Profiles To Match the Flight Data Errors

The monthly errors from this process are illustrated in Figures 9a and 9b. The match is approximate; in most cases, the model agrees with the average flight errors within the standard deviation of the flight data. During this exercise the following understanding was obtained. Efforts to raise the pitch errors relative to the LIMS profile renormalization scheme by raising and lowering the 60- and 80-deg latitude radiance profile intensities caused roll errors to exceed 0.5 deg. A profile renormalization scheme that would completely decouple the pitch error peak from the roll error peak was not found. This may be a result of not modeling the left and right scanner electronics separately. However, it was determined that the pitch error in the model is controlled by the radiance gradient between 20, 40, and 60 deg latitudes and the roll error is controlled by the gradient between 40, 60, and 80 deg latitudes. Thus, roll errors were moderated by flattening the gradients in the high latitudes, making 40- and 80-deg radiance differences smaller in the summer and winter seasons.

CONCLUSIONS

The results of this analysis contribute significantly to the understanding of the effect of horizon radiance-induced errors in IR scanner systems and complement earlier work performed in the GSFC/FDD. The flight data measurements of the effect provide an important comparison to and calibration of the Earth horizon radiance model. The results also answer questions about the significance of the horizon radiance effect for attitude determination pointing control accuracy and about the chances of improving the accuracy of attitude measurements and pointing performance using horizon radiance modeling techniques. They also demonstrate a modeling technique using spacecraft flight data. The measured horizon radiance effect for ERBS exceeds the original HRMU-modeled value by as much as a factor of 5. Measured values of 0.45 deg compare to 0.1 deg from the model of pitch errors for the midlatitudes in July. The source of part of this difference was known in 1984. It was reported then that the HRDB underestimated brightness contrast in the horizon radiance profiles for the polar latitudes in the winter and summer hemispheres. ERBS pitch and roll attitude information obtained from pure IR scanner telemetry thus requires the removal of average errors as large as 0.45 deg. The accuracy of the model of these errors can be assessed from the results in three ways:

1. Applying no correction for radiance effects to the IR scanner pitch and roll limits the average accuracy of the IR scanner attitude to 0.5 deg, ignoring all other errors due to biases, alignments, and electronic, optical, and mechanical noise.

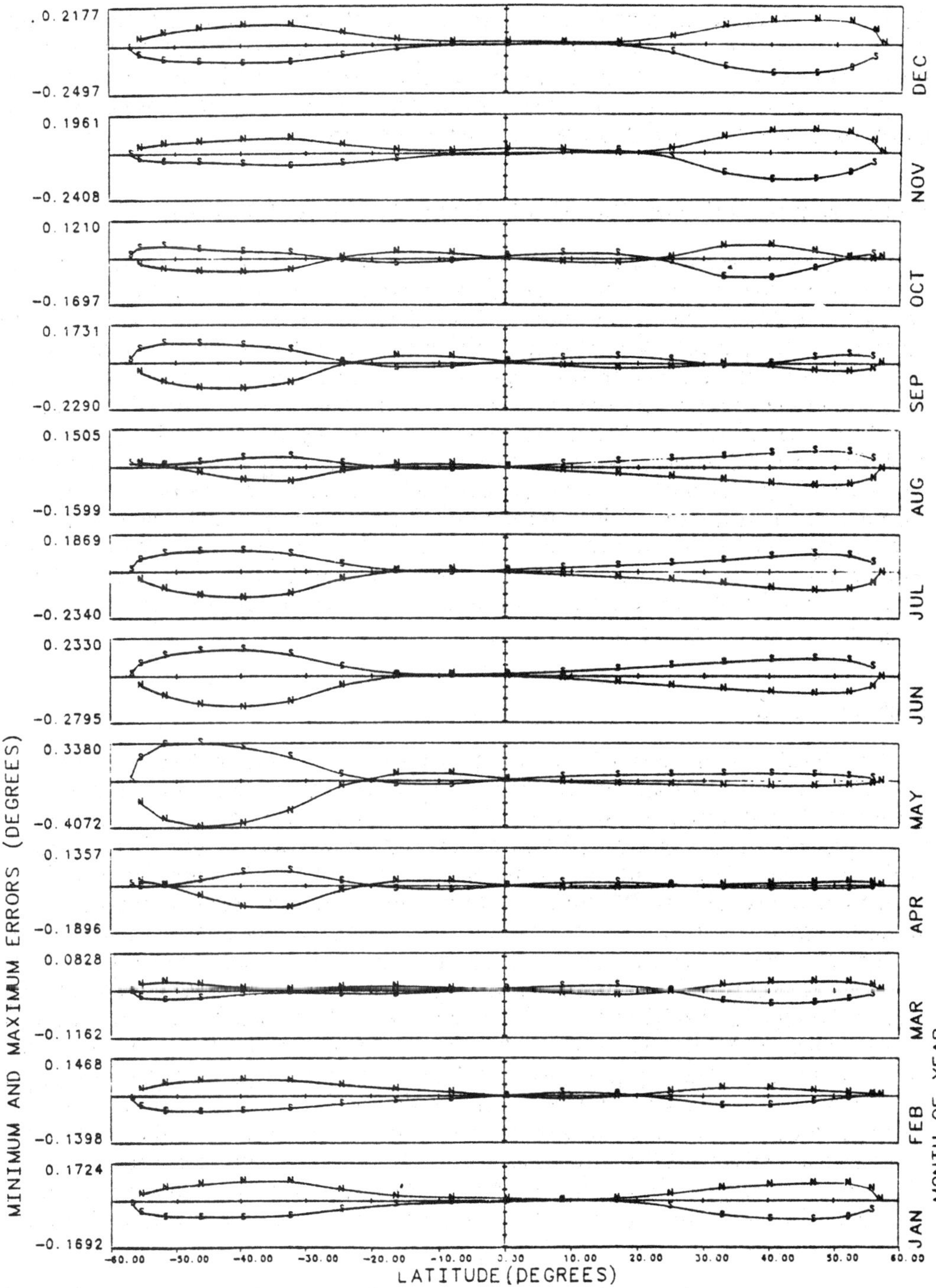

Fig. 8a Pitch Errors Versus Latitude and Month (LIMS Modified Model)

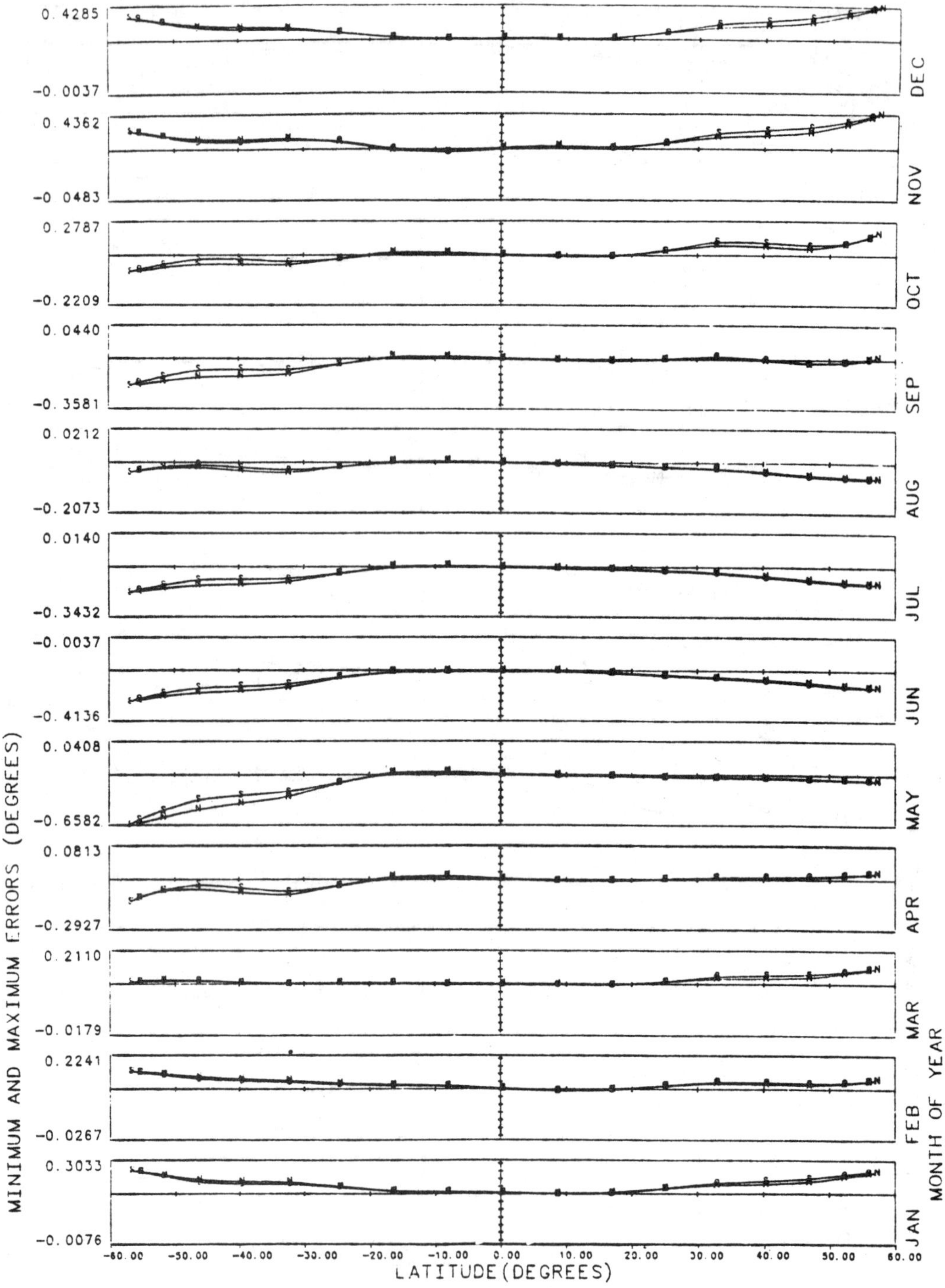

Fig. 8b Roll Errors Versus Latitude and Month (LIMS Modified Model)

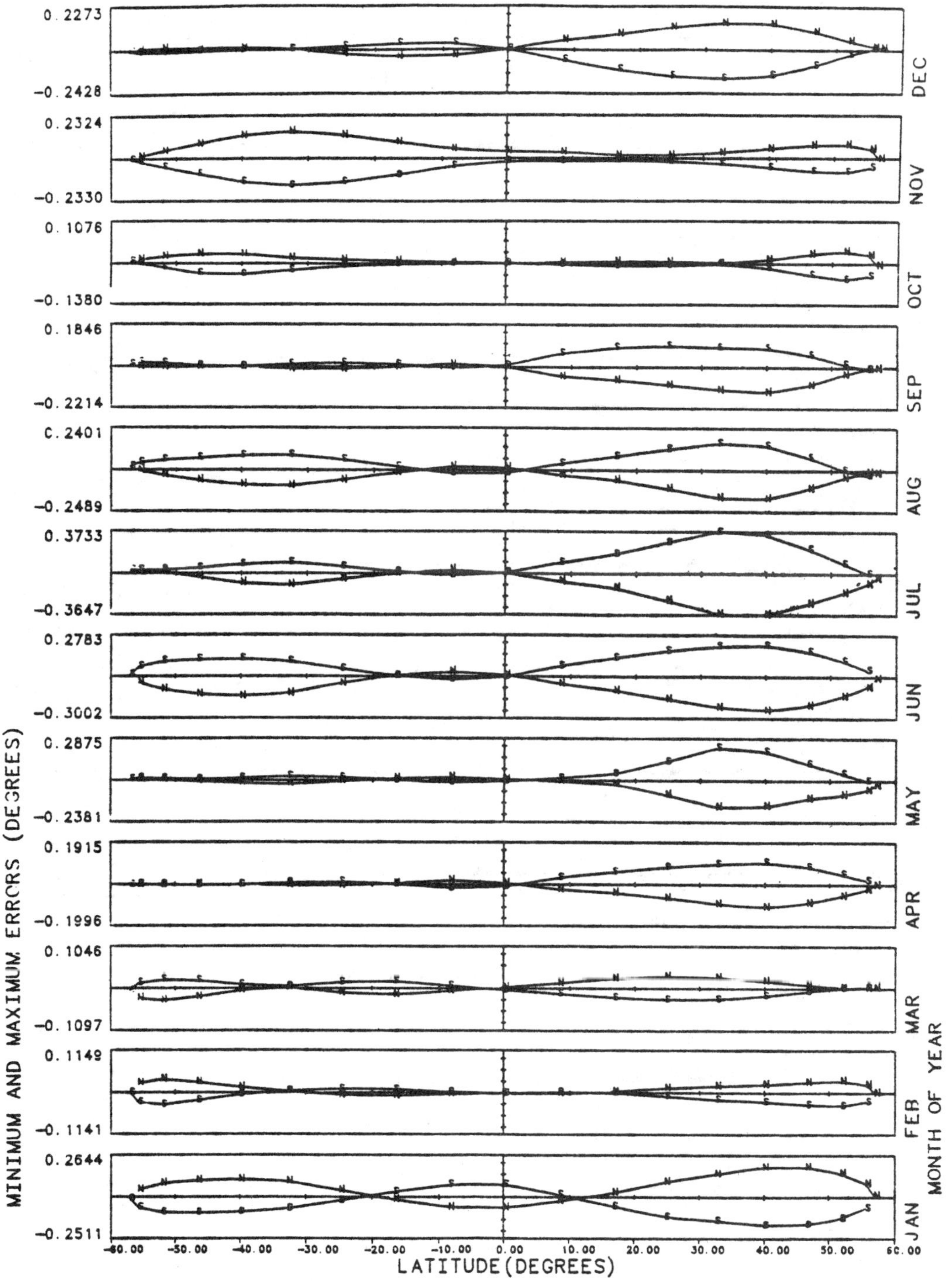

Fig. 9a Pitch Errors Versus Latitude and Month
(Flight Data Modified Model)

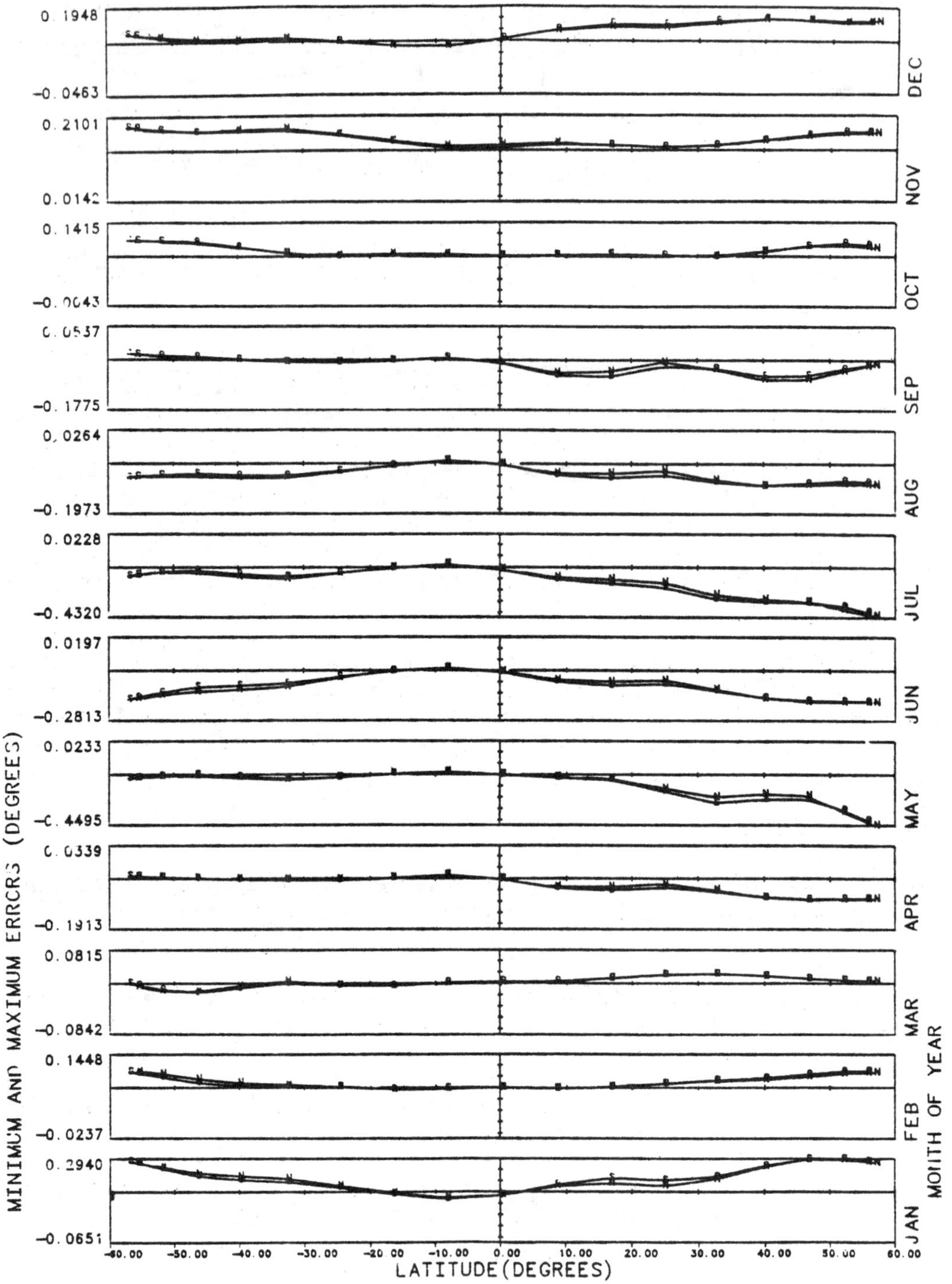

Fig. 9b Roll Errors Versus Latitude and Month
(Flight Data Modified Model)

2. Comparing the accuracy of pitch attitude results corrected using errors modeled from the original HRDB, there is a maximum 0.3-deg systematic error.

3. With a model based upon the flight data and with the extremes exhibited to compensate for average errors as large as 0.45 deg, the 3σ error due to variations in atmospheric radiance is approximately 0.3 deg.

4. From simulation analysis using the HRMU, it was determined that a significant reduction in the latitude-dependent error signal could be realized by moving the threshold voltage determination region closer to the edge of the Earth pulse.

5. An explanation for the inability to model lower roll errors (<0.2 deg) for the months where pitch errors approach 0.4 deg may be that the true stratosphere does not have a maximum brightness directly over the poles. This asymmetry would cause half of the orbits in 1 day to experience error effects characteristic of higher inclination orbits, which increase pitch errors and lower roll errors. The other half of the orbits would then experience lower pitch errors and lower roll errors, characteristic of lower inclination orbits. Thus, the average would enhance pitch errors and suppress roll errors relative to those from a stratospheric IR profile symmetric about the geographic poles.

The results of this analysis are preliminary in that a relatively small sample of the flight data was analyzed. For attitude determination support of ERBS in the future, a model based on this approach could be used as a replacement for the spacecraft attitude motion model currently provided by the IRU.

CITED REFERENCES

1. NASA Technical Memorandum 86181, *Infrared Horizon Sensor Modeling for Attitude Determination and Control: Analysis and Mission Experience*, M. Phenneger, S. P. Singhal, T. H. Lee, T. Stengle, March 1985

2. Computer Sciences Corporation, CSC/SD-84/6007, *Earth Radiation Budget Satellite (ERBS) Horizon Radiance Modeling Utility (HRMU) User's Guide and System Description*, E. Burgess, June 1984

3. --, CSC/TM-83/6177, *Evaluation of the Horizon Radiance Data Base (HRDB)*, M. Phenneger, S. P. Singhal, November 1983

4. Air Force Geophysical Laboratory, Report AFGL-TR-80-0067, *Atmospheric Transmittance/Radiance: Computer Code LOWTRAN-5*, F. X. Kneizys, E. P. Shuttle, et al., February 1980

5. Computer Sciences Corporation, CSC/TM-77/6133, *A Comparison of Radiosonde Temperature and Humidity Profile Data Bases*, R. A. Nieman, August 1977

6. --, CSC/SD-83/6118, *Earth Radiation Budget Satellite (ERBS) Attitude Analysis and Support Plan*, M. Phenneger, et al., September 1983

7. --, CSC/SD-82/6013, *Earth Radiation Budget Satellite (ERBS) Attitude Ground Support System (AGSS) Functional Specifications and Requirements*, G. Nair, et al., September 1982

UNCITED REFERENCES

Computer Sciences Corporation, CSC/SD-86/6756, *Earth Radiation Budget Satellite (ERBS) Pitch and Roll Error Analysis*, J. Kronenwetter, et al., November 1986

--, CSC/SD-84/6005, *Earth Radiation Budget Satellite (ERBS) Definitive Attitude Determination System (DADS) User's Guide and System Description*, Y. Teng, July 1984

--, CSC/SD-84/6004UD2, *Earth Radiation Budget Satellite (ERBS) Attitude Determination System (ADS) User's Guide and System Description*, Revision 1, Update 2, R. Boyer, June 1988

--, CSC/SD-86/6012, *Horizon Radiance Modeling Utility Requirements and Specifications*, J. Hashmall, March 1986

TEMPERATURE DEPENDENCE OF ATTITUDE SENSOR COALIGNMENTS ON THE SOLAR MAXIMUM MISSION (SMM)*

D. S. Pitone†, A. H. Eudell† and F. S. Patt‡

This paper presents results on the temperature correlation of the relative coalignment between the fine-pointing Sun sensor (FPSS) and fixed-head star trackers (FHSTs) on the Solar Maximum Mission (SMM). This correlation can be caused by spacecraft electronic and mechanical effects.

Routine daily measurements reveal a time-dependent sensor coalignment variation. The magnitude of the alignment variation is on the order of 120 arc-seconds (arc-sec), which greatly exceeds the prelaunch thermal structural analysis estimate of 15 arc-sec. Differences between FPSS-only and FHST-only yaw solutions as a function of mission day are correlated with the relevant spacecraft temperature. If unaccounted for, the sensor misalignments due to thermal effects are a significant source of error in attitude determination accuracy. Prominent sources of temperature variation are identified and correlated with the temperature profile observed on the SMM.

It has been determined that even relatively small changes in spacecraft temperature can affect the coalignments between the attitude hardware on the SMM and the science instrument support plate and that frequent recalibration of sensor alignments is necessary to compensate for this effect. An alternative to frequent recalibration is to model the variation of alignments as a function of temperature and use this to maintain accurate ground or onboard alignment estimates. These flight data analysis results may be important considerations for prelaunch analysis of future missions.

INTRODUCTION

This paper presents the Solar Maximum Mission (SMM) flight system measurement of the correlation between the spacecraft structure temperature and the coalignment of the fine attitude sensors, composed of two Adcole fine-pointing Sun

---

* This work was supported by the National Aeronautics and Space Administration (NASA)/Goddard Space Flight Center (GSFC), Greenbelt, Maryland, under Contract NAS 5-31500.

† Computer Sciences Corporation (CSC), 6565 Arlington Blvd., Falls Church, Virginia 22046.

‡ General Sciences Corporation (GSC).

sensors (FPSSs) and two Ball Aerospace CT401 fixed-head star trackers (FHSTs). An overview of the SMM, including mission history and configuration, is presented. Possible causes of the variation in the coalignment, subsequently referred to as the misalignment, are discussed, and the conclusion is drawn that the spacecraft temperature is the predominant factor affecting the FPSS-FHST misalignment. Two methods of compensating for this misalignment, frequent in-flight calibration and misalignment function modeling, are compared with regard to accuracy and impact to science data collection. This work was done by the Flight Dynamics Division (FDD) attitude determination and control ground support team, working in the Flight Dynamics Facility (FDF) at Goddard Space Flight Center (GSFC).

Mission History

The SMM was launched in February 1980 from the Eastern Test Range at Kennedy Space Center (KSC), into an approximately circular low-Earth orbit, with an inclination of nearly 28 degrees (deg) (Reference 1). The scientific objective of the mission was to study solar phenomena. The spacecraft attitude system provided three-axis stabilization and supported solar feature targeting. The spacecraft functioned normally until November 1980, when the standard reaction wheel (SRW) package that provides the controlling torques to the spacecraft began to fail. To preserve the mission, the SMM was put into a spin ($\approx$ 1 deg per second) about the minor principal axis in which it remained until April 1984. With the spacecraft in the spin mode, only minimal solar observation was possible. During April 1984, the spacecraft was repaired in-orbit by the Space Transportation System (STS). The entire attitude control system was replaced, and the spacecraft was returned to the nominal scientific observing mode.

Mission Configuration

The SMM was the first of the multimission modular spacecraft (MMS) series. The MMS were modular to facilitate mission repair and mission adaptation. The SMM basically consists of two parts as shown in Figure 1: the MMS itself and the experiment module. The modules that come with the MMS series are

- A Command and Data Handling (CD&H) system that handles all the communications between the ground and the spacecraft and includes the spacecraft onboard computer (OBC)
- A Modular Power System (MPS) that operates all the power systems, including the Solar Array System (SAS)
- A Modular Attitude Control System (MACS) that contains most of the sensors used in the attitude determination and control of the spacecraft. A High-Gain Antenna System (HGAS), for use in communicating with the Tracking and Data Relay Satellite System (TDRSS), is also attached to the end of the spacecraft.

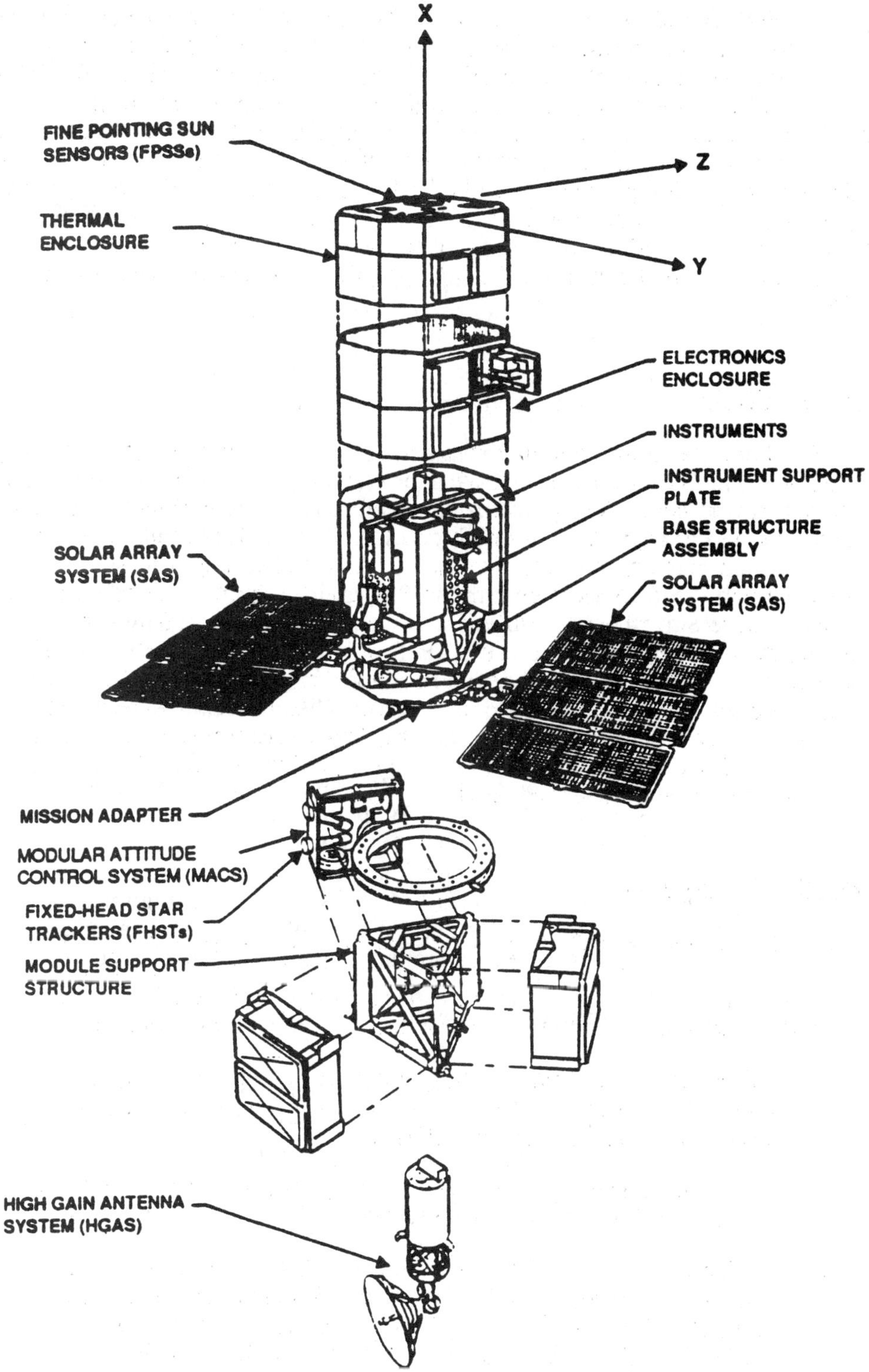

Fig. 1 SMM/MMS Structure

The experiment module contains all the SMM mission-specific components. The two main types of components are the SMM scientific instruments and the mission-specific attitude sensors. The instruments comprising the scientific payload mainly study the emissions of the Sun on several different wavelengths, including X-ray, ultraviolet, and gamma ray. The mission-specific attitude sensors are the coarse Sun sensor (CSS) and the FPSSs. The FPSSs are the primary sources of solar pointing information in the normal mission mode, so their colocation with the scientific instruments makes sense. The two separate modules are connected by a mission adapter ring as shown in Figure 1.

The FPSS, a vector sensor, outputs rotations, α and β, about two sensor axes. It has a 2-deg-by-2-deg field-of-view (FOV). The specified accuracy of the FPSS is 5 arc-seconds (arc-sec) (3σ) within a 1-deg square FOV (Reference 2). The FPSS is mounted on the instrument support plate (ISP). The normal to the ISP is parallel to the y-axis of the spacecraft body frame, as shown in Figure 1.

The spacecraft body coordinate frame is defined by the orientation of the FPSS (Reference 2). The x-axis (roll) of the body frame is defined as parallel to the boresight of the FPSS. The y-axis (pitch) and z-axis (yaw) of the body frame are parallel to the FPSS axes. This definition of the body frame facilitates the calculation of the spacecraft pitch and yaw by the FPSS. The spacecraft pitch and yaw attitude may be read directly from the FPSS readings. Thus, at nominal (zero) roll, the spot on the Sun where the spacecraft is pointing may be easily obtained. However, because of this orientation, the FPSS provides no resolution on the roll attitude of the spacecraft. Also, since the FPSS nominally defines the body frame, no alignment calibration of the FPSS is necessary. The only calibration of the FPSS is for the electronic angular response curve of the sensor.

The FHSTs are also vector sensors. They are mounted in the MACS in the MMS section of the spacecraft. The FHSTs have a two-axis sensor coordinate system, with an 8-deg-by-8-deg FOV. Star positions are output in the telemetry as projected angles in the FHST coordinate frame. These values are then converted to a vector and transformed to the body frame by the FHST alignment matrix, S. Since the FPSS defines the body frame, S represents the relative alignment of the FHST and the FPSS, called the coalignment. The position accuracy of a single FHST measurement is 30 arc-sec (3σ) (Reference 2). This noise in the observation is mainly due to instrument temperature and the varying magnetic field of the Earth.

Since there are two well-separated FHSTs, a full three-axis attitude may be obtained solely from FHST data. However, in August 1987, FHST 2 experienced a loss of power and became inoperable. Hence, the quality of the FHST-only attitudes became significantly degraded. Since star observations in the sensor FOV are only separated by a maximum of 11.3 deg, poor attitude resolution about the FHST boresight axis resulted; thus, after the failure of FHST 2, FHST-only attitudes were of minimal use. Consequently, the analysis in this paper is concerned only with the FHST misalignment behavior before August 1987.

The FHSTs collect data by tracking stars in the FOV. A single star will be tracked for several (N) observations. These N observations are combined to form a track group, which is the average position of the star in the FHST FOV. Thus, reduction in the uncertainty of the track group position can be averaged out by using

$$\sigma = 30/(N)^{1/2} \tag{1}$$

After a track group is formed, the FHST moves to another star and tracks the new star until another track group is formed. This process continues until the FHST becomes occulted.

The sampling rate of the star observations in telemetry and, therefore, the number of observations per track group, is set by the spacecraft telemetry mode. In science mode (the nominal operational mode) the number of observations that form a track group ranges from 6 to 20, while in engineering mode, 60 to 120 observations are available per track group. Engineering mode telemetry, however, does not contain experiment data. Thus, it is requested only for occasional calibration activities when the high FHST sampling rate can be justified. Because of this restriction, the accuracy of the attitude solutions computed from FHST data approximately ranges from 10 to 30 arc-sec (3σ), depending on the quality of the track groups and the degree to which the alignment matrix, S, is known.

Since, as stated previously, the FPSS is not able to determine the spacecraft roll, the FHSTs are the only source of fine roll determination. A coarse roll can be determined from magnetometer data; however, typical accuracies of the coarse-determined roll are between 1 and 2 deg, much larger than the specified roll accuracy of 0.1 deg. Thus, it is important to compute accurate coalignments of the FHSTs so that the computed roll is as accurate as possible. The nominal alignments of the FHSTs are the original design alignments. The calibrations are performed to calculate the misalignment from the nominal alignments.

The misalignment of the FHSTs has two components. The first component is the misalignment of the sensors with respect to the coordinate frame of the MACS. The second component is the misalignment of the MACS frame with respect to the body frame. Because of problems in observability of the orientation of the MACS frame, these components are combined into one set of values for the misalignment.

For the SMM, the misalignment matrices, M_i, represent the change in the alignment from the original design alignments, S_{oi}, where i = 1 or 2, depending on which FHST is being calibrated. The alignment matrices, S_i, of the FHSTs can then be represented by

$$S_i = M_i S_{oi} \quad i = 1,\ 2 \tag{2}$$

Since the S_{oi} are known, the purpose of the FHST alignment calibrations is to compute the M_i. The calibrations are performed by taking observed vectors from the three sensors, one FPSS and two FHSTs, and comparing them to the respective reference vectors. The theoretical aspects of the SMM alignment calibrations have been presented in Reference 3 and will not be repeated here. This alignment calibration scheme has typically yielded accuracies of 5 to 10 arc-sec in the relative alignment of the FHSTs if engineering mode telemetry is used in the calibration.

ANALYSIS

Possible Causes of the Misalignment Variation

Immediately after the SMM repair mission, the new FHST alignments were computed and used in the onboard and ground software. Thus, when pitch and yaw solutions were calculated by the FPSS and FHSTs separately, the resulting differences between the two solutions were initially very small, on the order of the position accuracy in the FHST. However, over the approximately 1-week period before the next alignment calibration was performed, a variation of the differences in yaw attitude, computed separately by the FHSTs and the FPSS, was exhibited that was clearly not random and was well in excess of the FHST solution accuracy (Reference 4). This initial variation is shown in Figure 2. Note that, as shown in Figure 3, the pitch differences show no such variation. Subsequently, to keep track of this variation, although the alignments were recalibrated, the original calibrated alignments were kept on hand and used to process a segment of data each day. The magnitude of the variation eventually reached 120 arc-sec, approximately 8 times greater than the original prerepair estimate for the alignment variation (Reference 5). Obviously, an unmodeled effect was causing this variation.

Several causes were proposed to account for the variation in misalignment:

- Spacecraft bending due to solar radiation pressure
- Electronic or mechanical changes in the sensors
- Uncertainty in the attitude solutions
- Mounting-plate expansion and contraction due to thermal effects

The first possible cause was eliminated because the effects would be too small to measure. Solar radiation pressure, while being a significant effect on spacecraft appendages, will not bend the body of a rigid spacecraft more than a fraction of an arc-second.

The next possibility is electronic or mechanical changes in the sensors themselves. Since the FPSS response was calibrated frequently and since the agreement between FHST readings did not exhibit measurable variations, the change in sensor response was ruled out. The FPSS response did show changes on the order of 1 to 2 arc-sec per month (Reference 4). This small change would not account for the 120-arc-sec differences being observed. An effect related to changes in the sensor, redefinition of the FPSS null, occurred once in December 1984. The null of the FPSS is defined as the output of the FPSS at the center of the Sun. Because of the manner of the FPSS angular response calibrations, the null of the FPSS is not

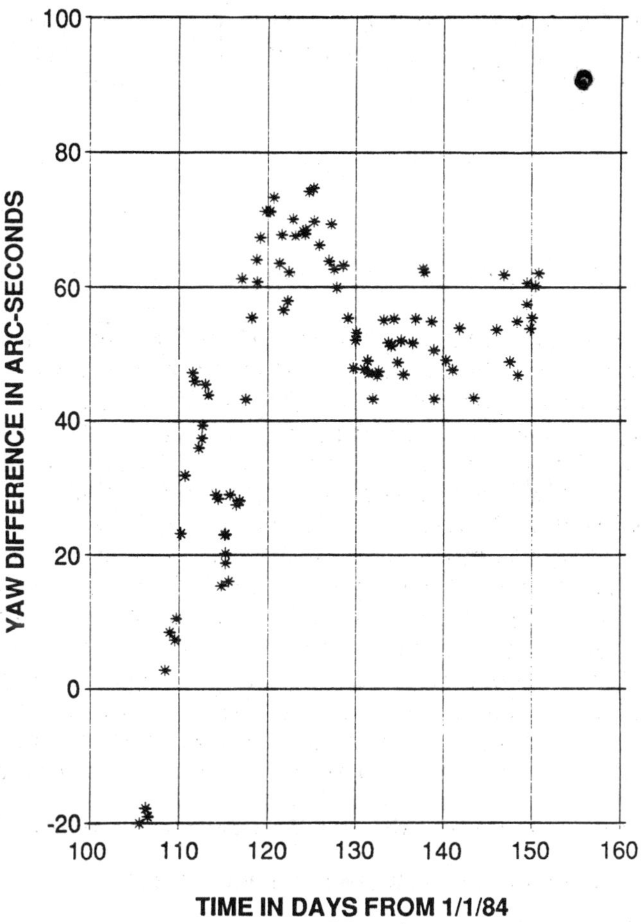

Fig. 2 Initial Yaw Misalignment Variations

determined directly. Some of the scientific instruments are capable of estimating the errors in their solar pointing. These readings have always been measured by the scientific personnel and communicated to the FDF personnel, who incorporated them into the FPSS response function in order to define the FPSS null. Because of changes in the electronic response of the FPSS, the null shows time-dependent shifts. Since scientists noticed a significant error in the FPSS null, it was redefined in December 1984 (Reference 4). This caused a significant (20 to 30 arc-sec) change in the misalignment of the FHSTs. This shift in the FPSS null was most likely accumulated over the time since the repair mission. However, because of the manner in which the FPSS calibrations are performed, it could not be accounted for until the discrepancy was noticed by the scientific instruments. The effect of the null shift can be most clearly seen in the pitch misalignment data, shown in Figure 3. Note the discontinuous drop at approximately day 350, which apparently

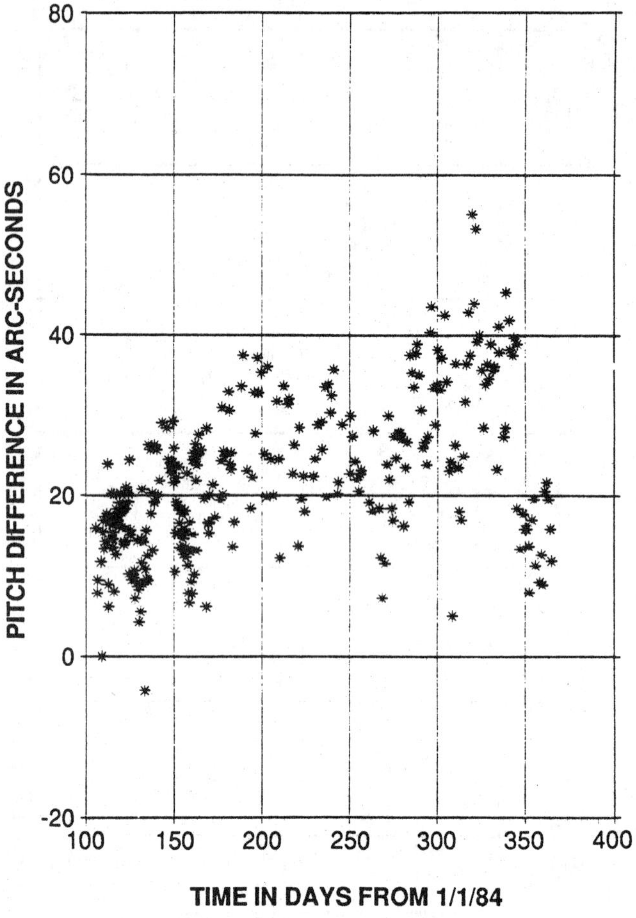

Fig. 3 Pitch Misalignment Variation

reversed the accumulated drift from day 100. This same effect occurred in the yaw misalignment data; however, since the effect was mainly in the pitch axis, the discontinuous change was not significant enough to show clearly on the plot. This, however, serves the purpose to show that the null variations are small compared to the misalignment variation.

The third possibility is the uncertainty in the attitude solutions. As mentioned previously, the misalignment variations were measured daily using FHST and FPSS attitudes computed during normal operations. This attitude comparison does not accurately determine the misalignments. More accurate results would have been obtained by performing frequent FHST calibrations; however, that approach was deemed infeasible because of the resources required and the need to interrupt scientific observation to increase the FHST sampling rate during calibration. Hence, the less accurate but much quicker method of subtracting the computed attitudes was devised. The disadvantage is that a noise level is introduced that is equal in magnitude to the uncertainty of the less accurate attitude solution, which is the FHST attitude. Thus, a noise level of 20 to 30 arc-sec was expected in the misalignment variation plot. However, this random noise would not explain the observed patterned variation, and its magnitude is approximately one-sixth of the magnitude of the observed variation. It is also worth noting that as full in-flight calibrations were performed every 2 to 3 months, the results were completely consistent with the daily attitude calculations.

The effect of temperature variation on attitude sensor alignments has been seen on previous missions, most notably the Magnetic Field Explorer (Magsat) (Reference 6). Thus, the possibility of bending due to thermal effects was explored by checking various temperatures in the spacecraft. It was found that temperatures on the ISP and the mission adapter ring showed a variation over time that was similar to the variation seen in the yaw misalignment. Six temperatures were initially monitored; however, since all six showed the same basic variation, only one was used for graphing and statistical purposes. Since the temperatures were available in the spacecraft telemetry, they could easily be recorded from all real-time station contacts. The temperatures varied slowly over time, taking at least several orbits to change measurably. After this determination was made, data were collected for yaw misalignment and temperature only once per day. The variation of the temperature superimposed on a graph of the variation of the yaw misalignment is shown in Figure 4. The relative scales in the plot were chosen by performing a fit of yaw differences to temperature. The correlation is obvious. Figure 4 shows only 8 months of data to accentuate the correlation. However, all data taken between the SMM repair and the FHST 2 failure exhibited this trend.

The similarity of the two variations points to spacecraft thermal bending as the cause of the misalignment variation. Thus, the changing spacecraft temperatures cause a temperature gradient which, in turn, causes the spacecraft structure to bend. A temperature gradient requires two temperatures. However, because of a lack of thermocouples on the spacecraft, no other temperature on the spacecraft, which when set up as a gradient with the ISP temperature, resulted in a variation similar to the yaw misalignment. However, it can be postulated that the second temperature, which is constant, is equal to the temperature in the laboratory when the original alignments were measured.

The pitch misalignment never showed any significant patterned variation similar to the yaw misalignment variation. As shown in Figure 3, the variation seems to have a noise level of 20 to 30 arc-sec, a slow drift over several months, and a major shift at the point of the FPSS null redefinition. Thus, it seems the dominant effects in the pitch misalignment variation are the uncertainty of the FHST attitude solutions and the shift in the FPSS null.

The fact that the thermal profile variations seemingly do not affect the pitch misalignment can possibly be explained by the orientation of the ISP and the mounting of the FPSS on the ISP. As shown in Figure 5, the ISP is aligned along the z-axis (yaw). The FPSS is mounted in the middle of the ISP, right on the y-axis (pitch) and off-set from the z-axis. Thus, bending of the ISP translates directly into

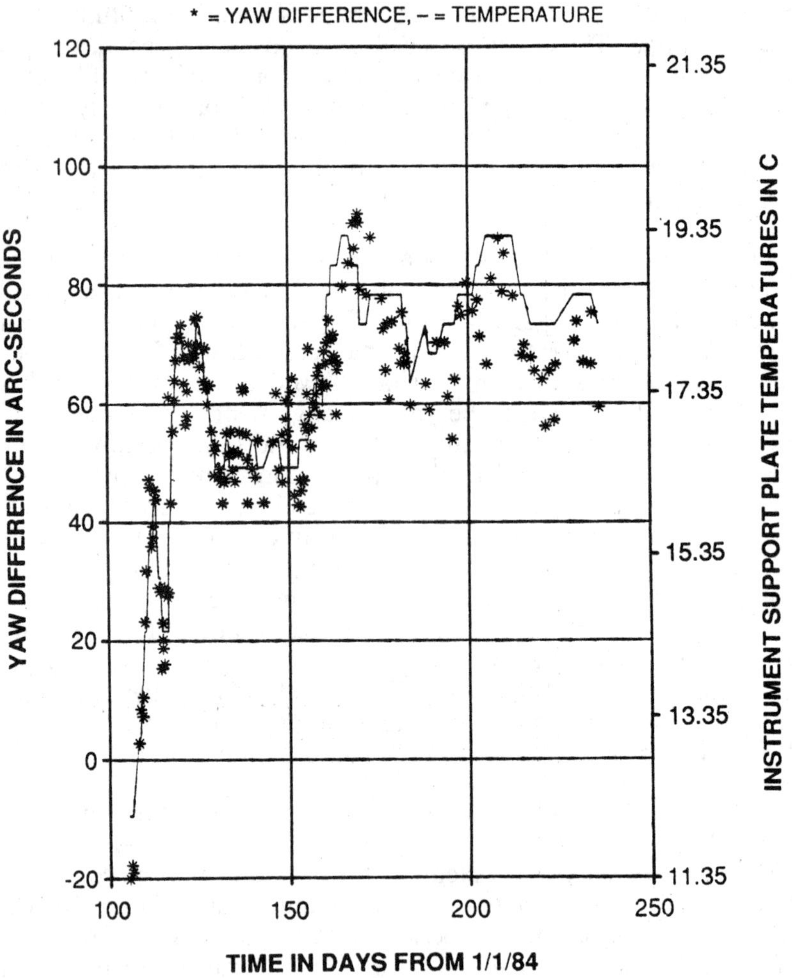

Fig. 4 Yaw Attitude Differences and Instrument Support Plate Temperatures

rotation of the FPSS about the yaw axis. Conversely, bending of the ISP will not cause any rotation about the pitch or roll axes.

Therefore, this analysis suggests that it is the FPSS that is actually moving with respect to the spacecraft. This causes a strange situation in that it is the movement of the FPSS that is causing the misalignment of the FHSTs. This peculiarity can be removed when it is realized that the FPSS defines the body-frame coordinates and,

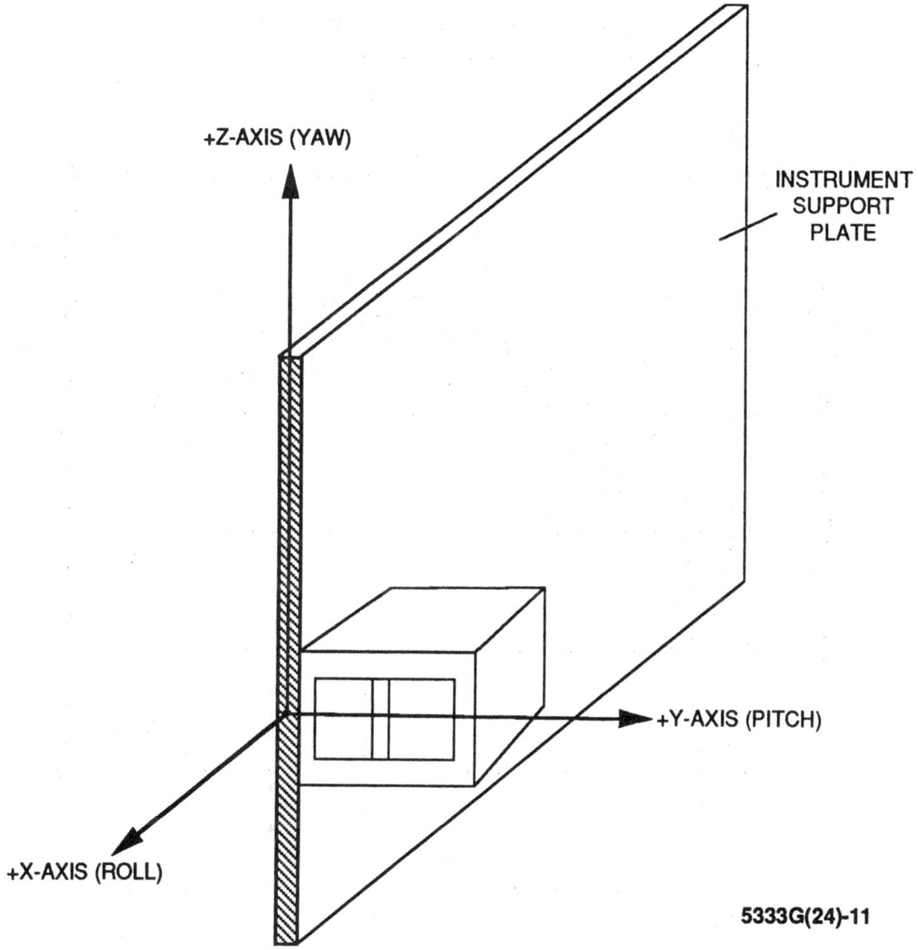

Fig. 5 Orientation of Instrument Support Plate and FPSS

in fact, provides the most critical pointing information for the experiments. Thus, the movement of the FPSS causes the body frame to rotate, which, in turn, causes the FHST misalignment to change. Therefore, the FHSTs never really experience any significant motion with respect to the spacecraft chassis; it is the moving body frame that causes the alignment of the FHSTs to change. Presumably, the other attitude sensors on the spacecraft, such as the three-axis magnetometer, experience this same phenomena. However, because of the low accuracy of these instruments, this phenomena cannot be seen in their alignments. This analysis is a possible explanation of the misalignment variation. However, the full structure of the spacecraft would need to be analyzed before any explanations of the variation could be proved.

Yaw Misalignment Estimation From the Temperature Profile

Since the dominating factor in the FHST misalignment variation about the yaw axis is the variation of the temperature, it would seem likely that the misalignment could be estimated from the observed temperature, thus reducing the need for in-flight FHST alignment calibration. The first step was to set up a scatter plot of the yaw misalignment versus the temperature. This scatter plot, shown in Figure 6, shows a nearly linear relationship between the yaw misalignment and the

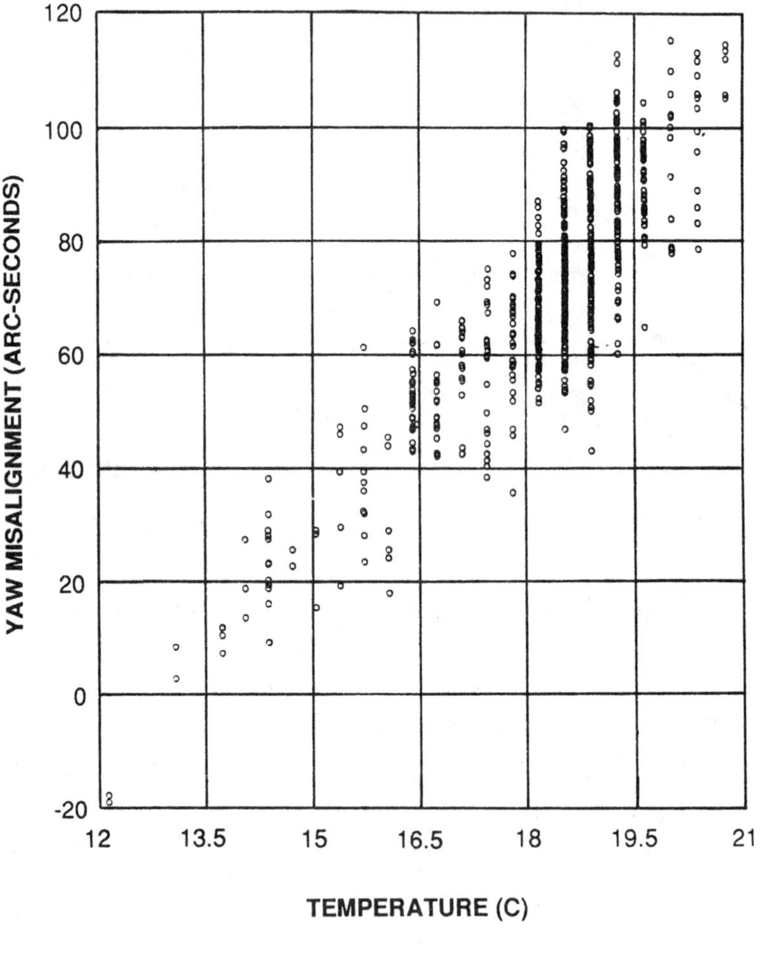

Fig. 6 Yaw Misalignment Versus Temperature

temperature. A linear least-squares fit was performed on the data, and the resulting equation was

$$M = 13.410 * T - 173.790 \tag{3}$$

where M is the scalar yaw misalignment and T is the temperature. The root mean square (RMS) residual of the straight line fit was approximately 10.5 arc-sec, thus 99.7 percent of the yaw misalignments calculated from the function in Equation (3) would be within 31.5 arc-sec (3σ). The error (1σ) in the slope is 0.3 arc-sec per deg celsius and the error (1σ) in the y-intercept is 5.6 arc-sec. This fit could be incorporated into the attitude determination system to estimate the misalignment matrix so that a more accurate attitude could be determined.

A more accurate fit may be determined if the effect due to the null shifts is taken into account. As shown in Figures 7 and 8, the data in Figure 6 can be broken down into two segments: data starting at the SMM repair mission and running until the date of the null shift in December 1984, and data starting at the null shift and running until the failure of FHST 2. Figures 7 and 8 show two different functions. Fits were done to the two functions and the equations were

$$M_1 = 10.801 * T - 130.029 \tag{4}$$

$$M_2 = 15.721 * T - 215.072 \tag{5}$$

The error (3σ) of the calculated misalignment in Equation (4) is 25.5 arc-sec, the uncertainty (1σ) in the slope is 0.3 arc-sec per deg, and the uncertainty (1σ) in the y-intercept is 5.9 arc-sec. The error (3σ) of the calculated misalignment in Equation (5) is 31.1 arc-sec, the uncertainty (1σ) in the slope is 0.5 arc-sec per deg, and the uncertainty (1σ) in the y-intercept is 7.6 arc-sec.

Equations (4) and (5) have different slopes, which poses an interesting question. Since the null shift is a change in the position of the boresight of the FPSS, one would think this would be reflected as a change in the constant term of the misalignment function only. While a significant change in the constant term of these equations exists, the change in the slope indicates that the dependence of the misalignment on the temperature is changing. This would indicate that the equation for the estimate of the alignment would need to be calibrated, seemingly undermining the estimate, since its prime use would be to replace FHST alignment calibration. However, since the variance improved by only about 2 arc-sec for the first segment, calibration of this equation would probably not exceed once per year.

Suggestions for Spacecraft Operations

It has been shown that as temperatures in the SMM structure vary the yaw misalignments of the FHSTs vary similarily, thus degrading the accuracy of the

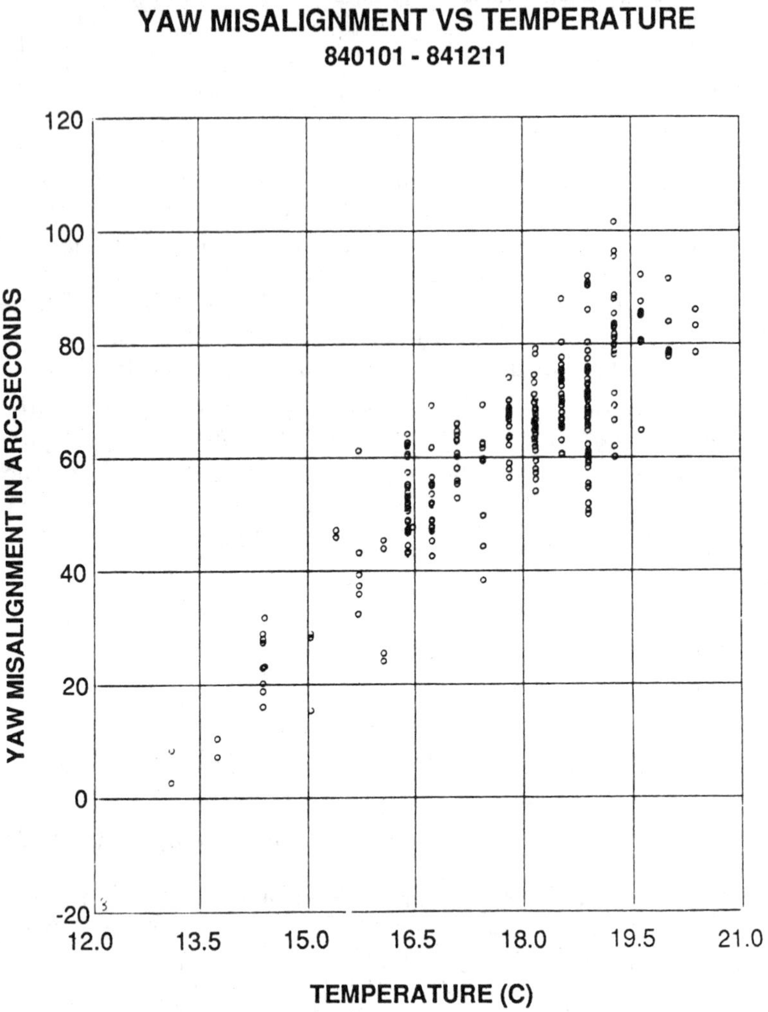

Fig. 7 Yaw Misalignment Versus Temperature Before Null Redefinition

attitude solutions. To compensate for this effect, two methods are available: (1) frequent in-flight alignment calibration and (2) misalignment function modeling. Both methods have their advantages and disadvantages.

The advantage of frequent in-flight calibration is that immediately following the calibration, the resulting alignments are known to a high degree of accuracy, better than 15 arc-sec. However, this accuracy will degrade over a couple of weeks as the temperature varies. This problem can be overcome by recalibrating the

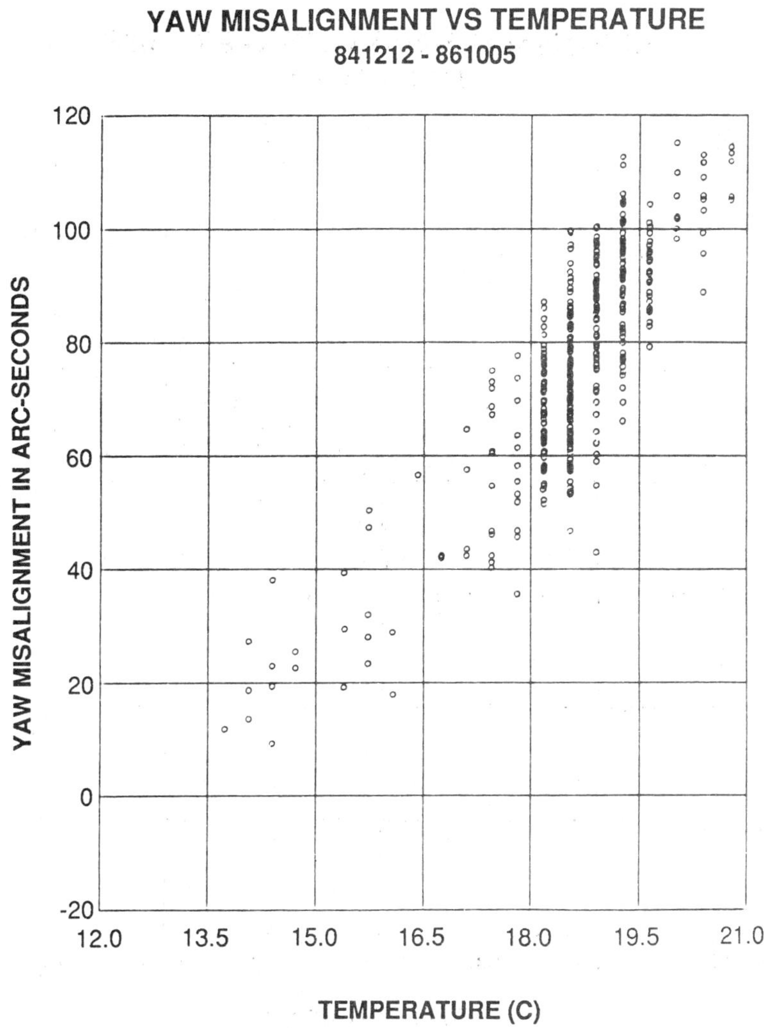

Fig. 8 Yaw Misalignment Versus Temperature After Null Redefinition

alignments every 2 to 3 weeks. However, this points out the major disadvantage of this scheme: in-flight calibration of the FHST misalignment requires operational time, typically three orbits, on the spacecraft and, in the case of the SMM, requires the use of engineering mode telemetry. Time spent calibrating the attitude sensors is lost to the scientists. Thus, ideally, these type of calibrations should be performed infrequently, perhaps no more often than every 2 months.

On the other hand, modeling of the misalignment function requires calibration of the function at most once a year. These calibrations of the function require no special operational time of the spacecraft; they can be completed during routine processing of the attitude data. However, the disadvantage of using the misalignment function is slightly degraded knowledge of the misalignments, between 5 and 15 arc-sec less accuracy than the in-flight calibration.

For best results, it is suggested that a combination of the two methods be used. The in-flight calibrations should be used initially to determine an accurate estimate of the misalignment matrix, M, and the misalignment function should be used to monitor the changes in M. Then, every 6 months to 1 year, the alignment calibration should be redone, using the in-flight method to maintain the most accurate estimate of M. This scheme combines the advantages of both methods by using as little operational time as possible and maintaining a high degree of knowledge of the FHST alignment matrix.

For example, the SMM could use the following scheme. The postrepair FHST alignments could be calibrated using the in-flight method. Then, the yaw misalignment function could be used to maintain the yaw misalignment accuracy. The pitch and roll misalignment would not need to be maintained because their misalignments are not affected by the varying temperatures as stated previously. The changes in these misalignments due to the other factors are much smaller and could be maintained by the in-flight calibrations every 6 months.

Modeling the Temperature Data

The relevant temperatures on the SMM are received from the spacecraft telemetry; thus, the misalignment can be easily calculated from the simple models presented previously. However, if the relevant spacecraft temperatures could be modeled so that they may be predicted, the misalignment could also be predicted. In addition, for missions where the spacecraft temperatures are not available in the telemetry or for missions in the planning stage, a model of the SMM temperature profile would be helpful.

For the SMM, the temperatures on the ISP and the mission adapter ring are a function of instrument activity, distance from the Sun, and the length of the spacecraft day. The attitude of the spacecraft, of course, plays a large part in the profile. However, the SMM maintains nearly the same attitude relative to the Sun so that the temperature does not vary due to this effect. Therefore, this effect is not taken into account in this analysis. However, for missions with attitudes that are dynamic relative to the Sun, i.e., Earth-pointing or astronomical missions, this effect should also be modeled.

As seen in Figure 9, the dominating effect on the spacecraft temperature is payload instrument activity. Operating the scientific instruments generates a great deal of heat that is dissipated and radiated to other parts of the spacecraft. The temperature profile experiences major changes along with changes in instrument activity. At the repair of the SMM, approximately day 100 of 1984, all the payload instruments were off, and, as seen in Figure 9, the temperature was very low. As the mission started scientific activities, the payload instruments were turned on, and the temperature rose rapidly. Another example is seen later near day 250 when the spacecraft went into safehold mode and all the scientific instruments were powered off. For the SMM, safehold mode is the spacecraft's response to a perceived dangerous situation. The MACS safehold electronics takes attitude control from the OBC and holds the spacecraft in a Sun-oriented attitude until the problem is resolved. To conserve spacecraft power, most of the instruments are, consequently, powered off. Thus, the temperature of the spacecraft drops quickly.

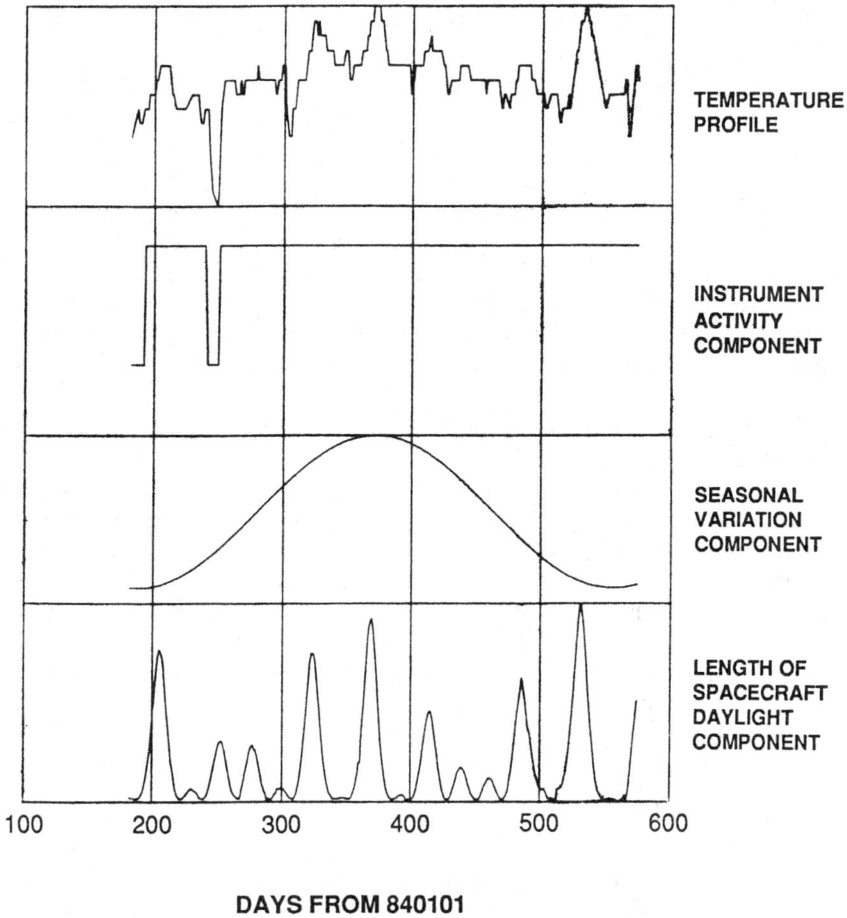

Fig. 9 SMM Temperature Profile and Components

Since the temperature changes in the case of powering on and off of the payload instruments occur relatively quickly, this effect can be modeled as a step function. The size of the step will vary with the number of instruments that are powered off. This component of the temperature profile is shown in Figure 9.

Another effect on the SMM temperature profile is the distance of the spacecraft from the Sun. This is a consequence of the Earth's orbit about the Sun. The Earth's orbit is not exactly circular, having an eccentricity of 0.016. Thus, the spacecraft is closer to the Sun in December at the Winter Solstice and farthest from the Sun in June at the Summer Solstice. This effect can be modeled by a sinusoidal pattern with a period of 1 year. The contribution of this effect on the SMM is shown in Figure 9.

The last effect on the temperature profile is the variation in the length of the spacecraft daylight period. The spacecraft daylight period is the amount of time per orbit that the spacecraft is in view of the Sun. This varies due to the geometry of the orbit. The length of the orbit daylight period varies inversely with the length of the Earth chord crossed by the spacecraft orbit, as viewed from the Sun. This chord depends entirely on the angle between the orbit normal and the Earth-to-Sun vector; the chord is longest when the vectors are close to perpendicular. The extreme of the angle between the vectors, and therefore of the orbit daylight period, occurs near the solstices as shown in Figure 9.

This effect is completely dependent on the orbit parameters and can be solved for exactly if the spacecraft ephemeris is known. If the mission is in the planning stages, the effect can be modeled from the preliminary knowledge of the orbital elements. This component of the temperature profile for the SMM is shown in Figure 9.

Once the effect of these parameters is taken into account, the total temperature profile can be formed. Then, after investigating the mission configuration and its response to temperature variation, an approximate function, analagous to Equations (3) through (5) can be determined. This function can be used in the premission attitude stability, determination, and control planning and the early mission operations. During the mission, data would be collected over a sufficiently long baseline to completely specify the model; if the SMM can be used as a guide, a minimum of 6 months would be required to account for the seasonal effects. Clearly, this approach requires substantially more intensive analysis than the use of direct temperature measurements.

SUMMARY AND CONCLUSION

In summary, flight data from the SMM attitude determination support demonstrates that spacecraft attitude sensor alignments vary with spacecraft temperature by up to 120 arc-sec over a 1-year period, with the majority of the variation occurring during the first few weeks as the temperatures stabilize. These levels are about eight times greater than were indicated in the currently available reports on the prelaunch thermal structural stability of the SMM. Methods have been proposed to incorporate flight measurement of the temperature-versus-alignment function and its variance to operational procedures with the benefit of reducing the spacecraft operations time required to support attitude sensor alignment calibration. Also, combining an approximate model of the temperature with the model of the alignment-versus-temperature could provide a significant reference for planning and analysis currently in progress for future missions. This prelaunch planning should also include incorporating the spacecraft structure temperatures in the attitude telemetry record.

ACKNOWLEDGMENTS

The authors wish to acknowledge the generous support of the following GSFC personnel: Mr. Frank Vaughn, Assistant Technical Representative, SMM Analytical Support Task; Mr. J. B. Joyce, Head, Mission and Network Support Section; and Mr. Gary Meyers, Head, Flight Dynamics Support Branch.

The authors also wish to thank Dr. Malcolm D. Shuster of the Johns Hopkins Applied Physics Laboratory for his technical advice and help in understanding and interpreting the data in this paper and Mr. Robert Shendock of the OAO Corporation for his help with the spacecraft hardware configurations. In addition, the authors would like to thank Dr. Milton Phenneger of Computer Sciences Corporation (CSC) for his help in preparing this work for publication. Acknowledgments for contributions to this paper are also extended to Mr. Bruce J. Twambly (CSC), Mr. Daniel R. Kulp (CSC), Ms. Deborah Roberts (CSC), Mr. Robert Feiertag (CSC), Mr. James R. Klein (CSC), and Mr. Joseph W. Dehen (CSC).

REFERENCES

1. Computer Sciences Corporation, CSC/TM-84/6131UD1, *Solar Maximum Mission Attitude Operations Handbook, Revision 4 - Update 1*, A. Eudell et al., July 1988

2. --, CSC/SD-78/6082, *Solar Maximum Mission (SMM) Attitude System Functional Specifications and Requirements*, R. Byrne et al., September 1978

3. Shuster, M. D., D. M. Chitre, and D. P. Niebur, "In-flight Estimation of Spacecraft Attitude Sensor Accuracies and Alignments," *Journal of Guidance, Control, and Dynamics*, Vol. 5, No. 4, July-August 1982, pp. 339-343

4. Computer Sciences Corporation, CSC/TM-85/6714, *Analysis of the Solar Maximum Mission (SMM) Fine Pointing Sun Sensor (FPSS) Response Changes*, F. Patt, May 1985

5. Hewitt, D. R., *Multimission Modular Spacecraft (MMS) Thermal System Specification*, S-700-12, February 1977

6. Computer Sciences Corporation, CSC/TM-81/6036, *High Precision Attitude Determination for Magsat*, F. VanLandingham, April 1981

AAS 89-392

REORIENTATION MANEUVER FOR SPINNING SPACECRAFT

C. Rahn* and P. Barba†

A spacecraft spinning about its minimum moment of inertia axis in the presence of energy dissipation is directionally unstable. Eventually, the spacecraft will reorient to a major axis spin. This passive reorientation maneuver is subject, however, to uncertainty in the final orientation of the spacecraft relative to the spin vector. The spacecraft can end up with either a positive or a negative major axis spin. This paper demonstrates that the maneuver, when augmented with two thruster firings based on gyro measurements, results in a desired final orientation.

Introduction

A single-body spacecraft spinning about its minor axis in the presence of energy dissipation is directionally unstable[1]. Although the spacecraft can be stabilized by active control, by the jet damping of a rocket motor[2], or by dampers on a controlled, despun platform[3], removing these stabilizing effects causes the spacecraft to reorient and rotate about its major axis due to the energy lost in fuel slosh and vibration. This passive reorientation maneuver is called spin transition.

Spacecraft may rotate about their minor axis for several reasons. First, launch vehicle fairing constraints often require that the long and narrow axis of the spacecraft be aligned with the longitudinal axis of the launch vehicle. Typically, the launch vehicle spins longitudinally prior to separation, resulting in a minor axis spin for the spacecraft after separation. Second, when a solid rocket motor is used for orbit raising, the rocket motor and spacecraft combination is spun about its minor axis to increase stability during the firing. When the firing is completed, the combination undergoes spin transition unless active control is utilized.

The spin transition maneuver is subject, however, to one drawback. The orientation of the spacecraft relative to the inertially fixed angular momentum vector at the end of the

* R&D Engineer, Ford Aerospace Corporation, 3939 Fabian Way, Palo Alto, California 94303-4697.

† Principal Engineer, Ford Aerospace Corporation, 3939 Fabian Way, Palo Alto, California 94303-4697.

maneuver cannot be determined a priori with any degree of accuracy. The spacecraft can end up with either a positive or a negative major axis spin. Physically, this corresponds to two final attitudes which are 180° apart. Many spacecraft have sensitive on-board instruments which must be shielded from the sun or directional communication equipment which must point towards the earth. In these cases it is desirable to ensure a final spin polarity.

There are techniques of optimally reorienting spacecraft[4], and methods of attitude acquisition utilizing momentum wheels have been developed[5], but they do not utilize the energy dissipation of the spacecraft to assist in the maneuver. In terms of fuel usage, the passive spin transition maneuver is, by definition, optimal. To make the maneuver truly useful, however, the final spin polarity must be controlled, requiring some fuel expenditure.

This paper presents a control system which guarantees a final orientation after spin transition. The control system uses gyros which determine when to fire thrusters, providing the desired orientation.

Dynamics

Two models of the spacecraft dynamics are used. For simulation purposes, the spacecraft is modeled as a rigid body with a spherical, dissipative fuel slug. The rigid body has three rates, ω_1, ω_2, and ω_3 about the major, intermediate, and minor body axes, respectively. The fuel is modeled as a spherical slug of mass J which is surrounded by a viscous layer. Designating the relative rates between the spacecraft body and the fuel slug as σ_1, σ_2, and σ_3, the equations of motion are written

$$\begin{aligned}
(I_1 - J)\dot{\omega}_1 &= (I_2 - I_3)\omega_2\omega_3 + \Delta\sigma_1 + T_1 \\
(I_2 - J)\dot{\omega}_2 &= (I_3 - I_1)\omega_1\omega_3 + \Delta\sigma_2 + T_2 \\
(I_3 - J)\dot{\omega}_3 &= (I_1 - I_2)\omega_1\omega_2 + \Delta\sigma_3 + T_3 \\
\dot{\sigma}_1 &= -\dot{\omega}_1 - \Delta\sigma_1/J - \omega_2\sigma_3 + \omega_3\sigma_2 \\
\dot{\sigma}_2 &= -\dot{\omega}_2 - \Delta\sigma_2/J - \omega_3\sigma_1 + \omega_1\sigma_3 \\
\dot{\sigma}_3 &= -\dot{\omega}_3 - \Delta\sigma_3/J - \omega_1\sigma_2 + \omega_2\sigma_1,
\end{aligned} \qquad (1)$$

where Δ is the viscous damping coefficient of the slug, I_1, I_2, and I_3 are the principal moments of inertia of the spacecraft, and T_1, T_2, and T_3 are the torques about the principal axes.

For analysis purposes, the spacecraft is modeled as a rigid body with an energy sink. The angular momentum, h, and rotational energy, T, of the rigid body can be expressed:

$$h^2 = I_1^2\omega_1^2 + I_2^2\omega_2^2 + I_3^2\omega_3^2 \qquad (2)$$

$$2T = I_1\omega_1^2 + I_2\omega_2^2 + I_3\omega_3^2, \qquad (3)$$

The path that the angular velocity vector traces in body axes coordinates is determined by the intersection of the momentum ellipsoid (Equation 2) and the energy ellipsoid (Equation 3) and is called the polhode. If there is no energy dissipation on-board the spacecraft,

h and T are constant and the polhode is a closed path corresponding to nutational motion of the spacecraft.

If there is energy dissipation, T decreases and the energy ellipsoid shrinks with time. This leads to an open polhode path that spirals outward from the minor axis and captures on the major axis as shown in Figure 1. The point at which the polhode switches from rotating about the minor axis to rotating about the major axis corresponds to crossing the "separatrix".

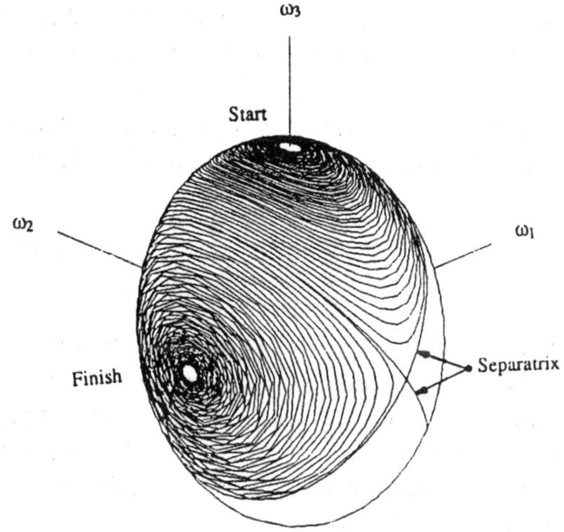

Figure 1: Polhode for a Typical Spin Transition

Simulation Inputs:

I_1	2000 kg.m^2
I_2	1500 kg.m^2
I_3	1000 kg.m^2
Initial Major Axis Rate	.1224 rad/s
Initial Intermediate Axis Rate	.0 rad/s
Initial Minor Axis Rate	2.99 rad/s

The exact point at which the polhode crosses the separatrix is dependent on the initial conditions and the energy dissipation characteristics of the spacecraft. Small changes in either of these parameters can change the side on which the polhode crosses the separatrix, changing the final spin polarity. There is a 50% chance of capturing on either side and it is impossible to predict with any degree of accuracy.

Figure 2 demonstrates the sensitivity of the final spin polarity to initial conditions. The horizontal axis is the initial major axis spin rate and the vertical axis is the initial intermediate axis spin rate. The initial minor axis spin rate is calculated to give a constant 3000 Nms angular momentum. The figure represents a 101 × 101 grid of simulations over a typical range of rates with the X's corresponding to a positive final spi The simulation in Figure 1 is marked with a •. It is clear that a small change in the initial conditions can lead to a change in the final spin polarity.

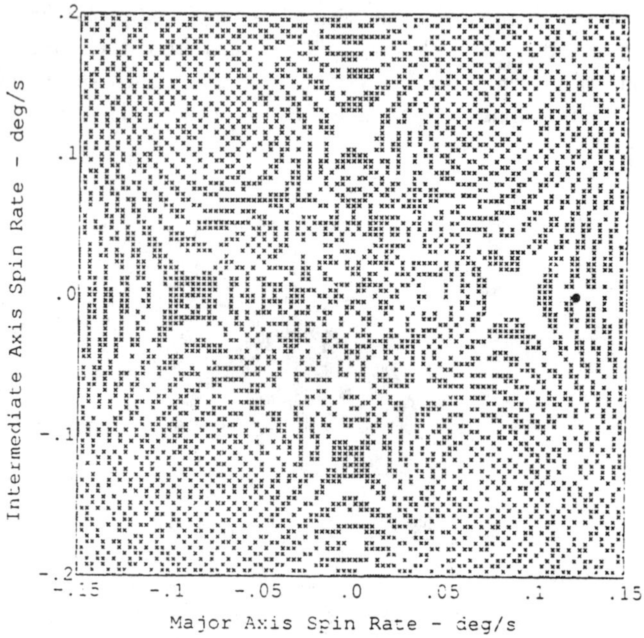

Figure 2: Spin Polarity from Initial Conditions, Constant Angular Momentum

Final Spin Polarity Control

The desired final spin polarity is guaranteed with the following control logic. Initially, the spacecraft spins about its minor axis with some small amount of nutation. The spacecraft begins to transit to flat spin, and, when the polhode crosses the separatrix, the spin direction is determined. If the spin direction is correct, no action is taken. If the spin direction is incorrect, a thruster fires to force the polhode to cross the separatrix. Then, a short time later, an opposite thruster fires, causing the polhode to recross the separatrix on the desired side.

To avoid the necessity for additional sensors on-board the spacecraft, the on-orbit rate gyros, nominally used for stationkeeping, are used for spin polarity control. Unfortunately, these gyros have limited range and are not designed to measure the relatively high rates which occur during spin transition. Therefore, they are only useful for determining the signs of the principal axis spin rates.

The Separatrix Crossing

Although predicting when the polhode will cross the separatrix is very difficult, it is relatively easy to determine that the crossing has occurred. Figure 3 shows the time response of the spacecraft rates during a spin transition. The spacecraft starts out spinning about its minor axis and ends up spinning about its major axis. When the minor axis spin rate changes sign, the polhode has crossed the separatrix. In all cases, the polhode crosses the separatrix before the minor axis spin rate changes. Therefore, the minor axis rate sign change is a conservative indicator of the separatrix crossing. Also shown are the angular momentum, h, which remains constant during the maneuver, the kinetic energy, T, which decreases during the maneuver; and the characteristic parameter, I.

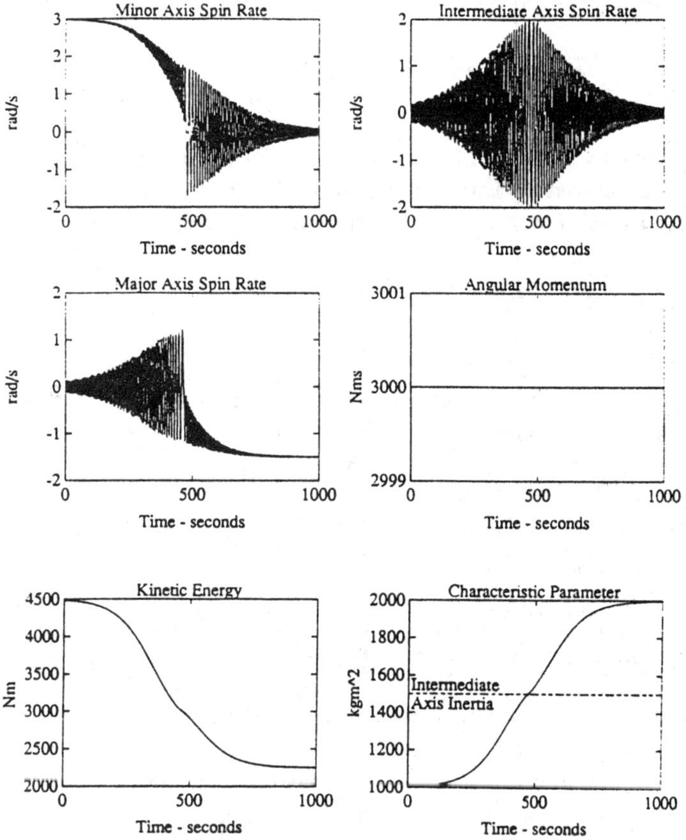

Figure 3: Time response of Spacecraft Rates, Angular Momentum, Kinetic Energy, and Characteristic Parameter During Spin Transition

Once the polhode has crossed the separatrix, the sign of the major axis spin rate can be determined from the gyros. If the sign corresponds to the desired spacecraft orientation, the maneuver is complete. If the sign is of opposite polarity, however, the spacecraft must be reoriented.

The Thruster Firings

If the spin polarity after the polhode has crossed the separatrix is correct, no thrusters are fired. If it is incorrect and there are no thruster failures, two thruster firings are required to change the spin polarity. The objective of each of the the two firings is to make the polhode cross the separatrix in a desired direction. This section discusses the effect of thruster firings on the characteristic parameter, I.

The "distance" between the polhode and the separatrix is measured by the characteristic parameter

$$I = \frac{h^2}{2T} \qquad (I_3 \leq I \leq I_1). \tag{4}$$

As the energy of the body decreases, I increases. At the separatrix crossing, $I = I_2$ as seen in Figure 3. The characteristic parameter qualitatively divides the motion into a regime of primarily minor axis spin ($I < I_2$) and primarily major axis spin ($I > I_2$).

Defining K as the equivalent gain of the thruster firing,

$$I^+ = KI, \tag{5}$$

where I^+ is the characteristic parameter after the firing. If $I > I_2$, and the objective of the firing is to cross the separatrix, then K is chosen less than 1 such that $I^+ < I_2$. For a separatrix crossing in the opposite direction, K is chosen greater than 1.

The equivalent gain is calculated from the thruster impulses i_1, i_2, and i_3, assuming the thruster firings are impulsive,

$$K = \frac{h^2 + 2\left(h_1 i_1 + h_2 i_2 + h_3 i_3\right) + i^2}{h^2 + 2\left(\frac{I_2}{I_1} h_1 i_1 + h_2 i_2 + \frac{I_2}{I_3} h_3 i_3\right) + \frac{I_2}{I_1} i_1^2 + i_2^2 + \frac{I_2}{I_3} i_3^2}, \tag{6}$$

where h_1, h_2, and h_3 are the components of the angular momentum prior to the firing.

For a given K, the timing and direction of the thruster impulse must be determined. Since the rate gyros provide only rate sign information the thrusters must be timed according to when the sign of a rate changes, indicating the rate is zero. Combined with the three possible impulse directions, there are nine combinations of impulse direction and firing time. For most practical cases, an impulse about the minor axis which is fired when $\omega_2 = 0$ uses the least propellant for a given K. The size of the impulse can be calculated by substituting $\omega_2 = 0$, $i_1 = 0$, and $i_2 = 0$ into Equation 6 and solving for i_3:

$$i_3 = -h_3 + \text{sign}(h_3)\sqrt{h_3^2 - \frac{I_3 h^2 (1-K)}{I_3 - K I_2}}. \tag{7}$$

The polhode is near the separatrix when each impulse is fired so

$$\omega_2 \approx s_2 \omega_{2m} \tanh\tau \quad \text{and} \quad \omega_3 \approx s_3 \omega_{3m} \text{sech}\tau, \tag{8}$$

where $s_2\omega_{2m}$, and $s_3\omega_{3m}$ are the signed amplitudes of the two angular rates and τ is time[6]. The intermediate axis spin rate is zero so $\tanh\tau = 0$ and $\mathrm{sech}\tau = 1$, resulting in

$$\omega_3 \approx s_3\omega_{3m} = s_3 h\sqrt{\frac{I_1 - I_2}{I_2 I_3 (I_1 - I_3)}}. \tag{9}$$

Substituting Equation 9 into Equation 7 yields

$$i_3 = h\begin{cases} \sqrt{C_1 - C_2} - \sqrt{C_1} & \omega_3 > 0 \\ \sqrt{C_1} - \sqrt{C_1 - C_2} & \omega_3 < 0 \end{cases}, \tag{10}$$

where

$$C_1 = \frac{I_3 (I_1 - I_2)}{I_2 (I_1 - I_3)} \quad and \quad C_2 = \frac{I_3 (1 - K)}{I_3 - K I_2}. \tag{11}$$

The Polhode Path

The polhode path affected by the spin polarity control logic is shown in Figure 4. The polhode is shown folded away from the inertia ellipsoid and projected on a plane. The separatrix resembles a figure eight and the angular rates are zero on vertical and horizontal lines. The horizontal line through the middle corresponds to $\omega_3 = 0$, and the vertical lines correspond to $\omega_1 = 0$ and $\omega_2 = 0$.

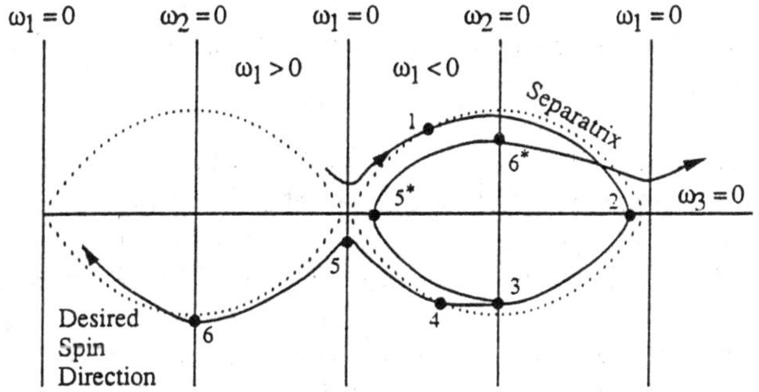

Figure 4: Diagram of the Polhode Path with Spin Polarity Control

A typical polhode path is shown crossing the separatrix at 1. The polhode passes through zero ω_3 at 2. The sign of ω_1 determines that the spin direction is negative, the opposite of the desired direction. At 3, when ω_2 is zero, a thruster fires to decrease I ($K < 1$) and the polhode goes back across the separatrix at 4. The polhode then crosses $\omega_1 = 0$ (5) and when $\omega_2 = 0$ (6) a thruster fires to increase I ($K > 1$) by the same amount as it was decreased in the previous firing, thus ensuring capture on the desired side.

Choosing K

The parameter K is used to calculate the thruster impulse that is required to push the polhode across the separatrix. When selecting this parameter, the rate of energy dissipation and the limits of the impulsive approximation must be considered.

To ensure that the first firing forces the polhode back across the separatrix, it must compensate for the energy lost since the polhode last crossed the separatrix. In the worst case, the polhode crosses the separatrix a half period from when ω_3 crosses zero, so by the time ω_2 is zero and the thruster is fired, 3/4 of a period has elapsed.

Designating Δt as the time from $\omega_3 = 0$ to $\omega_2 = 0$, the worst case I at the first thruster firing would be

$$I = I_2 + 3\Delta t \left(\frac{\partial I}{\partial t}\right)_{separatrix} . \tag{12}$$

The desired I^+ should be large enough to ensure the separatrix is not crossed before $\omega_1 = 0$ a quarter period later, or

$$I^+ = I_2 - \Delta t \left(\frac{\partial I}{\partial t}\right)_{separatrix} . \tag{13}$$

Substituting Equations 12 and 13 in Equation 5 and utilizing Equation 4 yields

$$K = \frac{I_2 + \Delta t \left(\frac{1}{T}\frac{\partial T}{\partial t}\right)_{separatrix}}{I_2 - 3\Delta t \left(\frac{1}{T}\frac{\partial T}{\partial t}\right)_{separatrix}} . \tag{14}$$

In order to calculate K from Equation 14, the energy dissipation characteristics of the spacecraft at the separatrix must be known. These characteristics can be determined experimentally[7] or by using telemetry data from previous spacecraft with similar energy dissipation characteristics. If there is some uncertainty in the energy dissipation, a larger impulse can be used.

Each of the two firings is the same in magnitude and opposite in direction. In practice, however, the second firing can be smaller than the first. After the first firing, the energy dissipation moves the polhode in the same direction as the desired separatrix crossing of the second firing. When the second thruster fires, the polhode is closer to the separatrix than it was for the first firing and a smaller impulse can be used.

Due to the limited torque capability of the thrusters, the actual firings are not impulsive. This results in a loss of efficiency, and the desired I^+ will not be achieved. If the thruster is fired when ω_2 changes sign, the firing must be finished by the time ω_3 changes sign, otherwise the firing will not produce the desired effect. If the required firing is too large for a given thruster configuration, the initial spin rate can be reduced, contracting the thruster firing time proportionally (See Equation 10).

Control Logic Description

Figure 5 shows the flow diagram for the control logic. Initially, the control system monitors the ω_3 rate from the gyro. When ω_3 changes sign, the sign of the saturated ω_1 gyro rate is measured. If this rate has the desired sign, the spin transition has been successfully completed. Therefore, there is a 50% chance that the controller will take no action. If the rate is of opposite sign, however, the controller waits for a change in sign of ω_2 and fires a thruster which reduces I according to Equation 10. The control system then waits for the zero crossing of ω_1 followed by a change in polarity of ω_2 and then fires a torque about the minor axis to guarantee capture.

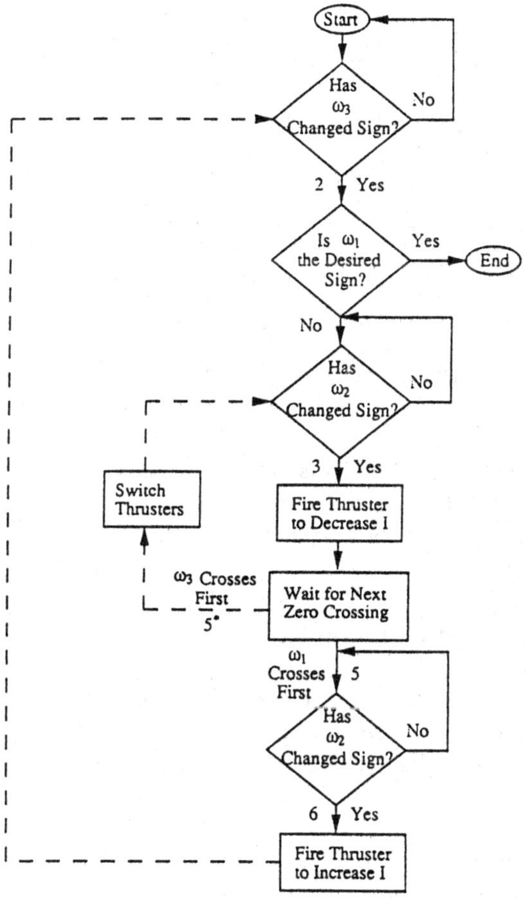

Figure 5: Control Logic Flow Diagram

Figure 6 show the polhode path for a controlled spin transition with the same initial conditions and mass properties as in Figure 1. The control system is configured to ensure a positive final spin rate. The polhode first crosses the separatrix at 1. When ω_3 is zero (2), ω_1 is less than zero. When ω_2 crosses zero (3), a thruster is fired and the polhode recrosses the separatrix at 4. Next, ω_1 crosses zero (5), signalling successful completion of the separatrix crossing. At (6), ω_2 crosses zero, the second firing increases I, and the polhode captures on the desired side.

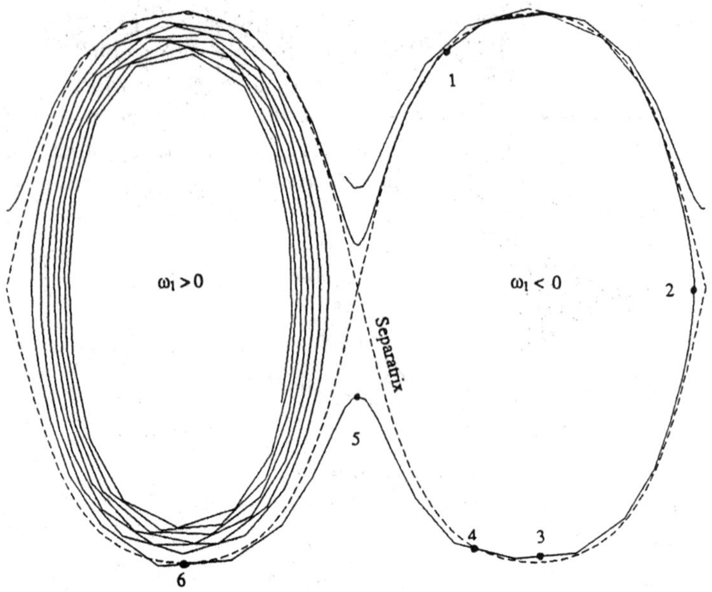

Figure 6: Polhode Path Simulation with Spin Polarity Control

The time response of the rates, h, T, and I is shown in Figure 7. The ω_1 response initially captures on the negative side and then is pushed by thruster firings to the positive side. The angular momentum and kinetic energy change each time a thruster fires with the parameter I decreasing and then increasing.

Autonomous Thruster Failure Detection

The dotted polhode path in Figure 4 would be followed if the thruster used for the first firing had failed. In that case, the ω_3 gyro rate will cross zero (5*) before the ω_1 gyro rate. If this happens, the control system switches autonomously to a backup string of thrusters and fires a larger impulse at 6*.

The dotted lines in Figure 5 show the autonomous thruster failure detection logic. After the first firing the controller monitors both the ω_1 and ω_3 rates. If ω_1 crosses first then the nominal logic is followed. If ω_3 crosses first then the thrusters are reconfigured and a larger impulse is fired. The dotted path after the second firing ensures that the second thruster has not failed. In practice, the number of passes through the flow diagram would be limited to preclude the possibility of limit cycles.

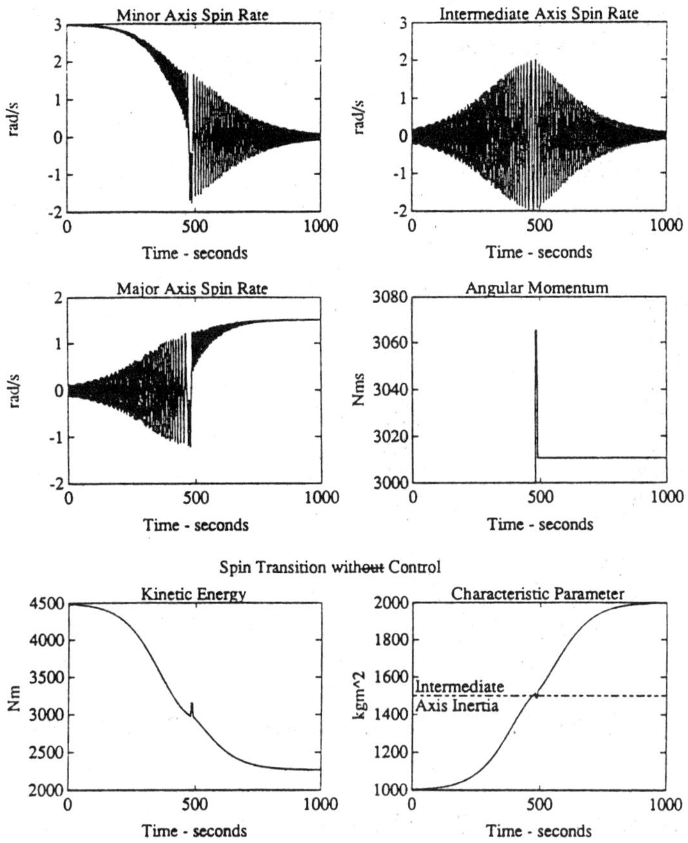

Figure 7: Controlled Time Response of Spacecraft Rates, Angular Momentum, Energy, and Characteristic Parameter During Spin Transition

Conclusions

The spin transition maneuver, when augmented with two thruster firings based on gyro measurements, provides a predetermined final spin polarity. The control logic utilizes rate sign changes to determine when the separatrix has been crossed and when thrusters must be fired. The sizing of the thruster firings is based on estimates of the energy dissipation in the spacecraft. Limitations in the torque capability of the thruster configuration may constrain the maximum initial spin rate which can be controlled.

Acknowledgements

The authors would like to thank Professor C. D. Mote, Jr. for the use of his computational facilities and Dr. Jonathan Wickert for his useful discussions on chaos.

References

[1]Likins, P. W.: "Effects of Energy Dissipation on the Free Body Motions of Spacecraft", NASA TR# 32-860, 1966.

[2]Thomson, W. T.: *Introduction to Space Dynamics*, John Wiley & Sons, Inc., New York, 1961, pp. 220-227.

[3]Leliakov, I. P. and Barba, P. M.: "Damping Spacecraft Nutation by Means of a Despun Antenna", AAS/AIAA Astrodynamics Conference, 1973.

[4]Junkins, J. L. and Turner, J. D.: *Optimal Spacecraft Rotational Maneuvers*, Elsevier, New York, 1986.

[5]Weissberg, J. and Ninomiya, K.: "Improved Method for the Initial Attitude Acquisiton Maneuver", *Journal of Guidance and Control*, Vol. 10, No. 3, May-June 1987, pp. 316-319.

[6]Hughes, P. C.: *Spacecraft Attitude Dynamics*, John Wiley & Sons, New York, 1986, pp. 104-108.

[7]Harrison, J. A., Garg, S. C., and Furumoto, N.: "A Free-Fall Technique to Measure Nutation Divergence, and Applications", AAS/AIAA 83-372, 1983.

AAS 89-393

VERY COARSE ATTITUDE DETERMINATION FOR THE THRUSTED VECTOR MISSION

M. D. Shuster*, H. L. Fisher† and K. L. Musser‡

Attitude estimation algorithms are presented for the Thrusted Vector Mission which determine attitude based on Sun sensor and very coarse albedo sensor measurements. On the basis of these measurements it has been demonstrated by comparison with more accurate gyro-based attitude that it is possible to estimate three-axis attitude with an average error per axis of 11 deg. Most of this error is about the Sun direction. Both deterministic quick-look and optimal estimates are examined.

INTRODUCTION

The Thrusted Vector mission (more commonly known by its launch vehicle, Delta-181) was the second mission to be launched by the Strategic Defense Initiative Organization. The Thrusted Vector mission was divided into two phases: a data-collection phase, which lasted approximately 12 hours, during which the experimental activities of the mission took place; and a data-retrieval phase, lasting up to several weeks, during which the experimental data stored on-board the spacecraft was telemetered to Earth. Power for the spacecraft was provided by two battery systems. One of these, whose lifetime was designed to be not much longer than the experiment phase, provided power for a very accurate set of three-axis gyros. During the data-retrieval phase, it was expected that passive stabilization of the spacecraft would maintain the down-link antenna Earth-pointing with sufficient accuracy to allow uninterrupted communication during passes over the mission ground stations. Therefore, no provision was made for attitude determination during the data-retrieval phase following

* Senior Professional Staff, Guidance and Control Group, Space Department, The Johns Hopkins University, Applied Physics Laboratory, Mail Stop 23-377, Johns Hopkins Road, Laurel, Maryland 20707-6099.

† Associate Engineer, Guidance and Control Analysis Section, Space Department, The Johns Hopkins University, Applied Physics Laboratory, Johns Hopkins Road, Laurel, Maryland 20707-6099.

‡ Associate Engineer, Guidance and Control Analysis Section, Space Department, The Johns Hopkins University, Applied Physics Laboratory. Currently, Graduate Student, Department of Electrical Engineering, University of Maryland, College Park, Maryland 20742.

the loss of gyro attitude. Shortly before launch, however, it was deemed advantageous to have some means of determining attitude during the data-retrieval phase in order to verify antenna pointing. In this way one could choose the optimal antenna or even avoid downlink during passes when the spacecraft attitude might be far from nominal and the transmission would be degraded. It is to the methods developed for attitude determination during the data-retrieval phase that this paper is devoted.

In addition to the gyro assembly, the Thrusted Vector spacecraft possessed two other sensors sensitive to the attitude. One of these was a two-axis Sun sensor with an accuracy of approximately one degree per axis, which was included to provide some attitude redundancy as a check on the gross operation of the gyro assembly. The other sensor was not intended as an attitude sensor at all but to measure Earth albedo as part of the mission experiments. It was on the basis of these two sensors that attitude would have to be determined during the data-retrieval phase.

The albedo sensor provides a very crude measurement of the direction of the Earth in the spacecraft reference frame and it is certainly possible to estimate attitude based on simultaneous measurement of the Sun and Earth directions. The important question, however, is whether attitude could be determined with useful accuracy given the crudity of the albedo sensor as an attitude sensing device. It turns out, fortunately, that the question can be answered affirmatively.

In the second section of this report we present measurement models for the attitude sensors. Following this, two attitude determination algorithms are presented: a "quick-look" attitude estimation algorithm, which constructs a rough estimate of the attitude based on lumped vectors representing the observed and predicted average directions of the Earth luminance; and an optimal algorithm which attempts to improve the "quick-look" attitude by fitting individual spatial components of the Earth albedo. Complete covariance analyses are carried out in each case. Finally, an absolute evaluation of the performance of these two algorithms is presented using actual sensor data collected during the experiment phase when the much more accurate gyro attitudes were available for comparison.

SENSOR MODELS

Two-Axis Sun Sensor

Two-axis Sun sensors have been flown on nearly every mission and their operation as attitude sensors has been extensively documented [1]. A simple model of the Sun sensor errors which has proved effective in many analyses [2] is

$$\hat{\mathbf{S}}_s = A\,\hat{\mathbf{S}}_r + \Delta\hat{\mathbf{S}}_s \quad ,$$

where $\hat{\mathbf{S}}_s$ is the observed Sun direction in the spacecraft frame and $\hat{\mathbf{S}}_r$ is the corresponding vector in the primary reference frame. The measurement error, $\Delta\hat{\mathbf{S}}_s$, is assumed to be Gaussian with zero mean and covariance matrix

$$E\{\Delta\hat{\mathbf{S}}_s\,\Delta\hat{\mathbf{S}}_s^T\} = \sigma_S^2\,[I_{3\times3} - (A\,\hat{\mathbf{S}}_r)(A\,\hat{\mathbf{S}}_r)^T] \quad .$$

The Albedo Sensor

The albedo sensor consists of 24 individual albedo detectors of approximately equal area, which together cover essentially the entire unit sphere. Geometrically, each detector is triangular. For purposes of the present analysis, however, it will be sufficient to model each detector as if the field of view were circular with a solid angle of $4\pi/24 = \pi/6$. Thus, the effective angular diameter of each detector in radians is nearly one radian. To this must be added the fact that the angular diameter of the Earth as seen from the spacecraft altitude of about 300 km, is approximately 2.5 radians. The albedo sensor is thus a very crude attitude sensing device. In addition, it is usually not possible to model the albedo of the Earth to better than 50%.

ATTITUDE ESTIMATION METHODS

Quick-Look Attitude

The quick-look algorithm uses the TRIAD algorithm [2], which computes a three-axis attitude from unit vector input. If the two observed unit vectors are $\hat{\mathbf{W}}_1$ and $\hat{\mathbf{W}}_2$ and the corresponding reference unit vectors are $\hat{\mathbf{V}}_1$ and $\hat{\mathbf{V}}_2$, then the TRIAD algorithm begins by computing the unit vectors

$$\hat{\mathbf{s}}_1 \equiv \hat{\mathbf{W}}_1 \quad , \quad \hat{\mathbf{s}}_2 \equiv \frac{\hat{\mathbf{W}}_1 \times \hat{\mathbf{W}}_2}{|\hat{\mathbf{W}}_1 \times \hat{\mathbf{W}}_2|} \quad , \quad \hat{\mathbf{s}}_3 \equiv \hat{\mathbf{s}}_1 \times \hat{\mathbf{s}}_2 \quad ,$$

$$\hat{\mathbf{r}}_1 \equiv \hat{\mathbf{V}}_1 \quad , \quad \hat{\mathbf{r}}_2 \equiv \frac{\hat{\mathbf{V}}_1 \times \hat{\mathbf{V}}_2}{|\hat{\mathbf{V}}_1 \times \hat{\mathbf{V}}_2|} \quad , \quad \hat{\mathbf{r}}_3 \equiv \hat{\mathbf{r}}_1 \times \hat{\mathbf{r}}_2 \quad ,$$

and the attitude matrix is given by

$$A \equiv [\, \hat{\mathbf{s}}_1 \vdots \hat{\mathbf{s}}_2 \vdots \hat{\mathbf{s}}_3 \,][\, \hat{\mathbf{r}}_1 \vdots \hat{\mathbf{r}}_2 \vdots \hat{\mathbf{r}}_3 \,]^T \quad .$$

The covariance matrix of the attitude estimate, defined as the covariance matrix of the incremental angles which parameterize the orthogonal transformation from the true attitude to the estimated attitude [2], is then given by $P_{\theta\theta}$ with

$$P_{\theta\theta}^{-1} = \frac{1}{\sigma_1^2} [I_{3\times 3} - \hat{\mathbf{s}}_1 \hat{\mathbf{s}}_1^T] + \frac{1}{\sigma_2^2} \hat{\mathbf{s}}_4 \hat{\mathbf{s}}_4^T \quad ,$$

with $\hat{\mathbf{s}}_4 \equiv \hat{\mathbf{W}}_2 \times \hat{\mathbf{s}}_2$. The covariance matrix will be smaller if $\hat{\mathbf{W}}_1$ is chosen to be the more accurately known of the two measured unit vectors. Thus, $\hat{\mathbf{W}}_1$ in this case is chosen to be the Sun unit vector.

The quick-look method consists of finding a quick method of computing a $\hat{\mathbf{W}}_2$. To do this we define the measured luminance direction $\hat{\mathbf{L}}_s$ to be

$$\hat{\mathbf{L}}_s \equiv \text{unit}\left(\sum_{i=1}^{24} N_i \mathbf{u}_i\right) \quad ,$$

where N_i is the intensity of the albedo radiance in detector i of the albedo sensor, $\mathbf{u}_i$ is the a characteristic outward normal unit vector of that detector, and unit($\cdot$) is the operation which generates a unit vector from a vector with arbitrary but non-vanishing normalization. In all the sums over individual albedo detector elements, the elements observing the Sun are not included. The corresponding reference vector is $\hat{\mathbf{L}}_r$, with

$$\hat{\mathbf{L}}_r \equiv \text{unit}\left(\int \Phi(\hat{\mathbf{n}})\,\hat{\mathbf{n}}\,d\Omega\right) \quad ,$$

where $\Phi(\hat{\mathbf{n}})$ is the radiance of the Earth albedo in the direction $\hat{\mathbf{n}}$, computed on the basis of Sun and spacecraft ephemeris and the reflectivity of the Earth.

We will assume that the luminance vector has an error representation similar to that of the Sun vector, namely

$$\hat{\mathbf{L}}_s = A\,\hat{\mathbf{L}}_r + \Delta\hat{\mathbf{L}}_s \quad ,$$

where the measurement error, $\Delta\hat{\mathbf{L}}_s$, is also assumed to be Gaussian with zero mean and covariance matrix

$$E\{\Delta\hat{\mathbf{L}}_s\,\Delta\hat{\mathbf{L}}_s^T\} = \sigma_L^2\,[I_{3\times3} - (A\,\hat{\mathbf{L}}_r)(A\,\hat{\mathbf{L}}_r)^T] \quad .$$

Optimal Attitude Estimation

Since the albedo sensor is so much less accurate than the Sun sensor, it makes sense to write the attitude matrix as

$$A = R\,A_{\text{quick-look}}$$

and optimize R, which was limited to rotations about the Sun vector. (Note that $\hat{\mathbf{S}}_s = A_{\text{quick-look}}\,\hat{\mathbf{S}}_r$ exactly, so this choice of R is the one which doesn't influence the agreement in the Sun direction.) Thus, the algorithm is sub-optimal. However, given a difference of an order of magnitude in the accuracy of the two sensors, little accuracy will be lost by this procedure. Thus,

$$R = R(\hat{\mathbf{S}}_s, \phi) \quad ,$$

and ϕ is the quantity over which we optimize. To carry out the optimization we consider the cost function

$$J(\phi) = \frac{1}{2}\sum_{i=1}^{24}\left[N_i^{\text{computed}}(\phi) - N_i^{\text{observed}}\right]^2 \quad ,$$

where the computed Earth albedo components are normalized so that

$$\sum_{i=1}^{24} N_i^{\text{computed}}(\phi) = \sum_{i=1}^{24} N_i^{\text{observed}} \quad .$$

The optimization was carried out by computing the cost function at five points equally spaced about $\phi = 0$, fitting a parabola to these points, and then finding the location of the minimum of the fitted parabola.

GRAPHICS INTERFACE

Figure 1 shows the graphical output that was obtained from the quick-look attitude determination. The attitude was calculated from the albedo and Sun sensor data and this was used to calculate the projection of the albedo detector boundaries on the visible Earth. The numbers in the figure label the albedo faces. The outlines of the southeastern United States, the Yucatan, and Cuba are visible. The Sun appears as a large black dot in the lower left. This does not indicate the Sun position but only the direction at which the plane determined by the spacecraft, the geocenter, and the heliocenter intersects the field of view. The terminator is visible in the right of the diagram. On the graphics screen the lit and unlit Earth were distinguished by different colors.

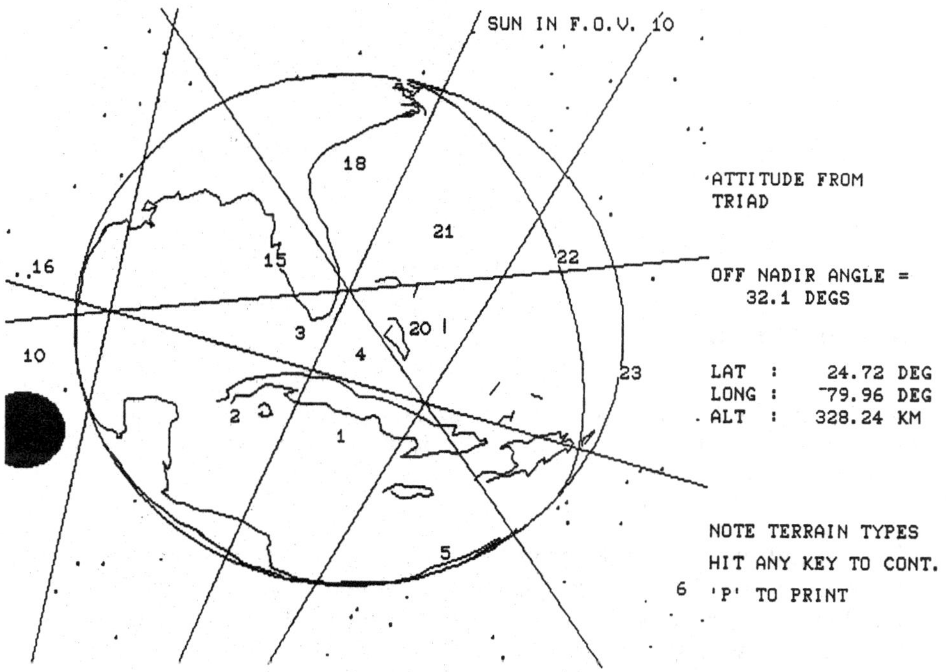

Fig. 1 Quick-Look Attitude Graphics

Figure 2 shows the cost function for the observed and simulated albedo profiles. The fitted curve is a parabola. The minimum of the parabola gives the optimal angle of the angular adjustment to the quick-look attitude. In this case the offset is only -1.5 deg. In cases where the five points are far from being fit by a parabola, one might choose not to compute an optimal attitude by this method.

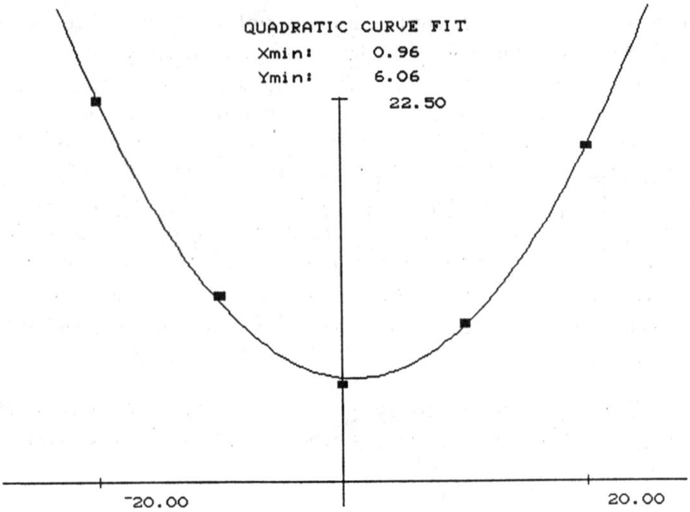

Fig. 2 Correction to Quick-Look Attitude

ERROR ANALYSIS

If we assume a uniform probability distribution for the position of the terminator or horizon in a partially illuminated albedo detector, then the sensor error standard deviations may be taken to be

$$\sigma_S = 1 \text{ deg} \quad , \quad \sigma_L = 7 \text{ deg} \quad .$$

On this basis, predictions for the quick-look attitude accuracy can be made based on the expression for the attitude error covariance given above. It should be pointed out that for the Sun directly overhead, the attitude will not be well determined, since the Sun vector and luminance vectors will be parallel. Likewise, for the Sun very near the horizon, the attitude will also be poorly determined because albedo data will be poor. If we assume for the sake of example that the Sun vector is along the x-axis and the luminance vector is in the xy-plane at an angle of 45 deg to the Sun vector, then the covariance matrix of the attitude (in rad^2) in this case using the assumed sensor variances is

$$P_{\theta\theta} = \begin{bmatrix} .03 & .0003 & 0 \\ .0003 & .0003 & 0 \\ 0 & 0 & .0003 \end{bmatrix}$$

so that two axes are determined with accuracies of about 1 deg and the remaining axis with an accuracy of about 10 deg. The average accuracy per axis (root-mean-square) for the quick-look method in this case is then about 6. deg.

We have computed a sampled covariance of the attitude errors from the quick-look method using the much more accurate gyro attitude, that was available during the experiment phase of the mission, as an absolute attitude reference. Based on sixteen samples, the estimated average covariance (averaged over different data geometries) was

$$P_{\theta\theta}^{\text{sampled}} = \begin{bmatrix} .062 & .028 & -.045 \\ .028 & .066 & -.040 \\ -.045 & -.040 & .062 \end{bmatrix}$$

corresponding to an average accuracy per axis of 14. deg. For 65 per cent of the data, however, the sampled covariance corresponded to a standard deviation per axis of 4.6 deg. The errors are, therefore, non-Gaussian. Examination of the data has shown, unfortunately, that these outlying cases could not have been identified and discarded using cosine checks of the angle between the Sun and luminance vectors. Most likely, these outliers are due to Sun glints from the spacecraft appendages and could be eliminated by suitable modeling. The accuracy needs of the mission for antenna pointing did not require that such modeling be carried out, however.

The analysis of the optimal method is more complicated but yields essentially similar results. The result for the sampled covariance is somewhat improved showing a standard deviation per axis of 11. deg.

DISCUSSIONS AND CONCLUSIONS

Simple, practical algorithms have been developed for estimating spacecraft attitude from sensors of extremely modest accuracy. Although these attitudes are less accurate certainly than those typical of even low accuracy missions, they are adequate for gross attitude estimation as might be necessary to monitor the health of the spacecraft or, as in the present instance, to verify antenna pointing. The general agreement of our error analyses with experience with actual flight data is good.

ACKNOWLEDGEMENTS

This work was presented under contract by the Strategic Defense Initiative Organization. The encouragement and support of Major Andrew J. Green, Thrusted Vector Program Manager, is gratefully acknowledged.

REFERENCES

1. Chen, L. C., and Lerner, G. M., "Sun Sensor Models," in J. R. Wertz (ed.), *Spacecraft Attitude Determination and Control,* D. Reidel Publishing Company, Dordrecht, the Netherlands, 1978.

2. Shuster, M. D., and Oh, S. D., "Three-Axis Attitude Determination from Vector Observations," *Journal of Guidance, Control and Dynamics,* Vol. 4, No. 1, Jan.–Feb. 1981 pp. 70–77.

AAS 89-394

ENERGY DISSIPATION AND VEHICLE ATTITUDE IN SPACE*

William T. Thomson†, Gabriel E. Arancibia‡ and Mitchell Thomas**

The case analyzed is an inflatable sphere with a hard structure attached at a pole. The system is initially spun around the axis of symmetry, but will eventually tumble around an orthogonal axis. The sphere is modeled as a stack of rings, and the motion of the rings are described by the forces and moments acting on each. The perturbations in the balloon geometry caused by the motion are assumed small so that integration over the surface can be done assuming the sphere remains a sphere.

The solution for the coning angle of the sphere as a function of time is obtained in closed form. The rate of growth of coning angle is directly proportional to the inflatable's pressure because of the resulting pre-stress in the balloon's skin. Most of the energy loss can be considered as a result of the stress variations in the balloon's skin caused by the bending of the axis of symmetry due to the gyroscopic moments and centripetal forces induced by the precessional motion.

NOMENCLATURE

A	Pitch/yaw moment of inertia
C	Roll moment of inertia
d	Thickness of spherical shell
E	Material elastic modulus
F	Force
f	"Form factor," Equation. (22)
h	Angular momentum
I	Area moment of inertia
K	Various groupings of physical parameters
ℓ	Radius of base plate
$\underline{M}$	Moment
$\overline{M}$	M/Q
M2	Centripetal-force moment
M4	Gyro-moment
m_s	Mass of spherical balloon
Q	$\frac{1}{2} m_s R^2 \dot{\psi}^2 \sin\theta \cos\theta$
R	Radius of sphere

* The appendix to this paper appears in the microfiche supplement to this volume, **AAS Microfiche Series**, Volume 59.

† Consultant for L'Garde, Inc. and Professor Emeritus of Engineering, University of California, Santa Barbara, California 93106.

‡ Member of Technical Staff; ** President; L'Garde, Inc., 15181 Woodlawn Ave., Tustin, California 92680-0771.

T Kinetic energy
t Time
t_0 Spin period
V Volume in material
$\bar{z}$ Offset of center of mass from sphere's center
α, β Azimuthal angle on sphere defining a ring around sphere
δ Azimuthal angle on sphere defining diameter of base plate, $\pi - \sin^{-1}(\ell/R)$
γ Hysteresis damping factor (fraction of energy stored in material dissipated during material cycling)
ω_0 Initial body fixed spin rate
ϕ Euler angle, rotation around spin axis
ψ Euler angle, rotation around momentum vector for steady precession
σ Tensile stress in material (σ_0 = pre-stress, $\sigma_\sim$ = cyclic stress)
θ Euler angle, coning half angle
τ Shear and bending stress

INTRODUCTION

Axisymmetric satellites are often spin stabilized around the axis of symmetry. If this axis is not the one with the largest moment of inertia, the satellite will eventually reorient itself to a position where the rotational energy is a minimum. The rate at which this reorientation takes place is determined by the rate at which rotational energy can be dissipated in the structure due to flexures caused by the motion.

Since the early use for inflatable structures by NASA (Echo I, Echo II, PAGEOS, Explorer IX, and Explorer XIX) the application of such lightweight devices has become increasingly important. Inflatables are being considered for a variety of space structures where their low packaged volume and weight make feasible missions otherwise impossible due to high launch costs. However, the ability to stabilize the orientation of hybrid devices consisting of part inflatable and part hard structure has not been previously investigated. The analysis in this paper shows that inflatables can rapidly dissipate rotational energy and thus change orientation.

The moment-free motion of a body of revolution with principal moments of inertia A,A,C is a steady precession of its spin axis about its angular momentum vector, h, through its center of mass. It is described by the well known relationship

$$\dot{\phi} = \dot{\psi}\left(\frac{A}{C} - 1\right)\cos\theta \qquad (1)$$

When there is dissipation of energy, this relationship no longer holds. Its moment-free motion takes place under the condition of constant angular momentum with decreasing kinetic energy, which leads to the equation

$$\dot{T} = \frac{h^2}{C}\left(\frac{C}{A} - 1\right)(\sin\theta\cos\theta)\,\dot{\theta} \qquad (2)^*$$

*See Reference 1

For a satellite consisting of a spherical balloon attached to a rigid cylindrical canister, as shown in Figure 1, gyroscopic moments and centrifugal forces will cause the balloon to distort as it spins and precesses. The resulting cyclic stress in the balloon material will therefore dissipate energy, and changes in the attitude angle θ will gradually take place towards its position of minimum kinetic energy at $\theta = 90°$.

The Ring Model

If the balloon of Figure 1 is absolutely rigid, the satellite will be subjected to the loads shown in Figure 2. The gyroscopic moment on each of the two components is clockwise, while the moment of the centrifugal forces about the system center of mass is counterclockwise. These loads tend to distort a flexible balloon into such a shape that its spin axis will bend as shown by the dotted line (drawn greatly exaggerated). This suggests a beam model where the inflated balloon is a short variable cross-section beam.

The inflated balloon, however, is a surprisingly stiff structure whose bending deformation is small compared to that of the displacement of its spin axis from the inertial axis through the cg of the system. We will therefore ignore the bending deformation and determine the load on the balloon as if it were absolutely rigid.

The problem of determining the maximum and minimum stress in the balloon is not a simple one, and several simplifying assumptions must be made. Any given point on the balloon undergoes variations in the stress as it rotates around the spin axis. This suggests the sphere to be composed of a series of concentric rings about the spin axis. The ring model is shown in Figure 3.

Since the rigid canister cannot contribute to the energy loss of the system, it is used only for establishing the location of the system cg, and the gyroscopic moment and force it exerts on the balloon at its juncture. We will therefore replace it with its moment and force at the juncture and focus our attention on the calculation of the energy loss of the balloon.

To determine the stress at any point p on the balloon, we will make an imaginary cut through a point perpendicular to the spin axis. The point p then moves in a circle of radius $R\sin\alpha$ in the cut surface. Forming free-body diagrams of the two sections made by the cut, we need first the forces and moments at that surface. We will call them $F(\alpha)$ and $M(\alpha)$. They act on each section with equal and opposite signs as shown in Figure 4.

Looking at the free-body section towards the free end of the balloon, we first examine a ring within this sector located by the angle β. The gyroscopic moment dM_β and the centripetal force dF_β necessary to maintain the motion of this ring are

$$dF_\beta = - dm_\beta \; \dot{\psi}^2 \, (\overline{z} + R\cos\beta) \sin\theta \tag{3}$$

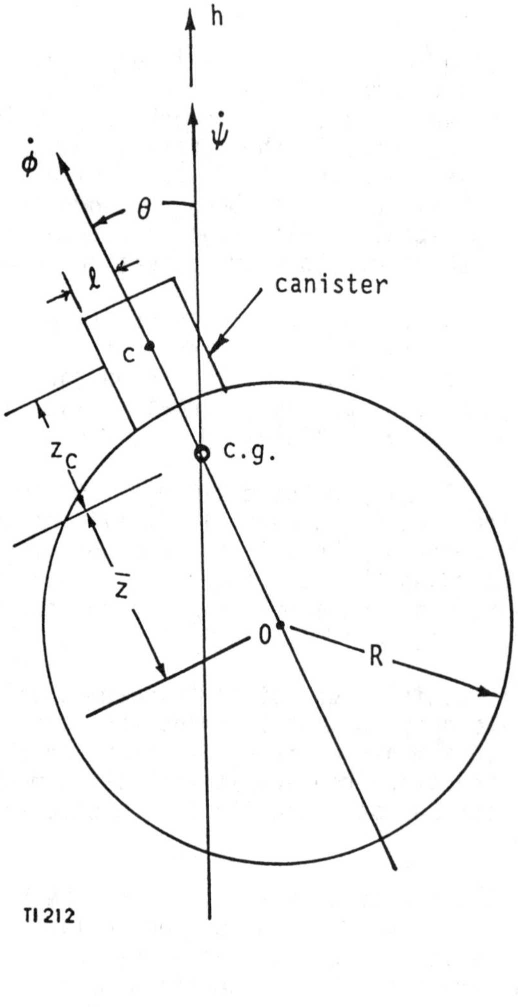

Figure 1. Inflatable Sphere with Attached Hard Canister

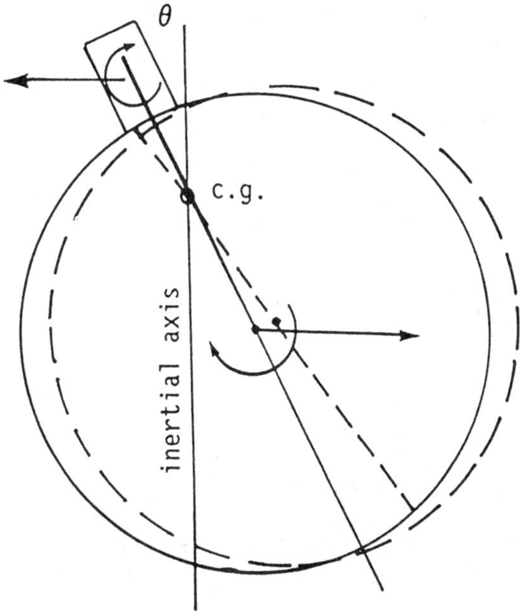

Figure 2. Distortion of Balloon (much Exaggerated) under Space Loads

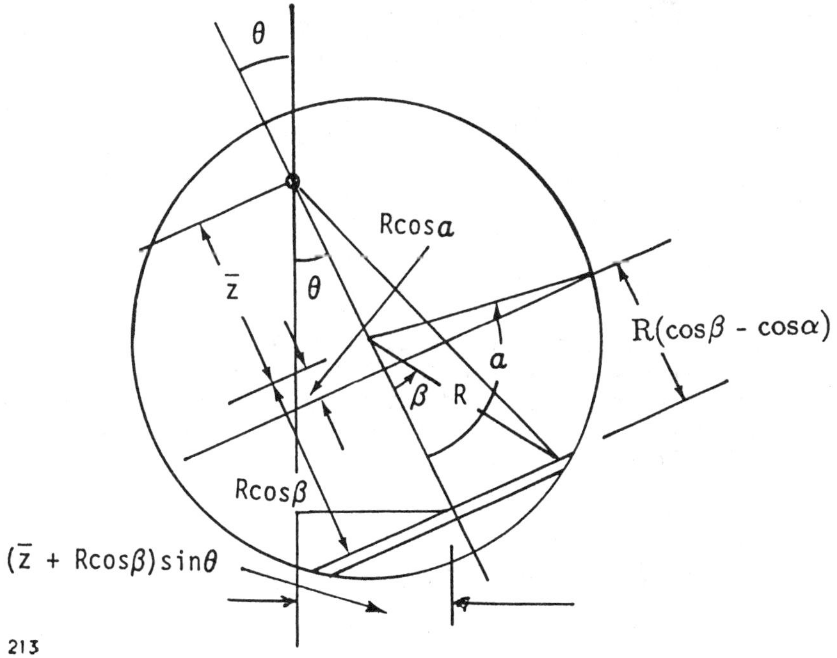

TI 213

Figure 3. Ring Model of Inflatable Sphere

$$dM_\beta = dC_\beta\,(\dot\phi + \dot\psi\cos\theta)\,\dot\psi\sin\theta - dA_\beta\,\dot\psi^2\sin\theta\cos\theta \tag{4}$$

where dm_β is the mass of the ring and dC_β and dA_β are the mass moments of inertia of the ring about its own ring cg.

The force $F(\alpha)$ is simply the sum of dF_β of all rings between $\beta = 0$ to $\beta = \alpha$. The moment at p of the single ring at β is, however, the difference of the gyro moment dM and the moment of the centripetal force dF_β, or

$$dM(\alpha) = dM_\beta - dF_\beta\,R(\cos\beta - \cos\alpha)\cos\theta \tag{5}$$

The total moment at p is then the sum of $dM(\alpha)$ for all rings. In the Appendix, the quantities in these equations are written in terms of the total mass m_s of the balloon, the radius R of the balloon, and the angle β, and integrated from $\beta = 0$ to $\beta = \alpha$. They are presented here as important load parameters for the determination of the stress at point p. For convenience of computation, the total moment at $M(\alpha)$ is presented in two parts - - M4 for the gyro-moment and M2 for the centripetal force moment.

$$F(\alpha) = -\frac{1}{2}\,m_s\,R\dot\psi^2\sin\theta\left\{\frac{1}{2}\sin^2\alpha + \frac{\overline{Z}}{R}(1-\cos\alpha)\right\} \tag{6}$$

$$M4 = Q\left(\frac{A}{C} - \frac{1}{2}\right)\left\{\frac{2}{3} - \frac{1}{3}\cos\alpha\,(\sin^2\alpha + 2)\right\} \tag{7}$$

$$M2 = Q\left\{\frac{1}{3}(1-\cos^3\alpha) - \left(\cos\alpha - \frac{\overline{Z}}{R}\right)\frac{1}{2}\sin^2\alpha - \frac{\overline{Z}}{R}\cos\alpha\,(1-\cos\alpha)\right\} \tag{8}$$

$$Q = \frac{1}{2}\,m_s\,R^2\,\dot\psi^2\sin\theta\cos\theta \tag{9}$$

$$M(\alpha) = M4 - M2 \tag{10}$$

Energy Dissipation

We will assume hysteresis damping in the balloon fabric as the source of energy loss. Over a broad frequency range, energy dissipation for this type of damping is independent of frequency and proportional to the square of the stress amplitude. It can be expressed by the simple equation

$$\text{energy loss} = \frac{\gamma\sigma^2}{2E}$$

where γ is the hysteresis damping factor.

With the stress varying between $\sigma_{max} = \sigma_0 + \sigma_\sim$ and $\sigma_{min} = \sigma_0 - \sigma_\sim$, where σ_0 is the prestress due to the internal pressure in the balloon and $\sigma_\sim$ the cyclic stress, the energy loss per unit volume of the material is

$$\frac{\gamma}{2E}\left[\sigma_{max}{}^2 - \sigma_{min}{}^2\right] \cong \frac{2\gamma}{E}\sigma_o\,\sigma_\sim \qquad (11)$$

Dividing by the time t_o per cycle of stress and integrating over the entire volume of the balloon, the energy dissipation can be found as

$$-\frac{2\gamma\sigma_o}{Et_o}\int \sigma_\sim dV \qquad (12)$$

For the determination of stress, we will consider the free-body diagram of a ring at α, as shown in Figure 4. We have already found the loads at its face as $M(\alpha)$ and $F(\alpha)$, the latter which can be resolved to $F(\alpha)\cos\theta$ parallel to its face and $F(\alpha)\sin\theta$ perpendicular to it.

Dissipation from Bending

We will first determine the stress due to $M(\alpha)$ and later those produced by $F(\alpha)$. We now make the assumption that the actual ring can be replaced by a ring of constant radius $R\sin\alpha$ and same width which is shown in Figure 5. From the beam theory, the bending stress is distributed as shown with maximum and minimum stress given as

$$\sigma_b = \pm\,\frac{M(\alpha)\,R\,\sin\alpha}{I(\alpha)} \qquad (13)$$

where the area moment of inertia is shown in the Appendix to be equal to

$$I(\alpha) = \pi d\,R^3 \sin^3 \alpha \qquad (14)$$

Thus σ_b at α becomes

$$\sigma_b(\alpha) = \frac{M(\alpha)}{\pi d\,R^2 \sin^2 \alpha} \qquad (15)$$

In this equation, the denominator goes to zero at $\alpha = 0$, but $M(\alpha)$ also goes to zero at this point, so that the equation is indeterminate in that region. This is to be expected since the application of the beam theory in this vicinity is far fetched. Regardless of the dilemma the contribution to the integral of $M(\alpha)$ in this region is quite small and could not influence the results of this problem very much.

To combine the maximum bending stress with the prestress σ_o which is tangent to the balloon surface, we multiply $\sigma_b(\alpha)$ with $\sin\alpha$. Then the maximum and minimum tensile stress in the ring due to the bending is $\sigma_o \pm \sigma_b\sin\alpha$ and the energy loss in the ring becomes

$$\frac{2\gamma}{E}\sigma_o\,\sigma_b(\alpha)\sin\alpha\,dV \qquad (16)$$

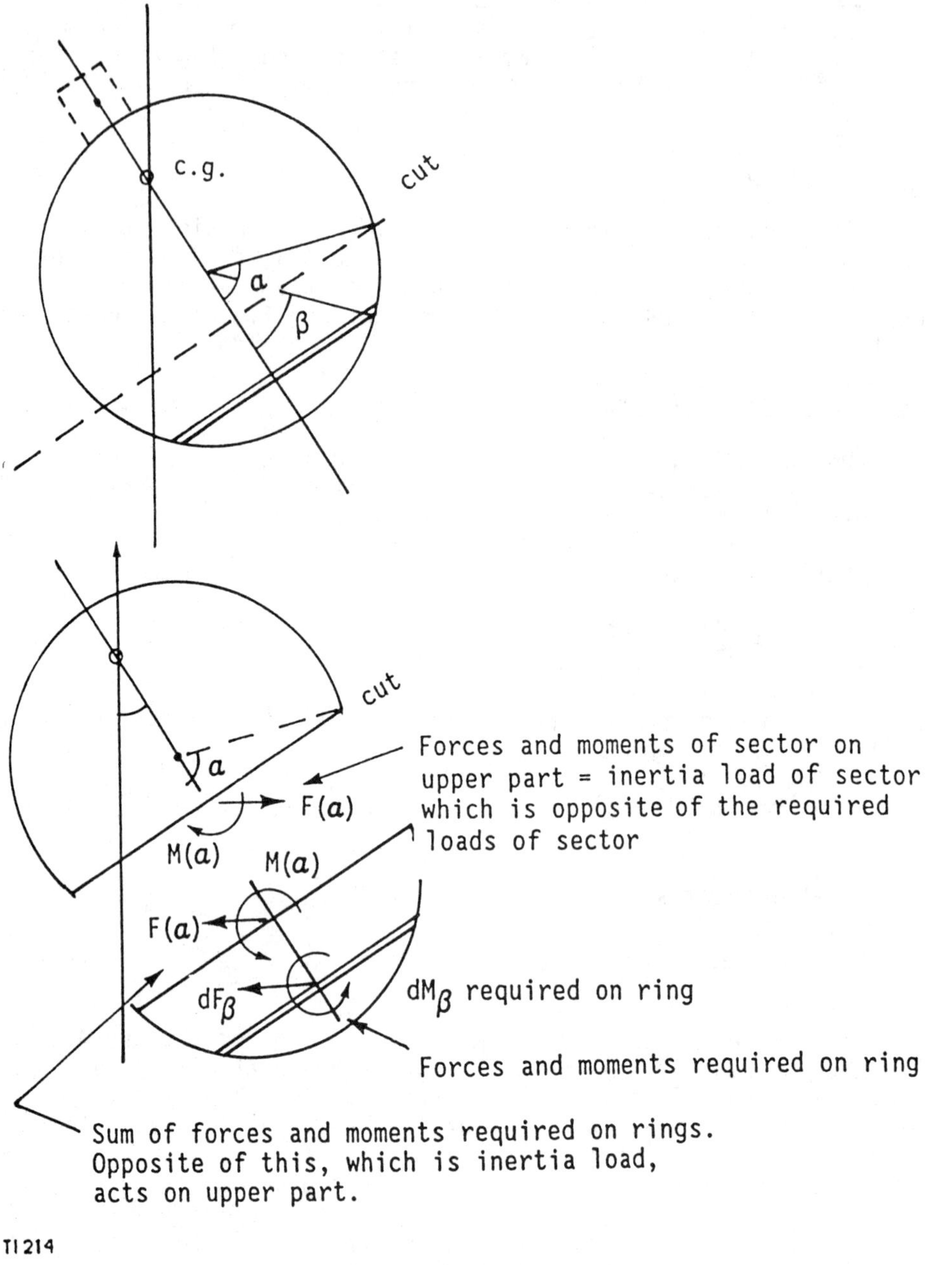

Figure 4. Free-body Diagram of Ring

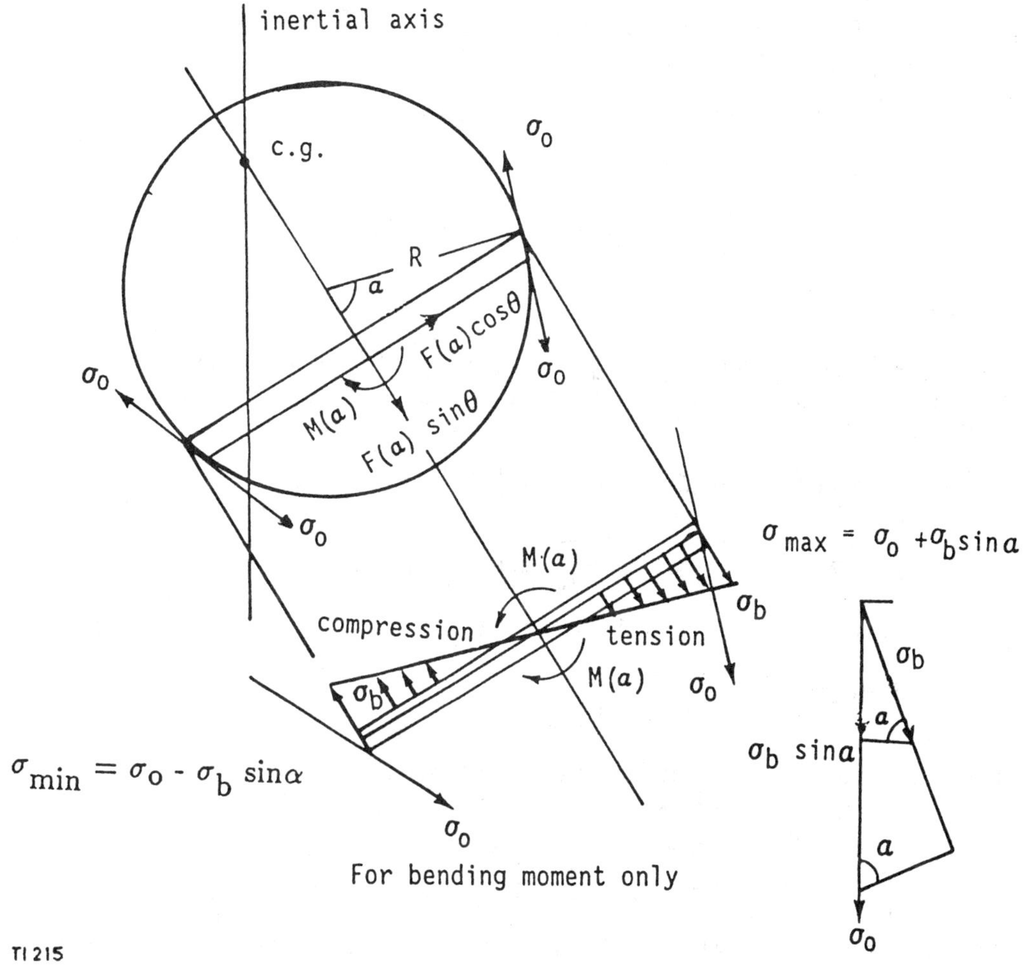

Figure 5. Distribution of Stress in Ring

where dV is the volume of the ring and equal to

$$dV = (2\pi\ R \sin\alpha)\ d\ R\ d\alpha \tag{17}$$

Putting all this together, dividing by the time $t_0 = 2\pi/\dot{\phi} = 2\pi/\dot{\psi}\left(\frac{A}{C} - 1\right)\cos\theta$ the rate of energy dissipation due only to the bending moment becomes

$$\dot{T} = \frac{\gamma m_s R^2 \sigma_0}{E\pi}\left(\frac{A}{C}-1\right)\left(\frac{C}{A}\right)^3\left(\frac{\omega_0}{\cos\theta_0}\right)^3 \sin\theta\cos^2\theta \int_0^{\delta} \bar{M}(\alpha)\, d\alpha \tag{18}$$

Combining equations (2), (7), (8), (10), and (18) we have

$$\dot{\theta} = \frac{\gamma m_s R^2 \sigma_o}{EC \cos\theta_o} \left(\frac{C}{A}\right)^2 \omega_o \cos\theta f(\delta) \tag{19}$$

for $\delta < a_c$, and

$$\dot{\theta} = \frac{\gamma m_s R^2 \sigma_o}{EC \cos\theta_o} \left(\frac{C}{A}\right)^2 \omega_o \cos\theta \left[2f(\alpha_c) - f(\delta)\right] \tag{19a}$$

for $\delta > a_c$

The crossover angle, a_c, is that position where the stress σ_b changes from one adding tension to one adding compression (about σ_o) or from Equation (7) and (8)

$$\left(\frac{A}{C} - \frac{1}{2}\right)\left[\frac{2}{3} - \frac{1}{3}\cos\delta\,(3 - \cos^2\delta)\right] = \frac{1}{3}(1 - \cos^3\delta) + \left(\frac{\bar{z}}{2R} - \frac{\cos\delta}{2}\right)(1 - \cos^2\delta) + \frac{\bar{z}}{R}(\cos\delta - 1)\cos\delta \tag{20}$$

It can be shown that the solution to Equation (20) of interest is

$$\cos\alpha_c = -2 + \frac{3\bar{z}}{2R\left(\frac{A}{C} - 1\right)}, \text{ and} \tag{21}$$

f(δ) is given by integrating the moment equations, viz.

$$f(\delta) = \frac{1}{\pi}\left(\frac{A}{C} - \frac{1}{2}\right)\left[\frac{2\delta}{3} - \frac{2}{3}\sin\delta - \frac{1}{9}\sin^3\delta\right] - \frac{1}{\pi}\left[\left(\frac{1}{3} + \frac{3\bar{z}}{4R}\right)\delta - \left(\frac{1}{3} + \frac{\bar{z}}{R}\right)\sin\delta + \frac{\bar{z}}{8R}\sin 2\delta - \frac{\sin^3\delta}{18}\right] \tag{22}$$

Integration of equation (19) or (19a) gives a coning angle history given by

$$K_2 t = \ell n \tan\left(\frac{\theta}{2} + \frac{\pi}{4}\right) \tag{23}$$

where, for instance, for the case of Equation (19)

$$K_2 = \frac{\gamma m_s R^2 \sigma_o}{EC \cos\theta_o}\left(\frac{C}{A}\right)^2 \omega_o f(\delta) \tag{24}$$

Other Mechanisms

We next examine the stress produced by the component of the force $F(a)\cos$ parallel to the face of the ring. With the area of the face equal to $(2\pi R \sin\alpha)\dfrac{d}{\sin\alpha} = 2\pi Rd$, the average shear stress is $\dfrac{F(a)\cos\theta}{2\pi Rd} = \tau_{a\nu}$ The stress $\tau_{a\nu}$ must then be distributed as shown in Figure 6 which is zero at the extreme fiber and a maximum at the neutral axis. It then cannot have any influence on the maximum bending stress σ_b.

The other component F(α)sinθ is perpendicular to the face of the ring and its tensile stress $\frac{F(\alpha)\sin\theta}{2\pi Rd} = \sigma_t$, is a constant around the circle. It therefore does add to the maximum bending stress and reduce the minimum bending stress which is compressive. So we now have for the maximum and minimum stress the equation

$$\sigma_{max} = [\sigma_o + (\sigma_b + \sigma_t) \sin \alpha] \qquad (25)$$
$$\sigma_{min} = [\sigma_o - (\sigma_b - \sigma_t) \sin \alpha]$$

and the difference of their squares is

$$4\sigma_o\sigma_b \sin \alpha \left[1 + \frac{\sigma_t}{\sigma_o} \sin \alpha\right] \qquad (26)$$

where σ_t is the tensile stress due to F(α)sinθ . Since $(\sigma_t/\sigma_o)\sin\alpha$ is small, the effect of the normal stress is not changed greatly.

Returning to the shear stress τ which is distributed as a sine function across the face of the ring, it had no effect on the maximum bending stress. At the neutral axis, however, the bending stress is zero and the shear stress is a maximum $\tau_m = \frac{2}{\pi} \tau_{a\nu}$. With $\tau_{a\nu} = \frac{F(\alpha) \cos \theta}{2\pi Rd}$, the maximum shear stress is

$$\tau_m = \frac{F(\alpha) \cos \theta}{\pi^2 Rd} \qquad (27)$$

This stress acts on an element of the ring with one side in its face as shown in Figure 6.

If we look at a plane making an angle 45° with the side, we find only tension σ_1. But this tension is found to be equal to the shear stress τ_m. On the other face of this 45° cube the normal force is also equal to τ_m but in compression. Since the prestress σ_o is the same in all directions, the maximum tensile stress at this point is

$$\sigma_o + \tau_m \sin \alpha \qquad (28)$$

The difference of the squares in the dissipation equation is then

$$(\sigma_o + \tau_{a\nu} \sin \alpha)^2 - \sigma_o^2 = \frac{4}{\pi} \sigma_o \tau_m \sin \alpha \left[1 + \frac{1}{2}\left(\frac{\tau_m}{\sigma_o}\right)\right]$$

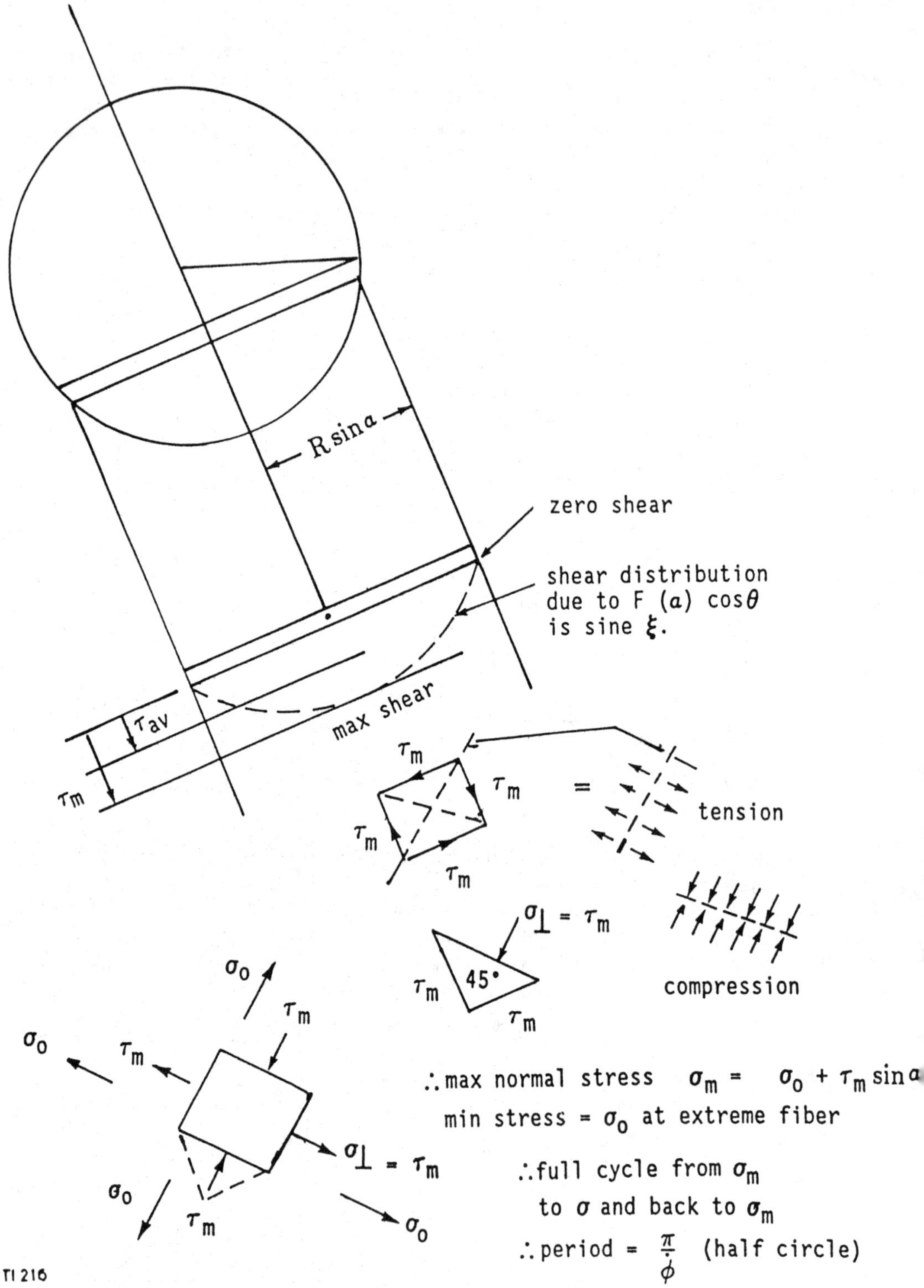

Figure 6. Stress Due to Shear Load F(α)cosθ

which can be rewritten as

$$\frac{4}{\pi}\sigma_o \frac{F(\alpha)\cos\theta\sin\alpha}{\pi^2 Rd}\left[1+\frac{1}{2}\left(\frac{\tau_m}{\sigma_o}\right)\right] \quad (29)$$

Following the procedure for the similar term in the case for the load $F(\alpha)\sin\theta$, it is possible to determine the fraction of the energy dissipation to be added to that of $M(\alpha)$. The period for the shear cycle is half that for the bending stress cycle.

Representative Calculations

To illustrate the energy dissipation effects we will consider a spherical balloon with the following properties:

A = 4520 lbm-in^2	γ = .25
C = 1792 lbm-in^2	E = 390000 psi
R = 43.3 in	ω_0 = 169°/sec
ℓ = 2.28 in	σ = 2598 psi
d = 0.001 in	m_s = 1.349 lbm
	θ_0 = 15°
$\bar{z}$ = 31.45 in	

First, we will look at the case of cyclic stresses due to the bending moment alone. The normalized moment (M/Q) profile is shown in Figure 7. Notice that for this case there is no crossover angle and thus equations (19) and (23) apply resulting in the coning angle history shown in Figure 8.

Now, we include the additional stresses due to the centripetal force. The normalized centripetal force profile is shown in Figure 9. These stresses representing the "other mechanisms" discussed above, add terms to equations (19) and (19a) thus requiring numerical solution. However, their effect on the coning angle history is small as can be observed in Figure 8. As expected, the bending moment induced stresses are the dominant factor and the principal energy dissipation mechanism.

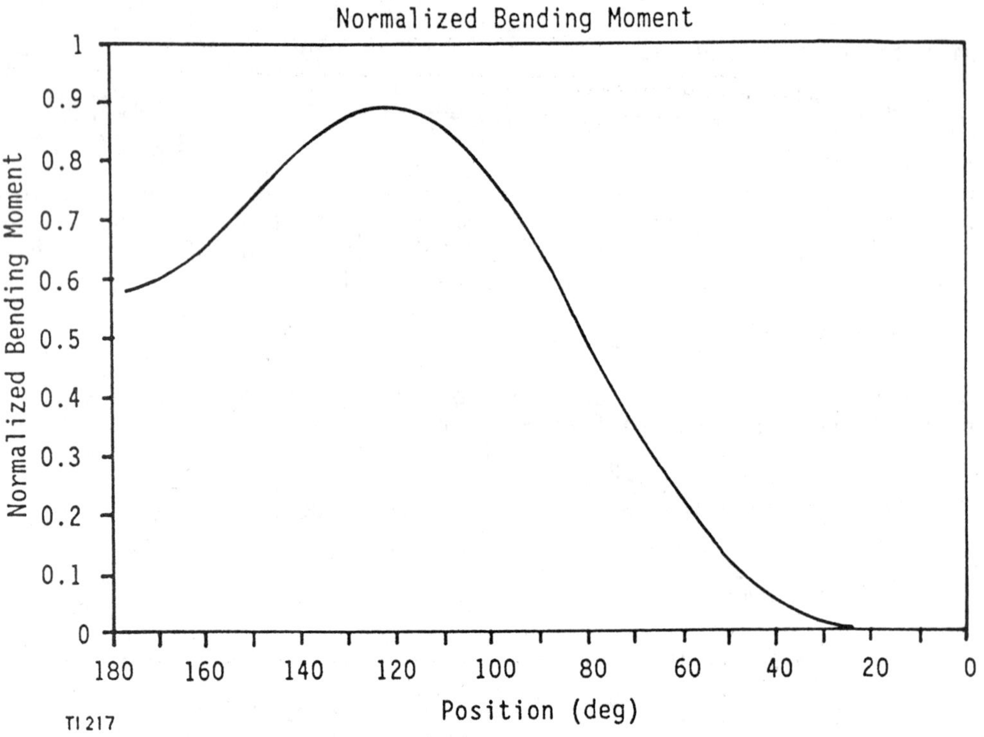

Figure 7. Normalized Bending Moment Profile

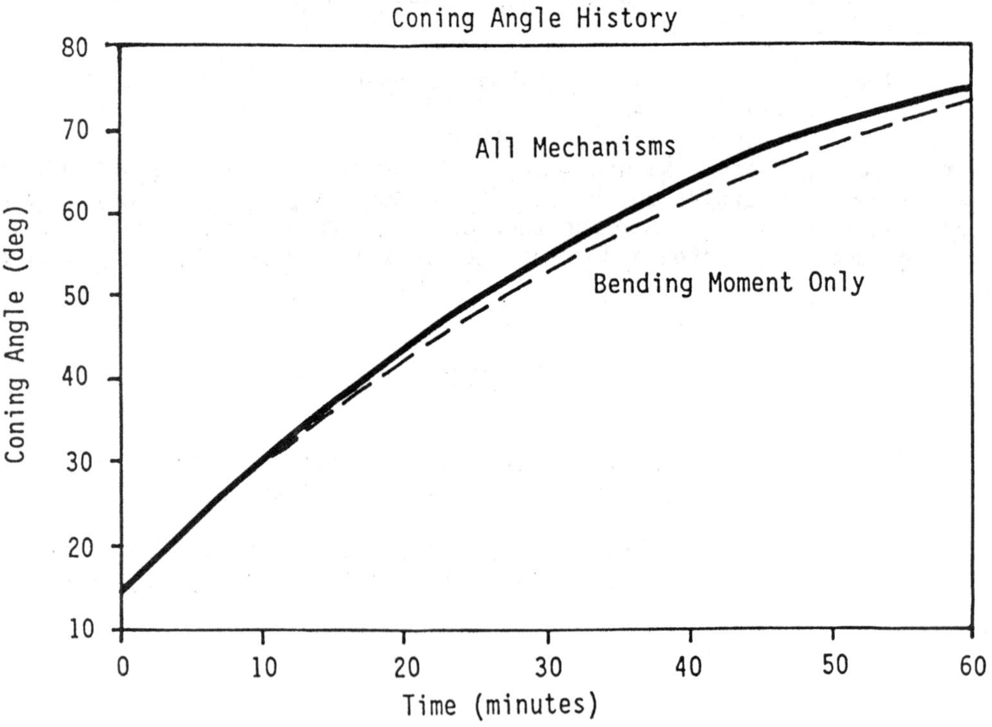

Figure 8. Comparative Coning Angle Histories

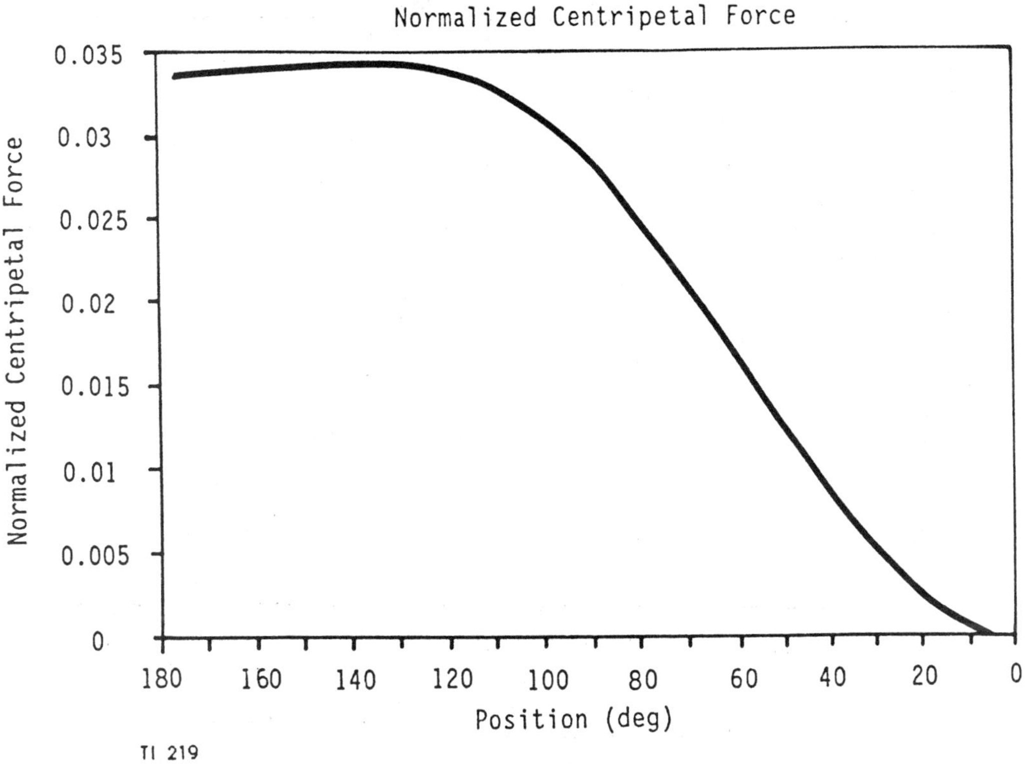

Figure 9. Normalized Centripetal Force Profile

ACKNOWLEDGMENTS

This work was sponsored partially by L'Garde, Inc. and partially by the USAF and Army under Contract F04704-85-C-0174, (Capt. Chris Hoskins, Project Engineer). Guidance was also provided by the L'Garde Program Engineer, Gil Friese. Participating in the early phase of the project were Chris J. Thompson and Dr. Rex Sinclair.

REFERENCES

1. Thomson, W. T., and G. S. Reifer, "Attitude Drift of Space Vehicles," J. Astronaut Sci., VII, No. 2 (1960), 29-34.

2. Thomson, W. T., "Introduction to Space Dynamics," John Wiley & Sons, 1960, reprinted by Dover Publications, 1986, page 214.

AAS 89-395

ELIMINATION OF ATTITUDE DETERMINATION ERROR CONSTRAINTS IN LAUNCH WINDOWS FOR GEOSYNCHRONOUS MISSIONS

J. J. Palimaka, A. J. Grisé, B. V. Burlton and J. M. Overduin*

In launch windows for missions to geosynchronous orbit it is common to find restrictions as a result of attitude determination error constraints. Based on the results of analytical work, simulations and mission experience, we argue that these restrictions are not necessary when sun-earth separation-angle data is included and the weighted-least-squares method is used to process the available attitude sensor data.

INTRODUCTION

Attitude determination accuracy is an important aspect of missions to geosynchronous orbit. The attitude determination process is critical when applied to determining spacecraft attitude prior to apogee motor firing (AMF). This maneuver is carried out to perform the transition from the transfer orbit state (into which the satellite is placed by the launch vehicle) to a near-synchronous orbit state referred to as drift orbit. An error in positioning of the spacecraft attitude of as little as 0.12° from the calculated optimum AMF attitude would require the expenditure of one month of stationkeeping fuel to correct.

Telesat Canada, which is responsible for all of Canada's domestic communication satellites, has conducted nine missions to geosynchronous orbit since 1972. Five of these missions

* Telesat Canada, 1601 Telesat Court, Gloucester, Ontario K1B 5P4, Canada.

have used the Delta launch vehicle and four have used the Shuttle Space Transportation System. All launches have taken place from the Eastern Space and Missile Centre (ESMC) at Cape Canaveral. In addition, we have provided mission support for other satellite operators worldwide through our consulting activities.

Two major changes have recently occurred in relation to the process of getting to geosynchronous orbit. Launch services are now available from facilities other than the ESMC and restartable liquid apogee engines (LAEs) are being used rather than solid-rocket apogee motors.

In the case of the Ariane launch vehicle, the latitude of the launch site at the Guiana Space Centre (GSC) is closer to the equator than the ESMC; therefore, the resulting transfer orbit is of lower inclination (7° rather than 25°). The lower inclination orbit leads to less geometrical variation in the relationships between the earth, satellite spin axis and sun vectors throughout the orbit and thus effectively less information with which to determine the attitude. This has been of some concern, particularly near the equinoxes where the geometrical variation is a minimum in any case due to the near-coplanarity of the earth, satellite spin axis and sun vectors. This concern extends for the same reasons to a mission with multiple LAE burns where the intermediate transfer orbits have a shallower inclination than the initial transfer orbit (which may itself have a relatively high inclination).

We have shown by analysis and simulations that, through the use of separation-angle data (defined below) and the weighted-least-squares (WLS) method, attitude determination of the required accuracy (usually 3σ uncertainties of less than 0.5°) can be attained for low-inclination transfer orbits and that attitude determination accuracy need not constrain the launch window during any time of the year. The conclusions of our analysis are similar to those reached by Van der Ha (Ref. 1) who examined the effect of measurement biases on attitude angle covariances. However, his conclusions concerning unacceptable attitude determination error under particular sun-earth geometries differ from ours in that he did not consider the effectiveness of the least-squares process in accurate attitude estimation.

ANALYTICAL RESULTS

In this paper we will express satellite attitude (defined as the orientation in space of the nominal angular velocity vector) using the angles right ascension (R) and declination (D) in a geocentric inertial reference frame. Satellite attitude, however, is not actually measured in terms of right ascension and declination but commonly by means of two onboard sensors: the sun sensor, which measures the sun angle (θ_s) between the satellite spin axis and the satellite-to-sun vector; and the earth sensor, which (with some processing of its output) measures the nadir angle (θ_e) between the spin axis and satellite-to-earth vector. In addition, a third sensor data type may be obtained by combining information from the two sensors: the separation (or dihedral) angle (α) between the plane defined by the spin axis and sun vector; and the plane defined by the spin axis and earth vector (Fig. 1).

One of the most critical parameters for accurate apogee motor firing is the spin axis declination. Our knowledge of this parameter is characterized by the variance in spin axis declination (σ_D^2). This quantity (together with the variance σ_R^2 in spin axis right ascension) occupies a diagonal position in the satellite's attitude state covariance matrix (P_s):

$$P_s = \begin{bmatrix} \sigma_R^2 & \sigma_{RD} \\ \sigma_{DR} & \sigma_D^2 \end{bmatrix} \qquad (1)$$

Correspondingly, the covariance matrix of the observations (sensor data) may be written:

$$P_0 = \begin{bmatrix} \sigma_s^2 & \sigma_{se} & \sigma_{s\alpha} \\ \sigma_{es} & \sigma_e^2 & \sigma_{e\alpha} \\ \sigma_{\alpha s} & \sigma_{\alpha e} & \sigma_\alpha^2 \end{bmatrix} \qquad (2)$$

Our goal is to find a relation between the diagonal elements (variances) of P_s and P_0. The least-squares estimation process, as applied to the attitude determination problem, can be represented by the non-linear regression equation:

$$y = h(X) + V \tag{3}$$

where y is the observation vector $(\theta_s, \theta_e, \alpha)$, X is the state vector (R, D), h is a known vector function (the measurement model) and V is a random noise vector characterized by the measurement covariance P_0.

This approach leads to the introduction of a diagonal weighting matrix (W) which has elements which are reciprocals of the diagonal elements of P_0. It can be shown (Ref. 2) that the state covariance is given by the covariance update equation

$$P_s = (H^T W H + P^{-1})^{-1} \tag{4}$$

where P is an a priori state covariance and H is a matrix made up of the partial derivatives of h with respect to the state variables:

$$H = \begin{bmatrix} \dfrac{\partial \theta_s}{\partial R} & \dfrac{\partial \theta_s}{\partial D} \\ \dfrac{\partial \theta_e}{\partial R} & \dfrac{\partial \theta_e}{\partial D} \\ \dfrac{\partial \alpha}{\partial R} & \dfrac{\partial \alpha}{\partial D} \end{bmatrix} \tag{5}$$

The exact nature of the partial derivatives in (5) depends on the geometry of the sun and earth sensors. It can be shown that:

$$\left.\begin{aligned} \frac{\partial \theta_s}{\partial R} &= -\hat{s} \cdot \hat{a}_R \csc\theta_s \cos D & \frac{\partial \theta_s}{\partial D} &= -\hat{s} \cdot \hat{a}_D \csc\theta_s \\ \frac{\partial \theta_e}{\partial R} &= -\hat{e} \cdot \hat{a}_R \csc\theta_e \cos D & \frac{\partial \theta_e}{\partial D} &= -\hat{e} \cdot \hat{a}_D \csc\theta_e \\ \frac{\partial \alpha}{\partial R} &= \vec{q} \cdot \hat{a}_R \cos D & \frac{\partial \alpha}{\partial D} &= \vec{q} \cdot \hat{a}_D \end{aligned}\right\} \tag{6}$$

where the "separation vector" ($\vec{q}$) has been defined by:

$$\vec{q} \equiv \frac{[(\hat{s}\cos\theta_e + \hat{e}\cos\theta_s)\sin\alpha + (\hat{s}\times\hat{e})\cos\alpha]}{D_s D_e} \tag{7}$$

and the unit vectors $\hat{a}_R$ and $\hat{a}_D$ are obtained from differentiating the spin axis unit vector ($\hat{a}$) with respect to right ascension and declination in polar coordinates:

$$\left.\begin{aligned} \hat{a}_R &\equiv \frac{\partial \hat{a}}{\partial R}\sec D = -\sin R\hat{\imath} + \cos R\hat{\jmath} \\ \hat{a}_D &\equiv \frac{\partial \hat{a}}{\partial D} = -\cos R\sin D\hat{\imath} - \sin R\sin D\hat{\jmath} + \cos D\hat{k} \end{aligned}\right\} \tag{8}$$

(The $x-y$ plane has been identified with the earth's equatorial plane, and the x-axis points toward Aries.) The unit vectors $\hat{s}$ and $\hat{e}$ lie along the satellite-to-sun and satellite-to-earth vectors respectively, and D_s and D_e are the magnitudes of those same vectors, projected onto the satellite's spin plane:

$$\left.\begin{aligned} D_s &\equiv |\hat{s} - \hat{a}\cos\theta_s| \\ D_e &\equiv |\hat{e} - \hat{a}\cos\theta_e| \end{aligned}\right\} \tag{9}$$

If we examine the least-squares estimation process at a single point in the orbit and ignore any previous knowledge about the state and covariance (i.e., the elements of P are large and the elements of P^{-1} are small) then the covariance update equation (4) becomes:

$$P_s = (H^T W H)^{-1} \tag{10}$$

Carrying through the matrix operations we obtain a state covariance matrix with the following elements:

$$P_s = \frac{\begin{bmatrix} \left(\frac{S_D^2}{\sigma_s^2}+\frac{E_D^2}{\sigma_e^2}+\frac{Q_D^2}{\sigma_\alpha^2}\right)\sec^2 D & -\left(\frac{S_R S_D}{\sigma_s^2}+\frac{E_R E_D}{\sigma_e^2}+\frac{Q_R Q_D}{\sigma_\alpha^2}\right)\sec D \\ -\left(\frac{S_D S_R}{\sigma_s^2}+\frac{E_D E_R}{\sigma_e^2}+\frac{Q_D Q_R}{\sigma_\alpha^2}\right)\sec D & \left(\frac{S_R^2}{\sigma_s^2}+\frac{E_R^2}{\sigma_e^2}+\frac{Q_R^2}{\sigma_\alpha^2}\right) \end{bmatrix}}{\left(\frac{C_\alpha^2}{\sigma_s^2\sigma_e^2}+\frac{C_e^2}{\sigma_s^2\sigma_\alpha^2}+\frac{C_s^2}{\sigma_e^2\sigma_\alpha^2}\right)} \tag{11}$$

where nine new factors have been introduced:

$$\left.\begin{aligned} C_\alpha &\equiv S_R E_D - E_R S_D \\ C_e &\equiv S_R Q_D - Q_R S_D \\ C_s &\equiv E_R Q_D - Q_R E_D \end{aligned}\right\} \tag{12}$$

$$\left.\begin{aligned} S_R &\equiv \hat{s}\cdot\hat{a}_R \csc\theta_s & S_D &\equiv \hat{s}\cdot\hat{a}_D \csc\theta_s \\ E_R &\equiv \hat{e}\cdot\hat{a}_R \csc\theta_e & E_D &\equiv \hat{e}\cdot\hat{a}_D \csc\theta_e \\ Q_R &\equiv \vec{q}\cdot\hat{a}_R & Q_D &\equiv \vec{q}\cdot\hat{a}_D \end{aligned}\right\} \tag{13}$$

It can be shown that C_α is equivalent to $\sin\alpha$. Variance in declination is then finally given by:

$$\sigma_D^2 = \frac{S_R^2\sigma_e^2 + E_R^2\sigma_s^2 + Q_R^2\left(\frac{\sigma_s^2\sigma_e^2}{\sigma_\alpha^2}\right)}{\sin^2\alpha + C_e^2\left(\frac{\sigma_e^2}{\sigma_\alpha^2}\right) + C_s^2\left(\frac{\sigma_s^2}{\sigma_\alpha^2}\right)} \tag{14}$$

This may be compared with the corresponding results of an analysis by Onstead and Tyler (Ref. 3), involving yaw and pitch rather than right ascension and declination, in which separation-angle data were not included:

$$\sigma_D^2 = \frac{S_R^2\sigma_e^2 + E_R^2\sigma_s^2}{\sin^2\alpha} \tag{15}$$

Equation (15) demonstrates that, under coplanar conditions ($\alpha \to 0$), the variance in declination will become very large without the benefit of separation-angle data, since S_R and E_R do not necessarily go to zero when $\sin\alpha$ does.

When separation-angle data is included, however, the variance in declination, as given by (14), will not grow arbitrarily large, even during coplanarity. This is because Q_R and Q_D, and hence C_e and C_s, are functions of the separation vector $\vec{q}$, which does not vanish as $\alpha \rightarrow 0$:

$$\vec{q}_{\alpha=0} = (\hat{s} \times \hat{e}) / D_s D_e \tag{16}$$

The presence of C_e and C_s in the denominator of (14) ensures that the variance in declination will remain acceptably small under coplanar conditions. Only in the event of an exact colinearity of the sun and earth vectors (so that $\hat{s} \times \hat{e} \rightarrow 0$) will single-point declination errors become very large.

ATTITUDE DETERMINATION METHOD

During a mission, the spin axis attitude is determined by the Telesat Flight Dynamics Software using the technique of weighted-least-squares with a priori statistics. The attitude vector is most conveniently defined in a sun-directed polar coordinate system (ϕ, θ) where ϕ is the angle between sun vector and attitude vector and θ is the angle between the two planes defined by the sun vector/earth spin axis and the sun vector/attitude vector. The mission software will handle five data types: sun angle, the north and south earth chord widths (from horizon crossing times), and the separation angles between the sun pulse and the north and south earth pulses. The estimated state includes: ϕ, θ, sun sensor cant angle bias (generally fixed), north and south earth sensor cant angle biases, and north and south earth sensor spin angle separation biases. Earth sensor models are used to account for the field of view of the earth sensor and for the delays introduced by the processing electronics. Overall weights for each data type are preset, but earth chord widths less than 10° are deweighted inversely proportional to their chord width. Automatic culling of bad data using a specified unacceptable residual level (in terms of a specified standard deviation of the data) is performed. Finally, chord width biases can be determined from the residuals obtained when the partial derivatives with respect to the attitude vector parameters are close to zero and used in subsequent re-processing of the data.

The results of using this system during recent missions will be presented in a later section. The technique is introduced at this point to provide background on the second simulation method described below.

SIMULATION METHODS

Two types of simulations were carried out to obtain an understanding of the behaviour of the attitude determination error. In one simulation approach, the errors at individual points in the orbit were determined from the earth, sun and separation angles using the analytical methods described previously and assuming sensor data errors on the order of 0.02°. The results of this analysis for declination error for a worst-case situation (where the colinear condition is coincident with the coplanar condition) are shown in Figure 2 for a GSC (Ariane) launch.

Referring to Figure 2, first note that for injection into transfer orbit at solstice, the single-point attitude determination error remains very low for a mean anomaly within 100° of apogee (180°). This is the time period over which attitude data collection would typically be carried out due to earth sensor data availability. For launches near equinox, when conditions of coplanarity and possibly colinearity are likely to occur, the errors are much larger over the nominal data collection interval. The other aspect to note in Figure 2 is that for the equinox case, the region where the error is large spans a much shorter time interval when separation data is included in the attitude determination calculation than when it is not. Similar trends in attitude determination error throughout the year, with and without separation angle data were shown by Parvez and Fox (Ref. 4).

Another approach we used to study attitude determination was to generate simulated data and to process it using algorithms similar to those used in processing mission data.

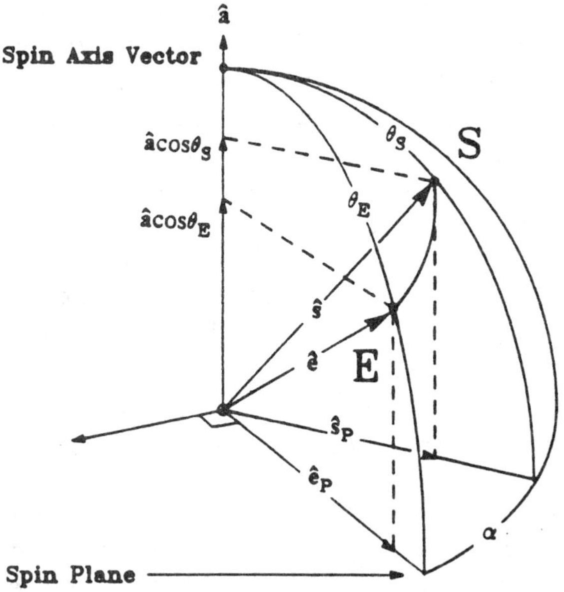

Figure 1. Sun, earth and spin axis vectors.

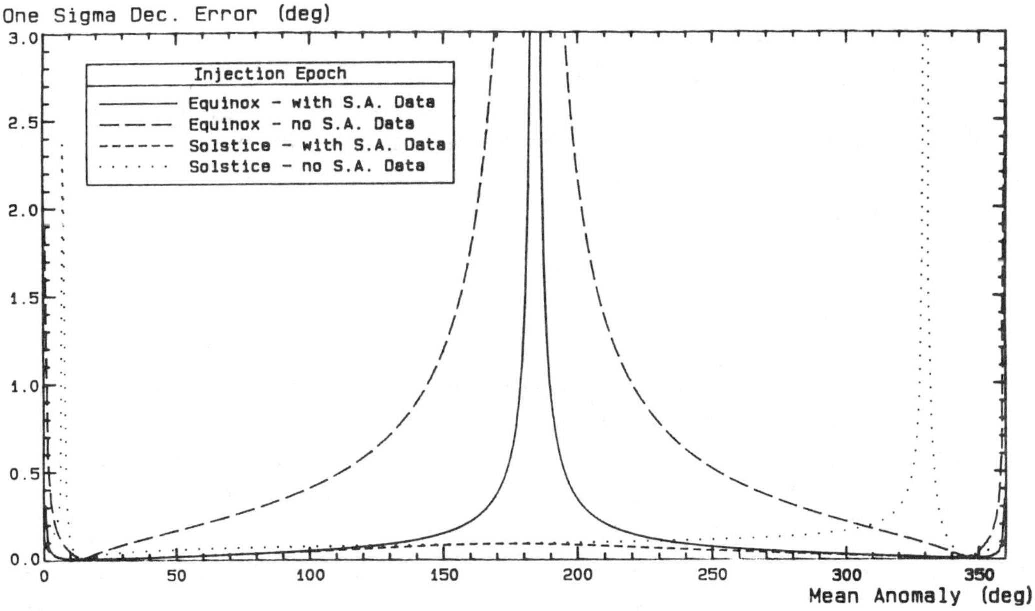

Figure 2. Single-point declination errors, with and without separation-angle (S.A.) data.

The technique used is to determine the attitude error for one pass through apogee using a weighted-least-squares estimation procedure with covariance analysis and using the injection state to provide an a priori state and covariance. The data was produced at a simulated sampling interval of 5 minutes, during the data collection interval of approximately 3.5 hours.

Various simulations have been performed as part of the mission planning process for missions which Telesat has supported. These simulations normally cover launch windows throughout the year for both midnight and noon injections into transfer orbit for both ESMC and GSC launches.

Contrary to the trends shown in the single point error runs (Fig. 2 and Ref. 4), the error was found to be nearly constant throughout the year, including the equinoxes as is shown in Figure 3 for Ariane launches throughout the midnight launch window. Generally, a 3σ attitude determination accuracy of less than 0.5° is attainable given typical sensor data errors. This was also found to be true for missions using LAEs where intermediate transfer orbit inclinations as low as 0.6° were simulated.

When the single-point error results are averaged over the data collection interval, using weighted averaging, it is found that the resultant error is comparable to the results of the mission data simulations. This consistency gives us some confidence in the conclusions of our analysis.

MISSION EXPERIENCE

Since our first satellite launch in November 1972, Telesat staff have performed attitude determination for our own nine missions. We have also provided substantial support to customers of our consulting services for eight other missions, of which five involved attitude determination. Specifically, in the last two years we have been involved in three Ariane missions: Aussat K3 on September 15, 1987; SBS F5 on September 8, 1988; and Skynet 4B on December 11, 1988. In the latter two missions, attitude determination was performed throughout

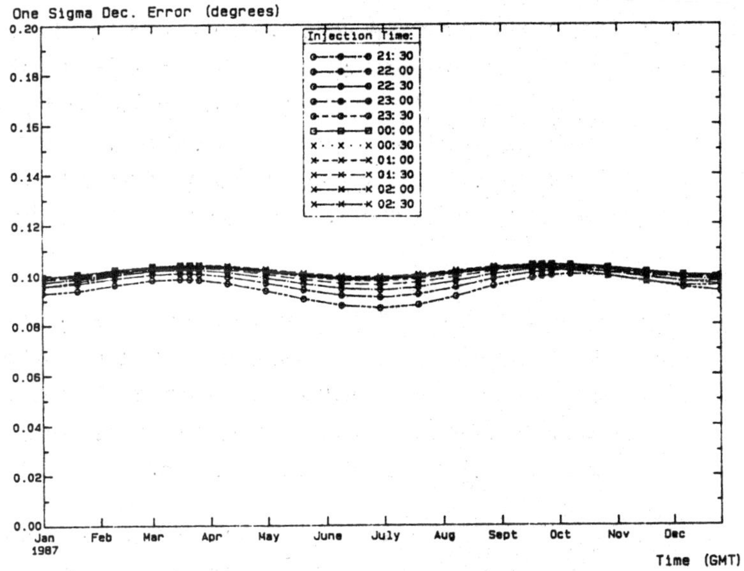

Figure 3. Spin axis declination error over one year, throughout an Ariane midnight launch window.

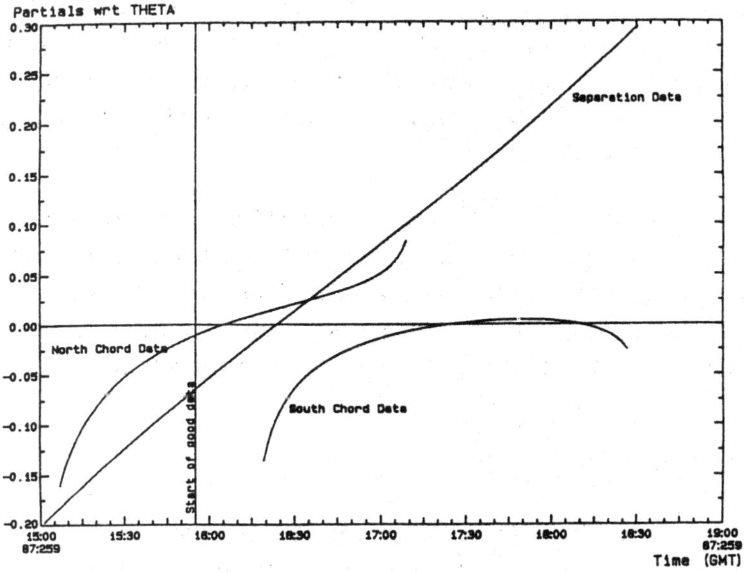

Figure 4. Partial derivatives of three of the data types with respect to θ over the data collection interval (Aussat K3 mission).

the mission and an AMF declination error of less than 0.05° (with respect to the actual AMF attitude determined from orbit matching) was obtained in both cases. This accuracy was obtained by combining attitude data over more than one revolution since the variation in declination between consecutive revolutions can be as much as 0.5°. In the case of the Aussat K3 mission, only attitude data from the second revolution was available to Telesat for data processing. Since a touch-up attitude maneuver was performed at the start of earth sensor data availability (which causes nutation afterward), almost all of the first half of the north earth sensor coverage had to be deleted. Unfortunately, this was the time when the earth chord provided the most valuable information toward determination of θ (which roughly corresponds to declination). The least-squares estimation process is most effective when the partial derivatives of the sensor data with respect to the attitude parameters is non-zero (Fig. 4). Using only 71% of the earth sensor coverage time, we determined an AMF attitude declination which was less than 0.24° from the actual attitude. It should be noted that during the Aussat K3 data coverage, the earth-satellite-sun line was within 3.2° of colinearity.

CONCLUSIONS

While the single-point errors near equinox may exceed the accuracy requirements of typical launch window criteria, the tendency toward excluding these regions from the launch window does not appear to be justified. Launch window cutouts due to attitude determination constraints are unnecessary in our opinion, given typical separation-angle data uncertainties and use of the weighted-least-squares method to process the attitude sensor data.

REFERENCES

1. Van der Ha, J. C., "Attitude Determination Covariance Analysis for Geostationary Transfer Orbits", *Journal of Guidance, Control and Dynamics,* Vol. 9, March-April 1986, pp.156-163.
2. Wertz, J. R., Ed., "Spacecraft Attitude Determination and Control", D. Reidel Publishing, Dordrecht, Holland, 1978.
3. Onstead, E. W. and Tyler, J. S., Jr., "Launch Constraints for a Synchronous Communications Satellite Mission", AIAA 2nd Communications Satellite Systems Conference, San Francisco, CA, 1968, AIAA Paper No. 68-445.
4. Parvez, S. and Fox, S., "Orbit Achievement of RCA Communication Satellites", CNES Conference on Space Dynamics for Geostationary Satellites, Toulouse, October 1985.

AAS 89-396

ERROR ANALYSIS FOR PULSED MANEUVERS OF A DUAL-SPIN SPACECRAFT

James M. Longuski*, Robert S. Campbell† and Eric W. Klumpe†

The Galileo is a dual-spin interplanetary spacecraft which is scheduled for launch towards Jupiter in October of 1989. The spacecraft has twelve thrusters, located on the rotor, which can be operated in continuous or pulsed modes. In this paper probabilistic error models are presented for the pulsed maneuvers which are used to change the spacecraft's velocity. These models take into account all error sources which might affect the maneuver accuracies, such as wobble, nutation, plume impingement, thrust level variation and misalignment, gyro drift, burn timing errors and mass property uncertainties. The analytic models are sufficiently general so that they may be applied to all spin and dual-spin stabilized spacecraft. When these models are applied to the case of the Galileo spacecraft, the numerical results demonstrate that the spacecraft can achieve the maneuver accuracies required in order to successfully reach Jupiter and perform its scientific mission.

Introduction

In October or November of 1989, the Galileo spacecraft will be lifted into low Earth orbit by the Shuttle and then injected into an interplanetary trajectory on its way toward Jupiter. Since the IUS (Inertial Upper Stage) system is insufficient for a direct trajectory to Jupiter, a circuitous route involving gravity assist from Venus and the Earth has been found by mission designers [1] in order to accomplish the goal (Fig. 1). While the VEEGA (Venus-Earth-Earth Gravity Assist) trajectory solves the problem of getting to Jupiter, it creates certain problems that affect the spacecraft design and operation. The closer distance to the sun required the addition of thermal protection (such as blankets and shades) in order to solve the heating problems. For the same reason, the spacecraft must maintain its antenna pointed toward the sun, instead of the Earth, whenever the vehicle is within one A.U. (Astronomical Unit) of the sun.

Another significant (but unrelated) change affecting the spacecraft's operation is the recent discovery that the thrusters (Fig. 2) may suffer thermal runaway if operated in the continuous mode. The current plan is to operate all twelve of the 10 Newton thrusters in a pulsed mode in which the thrusters are restricted to burn durations of less than a few seconds per half revolution.

* Assistant Professor, School of Aeronautics and Astronautics, Purdue University, West Lafayette, Indiana 47907.

† Members of Technical Staff, Jet Propulsion Laboratory, California Institute of Technology, 4800 Oak Grove Dr., Pasadena, California 91109.

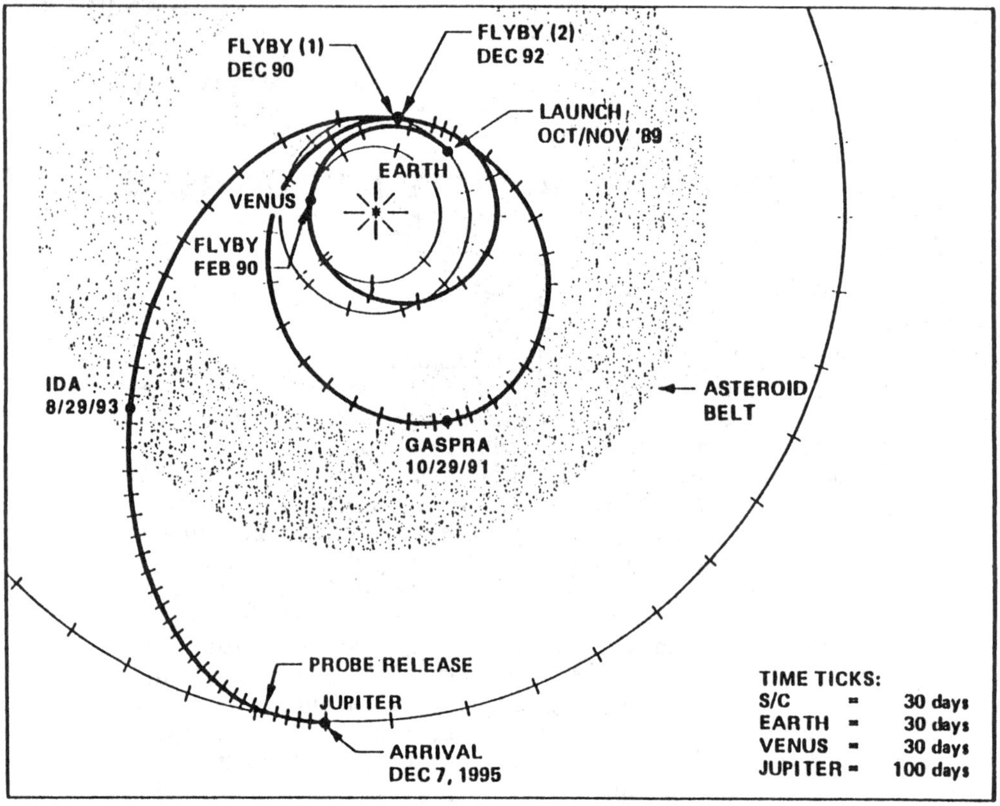

FIG. 1 Galileo VEEGA Trajectory.

Previous analyses included spacecraft turns [2,3], which are performed in pulsed mode, continuous delta-V maneuvers [2] and spin-change maneuvers [4,5]. Useful background material can be found in [6]. Related analyses of pointing performance and attitude control are provided in [7-9]. In this paper, an error analysis is presented for delta-V pulsed maneuvers of a dual-spin spacecraft. Probabilistic error models are developed for the velocity magnitude errors and the velocity pointing errors. The resulting models are applied to the lateral and axial maneuvers of the Galileo spacecraft. These models may be applied to all dual-spin and single-spin vehicles of similar design.

Pulsed Delta-V Maneuvers of a Dual-Spin Spacecraft

The Galileo Spacecraft

The twelve 10N thrusters of the Galileo spacecraft are located on the rotor, which always spins. The Retro Propulsion Module (RPM) is illustrated in Fig. 3. Note that the thrusters are named according to function (e.g. -SIA and -S1B for spin-down, P1A and P2A for pitch, L1B and L2B for lateral and -Z for velocity change in the -Z direction).

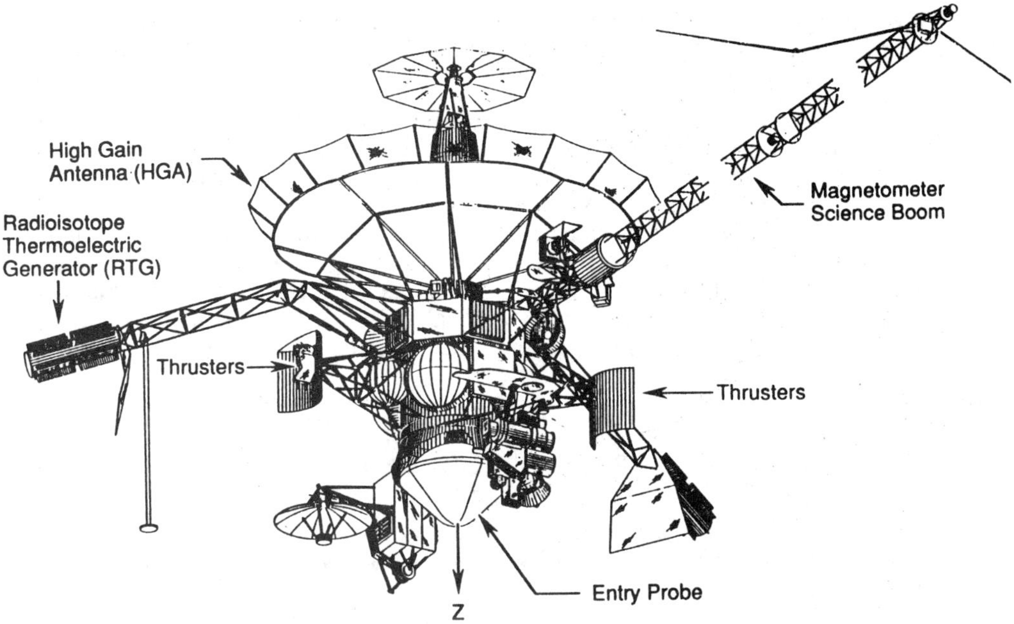

FIG. 2 The Galileo Spacecraft.

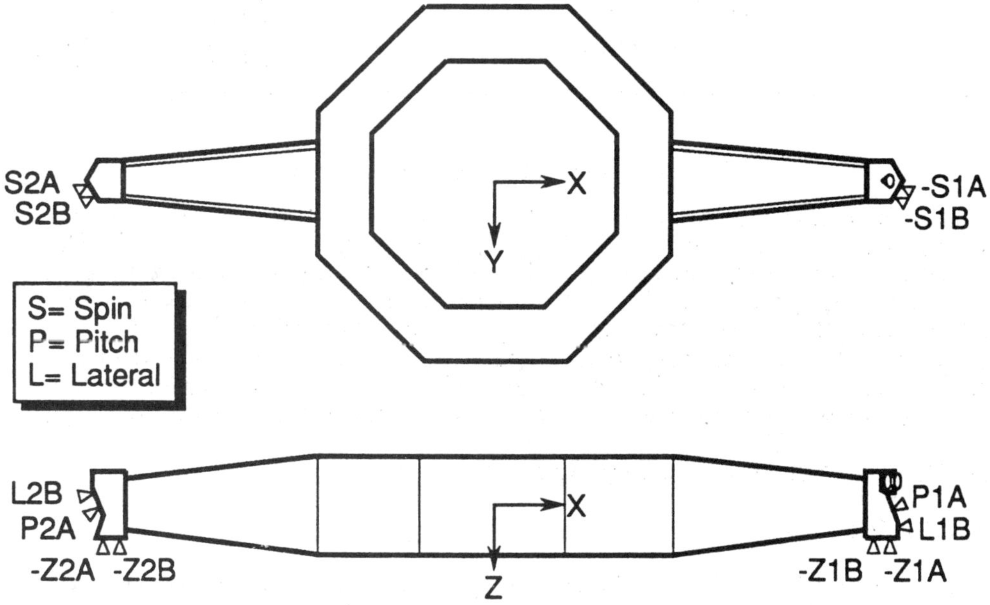

FIG. 3 The Retro Propulsion Module (RPM).

Figure 4 presents numerical information about the position and orientation of each of the 10N thrusters plus the 400N engine, which is only used at Jupiter for large delta-V maneuvers.

During a lateral maneuver, the L1B thruster is pulsed for up to about two seconds and after the rotor has rotated through a half rev, the L2B thruster is pulsed. Figure 5 illustrates the alternate pulsing with solid and dashed arrows. Note that the thrusters are canted by ± 10° from the spacecraft X axis. This allows moment compensation for the various positions that the spacecraft center of mass will take along the Z axis throughout the mission. The duration of the pulses are varied according to the estimated position of the center of mass. Since the center of mass is almost never located on the X axis in Fig. 5, the duration of these pulses are usually unequal. The resulting velocity component along the Z axis is, of course, taken into account during maneuver design.

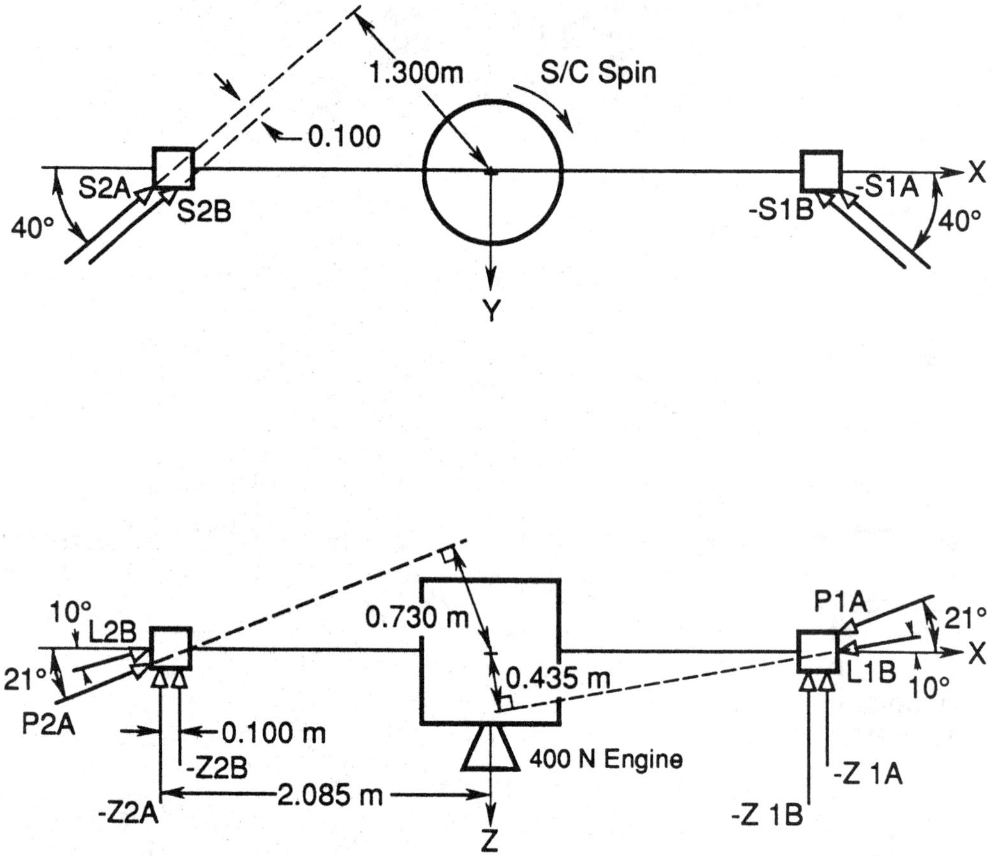

FIG. 4 Thruster and Engine Configuration.

Maneuvers along the -Z axis are performed by pulsing a -Z1 and a -Z2 thruster twice per rev (Fig. 5). Since A and B represent independent branches of the Retro Propulsion Module, either pair (-Z1A, -Z2A) or (-Z1B, -Z2B) may be used, but nominally the A pair will be used, with the B pair reserved for backup.

For maneuvers along the +Z direction, the P1A thruster is nominally used by pulsing the thruster every half rev. Because of the large angle between the Z axis and the thruster (69°), this is a very inefficient maneuver. The large angle is necessary to reduce plume impingement on the spacecraft (see Fig. 2). The backup thruster for +Z maneuvers is the L1B, which is even more severely canted from the Z axis. Note that the P thrusters also provide a backup system for the lateral maneuvers. The spacecraft is designed to perform all delta-V and turn maneuvers on either of the A and B branches. The cant angles have the dual-purpose of reducing plume impingement on the spacecraft (mainly to protect the instruments) and to allow most of the thrusters to provide backup capacity.

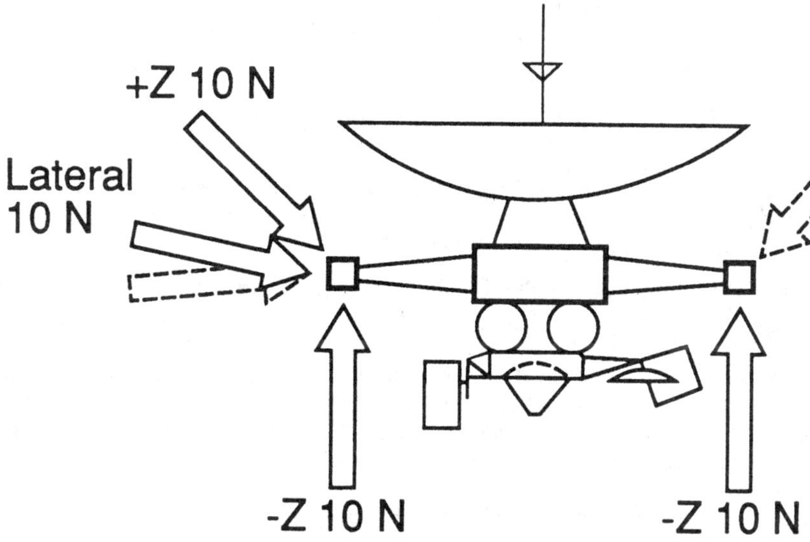

FIG. 5 Pulsed Maneuvers of the Galileo Spacecraft.

Maneuver Requirements

Maneuver requirements for delta-V accuracy are specified in terms of magnitude error and pointing error. In most cases, the requirements for the Galileo spacecraft are given in terms of proportional velocity magnitude error (in percent) and proportional velocity pointing error (in mrad) as illustrated in Fig. 6 and Fig. 7.

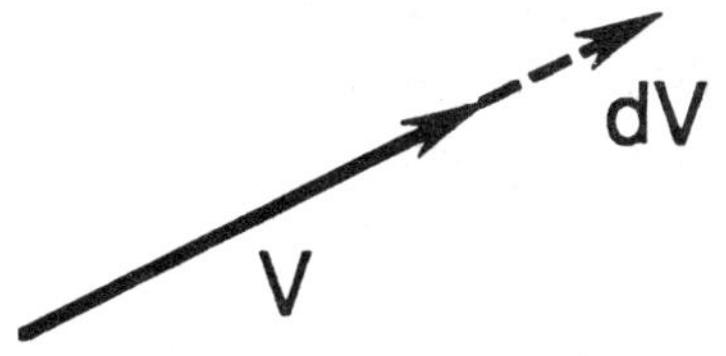

$$\varepsilon_V = \frac{dV}{V} \times 100 \quad \text{(percent)}$$

FIG. 6 Proportional Velocity Magnitude Error, ε_V.

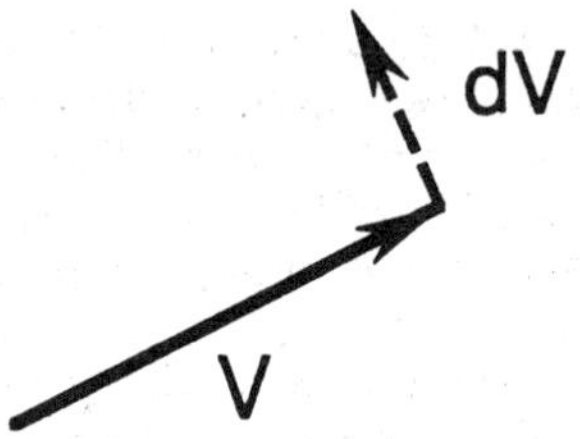

$$\varepsilon_p = \frac{dV}{V} \times 1000 \text{ (mrad)}$$

FIG. 7 Proportional Velocity Pointing Error, ε_p.

Clearly, these are errors which tend to remain proportional to the velocity. Another specification for maneuver accuracy is given in terms of fixed velocity errors in m/s (fixed velocity magnitude error and fixed velocity pointing error) but in the case of the Galileo spacecraft, this specification only applies to turns [2].

The specific requirements for the Galileo spacecraft are given in the sections on numerical results. Requirements are given in terms of "3σ accuracies." That is, the probability that the velocity accuracy of a specific maneuver remains within the 3σ value of its mean should be better than 99%. Two commonly used statistical distributions possess such a property, Gaussian and uniform. In the analysis which follows, almost all the error sources may be assumed to have Gaussian statistics based on the Central Limit Theorem. The 3σ value for these errors, then, will have an associated 99.73% probability.

Analysis for Proportional Velocity Magnitude Errors

All three pulsed maneuvers under consideration (lateral, +Z and -Z) are open loop maneuvers under timed control. For the simple case of a single timed pulse, the ΔV is governed by

$$\Delta V = (F/m)\Delta t \tag{1}$$

where

$$\begin{aligned} \Delta V &= \text{change of velocity (m/s)} \\ F &= \text{force (N)} \\ \Delta t &= \text{duration of burn (s)} \ . \end{aligned}$$

The variance (in percent^2) of equation (1) is easily shown to be (assuming the errors in F,m and Δt are small, have zero mean, and are uncorrelated)

$$\sigma_{\Delta V}^2 = \sigma_F^2 + \sigma_{\Delta t}^2 + \sigma_m^2 \tag{2}$$

where σ_F = standard deviation of thrust, including errors due to thrust level uncertainty, plume impingement and flexible body effects (percent),

$\sigma_{\Delta t}$ = standard deviation of burn time, including errors due to ignition and shutdown delays and quantization (percent), and

σ_m = standard deviation of mass, due primarily to uncertainty of propellant mass (throughout the mission) and injected mass (percent).

For a series of n pulses, the total velocity change is

$$\Delta V_T = \sum_{i=1}^{n} \Delta V_i = \sum_{i=1}^{n} (F_i \Delta t_i)/m \quad (3)$$

Assuming all errors have zero mean, the variance of equation (3) is the same as equation (2) if all the errors in F_i are all correlated and have the same variance, σ_F^2, and if all the errors in Δt_i are all correlated and have the same variance, $\sigma_{\Delta t}^2$. On the other hand, if all the errors in F_i (or Δt_i) are uncorrelated, but have the same variance, σ_F^2 (or $\sigma_{\Delta t}^2$), then the term σ_F^2 (or $\sigma_{\Delta t}^2$) is replaced by σ_F^2/n (or $\sigma_{\Delta t}^2/n$). An example of a correlated error is thruster bias, whereas an example of an uncorrelated error is thruster pulse-to-pulse error. The former error type is sometimes referred to as bias, while the latter is called noise.

Numerical Results for Proportional Velocity Magnitude Errors

Tables 1-3 present numerical results for the Lateral, +Z and -Z proportional velocity magnitude errors. The mass of the spacecraft is assumed to be 2556 kg which corresponds to the orbiter and probe with 100% propellant loading. Similar tables are used to evaluate the velocity magnitude errors of the orbiter, after the probe has been released into the Jovian atmosphere and with only 11% propellant loading remaining on the orbiter, which corresponds to a mass of 1387 kg. The requirements, capabilities and error sources in Tables 1-3 are given in percent and presented as 3σ values (3 standard deviations).

TABLE 1. Proportional Velocity Magnitude Error for Lateral Maneuver (%)

	Error Source	Short Burn	One Segment
1.	Mass Uncertainty	0.19	0.19
2.	Thrust Level Uncertainty	3.59	3.59
3.	Thrust Pulse-to-Pulse Uncertainty	0.71	0.13
4.	Timing Error	0.21	0.21
5.	Plume Impingement Uncertainty	0.25	0.25
6.	Thrust Vector Misalignment	0.12	0.12
7.	Flexible Body Effect	0.03	0.03
8.	Quantization	0.29	0.29
	Total (3σ)	3.68	3.64
	Requirement (3σ)	4.6	4.6

Note: Short Burn = One Rev (About 0.02 m/s)
One Segment = 0.488 m/s (31 Revs)

TABLE 2. Proportional Velocity Magnitude Error for +Z Maneuver (%)

	Error Source	Short Burn (n=14)	Long Burn (n=700)
1.	Mass Uncertainty	0.19	0.19
2.	Thrust Level Uncertainty	4.80	4.80
3.	Thrust Pulse-to-Pulse Uncertainty	0.27	0.04
4.	Timing Error	0.62	0.62
5.	Plume Impingement Uncertainty	1.95	1.95
6.	Thrust Vector Misalignment	2.41	2.41
7.	Flexible Body Effect	0.39	0.39
8.	Quantization	0.84	0.84
	Total (3σ)	5.83	5.82
	Requirement (3σ)	6.	6.

Note: Short Burn = About 0.02 m/s
Long Burn = 0.9765 m/s (10 segments, n=700)

TABLE 3. Proportional Velocity Magnitude Error for -Z Maneuver (%)

	Error Source	Short Burn (n=4)	Long Burn (n=1960)
1.	Mass Uncertainty	0.19	0.19
2.	Thrust Level Uncertainty	3.59	3.59
3.	Thrust Pulse-to-Pulse Uncertainty	0.50	0.02
4.	Timing Error	0.53	0.53
5.	Plume Impingement Uncertainty	0.26	0.26
6.	Thrust Vector Misalignment	0.00	0.00
7.	Flexible Body Effect	0.00	0.00
8.	Quantization	1.03	1.03
	Total (3σ)	3.82	3.79
	Requirement (3σ)	1.	1.

Note: Short Burn = About 0.02 m/s
Long Burn = 6.380 m/s (10 segments, n=1960)

In these tables, note that *short burns,* corresponding to a ΔV of about 0.02 m/s, and *long burns,* corresponding to the longest maneuver possible in one day, are presented. Long burns are composed of *segments* which are limited by the most stringent of the following conditions:

1.) That attitude excursions not exceed the ability of the HGA (High Gain Antenna) correction maneuver. This maneuver is a three burn scheme (using the P thruster couple) designed to yield zero nutation at the end of the third burn [9]. The maximum attitude excursion allowed by the HGA maneuver is less than 20 mrad.

2.) That spin rate excursions not exceed the two-pulse spin rate correction capability (using the S thrusters) which corresponds to approximately 0.005 rad/sec.

3.) That the RPM (Retro Propulsion Module) duty cycle capacity not be exceeded. The duty cycle capacity is the number of pulses that can safely be performed without risking thermal runaway. It can only be determined through testing and is affected by specific thruster characteristics and conditions of operation such as thruster "on" time and "off" time.

By these conditions, the lateral and -Z maneuvers are limited by duty cycle (3) and the +Z maneuver is limited by spin rate (2). After completion of a segment, spin rate and attitude corrections may be performed. In the case of the lateral maneuver (Table 1), after attitude correction, the errors for one segment and many segments are identical. Based on spacecraft memory limitations, up to 10 segments can be accomplished in one day.

Mass Uncertainty

The 3σ mass uncertainty is 0.19% for all three maneuvers. The mass uncertainty generally increases as the mission progresses, because the uncertainty in propellant loading increases as more propellant is used. Initially the mass uncertainty is based on the accuracy of weighing the fully loaded spacecraft and is very small. Near the end of the mission the uncertainty grows to 1.84% (3σ).

Thrust Level Uncertainty

The 3σ thrust level performance uncertainty of a single thruster for known temperature is 4.5%. The sensitivity to a 1° (3σ) temperature excursion is 1.7%. The combined effects of performance and temperature uncertainties is RSS (4.5%, 1.7%) = 4.8%. For two thrusters the effect of temperature is correlated, but the performance of each thruster is uncorrelated so that the combined uncertainties of two thrusters is RSS ($4.5\%/\sqrt{2}$, 1.7%) = 3.6%.

Thrust Pulse-to-Pulse Uncertainty

The thrust pulse-to-pulse errors are uncorrelated and have a 3σ value of 1%. For a single thruster the resulting error is $1\%/\sqrt{n}$ where n is the number of pulses. For a large number of pulses this error becomes small.

Timing Errors

Thruster tailoff duration uncertainty is 5 ms (3σ). Startup uncertainty is comparable but uncorrelated so that the total pulse duration uncertainty is $5\sqrt{2}$ ms = 7.07 ms. For a single thruster, this error source acts as a bias affecting all pulses alike. For two independent thrusters, the combined pulse duration uncertainty is RSS (7.07 ms, 7.07 ms) = 10 ms. Using the nominal pulse durations for the lateral maneuver ($\approx$ 2s/pulse), the +Z maneuver ($\approx$ 1s/pulse) and the -Z maneuver ($\approx$ 1s/pulse) the percent errors are easily calculated.

Plume Impingement Uncertainty

Predicted plume impingement effects are included in the maneuver design. The uncertainty in plume impingement is estimated to be ± 50% of the prediction. The lateral thrusters are predicted to lose about 0.05N in the X direction due to plume impingement. The predicted P1A thruster (used for the +Z maneuver) plume impingement loss in the Z direction is 0.134N and the predicted -Z thruster plume impingement loss in the Z direction is 0.05N. The plume impingement error acts as a bias.

Thrust Vector Misalignment

The 10N thrust vector misalignment uncertainty is 13 mrad total or 9.2 mrad (0.53°) per axis. For two independent thrusters, the combined misalignment uncertainty is 9.2 mrad/$\sqrt{2}$ = 6.5 mrad (0.37°). Thus, the lateral thrust uncertainty is (from cosine losses)

$$dF_L/F_L = [\cos(10.37°) - \cos(10°)]/\cos(10°) = 0.12\%$$

and the +Z thrust uncertainty in the Z direction is

$$dF_Z/F_Z = [\cos(69.53°) - \cos(69°)]/\cos(69°) = 2.41\ \%$$

The thrust vector misalignment has negligible impact on the -Z velocity magnitude because

$$dF_{-Z}/F_{-Z} = [\cos(0.53°) - \cos(0°)]/\cos 0° = 0.00\ \%$$

Nutation

The effect of nutation is taken into account by simulation during the maneuver design, so it provides no error source for velocity magnitude. Nutation is discussed in more detail in the next section.

Flexible Body Effects

Flexible body effects can usually be ignored when the lowest natural frequencies of the flexible components are at least an order of magnitude greater than the rigid body frequencies (such as nutation, control system or thruster pulse frequencies) [10, 6]. In the case of the Galileo spacecraft, the only flexible component that cannot be neglected is the magnetometer science boom (Fig. 2). The boom is purposely flexible at its hinged connection to the spacecraft rotor. It is attached to a spring and damper so that it provides passive nutation damping. The scientific purpose of the boom is to provide a constantly spinning platform for instruments which sense fields and particles. The magnetometer science boom has a natural frequency of about 0.1 - 0.2 hz and a time constant of approximately 5-10 min.

Analysis of the flexible body effect of such a boom on a dual-spin spacecraft has focused on the wobble angle, which has been defined as the angle between the spin axis (maximum axis of inertia) and the spin-bearing axis [7-9, 11]. When wobble is present, the spin-bearing axis assumes a coning motion about the spin axis at the spin frequency, effectively causing pointing errors in instruments [7] and misalignment of thrusters [2]. Wobble has three major sources:

1) POI wobble from Products of Inertia

2) BFT wobble from Body-Fixed Torques and

3) TAL wobble from Translational Acceleration Loading.

POI wobble results from a rotor imperfectly balanced about the bearing axis. Its correction is discussed in [12]. Wobble compensation is planned for the Galileo spacecraft [7] to reduce POI wobble by adjusting the positions of the RTG (Radioisotope Thermoelectric Generator) booms (Fig. 2). However, during thruster pulses, the body-fixed torques and acceleration will generate BFT and TAL wobble by causing the magnetometer boom to sag. These types of wobble are discussed in more detail in [13].

Since the wobble angle is ignored in the maneuver design, its effect can be considered to be an error source. For the purpose of velocity magnitude error analysis, the wobble angle will be considered constant, in which case it will act as a thruster misalignment. After wobble compensation, the total wobble angle for pulses of the lateral and +Z thrusters is approximately 1.5 mrad and 3 mrad for the -Z thrusters. Specific details for these calculations are provided in [14].

Quantization of Burn Duration

The Galileo thrusters are operated on a computer frame time of 11.11 ms. This means that each burn duration will last a quantized number of frame times. In addition, only two pulse durations can be chosen: one value for odd pulses and a second value for even pulses. In the maneuver design process, the maximum allowed pulse durations are computed, along with the corresponding number of rotor revs required to achieve the target velocity. The number of pulsing revs is then rounded up to the nearest integer, while the pulse durations are scaled down and then quantized to the nearest 11.11 ms. The pulse quantization has a uniform distribution between 0 and 11.11 ms with 5.55 ms mean. The 3σ for such a function is $5.55\sqrt{3}$ ms = 9.6 ms. This error acts as a bias from pulse to pulse. The percentage error is found by scaling the quantization with the nominal burn time as done in the section above on timing errors.

Analysis for Proportional Velocity Pointing Errors

In the following subsections, the major error sources affecting the velocity pointing of pulsed delta-V maneuvers will be briefly described and analyzed. Numerical results will be discussed in the next section.

Initial Attitude Uncertainty

The initial attitude uncertainty will result in a proportional velocity pointing error of equal magnitude. This error source is independent of the errors occuring during the actual maneuver and depends on the overall accuracy of sensors (star scanner), computer algorithms for attitude determination and attitude control.

Error Analysis of the Lateral Maneuver Deterministic Pointing Angle

During a lateral maneuver the L1B and L2B thrusters (see Figs. 4 and 5) cause precession, but in opposite directions. Since L1B and L2B are not symmetric with respect to the spacecraft center of mass (because it moves along the Z axis as a function of propellant loading), the pulse durations are trimmed so that their precession angles are equal and opposite. This is called moment compensation. The L1B and L2B pulse durations are chosen so that their ratio is inversely proportional to their effective moment arms, with respect to the spacecraft center of mass. The geometry is shown in Fig. 8 where *spacecraft* refers to the Galileo orbiter with probe attached and *orbiter* refers to the orbiter alone. Z_{cm} indicates the center-of-mass location along the Z axis.

Since the L1B and L2B pulses are of unequal duration, the resulting velocity has an axial component characterized by the deterministic pointing angle, θ_{DPA}, shown in Fig. 9. In this figure, the u axis corresponds to either +X or -X axes of the spacecraft.

The deterministic pointing angle is not a maneuver error, as it is accounted for in the maneuver design. By adding up the lateral and axial components of impulse, the deterministic pointing angle is found from

$$\tan\theta_{DPA} = V_Z/V_X = [(F_{L1}t_1m^{-1}\sin\alpha - F_{L2}t_2m^{-1}\sin\alpha)/(F_{L1}t_1m^{-1}\cos\alpha + F_{L2}t_2m^{-1}\cos\alpha)] \tag{4}$$

where t_1 = burn duration of L1B thruster
F_{L1} = thrust of L1B
α = cant angle of L thrusters (10°)
m = spacecraft mass

For $F_{L1} = F_{L2}$, equation (4) simplifies to

$$\tan\theta_{DPA} = [(t_1 - t_2)/(t_1 + t_2)]\tan\alpha \tag{5}$$

Clearly, there are two mechanisms that can lead to errors:

1.) errors which affect the size of θ_{DPA}, namely errors in F, t and α and

2.) errors which lead to precession (θ_{PREC}) which arise from errors in moment compensation due to uncertainties in F, t, α and r (moment arm).

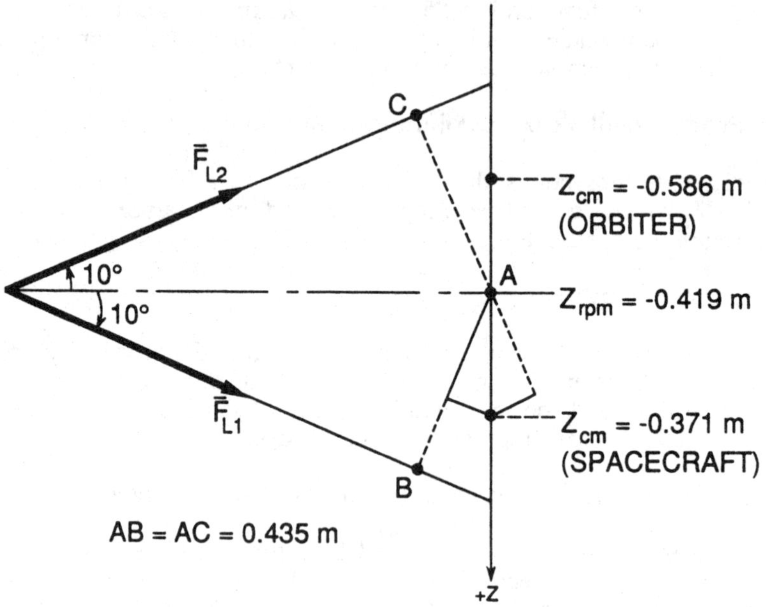

FIG. 8 Geometry of Moment Compensation for Lateral Maneuver.

The first error is analyzed by finding the variance of equation (4) in the same manner as equation (2) was derived from equation (1).

The analysis of the second error requires an assessment of the precession angle, θ_{PREC}. The precession angle is the total angle accrued from several pulses, each of which are governed by Euler's law

$$\mathbf{M} = \dot{\mathbf{H}} \quad (6)$$

where

$\mathbf{M}$ = moment vector

$\mathbf{H}$ = angular momentum vector .

For a pulse of duration Δt, the change of angular momentum is ΔH and the angular change, $\Delta\psi$, is given by

$$\Delta\psi = (\Delta H/H) = M\Delta t/I_z\omega_z \quad (7)$$

where

I_z = rotor moment of inertia($\approx$ 5000 kg–m^2)
ω_z = spin rate ($\approx$ 0.33 r/s)

The total precession angle is

$$\theta_{PREC} = \sum_{i=1}^{n} \Delta\psi_i \quad (8)$$

where n is the number of pulses. Moment compensation attempts to keep the precession angle near zero. If the initial thrust vector is pointed along V_{LAT} as shown in Fig. 9, and

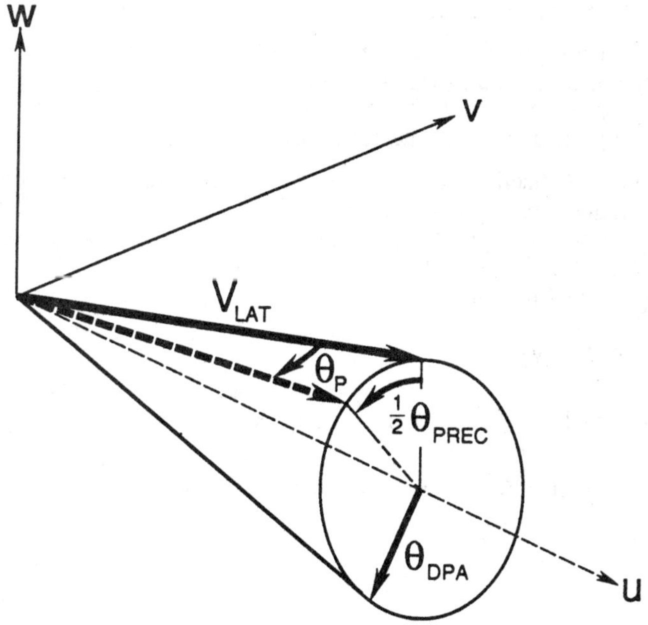

FIG. 9 Precession About the Lateral Maneuver Velocity Axis.

the final thrust vector is rotated through the precession angle, θ_{PREC}, then the average velocity pointing direction, θ_p, occurs at ½ θ_{PREC} and is found by

$$\sin(\tfrac{1}{2}\theta_p) = \sin\theta_{DPA}\,\sin(\tfrac{1}{4}\theta_{PREC}) \tag{9}$$

For θ_p, θ_{DPA} and θ_{PREC} assumed small,

$$\theta_p = \theta_{DPA}\theta_{PREC}/2 \tag{10}$$

gives a convenient approximation. A simple approach to the error analysis is to take the variance of equation (10) with respect to θ_{PREC}. The result is that two types of errors arise for the variance of θ_{PREC}: a bias error

$$\sigma^2_{PREC} = n^2\sigma^2_{\Delta\psi} \tag{11}$$

and noise

$$\sigma^2_{PREC} = n\sigma^2_{\Delta\psi}$$

The correct interpretation of errors resulting from the variance of equation (7) is necessary. Naturally, bias errors are much more important and the most significant error source is usually from thrust level variation.

A generalization of these concepts to any pulsed maneuver is presented in the next subsection.

General Model for Proportional Velocity Pointing Errors

In this subsection a general mathematical model for the three different pulsed ΔV maneuvers (lateral, +Z and -Z) is discussed. These maneuvers are similar in that each is performed in inertial mode with thrusters pulsed at every 180 degrees of rotation of the rotor. It is this similarity that allows one set of general equations to be developed that can describe any one of the above maneuvers.

For the sake of simplicity it will be assumed that a thruster pulse occurs when the positive X-axis of the rotor is aligned with the positive X-axis or the negative X-axis of the stator. It will also be assumed that the ΔV imparted to the spacecraft as a result of one pulse will occur at a point located at the center of the burn arc.

The thrusters to be used for these maneuvers have thrust components strictly in the XZ-plane which create torque about the Y-axis causing the Z-axis of the spacecraft to precess about the X-axis. The precession angle associated with a single thruster pulse will be referred to as $\Delta\psi$. For "n" thruster pulses, where "n" is an even integer, the total ΔV imparted to the spacecraft in inertial space is:

$$\Delta V_{XT} = \sum_{i=1}^{n/2}\Delta V_{X1} + \sum_{i=1}^{n/2}\Delta V_{X2} \tag{13}$$

$$\begin{aligned}\Delta V_{YT} = {} & \sum_{i=1}^{n/2}\Delta V_{Y1}\cos[(i-\tfrac{1}{2})\Delta\psi_1 + (i-1)\Delta\psi_2] + \sum_{i=1}^{n/2}\Delta V_{Y2}\cos[(i)\Delta\psi_1 + (i-\tfrac{1}{2})\Delta\psi_2] \\ & - \sum_{i=1}^{n/2}\Delta V_{Z1}\sin[(i-\tfrac{1}{2})\Delta\psi_1 + (i-1)\Delta\psi_2] - \sum_{i=1}^{n/2}\Delta V_{Z2}\sin[(i)\Delta\psi_1 + (i-\tfrac{1}{2})\Delta\psi_2]\end{aligned} \tag{14}$$

$$\Delta V_{ZT} = \sum_{i=1}^{n/2} \Delta V_{Y1} \sin[(i - \tfrac{1}{2})\Delta\psi_1 + (i-1)\Delta\psi_2] + \sum_{i=1}^{n/2} \Delta V_{Y2} \sin[(i)\Delta\psi_1 + (i - \tfrac{1}{2})\Delta\psi_2]$$
$$+ \sum_{i=1}^{n/2} \Delta V_{Z1} \cos[(i - \tfrac{1}{2})\Delta\psi_1 + (i-1)\Delta\psi_2] + \sum_{i=1}^{n/2} \Delta V_{Z2} \cos[(i)\Delta\psi_1 + (i - \tfrac{1}{2})\Delta\psi_2] \quad (15)$$

where:

$\Delta V_{X1}, \Delta V_{Y1}, \Delta V_{Z1}$	=	the components of the ΔV imparted to the spacecraft on the odd thruster pulses.
$\Delta V_{X2}, \Delta V_{Y2}, \Delta V_{Z2}$	=	the components of the ΔV imparted to the spacecraft on the even thruster pulses.
$\Delta\psi_1, \Delta\psi_2$	=	the precession angles associated with the odd and even thruster pulses, respectively.

The series in equations (13) - (15) can be summed to provide

$$\Delta V_{XT} = n(\Delta V_{X1} + \Delta V_{X2})/2 \quad (16)$$
$$\Delta V_{YT} = [\Delta V_{Y1} \cos \tfrac{1}{2}(n\Delta\bar{\psi} - \Delta\psi_2) + \Delta V_{Y2} \cos \tfrac{1}{2}(n\Delta\bar{\psi} + \Delta\psi_1) - \Delta V_{Z1} \sin \tfrac{1}{2}(n\Delta\bar{\psi} - \Delta\psi_2) - \Delta V_{Z2} \sin \tfrac{1}{2}(n\Delta\bar{\psi} + \Delta\psi_1)] \cdot \sin \tfrac{1}{2}(n\Delta\bar{\psi})/\sin(\Delta\bar{\psi}) \quad (17)$$

$$\Delta V_{ZT} = [\Delta V_{Y1} \sin \tfrac{1}{2}(n\Delta\bar{\psi} - \Delta\psi_2) + \Delta V_{Y2} \sin \tfrac{1}{2}(n\Delta\bar{\psi} + \Delta\psi_1) + \Delta V_{Z1} \cos \tfrac{1}{2}(n\Delta\bar{\psi} - \Delta\psi_2) + \Delta V_{Z2} \cos \tfrac{1}{2}(n\Delta\bar{\psi} + \Delta\psi_1)] \cdot \sin \tfrac{1}{2}(n\Delta\bar{\psi})/\sin(\Delta\bar{\psi}) \quad (18)$$

where

$$\Delta\bar{\psi} = (\Delta\psi_1 + \Delta\psi_2)/2$$

During moment compensation, the thruster "on times" are adjusted such that the net precession angle is zero. Thus

$$\Delta\psi_1 = -\Delta\psi_2 = \Delta\psi; \quad \Delta\bar{\psi} = 0. \quad (19)$$

Equations (19) are substituted into equations (16) - (18) to form simpler expressions as follows:

$$\Delta V_{XT} = n(\Delta V_{X1} + \Delta V_{X2})/2 \quad (20)$$
$$\Delta V_{YT} = n[(\Delta V_{Y1} + \Delta V_{Y2}) \cos \tfrac{1}{2}\Delta\psi - (\Delta V_{Z1} + \Delta V_{Z2}) \sin \tfrac{1}{2}\Delta\psi]/2 \quad (21)$$
$$\Delta V_{ZT} = n[(\Delta V_{Y1} + \Delta V_{Y2}) \sin \tfrac{1}{2}\Delta\psi + (\Delta V_{Z1} + \Delta V_{Z2}) \cos \tfrac{1}{2}\Delta\psi]/2 \quad (22)$$

Since the thrusters to be used for these maneuvers have no thrust component in the Y direction, equations (20)-(22) can be further simplified to:

$$\Delta V_{XT} = n(\Delta V_{X1} + \Delta V_{X2})/2 \quad (23)$$
$$\Delta V_{YT} = (-n/2)(\Delta V_{Z1} + \Delta V_{Z2}) \sin \tfrac{1}{2}\Delta\psi \quad (24)$$
$$\Delta V_{ZT} = (n/2)(\Delta V_{Z1} + \Delta V_{Z2}) \cos \tfrac{1}{2}\Delta\psi \quad (25)$$

which completes the development of the nominal ΔV equations. Equations (23) - (25) govern the behavior of all three maneuver types. Equation (23) indicates a linear accumulation along the X axis, while equations (24) and (25) indicate circular motion in the YZ plane. Thus, these equations explain the behavior of the lateral maneuver depicted in Fig. 9. In the case of an axial maneuver, the contribution of velocity along the X axis is expected to cancel, so that equation (23) is equal to zero, but the velocity circle, provided by equations (24) and (25), still remains. Note that equations (23) - (25) provide the deterministic pointing angle, θ_{DPA}, for the lateral maneuver in which the desired velocity is along the X direction:

$$\tan\theta_{DPA} = (\Delta V_{YT}^2 + \Delta V_{ZT}^2)^{1/2}/\Delta V_{XT} = (\Delta V_{Z1} + \Delta V_{Z2})/(\Delta V_{X1} + \Delta V_{X2}) \qquad (26)$$

Since:

$$\Delta V_{ji} = K_j F_j \Delta t_i / m \qquad (27)$$

where

i	=	1, 2
j	=	X, Y, Z
K_j	=	efficiency of the burn arc along the j-axis = $(\sin\gamma)/\gamma$ where γ is half of the total burn arc [15]
Δt_i	=	duration of the ith thruster pulse
m	=	mass of the spacecraft
F_j	=	thrust applied in the jth direction.

When $K_j = 1$ and $F_j = F_1 = F_2$, equation (5) is obtained.

The desired ΔV associated with +Z and -Z pulsed maneuvers is along the Z-axis, therefore the pointing error, θ_{PE}, is:

$$\tan\theta_{PE} = (\Delta V_{XT}^2 + \Delta V_{YT}^2)^{1/2}/\Delta V_{ZT} \qquad (28)$$

However, in these cases the thrusters fire every 180 degrees of rotation of the rotor, resulting in the following expression

$$\Delta V_{X1} = -\Delta V_{X2} \;, \qquad (29)$$

which causes equation (28) to reduce to:

$$\tan\theta_{PE} = \tan \tfrac{1}{2}\Delta\psi \approx \tfrac{1}{2}\Delta\psi \qquad (30)$$

since $\Delta\psi$ is usually on the order of milliradians. The result is that the axial maneuvers do not have a velocity component along the u axis of Fig. 9, but precession still occurs and creates a velocity circle as shown in Fig. 9, parallel to the wv plane. Similar velocity circles arise in the analysis of spacecraft turns [2]. Another component of the pointing error arises from the pulse-to-pulse impulse uncertainty of a single thruster, in which equation (29) is not precisely satisfied.

With the basic behavior of the pulsed maneuver velocities contained in equations (23) - (25), the perturbation of these and related equations provide the error models.

In the case of *lateral maneuver,* which has already been discussed in the previous subsection, the velocity error model consists mainly of the uncertainty in the pointing angle, θ_{DPA}, and in the precession angle, θ_{PREC}.

In the case of the *axial maneuvers,* the pointing error, θ_{PE}, is nominally zero. However, the precession angle, θ_{PREC}, still is an important error source, since it governs the velocity circles which are parallel to the YZ plane. In addition, any thruster misalignment (or other error source) resulting in a component of thrust along the X axis will be an error source.

In all cases the precession angle, θ_{PREC}, must be analyzed as an error source and appropriately categorized as a bias error (equation 11) or noise (equation 12). These interpretations are discussed below, in the section on numerical results.

Burn Timing Delay

As mentioned above, burn timing can affect both the proportional velocity magnitude error (equation (2)) and proportional velocity pointing error related to the deterministic pointing angle and precession (equations (4) and (8)). Figure 10 illustrates the effect of *burn timing delay* as it affects the lateral maneuver.

It is clear from the figure that if there is a delay in the actual ignition and shutdown times from the desired values then the actual velocity vector, $\mathbf{V_A}$, will have a proportional velocity pointing error relative to the desired velocity, $\mathbf{V_D}$. The pointing error, δ, due to delays corresponding to the angles δ_I and δ_S is

$$\delta = (\delta_I + \delta_S)/2 \tag{31}$$

The standard deviation is

$$\sigma_\delta = \sigma_t \omega_z / \sqrt{2} \tag{32}$$

where it is assumed that ignition and shutdown delays have the same standard deviation, σ_t, and are independent.

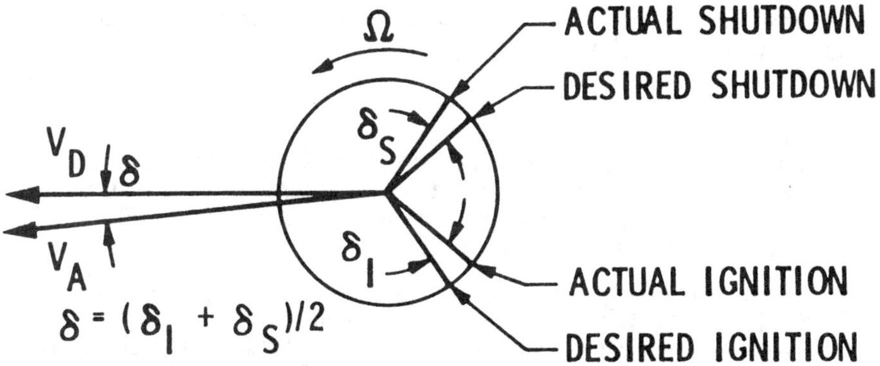

FIG. 10 Burn Timing Delay (Showing Error in Clock Orientation).

Nutation

Maneuver torques induce a precession of the spin axis (the maximum axis of inertia) about the spacecraft angular momentum vector. The angle between these two vectors is called the *nutation angle*. Usually an attempt is made to control nutation by active or passive means. In the case of the Galileo, a passive nutation damper is used, which (as mentioned above) has a time constant of 5-10 min.

The effect of nutation has been ignored in the derivation of the nominal velocity equations (equations (23) - (25)). Nutation could potentially be a serious problem, destabilizing the spacecraft if the thruster pulses occur at the same frequency or integer values of the nutation frequency. Naturally this problem is avoided at the early stages of spacecraft design so that the mass properties of the vehicle are selected to avoid nutation instability.

A simple approach to error analysis of the nutation effect is to consider that the first pulse of a pulse pair causes a precession angle equal to θ_{ppp} (precession per pulse). In the worst case phasing (the destabilizing case which is avoided by design) the second pulse occurs at a nutation angle of $2\theta_{ppp}$. Both pulses have an average pointing error of one half the nutation angle, for a total of θ_{ppp}. Note that θ_{ppp} has the same value as $\Delta\psi$ of equation (7)

$$\theta_{ppp} = \Delta\psi = M\Delta t/I_z\omega_z \tag{33}$$

Attitude Estimate Error

During the maneuvers, the spacecraft attitude may drift, leading to a proportional velocity pointing error. This error source is highly dependent on the spacecraft attitude control system.

In the Galileo spacecraft, the star scanner cannot be used during pulsed burns, so the attitude estimate is obtained from gyros. An error in the Z axis attitude estimate has no effect on pulsed maneuvers, since the spacecraft takes no action during the burn to correct the supposed error. The attitude estimator is used only to control the clock orientation (Fig. 10) of the burn arcs. The gyro-based attitude estimate is subject to gyro drift rate and quaternion integration errors, which cause the clock orientation of the burn arc centers to slowly drift from the desired inertial plane.

This error source has no effect on the +Z or -Z maneuver performance, but directly affects the lateral maneuver pointing accuracy.

Spin Rate Corrections

During pulsed maneuvers, small undesired torques about the spin axis cause the spin rate to change. This is a general effect on all spinning and dual-spin spacecraft which have thrusters that are not located near or at the center of mass (Fig. 4). Thruster misalignment will result in spin-changing torques.

This is one reason why the Galileo spacecraft has spin correcting thrusters (-S1A, -S1B, S2A, S2B). After each burn segment is completed there will be a spin-rate correction opportunity. Since the Galileo spin-rate maintenance algorithm divides the spin-change maneuvers into two burns, 180° apart, the ΔV accrued during the correction is negligible.

Flexible Body Effects

Flexible body effects have already been discussed in connection with proportional velocity magnitude error. The wobble angle has effects similar to thruster misalignment and nutation.

Quantization

Quantization of thruster burn duration results in errors similar to burn timing delay. If the mean time delay, μ_t, is not taken into account, then the velocity pointing error will contain a bias angle, δ_μ

$$\delta_\mu = \mu_t\omega_z \tag{34}$$

The standard deviation of the quantized velocity pointing error, σ_q, is

$$\sigma_q = (\sqrt{3}/3)\mu_t\omega_z/\sqrt{2n} \qquad (35)$$

based on a uniform distribution and n thruster pulses.

Numerical Results for Proportional Velocity Pointing Errors

Tables 4-6 present numerical results for the Lateral, +Z and -Z proportional velocity pointing errors. The mass properties of the fully loaded spacecraft (orbiter plus probe) are used for the calculations. The moments of inertia are 5456 kg-m^2 for the rotor and 5949 kg-m^2 for the entire vehicle. Test values for the thrusters are used, corresponding to averages of 8.914 N, 8.857 N and 9.005 N for the Z, P and L thrusters, respectively. (Thrusters are not expected to be operated at the optimal levels of 10 N, but rather at levels which will improve longevity.) Figs. 4 and 8 provide the moment arms.

TABLE 4. Proportional Velocity Pointing Error for Lateral Maneuver (mrad)

	Error Source	Short Burn (n=2)	One Segment (n=62)
1.	Initial Attitude	0.5	0.5
2.	Deterministic Pointing Angle		
	2a. Impulse Uncertainty	5.6	5.5
	2b. Thrust Misalignment	9.35	9.35
3.	Precession	0.0	0.20
4.	Burn Timing Delay	0.8	0.8
5.	Nutation	0.0	0.0
6.	Attitude Estimate	0.0	.85/.06
7.	Spin Rate Correction	0.0	0.0
8.	Flexible Body Effect	1.5	1.5
9.	Quantization	3.4	2.1
	Total (3σ)	11.6	11.2/11.2
	Requirement (3σ)	9.	41.

Note: Short Burn = One Rev (About 0.02 m/s)
One Segment = 0.488 m/s (31 revs)

TABLE 5. Proportional Velocity Pointing Error for +Z Maneuver (mrad)

	Error Source	Short Burn (n=14)	Long Burn (n=700)
1.	Initial Attitude	0.5	0.5
2.	Precession	0.07	0.48
3.	Pulse-To-Pulse Impulse Uncertainty	6.96	0.98
4.	Burn Timing Delay	0.0	0.0
5.	Nutation	0.0	0.0
6.	Attitude Estimate	0.0	0.0
7.	Spin Rate Correction	0.0	0.0
8.	Flexible Body Effect	1.5	1.5
9.	Quantization	1.56	0.22
	Total (3σ)	7.31	1.9
	Requirement (3σ)	10.	10.

Note: Short Burn = About 0.02 m/s
Long Burn = 0.9675 m/s (10 Segments, n=700)

TABLE 6. Proportional Velocity Pointing Error for -Z Maneuver (mrad)

	Error Source	Short Burn (n=8)	Long Burn (n=1960)
1.	Initial Attitude	0.5	0.5
2.	Precession	0.1	2.0
3.	Burn Timing Delay	0.0	0.0
4.	Nutation	0.0	0.0
5.	Attitude Estimate	0.0	0.0
6.	Spin Rate Correction	0.0	0.0
7.	Flexible Body Effect	3.0	3.0
8.	Quantization	0.0	0.0
	Total (3σ)	3.0	3.6
	Requirement (3σ)	29.6	10.

Note: Short Burn = About 0.02 m/s
Long Burn = 6.38 m/s (10 Segments, n=1960)

Initial Attitude Uncertainty

The Galileo spacecraft uses a star scanner to estimate the angular momentum vector attitude. The uncertainty in the vector is 0.5 mrad (3σ) which translates directly into a proportional velocity pointing error of the same magnitude.

Deterministic Pointing Angle Uncertainty (Lateral Maneuver)

This error is found by perturbing equation (4) with respect to thruster impulse and thruster misalignment.

Precession

For the *lateral maneuver,* uncertainties in thrust level, thruster alignment, burn timing, center-of-mass (CM) offset and plume impingement contribute to an uncertainty in the precession per pulse pair, which is nominally zero. The following values are found for the 3σ errors in total precession:

Error Source	Precession Uncertainity (3σ) in mrad
Thrust level	0.153n
Burn timing	0.011n
Plume impingement	0.053n
Thruster Misalignment	0.162n
CM offset	0.092n
RSS	0.247n

For the fully loaded Galileo spacecraft the deterministic pointing angle, θ_{DPA}, is 24.8 mrad. From equation (10) the resulting proportional velocity pointing error due to precession is given by

$$3\sigma_p = 0.0031n$$

in mrad (3σ).

Since the *+Z maneuver* is performed with a single thruster, the fixed errors associated with thrust level uncertainty, thruster misalignment, center-of-mass offset and plume impingement all cancel and therefore do not contribute to precession. The chief error source remaining is the 1% (3σ) pulse-to-pulse impulse uncertainty, which will cause uncertainty in the magnitude of the precession angle. Since the error acts as noise, the 3σ value for the precession angle is

$$3\sigma_{PREC} = 0.036\sqrt{n}$$

in mrad. From equation (30), note that the final pointing error is half the value of the precession angle.

For the *-Z maneuver,* the same errors cancel as in the +Z maneuver, leaving only the pulse-to-pulse impulse uncertainty which gives a precession angle of

$$3\sigma_{PREC} = 0.09\sqrt{n}\ .$$

Pulse-to-Pulse Pointing Error

This error source creates an additional pointing error for the special case of the +Z maneuver, in which equation (29) is not precisely satisfied. Since most of the +Z thrust is normal to the intended velocity direction (see thruster P1A in Fig. 4), the pointing error is very sensitive to pulse-to-pulse impulse variation. For a 1% (3σ) pulse-to-pulse impulse uncertainty, the ratio of lateral velocity error to axial velocity is

$$\Delta V_L/V_Z = 0.01 \cos 21°/1.01 \sin 21° = 0.01/\tan 21°$$

The impulse uncertainty behaves as noise so that the velocity pointing error is

$$\sigma_p = 26.05/\sqrt{n} \text{ mrad}.$$

Burn Timing Delay

For the Galileo spacecraft, the thruster tailoff duration uncertainty and the thruster startup uncertainty are each assumed to be 5 ms (3σ).

For the *lateral maneuver,* since two thrusters are used for each rev and since these thruster pulses are uncorrelated, equation (32) is modified to

$$3\sigma_\delta = 3\sigma_t\omega_z/2 = (5\text{ms})(.33 \text{ r/s})/2 = 0.8 \text{ mrad}$$

Burn timing delay has no effect on the *+Z or -Z maneuvers.*

Nutation

Nutation, if ignored, is a possible source of velocity pointing error. This source is eliminated in the Galileo spacecraft by accounting for the effect of nutation in the maneuver design (by simulation).

Attitude Estimate Error

As noted above, the attitude estimate error only affects the lateral maneuver. The lateral maneuver proportional velocity pointing error due to attitude estimate error is given by

$$\sigma_p = \tfrac{1}{2} \text{ RSS } (\omega_d t, \omega_q t, \tfrac{1}{2} a_d t^2)$$

where

ω_d = gyro drift rate uncertainty (3σ) = 9.5 mrad/hr (uncalibrated); = 0.5 mrad/hr (calibrated)
ω_q = quaternion integration error (3σ) = 0.04 mrad/hr
a_d = gyro drift rate instability (3σ) = 0.1 mrad/hr^2
t = time from start of the first segment to the end of the last segment (hours)

The gyro drift rate and instability values come from extensive observation of the six Voyager gyros, which indicate a start-up to start-up gyro drift rate variation of 0.03 deg/hr or less and drift rate stabilities over a six hour period of about the same magnitude.

Spin Rate Corrections

This error source is expected to be negligible for the Galileo spacecraft.

Flexible Body Effects

For the lateral and +Z maneuvers, the wobble angle is expected to be about 1.5 mrad. For the -Z maneuver it is expected to be about 3 mrad. A conservative approach to the flexible body effect is to assume that the proportional velocity pointing error will be equal to the wobble angle for short and long burns.

Quantization

The effect of quantization on the *lateral maneuver* pointing error is found by substituting $\mu_t = 5.55$ ms into equations (34) and (35).

For the *+Z maneuver,* a similar error occurs when timing errors result in a velocity component along the Y direction from equation (21) (the inexact cancellation of the X component of velocity of equation (29) is less important). The timing uncertainty causes uncertainty in pulse clock orientation, σ_ϕ

$$\sigma_\phi = \omega_z \sigma_t$$

so that the velocity pointing error per pulse is

$$\sigma_v = (\sigma_{VY}/V_Z) = [(F/m)\Delta t \cos 21^\circ \sigma_\phi] / [(F/m)\Delta t \sin 21^\circ]$$
$$= \omega_z \sigma_t/\tan 21^\circ = \omega_z(\sqrt{3}/3)\mu_t/\sqrt{n}\ \tan 21^\circ$$

Note that mean delays do not result in an error because they cancel.

Quantization does not affect the *-Z maneuver* proportional velocity pointing error.

Conclusions

The maneuver analysis of the delta-V pulsed maneuvers of the Galileo spacecraft has resulted in the development of useful probabilistic error models which, due to their generality, contribute to understanding single-spin and dual-spin spacecraft. The analytic models are simple in form and yet take into account the most important error sources affecting velocity magnitude and pointing errors. The results indicate that the Galileo spacecraft achieves or nearly achieves all of its maneuver accuracy requirements with the exception of magnitude accuracy of an axial maneuver. Negotiation between the orbiter engineering team, the navigation team and the mission design team may be necessary to resolve this remaining problem.

Acknowledgments

The authors thank D.E. Bernard, W.G. Breckenridge, L.A. D'Amario, F.O. Eke, S.A. Hayati, M.H. Jahanshahi, E.P. Kan, T. Kia, G. Macala, G.K. Man, and H. Otake for providing valuable information.

Part of the research described in this paper was carried out at the Jet Propulsion Laboratory, California Institute of Technology, under contract with the National Aeronautics and Space Administration.

References

[1] D'Amario, L. A., Bright, L. E., Byrnes, D. V., Johannesen, J. R. and Ludwinski, J., "Galileo 1989 VEEGA Mission Description," AAS Paper No. 89-431, August 7-10, 1989, Stowe, Vermont.

[2] Longuski, J. M., "Galileo Maneuver Analysis," AAS/AIAA Astrodynamics Conference, AAS Paper No. 81-137, August 3-5, 1981, Lake Tahoe, Nevada.

[3] Hintz, G. R. and Longuski, J. M., "Error Analysis for the Delivery of a Spinning Probe to Jupiter," *AIAA Journal of Guidance, Control and Dynamics,* Vol. 8, No. 3, May-June 1985, pp. 384-390.

[4] Longuski, J. M. and Kia, T., "A Parametric Study of the Behavior of the Angular Momentum Vector During Spin Rate Changes of Rigid Body Spacecraft, *AIAA Journal of Guidance, Control and Dynamics,* Vol. 7, No. 3, May-June 1984, pp. 295-300.

[5] Longuski, J. M., Kia, T., and Breckenridge, W. G., "Annihilation of Angular Momentum Drift During Spinning-Up and Thrusting Maneuvers of Rigid Bodies," AAS Paper No. 83-408, AAS/AIAA Astrodynamics Conference, August 22-25, 1983, Lake Placid, New York. To appear in *The Journal of the Astronautical Sciences*.

[6] Noll, R. B., Zvara, J. and Deyst, J. J., "Spacecraft Attitude Control During Thrusting Maneuvers," NASA Space Vehicle Design Criteria (Guidance and Control), NASA SP-8059, February, 1971.

[7] Hayati, S. A. and Jahanshahi, M. H., "Pointing Performance of a Dual-Spin Spacecraft," *The Journal of Astronautical Sciences,* Vol. 31, No. 4, October-December, 1983, pp. 545-559.

[8] Ward, R. S., "Galileo Dual-Spin Attitude and Articulation Control System," AIAA Paper No. 78-1286, AIAA Guidance and Control Conference, Palo Alto, California, August 1978.

[9] Ward, R. S. and Tauke, G. J., "Dual-Spin Attitude Control for Outer Planet Missions," JPL Publication 77-74, Jet Propulsion Laboratory, Pasadena, California, December 15, 1977.

[10] Wertz, J. R. (editor), *Spacecraft Attitude Determination and Control,* D. Reidel Publishing Co., Boston, 1980.

[11] McIntyre, J. E. and Gianelli, M. J., "Bearing Axis Wobble for a Dual-Spin Vehicle," *Journal of Spacecraft and Rockets,* Vol. 8, No. 9, September 1971, pp. 945-951.

[12] Wright, W. H., "Wobble Correction for a Dual-Spin Vehicle," *Journal of Spacecraft and Rockets,* Vol. 11, No. 4, April, 1974, pp.255-257.

[13] Bernard, D. E., "Passive Nutation Damping Using Large Appendages with Application to Galileo," *Journal of Guidance and Control,* Vol. 5, No. 2, March-April, 1982, pp. 174-180.

[14] Longuski, J. M., "Galileo Maneuver Execution Analysis Part III: Proportional Delta-V Magnitude Errors," Jet Propulsion Laboratory Memorandum 343-537 (internal document) July 1, 1981.

[15] Kaplan, M. H., *Modern Spacecraft Dynamics & Control,* John Wiley & Sons, Inc., New York, 1976.

AAS 89-362

COMPARISON OF GENERAL AND SPECIFIC SIMULATION PROGRAMS FOR SPACECRAFT DYNAMICS AND CONTROL*

J. E. Cochran, Jr.†, T. S. No‡ and N. G. Fitz-Coy†

Three methods for deriving equations used in simulating spacecraft attitude dynamics and control are discussed and results from several digital computer codes which are based on the different equations are compared. The first code is a generic, multi-body code, called TREETOPS, which is based on equations of motion derived using Kane's method. A second is a fairly general code for multi-rigid-body systems based on equations derived using Lagrange's method and quasi-coordinates. A third code is based on equations obtained in an *ad hoc* manner from Newton-Euler equations for a specific multi-body system. Finally, a fourth code is a specific one based on Lagrange's method and quasi-coordinates. A comparison of the methods is followed by example results for a three-body spacecraft. The performances of the programs and in the results obtained are discussed.

---

* This paper appears in full in the microfiche supplement to this volume, **AAS Microfiche Series**, Volume 59.

† Professor, Department of Aerospace Engineering, Auburn University, Alabama 36849-5338. Associate Fellow, AIAA; Senior Member, AAS.

‡ Graduate Student, Department of Aerospace Engineering, Auburn University, Alabama 36849-5338.

AAS 89-363

ENVIRONMENTAL DISTURBANCE TORQUE REJECTION FOR MOMENTUM BIAS STABILIZED SPACECRAFT*

C. Rahn and B. Gretz†

Many momentum bias stabilized spacecraft (Insat-1, Superbird-1, and Intelsat VII) rely on quarter orbit coupling to ensure that yaw error can be controlled through the use of an Earth sensor which measures only roll error. The yaw accuracy achieved depends on the nature of the environmental disturbance torques which act on the spacecraft. This paper presents a method of improving the disturbance torque rejection of momentum bias spacecraft utilizing a roll sensor, yaw estimation, disturbance torque estimation, and magnetic torquers.

---

* This paper appears in full in the microfiche supplement to this volume, **AAS Microfiche Series**, Volume 59.

† R&D Engineer, Ford Aerospace Corporation, 3939 Fabian Way, Palo Alto, California 94303-4697.

SATELLITE DYNAMICS

SESSION 4

Chairman: Bernard Kaufman
U. S. Naval Research Laboratory

AAS 89-369

SLEW DISTURBANCE COMPENSATION FOR MULTIPLE SPACECRAFT PAYLOADS

Peter Y. Chu*

Multiple slewing spacecraft payloads such as mirrors and antennas cause dynamic disturbances to the spacecraft and themselves. A centralized compensation logic uses the payload slew commands to calculate these interactions, or essentially runs a mini-simulation of spacecraft response in real-time, and modifies the lines-of-sight accordingly. No sensor is used. From the payload designers and users point of view, the spacecraft bus appears not disturbed. The scheme is readily expanded to any number of payloads. The implementation on the new geo-stationary GOES weather satellites with two double-gimballed scanners is described. The pointing error reduction ratrio is about 15:1. Together with the baseline spacecraft feedback control system, the effective spectral band is from DC to about 1 Hz.

INTRODUCTION

The scanning or slewing motions of a payload cause reaction torques and forces which disturb the whole spacecraft. The resulting spacecraft motion in turn disturbs the pointing of this and all other payloads. The magnitude and duration of the resulting pointing error depends on the slew pattern, gimbal geometry, inertia ratios, and reactions of the spacecraft control system. With the trend of putting more independently-operated payloads on a single spacecraft in order to reduce the cost per payload, and the trend of tighter pointing requirements, the payload interaction problem is increasingly important.

The have been a variety of approaches to handle this pointing error:

(1) Minimize the size of payload motion and the inertias of the moving parts, and accept the error in the pointing budget. The images can be enhanced by ground processing using landmarks. A very large spacecraft bus does not necessarily make this approach a trivial choice, because it probably carries more payloads or larger aperture instruments requiring tighter pointing.

* Ford Aerospace & Communications Corporation, 3939 Fabian Way, Palo Alto, California 94303.

(2) Incorporate the anticipated spacecraft motions in the the scan profiles stored in the payload controller. The exact scan sequence must be determined before the generator is programmed, and the method breaks down if there are other payloads operating independently.

(3) Use counter-rotating inertia or local momentum devices to make payloads reactionless[1]. This could simplify system issues but requires additional weight, power, and cost.

(4) Measure spacecraft attitude errors, regard them as line-of-sight errors, and correct the images accordingly.

(5) Measure the line-of-sight error using inertial and/or stellar sensors local to the payload or located at the spacecraft navigation package. The line-of-sight is corrected by steering a gimballed platform or mirrors in real-time[2]. If the payload is pointed to a fixed target, a focal plane detector can be the best sensor for this purpose[3].

Approaches (4) and (5) can handle any disturbance whether it is stochastic or deterministic. It is not necessary to know the disturbance sources. If the sensors and attitude determination system is local to the payload and have sufficient bandwidth, it can even handle spacecraft flexible structure response and high frequency jitter. However, they require sensors which may not be otherwise needed, thus increase cost.

(6) Use synthetic scan with detector arrays separated spatially or temporally, and abolish mechanical scan altogether. This is not really a compensation method since it uses an entirely different payload.

This paper describes another approach which needs no sensor, does not require attitude determination, allow totally independent operation of each

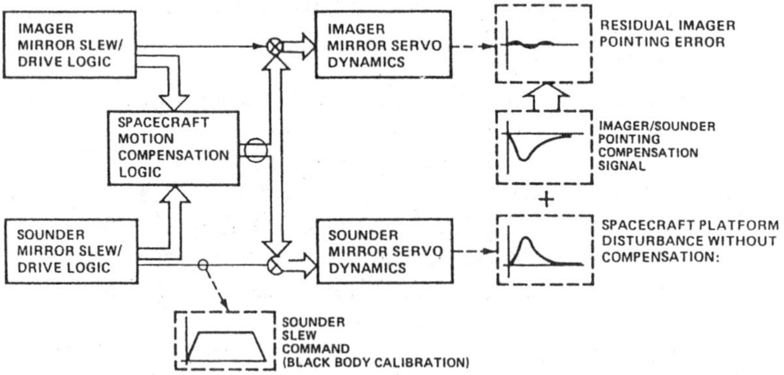

Figure 1: Principle of slew disturbance compensation

payload, the images are corrected in real time at the source, and the logic is readily expandable to any number of payloads. The basic concept is illustrated in Fig. 1. A spacecraft central control algorithm uses the gimbal slew commands from all important payloads to calculate the disturbance effects, or essentially runs a mini-simulation of spacecraft response in real-time, and sends compensation signals to each payload to modify the lines-of-sight accordingly. The overall block diagram is shown in Fig. 2. It shows the input of the slew disturbance compensation logic from the payload slew commands and the output to the payload steering servo loops. Its sub-blocks are further explained in the subsequent sections. A particular method of integrating the compensation signals in the payload servo loop is also shown. However, the details of the payload control system and the baseline spacecraft control system are not the main subject of this paper.

The approach described in this paper is best suited where deterministic payload motions are the dominant disturbances, and the dynamic interaction among spacecraft and payloads can be accurately modeled (with the possible help of after-launch parameter tuning). It is inappropriate to

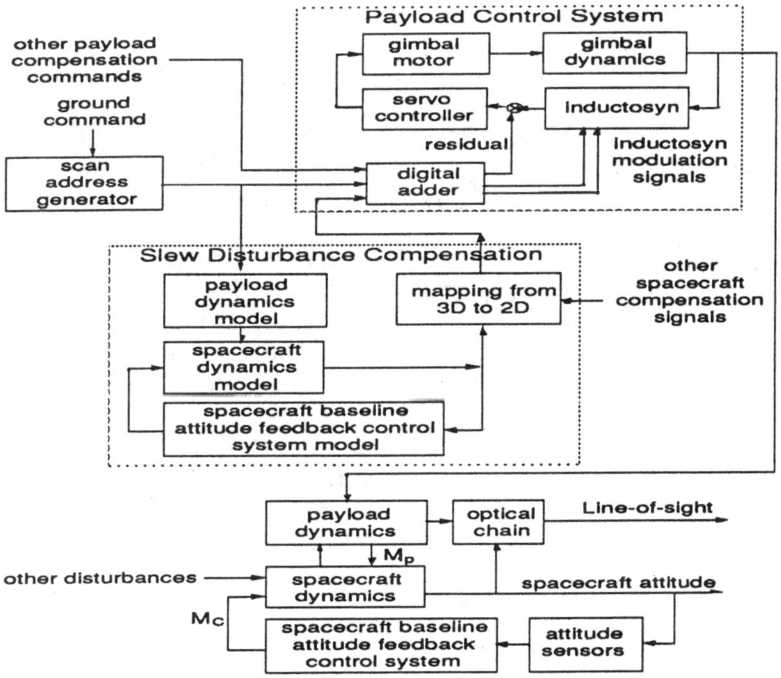

Figure 2: Payload slew disturbance compensation logic block diagram. Also shown is the interface from/to the payload electronics. Similarity between the middle and lower sections of the figure reflects the idea of running a mini-simulation onboard in real time.

extend the idea to handle flexible structural response because of modeling uncertainties. However, this limitation is often not important because (1) only the starts and stops of the slews excite structural modes, and (2) the magnitude is only a small fraction of the rigid body response. Since the speeds of spacecraft computers are quite limited, only the important portions of the interaction model should be implemented. Algorithm simplifications are taken whenever possible based on geometry, alignment, inertia ratios, and scan patterns and their spectral contents. Those applied to the GOES satellite are explained in the following section.

The Geo-stationary Operational Environmental Satellite (GOES) has been the major space system for weather forecast in the United States. The next-generation GOES is momentum-biased and 3-axis stabilized. The major payloads are an Imager and a Sounder of similar design operating in visible and infrared spectrums. Each instrument has a 31 cm mirror mounted on a double-gimbal. The scheduled launch is in 1990.

DYNAMIC INTERACTIONS AMONG PAYLOADS & SPACECRAFT

The equation of motion of the spacecraft can be written in the spacecraft body coordinates as

$$I_s\dot{\omega}_s + \omega_s \times H_s + M_c + \sum_i M_{p_i} = 0 \qquad (1)$$

where I_s is the spacecraft inertia, ω_s is the spacecraft attitude rate, M_c is the torque from the spacecraft control actuators, H_s is the bias momentum, and M_{pi} is the reaction torques from each moving payload.

A desirable and common design of the payload gimbal/mirror assembly has all moving parts statically balanced, i.e., the center of mass of the mirror resides on the inner gimbal axis, and the center of mass of the mirror and inner gimbal, regardless of the inner gimbal angle, resides on the outer gimbal axis. In other words, the center of mass of the moving parts of a payload, denoted as point P, is at the intersection of the two gimbal axes and is fixed relative to the rest of the payload. The center of mass of the spacecraft therefore is fixed in the spacecraft body. M_p can be written with respect to P as

$$M_p = I_p\,\dot{\omega}_p + \omega_p \times \omega_p I_p \qquad (2)$$

where I_p is the moving part inertia, and ω_p is the inertial payload angular rate.

Usually the inertias of the payload moving parts are much smaller than the spacecraft inertia, which include the payload non-moving parts. The ratios for GOES are in the order of 10^{-5}. The induced motion of spacecraft is so small that the inertial rate of the payloads ω_p can be regarded as the rate relative to the spacecraft. The payload motions are prescribed by gimbal address command generators and enforced by the gimbal servo loops. The gimbal address commands are generated by the payload electronics based on ground commands and built-in profiles and priorities. The actual maximum acceleration is limited by gimbal servo bandwidth.

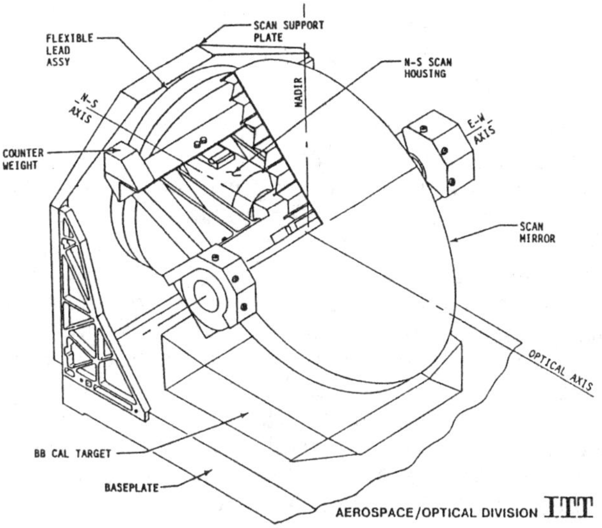

Figure 3: GOES Imager scanner

Fig. 3 shows the scanner of the GOES Imager which is a representative Earth-observation scanning instrument for the geo-stationary orbit. The Sounder is identical mechanically and optically except for the additional filter wheel. The Earth disk radius as seen from the geo-stationary orbit is 8.7°, thus the range of the line-of-sight is about ±10° to accommodate full disk scans, cold space look for calibration, and sighting stars near the Earth disk.

The scanner has an outer gimbal for north-south scans and an inner gimbal for east-west scans. The motor and the inductosyn for the inner gimbal are located at opposite ends of the mirror shaft. The outer gimbal axis is aligned with the spacecraft x axis for north-south scan. Periodic black-body calibrations (BBC) require the outer gimbal be slewed 180° to dwell on a black-body cavity and be slewed back. The inner gimbal and the mirror shaft are aligned with the spacecraft y axis when the outer gimbal is at its null position. The optical scan angle is double of the mechanical scan angle, therefore the east-west mechanical scan is only ±5°. The inner gimbal stays at the null position when the outer gimbal angle undergoes the 180° slew. Based on inspection of the scanner/mirror geometry and mechanical design data, it is noted that the cross-coupling inertia I_{xz} is significant, I_{xy} is small in this design, and I_{yz} is negligible. M_p can then be approximated in the gimbal body coordinates as

$$M_p = \begin{bmatrix} I_x \dot{\omega}_x \\ -I_{xy} \dot{\omega}_x + I_{xz} \omega_x^2 \\ -I_{xz} \dot{\omega}_x - I_{xy} \omega_x^2 \end{bmatrix} + \begin{bmatrix} 0 \\ I_y \dot{\omega}_y \\ 0 \end{bmatrix} \quad (3)$$

where I_x, I_{xz}, I_{xy} are the inertias of the north-south moving parts (outer gimbal, inner gimbal and mirror at the nominal position), I_y is the y inertia of east-west moving parts (inner gimbal and mirror) about the mirror shaft, and subscripts x, y, z refer to roll, pitch, yaw. The terms $I_x\dot{\omega}_x$ and $I_y\dot{\omega}_y$ are direct momentum exchanges, $I_{xy}\dot{\omega}_x$ and $I_{xz}\dot{\omega}_x$ are cross-axes momentum exchanges, and $I_{xy}\omega_x^2$ and $I_{xz}\omega_x^2$ represent the natural tendency of the moving parts to align with the principal axes. M_p along the gimbal axis is driven by the gimbal motor and causes torsion of the axle, while M_p normal to the axis is carried by the gimbal bearing and cause bending of the gimbal axis and the gimbal yoke. Since the east-west motion is less than 5°, the difference between the mirror body coordinates and the inner gimbal body coordinates is ignored. The difference between the gimbal coordinates and the spacecraft body coordinates is the angle ϕ, therefore M_p in the spacecraft body coordinates is

$$M_p = \begin{bmatrix} I_x\ddot{\phi} \\ I_y\ddot{\theta} + I_{xz}(\dot{\phi}^2\cos\phi + \ddot{\phi}\sin\phi) + I_{xy}(\dot{\phi}^2\sin\phi - \ddot{\phi}\cos\phi) \\ I_{xz}(\dot{\phi}^2\sin\phi - \ddot{\phi}\cos\phi) + I_{xy}(-\dot{\phi}^2\cos\phi - \ddot{\phi}\sin\phi) \end{bmatrix} \quad (4)$$

where ϕ, θ are the gimbal angles of the outer and inner gimbals, respectively. Notice that M_p is not zero when $\ddot{\phi}$ is zero but ϕ has a constant rate. Because of geometry, centrifugal torque, cross-axis inertia, spacecraft momentum bias, and spacecraft control system, parts of interactions from the forward and return slews do not cancel and have to be slowly attenuated by the spacecraft attitude control system.

RELATION WITH BASELINE SPACECRAFT ATTITUTDE CONTROL SYSTEM

After the term M_p in Eq. (1) has been derived, the next step is to obtain M_c. The actual command sent to the actuators from the baseline control logic is readily available in the spacecraft computer. If it is used in the slew compensation logic, however, its noise components ignored by the low-bandwidth spacecraft actuators will be followed by the high-bandwidth payload servo loops and degrade line-of-sight pointing. A better alternative is to compute a special version of M_c attributed only to payload disturbances. This term gradually attenuates the payload correction signal as the spacecraft attenuates the physical attitude error. This leads to the interesting feature that the lower bandwidth limit of this compensation scheme fits *exactly* with the higher bandwidth limit of the spacecraft control system, as shown in Fig. 4. The compensation signals are not fed back into the compensation logic because their effects are second order. The closed loops in the slew disturbance compensation logic are always identical to those of the baseline spacecraft control system. Consequently, the slew compensation logic has no stability issues by itself.

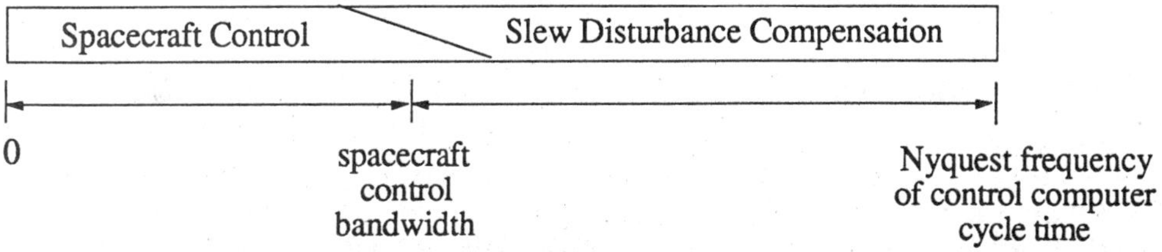

Figure 4: Effective spectral bands of slew disturbance compensation and the baseline spacecraft pointing control system. The transition between the two control systems is gradual and without overlap or gap.

CORRECTION OF SLEW DISTURBANCE EFFECTS

Based on small angle assumption, individual components of ω in Eq. (1) are integrated to obtain spacecraft attitude errors attributed to payload slews. The bias momentum can be calculated from onboard information such as the speeds of the momentum or reaction wheel set.

The pointing error is proportional to the inertia ratio between the moving parts and the spacecraft. The absolute pointing error is proportional to the slew angle. The pointing stability error (i.e. pointing error in pixel integration time) is proportional to the slew rate, which is often high during calibration slews to minimize service outage.

For payloads pointed to the nadir, the spacecraft roll/pitch/yaw errors are mapped to the pixel location errors as

$$\Delta EW = -\sin(NS)\,\psi - \cos(NS)\,\theta, \tag{5A}$$

$$\Delta NS = \frac{\sin(EW)*\cos(NS)}{\cos(EW)}\,\psi - \frac{\sin(EW)*\sin(NS)}{\cos(EW)}\,\theta - \phi. \tag{5B}$$

where EW is along east-west, NS is along north-south, and Δ denotes deviations on that direction. Since the gimbals are aligned with the EW and NS directions, the pixel location errors ΔEW and ΔNS are the correction gimbal angles to be subtracted from the scan pattern commands. The payload gimbal servo usually has much higher bandwidth than the update rate of the spacecraft compensation commands. The resulting physical lines-of-sight achieves the intended patterns which has been programmed without considering the dynamic interaction effect.

IMPLMENTATION ISSUES

The signal interface between the payloads and the spacecraft electronics needs coordination in the earliest possible time. If the payload electronics is digital, the interface should be digital. The implementation on the GOES weather satellite has taken further simplifications for the similarities between the two scanners, adjusted for the different mounting geometry. Since the gimbal servos have high bandwidths, the software code must not experience overflow when mirrors move abruptly. The computation delays between the slew command input and the correction signal output must be minimized. The gain switches and mode changes of the spacecraft attitude control logic must also be propagated to the code for slew disturbance compensation. A more sophisticated version of Eq. (5) is implemented as part of another GOES fine-pointing logic dealing with longer-term error sources.

Finally, ground-commandable dimensionless scale factors have been included in the software code such that the parameters related to spacecraft and payload mass properties can be adjusted after launch and a scheme for parameter identification has been designed. The most sensitive parameter is nutation frequency, but it is also the simplest to tune.

PERFORMANCE RESULTS

Fig. 5 shows the performance of the GOES slew disturbance compensation for a worst case payload activity schedule. The spacecraft inertia matrix is

$$I_s = \begin{bmatrix} 3426 & \text{small} & \text{small} \\ & 992 & \text{small} \\ & & 3507 \end{bmatrix} \text{kgm}^2$$

and the scanner mass properties are

gimbal and mirror:

$$I_x = 0.159 \text{ kgm}^2$$
$$I_{xz} = 0.0186 \text{ kgm}^2,$$
$$I_{xy} = \text{negligible},$$

mirror only:

$$I_{yy} = 0.0305 \text{ kgm}^2.$$

The left column is the uncompensated line-of-sight errors in roll, pitch, and yaw. They range between 50 and 125 μrad and are not acceptable. The Imager outer gimbal exercises black-body calibration slews every 10 minutes, consisting of a 180° rotation at a rate of 10°/s, a dwell for 4 seconds, and a symmetric return rotation. The Sounder exercises similar maneuvers to the opposite direction every 20 minutes. The spacing between the Imager and the Sounder slews is 10 seconds, reflecting realistic conditions. Roll and yaw are coupled through the bias momentum. The size of the roll motion is primarily affected by I_x. The pitch motion is primarily affected by I_{xz} (and I_{xy} if it is significant). The yaw motion is driven by both nutation coupling and the I_{xz} term. All three errors are linearly affected by the total slew angle. Pedagogical cases can be made by having one payload slew 1/2 nutation cycle later than the other to maximize the disturbances. Then the errors will be about 50% higher. The disturbance effects of regular scans are very small compared to those of the calibrations.

The right column of Fig. 5 is the residual errors after compensation. Not shown is the structural flexibility effect of about 5 μrad which is separately accounted for in the system error budget. The error reduction is about 15:1, when the inertia knowledge is good to 1% and the nutation period knowledge is good to 0.2%. The magnitudes of the uncompensated errors affect the residual errors linearly. The knowledge of inertia ratios affects the residual errors linearly. The roll and yaw errors are affected much more than linearly by the phase mismatch of actual and calculated nutation motions. If the mismatch is large (say 3 %) and the spacecraft control system time constant is considerably slower than the nutation period, the largest roll/yaw residual errors could approach the uncompensated error. Notice that the peaks of the roll/yaw errors occur long after the slew which initiated the errors have finished. Other error sources such as time delays and modeling simplifications are difficult to reduce. The details of the residual error profile depends on the phasing between spacecraft transients and successive scanner maneuvers, but the magnitude of the residual errors stay in the same range. Any repetitive activity on the spacecraft or payloads should avoid resonating with the nutation cycle.

CONCLUSION

A general payload disturbance compensation scheme and its onboard implementation on the GOES satellite with two scanning instruments have been described. The logic is essentially a simplified simulation of the payload/spacecraft interaction in real time. No sensor is used, and the algorithm is readily expandable to any number of payloads.

ACKNOWLEDGEMENT

Carl Plescia and Don Gamble of Ford Aerospace conceived the compensation scheme illustrated in Fig. 1 (Ref 4). Carl also jointly designed the GOES implementation. NASA Goddard Space Flight Center and NOAA are the sponsors of the GOES project.

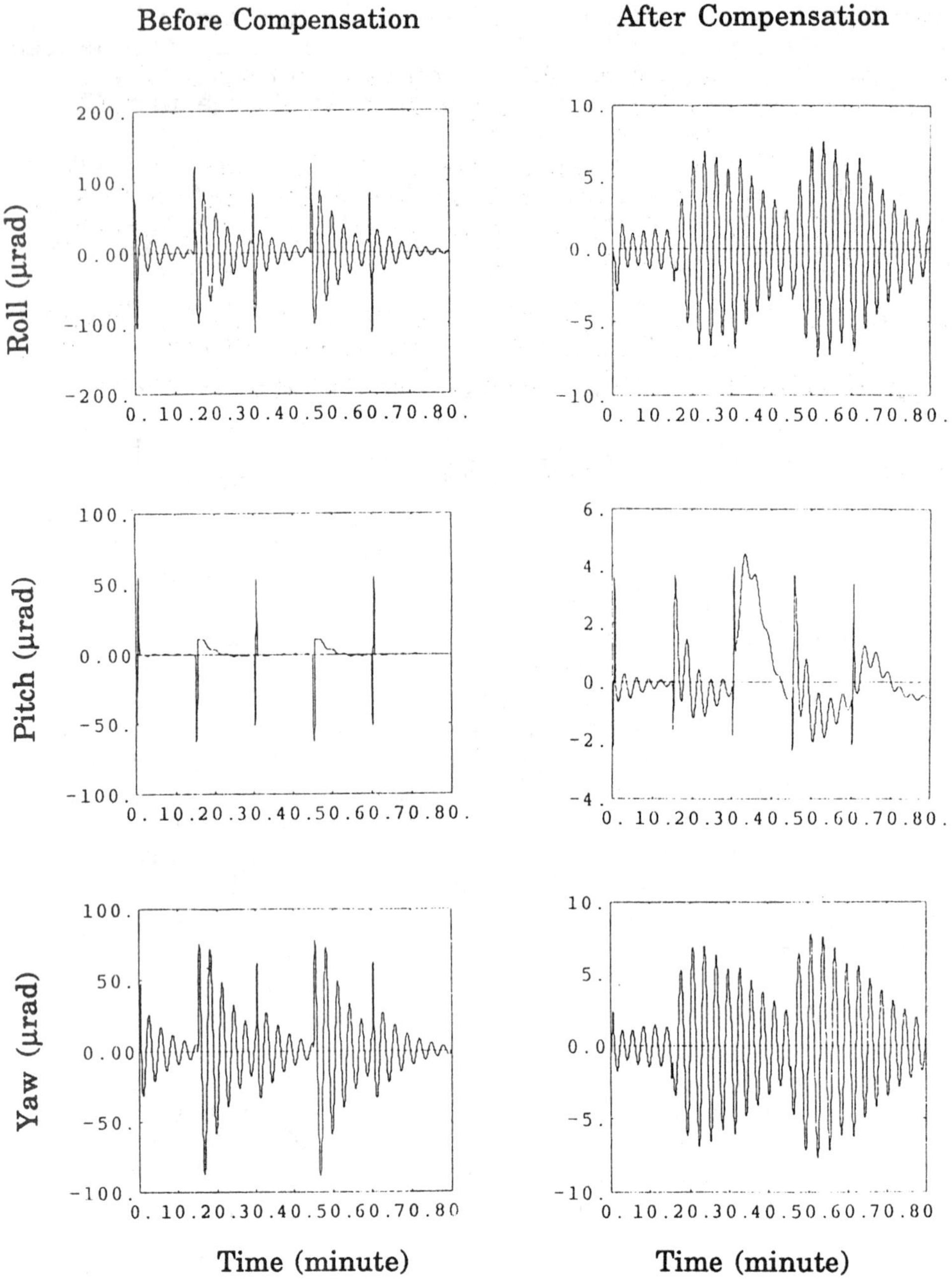

Figure 5: Slew disturbance compensation performance of GOES blackbody calibrations. Left column is uncompensated pointing error, right column is compensated residual error.

REFERENCE

1. Marchetto, C.A., “Reactuator”, JPL exhibit at 11-th Annual AAS Guidance and Control Conference, Keystone, Colorado, Jan 30, 1988

2. Bauer, F.H., "Ultraviolet Imaging Telescope (UIT) Image Compensation Actuator (IMCA) Control Systems Analysis and Compensation Design," NASA Goddard Space Flight Center, Guidance and Control Branch, Branch Report No. 331, July 13, 1983

3. Tarbell, T.D., Duncan, D.W., Finch, M.L., "Image Motion Compensation on the Spacelab 2 Solar Optical Universal Polarimeter (SL2 SOUP), SPIE Vol 265 Supplement - Shuttle Pointing of Electro-optical Experiments (1981)

4. Plescia, C., Gamble, D., U.S. Patent 4687161.

AAS 89-370

SENSITIVITY STUDIES OF MARS ORBITERS FOR MARS GRAVITY RECOVERY

George W. Rosborough* and Frank G. Lemoine†

Knowledge of the Mars gravity field must currently be derived solely from tracking of Mars orbiting satellites. The existing data consist of S-band Doppler tracking of three spacecraft, Mariner 9 and Viking Orbiters 1 and 2, in highly eccentric orbits at effectively five different inclinations. There is also optical tracking of the two natural satellites, Phobos and Deimos, which can be utilized. The Mars Observer spacecraft, scheduled for launch in September 1992, will be in a near polar orbit and will be tracked by X-band Doppler. A sensitivity study for each of these orbiters has been conducted to determine their potential contributions to a new model of the Mars gravity field. The results show that the Mariner 9 data will contribute to the determinations of the even orders out to degree and order 10. Viking data from the near resonant and deeply resonant one day orbits will contribute to all orders at least up to degree and order 18. Phobos and Deimos will only contribute to the lowest degrees of the gravity field, but in a valuable manner as a consequence of their low inclination orbits. In addition to the lower degree and order harmonics, the Mars Observer data will be sensitive to coefficients potentially as high as degree and order 50. In all cases studied, it was found that the radial velocity perturbations due to the Mars gravity field are significantly larger than the velocity perturbations both along-track, and cross-track to the orbit. The results definitively show that the sensitivity to the gravitational signal is enhanced when the spacecraft orbit plane is oriented in such a way that the radial velocity perturbations are directly measured by the Doppler data. When the spacecraft orbit plane is perpendicular to the Mars-Earth line, only cross-track velocity perturbations are directly sensed. As a consequence, this orientation degrades the gravity signal content of the Doppler data.

INTRODUCTION

The tracking of Earth orbiting satellites provides one of the most direct means for determining models of the planet's gravity field. This approach has been used since the start of the space-age and continues to be the most effective method for obtaining accurate representations of the Earth's global gravity field[1]. This approach has extended naturally to other planets where orbiters have been placed and tracked through the Deep Space Network (DSN).

The accuracy with which a gravity field may be determined from tracking orbiting satellites is a complicated question which depends on the orbital geometries and the accuracy and coverage of the tracking data. The primary difficulty is that given noisy observations of a single satellite, there are many possible representations of the gravity field which could reproduce the accelerations which the

* Assistant Professor, Department of Aerospace Engineering Sciences, Colorado Center for Astrodynamics Research, University of Colorado, Campus Box 429, Boulder, Colorado 80309-0429.

† Graduate Research Assistant, Department of Aerospace Engineering Sciences, Colorado Center for Astrodynamics Research, Campus Box 429, University of Colorado, Boulder, Colorado 80309-0429.

satellite is sensing. To help distinguish among these representations other information is needed, such as tracking data of another satellite with a different orbital geometry. The more satellites which can be observed, the better constrained and more accurate the gravity field becomes. The most recent solution for the Earth's gravity field used tracking data on 17 satellites, and a follow-up to that solution will incorporate 31 satellites[2].

The intent of the current paper is to examine the Mars orbiters and the planned Mars Observer with regard to their sensitivity to the Mars gravity field. Due to the different orbital geometries of these orbiters they each provide unique information which will contribute to a global solution for the Mars field. The sensitivities have been evaluated in terms of the expected position and velocity disturbances that will be produced by the Mars gravity field. The various disturbances have been quantified in terms of their magnitude as well as their frequency. These results are compared to the accuracy of the tracking data and to the coverage of the Mars orbiters to ascertain which components of the Mars gravity field may be determined accurately.

MARS ORBITERS

The spacecraft and satellites whose data will contribute to a global solution for a Mars gravity field include Mariner 9, Viking Orbiters 1 and 2, Mars Observer and the natural satellites of Mars, Phobos and Deimos. These objects and their nominal orbital elements are listed in Table 1. The Viking Orbiters changed orbits many times, as summarized by Hildebrand et al.[3], Snyder[4], and Snyder and Evans[5]. Thus, only a representative sampling of these orbits is listed in Table 1. For the purposes of this study, two Viking 1 orbits were selected for analysis. These include an orbit in deep resonance with the Mars gravity field, with a period of 24.65 hours, and an eccentricity of 0.76. And an orbit in shallow resonance was also chosen with a period of 23.05 hours and eccentricity of 0.75. These orbits were selected to compare and contrast the sensitivity of the Viking orbits to the Mars gravity field, with a special emphasis on their resonance characteristics.

Table 1

MARS ORBITERS CHARACTERISTICS

Satellite	Period (hrs)	Semimajor Axis (km)	Inclination (deg)	Eccentricity
Mariner 9	11.97	12629.5	64.6	0.62
Viking 1	24.65	20442.6	38.2	0.76
	23.05	19548.1	39.1	0.75
Viking 2	24.62	20426.0	55.4	0.76
	26.79	21609.3	74.9	0.77
	26.48	21442.3	80.1	0.81
Mars Observer	1.95	3758.3	92.8	0.0071
Phobos	7.66	9378.5	1.0	0.015
Deimos	30.29	23457.7	2.8	0.0008

SENSITIVITY ANALYSIS

The gravitational field of Mars will be represented by a spherical harmonic expansion of the form,

$$U = \frac{\mu}{r} \sum_{l=0}^{\infty} \left(\frac{a_e}{r}\right)^l \sum_{m=0}^{l} \bar{P}_{lm}(\sin\phi)\left(\bar{C}_{lm}\cos m\lambda + \bar{S}_{lm}\sin m\lambda\right) \tag{1}$$

where

μ	$\equiv$	product of the universal constant of gravitation and the mass of Mars
a_e	$\equiv$	semi-major axis of the Mars reference ellipsoid
r, ϕ, λ	$\equiv$	satellite distance, latitude and longitude, respectively, in a body-fixed coordinate system
$\bar{C}_{lm}, \bar{S}_{lm}$	$\equiv$	normalized geopotential harmonic coefficients of degree l and order m
$\bar{P}_{lm}$	$\equiv$	normalized associated Legendre functions

Of general interest will be the perturbations in the orbital motion of the various orbiters due to $\bar{C}_{lm}$ and $\bar{S}_{lm}$. In the remainder of the paper the notation J_{lm} will be used to denote a $\bar{C}_{lm}$ and $\bar{S}_{lm}$ coefficient pair. The most recent Mars gravity field, complete to degree and order 18, developed by Balmino et al.[6], was used as the baseline in this study. In addition, the power law of the Mars gravity field as derived by Balmino, $13.0 \times 10^{-5}/l^2$, was used to simulate coefficients above degree 18.

The perturbations in the orbital motion can be determined most easily using the linear results derived by Kaula[7]. However, for those cases where the linear analysis is not applicable (most notably for resonance) the sensitivities can be calculated using numerical integration. Each of these approaches as applied in this study are described in the following subsections.

In either approach it is not simply the variations of the orbital elements which are desired but the sensitivity of the tracking data to those variations. The data types of interest here are optical tracking of the natural satellites and Deep Space Network (DSN) tracking of the orbiters. Thus the orbital perturbations will be given in terms of the radial, transverse and normal components of the position and velocity variations in the orbital motion. The transverse component is in the orbital plane and perpendicular to the radius vector, and the normal component is orthogonal to the orbital plane. Given this information and the relative orientation of the tracking geometry, it is possible to infer the sensitivity of the tracking data to the various components of the gravity field.

For an aspect angle of 0° (the angle between the observer and the orbital plane) the DSN Doppler data will most directly sense the radial and transverse variations of the satellites velocity. When the aspect angle reaches 90°, then the normal perturbations in the velocity will be measured.

Linear Analysis

Kaula's results provide the perturbations in each of the Keplerian orbital elements for each degree and order of the gravity field. Linear coordinate transformations can be used to recast the Kepler orbital perturbations in terms of orbital position and velocity. For example, the position perturbations in the radial, transverse and normal directions are:

$$\Delta r = \frac{\partial r}{\partial a}\Delta a + \frac{\partial r}{\partial e}\Delta e + \frac{\partial r}{\partial M}\Delta M \tag{2}$$

$$\Delta\tau = r(\Delta\omega + \Delta f + \Delta\Omega\cos i), \tag{3}$$

$$\Delta\eta = r(\Delta i\sin(\omega + f) - \Delta\Omega\sin i\cos(\omega + f)) \tag{4}$$

Substituting Kaula's solution for the orbital element perturbations into these expressions then provides a direct means for evaluating the orbital position perturbations. Such general expressions are given in Rosborough and Tapley[8] and were used in this analysis.

Similar transformation expressions exist for the velocity components but have not as of yet been implemented in a general manner. Wagner[9] provides expressions for the components of the velocity perturbations, but only for low eccentricity orbits. In view of the high eccentricity of several of the Mars orbiters, these expressions are not applicable, so an alternate procedure was used. This consisted of generating in time both the reference trajectory (a secularly precessing ellipse) and the perturbed trajectory, which is defined by adding Kaula's linear perturbations to the reference orbit. These trajectories were then converted and differenced in cartesian space and the resulting differences were rotated into the radial, transverse, and normal directions. Thus, a time series of the differences in position and velocity are obtained directly, from which the Root Mean Square (RMS) of the differences may be computed for each component.

This procedure is completely rigorous as long as it can be assumed that Kaula's solution is valid. To avoid unduly biasing the result, it is necessary to ensure that the time series be long enough to encompass multiple periods of the lowest frequency term present in Kaula's solution. Further, the time series must have adequate resolution to resolve the nonlinear mapping of the long period orbit element perturbations into position and velocity perturbations with a frequency of one or more cycles per satellite revolution. This procedure can be time consuming and computationally intensive. In practice, it was found that a time series with a length 20 to 100 times the period of the lowest frequency term in Kaula's solution, and a time step of 0.005 days was adequate. Use of the University of Colorado's Alliant FX/8 mini-supercomputer[10] facilitated the determination of the sensitivities of the spacecraft orbiters to the Mars gravity field. As will be shown in detail subsequently in this paper, the position and velocity RMS's computed via the procedure just outlined, compare favorably with results from straightforward numerical integration.

High Eccentricity. Kaula's linear theory which describes the variations of the orbital elements caused by the constituents of the gravity field has been applied extensively to study orbits of low eccentricity. However, direct application of Kaula's solution to the high eccentricity orbits of Mariner 9 and Viking Orbiters 1 and 2 poses some difficulty. The eccentricity functions in Kaula's solution, $G_{lpq}(e)$, and their derivatives with respect to eccentricity, $G'_{lpq}(e)$, are formally defined as finite sums of factorials and binomial coefficients. This analytic formulation remains mathematically rigorous at the higher eccentricities. However these formulae become unwieldy and unstable when evaluated numerically, due to the inherent finite precision of any computer. As shown by Gooding and King-Hele[11], the analytic formulation can be replaced by an integral such as,

$$G_{lpq} = \frac{1}{\pi}\int_0^{\pi} \frac{1}{(1 - e\cos E)^l}\cos[(l-2p)f - (l-2p+q)M]dE, \tag{5}$$

where e, is the eccentricity, f, M, and E are the true, mean, and eccentric anomalies, and l, p, and q are the indices from Kaula's solution.

This integral can be evaluated by numerical quadrature, and remains stable even at high eccentricities. In this study, a 4 step Newton-Cotes formula with truncation error $O(h^7)$ was used[12]. The derivative of the eccentricity function was evaluated numerically, using the central difference form of the derivative,

$$G'_{lpq} = [G_{lpq}(e+\delta e) - G_{lpq}(e-\delta e)]/2\delta e. \tag{6}$$

The optimum value of δe was found to be $\eta^{1/3}$, where η is machine epsilon, rather than $\eta^{1/2}$, as suggested by Dennis and Schnabel[13]. Use of these equations facilitated the evaluation of the orbit element perturbations for the high eccentricity orbiters.

Nonlinear Analysis

Nonlinearities that potentially invalidate Kaula's solution include resonance, as well as the effects of the frozen orbit geometry. Resonance between the tesseral components of the spherical harmonic expansion and the spacecraft orbit, occurs whenever the spacecraft completes an integral number of rotations with respect to the line of nodes, in the same time it takes the planet to undergo an integral number of rotations. These are the conditions for a ground track to repeat. The proximity of the repeat controls the depth and strength of the resonance. The perturbations are long period, compared to the spacecraft mean motion. Resonance results in large deviations from the mean orbit and the secularly precessing ellipse assumed in Kaula's solution. Physically, these large changes in the mean elements drive changes in the secular rates, which feed back into further changes in the orbit parameters. This sort of nonlinear interaction, characteristic of deep resonance, means that the sensitivity to the gravity field can only be assessed by direct integration of the equations of motion. For the case of shallow resonance, the use of Kaula's solution remains adequate to study the effect of the resonant orders on the spacecraft orbit.

A frozen orbit arises from the balancing of the effects of the even and odd zonal harmonics. For appropriate initial values of eccentricity and argument of periapsis, these elements undergo a libration[14]. As a consequence, periapsis is not free to precess, and remains constrained to small excursions from $\omega = \pm 90°$ (depending on the sign of the third degree zonal). Mars Observer will be in such an orbit ($\omega = -90°$). In performing a sensitivity analysis, using Kaula's solution, the frozen orbit can be accommodated by redefining the conventional secularly precessing reference orbit to have a null periapse rate. The degree to which this analysis will reflect the actual orbit perturbations will be determined by how well the frozen orbit geometry is maintained in the presence of other perturbations such as higher degree and order gravity terms, drag and solar radiation pressure.

Software

In this study, the linear analysis based on Kaula's solution and the linear transformations to position and velocity were carried out by a computer program called SLOP (Satellite Linear Orbit Perturbations). The nonlinear analysis was conducted using the programs, GEODYN[15], UTOPIA[16] (University of Texas Orbit Processor), and POLOP[17] (Planetary Observer Long Orbit Predictor), developed by the Jet Propulsion Laboratory. These programs were used to evaluate the cases of resonance and also to corroborate the results predicted from Kaula's solution computed by SLOP.

GEODYN and UTOPIA are least squares orbit determination batch processors that perform a Cowell integration. POLOP integrates the Lagrange Planetary Equations, propagating only the secular, long period, and resonant portions of the disturbing function. The UTOPIA program was designed to process tracking data from Earth orbiters. For the purposes of this study modifications were introduced to allow the program to propagate Mars orbiters. GEODYN is used by the Goddard Space Flight Center (GSFC) to process actual and simulated Mars Orbiter tracking data.

RESULTS

Phobos and Deimos

Introduction. The availability of Earth based and spacecraft based optical data of the natural satellites of Mars motivated studies to determine their sensitivity to the Mars gravity field. Further, prior to its demise, the Soviet Phobos II spacecraft was scheduled to deposit a transponder on Phobos[18]. This would have permitted high precision tracking of the satellite. A final motivation for this present study is that both satellites are at low inclinations. Even current advanced Earth gravity models, such as GEM-T1[1], lack sufficient data from low inclination satellites. This conclusion is based on analyses of the radial orbit error predicted at the TOPEX altitude from the GEM-T1 covariance at various inclinations[19]. Therefore, tracking data of these satellites or of spacecraft in similar orbits are expected to contribute useful information to models of the global Mars gravity field.

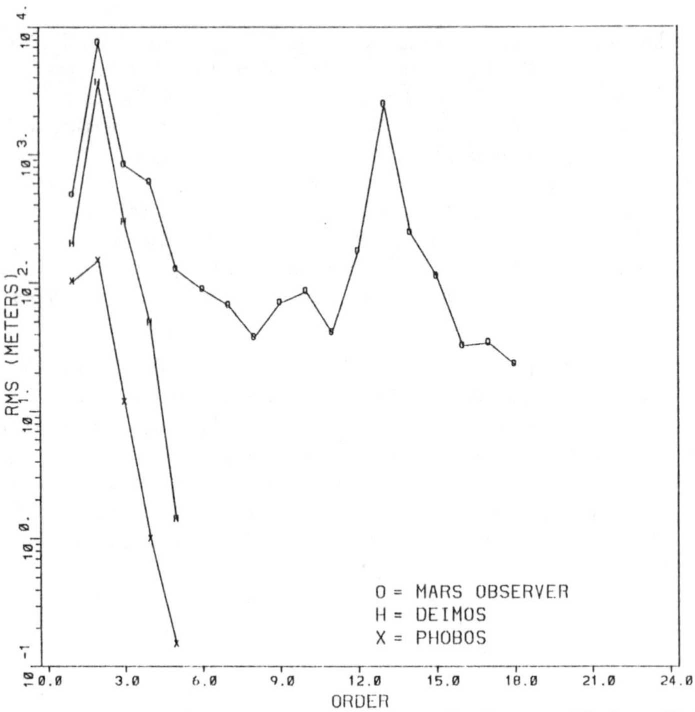

Figure 1: Comparison of RMS of Transverse Orbit Perturbations on Phobos, Deimos and Mars Observer Using Balmino (18 x 18) Field.

Sinclair[20] analyzed all available Earth based observations of the positions of the satellites of Mars from 1877 to 1969. His results included estimates of the gravitational constant (GM), and the oblateness of Mars (J_2). Born[21] determined the spherical harmonic coefficients of the Mars gravity field to degree 6, the Mars gravitational constant, as well as other parameters from a combination of Mariner 9 Doppler data and Phobos and Deimos optical data. Born reported that his estimates for the gravity field were determined principally by measurements of the motion of the Mariner 9 spacecraft, although Phobos and Deimos were perturbed strongly by the Mars oblateness (J_2).

It is clear from the work of previous investigators, that the optical data are sensitive to GM and J_2 of Mars. The objective of this study was to quantify the sensitivity of Phobos and Deimos to the Mars gravity field. Nonlinear or mutual interactions between Phobos and Deimos and Mars were not considered. This is a reasonable approximation for the purpose of determining the sensitivity of these satellites to the Mars gravity field since the mass ratios of Mars to Phobos and Mars to Deimos are 51×10^6 and 357×10^6 respectively.

Analysis. Due to their distance from the planet, and due to the low eccentricities of their orbits, these satellites are sensitive primarily to the low degree and order terms of the spherical harmonic expansion, as summarized in Figur. 1. The corresponding sensitivity of Mars Observer, which is discussed in a later section, is also shown for comparison. Although Deimos is farther from Mars than Phobos, the Deimos perturbations are larger due to resonance (Table 3).

It is not only the relative size of the perturbations that are important, but their frequencies as well. Phobos is in shallow resonance with the order 3 terms of the gravity field. These terms produce a resonance period of approximately 4.7 days, although the sensitivity to the order three coefficients is rather small. The various periods of the near resonances for both Phobos and Deimos are listed in Table 2. The effect of individual harmonic coefficients on the motion of the two satellites is given in

Table 3. The J_3 harmonic produces the largest perturbation on Phobos, with a period proportional to the rate of periapse circulation, or 412.5 days. Most of the signal strength of the J_{31} harmonic results from near resonant perturbations with a period of 0.463 days. Short period perturbations with a frequency of 1 or more cycles per satellite revolution produce the dominant periodic effects on the J_{22} and J_2 harmonics.

Deimos is beyond the synchronous orbit height for Mars. It is in primary resonance with the order 1 terms and in secondary resonance with all higher orders (see Table 2). In addition to the resonance condition, the unusually large equatorial ellipticity, manifested by the bulge in the Tharsis region of Mars, produces the kilometer level radial and along-track perturbations induced by the J_{22} harmonic (Table 3).

Table 2

PHOBOS AND DEIMOS RESONANT PERIODS

Phobos			Deimos		
Order	Period (Days)	Type	Order	Period (Days)	Type
1	0.463	Near	1	5.482	Primary
2	0.843 0.842	Near Near	2	2.741 2.334	Secondary
3	4.781 4.726 4.673	Primary	3	1.827 4.081	Secondary
			4	15.944 15.969	Secondary

The rate of sampling, or the frequency of the optical observations is much less than what would be theoretically adequate (assuming a Nyquist criterion) to resolve properly the effects of certain coefficients on the motion of the satellite. Frequently, though not always, multiple observations were obtained as the Viking Orbiters made a single flyby of Deimos or Phobos. The next sequence of observations might take place in a matter of days or weeks. The sparsity of the data relative to certain frequencies of perturbations suggests that only those coefficients that produce long period perturbations might be recoverable from the optical data. In the case of Phobos, this suggests that J_3 might be recovered easily from the optical data but that the J_{31}, and J_{22} harmonics would not.

Table 3

PHOBOS AND DEIMOS RMS POSITION PERTURBATIONS DUE TO MARS GRAVITY FIELD

(all values are in meters)

	Phobos			Deimos		
Coefficient	Radial	Transverse	Normal	Radial	Transverse	Normal
J_3	347.85	696.30	287.89	929.30	1837.77	15.59
J_{31}	26.90	105.30	1.67	27.35	198.74	1.69
J_{41}	0.23	1.46	8.60	0.09	0.11	0.30
J_{22}	18.36	147.25	6.97	974.83	3610.27	5.63
J_{32}	0.42	0.29	32.96	1.28	3.21	5.98
J_{42}	4.39	4.38	0.37	2.07	7.87	0.61
J_{33}	9.60	11.68	11.24	110.56	291.54	0.36
J_{43}	0.41	0.71	1.54	0.18	0.84	0.02
J_{44}	1.05	0.98	0.04	19.39	40.45	0.02

In order to make an accurate appraisal of the potential for use of the optical data to recover

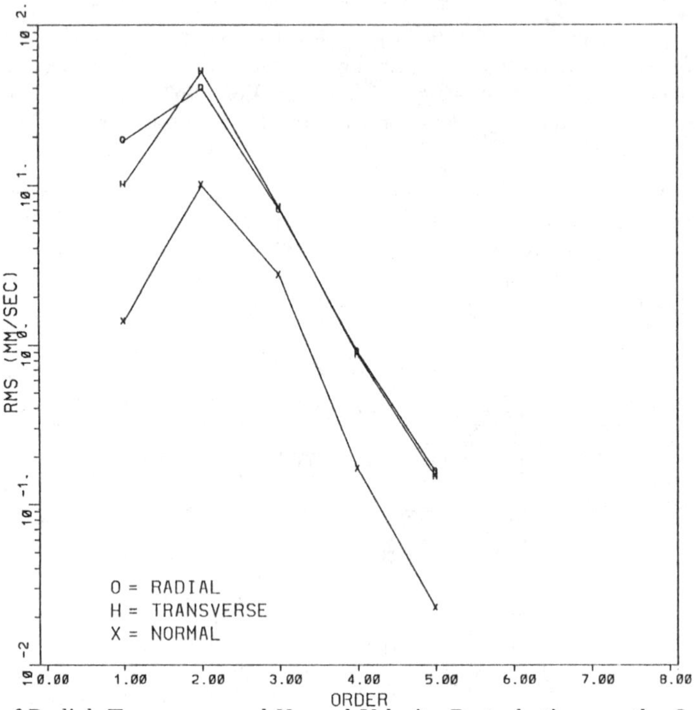

Figure 2: RMS of Radial, Transverse, and Normal Velocity Perturbations on the Orbit of Phobos.

coefficients of the Mars gravity field other than the J_2 harmonic, a discussion of the precision of the optical observations is essential. Sinclair[20], and others[22] have processed this data to derive an ephemeris of the satellites, as well as to estimate their secular acceleration. They found that the RMS of the residuals of the Earth based observations range from 0.15 to 0.95 arcseconds[22]. If one accepts these residuals as an indication of the observation uncertainty, then even at the most favorable oppositions, when Mars is at only 0.395 AU, these uncertainties translate into large position errors, from 43 to 270 km. Clearly, the ability to recover J_2 from this data results from the long-term measurement of the satellite's secular motion. Given the sparsity of the data relative to the frequency of the principal periodic perturbations induced by the Mars gravity field, and the observational uncertainties inherent in the Earth based data, it is apparent that only the spacecraft based observations have the potential to contribute significantly to a comprehensive Mars gravity solution.

The Mariner 9 optical data consist of 80 pictures of the satellites acquired during the approach and orbital phases of the mission[23]. Uncertainties in camera pointing and in the spacecraft position at shutter times map into position uncertainties for the satellites of 3 to 10 km. The Viking astrometric data consist of 275 spacecraft centered observations spanning four years, 1976 to 1980[24]. Duxbury and Callahan attribute the dominant errors in this data as uncertainties in knowledge of the spacecraft position, and center of figure errors. The latter derives from an inability to discriminate accurately the center of mass from the center of figure in these irregularly shaped natural satellites. Duxbury and Callahan maintain that the effects of Orbiter position errors can be mitigated by adjusting the Viking orbits while solving for the Phobos/Deimos ephemerides This results in a limiting accuracy for this data of 1 km (1σ). If a global control network were established, the center of finding errors would be reduced an order of magnitude. This would result in a limiting accuracy of a few hundred meters for the optical data, if the errors in the Viking spacecraft orbits are modeled properly.

Even making the most favorable assumption, that the Viking astrometric data will ultimately have a precision of a few hundred meters, it is extremely difficult to envisage how these observations could be used to estimate more than a few coefficients of the Mars gravity field. From Table 3, only the J_3, J_{22}, and J_{31} harmonics perturb the orbit of Phobos by more than 100 meters. Also from Table 3, only the J_{22}, J_{31}, and J_{33} harmonics produce effects of this magnitude on the orbit of Deimos. The Deimos J_3 perturbation is extremely long period (10276.9 days). In summary, the optical data of Phobos and Deimos are sensitive only to the lowest degree and order terms of the Mars gravity field.

The velocity perturbations due to the Mars gravity field on the orbit of Phobos were also calculated using Kaula's theory. These results, summarized in Figure 2, are representative of the sensitivity of the Doppler tracking data to a transponder on Phobos or of a spacecraft in an orbit similar to that of the natural satellite. The in-track perturbations dominate, and order 3 is the highest order of the gravity field that can be directly sensed, assuming DSN tracking with a precision of 1 mm/sec. These results should also reflect the information content of any Doppler tracking of the Soviet Phobos II spacecraft as it neared the satellite Phobos.

Mariner 9

Introduction. Mariner 9 entered Mars orbit on November 14, 1971, and remained operational until October 27, 1972, when the attitude control propellant was exhausted[25]. In order to maximize data returned through the Goldstone DSN site, two orbit trims were performed on November 16, and December 31, 1971[26]. Throughout its mission, Mariner 9 was in a near 12 hour orbit. This results in Mariner 9 being in resonance with the even orders of the Mars gravity field, and in near resonance with all orders (see Table 4).

Table 4

PERIODS OF MARINER 9 RESONANT AND NEAR RESONANT PERTURBATIONS

Resonant		Near Resonant	
Order	Period (days)	Order	Period (days)
2	18.40	1	1.03
4	9.20	3	1.09
6	6.13	5	1.15
8	4.60	7	1.22
10	3.68	9	1.32
12	3.07	11	1.42
14	2.63	13	1.54
16	2.30	15	1.68
18	2.04	17	1.85

From orbit insertion through March, 1972, the spacecraft was tracked on a near continuous basis using the 26 meter antennae at the DSN stations of Goldstone, Madrid, and Woomera (Australia). Although the data were contaminated by gas leaks from the attitude control system, producing accelerations on the levels of 1.0×10^{-12} km/sec^2 the near continuous tracking permitted Reasenberg et al.[27] to process data from six 19 day spans to construct a complete 6 by 6 gravity field for Mars. These long data arcs allowed complete sampling of the resonance effects in addition to taking advantage of the short period effects manifested near periapsis.

As the Earth-Mars distance reached 2.0 AU in March, 1972 (compared to 0.80 AU at the start of the mission), the noise from the 26 meter antennae increased dramatically. From March 1972, through the end of the Mariner 9 mission, the spacecraft was tracked with only the 64 meter antenna at Goldstone, resulting in the acquisition of only one pass of tracking data per day[25]. Using data gathered in June, and July, 1972, when the Mariner 9 orbit aspect angle was around zero degrees, Sjogren, et al.[28] reduced 28 independent 2 hour arcs near periapsis to estimate 92 surface mass points on Mars. These are comparable to a field complete to degree 4 with zonals through J_9, when compared to spherical harmonics. Over the course of the mission, the Mariner 9 Navigation Team was able to refine a field complete to degree 10 from the tracking data[26]. These results from various investigators suggest that the tracking data are sensitive to coefficients of the Mars gravity field at least up to degree and order 10. The sensitivity analysis performed using Kaula's solution with SLOP, as well as numerical integration with GEODYN and UTOPIA agree with these results.

Analysis. Figure 3 summarizes the RMS of the position perturbations induced by the Mars gravity field on the orbit of Mariner 9. The high eccentricity of the Mariner 9 orbit necessitated the evaluation of Kaula's solution out to $q = \pm 40$. The peaks induced by the resonances at the even orders are evident. Figure 4 summarizes the velocity perturbations induced on the spacecraft orbit by the Mars gravity field.

The sensitivity of Mariner 9 to the higher orders of the Mars gravity field is limited. In a simulation, GEODYN was used to generate two trajectories for seven days with the full 18 by 18 field, and the same field truncated at degree and order 12. Adjusting the initial conditions and performing a least squares fit between the two trajectories resulted in a total RMS difference in position of 6.02 meters and in velocity of 0.166 m/s.

The S-band tracking data at 2200 MHz had a nominal precision of 1 to 3 mm/sec. The data noise rose with proximity to solar conjunction, reaching a peak of 15 to 20 cm/sec in late August, and early September, 1972[26]. Under optimum conditions, this figure indicates that Mariner 9 orbit should be sensitive to coefficients out to degree and order 12. Figure 4 also underscores the importance of the orientation (aspect angle) of the spacecraft with respect to the Mars-Earth line. If the aspect angle is small, then the in-plane radial and transverse perturbations are emphasized. If the aspect angle approaches 90 degrees, normal (out of plane) velocity perturbations are emphasized. If only normal and transverse velocity perturbations are sensed then, then the limit of the sensitivity to the Mars gravity field will be degree and order 8. Processing of Mariner 9 tracking data at GSFC has shown that the aspect angle ranged from 37°.48 to 41°.74 from January 13, 1972 to February 10, 1972. Over this time span, the range of orbit orientations degrades the sensitivity to the radial velocity perturbations by 20%.

Software Comparison. The high eccentricity Mariner 9 orbit was used as a test case for evaluating the results based on Kaula's linear solution computed by SLOP. These results were compared against numerical integration with both GEODYN and UTOPIA. The linear results for the resonance coefficients were further compared with the POLOP program.

The procedure for GEODYN and UTOPIA involved generating trajectories with and without a coefficient or a set of coefficients, and performing a least squares fit between the reference and the perturbed trajectory. The resulting RMS differences between the fitted and the baseline trajectories represents the sensitivity of the orbit to a coefficient or a set of coefficients of the gravity field. The RMS of these orbit differences should agree with those derived from Kaula's linear theory.

The results from GEODYN and UTOPIA were generally found to agree with the SLOP results to within 5%. A sampling of the comparison versus GEODYN is shown in Table 5.

In a separate analysis, the amplitude of the periodic resonant perturbations as predicted by SLOP and by POLOP were compared. The amplitude of the semimajor axis, eccentricity, and inclination perturbations predicted by linear theory and by numerical integration agree to within 5%. The close agreement indicated in Tables 5 and 6 validates the application of Kaula's theory to the analysis of the orbital perturbations of the high eccentricity Mariner 9 orbit.

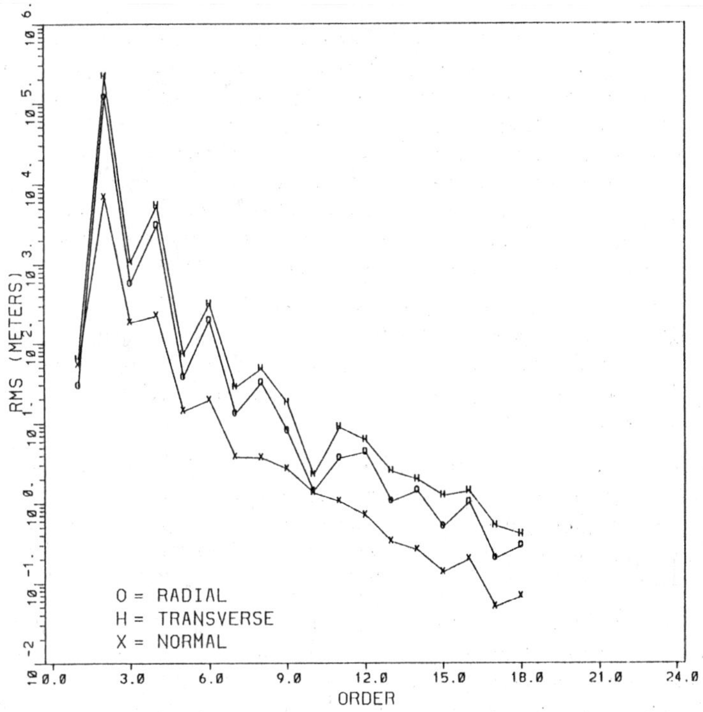

Figure 3: RMS of Radial, Transverse, and Normal Position Perturbations on the Mariner 9 Orbit.

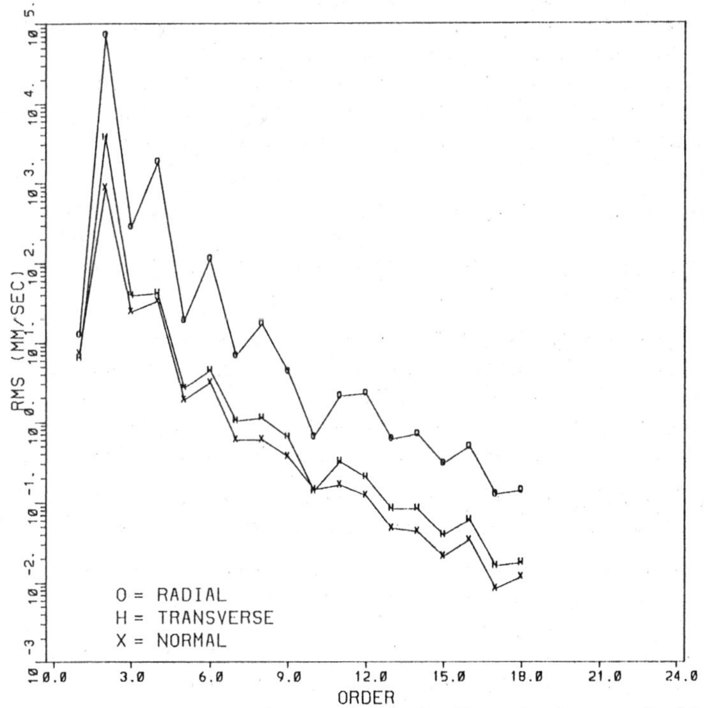

Figure 4: RMS of Radial, Transverse, and Normal Velocity Perturbations on the Mariner 9 Orbit.

Table 5

MARINER 9 SENSITIVITY TO THE GRAVITY FIELD SOFTWARE INTERCOMPARISON: GEODYN VS SLOP

		Position RMS (m)			Velocity RMS (mm/s)		
Coefficient	Program	Radial	Transverse	Normal	Radial	Transverse	Normal
J_{31}	SLOP	27.26	65.56	56.48	11.02	6.62	7.73
	GEODYN	24.44	58.13	61.27	10.03	6.35	8.32
J_{33}	SLOP	599.15	1111.71	182.79	301.19	42.27	25.23
	GEODYN	601.00	1130.18	182.67	305.85	42.83	25.52
J_{43}	SLOP	47.45	108.29	18.35	24.74	4.81	2.46
	GEODYN	48.60	109.25	18.62	25.54	4.81	2.52
J_{55}	SLOP	47.77	96.03	15.26	24.58	3.65	2.16
	GEODYN	48.75	99.61	15.48	25.58	3.74	2.21
J_{65}	SLOP	12.89	29.13	4.13	6.78	1.21	0.59
	GEODYN	12.94	30.20	4.19	6.95	1.26	0.60
J_{75}	SLOP	2.83	7.04	2.56	1.50	0.32	0.31
	GEODYN	2.85	7.09	2.61	1.54	0.32	0.32
J_{77}	SLOP	11.81	24.94	3.70	6.25	0.92	0.54
	GEODYN	12.29	26.36	3.82	6.65	0.95	0.56
J_{87}	SLOP	6.05	13.90	1.76	3.26	0.56	0.26
	GEODYN	6.23	14.66	1.81	3.44	0.58	0.28

Table 6

COMPARISON OF AMPLITUDE OF ORBIT ELEMENT PERTURBATIONS ON MARINER 9 ORBIT DUE TO RESONANCE SOFTWARE INTERCOMPARISON: POLOP VS. SLOP

Coefficient or order	Program	semimajor axis (m)	eccentricity	inclination (deg)
J_{22}	SLOP	6459.8	4.53×10^{-4}	4.63×10^{-2}
	POLOP	6552.0	4.61×10^{-4}	4.70×10^{-2}
J_{32}	SLOP	549.5	3.43×10^{-5}	4.02×10^{-3}
	POLOP	595.2	3.48×10^{-5}	4.08×10^{-3}
J_{44}	SLOP	436.1	8.60×10^{-6}	2.54×10^{-3}
	POLOP	445.3	8.79×10^{-6}	2.58×10^{-3}
J_{54}	SLOP	105.2	4.05×10^{-6}	6.65×10^{-4}
	POLOP	105.0	4.05×10^{-6}	6.64×10^{-4}
Order 2	SLOP	6537.5	4.61×10^{-4}	4.70×10^{-2}
	POLOP	6664.1	4.68×10^{-4}	4.78×10^{-2}
Order 4	SLOP	360.6	6.53×10^{-6}	2.08×10^{-3}
	POLOP	362.8	6.65×10^{-6}	2.09×10^{-3}
Order 6	SLOP	31.1	7.63×10^{-7}	1.84×10^{-4}
	POLOP	34.2	8.94×10^{-7}	2.02×10^{-4}
Order 8	SLOP	6.8	2.34×10^{-7}	4.20×10^{-5}
	POLOP	7.4	2.22×10^{-7}	4.53×10^{-5}

Viking Orbiters

Introduction. Viking Orbiters 1 and 2 entered Mars Orbit on June 19, 1976 and August 7, 1976. Their primary task consisted of deployment of the Viking Landers. During the course of their missions, which lasted until July 25, 1978 for Viking Orbiter 2 (VO-2), and until August 7, 1980 for Viking Orbiter 1 (VO-1), the spacecraft served as relays for the landers, collected hundreds of images of the planet, and conducted experiments with other scientific instruments.

The Viking orbiters performed numerous orbit trims and maneuvers in order to maintain relay with the landers, to optimize coverage and data returned from the orbiter science instruments, and to enable close flybys of the satellites, Phobos and Deimos. The sequence of mission operations and the resulting orbit changes are summarized in detail by Hildebrand et al.[3], Snyder[4], Snyder and Evans[5] and O' Neill et al.[29]

The orbit of VO-1 was synchronized over that of the Viking-2 Lander in September, 1976, at an altitude of 1500 km. This resulted in an orbit in deep resonance with the Mars gravity field. Orbit trims in January, and February, 1977 permitted frequent Phobos flybys in February, 1977. In March, 1977, the periapsis of VO-1 was lowered to 300 km, providing highly localized gravity information during each periapsis pass. On December 2, 1978 the orbit was again changed to make it synchronous with the Mars rotation rate.

The inclination of VO-2 was changed from 55°, to 75°, and then to 80° during the first 5 months of the mission, in order to provide better visibility of the entire planet. In addition periapsis was lowered to an altitude of 800 km on December 20, 1976, and to 300 km in altitude in October, 1977.

An important consideration in the recovery of a global gravity field for Mars is an adequate geographical distribution of low altitude periapsis passes. The periapsis of VO-2 remained exclusively in the Northern Hemisphere of Mars. Periapsis of the VO-1 spacecraft occurred in the Northern Hemisphere through the end of 1977, and then remained in the Southern Hemisphere through the end of 1979. This contrasts with Mariner 9, whose periapsis remained close to 23° S throughout its mission, due to its proximity to the critical inclination. The combined Viking and Mariner tracking data did not sense the local gravity at the higher southern latitudes or near the poles. This is the reason Balmino[6] mentions that the zonals higher than degree 10 in his field are poorly determined.

The S-band tracking data of the Viking orbiters are less complete than the near continuous tracking provided by the Mariner 9 data. The relative sparsity of the data, the frequent orbit maneuvers, and gas leaks in the attitude control system limit the length of the arcs of data that can be used. Formal computed total initial position uncertainties for six day arcs of Viking tracking data, recently processed with GEODYN, range from 360 to 1500 meters[30]. These are an order of magnitude larger than the computed uncertainties obtained from GEODYN processing of Mariner 9 data, and this reflects the smaller quantity of available tracking data.

Numerous investigators have analyzed the Viking Orbiter tracking data in combination with the Mariner 9 data to derive estimates of the Mars gravity field. Gapcynski, et al.[31] used a short arc technique, utilizing four hours of data centered on periapsis to derive a sixth degree and order gravity field. This technique minimized computation time, and aliasing due to unmodeled accelerations. Christensen and Williams took advantage of the orbit element variations commensurate with the orbit period. Analyzing data from the 1500 km phases of the VO-1, and VO-2 missions, they estimated a sixth degree and order gravity field[32]. Christensen and Balmino[33] processed both Mariner 9 and VO-1, and VO-2 data to obtain a gravity field complete to degree and order 12. Balmino et al.[6], later extended this work, adding further arcs of Viking tracking data and derived an 18 by 18 field. Sjogren[34] processed data from the low altitude (300 km) phases of VO-2 in 2 hour arcs centered on periapsis. He computed the line of sight accelerations that revealed gravity anomalies associated with distinct topographic features.

The analyses by Sjogren, Balmino, and Gapcynski demonstrated the information content of low altitude passes of tracking data. The analysis in this paper will demonstrate that the effects due to resonance also contribute to the sensitivities of the Viking Orbiters to the Mars gravity field. This consideration is important since both spacecraft were maintained in orbits that were nearly synchronous with the Mars rotation rate for extended periods of time.

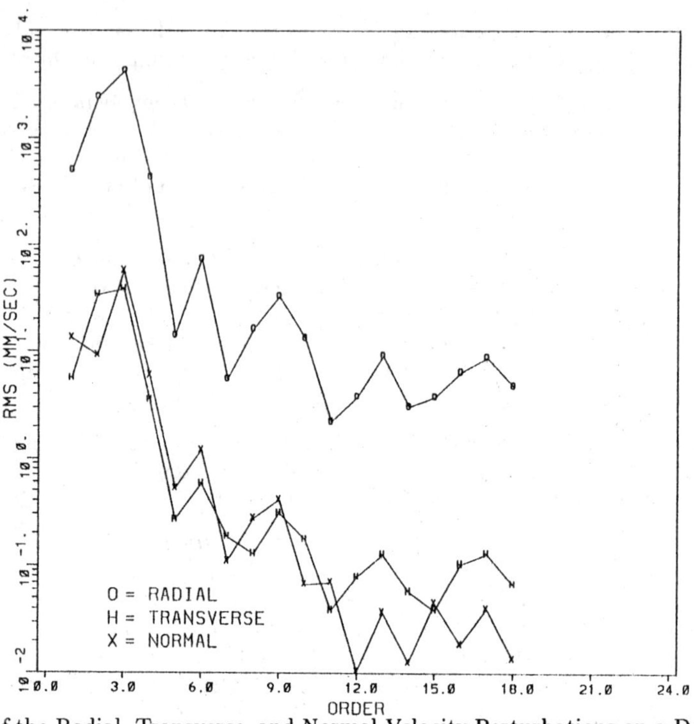

Figure 5: RMS of the Radial, Transverse, and Normal Velocity Perturbations on a Deeply Resonant Viking Orbit over 10.0 Days.

Analysis. It is not possible to apply Kaula's linear solution to the analysis of the Viking deeply resonant orbits due to the magnitude of the nonlinear effects. Instead POLOP was used to propagate the trajectories of a VO-1 orbit for 10.0 and 44.0 days using actual initial conditions obtained from an orbit fit on September 29, 1976. A trajectory was produced to study the effect of each order of the Mars gravity field. A mean set of orbit elements and secular rates was estimated for each trajectory. The RMS differences between the mean orbit and the actual POLOP trajectory were then computed. This process should simulate the adjustment of initial conditions that would occur in an actual orbit fit. The resonance perturbations are long period, with periods on the order of hundreds of days, so that changes in the orbit elements due to constituents of the gravity field will appear as secular trends. These will map into once or more cycles per revolution position and velocity differences between the mean orbit and the actual trajectory. The velocity perturbations are summarized in Figure 5. The radial velocity perturbations dominate the spectrum, with a signal well in excess of 1 mm/s at all orders. Order 1 is the primary resonance for this VO-1 orbit, yet its perturbations do not dominate the spectrum. The order 1 perturbations are indeed large amplitude, but also extremely long period, and have not had a chance to build up significantly over only ten days. Longer arc lengths amplify the magnitude of the resonance effects as shown in Fig. 6. The 44.0 day arc corresponds to the longest period for which a near continuous set of tracking data is available for VO-1.

These results underscore the fact that information about the Mars gravity field can be derived from analysis of the long period resonance effects. Nevertheless, it may be extremely difficult to obtain higher degree and order terms from the Viking data alone, even though the signature is potentially present in the data. The full potential of the data may not be realized until it is combined with data from other sources, such as from Mars Observer.

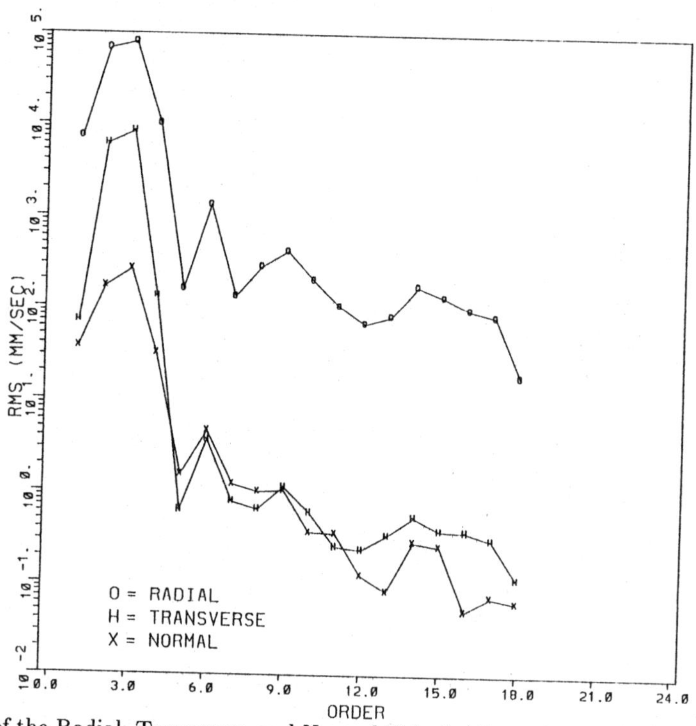

Figure 6: RMS of the Radial, Transverse, and Normal Velocity Perturbations on a Deeply Resonant Viking Orbit Over 44.0 Days.

Movement away from resonance reduces the sensitivity to the higher orders. To illustrate this, an orbit whose period was synchronized 3:1 with the period of Phobos was studied. This orbit (a = 19553 km, e = 0.7507, i = 39°.10) was located sufficiently off resonance so that its analysis remained tractable using Kaula's solution. The periods of the resonances ranged from 1 to 15 days from orders 1 to 14. Orders 15, 16, and 17 have resonance periods of 21, 50 and 12 days respectively. As illustrated in Figure 7, moving away from the deep resonance with the Mars rotational period diminishes the sensitivity to the higher orders. Measurement of the radial velocity perturbation would provide resolution to order 9. If only the cross track component of the velocity perturbations contribute to the Doppler signal then the limit of resolution would be order 5. Orbits whose periods are between 24.6 hours and 23.05 hours, would have sensitivities to the gravity field that are intermediate to the two cases cited above.

The evolution of the orbit aspect angle further reveals the information content of the tracking data of VO-1. Figure 8 illustrates the evolution of the orbit aspect angle for nine months, from September 29, 1976, through June, 1977. The sensitivity to the radial velocity perturbation induced by the Mars gravity field will be maximum on October, 1976, and May, 1977. From, December 21, 1976 through March 15, 1977, the Doppler sensitivity to the radial velocity perturbation was diminished by more than 20%.

Analysis of the Viking orbits is still in progress. These results suggest that exclusive of the local gravity information obtained by low altitude periapsis passes, resonance effects alone provide sensitivity to the Mars gravity field beyond the nominal 18 by 18 field computed by Balmino et al.[6]

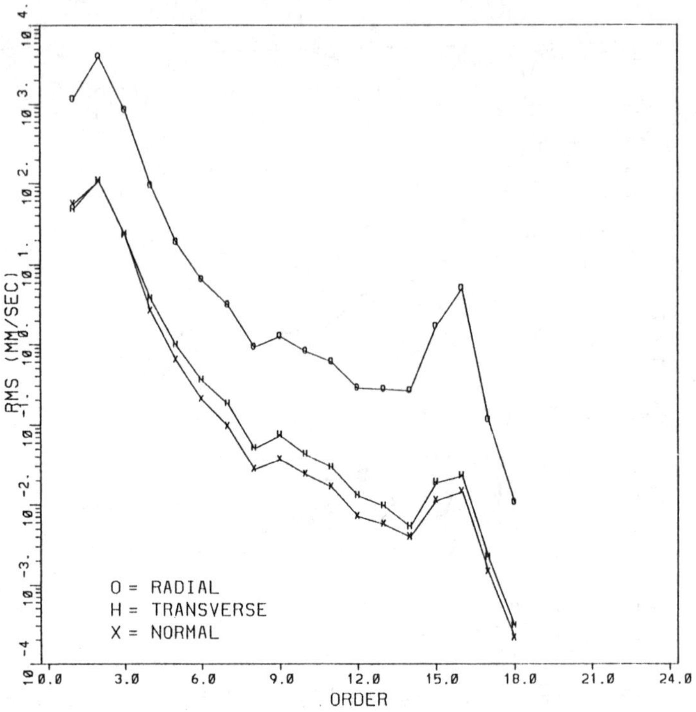

Figure 7: RMS of the Radial, Transverse, and Normal Velocity Perturbations on a Viking Near Resonant Orbit.

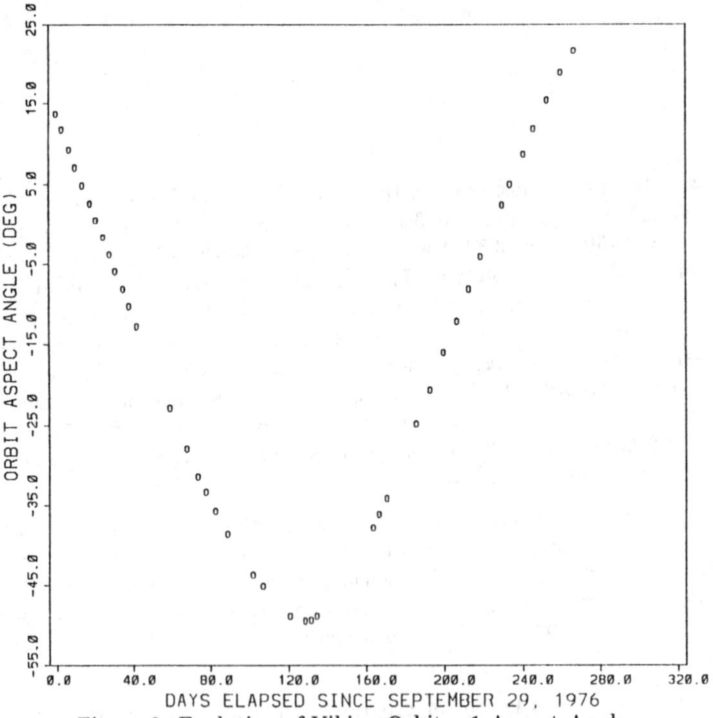

Figure 8: Evolution of Viking Orbiter 1 Aspect Angle.

Mars Observer

Introduction. The Mars Observer (MO) spacecraft, scheduled for launch between September 10, and October 9, 1992, will map the planet Mars, and study the atmosphere with a variety of instruments, as well as define globally the topography and gravity field[35]. The baseline mission plan requires the spacecraft to be inserted into a sun-synchronous, nearly circular orbit, with an inclination of 92°.8, and periapsis frozen near the South Pole. During the mapping mission, which is scheduled to last one Martian year, from December 1993 to November 1995, the satellite will be in a near three *sol* (Martian day) repeat ground track. In this orbit the ground track shifts 30 km to the west every three sols, or after 38 orbits. The ground track repeats exactly every 56 sols or 709 orbits[36].

Prior to insertion into the mapping orbit, the spacecraft will be placed in an interim orbit to facilitate the derivation of an improved gravity field for the Mars Observer mission. The spacecraft will remain in this interim orbit, or Gravity Calibration Orbit (GCO) for seven days, during which the spacecraft will be continuously tracked except for occultations. The orbital characteristics for both the mapping orbit and GCO are summarized in Table 7.

Table 7

MARS OBSERVER ORBITS

Parameter	Mapping Orbit (Baseline)	Gravity Calibration Orbit (GCO)
Semi-major axis	3758.3 km	3776.7 km
Eccentricity	0.00724	0.00720
Inclination	92°.816	92°.865
Argument of periapse	−90°.0	−90°.0
Nodal Orbital Period	7011.87 s	7063.24 s
Near Repeat Cycle		
• Mars Rotations	3	7
• Satellite Orbits	38	88
Mapping Cycle		
• Mars Rotations	56	51
• Satellite Orbits	709	641

Growth in excess of the total mass allotted to the spacecraft, as well as concerns about the potentially high drag environment in the mapping orbit have prompted consideration of moving the mapping orbit from the current 3 day repeat to the 7 day repeat of the GCO. The 33% reduction in mean atmospheric density, and the elimination of the need to reboost the MO spacecraft to a higher altitude at the end of the mission, reduce the propellant requirements by 13 kg[36].

Analysis. Both the original mapping orbit and the GCO were analyzed in this sensitivity study. These orbits have differing resonance characteristics. For the mapping orbit, the periods of the resonances range from 0.60 to 3.58 days (Table 8). The strong resonance at order 38 is driven by the 3 day near repeat of the mapping orbit ground trace to within 0°.51 (30.1 km) after 38 orbits. For the GCO, the periods of the main resonances are shorter than for the 3 day mapping orbit. The single exception is the resonance at order 25. For the GCO, this resonance is driven by a near repeat of the ground trace to within 3°.93 (232.8 km), as opposed to a repeat within 9°.14 (541.5 km) at this order for the 3 day mapping orbit. The geometry of the ground traces for the mapping and GCO orbits are summarized in Table 9.

Table 8 MARS OBSERVER RESONANCE AND NEAR RESONANCE PERIODS		
	Baseline Mapping Orbit Three Sol Near Repeat	GCO Seven Sol Near Repeat
Order	Period (days)	Period (Days)
12	1.51	1.75
13	3.19	2.48
14	0.78	0.73
24	0.76	0.88
25	2.88	5.77
26	1.60	1.24
37	0.99	1.36
38	29.67	4.24
39	1.06	0.83
49	0.60	0.76
50	1.44	2.99
51	3.58	1.56
52	0.80	0.62

Kaula's solution was used to compute the linear orbit element perturbations for both the original mapping orbit and the GCO. All coefficients were simulated using the power law for Mars, as derived by Balmino in his 18 by 18 Mars gravity model[6]. The computed orbit element perturbations were mapped into position and velocity components and are summarized in Figures 9 and 10. The resonances at orders 13, 25, and 38 are clearly evident when the perturbations are grouped and mapped by order. The order 38 terms produce long period (29.67 days) perturbations that have a total RMS amplitude of 4989 meters. For comparison, the other resonances produce perturbations with an RMS amplitude of 3073 meters for order 13, and 252 meters for order 25.

Table 9 SEPARATION OF GROUND TRACES FOR MARS OBSERVER MAPPING ORBIT AND GCO		
	Track Separation	
Number of Revolutions[a]	Mapping Orbit	GCO
13	9°.64 W	12°.36 W
25	9°.14 E	3°.93 E
38	0°.51 W	8°.84 W
50	18°.28 E	7°.86 E
51	10°.15 W	20°.78 W
63	8°.63 E	4°.49 E
76	1°.02 W	16°.85 W
88	17°.77 E	0°.56 W
89	10°.66 W	29°.20 W

[a] The number of revolutions completed before a near repeat of the ground trace corresponds to the order (m) of the gravity field with which the satellite will be in resonance. The closer the repeat, the stronger the resonance.

Smith et al.[37] suggest that the projected orbit error on Mars Observer for a ten day arc, with one DSN station tracking will be 11.2 to 19.1 meters. Figure 9 shows that coefficients out to order 30 will

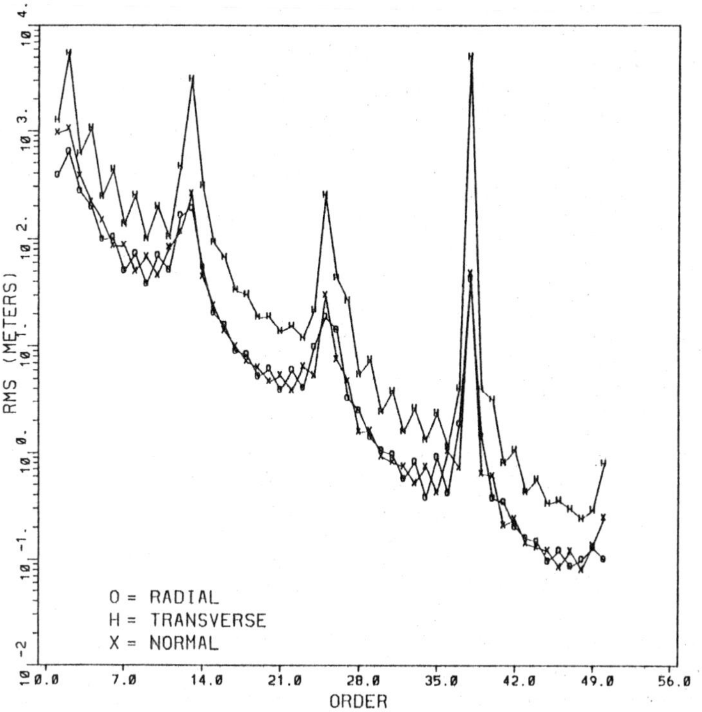

Figure 9: RMS of Position Perturbations for the Mars Observer 3-Day Mapping Orbit.

produce perturbations of this magnitude. This should not be taken as an indication of the size of the field that can be obtained from MO, since it is the precise measurement of the of the spacecraft velocity perturbations by the Doppler tracking data that provides the most sensitivity to the gravity field. Nominal precision for the X-band Doppler to be used on Mars Observer is expected to be 0.1 mm/sec. This is an order of magnitude improvement over the S-band Doppler tracking available for the Mariner 9 and the Viking Orbiters. At this level of precision, it is apparent that the Doppler data will provide sensitivity to the gravity field out to order 50 (Figure 10). In this spectrum, it is the radial velocity perturbation that has the highest sensitivity to the gravity field. For Mars Observer the sensitivity to the higher orders ($m > 42$) comes exclusively from the measurement of the radial velocity perturbation. This occurs because the transverse and normal perturbations are at the same level as the precision of the X-band Doppler data.

As discussed previously, the extent to which the Doppler data will sense the in-track (radial and transverse) perturbations versus the cross-track (normal) perturbations is determined by the aspect angle. For MO, the orientation of the orbit is known in advance, since the mission requires the satellite to be maintained in a 2 P.M. sun-synchronous orbit. As shown in Figure 11 the aspect angle approaches zero at two epochs during the nominal mission, 160 days, and 405 days after the initiation of the mapping mission on December 6, 1993. The first epoch follows perihelion and the onset of summer in the Southern Hemisphere of Mars. At this time, the bulk of the South Polar Cap has disappeared, and the surface pressure reaches its maximum value of approximately 7 mbars. The mean atmospheric density expected at the MO altitude at this time is 7 to 8×10^{-13} kg/m^3. During the second passage through the zero aspect angle, the mean atmospheric density expected at the MO altitude is 7 to 8×10^{-14} kg/m^3 [38]. This factor of 10 reduction in the mean atmospheric density will reduce the frequency of the drag compensation maneuvers and momentum dumps. These phenomena have deleterious effects on the recovery of coefficients of the gravity field, if not modeled correctly. In particular, these effects may become aliased with the large long period effects due to resonance, corrupting the precision with which the resonant orders can be recovered.

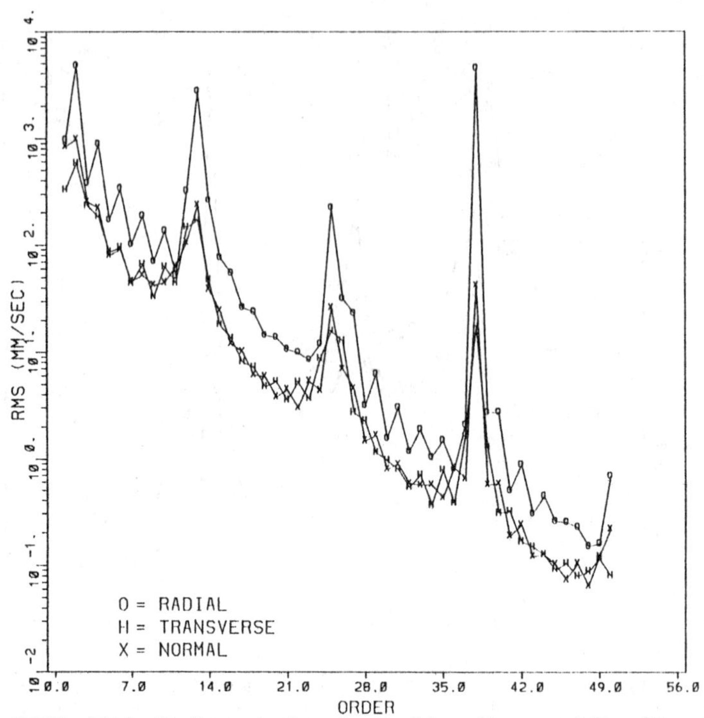

Figure 10: RMS of Velocity Perturbations for the Mars Observer 3-Day Mapping Orbit.

Tracking data obtained 405 days after the initiation of the mapping mission, around January 15, 1995, should be more pristine, and have maximum sensitivity to the Mars gravity field.

Figure 12 shows that the GCO manifests less dramatic resonance effects. The near seven day repeat diminishes the power of the resonance at orders 13, 38, and 51, but accentuates the power of the resonance at order 25 . The total RMS of the position perturbations is 1823 meters for order 13, 1093 meters for order 25, and 98 meters for order 38. Each one of the resonances manifested in the GCO or the original mapping orbit is driven by a close ground track repeat (Table 9). For a given orbit, the strength of the resonance and the period of the induced perturbations are proportional to the proximity of the repeat. From Table 9, a powerful resonance at order 76 is indicated for the 3 day mapping orbit. The change from a 3 day mapping orbit to the 7 day repeat as the baseline mapping orbit for the MO mission results in increased power for the resonances at orders 63, and 88. Whether the resonances at the higher orders (63, 76, and 88) produce observable signals on the spacecraft orbit has not been assessed. Their potential effects should be investigated in future research.

A pertinent question is whether a 50 by 50 gravity field will adequately describe the motion of Mars Observer. The effects of truncating the field at degree and order 50 can be assessed by studying the effects of higher degree terms, such as those from degrees 51 to 60. In this case, the coefficients beyond degree 50 are also simulated using the Mars power law as derived by Balmino[6]. The total velocity and position perturbations for selected orders are summarized in Table 10, for the baseline mapping orbit and the GCO. Note that the transverse position and the radial velocity perturbations will be the largest contributors to the total position and velocity perturbations.

For both the 3 sol mapping orbit and the GCO, there remains considerable residual information in the coefficients beyond degree 50, almost exclusively at the resonant orders. The power of the order 38 resonance for the baseline mapping orbit is manifest. It is extremely difficult to determine, *a priori*, how many terms at beyond degree 50 might be required to model this residual signal.

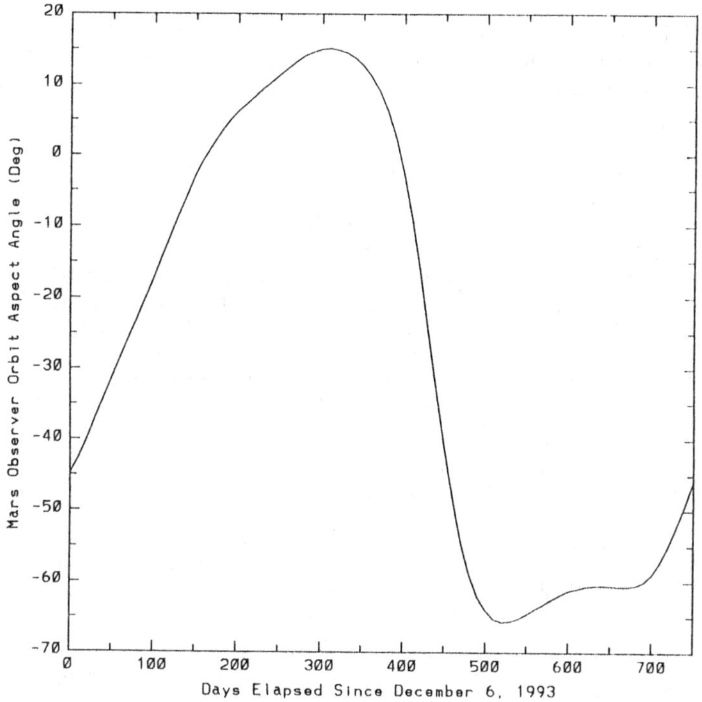

Figure 11: Variation of Mars Observer Orbit Aspect Angle Over the Duration of the Mapping Mission.

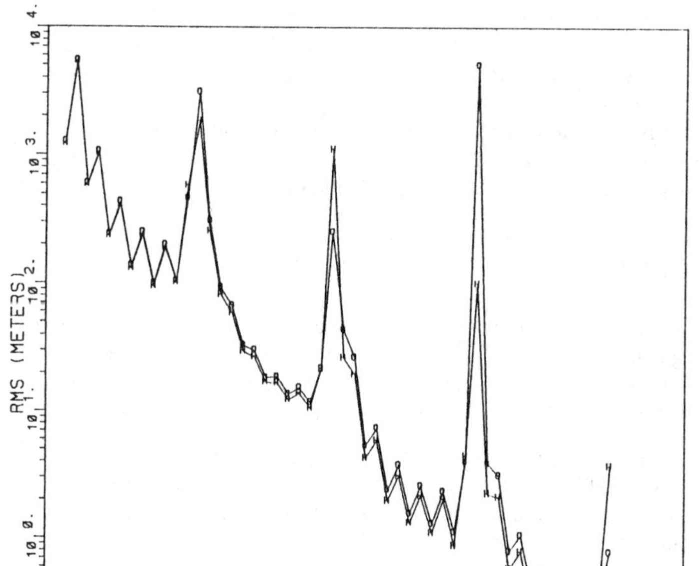

Figure 12: Comparison of Transverse Orbit Perturbations on Mars Observer: 3-Day Mapping Orbit vs. 7-Day GCO.

A portion of this signal will be aliased into the lower degree terms at the resonant orders. Only additional data from other low altitude satellites could enhance the separability of these coefficients, reduce the intra-order correlations and improve the accuracy of the estimates for these constituents of the gravity field.

Table 10

**MARS OBSERVER
TOTAL RESIDUAL POSITION AND VELOCITY PERTURBATIONS DUE TO TRUNCATION OF GRAVITY FIELD AT DEGREE AND ORDER 50**

	3 sol Near Repeat (Baseline Mapping Orbit)		7 sol Near Repeat (GCO)	
order	Position (m)	Velocity (mm/s)	Position (m)	Velocity (mm/s)
1	0.75	0.56	0.66	0.51
2	1.73	1.40	1.34	1.13
13	4.05	3.49	1.99	1.68
25	2.48	1.93	9.86	8.69
38	689.53	619.47	13.84	12.39
50	1.55	1.22	5.88	5.19
51	13.54	12.14	2.39	2.13
total (all orders)	689.69	619.60	18.43	16.35

Finally, the sensitivity study using Kaula's solution demonstrates that the increased altitude of the GCO with respect to the baseline mapping orbit diminishes the total sensitivity of the Mars gravity field by 10 to 25%. This is negligible for the lower degrees, where there is considerable signal. But at the higher degree terms this is quite significant. Since the radial velocity perturbation is the dominant component of the total velocity signal, when the aspect angle deviates from zero degrees, the loss in sensitivity for the higher degree terms will be more acute for the 7 day near repeat orbit (GCO) than for the baseline mapping orbit. A precise quantitative assessment requires a detailed numerical simulation, such as the analysis performed by Smith et al.[37]

CONCLUSIONS

The sensitivity of the Mars orbiters to the Mars gravity field has been evaluated and quantified in terms of the magnitude and frequency of their position and velocity perturbations. This information, in consort with the coverage and accuracy of the tracking data, provides insight into the strengths and weaknesses of each orbiter when treated as a sensor of the Mars gravity field.

The present analysis is predicated on the assumption that the Mars gravity field and power law derived by Balmino et al.[6] is an adequate representation of the actual behavior of the Mars gravity field. In the absence of other information, use of this gravity field and the power law is the only reasonable alternative.

A further assumption implicit in this analysis is that Mars is a smooth planet gravitationally, meaning that it has a smooth equipotential surface (areoid, or Martian geoid) that is well suited to modeling by a spherical harmonic expansion. Previous investigators, however, have determined that Mars has sharp and powerful gravity anomalies associated with distinct features on the planet's surface (e.g. volcanos and impact basins). Since Mars is known to be gravitationally rougher than the Earth, it may be reasonable to expect that a combination of spherical harmonics and mascons would be the most effective means of modeling the global gravity field of the planet.

The natural satellites, in addition to providing information for the Mars gravitational constant and J_2, also provide sensible information on a few low degree coefficients. While the other orbiters are much more sensitive to these low degree coefficients, the low inclinations of Phobos and Deimos can provide a valuable constraint in solving for the low order terms.

Mariner 9 and Viking Orbiter data provide strong information on the low degree and order coefficients. Mariner 9 will principally contribute to the even ordered terms out to about degrees 10 to 12. Low altitude periapsis passes by Viking have been shown to contain highly valuable local gravity information. In addition resonance effects provide sensitivity to the Mars gravity field beyond degree 18 whenever the Viking Orbiters were in orbits synchronous with the Mars rotation.

Under optimum viewing geometries, and during periods of reduced drag, Mars Observer tracking data should sense the Mars gravity field out to degree and order 50. The sensitivity at other epochs may be degraded (at the higher degrees) due to unfavorable orientations. This is a consequence of the in-plane velocity perturbations being of significantly higher amplitude than the out-of-plane velocity perturbations.

In the future there are plans and prospects for a number of further missions to Mars. Although, none of the missions are planned specifically for improving the Mars gravity field, the precise tracking of additional Mars orbiters can always be used for this purpose. It is plausible to expect that models of the Mars gravity field will evolve and improve in the same way that the series of Goddard Earth Models and other gravity models have improved our knowledge of the Earth's geopotential since the 1960's.

As refinements in the Mars gravity field are made through the use of MO and other orbiters, a better understanding of the planet's outer structure will follow. This in turn will contribute to resolving key questions about the planet's origin and evolution, including the nature of the Mars crustal dichotomy and the distribution of tectonic features associated with the Tharsis uplift.

ACKNOWLEDGEMENTS

H. B. Iz (ST Systems Corp., Lanham, Md.) provided the data for Figure 11. M. Zulauf (University of Colorado, Boulder) generously provided programming support for many of the computations. E. J. Christensen (JPL), and G. H. Born (University of Colorado, Boulder) provided valuable information and documentation on orbit determination for the Mariner 9 and Viking Orbiter spacecraft. It is our pleasure to acknowledge numerous helpful discussions on the subject of Mars gravity with F. J. Lerch, and D. E. Smith of NASA/GSFC. B.H. Putney's (NASA/GSFC) advice and expertise facilitated data processing with GEODYN.

This research was supported by NASA Graduate Student Researcher's Program Grant NGT-50329.

REFERENCES

1 Marsh, J. G., F. J. Lerch, B. H. Putney, D. C. Christodoulidis, D. E. Smith, T. L. Felsentreger, B. V. Sanchez, S. M. Klosko, E. C. Pavlis, T. V. Martin, J. W. Robbins, R. G. Williamson, O. L. Colombo, D. D. Rowlands, W. F. Eddy, N. L. Chandler, K. E. Rachlin, G. B. Patel, S. Bhati and D. S. Chinn, "A New Gravitational Model for the Earth From Satellite Tracking Data: GEM-T1," *J. Geophysical Research*, Vol. 93, pp. 6169–6215, June 10, 1988.

2 Marsh, J. G., "An Enhanced 'Satellite Only' Gravitational Model: GEM-T2," *EOS, Transactions, American Geophysical Union*, Vol. 69, p. 329, 1988.

3 Hildebrand, C. E., E. J. Christensen, D. H. Boggs, G. H. Born, H. Hokikian, and J. F. Jordan,

"Viking Satellite Orbit Determination," AIAA Paper 77–70, AIAA 15th Aerospace Sciences Meeting, Los Angeles, California, January 1977.

4 Snyder, C. W., "The Extended Mission of Viking," *J. Geophysical Research*, Vol. 84, pp. 7917–7933, December 1979.

5 Snyder, C. W. and N. Evans, "The Final Phases of the Viking Mission to Mars," *Icarus*, Vol. 45, pp. 2–24, 1981.

6 Balmino G., B. Moynot, and N. Valès, "Gravity Field of Mars in Spherical Harmonics Up to Degree and Order Eighteen," *J. Geophysical Research*, Vol. 87, pp. 9735–9746, November 1982.

7 Kaula, W. M., *Theory of Satellite Geodesy*, Blaisdell Publ. Co., Waltham, MA, 1966.

8 Rosborough, G. W. and B. D. Tapley, "Radial, Transverse and Normal Satellite Position Perturbations Due to the Geopotential", *Celestial Mechanics*, Vol. 40, pp. 409–421, 1987.

9 Wagner, C. A., "Geopotential Orbit Variations: Applications to Error Analysis," *J. Geophysical Research*, Vol. 92, pp. 8136–8146, July 1987.

10 *FX/Fortran Programmers Handbook, Version 4.0*, Alliant Computer System's Corp., Littleton, MA, July 1988.

11 Gooding, R. H., and D. G. King-Hele, "Explicit Forms of Some Functions Arising in the Analysis of Resonant Satellite Orbits," *Proc. R. Soc. London, A*, Vol. 422, pp. 241–259, 1989.

12 Burden, R. L., and J. D. Faires, *Numerical Analysis*, Prindle, Weber, and Schmidt, Boston, 1985.

13 Dennis, J. E., Jr., and R. B. Schnabel, *Numerical Methods for Unconstrained Optimization and Nonlinear Equations*, Prentice-Hall, Inc., 1983.

14 Cutting E., G. H. Born, and J. C. Frautnick, "Orbit Analysis for SEASAT-A," *J. Astronautical Sci.*, Vol. 26, pp. 315–342, 1978.

15 Martin, T. V., W. F. Eddy, D. D. Rowlands, and D. E. Pavlis, "GEODYN System Description," *E.G.G. Contractor Report*, 5 Volumes, E.G.G. WASC, Lanham, MD, 1985.

16 Schutz, B. E. and B. D. Tapley, "UTOPIA: University of Texas Orbit Processor," *IASOM TR-80-1*, Center for Space Research, The University of Texas at Austin, 1980.

17 Vincent, M. A., "The Planetary Observer Planning Software," *JPL EM 312/89-157*, Jet Propulsion Laboratory, Pasadena, CA, June 1989.

18 Academy of Sciences of the USSR, "Phobos: Exploration of Phobos, Mars, Sun, and Interplanetary Space," Space Research Institute, Moscow, 1987.

19 Rosborough, G. W., "Orbit Error Due to Gravity Model Error," *Proceedings of the AAS/AIAA Astrodynamics Specialist Conference*, pp. 1417–1425, Kalispell, Montana, August 1987.

20 Sinclair, A. T., "The Motions of the Satellites of Mars," *Mon. Not. R. Astr. Society*, Vol. 155, 249, 1972.

21 Born, G. H., "Mars Physical Parameters as Determined From Mariner 9 Observations of the Natural Satellites and Doppler Tracking," *J. Geophysical Research*, Vol. 79, November 1974.

22 Jones, D. H. P., A. T. Sinclair, and I.P. Williams, "Secular Acceleration of Phobos Confirmed From Positions Obtained at La Palma," *Mon. Not. R. Astr. Society*, Vol. 237, pp. 15–19, 1989.

23 Born, G. H., and T. C. Duxbury, "The Motions of Phobos and Deimos From Mariner 9 TV Data," *Celestial Mechanics*, Vol. 12, pp. 77–88, 1975.

24 Duxbury, T. C., and J. D. Callahan, "Phobos and Deimos Astrometric Observations from Viking," *Astronomy and Astrophysics*, Vol. 201, p. 169, 1988.

25 Born, G. H., E. J. Christensen, A. J. Ferrari, J. F. Jordan, and S. J. Reinbold, "The Determination of the Satellite Orbit of Mariner 9," *Celestial Mechanics*, Vol. 9, pp. 395–414, 1974.

26 O' Neil, W. J., J. F. Jordan, J. W. Zielenbach, S. K. Wong, R. T. Mitchell, W. A. Webb, P. E. Koskela, et al., "Mariner 9 Navigation," JPL Technical Report 32–1586, Jet Propulsion Laboratory, Pasadena, CA, November 1973.

27 Reasenberg, R. D., I. I. Shapiro, and R. D. White, "The Gravity Field of Mars," *Geophysical Research Letters*, Vol. 2, pp. 89–92, March, 1975.

28 Sjogren, W. L., J. Lorell, L. Wong, and W. Downs, "Mars Gravity Field Based on a Short Arc Technique," *J. Geophysical Research*, Vol. 80, July 1975.

29 O' Neil, W. J.,R. P. Rudd, D. L. Farless, C. E. Hildebrand, R. T. Mitchell, K. H. Rourke et al., and E. A. Euler, "Viking Navigation," JPL Publication 78–38, Jet Propulsion Laboratory, Pasadena, CA, November 1979.

30 Smith, D. E., Personal Communication, June 1989.

31 Gapcynski, J. P., R. H. Tolson, and W. H. Michael, Jr., "Mars Gravity Field: Combined Viking and Mariner 9 Results," *J. Geophysical Research*, Vol. 82, pp. 4325–4327, September 1977.

32 Christensen, E. J., and B. G. Williams, "Mars Gravity Field Derived From Viking-1 and Viking-2: The Navigation Result," *J. Guidance and Control*, Vol. 2, pp. 179–183, May-June, 1979.

33 Christensen, E. J., and G. Balmino, "Development and Analysis of a Twelfth Degree and Order Gravity Model for Mars," *J. Geophysical Research*, Vol. 84, pp. 7943–7953, December 1979.

34 Sjogren, W. L., "Mars Gravity: High Resolution Results From Viking Orbiter 2," *Science*, Vol. 203, pp. 1006–1010, March 1979.

35 Albee, A., "Mars Observer: Science," *Proceedings of the NASA Mars Conference*, AAS Paper 86–135, July 21–23, 1986, Washington D. C., Univelt, San Diego, CA, 1988.

36 Albee, A., Attachment to letter to MO Project Staff and MO Experiment Representatives, "Description of 7-Sol Repeating Orbit at 379-Km Altitude and Comparison to the 361-Km Mapping Orbit," May 1989.

37 Smith, D. E., F. J. Lerch, J. C. Chan, D. S. Chinn, H. B. Iz, A. Mallama and G. B. Patel, "Error Analysis of the Mars Gravity Field From Radio Tracking of Mars Observer," Submitted to *J. Geophysical Research*, July 1989.

38 Esposito, P. B., "Mars Observer Navigation Plan: Update For 1992 Launch: Preliminary," *JPL Document D-3820*, Jet Propulsion Laboratory, Pasadena, CA, December 1988.

AAS 89-368

THE SPACE STATION REBOOST PROBLEM: MOTION OF AN ORBITING SPACECRAFT UNDER CONSTANT THRUST*

W. D. McClain†

Two analytical models are developed for the motion of an orbiting spacecraft under the influence of continuous low level thrust in the transverse direction. The models are developed in orbit elements using a perturbation method with the magnitude of normalized thrust acceleration as the small parameter. Formulations are presented for the equinoctial elements and for the classical elements. The equinoctial element formulation is closed form in the eccentricity at first order and a small eccentricity approximation is used at second order. The formal development of the classical element model is similar to the equinoctial model at first and second orders, however, due to the singularity in the equations of motion caused by a vanishing eccentricity the formalism breaks down for small eccentricity cases. Consequently, for small eccentricity cases it is not possible to obtain a complete zeroth order model for the eccentricity. The performance of each model is evaluated by comparison with numerical integration.

---

* This paper appears in full in the microfiche supplement to this volume, **AAS Microfiche Series**, Volume 59.

† The Charles Stark Draper Laboratory, Inc., 555 Technology Square, Cambridge, Massachusetts 02139.

MINIMUM-FUEL RESCUE TRAJECTORIES FOR THE EXTRAVEHICULAR EXCURSION UNIT*

J. M. Neff and W. T. Fowler†

The problem of determining minimum-fuel trajectories for rescuing astronauts or equipment which become separated from a space station is addressed. Using the Clohessy-Wiltshire equations of relative motion and assuming impulsive ΔVs, the minimum-fuel rescue problem is shown to be a parameter optimization problem. Minimum-fuel rescue trajectories are found for seventeen test cases using a recursive quadratic programming algorithm. The results are analyzed and general rules for astronaut rescue and equipment retrieval are developed.

---

* This paper appears in full in the microfiche supplement to this volume, **AAS Microfiche Series**, Volume 59. This work is sponsored by NASA under NGT-50222.

† Department of Aerospace Engineering and Engineering Mechanics, The University of Texas, Austin, Texas 78712-1085.

AAS 89-372

DYNAMICS OF STRAP-ON SEPARATION*

Rajeev Lochan†, K. Kumar‡ and V. Adimurthy**

The dynamics of separating strap-ons is the subject-matter of this paper. Euler's formulation is employed for description of motion of this multi-body system. Different models for ejection mechanisms are developed. On-going stage is assumed perfectly controlled whereas aerodynamics is ignored.

The geometry of collision is analyzed. Different schemes are proposed for "cleanliness" of separation. The assumption of perfect control for on-going stage is scrutinized next. The allowable limits on the imperfections of control are constructed. The methodology is demonstrated using two typical examples.

---

* Only an abstract for this paper was available for publication.

† Senior Scientist, Aerodynamics Division, VSSC, Trivandrum, India.

‡ Professor, IIT Kanpur, India.

** Head, Aerodynamics Division, VSSC, Trivandrum, India.

ORBITING DEBRIS

SESSIONS 5 & 9

Chairman: Darren S. McKnight
U. S. Air Force Academy

The following paper numbers were not assigned:

AAS 89-379

AAS 89-401

AAS 89-374

A NEW APPROACH IN TREATING THE BALLISTIC COEFFICIENT IN THE DIFFERENTIAL CORRECTION FITTING PROGRAM

William N. Barker*, Thomas J. Eller† and Leland E. Herder‡

This paper describes the results of an Air Force Space Command Space and Warning Systems Center sponsored study†††† to improve the accuracy of Tracking and Impact Prediction Special Perturbations software in use at the Space Surveillance Center. First, we consider historical data for a wide range of decayed satellites. In general, this data indicates that the satellite ballistic coefficient (a model parameter in the differential correction process) varies as a function of time in the last days prior to decay. At present, this effect is not modeled and the ballistic coefficient is held constant over the differential correction observation span. The new approach, which we present here, is a parameterization of the ballistic coefficient in the form of a simple linear function with time. The slope of this function is the time derivative of the ballistic coefficient which is treated as a new model parameter. Numerical results obtained from processing two important historical satellite decay cases are presented.

INTRODUCTION

At the Space Surveillance Center one of the primary missions is the accurate prediction of orbital decay of artificial Earth satellites. This process begins some fifteen to twenty days prior to a General Perturbations estimate of the decay date. At that time, Special Perturbations processing is initiated along with increased tasking to radar

* Astrodynamicist, Kaman Sciences Corporation, 1500 Garden of the Gods Rd., P.O. Box 7463, Colorado Springs, Colorado 80933-7463.

† Astrodynamics Project Manager, Kaman Sciences Corporation, 1500 Garden of the Gods Rd., P.O. Box 7463, Colorado Springs, Colorado 80933-7463.

‡ Physicist, Kaman Sciences Corporation, 1500 Garden of the Gods Rd., P.O. Box 7463, Colorado Springs, Colorado 80933-7463.

†††† Contract F05603-87-C-0001.

sensors for metric observational data. Orbit determination is accomplished using a first order linear, weighted, least square curve fitting process often called differential correction[11]. Sliding fits are used to process both old and new metric data until the satellite is no longer in orbit.

The force models used for orbit representation and ephemeris generation are those of the Jacchia dynamic atmosphere of 1965[1] and the World Geodetic System Earth gravity model of 1972[2]. The latter is truncated to sixth order and degree for satellite decay processing.

A direct step-by-step numerical integration of the total acceleration acting on the decaying satellite is employed. This is generally referred to as Cowell's Method[3]. The Gauss-Jackson[4] eighth order predictor-corrector formulas in ordinate form are used to integrate the equations of motion. Because of computer run-time constraints, the partial derivatives necessary for differential correction are computed analytically except for the secular variations due to atmospheric drag of orbit semi-major axis and eccentricity. These are integrated numerically using a low order (trapezoidal rule) integration.

Currently, the differential correction solution state consists of the equinoctial elements † and the satellite ballistic coefficient model parameter, B_0. After each solution, the new state is used to predict the time of decay. At specific intervals prior to predicted decay, messages are sent from the Space Surveillance Center alerting various agencies. The decay times contained in these messages are highly variable from prediction to prediction as decay approaches. The purpose of this paper is to present a new method which removes a significant amount of this variability and improves the overall prediction quality.

Analysis of historical data shows that during the days prior to decay, the ballistic coefficient, B, exhibits a pronounced variation with time. This variation is due to one or more of several causes: tumbling or weather vaning of the satellite, loss of mass, and variation of the drag coefficient, C_D. At present, none of these effects are modeled explicitly since this would require precise knowledge of the satellite size, shape, mass distribution and orbital attitude. Generally, these parameters are not known for most satellites tracked by the Space Surveillance Center. In addition, any unmodeled forces acting in the in-track direction, such as those caused by inaccuracies in the atmospheric density model, can manifest themselves in the solution ballistic coefficient.

In spite of the lack of specific knowledge of unknown forces acting on a decaying satellite, an improvement can be made by the introduction of a new model parameter. This parameter accounts for the fact that for many decaying satellites, the variation in B appears to be linear during the last several days prior to decay. The new model

---

† The set of orbital elements adopted for the Space Surveillance Center which are free from both low eccentricity and low inclination difficulties (see NOTATION).

parameter, $\dot{B}_0$, is the time derivative of the ballistic coefficient, B, which is parameterized as

$$B = B_0 + \dot{B}_0 (t - t_0) \quad (1)$$

where the subscript "0" indicates the quantities at the solution state epoch, t_0. The introduction of $\dot{B}_0$ requires that new semi-analytic partial derivatives for B_0 and $\dot{B}_0$ be derived.

The historical data analysis, development and verification of the partial derivatives, application of the new method to historical cases and conclusions are presented in the following sections.

HISTORICAL DATA ANALYSIS

The proposed new model requires an average ballistic coefficient, B_0, and the time rate of change of the ballistic coefficient, $\dot{B}_0$. These are to be determined over a finite time span using a differential correction fitting process. Several such differential corrections must be made over the history of an object's orbital decay, with a new B_0 and $\dot{B}_0$ determined for each differential correction. However, it is necessary to start the fitting process with close approximations of B_0 and $\dot{B}_0$ for the first differential correction. It is also important for the purposes of this paper to determine if B is linear in the last day or two before final decay. Good initial guesses at the values of these two parameters can be obtained by studying historical orbital decay data. The data will also show how nearly linear B is. In particular, we have at our disposal, the orbital decay data of 264 objects re-entering the Earth's atmosphere from Jan. 1977 to Mar. 1979, as well as a few cases outside this range. This period of time happens to contain those decays just prior to the previous solar maximum activity. The data come from Tracking and Impact Prediction log records from the Space Surveillance Center.

Table 1 shows by month and year, the number of objects which decayed during our data sample period. As mentioned above, the total number of orbitally decaying objects in our sample was 264. However, a sizeable number of these were impractical for study since they included rocket parts and "debris" which could be of any size, shape, or mass. These were excluded from the data for this study. Once these final cuts were made, there were still 156 useful configurations such as rocket bodies and payloads which presumably have well-known physical attributes. Because most of the data were made up of Soviet vehicles, we decided, for now, to confine our study to only Soviet rocket bodies and payloads which are believed to be uniform in shape, composition, and mass (within their configuration type: A-2, B-1, RORSAT, etc.). The behavior of B could be further studied according to configuration type (shape, mass) but then we

TABLE 1

DISTRIBUTION OF 264 TIP LOG DECAYED ORBITAL OBJECTS

YEAR:	1970											
MONTH:	J	F	M	A	M	J	J	A	S	O	N	D
NUMBER:								2				

YEAR:	1977											
MONTH:	J	F	M	A	M	J	J	A	S	O	N	D
NUMBER:	8	8	11	6	15	8	8	11	14	9	8	7

YEAR:	1978											
MONTH:	J	F	M	A	M	J	J	A	S	O	N	D
NUMBER:	11	9	9	13	6	12	7	8	11	13	8	18

YEAR:	1979											
MONTH:	J	F	M	A	M	J	J	A	S	O	N	D
NUMBER:	6	10	7									

YEAR:	1983											
MONTH:	J	F	M	A	M	J	J	A	S	O	N	D
NUMBER:	1											

- Selected 156 Soviet orbitally decayed object distribution:[6,7,8]
  - **ROCKET BODIES**

				DIA(m)	LNG (m)
21	A2	VOSTOK	(440 kg)	2.6	7.5
89	A2	VOSTOK	(2500 kg)	2.0	2.0
4	B1	COSMOS	(1500 kg)	1.65	8.0
8	C1	SKEAN	(2000 kg)	2.0	6.0
3	D1	PROTON	(1900 kg)	3.9	4.0
3	D1	PROTON	(4000 kg)	4.0	12.0
3	F1M	SCARP	(700 kg)	2.0	5.0
1	F1M	SCARP	(1500 kg)	2.5	6.0

  - **PAYLOADS**[§]

3	(875 kg)	?	?
1	(325 kg)		
1	(3640 kg)		
1	(680 kg)		
8	(400 kg)		
2	(550 kg)		
4	(4500 kg)		
2	(900 kg)		
1	(1120 kg)		
1	(3800 kg)		

§ The exact size and shape of the payloads are unknown at this point.

would not have enough statistics in any one subsample for a meaningful study. So, for this paper we have lumped the Soviet rocket body and payload data together in one large, conglomerated data sample. The distribution of these Soviet decay objects is given at the bottom half of Table 1 and consisted of 132 rocket bodies (particularly the A-2 series which included Vostok) and 24 payloads.

This uniformity is good for research, but on the other hand, it is contrary to operational experience where most of the physical characteristics of space objects are unknown to the Space Surveillance Center. By confining our study to the Soviet data subsample, we reduce many uncertainties in the problem.

To get an appreciation of why we pursued this study, one only needs to examine the results in Figures 1 and 2. Figure 1 is a scatter plot of the great circle length difference between the decay location predicted using the present astrodynamic theory employed by the Space Surveillance Center and the final location of impact on the Earth's surface. Each point in the scatter plot is an individual prediction derived from each differential correction for a given object. The decay histories of all 156 Soviet objects have been plotted. We will refer to an object's "history" as all differential corrections (on average about 7) listed in the Tracking and Impact Prediction Log as made in the final days of orbital decay. This is over a span of time less than 25 days from final decay and usually less than about 13 days for most of the data. The length of the Tracking and Impact Prediction history data span depended upon how important an object was (whether it had a nuclear reactor on board or was massive enough to reach the ground), how often the differential correction code was run, and how near to final decay the object actually was. Note that even as little as 1 day out, the predicted location could have been off by as much as 20,000 km (half the Earth's circumference). Keep in mind that this is absolute distance between predicted and actual impact locations. Beyond 1 day out from final decay, the error in predicted impact location was essentially the same as at 1 day out. An error between 10,000 and 20,000 km was typical. Obviously, we would like to improve the decay prediction location.

Figure 2 is a scatter plot of prediction time error versus time to decay for all histories. Beyond about 10 days out there appears to be a slight tendency to under predict the time. At three days out most predictions fall within $\pm$ 1 day. As shown in the figure, even at one day out, the prediction can be off by as much as 3 hours (approximately 2 revolutions). Generally speaking, as the fitting process got closer to final decay, there was a tendency to over-predict the time, in contrast to the under-predicting of final decay time several days out. Clearly, there is room for improvement in the decay time prediction also.

In order to improve the decay prediction further, we need to make improvements over the way that the ballistic coefficient, B, has been handled in the past. One particular behavior that we have noted in the data is that, in some cases, particularly several days out, B exhibits a periodic behavior over a typical history (with a period of about 12 days or so) usually followed by a linear trend 3 to 5 days before final decay. A short section of a sinusoidal curve (less than a quarter period) can be approximated by a straight line. In our data, during the last 3 to 5 days of decay, the B curve tends to

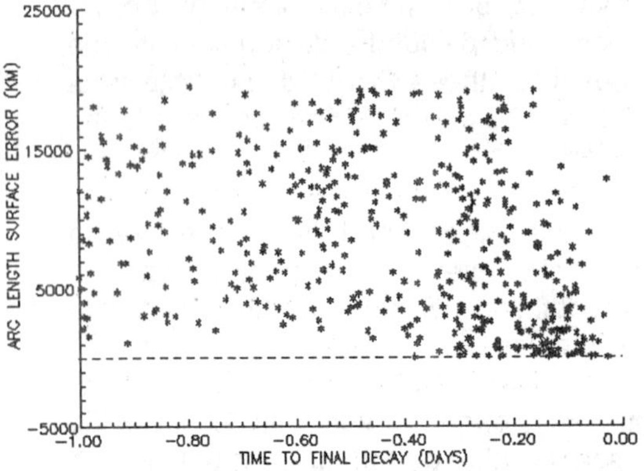

Fig. 1 The Arc Surface Length Error Between Predicted and Final Decay Location

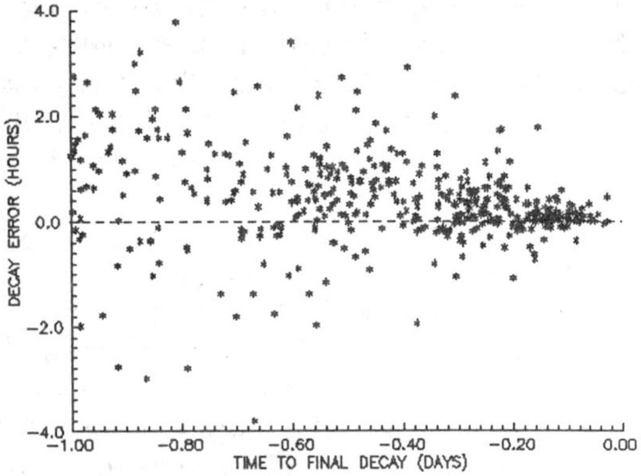

Fig. 2 The Difference in Predicted Time of Final Decay and the Actual Time of Final Decay, One Day Out

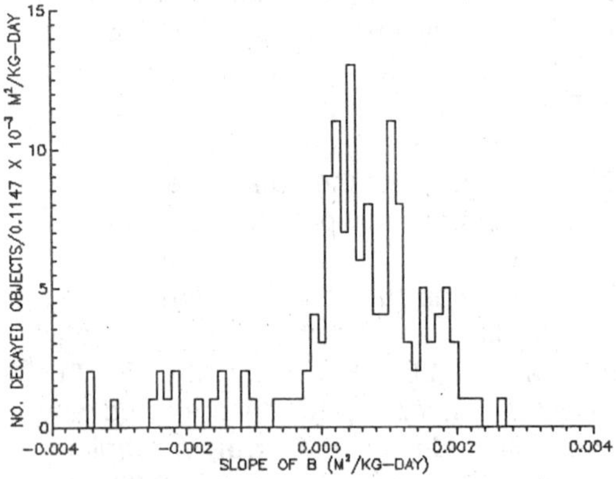

Fig. 3 Least Squares Fitted Slope of B for Linear Region of B

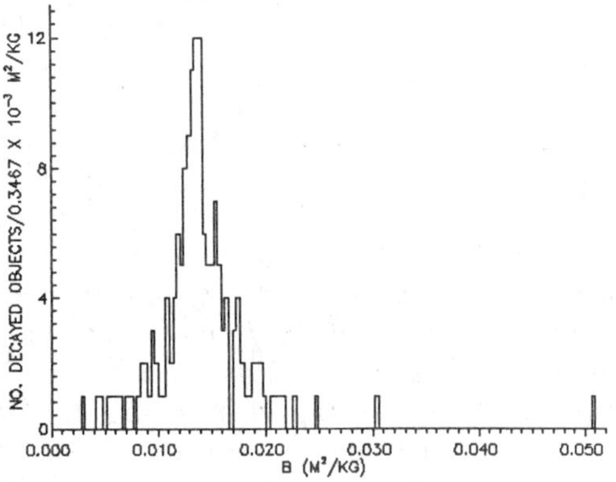

Fig. 4 Value of B when B Becomes Linear

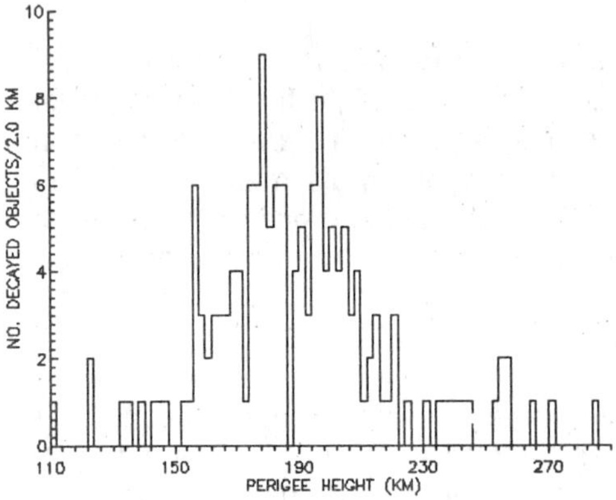

Fig. 5 Value of Perigee when B Becomes Linear

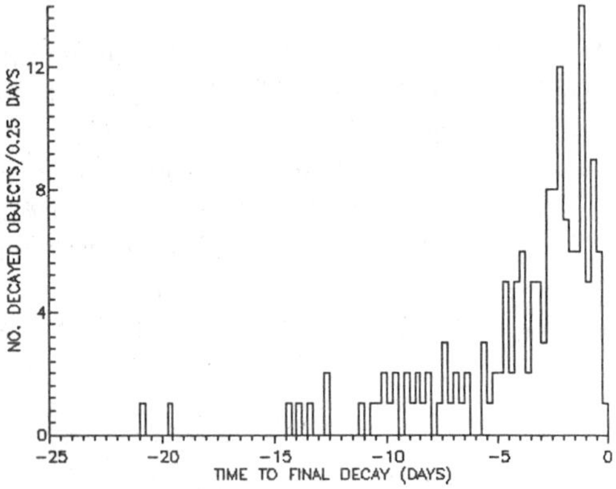

Fig. 6 Time to Final Decay when B Becomes Linear

shoot linearly upwards (or downwards) more quickly than the periodic trend of B several days out. Some of the data have a linear trend throughout the object's Tracking and Impact Prediction history of several days. Of the 156 decayed objects, 66 show a linear trend with time throughout their decay history (usually 5 days or less) while the rest show the hybrid behavior: part periodic 5 to 25 days out, followed by a steeper linear trend in the final 3-5 days. It is this final 3 to 5 day linear trend that we want to concentrate on for now.

Figure 3 is a histogram of the number of decaying objects versus the fitted slope of B at the point when B begins a linear trend in the object history. This value of B will be called, B_C. Only 17 of the 156 Soviet objects have slopes outside the histogram range. The average value of the slope is about 0.0008 m^2/kg-day (9.26 X 10^{-9} m^2/kg-s). The slope is positive, indicating that in the final days of decay, the current differential correction code over-predicts the final decay time.

The value of B_C for each object is shown collectively in the histogram of Figure 4. In the figure, B_C peaks at about 0.01375 m^2/kg. We have not shown the average B coefficient per object for the sake of brevity, but the distribution looks very similar. B_C can be used as an initial guess to B_0 in our model for the very first differential correction. Thereafter, the initial value of B_0 is the result of the previous differential correction.

There are other interesting data as well. Figure 5 is a histogram of number of objects versus the perigee height at B_C. As we can see, the distribution peaks at about 187 km. There is a depletion of data right at this value: this is an artifact of the bin size that we have chosen and nothing more. There are 155 decays within the limits of this plot. What the data suggest are that our linear form for B can be employed, perhaps, when the object has decayed to an altitude of 220 km or lower. Most of the decays at this point are within 5 days of final decay.

Lastly, the time to decay when B becomes linear is displayed in Figure 6. This is defined in the Tracking and Impact Prediction log data as the time from the end of the B_C fit interval to actual decay. Note in the histogram that we don't actually have data at 0 days out because there is a time interval from the end of the last data point to the final decay time. The final decay time must be projected from the last observational set of data (i.e., the last differential correction). The data is offset by this time interval.

We note, though, that most of the decay times predicted at B_C are within 5 days of the final decay date. With this information in mind, we can proceed with the discussion of our new method of handling the B term. Good starting values for B_0 and $\dot{B}_0$ have been determined, and we have found that the data tends to be linear in the last 5 days prior to decay. Our model should work well for this period of time.

DEVELOPMENT AND VERIFICATION OF THE PARTIAL DERIVATIVES

In order to solve for B_0 and $\dot{B}_0$ in the differential correction, new partial derivatives[5] must be developed. These "partials", used in the state transition matrix, relate elements at time t to elements and model parameters at the epoch time, t_0. Only first-order partials are used. The predominant effect is that of Keplerian motion but in many cases, such as a decaying satellite, perturbations are significant. In the Space Surveillance Center, the J_2 term of the geopotential perturbations is accounted for analytically in the partial derivatives. The atmospheric drag effect is too complicated for this and is handled by a low order (trapezoidal rule) numerical method during the step-by-step integration of the orbit.

The partials for drag are based on secular variations[5] in orbit semi-major axis, a, and eccentricity, e. The method of variation of parameters[10] leads to the following expressions for the equinoctial elements at time, t:

$$a_f = a_{fo} + a_{fo} \int_{t_0}^{t} \frac{e`}{e}\, dt \tag{2}$$

$$a_g = a_{go} + a_{go} \int_{t_0}^{t} \frac{e`}{e}\, dt \tag{3}$$

$$n = n_0 - \frac{3}{2} \int_{t_0}^{t} \frac{n}{a}\, a`\, dt \tag{4}$$

$$L = L_0 + n_0 (t - t_0) - \frac{3}{2} \iint_{t_0}^{t} \frac{n}{a}\, a`\, dt^2 \tag{5}$$

$$\chi = \chi_0 \tag{6}$$

$$\psi = \psi_0 \tag{7}$$

where the "grave" accent symbol (`) will be used to denote perturbative variations, i.e., variations with respect to time (t) that are caused by perturbing forces.

The integrands[5,9] are given by the following expressions:

$$a` = \frac{2a^2}{\mu} \dot{\mathbf{r}} \cdot \dot{\mathbf{r}}` \tag{8}$$

$$\frac{e`}{e} = \frac{1}{e^2} \left\{ \frac{a`}{a} \left[\frac{r}{a} (e \cos E) - \frac{1}{2} (e \sin E)^2 \right] + \frac{e \sin E}{\sqrt{\mu a}} \mathbf{r} \cdot \dot{\mathbf{r}}` \right\} \tag{9}$$

where boldface type indicates vector quantities.

First, we will consider the current method where the acceleration acting on a satellite due to atmospheric drag is given by

$$\dot{\mathbf{r}}` = -\frac{1}{2} B_0 \rho v_a \dot{\mathbf{r}}_a \tag{10}$$

and where the ballistic coefficient is constant with respect to time and given by

$$B_0 = C_D \frac{A}{m} \tag{11}$$

Substituting equation (10) into equations (8) and (9) and then factoring out B_0 gives

$$a` = B_0 \left(\frac{2a^2}{\mu} \right) \left(-\frac{1}{2} \rho v_a \right) \dot{\mathbf{r}} \cdot \dot{\mathbf{r}}_a = B_0 \zeta(t) \tag{12}$$

and likewise, we can write

$$\frac{e`}{e} = B_0 \xi(t) \tag{13}$$

where $\zeta(t)$ and $\xi(t)$ are functions of time, t.

Numerical studies[5] have shown that the following approximations can be made:

$$\frac{n}{a} \approx \frac{n_0}{a_0} \tag{14}$$

and

$$\iint_{t_0}^{t} a` \, dt^2 \approx \frac{1}{2} (t - t_0) \int_{t_0}^{t} a` \, dt \tag{15}$$

Using these, substituting equations (12) and (13) into equations (2) through (5) and taking partial derivatives with respect to B_0, we get the following partials necessary to solve for the current model parameter, B_0,

$$B_0 \frac{\partial a_f}{\partial B_0} = a_{f0} \int_{t_0}^{t} \frac{e`}{e} \, dt \tag{16}$$

$$B_0 \frac{\partial a_g}{\partial B_0} = a_{g0} \int_{t_0}^{t} \frac{e`}{e} \, dt \tag{17}$$

$$B_0 \frac{\partial n}{\partial B_0} = - \frac{3}{2} \frac{n_0}{a_0} \int_{t_0}^{t} a` \, dt \tag{18}$$

$$B_0 \frac{\partial L}{\partial B_0} = - \frac{3}{4} \frac{n_0}{a_0} (t - t_0) \int_{t_0}^{t} a` \, dt \tag{19}$$

$$B_0 \frac{\partial \chi}{\partial B_0} = B_0 \frac{\partial \psi}{\partial B_0} = 0 \tag{20}$$

Equations (16) through (20) represent the current method. Now we will introduce the notion that the ballistic coefficient is not a constant in equation (10) but is a function of time as shown in equation (1).

This leads to the following new expression for the drag acceleration:

$$\dot{\mathbf{r}}` = - \frac{1}{2} B \rho v_a \dot{\mathbf{r}}_a \tag{21}$$

Simply replacing B_0 by B in equations (12) and (13) gives:

$$a` = B \left(\frac{2a^2}{\mu} \right) \left(- \frac{1}{2} \rho v_a \right) \dot{\mathbf{r}} \cdot \mathbf{r}_a = B \, \zeta(t) \tag{22}$$

and,

$$\frac{e'}{e} = B\,\xi(t) \tag{23}$$

The partial derivatives of the equinoctial elements with respect to B_0 must be rederived by substituting equations (22) and (23) in equations (2) through (5) and taking the partial derivatives with respect to B_0 as before.

Differentiating equation (1) with respect to B_0 gives

$$\frac{\partial B}{\partial B_0} = \frac{\partial B_0}{\partial B_0} + \frac{\partial}{\partial B_0}\left[\dot{B}_0\,(t - t_0)\right] = 1$$

Using this result and the approximations of equations (14) and (15), we get the following new partials for B_0,

$$B_0 \frac{\partial a_f}{\partial B_0} = B_0\, a_{f0} \int_{t_0}^{t} \frac{1}{B} \frac{e'}{e}\, dt \tag{24}$$

$$B_0 \frac{\partial a_g}{\partial B_0} = B_0\, a_{g0} \int_{t_0}^{t} \frac{1}{B} \frac{e'}{e}\, dt \tag{25}$$

$$B_0 \frac{\partial n}{\partial B_0} = -\frac{3}{2} B_0 \frac{n_0}{a_0} \int_{t_0}^{t} \frac{1}{B}\, a'\, dt \tag{26}$$

$$B_0 \frac{\partial L}{\partial B_0} = -\frac{3}{4} B_0 \frac{n_0}{a_0} (t - t_0) \int_{t_0}^{t} \frac{1}{B}\, a'\, dt \tag{27}$$

The partials of χ and ψ with respect to B_0 are still as in equations (20).

The new model parameter is $\dot{B}_0$. The partial derivatives of the equinoctial elements with respect to $\dot{B}_0$ can be obtained by substituting equations (22) and (23) in equations (2) through (5) and taking the partial derivatives with respect to $\dot{B}_0$.

Differentiating equation (1) with respect to $\dot{B}_0$ gives

$$\frac{\partial B}{\partial \dot{B}_0} = \frac{\partial B_0}{\partial \dot{B}_0} + \frac{\partial}{\partial \dot{B}_0}\left[\dot{B}_0\,(t - t_0)\right] = t - t_0$$

Using this result and the approximations of equations (14) and (15), we get the following partials for $\dot{B}_0$,

$$\dot{B}_0 \frac{\partial a_f}{\partial \dot{B}_0} = \dot{B}_0\, a_{f0} \int_{t_0}^{t} \left[(t - t_0)\,\frac{1}{B}\,\frac{e^`}{e}\right] dt \qquad (28)$$

$$\dot{B}_0 \frac{\partial a_g}{\partial \dot{B}_0} = \dot{B}_0\, a_{g0} \int_{t_0}^{t} \left[(t - t_0)\,\frac{1}{B}\,\frac{e^`}{e}\right] dt \qquad (29)$$

$$\dot{B}_0 \frac{\partial n}{\partial \dot{B}_0} = -\frac{3}{2}\,\dot{B}_0\,\frac{n_0}{a_0} \int_{t_0}^{t} \left[(t - t_0)\,\frac{1}{B}\,a^`\right] dt \qquad (30)$$

$$\dot{B}_0 \frac{\partial L}{\partial \dot{B}_0} = -\frac{3}{4}\,\dot{B}_0\,\frac{n_0}{a_0}\,(t - t_0) \int_{t_0}^{t} \left[(t - t_0)\,\frac{1}{B}\,a^`\right] dt \qquad (31)$$

$$\dot{B}_0 \frac{\partial \chi}{\partial \dot{B}_0} = \dot{B}_0 \frac{\partial \psi}{\partial \dot{B}_0} = 0 \qquad (32)$$

A differential correction program was modified to include the time dependent ballistic coefficient formulation just derived. In order to verify the partial derivatives, simulated observations for an arbitrary sensor were generated from a special perturbations ephemeris. No systematic or random errors were added, therefore keeping the problem deterministic. Offsets were added to the equinoctial elements and model parameters on which the simulated observations were based. Using the simulated observations and the corrupted state, differential corrections were performed to estimate the original values. These offsets were as large as 40% for B_0 and 400% for $\dot{B}_0$. In all cases the fitting process converged to within 0.1% of the original values thereby validating the new partial derivatives.

APPLICATION OF THE NEW METHOD

The Soviet satellites COSMOS 954 and COSMOS 1402 were chosen for application of the new method. These were selected because they represent a type of satellite which has on board a small nuclear reactor capable of surviving re-entry and posing hazards to the ground. As such, these types are of great interest to the Space Surveillance Center and any improvement in the ability to predict their decay is deemed important. Another factor governing their selection was the availability of historical data with which to do analysis. The decay times and approximate locations for both are known and observational data for the ten days prior to decay was available. Using this data, two sets of twenty-five Special Perturbations differential corrections were performed on COSMOS 954 and two sets of twenty-six on COSMOS 1402. The first set on each used the current method of solving for the six equinoctial elements and the ballistic coefficient, B_0. The second set solved for the six elements plus B_0 and $\dot{B}_0$. This section of the paper compares the results of these differential corrections and their resulting decay time predictions. For purposes here, decay time and decay location are synonymous.

First, we will consider Figures 7 through 10 for satellite COSMOS 954. Figure 7 shows B_0 history over the twenty-five sliding-fit runs during the nine days prior to decay. The horizontal axis represents the prediction interval to the known decay time. Note the point at about minus three-quarters of a day where B_0 increases rapidly in an approximately linear fashion. The value of B_0 (marked B_C on the plot) at this time is approximately 0.006 m^2/kg. Figure 8 shows a comparison between the decay time error produced by the current method and the new method. For this plot, $\dot{B}_0$ was a solution parameter in the differential corrections but then was but set to zero for the decay predictions. The reason for this is that the linear model for B of equation (1) does not hold for long term predictions where B varies in a non-linear fashion. Fortunately, solving for $\dot{B}_0$ over the differential correction span and setting it to zero for prediction yields good results. As we will see presently, this procedure will become important in actual application. From Figure 8, the variability of decay time error suffers somewhat with the new method until the point where B_0 equals B_C. Figure 9 shows decay time error for this last phase on an expanded scale and includes predictions using the non-zero $\dot{B}_0$. Both procedures yield good results here but setting $\dot{B}_0$ equal to zero is preferred. This is because, in practice, a priori knowledge of the point where B_0 equals B_C will not be known. Figure 10 shows the best fit differential correction root mean square (RMS) of residuals for each of the twenty-five runs. As can be seen, the RMS history is considerably better for the new method, thereby indicating a better overall model.

Figures 11 through 14 are for COSMOS 1402. The same comments as those for COSMOS 954 apply except that the point at which B_0 equals B_C occurs some three and one half days from decay.

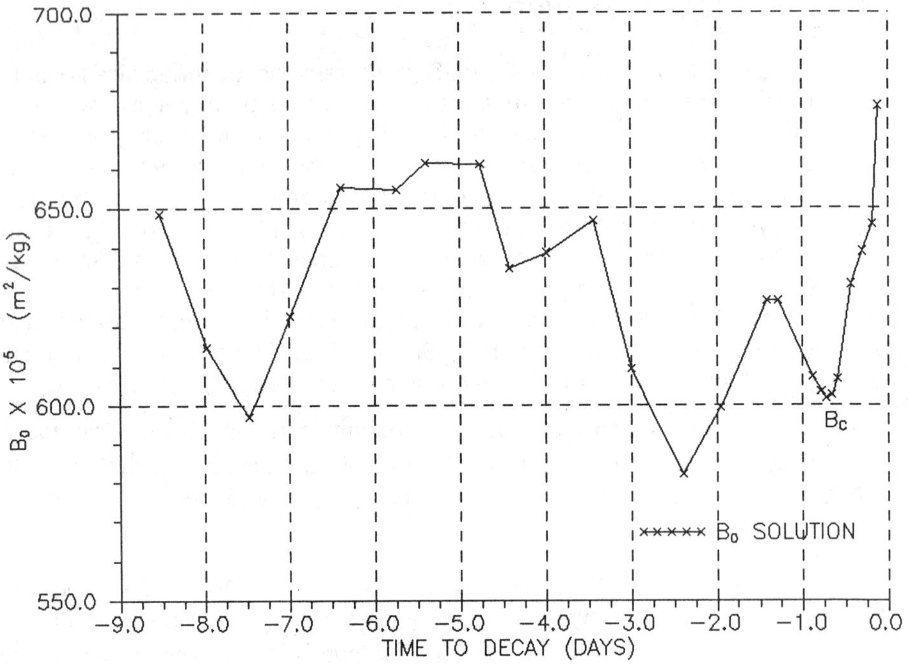

Fig. 7 Ballistic Coefficient (B_0) For COSMOS 954

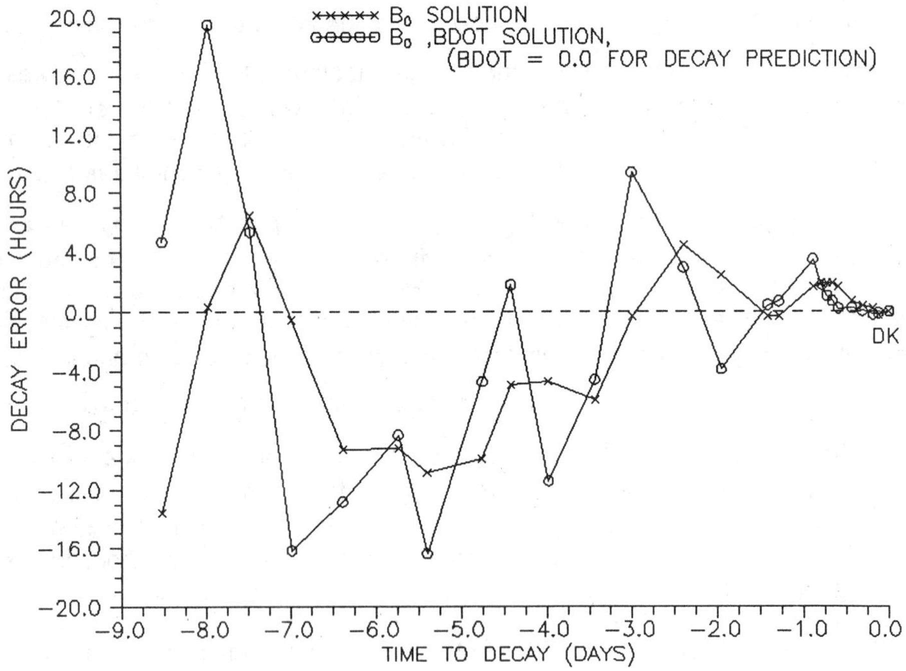

Fig. 8 Decay Prediction Error For COSMOS 954

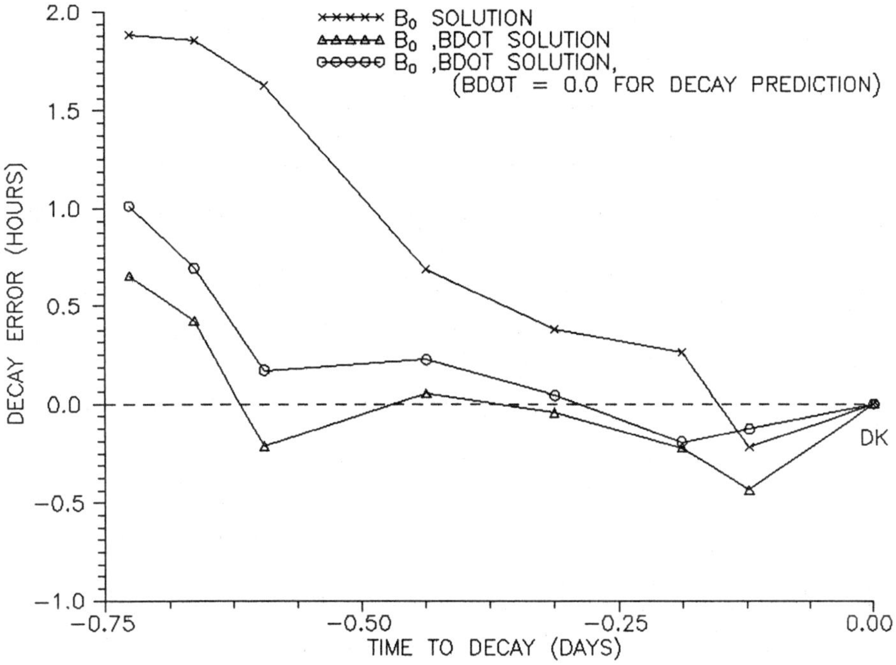

Fig. 9 Decay Prediction Error (Last .75 Day) For COSMOS 954

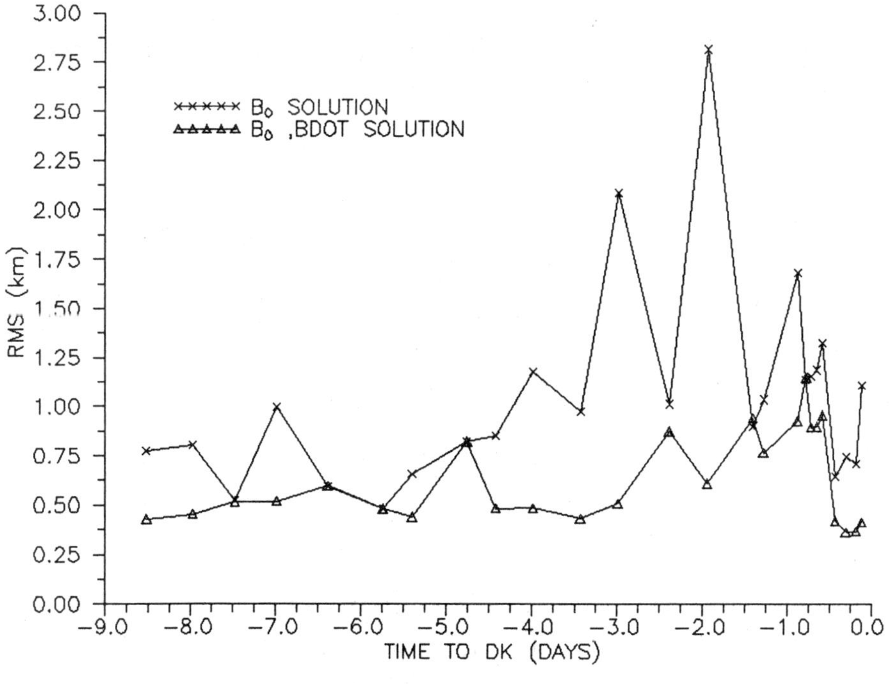

Fig. 10 Best Fit Differential Correction Residual Root Mean Square (RMS) For COSMOS 954

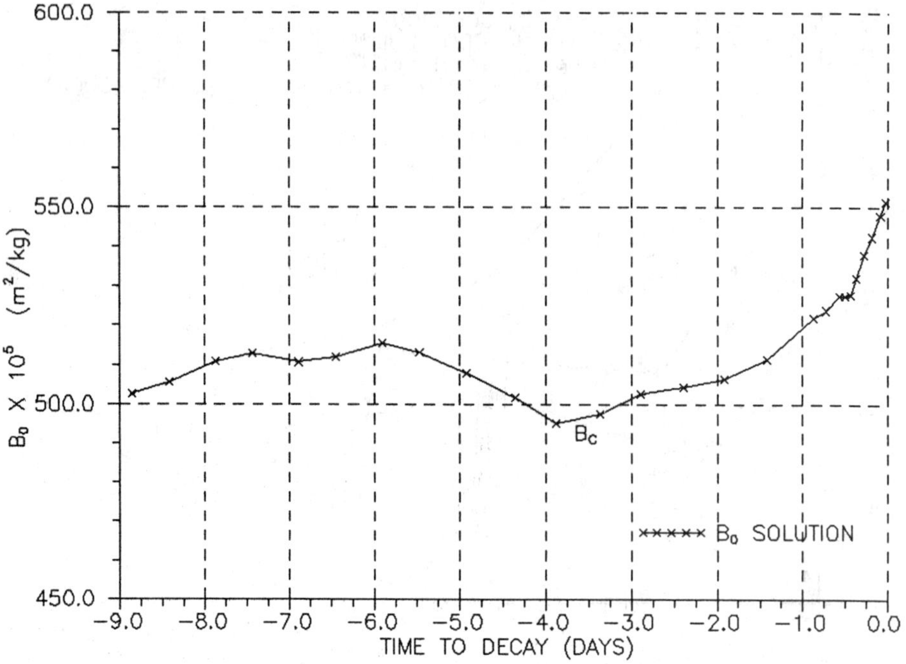

Fig. 11 Ballistic Coefficient (B_0) For COSMOS 1402

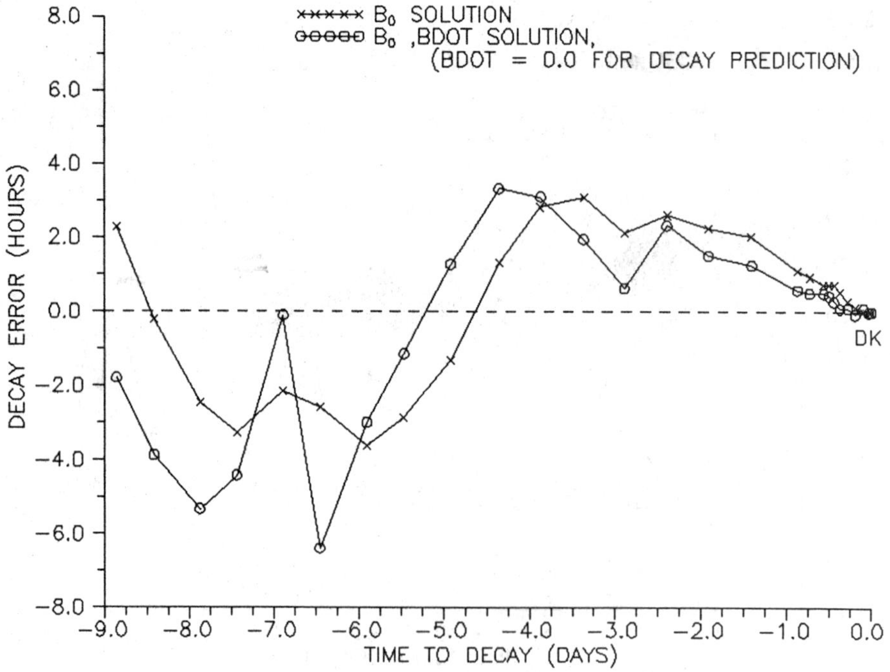

Fig. 12 Decay Prediction Error For COSMOS 1402

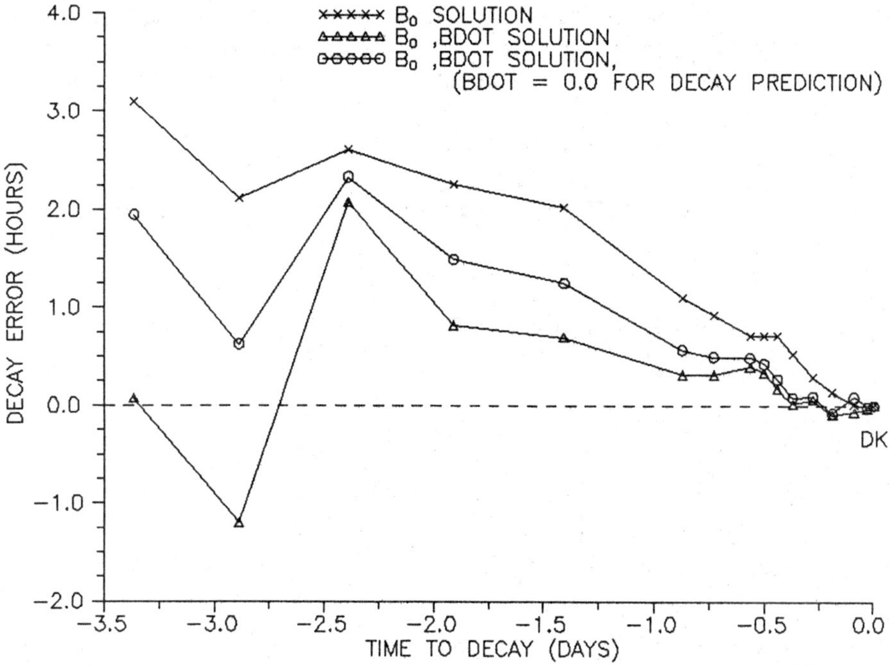

Fig. 13 Decay Prediction Error (Last 3.5 Days) For COSMOS 1402

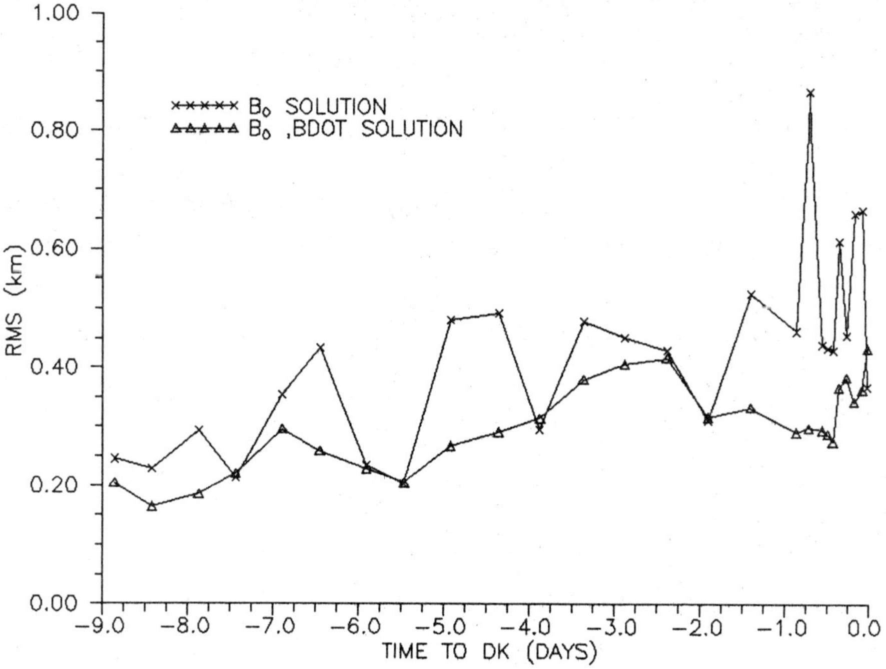

Fig. 14 Best Fit Differential Correction Residual
Root Mean Square (RMS) For COSMOS 1402

Since the historical analysis indicates that the linear trend in B tends to occur during the final 3 to 5 days prior to decay, only that time period will be considered in the following tables. Table 2 shows the decay time error and percent improvement for this period for both satellites.

TABLE 2
DECAY TIME ERROR ROOT MEAN SQUARE (RMS) OVER LAST 5 DAYS

Satellite	Decay Error RMS(hrs) B_0 Solution	Decay Error RMS(hrs) B_0, $\dot{B}_0$ Solution, $\dot{B}_0$=0.0 for Prediction	Improvement(%)
COSMOS 954	3.51	4.11	-17.09
COSMOS 1402	1.59	1.43	10.06

As expected, for this time period, COSMOS 954 did not improve since the linear trend in B did not begin until the last day. The degradation is not considered significant, however, in view of the improvement of the final day. Table 3 shows the decay error and percent improvement for both satellites after B becomes linear. As can be seen, the improvement during this phase is significant.

TABLE 3
DECAY TIME ERROR ROOT MEAN SQUARE (RMS) AFTER B BECOMES LINEAR

Satellite	Decay Error RMS(hrs) B_0 Solution	Decay Error RMS(hrs) B_0, $\dot{B}_0$ Solution, $\dot{B}_0$=0.0 for Prediction	Improvement(%)
COSMOS 954	1.22	0.48	60.66
COSMOS 1402	1.51	0.98	35.10

CONCLUSION

A method has been presented which improves the accuracy of decay time predictions during the final phase of the lifetime of an Earth orbiting artificial satellite. The method is based on some 156 historical cases where the behavior of the satellite's ballistic coefficient was analyzed. While helpful during this research, a priori knowledge of the satellite's ballistic coefficient behavior is not required for application of the method. Two important historical decay cases were processed with the new method yielding improvements in decay time prediction of 61% and 35% respectively.

ACKNOWLEDGEMENT

The authors wish to acknowledge the contributions of the following people: Bob Morris and Paul Major of AFSPACECOM/DOA; Major Darryl Boden, USAFA/DFAS; Dr. Dan DeBra, Stanford University; Dr. Bob Schutz, University of Texas; Dr. Angelo Miele, Rice University, Dr. Richard Battin, MIT; Hewett Phillips, NASA Langley. The manuscript was carefully prepared by Edmond D. Van Doren and Mary Rodgers of Kaman Sciences Corporation.

NOTATION†

a = orbit semimajor axis
$\grave{a}$ = time rate of change of a due to drag perturbations
a_f = equinoctial element; $a_f = e \cos(\Omega + \omega)$
a_g = equinoctial element; $a_g = e \sin(\Omega + \omega)$
A = effective cross-sectional area of satellite to atmospheric drag
B_0 = ballistic coefficient of a satellite defined at epoch; $B_0 = C_D A/m$
B = time variable ballistic coefficient
$\dot{B}_0$ = time rate of change of B defined at epoch
B_C = value of B when B trend becomes linear
C_D = drag coefficient (dimensionless)
e = orbit eccentricity
$\grave{e}$ = time rate of change of e due to drag perturbations
E = eccentric anomaly
i = orbit inclination
J_2 = coefficient of the second zonal harmonic of the Earth's gravitational potential
L = mean longitude; $L = M + \Omega + \omega$, (part of equinoctial set of elements)
m = satellite mass
M = mean anomaly
μ = Earth gravitational parameter
n = mean angular motion (part of equinoctial set of elements)

† Bold face type denotes a vector quantity; "grave" (`) denotes perturbative variation with time.

0 = subscript denoting epoch time
ρ = atmospheric density at satellite's altitude
$\mathbf{r}$ = inertial radius rector from the center of the Earth to the satellite
$\dot{\mathbf{r}}$ = inertial velocity vector
$\dot{\mathbf{r}}_a$ = velocity vector relative to the Earth's atmosphere
$\dot{\mathbf{r}}$` = inertial acceleration vector due to drag perturbations
t = time; when unsubscripted t means current time
v_a = magnitude of relative velocity vector; $v_a = |\dot{\mathbf{r}}_a|$
χ = equinoctial element for orbit-plane orientation; $\chi = \tan(i/2)\sin\Omega$
ψ = equinoctial element for orbit-plane orientation; $\psi = \tan(i/2)\cos\Omega$
ω = argument of perigee
Ω = longitude of the ascending node

REFERENCES

1. Jacchia, L.G., "Static Diffusion Models of the upper Atmosphere with Empirical Temperature Profiles", Smithsonian Astrophysical Observatory Special Reports SR-170, 1964, and SR-173, 1965.

2. Sepplin, T.O., "The Department of Defense World Geodetic System 1972", Technical Paper, Headquarters, Defense Mapping Agency, Washington, DC, May 1974.

3. Danby, J.M.A., *Fundamentals of Celestial Mechanics*, The Macmillan Company, New York, 1962, pp. 234-235.

4. Herrick, S., *Astrodynamics Volume 2*, Van Nostrand, London, 1972, Chapter 14.

5. Arsenault, J.L.; Ford, K.C. and Koskela, P.E., "Orbit Determination Using Analytic Partial Derivatives of Perturbed Motion", AIAA JOURNAL, Vol. 8, No. 1, January 1970, pp. 4-12.

6. TRW Space Log, TRW Systems Group, TRW Inc., Winter 1968-69, pp. 11-17.

7. TRW Space Log, TRW Space and Technology Group, TRW Inc., Vol. 23, 1957-1987, pp. 42-239.

8. TRW Space Log, TRW Systems Group, TRW Inc., 1974, pp. 6-7.

9. Moulton, F.R., *An Introduction to Celestial Mechanics*, Dover Publications, Inc., New York, 1970, pp. 404-405.

10. Herrick, S., *Astrodynamics Volume 2*, Van Nostrand, London, 1972, Chapter 16.

11. Baker, R.M.L. Jr., *Astrodynamics*, Academic Press, New York, 1967, pp. 77-95.

AAS 89-376

ESTIMATING AND MODELING THE BACKGROUND ORBITAL DEBRIS POPULATION

Robert D. Culp*, Timothy D. Maclay† and Ron Madler‡

One of the most difficult problems of quantifying the hazards from artificial space debris arises from the fact that most of the threatening particles are untrackable. The dangers from these particles can be characterized by modeling statistically the untrackable orbital debris population.

Two methods are presented for constructing a realistic model of the 1 mm - 10 cm space debris population. One method uses the larger, tracked and catalogued objects to create a statistical representation of the untrackable debris. The second method simulates the 97 known satellite breakups by generating the resulting debris fragments and produces a spatial density matrix of debris.

Samples from the databases generated by the two methods are presented and comparisons between them are made. It is concluded that both methods result in valid and useful statistical orbital debris models.

Introduction

As the number and duration of space activities increase, so does the concern about their safety and reliability. One issue that has recently received greater attention is that of orbital space debris. How to track it, model it, and avoid it have all become growing areas of interest to the space community. An ability to avoid debris during space operations is directly dependent on our ability to track and model it. While debris larger than 10 cm can be both tracked and modeled, a description of the smaller, untrackable particles must rely almost exclusively on mathematical models.

* Professor, Department of Aerospace Engineering Sciences, University of Colorado, Campus Box 431, Boulder, Colorado 80309-0431. Fellow AAS.

† Century XXI Graduate Fellow, Doctoral Student, Department of Aerospace Engineering Sciences, University of Colorado, Campus Box 431, Boulder, Colorado 80309-0431.

‡ NASA Graduate Fellow, Doctoral Student, Department of Aerospace Engineering Sciences, University of Colorado, Campus Box 431, Boulder, Colorado 80309-0431.

Two methods of generating a spatial density matrix describing the untrackable 1 mm-10 cm background debris population are presented here. One method bases an estimate on the present trackable population while the other method models the 97 known satellite fragmentations to produce another estimate.

Method 1 - Using Trackables

This method is designed to be a fairly easy way to estimate the untracked orbital debris by associating it with the catalogued debris. The rationale behind this is that there should be smaller untracked pieces of debris associated with any kind of tracked object. The tracked objects are separated into four classes: satellites, rocket bodies, debris not associated with a fragmentation, and fragmentation debris.

The principal assumption behind this method is that functioning satellites and rocket bodies will shed almost no untrackable debris larger than 1 mm in diameter. If, however, one was interested in objects smaller than 1 mm, this would have to be considered as a source of these objects. The tracked operational and fragmentation debris, on the other hand, surely have untrackable objects larger than 1 mm in very similar orbits. For example, a shroud will have pieces associated with its explosive bolts and other hardware.

The most important source of untrackable objects would have to be the 97 fragmentations which have occurred. Fragmentation debris accounts for close to 50% of the tracked objects in orbit now, and that figure only includes objects down to approximately 10 cm in diameter - the limit of present tracking radars for U.S. Space Command. The distribution of satellites can be seen in Fig. 1. Other researchers have done studies which indicate that the number of objects 1 cm and greater is actually between two and ten times the number of catalogued objects presently being tracked (Refs. 1 and 2), and there is probably an equal number of untracked particles between 1 mm and 1 cm. NASA is currently planning a debris radar which will be capable of monitoring cm size objects. It will eventually be used to evaluate the present debris background models (Refs. 3 and 4).

In order to estimate the number of pieces that are associated with each tracked object, the objects in orbit have been analyzed according to their number and size. These distributions are then modeled and an estimate of the number of smaller pieces associated with the larger pieces is obtained. This task of finding a multiplying factor for the debris is a most difficult one because different breakups have different characteristics. Violent fragmentations produce a large number of fragments, while others only produce a small number of pieces and decay rapidly. In this method, fragmentations which only produce a few pieces will be modeled as operational debris so that there will be fewer pieces associated with them.

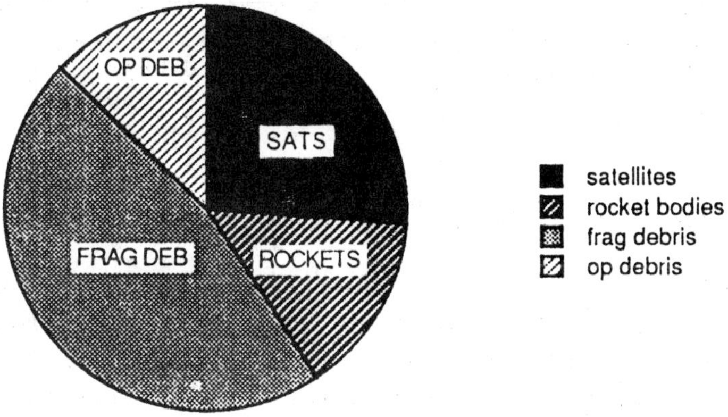

Figure 1: Distribution of Catalogued Satellites in Orbit

Object Classification

Determining the class of a tracked object is more involved than was originally expected because the datafiles used in this modeling did not contain any information on the type of the object. For this reason, another method was used to estimate the class of the object in question. This technique uses the piece number of the object associated with its launch.

The piece number is a number assigned by the U.S. Space Command when it tracks and catalogues an object. Each separate piece that is identified is assigned a piece number sequentially. For this reason, the satellite is assigned the number 1, the rocket body 2, and later debris is numbered as it is catalogued. Although there is some variation from this norm, it generally holds true and is used to categorize the objects in this modeling effort.

Fig. 2 shows that the ranges are quite distinct except for the region in which the operational debris and the fragmentation debris overlap. From this figure it can be seen that objects with numbers 1 and 2 are satellites and rocket bodies, respectively, while larger numbers are generally some form of debris. Distinguishing between the operational and fragmentation debris is more difficult. In this analysis, those objects with numbers greater than 12 are assumed to be fragmentation debris. This technique, though not always accurate, gives a very good estimate of the nature of the object from the information available.

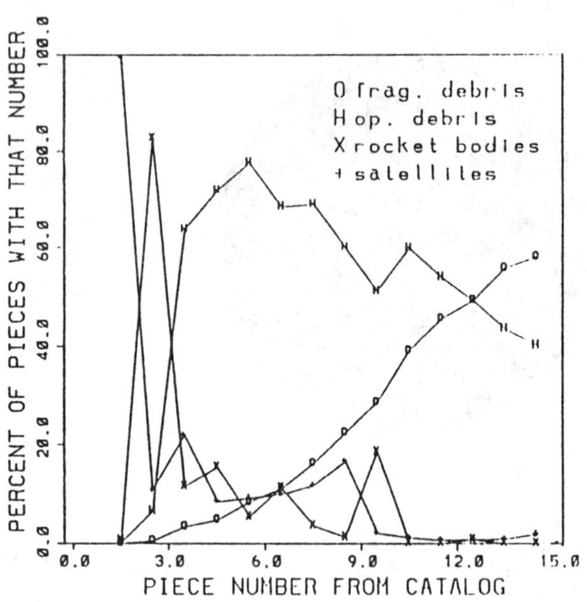

Figure 2: Comparison of Different Classes of Space Objects with Piece Number

Number Distribution

According to theories on explosions of satellites, the number of fragments approximates an exponential distribution (Refs. 5 and 6), normally written as in Eq. (1).

$$N = N_0 \exp\left(-c\sqrt{M}\right) \tag{1}$$

N is the number of fragments of mass M or larger, and N_0 and c are constants defining the total number of fragments and the slope of the distribution curve, respectively.

In this method the value of radar cross section (RCS) is used instead of mass because this is the value which is available for satellites in orbit. Eq. (2) (Ref. 7) is based on the densities of common materials found in spacecraft and rocket body construction and provides a relationship between debris mass, M, and RCS, A_c. The units are grams and meters. This equation was further verified in Refs. 8 and 9.

$$M = 62,000 A_c^{1.13} \tag{2}$$

Weighting Factors

The catalogued satellites were analyzed by fitting Eq. (1) to the data and then extrapolating down to the smaller sizes. The difference between this extrapolation of the untrackable objects and the trackables leads to the weighting factor, which is then applied to all fragments in that class. From this a total background population and its resulting spatial density is computed. The weighting factors from this analysis are presented in Table 1 as case A.

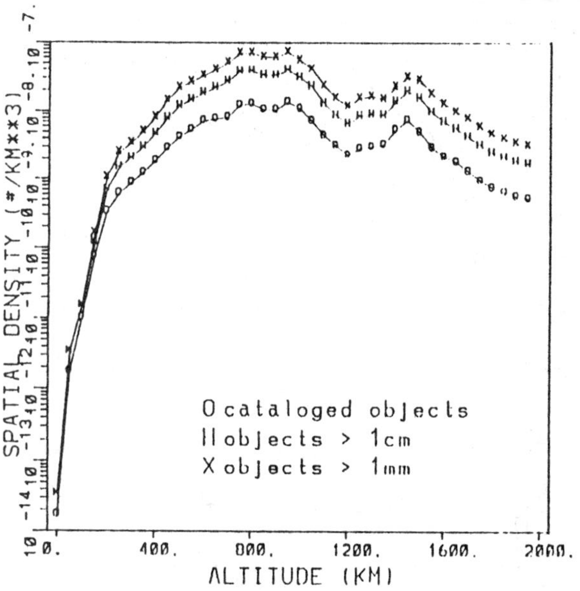

Figure 3: Spatial Density Averaged Over Latitude - Method 1: Case A - Nov. '88 Sat Data

TABLE 1 - CASE A - WEIGHTING FACTORS

class of object	factor for > 1 cm	factor for > 1 mm
satellites	1	1
rocket bodies	1	1
op. debris	2	2
frag. debris	4	8

Spatial Density

Once the weighting factor is known for each class of object, the next step is to find out how this affects the spatial density of the near Earth environment. To do this, a method developed by Kessler is used to find the spatial density for all the objects as a function of altitude above the Earth (Refs. 10 and 11). Fig. 3 shows the spatial density versus altitude for 1988 data using the weighting parameters in Table 1. It can be seen from this figure that the spatial density goes up by about a factor of three for the 1 cm objects, which is consistent with projections of the environment.

When this method is compared with the results of Su and Kessler (Ref. 12) for the 1983 satellite data, it is found that the projected densities differ by approximately a factor of three. Another case was run in which the new method was calibrated to the results of Su and Kessler; these new weighting factors are in Table 2. These new factors were then applied to the 1988 database with the resulting spatial density shown in Fig. 4.

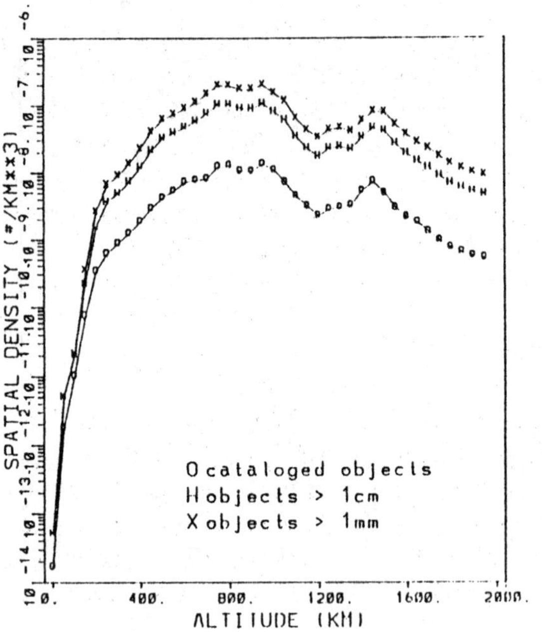

Figure 4: Spatial Density Averaged Over Latitude - Method 1: Case B - Nov. '88 Sat Data

TABLE 2 - CASE B - WEIGHTING FACTORS

class of object	factor for > 1 cm	factor for > 1 mm
satellites	1	1
rocket bodies	1	1
op. debris	3	3
frag. debris	12	24

Method 2 - Modeling Fragmentations

This approach is based on the assumption that virtually all of the orbital debris in the 1 mm-10 cm size range has been generated by the 97 satellite fragmentations recorded to date. This assumption is validated by ground based explosion and collision experiments (Refs. 5, 6, and 13) which lead to predictions that between tens of thousands and hundreds of thousands of 1 mm-10 cm fragments are produced in each breakup event. In fact, the uncertainties alone that are inherent in these predictions are believed to overshadow any operational debris produced by explosive bolts, shrouds, etc. It is therefore the goal of this method to arrive at a reasonable estimate of the untrackable background debris population by modeling only the known on-orbit fragmentation events.

Each event's contribution to the background environment is determined from its fragmentation type and the parent satellite's mass and orbital elements (Refs. 14 - 17), and is added to a 3-d spatial density matrix with dimensions of particle size, latitude, and altitude. First, the breakup date is compared to the present date and a time-since-breakup cutoff to determine whether the debris has effectively become part of the background environment. If not, then the breakup is left to be overlaid on the final matrix by another method. An amount of debris dependent on the satellite's mass and fragmentation mechanism is then generated via empirical distribution equations, and the effect of atmospheric drag is computed. Lastly, the debris is uniformly spread over the entire volume it encounters and is added to the spatial density matrix.

Time Cutoff

After a satellite fragments, the resulting debris continues orbiting as a cloud of individual, smaller satellites. The shape of this cloud evolves as the perigees and ascending nodes of the different orbits precess at different rates. The precession is due primarily to the J2 term of the geopotential, and typically takes on the order of several years for the fastest precessing node to drift 180 degrees from the slowest precessing node (Ref. 18). At this point, the breakup is considered to have become part of the background environment.

Only those debris clouds which have reached this state can be considered to be longitudinally independent and uniformly distributed over some static volume. Therefore, a time-since-breakup cutoff must be chosen to exclude the more recent breakups. Presently, the model uses a constant cutoff of one year which is applied to each event. For a more realistic choice, however, it is possible to estimate a cutoff for each event based on the delta velocity distribution imparted to its fragments as Chobotov has done in Ref. 18.

Debris Generation

Once an event has been deemed a contributor to the background environment, debris pieces are generated from empirically derived distribution equations according to the satellite's mass and fragmentation mechanism. The cause of each satellite's destruction is classified as either a collision with some projectile, an explosion in which charge is not in direct contact with the structure (low intensity), or an explosion in which charge is in direct contact with the structure (high intensity).

The distribution equation used to describe the results of a hypervelocity collision is the power law given by Eq. (3) (Ref. 12).

$$N = 0.4478 \left(\frac{M}{M_e}\right)^{-0.7496} \tag{3}$$

N is the number of fragments of mass M or larger, and M_e is the total ejecta mass. Since all of the collision events considered here are assumed to be catastrophic, the ejecta mass is the entire spacecraft mass in each case. Projectile mass is neglected. The constants in this equation were determined by Su and Kessler (Ref. 12) by reanalyzing Bess's work.

Eq. (2) is rewritten in Eq. (4) (Ref. 19) to provide a relationship between debris mass, M, and characteristic length, d. The units are again grams and meters.

$$M = 4.72 \times 10^4 d^{2.26} \tag{4}$$

Eq. (5) is also taken from Su and Kessler (Ref. 12) and is used to produce debris from a low intensity explosion.

$$N = 8.6921 \times 10^{-4}\ M_t \exp\left(-0.05756\sqrt{M}\right), for M < 1936g \tag{5}$$

M_t is the total mass of the original satellite. It should be noted that although a mass limit is imposed on this distribution, the untrackables under consideration range in diameter from 1 mm to 10 cm, which, by Eq. (4), are well within this limit.

The high intensity explosion distribution used in this model is Eq. (1), repeated here for convenience as Eq. (6).

$$N = N_0 \exp\left(-c\sqrt{M}\right) \tag{6}$$

The values of the constants in Eq. (6) range empirically from 6,921 to 33,880 for N_0 and 0.23 to 0.44 for c (Ref. 20), and they depend on the individual event's structural and detonation characteristics. However, since these characteristics are often unknown and since this is a statistical treatment of a large number of satellites, it is desirable to find one pair of representative values that could be used for all high intensity explosion simulations. After several unsuccessful attempts to calibrate Eq. (6), the representative values of 25,000 for N_0 and 0.4 for c were chosen.

Distribution Range

The next step in the modeling process is to select orbits for the debris that have been generated. Since the final background model is to be static and statistical in nature, the position of a fragment within its orbit is not of primary interest. More important is the determination of the spatial distribution range for each cloud of debris. Once this range is chosen, the debris is spread uniformly over the encountered volume.

It is only necessary to define limits in latitude and altitude, since the application of the time cutoff assures that the debris has spread over 360 degrees in longitude.

The latitude range of a debris cloud is symmetric about the equator and is assumed to be bounded by the inclination of the parent satellite. The reason for this selection stems from the fact that only a significant out of plane Δv directly affects a satellite's inclination, and therefore its maximum latitude. Although this choice is not entirely justifiable for the more violent breakups which can impart significant out of plane Δv's to fragments, its effect on the final background debris matrix is small.

An assumption must be made in choosing altitude limits as well. It is assumed in this model that the volume encountered by a debris cloud is bounded in altitude by the parent satellite's perigee and apogee. This assumption is also not altogether valid, since the Δv's resulting from satellite breakups can be upwards of several km/sec. Impulses of this magnitude can easily throw fragments into extremely high orbits or into the atmosphere within a single orbital revolution. It may be possible to incorporate a delta velocity distribution in a manner that would allow more accurate estimates of altitude limits, but until then, this choice will be used.

The main effect of these assumptions is that the final spatial density matrix has slightly more severe peaks and valleys than would otherwise be expected.

Atmospheric Decay

Besides allowing the user to predict collision hazards to a spacecraft due to different size particles, size is included as a dimension of the spatial density matrix to allow a more accurate calculation of the effect of atmospheric drag on particles that decay at different rates.

Atmospheric decay calculations are performed analytically and are based on the theory and equations developed by King-Hele (Ref. 21). The assumptions made in deriving his method are:

- the atmosphere is spherically symmetric;
- air density does not vary with time;
- air density varies exponentially with height;
- only tangential drag is considered (no lift);
- the atmosphere rotates with constant angular velocity;
- the unperturbed orbit is an exact ellipse;
- during one revolution, the action of air drag changes the orbit a small amount, and its square is negligible;
- lunisolar perturbations are ignored (justifiable except for orbits of large eccentricity or satellites of large area-to-mass ratio).

Although these computations are done analytically, they compare very well with results from integrating routines using the same assumptions. Fig. 5 is taken from Ref. 22 and shows the altitude history of a satellite originally in a 475 km altitude circular orbit as computed by the analytic and integrating methods.

Discussion

Results from Method 2 are plotted in Figs. 6 and 7. Fig. 6 shows the latitude-averaged estimates of two size ranges of background debris up to 2400 km in altitude, while Fig. 7 shows the full spatial density matrix for 1 mm - 1 cm debris. Increments of 50 km in altitude and 5 degrees in latitude were chosen for this example.

It should be noted in comparing Figs. 4 and 6 that results from both methods are highly sensitive to various model parameters. In order to avoid any artificial correlation between estimates, an effort was made to isolate one method's intermediate results from the other's. This isolation leads to a truer comparison of the two methods.

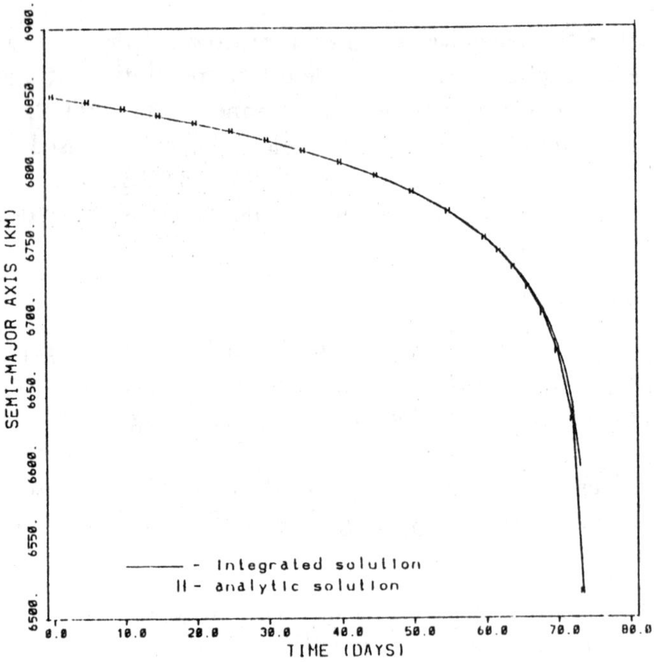

Figure 5: Comparison of Analytic and Integrated Drag Computations on a 1 mm Diameter Debris Fragment in Circular Orbit

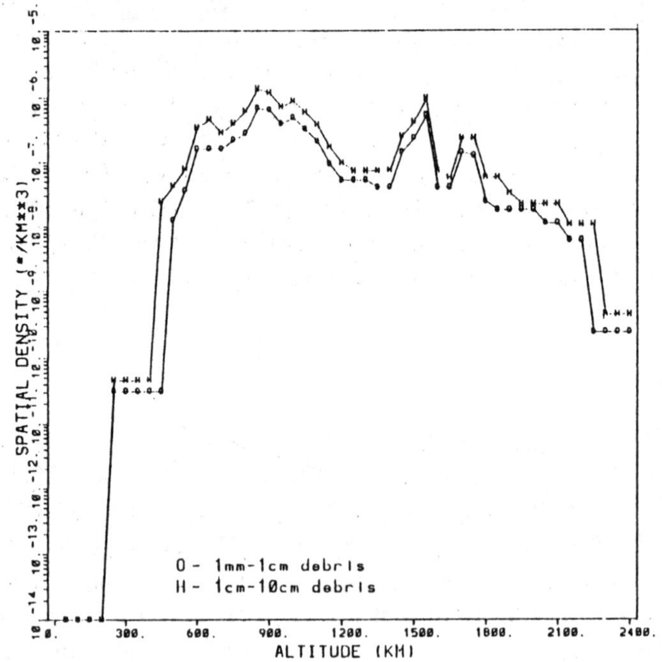

Figure 6: Spatial Density Averaged Over Latitude - Method 2

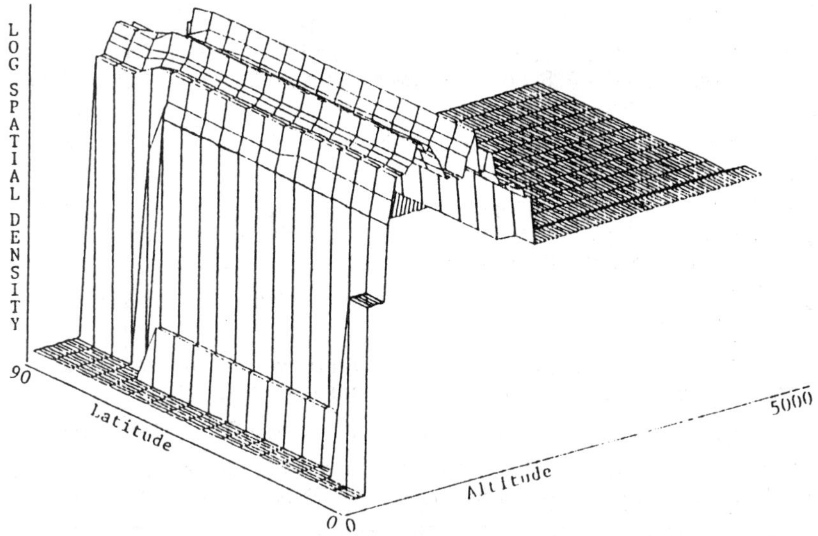

Figure 7: 3-D Representation of 1 mm - 1 cm Debris Matrix, Method 2

The methods are perhaps most effectively evaluated by comparing the shapes and magnitudes of Figs. 4 and 6. The shapes of the two estimates are very similar. In both cases, a wide peak in spatial density between 800 km and 1000 km and a narrower one at approximately 1500 km are present. One very noticeable difference is that Method 2 produces a much more jagged plot than Method 1. This roughness is due to the unrealistic choice of fragment distribution limits and would undoubtedly become smoother if debris were allowed to be thrown into orbits with perigees, apogees, and inclinations different from those of their parent satellite. With the exception of the very low altitude end of the plots, the estimates consistently differ by about one order of magnitude. This result is not surprising, nor is it disturbing. These methods are based on different assumptions, and their results are not only highly dependent on model methodology, but are also not presently verifiable.

Conclusion

The steadily increasing amount of orbital debris in low Earth orbits has prompted us to take a closer look at the hazards it poses to space operations. Modeling the untrackable background environment will help us to quantify these hazards. Presented here are two methods of generating a statistical background debris matrix to describe the small particle environment, and although a few specific areas merit additional attention, both approaches to statistically describing the untrackable debris population yield promising estimates.

References

1. Jonuskis, D. M., and Taff, L. G. "Debris Search at Linclon Laboratory's Experimental Test System", Presented at SDIO/IST Second Pathways to Space Experimentation Workshop, Orlando, Florida, June 17-19, 1986.
2. Taff, L. G., *The Analysis of Near-Earth Satellite Astrometric Data at the ETS,* Linclon Laboratory Project Report ETS-76, (1985).
3. Kessler, D. and Stansbery, G. "NASA's Planned and Proposed Techniques for Monitoring 1 cm to 1 mm Space Debris", Presented by Gene Stansbery at the SMART Catalogue Working Group - Session II, Colorado Springs, CO, May 25-26, 1988.
4. Porter, J. W. *Orbital Debris Radar Parameter Definition,* Teledyne Brown Engineering CS87-LKD-006, May 15, 1987.
5. Bess, T. Dale, *Size Distribution of Fragment Debris Produced by Simulated Meteoroid Impact of Spacecraft Wall,* Eighth Conference on Space Simulation, NASA SP-379, 1975, pp.575-587.
6. Bess, T. Dale, *Mass Distribution of Orbiting Man-Made Space Debris,* NASA TND-8108, 1975.
7. Kessler, D. J. and Cour-Palais, B. G. "Collision Frequency of Artificial Satellites: The Creation of a Debris Belt", *Journal of Geophysical Research,* Vol. 83, No. A6, June 1, 1978, pp.2637-2646.
8. Dickey, M. R. and Culp, R. D. "Determining Characteristic Mass for Low Earth Orbiting Debris Objects", *Journal of Spacecraft and Rockets,* Sept./Oct. 1989.
9. Madler, R., Maclay, T. D. and Hinga, M. *Preliminary Analysis of Collision Fragments,* IRR Report No. SD-89-02r, CCAR, University of Colorado, Boulder, May 1989.
10. Kessler, D. J. "Derivation of the Collision Probability between Orbiting Objects: The Lifetimes of Jupiter's Outer Moons", *ICARUS,* 48 (1981), pp.39-48.
11. Madler, R. *A Simple Method for Estimating the Background Population from the Tracked Population,* IRR Report No. SD-88-03r, CCAR, University of Colorado, Boulder, July 1988.
12. Su, S.-Y., and Dessler, D. J. "Contribution of Explosion and Future Collision Fragments to the Orbital Debris Environment", *COSPAR Advances in Space Research,* Vol. 5, No. 2, 1985.
13. Maclay, T. D., Hinga, M. and Madler, R. *Analysis of Shot 6470 Fragments,* IRR Report No. SD-89-02T, CCAR, University of Colorado, Boulder, July, 1989.
14. Maclay, T. D. *A Statistical Model of Background Orbital Debris,* IRR Report No. 153-6465-88-03, CCAR, University of Colorado, Boulder, July, 1988.
15. Correspondence with Nick Johnson, Teledyne Brown Engineering, Colorado Springs, CO, June, 1989.
16. Johnson, N. L. and Nauer, D. J. *History of On-Orbit Satellite Fragmentations,* Third Ed., Teledyne Brown Engineering Tech. Rept. CS88-LKD-001, Colorado Springs, CO, Oct. 4, 1987.
17. King-Hele, D. G., Pilkington, J. A., Walker, D. M. C., Hiller, H. and Winterbottom, A. N. *The RAE Table of Earth Satellites 1957-1982,* Second Ed., John Wiley and Sons, New York, 1983.
18. Chobotov, V. A., Spencer, D. B., Schmitt, D. L., Gupta, R. P., Hopkins, R. G. and Knapp, D. T. *Dynamics of Debris Motion and the Collision Hazard to Spacecraft Resulting from an Orbital Breakup,* Report No. SD-TR-88-96, The Aerospace Corporation, January 1988.
19. Anz-Meador, P. D., Rast, R. H. and Talent, D. L. *Hazard Analysis for the Breakup of Satellites 16937 and 16938,* Lockheed Engineering Report LEMSCO 23613, Feb. 27, 1987.
20. Johnson, N. L. and McKnight, D. S. *Artificial Space Debris,* Orbit Book Co., Malabar, FL, 1987.
21. King-Hele, D. *Satellite Orbits in an Atmosphere: Theory and Applications,* Blackie and Son Ltd., London, 1987.
22. Maclay, T. D. *An Analytic Method for Calculating Drag Effects,* IRR Report No. SD-89-01T, CCAR, University of Colorado, Boulder, July 1989.

AAS 89-378

DISPOSAL OF SPACECRAFT AT END-OF-LIFE IN GEOSYNCHRONOUS ORBIT

V. A. Chobotov*

A number of national and international organizations have adopted a policy of removing spacecraft at end of life from the geosynchronous ring. However, a resolution of several technical issues still remains before an international policy can be developed and implemented by all users of geosynchronous orbits.

The present study reviews the cataloged population of objects in the geosynchronous ring. The probability of collision is computed as a function of range and orbital inclination of an object. Long-term stability of super-synchronous disposal orbits and of the geopotential stable points is examined. Several disposal options are considered.

The results of the study show that the use of the supersynchronous orbits for disposal purposes is economical and effective in significantly reducing the collision hazard. The use of the geopotential stable points for disposal of spacecraft at end of life, on the other hand, was found to be impractical.

INTRODUCTION

Because of the uniqueness and usefulness of the geosynchronous equatorial orbit (GEO), the population of objects in this orbit has increased continuously. Reference 1 shows, for example, that at the beginning of 1988, there were 286 cataloged satellites residing in this orbit, not including spent upper stages. Of these, 110 to 130 operating satellites were on station along with 150 that are nonfunctional or abandoned.

Reference 2, on the other hand, shows that of the approximately 453 cataloged objects traversing GEO altitudes, only about 150 are geostationary. The others are either geosynchronous or in highly elliptical ("Molniya") semi-synchronous orbits. There are also a number of military spacecraft, the orbits of which are not generally available in published form.

To reduce the collision hazard within GEO, approximately 30 to 35 satellites have been removed from GEO at end of life. Several strategies are available for disposing of GEO satellites, such as moving them to stable longitudes or raising them to orbits above geostationary altitude.

---

* Manager, Space Hazards Section, The Aerospace Corporation, P.O. Box 92957, Los Angeles, California 90009-2957.

Table 1. 1988 Geosynchronous Catalog

Owner	Spacecraft	Rocket Bodies	
AUST	3		
BRAZIL	2		
CANADA	10		
ESA	8		
FRANCE	4		
GR. BRIT	1		
INDIA	3		
INDONESIA	4		
INTLSAT	20		
ITALY	1		
ITSO	4		
JAPAN	12		
MEXICO	2		
NATO	7		
PRC	2		
SAUD	2		
US	91	24	
USSR	64	65	
UNKNOWN	50		
TOTAL	290	89	379

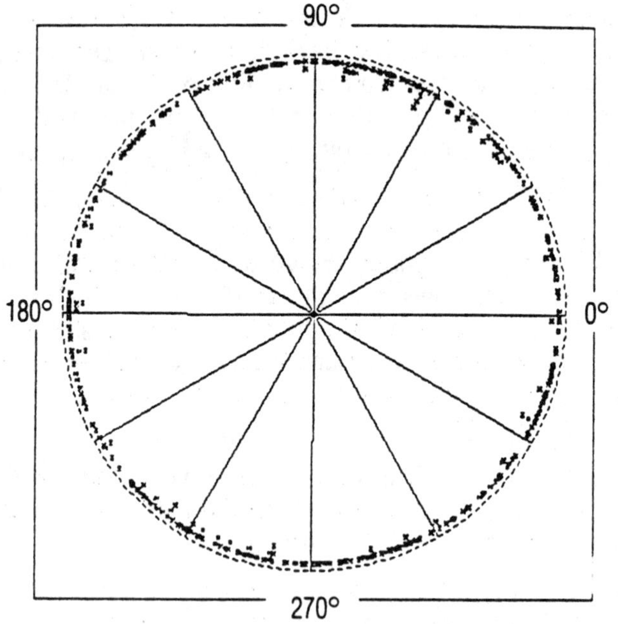

Figure 1. Longitudinal Distribution of Object in the Geosynchronous Ring (March 1989)

The benefits or removing old spacecraft from GEO are: (a) the freeing of desirable longitudinal positions for replacement spacecraft, and (b) reduction of collision damage potential which can become high for collocated satellites (Ref. 3). The primary issues involved are: (a) the protection of GEO as an unique natural resource, and (b) operational considerations such as: cost-effectiveness of the maneuver, risks of explosion, structural breakup, or other malfunctions in carrying out the maneuver, and national and international policy considerations. Other issues include the potential impact on future traffic to GEO and the perturbative effects on disposal orbits (e.g., solar radiation and the Sun/Moon gravitational perturbations). In other words, abandoned rocket stages or satellites should not be a hazard to future missions, and their orbits should not intersect GEO over long periods of time.

The purpose of the present study is to review the population environment in GEO, examine average probabilities of collision as a function of altitude, and consider the feasibility of removing spacecraft to supersynchronous (higher altitude) orbits or to the geopotential stable points. Distributions of the GEO population are described, average spatial density and collision probability are computed, and the rationale for removing spacecraft from GEO is reviewed.

GEOSYNCHRONOUS RING

Orbits in the geosynchronous ring may be classified as: (a) low or high inclination, and (b) sub- or supersynchronous altitude drifting orbits. The ideal geostationary orbit has an altitude of 35,787 km and a period of 1436.2 min. A satellite whose period of revolution is equal to the period of rotation of the Earth about its axis is called a geosynchronous satellite. A geostationary satellite is a geosynchronous satellite in a circular orbit which lies in the plan of the Earth's equator and rotates about the Earth's polar axis in the same direction. Consequently, it remains "fixed" over a point on the equator.

The owners of the 379 GEO objects in the United States Space Command Catalog of February 1988 are identified in Table 1. An illustration of the GEO objects in the equatorial plane is shown in Figure 1, where the objects located at or near the same longitude are shown displaced radially inward. The distribution in right accession of the ascending node (RAAN) is given in Figures 2 and 3, which illustrate the clustering (bunching) of orbits in the 60° to 100° range. A view of the GEO population is also shown in Figure 4, where the apparent sinusoidal distribution in latitude is the result of the orbital bunching of orbit nodes of aging spacecraft and rocket bodies.

Distributions of the GEO objects with orbital inclination and eccentricity (apogee-perigee difference) are shown in the Figures 5 and 6, respectively, for the February 1988 and the October 1979 catalogs.

A significant increase in the number of objects is apparent in the time interval considered. The drift rate distribution for the 1988 catalog is also shown in Figure 7. It can be seen that only about 70 objects have drifts rates of less than

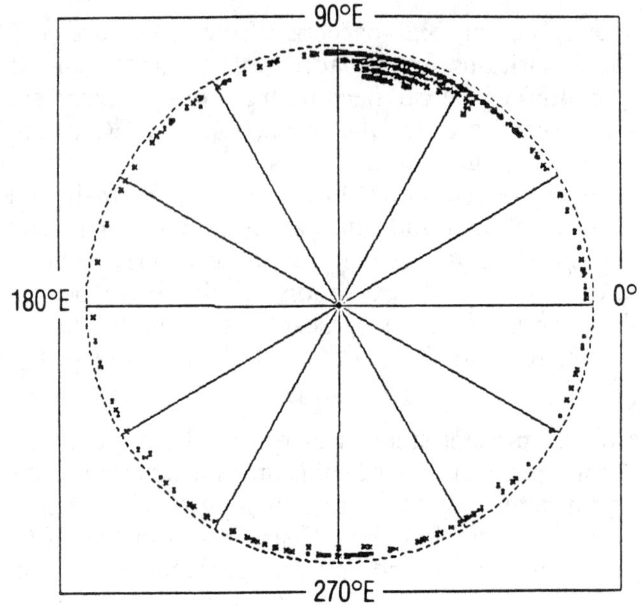

Figure 2. Right Ascension of Ascending Node (RAAN) Distribution of Objects in the Geosynchronous Ring (March 1989)

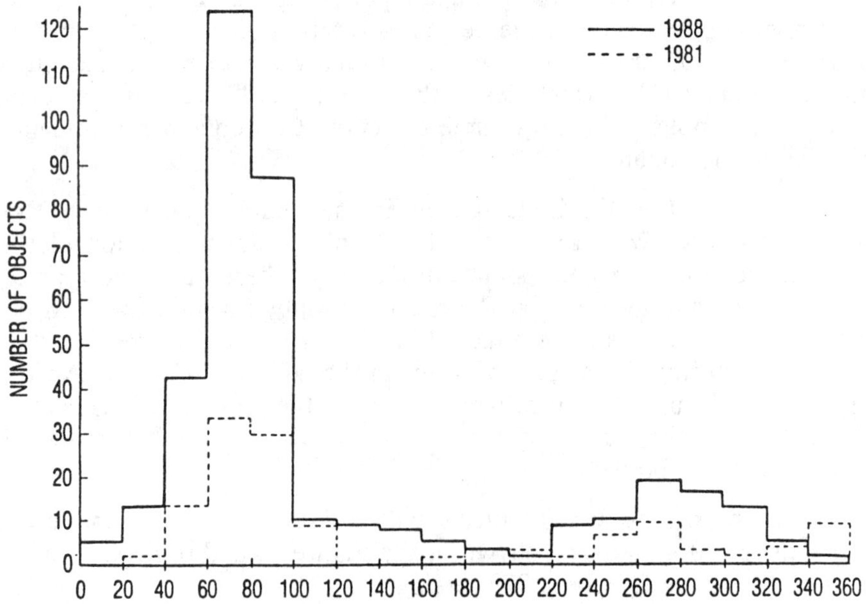

Figure 3. Right Ascension of Ascending Node (deg)

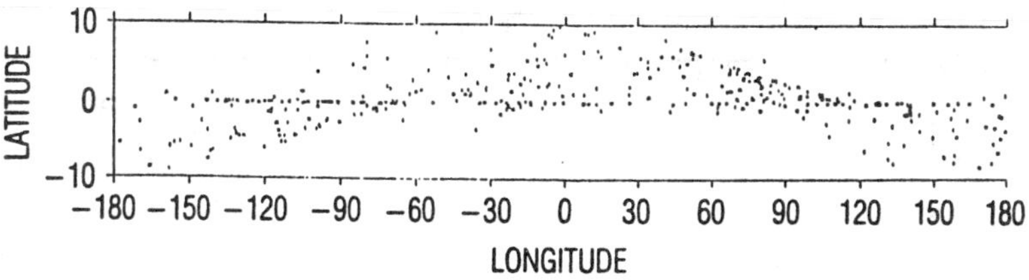

Figure 4. View of Geosynchronous Objects

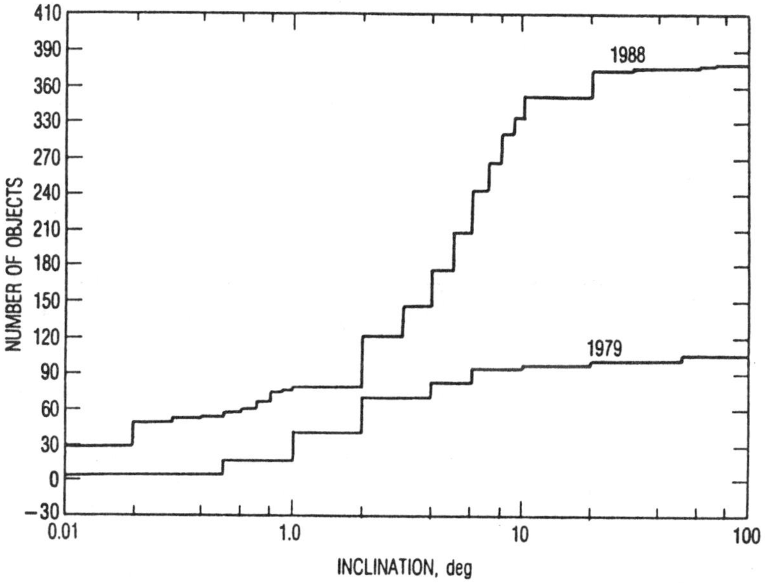

Figure 5. Geosynchronous Orbit Inclination Distribution

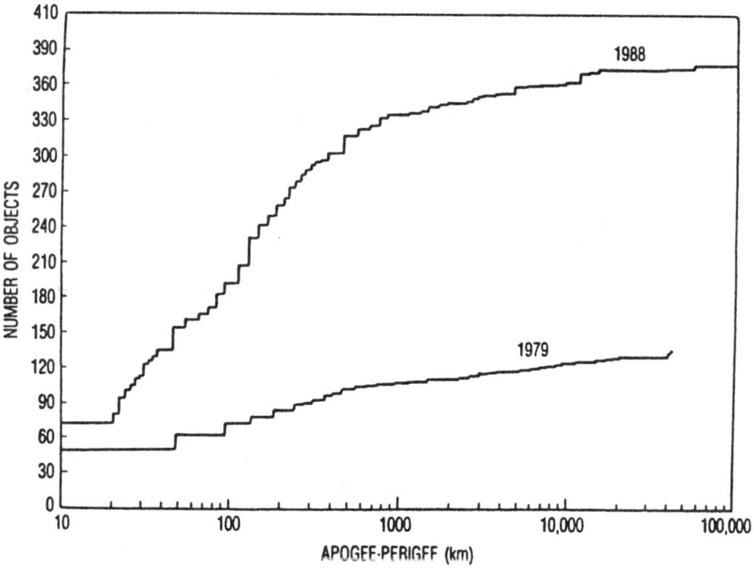

Figure 6. Geosynchronous Eccentricity Distribution

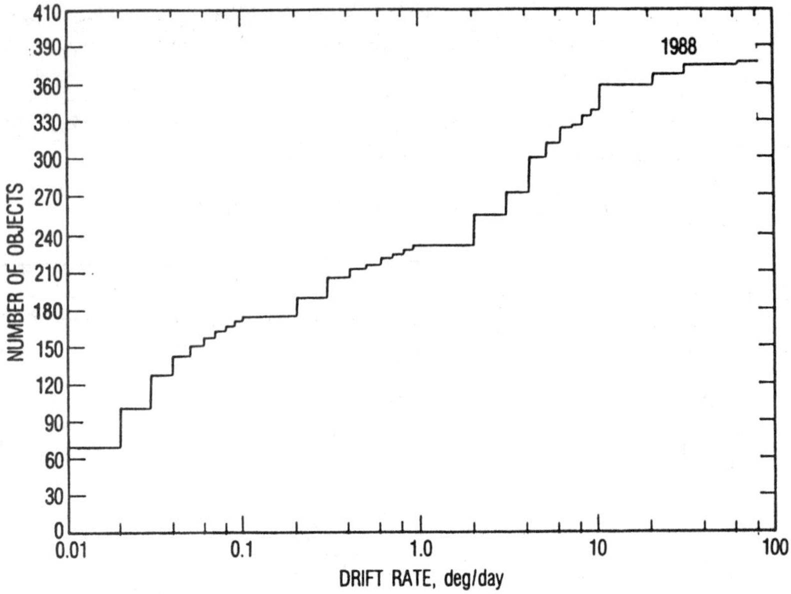

Figure 7. Geosynchronous Drift Rate Distribution

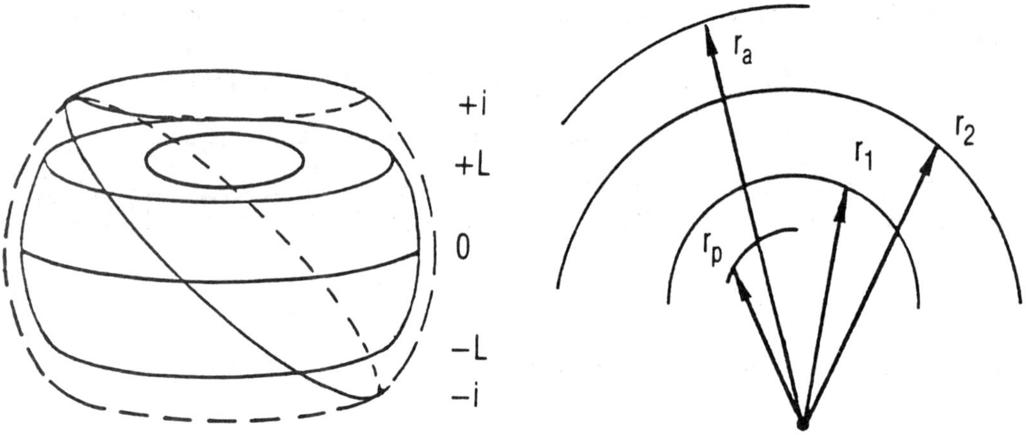

(a) Toroidal Volume (b) Radial Parameters

Figure 8. Orbital Geometry

0.01°/day and fewer than 180 objects less than 0.1°/day. These are "fixed" or active satellites, and the remaining objects are probably uncontrolled or abandoned debris.

SPATIAL DENSITY AND COLLISION HAZARD

Spatial Density

The spatial density as a function of radius r and latitude L of the N objects in the geosynchronous ring, defined by a toroidal volume, can be expressed approximately as

$$\rho(r, L) \approx \sum_{k=1}^{N} \rho_k \qquad (1)$$

where

$$\rho_k = \frac{P_k(r_1, r_2)\ P_k(-L, L)}{\Delta V_k} \qquad (2)$$

with the probability of the k-th object being within the toroidal volume

$$\Delta V_k = \frac{4\pi}{3}(r_2^3 - r_1^3) \sin L \qquad (3)$$

defined by the arbitrary orbital radii r_1, r_2, and latitude L as shown in Figure 8. Here

$P_k(r_1, r_2)$ = probability that the k-th object is within the radial range $\Delta r = r_2 - r_1$

$P_k(-L, L)$ = probability that the k-th object is between the latitudes -L and L.

For example, if $r_a \geq r_2 > r_1 \geq r_p$ where r_a and r_p are the orbit apogee and perigee, respectively, and if $i > L$ where i is the orbit plane inclination of the object

$$P_k(r_1, r_2) = P_k(r_p, r_2) - P_k(r_p, r_1) \qquad (4)$$

where the probability that the object lies between perigee (r_p) and some radius (r), is derived in Reference 4 as

$$P(r_p, r) = \frac{1}{2} + \frac{1}{\pi}\sin^{-1}\left[\frac{2(r-a)}{(r_a-r_p)}\right] - \frac{1}{a\pi}\sqrt{(r_a-r)(r-r_p)} \qquad (5)$$

and the probability that the object is between the latitudes -L and L is

$$P(-L, L) = \frac{2}{\pi}\sin^{-1}\left(\frac{\sin L}{\sin i}\right) \qquad (6)$$

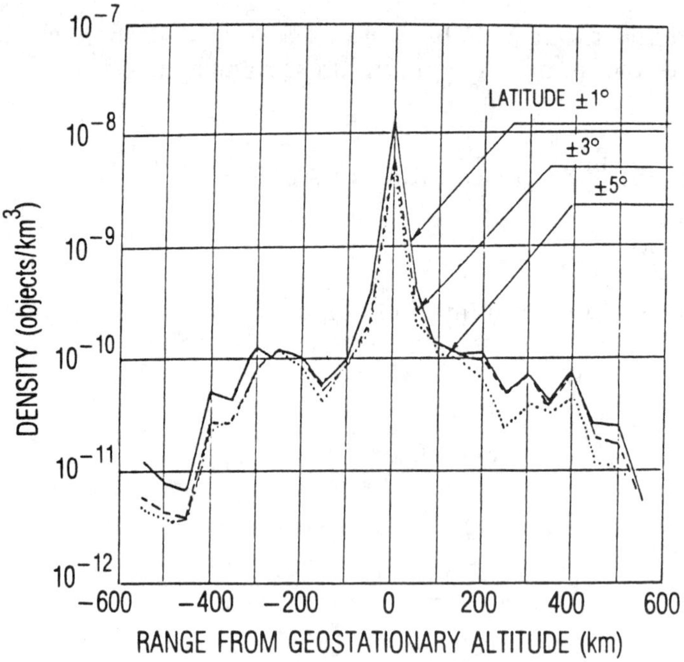

Figure 9. Population Density as a Function of Range and Latitude from GEO

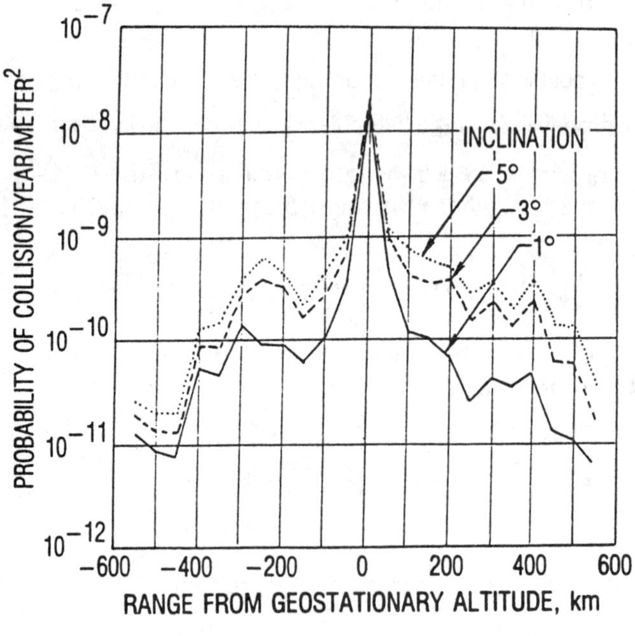

Figure 10. Probability of Collision for a Circular Orbits as a Function of Inclination and Range from GEO

where a and i are the object orbit semi-major axis and inclination, respectively.

Equation (1) is plotted in Figure 9 for the 379 objects of the GEO catalog considered. The results show that the spatial density is maximum (about 10^{-8} sats/km^3) in a narrow range of the geosynchronous altitude. It decreases by about two orders of magnitude at ±100 km above or below GEO. The subsequent decrease with altitude is not as high as in the first 100 km range, being about three orders of magnitude at ±600 km range.

Probability of Collision

The probability per revolution that a satellite in a circular orbit will collide with an object can be computed from the relation (Ref. 5)

$$P(col/rev) = A\bar{V}_r T \int_{r_1}^{r_2} \int_{-L}^{L} \rho(r,L)\, f(L) dL dr \qquad (7)$$

where

p(r, L) = object density function

f(L) = weighting function to indicate the time fraction spent by a satellite in a latitude band, -L, L

T = satellite period

$\bar{V}_r$ = mean relative velocity between the satellite and the objects in GEO

A = satellite projected area normal to $\bar{V}_r$

The probability of collision for a a spacecraft in a circular orbit with inclination i, for a given range above or below GEO, has been computed and plotted in Figure 10. The results show that the probability of collision is greatest for GEO satellites and decreases as a function of range above or below GEO. It also increases with inclination due to higher relative velocities between the satellite and the objects in GEO. For example, a decrease of two orders of magnitude is apparent for a spacecraft in an orbit with an inclination of 1° at ±100 km above or below GEO.

DISPOSAL OPTIONS

Some of the satellites launched into the geosynchronous orbit have been removed at end of life to either higher or lower altitudes. The primary benefits for removing old spacecraft from the geosynchronous orbit are:

a. Freeing of desirable longitudinal positions for replacement spacecraft to avoid interference from old spacecraft

b. Reduction of collision damage potential for other spacecraft

The removal of inactive satellites has evidently become a practice at Intelsat and the European Space Agency (ESA). The first case of end-of-life removal occurred in May 1977, when three Intelsat 3 satellites F2, F3, and F4 were moved to disposal orbits at 3580, 3700, and 4000 km altitudes beyond the altitude of the geosynchronous orbit (Ref. 6). ATS 6, Raduga 5, Anik 1, SMS 1, SMS 2, Intelsat 4 F2 and F4 were removed later to disposal orbits. Similarly, in January 1984, GEOS 2 was placed 269 km above GEO by ESA using 2 kg of propellant in three impulses.

Similar requirements are placed on future spacecraft by the U.S. Navy UHF Follow-On, DSCS-III, and so forth, which require removal of satellites to supersynchronous orbits by at least 370 km from GEO. Also, AUSSAT, the Australian Agency, requires removal capability of up to 1110 km above GEO at the end of life.

The practice of removing satellites from GEO is not without some risk and cost to the spacecraft operator. The risk is the probability of failure during the maneuver, which may result in an orbit with higher relative velocities or even a collisional breakup. The cost is the slightly reduced life and the potential loss of the satellite as a spare or a decoy. (The latter may sometimes be of benefit to military spacecraft).

In view of this, decisions concerning removal of a spacecraft at the end of life should be made only after careful analysis of all factors. To minimize the explosion hazard from fuel decomposition, a burn to depletion of propellants should also be considered.

Supersynchronous Orbit Disposal

The potential for reduction of the probability of collision is illustrated in Figure 10, which shows that a two or three-order of magnitude decrease in the collision hazard results by placing a spacecraft about 300 km above the geosynchronous orbit. The velocity impulse for such a maneuver is only 10.8 m/s, requiring, for example, only 5 to 10 kg of propellant for a typical spacecraft. This translates to a month or two of life reduction and can be accomplished by a two-impulse maneuver.

The technique of initially deploying satellites to a supersynchronous orbit can also be used to ensure that the apogee-injection stage remains above geosynchronous. This procedure is attractive in that it effectively removes the threat posed by the apogee-injection stage to objects in the geosynchronous corridor.

The long-term stability of supersynchronous orbits has been examined numerically using program ELEMENT (Ref. 7) and found to be satisfactory. For example, the first order, nearly resonant perturbation for circular orbits displaced in an altitude ±150 km from a typical synchronous, nearly equatorial orbit results in the maximum radial excursion on the order of only ±2.31 km almost entirely due to the lowest order tesseral harmonic $J_{2,2}$.

The effects on the Sun/Moon perturbation, however, will cause an orbital inclination growth from zero to nearly 15° in 27.4 years, as illustrated in Figure 11. The in-plane motion, on the other hand, is caused mostly by solar radiation pressure. For example, for an object with an area-to-mass ratio of 0.014 m^2/kg and a

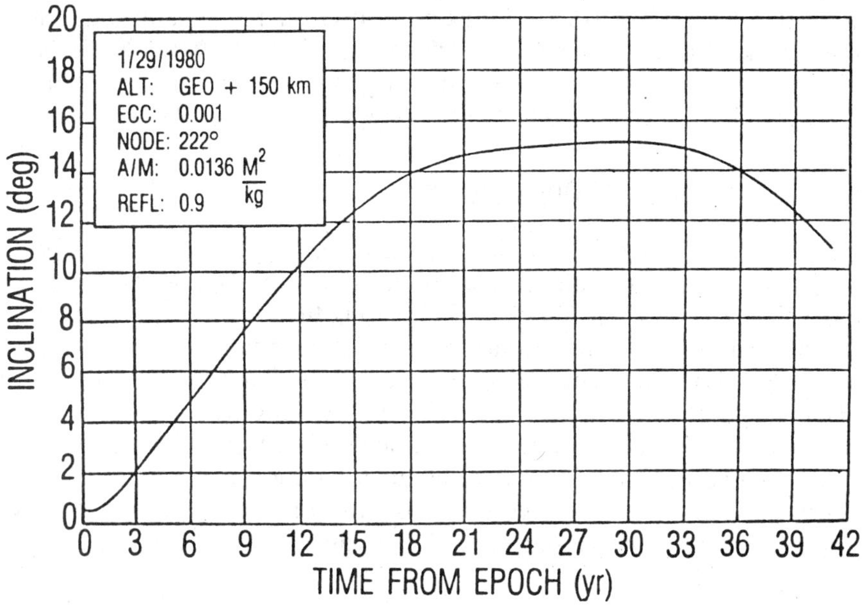

Figure 11. Inclination Drift in Supersynchronous Orbit

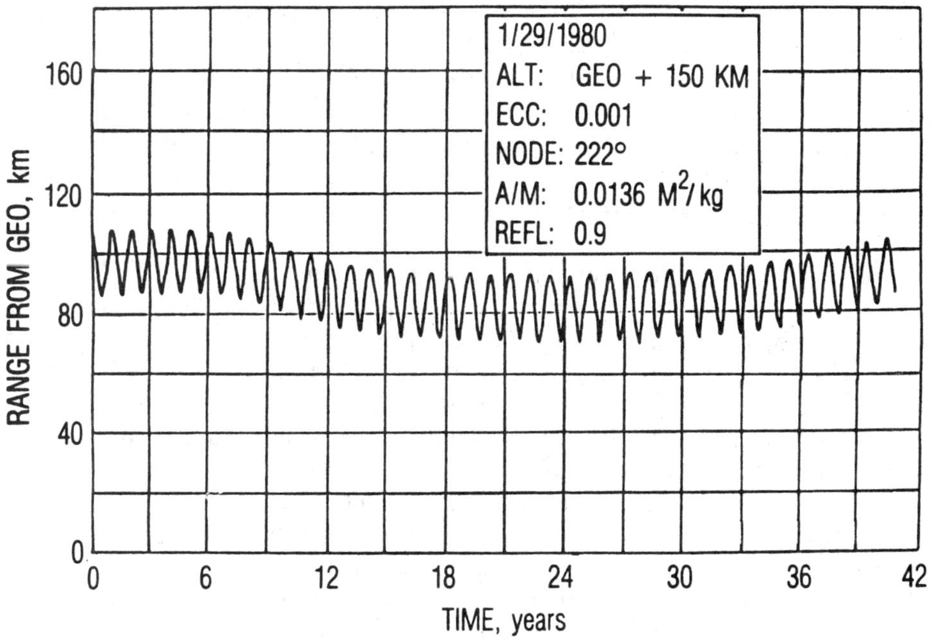

Figure 12. Perigee Drift in Supersynchronous Orbit

coefficient of reflection of 0.9, the maximum cyclical perigee altitude variation is only about 24 km/yr. A longer-term perigee variation of about 20 km is also apparent in Figure 12, which illustrates the perigee drift due to solar radiation pressure. For a detailed description of the general inclination history, see Reference 8.

Geopotential Stable Point Deployment

Another possibility exists, at least in theory, of placing spent satellites and rocket stages at the geopotential stable points. There are two such points located in the equatorial plane at 75.3°E and 255.3°E longitude, as illustrated in Figure 13. Objects placed a given distance away will oscillate about the stable point (on the Earth's minor axis) with a period of several years. Therefore, if objects can be placed at such points with negligible position and velocity errors, they will tend to remain in the longitudinal vicinity of such points. However, they will still be subject to the perturbative effects of the solar/lunar gravitation and radiation pressure. A velocity error (between nominal geosynchronous velocity and true orbital velocity) of only a fraction of a meter per second can cause large oscillations of an object about the stable point.

Figures 14, 15, and 16 illustrate the longitudinal displacement, inclination, and eccentricity changes in 1000 days for an object initially at 75.3°E longitude with a velocity error of 0.6 m/s and an area-to-mass ratio of 0.05 m^2/kg obtained using program GEOSYN (Ref. 9). The coefficient of reflectivity is 0.5. The results show the extreme sensitivity of the longitudinal position of an object to an initial velocity error. It is also clear that the Sun/Moon gravitational perturbation is present. The placement of spent satellites and rocket stages at the geopotential stable points would therefore be ineffective in reducing the collision hazard. The resultant oscillations of the objects would continue to pose a collision threat to each other, to active spacecraft, and to drifting bodies in the vicinity or passing near such points.

SUMMARY AND CONCLUSIONS

The tracked satellite population in the geosynchronous orbit has been examined and its spatial density evaluated. The probability that a spacecraft will collide with an object in the ring has been computed and found to be on the order of $10^{-8}/m^2$ of spacecraft area per year, not including the effects of longitudinal bunching of objects (e.g., collocation of satellites) or of smaller debris that cannot be tracked. The maximum probability of collision for spacecraft in circular, inclined orbits was found to be at geosynchronous altitude, decreasing rapidly with distance from this altitude. Thus, the removal of spacecraft at the end of life to supersynchronous orbits 300 to 600 km above geosynchronous would free longitudinal locations for future satellites (avoiding bunching effects) and would decrease the collision hazard significantly (up to three orders of magnitude) with the current population of objects. The velocity impulse for such maneuvers is modest (on the order of 10 m/s) with a typical propellant requirement of less than 10 kg. The long-term stability of supersynchronous orbits was found to be adequate.

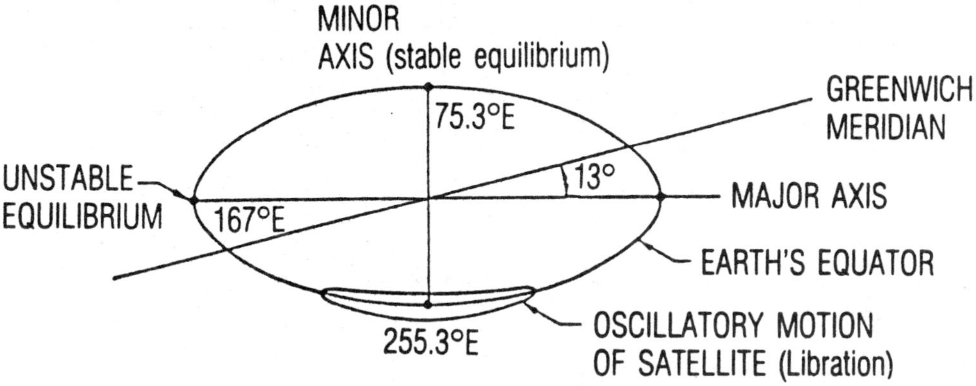

Figure 13. Geopotential Stable Points

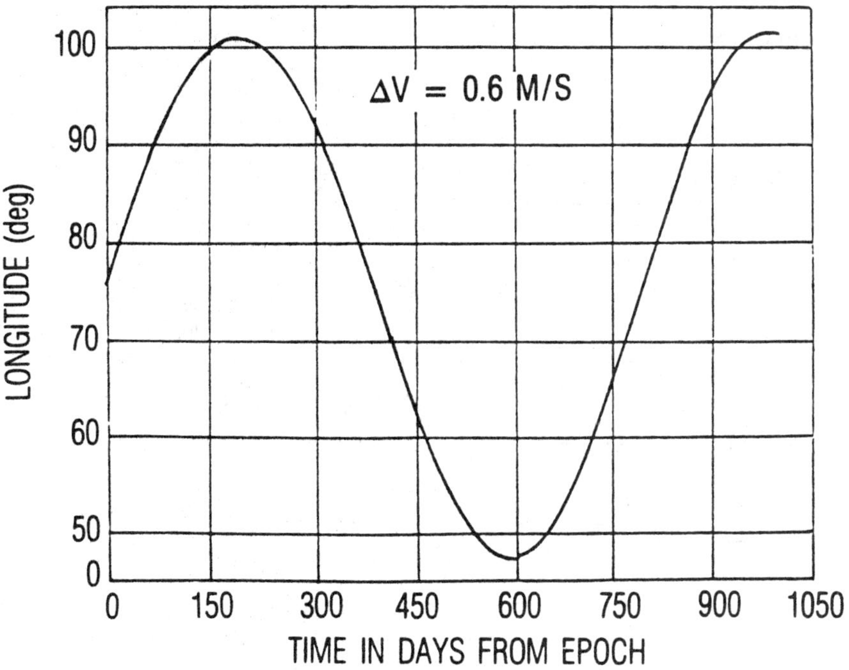

Figure 14. Longitude Time History Near Stable Point

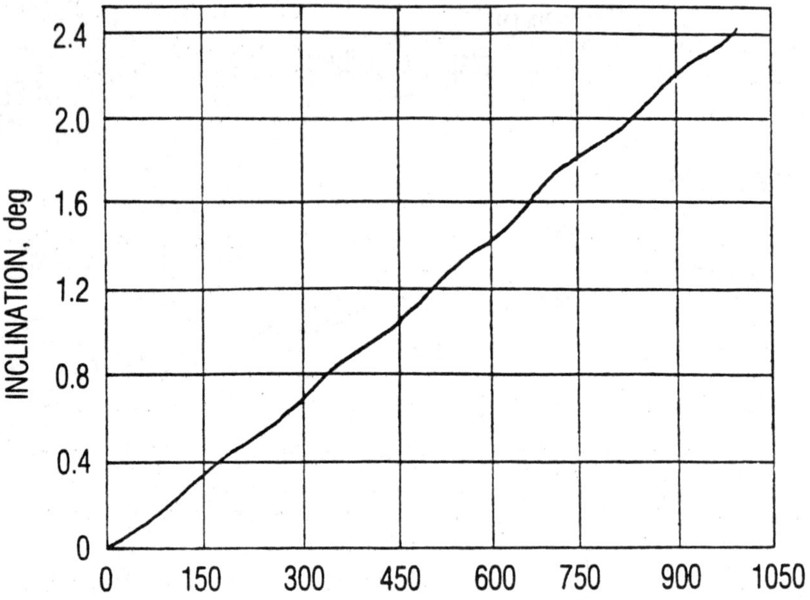

Figure 15. Inclination Time History Near Stable Point

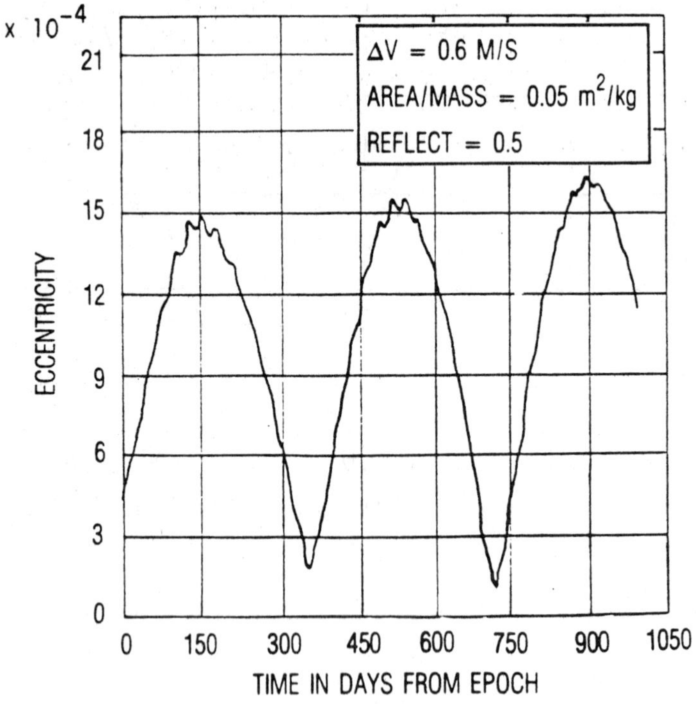

Figure 16. Eccentricity Time History Near Stable Point

The feasibility of using the geopotential stable points for end-of-life disposal was also examined. However, the high sensitivity to initial velocity errors at deployment and the Sun/Moon and the solar radiation pressure perturbation effects would make this approach impractical.

ACKNOWLEDGEMENT

The author wishes to thank M. R. Rakoczy for her assistance in the evaluation of the collision hazard in the geosynchronous orbit.

REFERENCES

1. Space Debris Report, ESA SP1109, ESTEC, Noordwijk, The Netherlands, November 1988.
2. Report on Orbital Debris, Interagency Group (Space) for National Security Council, Washington, D.C., February 1989.
3. Harting, A., et al., "On the Collision Hazard of Collocated Geostationary Satellites," AIAA/AAS Conference, Paper No. 88-4239-CP, August 1988.
4. Dennis, N. G., "Probabilistic Theory and Statistical Distribution of Earth Satellites," *Journal of British Interplanetary Society,* Vol. 25, pp.333-376, 1972.
5. Chobotov, V. A., "The Collision Hazard in Space," *The Journal of the Astronautical Sciences,* Vol. 30, No. 3, July-September 1982.
6. Perek, L., "The Scientific and Technical Aspects of the Geostationary Orbit," 38th IAF Congress, 10-17 October 1987, Brighton, UK.
7. Baxter, B. E., "Notes on Program ELEMENT (Special Perturbations via An Averaged Disturbing Function)," The Aerospace Corporation, Report No. TOR-0172(2441-02)-18, September 1971.
8. Allan, R. R. and G. E. Cook, "The Long Period Motion of the Plane of a Distant Circular Orbit," *Proceedings of the Royal Society (London),* Vol. 280, No. 1380, 7 July 1964.
9. Chao, C. C. and J. M. Baker, "On the Propagation of Geosynchronous Orbits," *The Journal of Astronautical Sciences,* Vol. 31, No. 1, pp.99-115, January 1983.

AAS 89-397

A NONLINEAR DYNAMICAL MODEL OF RELATIVE MOTION FOR THE ORBITING DEBRIS PROBLEM*

Richard S. Hujsak†

A nonlinear dynamical model of relative motion is derived for application to the orbital breakup problem. This model is designed for analysis of debris density, beginning at the time of breakup, and continuing until after the cloud "rings" the Earth. A linear relative motion model is derived, which generalizes the Clohessy Wiltshire equations to all eccentricity and secular J_2. Then selected higher order terms are added, yielding generalized equations which are useful at large displacements from the reference ephemeris. A significant improvement over linear relative motion techniques is demonstrated for the expanding debris cloud. The inverse of the relative motion transformation is then shown to be an approximate solution to Lambert's problem for multiple orbits in a perturbed force field. This model enables an analysis of debris density differentially at a point in space as a function of initial breakup spread velocity statistics, and eliminates the need to assume cloud size or shape. Analysis is provided showing that particle density can vary by four orders of magnitude within the instantaneous cloud. Density extrema always occur at the "pinched" points of the cloud and, in a perturbed force field, also occur at discrete locations within the envelope of the expanding cloud.

INTRODUCTION

Consider the problem of planning a space test where one or more vehicles will break up during collision or explosion. The test must be scheduled such that test debris does not pose a serious threat to other operational satellites. One may calculate the probability of collision between debris and each operational satellite, establish an acceptable safety threshold, and permute test conditions until all satellites are deemed safe. To

* Developed, in part, under Air Force contract F04690-86-C-0006.

† Senior Engineer, Applied Technology Associates, Inc., 400 Gordon Drive, Suite #504, Lionville, Pennsylvania 19353.

compute probability of collision one needs to know the probable distribution of debris particles along the satellite's orbit path.

There have been a few attempts[1,2,3] to solve the near term space safety problem. Chobotov et al[1,2] have developed a fragmentation model which computes the spread velocity, mass, and size distribution of particles at the time of breakup as a function of the size(s) of the vehicle(s) and mass(es), while conserving momentum, mass, and energy. Fragmentation model outputs are used to find the rates of expansion of two fixed cloud shapes, one of three concentric spheres, and one of concentric, pinched, truncated tori. An averaged particle density is assigned to each cloud shape and probability of collision is measured along the path of a satellite intersecting the cloud shape.

McKnight[3] uses a fragmentation model which does not conserve momentum, mass, or energy. The fragmentation model is used to scale three fixed shapes, a hyperellipsoid, a torus, and a band about the earth. Again averaged particle densities (i.e., homogeneous distributions) are assigned to each shape. McKnight does not actually find satellite-to-cloud intersections, but computes worst case and average probabilities of collision.

These works suffer from a common problem. The authors need to predict spatial particle density from fragmentation model outputs. But fragmentation outputs are stated as probable spread velocity distributions. Techniques for rigorously predicting spatial distributions from initial spread velocity distributions do not exist. So the authors compute spatial density by computing total particles per total volume, and an empirical (geometric) shape must be assigned to determine volume.

A debris cloud is defined by the ensemble of instantaneous positions of all of the orbiting particles from a single breakup. A cloud does not conform to any specific geometric shape. Errors introduced by assuming an empirical shape lead to errors in volume and hence in density. Moreover, the chosen shape may not always contain the real cloud, which is unacceptable for test safety analysis. Finally, the forced assumption of average density is seriously in error. As we will show, particle density within the instantaneous cloud can vary by orders of magnitude. Given that these problems cause errors of orders of magnitude in the probability of collision, there is a need for an improved method of calculating particle density. If the method we seek is to be verifiable it is appropriate to seek a functional relationship between debris density and initial spread velocity which is rigorously based in dynamics.

Relative motion models have been explored[2] as a natural means of establishing the dynamical relationship we seek. *Linear* equations of relative motion expressed in rectangular coordinates[4] were found to be inadequate for the breakup problem. Attempts to add *linear* perturbations have been unsuccessful in resolving the problem.

The following discussion develops a *nonlinear* relative motion model and evaluates the model with a simple breakup simulation. The model is shown to provide the missing dynamical relationship between spread velocity distributions and spatial density. Once that claim is established, we use the model to demonstrate that the averaged particle density can differ with the actual particle density by orders of magnitude.

A LINEAR RELATIVE MOTION MODEL

Model Form

A relative motion model defines motion relative to a reference ephemeris. The reference ephemeris is assumed to be generated by known analytic or numerical techniques, passing through the position of breakup at the time of breakup. The velocity of the reference ephemeris at the time of breakup can be conveniently chosen so that the reference ephemeris will always lie within the orbiting cloud. The inputs and outputs of a relative motion model normally include relative position and velocity components. Because we need a metric to define distance, and hence volume, we constrain inputs and outputs to rectangular Gaussian coordinates.

Linear Model Development

We seek a 6×6 transformation Φ_{t,t_0} which computes relative radial (ΔY_1), circumferential ("intrack") (ΔY_2), and crosstrack (ΔY_3) position and velocity components in a cloud frame (c) at a time of interest (t) from relative position and velocity components in another cloud frame (b) at a reference time (t_0):

$$[\Delta Y_1(t) \;\; \Delta Y_2(t) \;\; \Delta Y_3(t) \;\; \Delta\dot{Y}_1(t) \;\; \Delta\dot{Y}_2(t) \;\; \Delta\dot{Y}_3(t)]_c^T =$$
$$\Phi^{\#1}_{t,t_0} \, [\Delta Y_1(t_0) \;\; \Delta Y_2(t_0) \;\; \Delta Y_3(t_0) \;\; \Delta\dot{Y}_1(t_0) \;\; \Delta\dot{Y}_2(t_0) \;\; \Delta\dot{Y}_3(t_0)]_b^T \, .$$

Define a linear transformation, $\Phi^{\#1}_{t,t_0}$, using the Chain Rule, as the product of three 6×6 transformations:

$$\Phi^{\#1}_{t,t_0} = \underbrace{\left[\frac{\partial Y_i(t)}{\partial F_j(t)}\right]}_{G_1(t)} \underbrace{\left[\frac{\partial F_j(t)}{\partial F_k(t_0)}\right]}_{G_2(t,t_0)} \underbrace{\left[\frac{\partial F_k(t_0)}{\partial Y_m(t_0)}\right]}_{G_3(t_0)}$$

where F_j are orbital elements $j = 1,2,\cdots,6$. If F_j are equinoctial $(a_f, a_g, n, L, \chi, \psi)$, then $G_1(t)$ is a well-known[5] linear transformation from orbital elements to position and velocity. $G_1(t)$ is complete in eccentricity and nonsingular for all eccentricity $(0 \leq e < 1)$ and inclination $(0 \leq i < \pi)$. $G_3(t_0)$ is simply $G_1(t_0)^{-1}$, and likewise nonsingular and valid for all eccentricity and inclination. $G_2(t,t_0)$ is exact for two body dynamics as:

$$G_2(t,t_0) = \begin{bmatrix} 1 & 0 & 0 & 0 & 0 & 0 \\ 0 & 1 & 0 & 0 & 0 & 0 \\ 0 & 0 & 1 & 0 & 0 & 0 \\ 0 & 0 & \tau & 1 & 0 & 0 \\ 0 & 0 & 0 & 0 & 1 & 0 \\ 0 & 0 & 0 & 0 & 0 & 1 \end{bmatrix}, \quad \tau = t - t_0 .$$

Formulations for $G_2(t,t_0)$ which are complete in secular J_2 already exist[5]. By adopting these existing formulations, and evaluating G_i at reference ephemeris values at times t and t_0 , we claim that:

$\Phi^{\#1}_{t,t_0}$ is a generalization of the Clohessy Wiltshire "rendezvous equations", valid for all eccentricity and for secular J_2 .

<u>Restriction to Time of Breakup</u>

By placing t_0 at the time of breakup, input relative positions can be set to zero. To compute debris density, the relative velocity outputs can be neglected. As a result, we are only interested in a 3×3 subset of $\Phi^{\#1}$, denoted as $\Phi^{\#2}$:

$$\Phi^{\#2}_{t,t_0} = G_1'(t)\ G_2(t,t_0)\ G_3'(t_0)$$

where $G_1'(t)$ is a 3×6 subset of $G_1(t)$ and $G_3'(t_0)$ is a 6×3 subset of $G_3(t_0)$.

Linear Model Shortcomings

The linear transformation $G_1'(t)$ is a 3×6 matrix of dot products of the form:

$$G_1'(t) = [\ \underline{\hat{u}}\ \ \underline{\hat{v}}\ \ \underline{\hat{w}}\ ]^T \cdot \begin{bmatrix} \dfrac{\partial \underline{r}}{\partial a_f} & \dfrac{\partial \underline{r}}{\partial a_g} & \dfrac{\partial \underline{r}}{\partial n} & \dfrac{\partial \underline{r}}{\partial L} & \dfrac{\partial \underline{r}}{\partial \chi} & \dfrac{\partial \underline{r}}{\partial \psi} \end{bmatrix},$$

where $\underline{r}$ is position and $\underline{\hat{u}}$, $\underline{\hat{v}}$, $\underline{\hat{w}}$ are unit vectors in radial, intrack, and crosstrack directions. $G_1'(t)$ maps relative orbital elements into relative cartesian position (relative to the reference ephemeris at t):

$$[\ \Delta Y_1(t)\ \ \Delta Y_2(t)\ \ \Delta Y_3(t)\ ]^T = G_1'(t)\ [\ \Delta a_f(t)\ \ \Delta a_g(t)\ \ \cdots\ \ \Delta\psi(t)\ ]^T .$$

As the cloud expands, the relative orbital elements change slowly, all except relative mean longitude $\Delta L(t)$. (When the "leading edge" of the cloud catches the center of mass, $|\Delta L(t)| \approx 2\pi$, while other ΔF_j are small.)

If one examines higher order partial derivatives in mean longitude of the form:

$$\frac{1}{k!}\ \frac{\partial^k \underline{r}}{\partial x L^{k-1}}\ \Delta L^{k-1} \quad , \quad x = a_f\ ,\ a_g\ ,\ \cdots$$

one can see that the linear transformation $G_1'(t)$ is only valid in a small neighborhood of the reference ephemeris ($|\Delta L| \ll 1$ radian). This explains why Spencer[2] reports that the linear rendezvous equations are inadequate for debris cloud modeling, except for very small spread velocities.

NONLINEAR EXTENSION

Derivation

To remove the error due to neglected higher order partial derivatives in L , we seek a transformation Φ such that:

$$\Phi_{t,t_0} = \left[G_1'(t) + G_1^*(t,\Delta L)\right] G_2(t,t_0)\, G_3'(t_0)$$

and:

$$\begin{bmatrix} \Delta Y_1(t) \\ \Delta Y_2(t) \\ \Delta Y_3(t) \end{bmatrix}_c = \left[G_1'(t) + G_1^*(t,\Delta L) \right] G_2(t,t_0)\, G_3'(t_0) \begin{bmatrix} \Delta \dot{Y}_1(t_0) \\ \Delta \dot{Y}_2(t_0) \\ \Delta \dot{Y}_3(t_0) \end{bmatrix}_b$$

where $G_1^*(t,\Delta L)$ will have the form:

$$[\,\hat{\underline{u}} \quad \hat{\underline{v}} \quad \hat{\underline{w}}\,]^T \cdot \begin{bmatrix} \displaystyle\sum_{k=2}^{\infty} \frac{\partial^k \underline{r}}{\partial x L^{k-1}} \frac{1}{(k-1)!} \Delta x\, \Delta L^{k-1} & x = a_f,\ a_g,\ n,\ \chi,\ \psi \\ \text{or} & \\ \displaystyle\sum_{k=2}^{\infty} \frac{\partial^k \underline{r}}{\partial L^k} \frac{1}{k!} \Delta L^{k-1} & x = L \end{bmatrix}$$

and is an explicit function of ΔL .

Omitting laborious details, each series, evaluated at $e = 0$, can be determined to be an expansion of a closed form trigonometric function of ΔL . For example,

$$\left[\sum_{k=1}^{\infty} \frac{\partial^k \underline{r}}{\partial L^k} \left[\frac{1}{k!} \right] \Delta L^k \right]_{\epsilon=0} = a \left\{ \hat{\underline{u}} \left[-\frac{\Delta L^2}{2!} + \frac{\Delta L^4}{4!} - \frac{\Delta L^6}{6!} + - \cdots \right] + \hat{\underline{v}} \left[\Delta L - \frac{\Delta L^3}{3!} + \frac{\Delta L^5}{5!} - + \cdots \right] \right\} .$$

Therefore:

$$\left[\sum_{k=2}^{\infty} \frac{\partial^k \underline{r}}{\partial L^k} \left[\frac{1}{k!} \right] \Delta L^{k-1} \right]_{e=0} = \frac{a}{\Delta L} \left\{ \hat{\underline{u}} \, (-1 + \cos \Delta L) + \hat{\underline{v}} \, (-\Delta L + \sin \Delta L) \right\}$$

Let $G_1^*(t,\Delta L) \equiv [g_{ij}]$, $i = 1, 2, 3$, $j = 1,\cdots, 6$. Then:

$$g_{11} = -g_{22} = a\,(-\cos L \sin^2\Delta L - \sin L \sin \Delta L \cos \Delta L)$$

$$g_{12} = g_{21} = a\,(-\sin L \sin^2\Delta L + \cos L \sin \Delta L \cos \Delta L)$$

$$g_{13} = \frac{-2}{3}\frac{a}{n}\,(\cos \Delta L - 1)$$

$$g_{14} = a\,(\cos \Delta L - 1) \,/\, \Delta L \qquad (= 0 \;\; \text{for} \;\; \Delta L = 0)$$

$$g_{15} = -a\, w_y \sin \Delta L$$

$$g_{16} = -a\, w_x \sin \Delta L$$

$$g_{23} = \frac{-2}{3}\frac{a}{n} \sin (\Delta L)$$

$$g_{24} = a\,(\sin \Delta L - \Delta L) \,/\, \Delta L \qquad (= a \;\; \text{for} \;\; \Delta L = 0)$$

$$g_{25} = a\, w_y\,(\cos \Delta L - 1)$$

$$g_{26} = a\, w_x\,(\cos \Delta L - 1)$$

$$g_{31} = g_{32} = g_{33} = g_{34} = 0$$

$$g_{35} = a\,(1 + w_z)\,(\cos L - \cos (L + \Delta L))$$

$$g_{36} = a\,(1 + w_z)\,(-\sin L + \sin (L + \Delta L))$$

where w_x , w_y , and w_z are components of $\hat{\underline{w}}$ resolved on an inertial basis, and a is semimajor axis.

A Simple Test of Nonlinear Extension

Consider a 97 minute period, circular orbit, and a fictitious explosion at $t = 0$, with all pieces ejected at 300 m/sec, isotropically. After 97 minutes the cloud extends $\pm 43^{o}$ in the orbit plane from the reference point. Assume there are 18 pieces of debris, use isotropic spread velocity directions to construct 18 orbital element sets and propagate each (two body) separately. Relative positions of each are calculated and plotted as "truth" in the attached figures (x's). Figure 1 contains three projections of the same cloud and depicts the 18 locations (o's) predicted by a similar subset of the Clohessy Wiltshire rendezvous equations[4]. In Figure 2 the experiment is repeated using the nonlinear transformation Φ (with $J_2 = 0$).

> ***A nonlinear extension to the linear rendezvous equations allows relative motion techniques to be applied at very large relative displacements from the reference ephemeris.***

A GENERALIZATION OF LAMBERT'S PROBLEM

For purposes of this paper we refer to the Generalized Lambert Problem as Lambert's problem stated for multiple orbits in a perturbed force field, restricted to elliptic motion.

ACCURACY OF SOLUTION TO THE GENERALIZED LAMBERT PROBLEM

We claim that Φ^{-1} is an approximate solution to the Generalized Lambert Problem. To demonstrate this claim, consider a single piece of debris, ejected from a breakup with relative velocity magnitude = 100 m/s, assigned as $\Delta\dot{Y}_1$ = -42.1 m/s, $\Delta\dot{Y}_2$ = -7.4 m/s, $\Delta\dot{Y}_3$ = 90.4 m/s. Compute orbital elements for this piece and generate a "truth" ephemeris (two body + J_2 secular). At each position in the truth ephemeris, apply Φ^{-1} to reconstruct $[\Delta\dot{Y}_i(t_0)]$. Figure 3 provides the values determined, graphed versus cloud revs since breakup.

Figure 3 gives evidence that Φ^{-1} is accurate to within a few percent in all three coordinates over much of space, but "loses" accuracy in one or two coordinates at multiple half revs of the piece. Chobotov et al[2] call these half rev and integer rev points "pinch wedges" and a "pinch points", places where the cloud is actually pinched through a small volume of space.

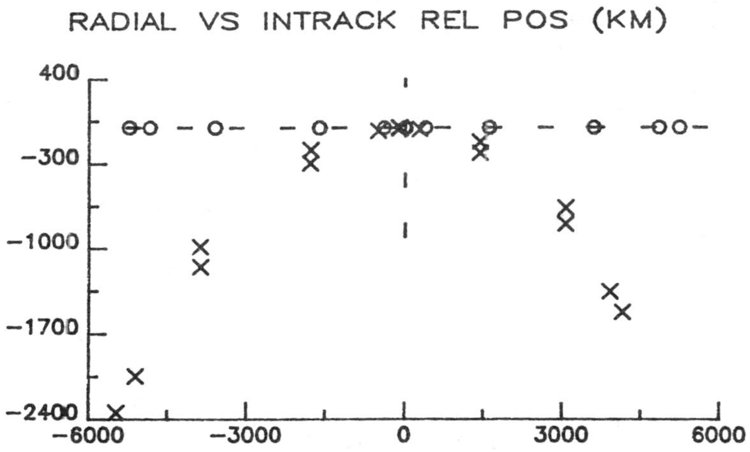

FIG 1. Clohessy Wiltshire Analysis

time predicted = 97 min
delta-v = 300 m/s
period = 97 min
eccentricity = 0
inclination = 65 deg

x = true cloud (two-body)
o = Clohessy Wiltshire transform

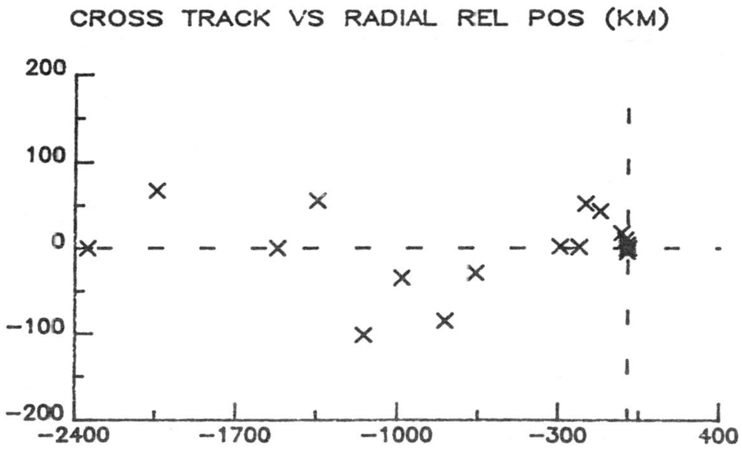

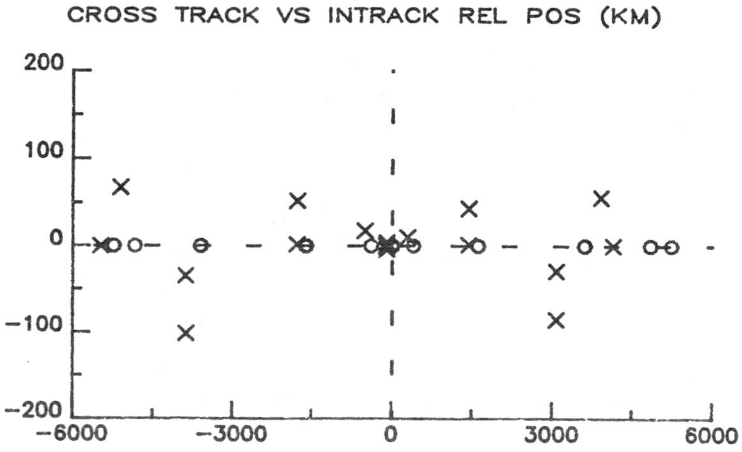

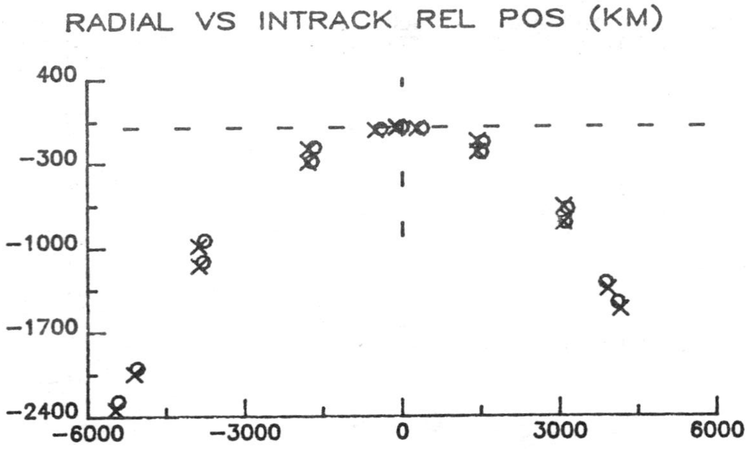

FIG 2. Nonlinear Transform Analysis

time predicted	=	97 min
delta-v	=	300 m/s
period	=	97 min
eccentricity	=	0
inclination	=	65 deg

x = true cloud (two-body)
o = nonlinear transform

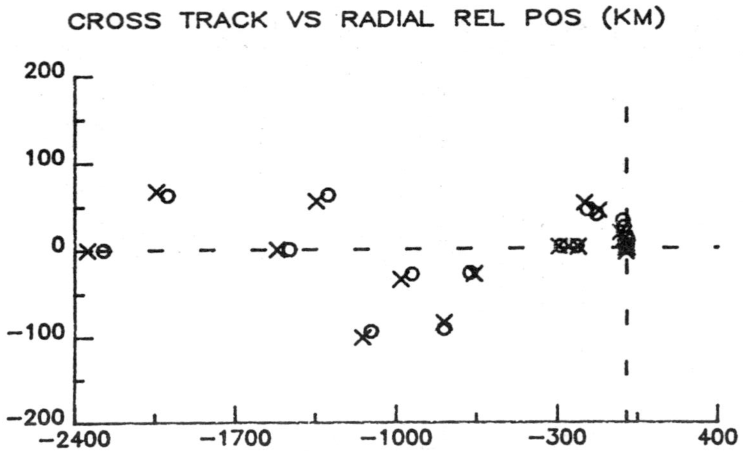

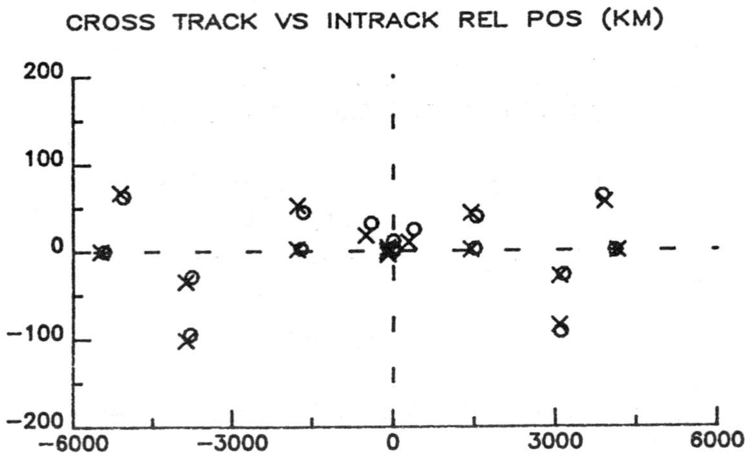

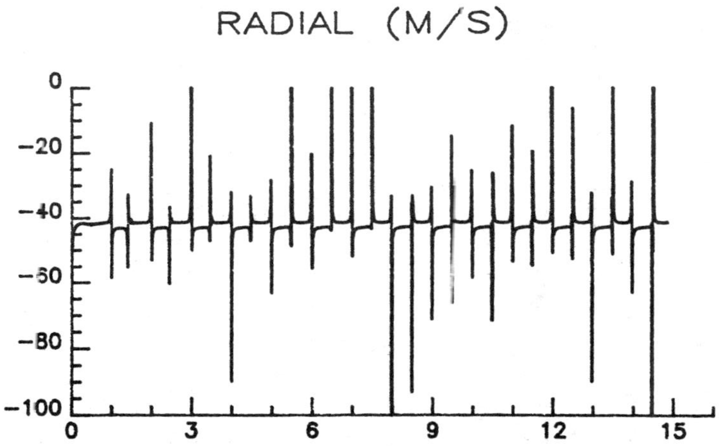

FIG 3. Reconstructed Spread Velocities
vs
Revs Since Breakup

cloud period	=	97	min
eccentricity	=	0	
inclination	=	65	deg

actual r.i.c. spread velocities =
−42.1, −7.4, 90.4 m/s

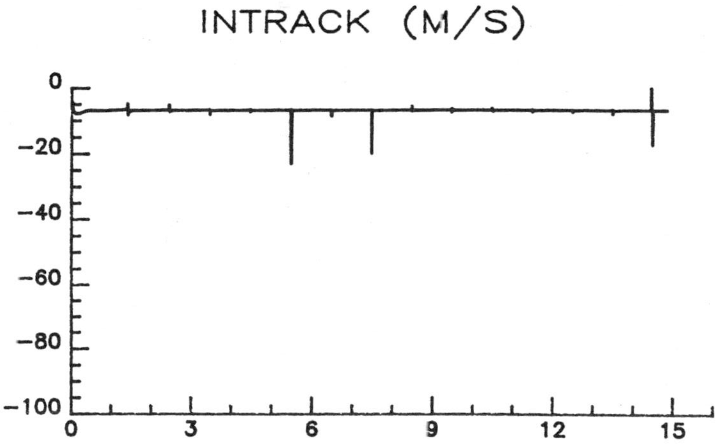

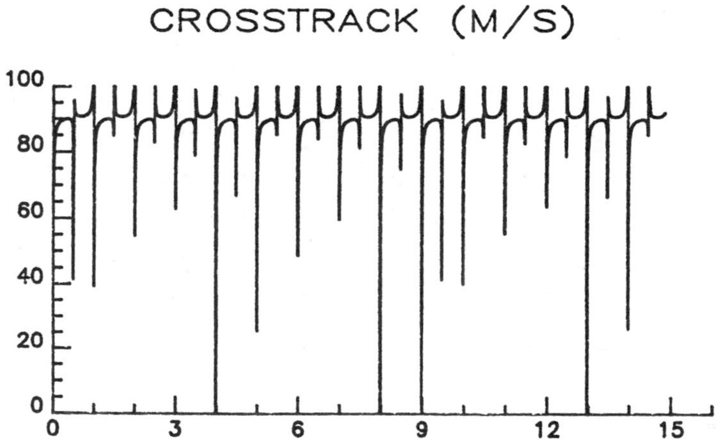

This solution to the Generalized Lambert Problem is singular at pinched points, just as Gauss's solution to Lambert's problem is singular at π [6]. As noted by Sun et al[7], the singularity at each half rev occurs simply because the solution to Lambert's problem is not unique. There are an infinite number of solutions to the Generalized Lambert Problem at the pinched points, which are not readily characterized by the transformation Φ , which is one-to-one. Thus:

> *When the solution to the Generalized Lambert Problem is unique, Φ^{-1} provides a non-iterative approximate solution which is accurate to within a few percent of the spread velocity magnitude.*

USING THE NONLINEAR MODEL TO FIND DEBRIS DENSITY

The power of the nonlinear model of relative motion is apparent when it is used to characterize debris density differentially, at some point of interest, in terms of the initial conditions of the breakup. Given the position of a satellite $\underline{r}(t)$, compute $\Delta Y_i(t)$. Then apply the inverse transformation:

$$[\Delta\dot{Y}_1(t_0) \quad \Delta\dot{Y}_2(t_0) \quad \Delta\dot{Y}_3(t_0)]^T = \Phi^{-1}_{t,t_0} \; [\Delta Y_1(t) \quad \Delta Y_2(t) \quad \Delta Y_3(t)]^T$$

to find an approximate solution to the Generalized Lambert Problem for multiple orbits in a perturbed gravity field. By adding differential increments to $\Delta Y_i(t)$ and repeating the process, we can construct a differential volume of spread velocity which maps into a differential position volume in a neighborhood of $\underline{r}(t)$. Computing the probable number of particles in the velocity volume, we obtain the probable number of particles in the position volume. This yields density at one point in space. Summing along the satellite path yields the probable number of particles encountered by the satellite as it passes through the cloud.

An Example of Density Computation

For purposes of illustration, we simulated a single piece ejected from a breakup with 100 m/s in the crosstrack direction. The cloud reference ephemeris was assigned a period of 97 minutes and assumed to be circular. The piece was found to pass through the pinch point after 97.0304 minutes. The nonlinear transformation (inverse) was applied to find the differential volume of spread velocity which maps into a 10m × 10m × 10m position volume around the piece at various times and places during the evolution of the debris cloud. We were particularly interested in the utility of the nonlinear transformation in determining particle density near the pinched points.

First the model was tested in a region where the solution to the Generalized Lambert Problem is unique (at t = 80 min). The model predicted that a quasi-hyperellipsoid of spread velocities with principle semi-axes of .0113 m/s × .0005 m/s × .012 m/s maps into the differential position volume. To confirm these results an interactive search was performed, where candidate spread velocities were used to generate orbital elements, which were propagated, converted to positions, and differenced with the piece location. The confirmation analysis found that actual spread velocity differential volume had dimensions .012 m/s × .0004 m/s × .012 m/s, and the differential volume was correctly predicted to within 6 percent.

The analysis was repeated at t = 96 min, much closer to the model singularity. The inverse transformation predicted a differential spread velocity volume with semi-axes of .16 m/s × .0005 m/s × .14 m/s mapping into the differential position volume. The confirmation analysis found the actual semi-axes to be .155 m/s × .0005 m/s × .155 m/s, accurate to within 7 percent.

As the analysis approached the pinch point the capability of the nonlinear transformation became less reliable. At t = 96.5 minutes the model predicted .31 m/s × .001 m/s × .24 m/s, while confirmation analysis yielded .30 m/s × .0005 m/s × .29 m/s. The volume calculated by the model is too large by a factor of 1.7.

At t = 97 min, the model predicted 1.6 m/s × .027 m/s × 5.4 m/s while confirmation analysis yielded 2.5 m/s × .0005 m/s × 2.5 m/s. The error in the volume calculation near the pinch point is a factor of 75.

These analyses indicate that this method requires special case models near the pinched points. It has been suggested that if one could determine the persistent intrack spread velocity semi-axis value (.0005 m/s in this case) that the volumes predicted at 96.5 and 97 minutes could be substantially corrected. Even with the present limitations, this model is an improvement over current methods which assume an averaged density.

Our confirmation analysis proves that the differential velocity volume which maps into a differential position volume can change by four orders of magnitude, depending upon the location in the cloud. Consequently:

> *Particle density can vary by at least four orders of magnitude, depending on location within the cloud, and an averaged density assumption can be in error by at least two orders of magnitude.*

OTHER LOCATIONS OF INCREASED DENSITY

The pinched locations are not the only regions of the cloud which exhibit increased particle density, although they are the most obvious. There are other locations, which occur within the envelope of the cloud, which also exhibit increased density. These were detected by repeating the experiment for Figure 3, but assigning $\Delta\dot{Y}_1$ = 26.6 m/s, $\Delta\dot{Y}_2$ = 68.8 m/s, and $\Delta\dot{Y}_3$ = 67.5 m/s. The ability of Φ^{-1} to solve the Generalized Lambert Problem can be seen in Figure 4. In addition to the singularities at the pinched locations, additional spikes occur every half rev, well removed from the pinch point.

Investigating this new phenomenon, the transformation was found to be nonsingular. The location of the spike was found to be quite distant from pinched features. Differential volume calculations were performed in a neighborhood of the new spikes using Φ^{-1} on a differential position volume (and confirmed interactively). It was found that the differential velocity volume mapping into these points is also as much as four orders of magnitude greater than where Φ^{-1} yields a unique solution. Increased differential velocity volume can yield a proportional increase in predicted debris density.

This increased density in the debris cloud has not been detected in two body analysis, and it is assumed that this is a manifestation of the perturbations. Apparently perturbations yield new conditions under which the solution to the Generalized Lambert Problem is not unique. One can only conclude that:

> *Perturbations introduce new density structures into the debris cloud, structures which are significant in terms of probability of collision analysis.*

REMARKS

1. The reference ephemeris can be generated with any force model appropriate to the accuracy desired, including numerical integration of a robust perturbation model.

2. $\Phi^{\#2}$ can be generated to any accuracy (e.g., by using variational equations).

3. The function $G_1^*(t,\Delta L)$ can be extended to higher order terms in eccentricity using the Taylor Series.

4. As the debris cloud "rings" the Earth, a single point in space $\underline{r}(t)$ may be achieved by a finite number of particles, simultaneously. $G_1^*(t,\Delta L)$ is a closed form function of sines and cosines of ΔL , and can be repeatedly evaluated for $\Delta L^* = \pm n(2\pi) + \Delta L$. Particle density at $\underline{r}(t)$ is the sum of differential densities obtained by repeated application of Φ for various ΔL^* . Φ effectively provides a

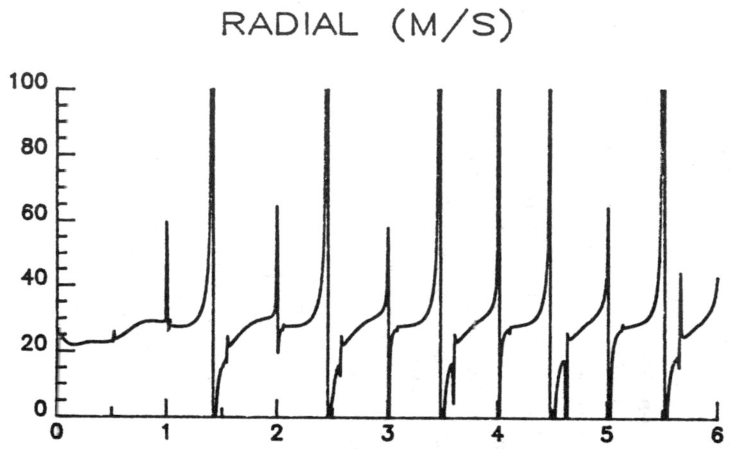

FIG 4. Reconstructed Spread Velocities
vs
Revs Since Breakup

cloud period = 97 min
eccentricity = 0
inclination = 65 deg

actual r.i.c. spread velocities =
26.6, 68.8, 67.5 m/s

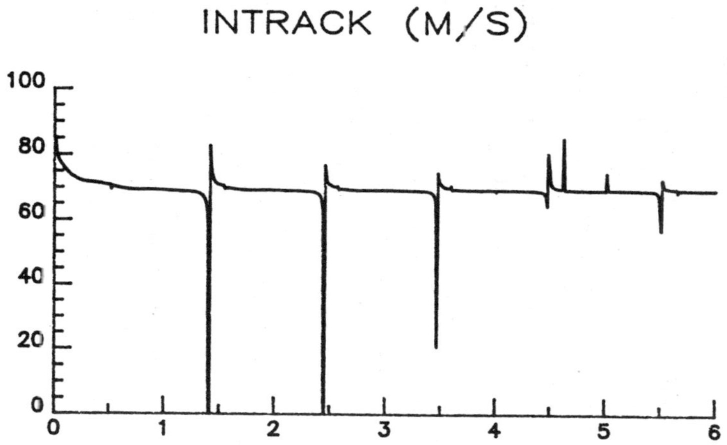

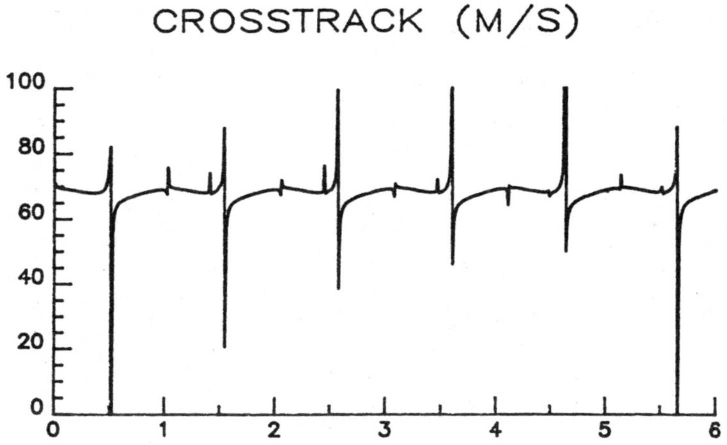

single dynamical model of differential cloud density, valid from the time of breakup until after the cloud "rings" the earth.

5. The choice of orbital elements in G_j depends upon the application. Classical elements result in singularities for zero eccentricity and inclination, and should be avoided[8]. Nodal elements can not be used for equatorial breakups, while equinoctial elements are nonsingular for all elliptic motion. Universal variables are recommended for a complete analysis of cases where elliptic and hyperbolic debris trajectories are of interest.

6. In the two body debris problem the pinched cloud features are fixed in inertial space at $n\pi$. For the perturbed problem, however, the pinched locations precess. The singularities in Φ occur at the locations of the precessed pinched cloud features.

CONCLUSIONS

A nonlinear dynamical model of relative motion has been derived for application to the orbital debris problem. The model is valid for large displacements for the reference ephemeris. The linear expressions are complete in eccentricity and secular J_2, while the nonlinear terms are defined for zero eccentricity.

This model of relative motion can be successfully applied to compute particle density in the space safety problem. Doing so eliminates the need to assume a cloud shape, eliminates the need to use averaged particle density, and allows spatial density to be computed differentially. By treating the debris problem differentially, as an Extended Lambert Problem, spatial density can be computed directly from the spread velocity distribution generated by a fragmentation model.

We have also gained considerable insight into the structure of a debris cloud. An average particle density can seriously misrepresent actual particle density. Debris density increases by four orders of magnitude near the well known pinched features of the cloud. Density also increases by four orders of magnitude, due to perturbations, at locations internal to the cloud. Space safety analysis based upon averaged particle densities may be unacceptable in light of this new result.

This model is singular at each half rev from the breakup point and requires a complimentary set of special case models to properly compute density everywhere. The required special case models do not yet exist. In the absence of these models the relative motion model correctly identifies the locations of increased density, even though the computed densities are incorrect. While the need for more work in this area is apparent, it is also clear that significant improvements in debris density calculations have been achieved with the Chain Rule, the Taylor Series, and a few analytic equations from celestial mechanics. It is our fervent hope that future developments in this area attempt to improve upon this approach.

ACKNOWLEDGEMENT

I wish to thank Jeannine Arsenault for observing that the nonlinear equations can be evaluated at the local value of ΔL , she provided the key breakthrough in this model. Jeannine also verified all of the equations during model development. Thanks go to Dr. Lou Walters, Dr. Val Chobotov, and Bill Sampson for review and interpretation and to the Air Force for initial funding of this effort.

REFERENCES

1. Chobotov, V.A., "Dynamics of Orbiting Debris Clouds and the Resulting Collision Hazard to Spacecraft", International Astronautical Federation IAA-87-571, October 1987.

2. Chobotov, V.A., Spencer, D.B., Schmitt, D.L., Gupta, R.P., Hopkins, R.G., and Knapp, D.T., "Dynamics of Debris Motion and the Collision Hazard to Spacecraft Resulting From an Orbital Breakup", Aerospace Report SD-TR-88-96, January 1988.

3. McKnight, D., "A Phased Approach to Collision Hazard Analysis", COSPAR III.2.4, undated.

4. Clohessy, W.H. and Wiltshire, R.S., "Terminal Guidance System for Satellite Rendezvous", J. Aerospace Sci, Vol. 27, 1960.

5. Koskela, P.E., "Astrodynamic Analysis for the Advanced Orbit/Ephemeris Subsystem", Aeronutronic Publication No. U4180, September 1967.

6. Battin, R.H., ***An Introduction to The Mathematics and Methods of ASTRODYNAMICS***, AIAA Education Series, 1987.

7. Sun, F., Vinh, N.X., Chern, T., "Analytic Study of the Solution Families of the Extended Godal's Time Equation for Lambert's Problem", AAS85-351, August 1985.

8. Badhwar, G.D., Tan, A., Reynolds, R.C., "Velocity Perturbation Distributions in the Breakup of Artificial Satellites", NASA Johnson Space Center, undated.

AAS 89-398

STUDY OF POSSIBLE COLLISION-INDUCED BREAKUPS

Darren S. McKnight*

The near Earth environment has been strewn with debris from thousands of space launches and nearly 100 satellite breakups. Some of these may be the result of on-orbit collisions. This study outlines a series of tests which are used to classify characteristics of a breakup event. This group of tests is called the Satellite Fragmentation Event (SAFE) Test III, which is an improvement on the earlier SAFE Test II. The results from this paper reinforce earlier analyses which have concluded that possibly three of the unknown breakups have characteristics of a hypothetical collision-induced fragmentation. To provide more accurate predictions of breakup cause the phenomenology of explosions/collisions of satellites should be studied further and a greater percentage of debris fragments must be able to be sensed.

INTRODUCTION

The threat of space debris to operational space systems has received much more attention over the last few years [1,2,3]. If the debris is massive enough and the relative velocity between it and a satellite large enough the parent satellite may be completely destroyed. Of the nearly 100 on-orbit fragmentation events about half have no known cause but a few of these are candidates for being collision-induced [4,5]. In 1985, the Satellite Fragmentation Event (SAFE) Test was created for this purpose. It was found to be too subjective and neglected to use all the information available about each breakup [6]. The SAFE Test II was formed in 1987 which corrected the discrepancies of the original SAFE Test [5]. This paper develops a follow-on investigative tool, the SAFE Test III.

---

* Captain, U.S. Air Force, U.S. Air Force Academy, Department of Physics, Colorado Springs, Colorado 80840.

To formulate a test to determine if a satellite breakup was caused by a collision with space debris two phases of analysis are necessary. First, the prebreakup parameters for the scenario are covered. This includes mission duration, collision hazard, fragment lethality and past history. Secondly, postbreakup debris analysis is required. The velocity and mass distributions of the resulting debris are covered in detail. The postbreakup investigations are aided by access to unpublished hypervelocity impact data which has provided greater insight into the fragmentation process. A number of efforts have been made in the past to determine the cause of satellite breakups [7,8,9]. The study accomplishes this difficult task by using information available to all space users which should encourage future use and validation by others.

PREBREAKUP PARAMETERS

The probability of collision from the trackable debris environment can be calculated using spatial density values derived from a catalog of trackable objects [10].

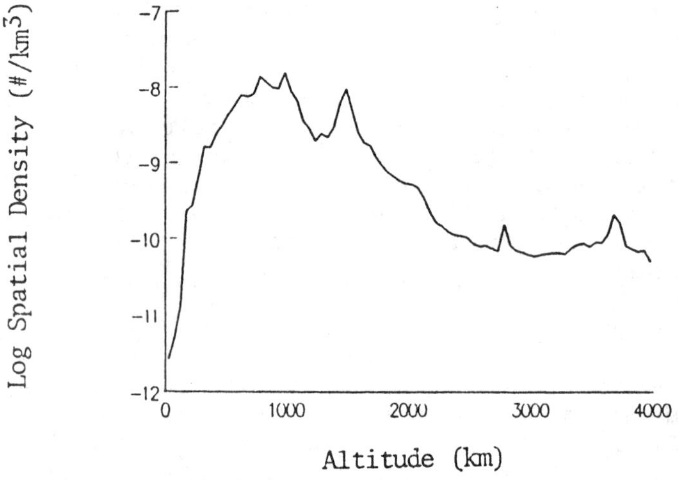

Fig. 1 Spatial Density Values

The probability of collision per year times the satellite's mission duration provides the starting point for this study. This number will be adjusted for each breakup to result in a "probability" of being collision-induced. Two corrections for nontrackable fragments are made. First, the probability of collision term is multiplied by a factor to correct for the greater number of nontrackable fragments at a given altitude that might still cause a catastrophic collision. Secondly, the lethality of the nontrackable fragments is factored into the analysis. A collision will not necessarily result in a complete fragmentation of the parent satellite; the satellite will fragment totally only if the fragment possesses enough kinetic energy relative to the satellite. A fragment's lethality is a function of its relative size and relative velocity with respect to the target satellite.

The last prebreakup adjustment accounts for the history of explosion events. If similar satellites have experienced explosions or deliberate fragmentations in the past then the possibility of this breakup being collision-induced is assumed less likely. All these adjustments are made to produce the Total Prebreakup Probability.

POSTBREAKUP PARAMETERS

Now this term is modified by categorizing characteristics of the resulting debris as being representative of a collision or explosion. These postbreakup tests are similar to those used in a previous analyses by the author but have been modified to account for new hypervelocity impact data [11].

Three of the postbreakup criteria are concerned with the velocity distribution of the debris from a breakup event. The magnitude of the average velocity imparted (delta velocity) to the larger fragments is usually less than 100 m/s for a debris-satellite encounter. Additionally, in a collision-induced fragmentation the smaller fragments have a tendency to have larger delta velocity values. Lastly, the more massive fragments are usually symmetrically dispersed from a collision-induced breakup. These characteristics take the form of three tests.

The last postbreakup correction results from analysis of the mass distribution the fragmentation debris. Three different equations were investigated to find the best representation to characterize the mass distribution. Power, exponential and hyperbola equations are investigated as outlined below.

$$CN = A\ (m/M_e)^{-B} \quad (1)$$

$$CN = A \exp(-B * \sqrt{m}) \quad (2)$$

$$-CN^2 = A + Bm + C(CN) + Dm(CN) + E(m^2) \quad (3)$$

where CN = cumulative number of fragments of mass greater than m

A,B,C,D,E,F = constants

m = fragment mass

M_e = mass of fragmentation debris

The power law representation is a good "order of magnitude" approximation [10]. The exponential distribution performed slightly better with three tests having significant confidence levels (greater than 95%). These two mathematical representations have both been used but neither truly depict the empirical results from the events analyzed as well as hoped [13]. The last curve, the hyperbola, performed much better. This five-term equation fit curves to the data at a +99% confidence level for all events. The new use of this representation has its liabilities. The physical interpretation of the constants in the hyperbola equation is in its infancy. These concerns are addressed in Reference 12. For the time being, parameters from events of known cause are simply compared to the terms of the events of unknown cause to assign a score for the last postbreakup test.

RESULTS

The compilation of all of these corrections to the original probability of collision value is called the Satellite Fragmentation Event (SAFE) Test III. Figure 2 is the form for the SAFE Test III. Fourteen different satellite breakups are analyzed. Preliminary results suggest that at least one breakup, and possibly three, events have characteristics assigned to a collision-induced event. Table 1 lists the breakups studied in order from most likely collision-induced to least likely.

Table 1
SAFE TEST III RESULTS

International Designator	Common Name	Previous Breakup Classification
1981-53	Cosmos 1275 P/L	Unknown
*1979-17	P-78/1 P/L	Deliberate-ASAT
1976-67	Cosmos 839 P/L	Unknown
1982-55	Cosmos 1375 P/L	Unknown
1970-25	Nimbus 4 R/B	Explosion
1972-58	Landsat 1 R/B	Explosion
1978-26	Landsat 3 R/B	Explosion
1976-77	NOAA-5 R/B	Explosion
1973-86	NOAA-3 R/B	Explosion
1986-19	Viking/Spot R/B	Unk./Expl.
1965-82	Titan 3C R/B	Explosion
1970-91	Cosmos 375 P/L	Unknown
1961-OMI	Transit 4A R/B	Explosion

*Partial Analysis

SAFE Test III

Satellite____

PREBREAKUP

Probability of Collision/Year ____________________________
Years in Orbit X____________________________
Nontrackable Fragments X____________________________
Lethality [Yes:0.1]
[No:1.0] X____________________________
History of Explosion Events
[Yes:0.1]
[No:1.0] X____________________________

TOTAL PREBREAKUP PROBABILITY ____________________________
____

POSTBREAKUP

Average Delta Velocity
[V > 100 m/s : 0.1]
[v < 100 m/s :10.0] X____________________________
Delta Velocity Inversely Proportional to Mass
[Yes:10.0]
[No: 0.1] X____________________________
Symmetry of Large Fragments by Delta Velocity
[Yes:10.0]
[No: 0.1] X____________________________
Hyperbola Curve
[Collision:10.0]
[Explosion: 0.1]
[Unknown: 1.0] X____________________________

"PROBABILITY OF BEING COLLISION-INDUCED" __________________

Fig. 2 SAFE Test III Worksheet

CONCLUSIONS

Despite significant efforts in the area of breakup-cause technology it is still very difficult to ascertain the reason why a satellite has fragmented. The SAFE Test III has placed a number of unknown breakups into the possible collision-induced category while leaving most explosions away from this trend. The analysis is very subjective and suffers from two major drawbacks. First, the limited number of observable fragments makes it difficult to statistically represent the breakup's debris cloud accurately. Much of the postbreakup information resides in objects too small (optically and/or physically) to be detected. Second, the hyperbola curve appears to be more appropriate to model the debris produced from fragmentation events. More research is needed to quantify the physical significance of each of the parameters associated with the hyperbola model. The goal of this ongoing work is to be better able to distinguish between different types of breakup events. Only through the pursuit of this basic research will directly applicable findings result.

ACKNOWLEDGEMENTS

The research conducted by Kenton Leitzau during an independent study project at the Air Force Academy has been invaluable. His fresh insight was a useful addition to our research project. Captain Christopher Brechin has helped to design the course of our research. A special thanks is extended to Pat Ridley for typing this paper.

REFERENCES

1. Report on Orbital Debris by Interagency Group (Space) for National Security Council, February 1989.
2. "Space Debris", The Report of the ESA Space Debris Working Group, ESA SP-1109, November 1988.
3. McKnight, D. S. and Johnson, N. L., "Error Found in Orbital Debris Projections", *Aerospace America,* April 1989.
4. Johnson, N. L., "Hazards of the Artificial Space Debris Environment", A paper presented to the Legal Subcommittee of the Committee on the Peaceful Uses of Outer Space, 30 March 1989.
5. McKnight, D. S., "Determining the Cause of a Satellite Fragmentation: A Case Study of the Kosmos 1275 Breakup", International Astronautical Federation Congress, Brighton, England, 10-17 October 1987.
6. Culp, R. D. and McKnight, D. S., "Distinguishing Between Collision-Induced and Explosion-Induced Satellite Breakup Through Debris Analysis", Astrodynamics 1985, Vol. 58 Part 1, *Advances in the Astronautical Sciences,* AAS Paper Number 85-370, Univelt Inc., San Diego, CA, 1986, pp.739-758.
7. Badhwar, G. D., et al., "Characteristics of Satellite Breakups as Derived from the Radar Cross Section and the Plane Change Angle", Presented at IAF Congress in Bangalore, India, October 1988.
8. Badhwar, G. D. and Anz-Meador, "Determination of the Area and Mass Distribution of Orbital Debris Fragments", Presented at IAF Congress in Bangalore, India, October 1988.
9. Orbital Debris Monitor, quarterly Newsletter, P.O. Box 136, USAFA, CO, April 1989.
10. Brechin, C. B. and McKnight, D. S., "Mass Distributions from Recent Hypervelocity Impact Tests", Astrodynamics 1989, Vol. 59, *AAS Microfiche Series,* AAS Paper Number 89-375, Univelt Inc., San Diego CA, 1990.
11. McKnight, D. S., et al., "Mass Distribution from Hypervelocity Impact Tests of Spacecraft Structures", to be presented at the 1989 Hypervelocity Impact Symposium, San Antonio, TX, 12-14 December 1989.
12. Leitzau, K. R., "Physics 499 - Orbital Debris Analysis", Internal Final Report, USAFA/Dept. of Physics, 21 May 1989.

AAS 89-399

AVOIDING COLLISIONS WITH RESIDENT SPACE OBJECTS*

Ralph G. Schinnerer†

A method is presented for determining the minimum velocity impulse maneuver which may be applied to a satellite at a specified time to permit safe passage with all other known satellites for a specified time into the future. The process may be initiated after it is determined that a high–valued object will come dangerously close to another object. The minimum velocity impulse determined by the technique will be that which not only avoids the *anticipated* conflict but also ensures that conflicts will not be created with any other resident objects as a result of the proposed maneuver.

The essence of the technique is that of linearizing the problem about the nominal "non–maneuvered" orbit. To do this, it is assumed that the required maneuver impulse will be small enough to permit the relationship between the components of maneuver impulse and the position at potential collision times to be expressed as a linear Taylor's series.

An analysis is presented which aids in an understanding of the problem; a workable approach is described that makes reasonable use of computer resources; and some typical results are presented wherein real data are used.

The technique is limited to considering only cataloged resident space objects for which orbital elements are known (ie., clouds of debris too small to be tracked as discrete objects are not considered). The technique also makes use of the common simplification that a collision can be safely avoided if a specified "stand–off" radius is ensured at the point of closest approach between two objects. Hence, considerations of object size and orbital element accuracy are sophistications not included in determining probabilities of collision.

BACKGROUND

The technique presented in this paper makes use of some capabilities which have been available for many years in typical large-scale space surveillance systems. References are provided which describe these capabilities, but the capabilities are used without supporting analyses included herein as to their accuracy. (Extensive such analyses *have* been undertaken over the years but were documented as internal memoranda.) It is helpful to review these capabilities in order to more easily understand the approach presented herein.

---

* Work performed under Government contract F19628-83-C-0019.

† Principal Engineer, Ford Aerospace Corporation, 9970 Federal Drive, Colorado Springs, Colorado 80921. Associate Fellow, AIAA. Member, AAS.

The first such capability is a computer program for determining the points of closest approach between any cataloged space objects. Such a program is typically used to warn of potential collisions in space. This capability will henceforth be referred to as "COMBO" (Computation of Miss Between Orbits). An excellent reference for the COMBO capability is a paper by Hoots, et al[1], wherein many improvements are described. In spite of these improvements, using COMBO to determine close approaches between any object and the rest of the entire catalog is computationally intensive. The COMBO capability is used by the technique described herein to establish a baseline about which analyses of proposed maneuvers can be made using linear techniques. It is also used as a final check of the post-maneuver elements to ensure that no new close approaches are created.

The second important capability used is the computation of the matrix of partial derivatives relating changes in epoch elements to changes in position and velocity at a prediction time[2,3]. While these partial derivatives are used in a typical surveillance system primarily for differential correction, they are used in the technique described herein to provide a linear relationship between the small components of an impulsive maneuver and departures in position and velocity at potential collision times.

Other capabilities used include various general perturbations orbit representation packages[4,5], and various mathematical routines for vector and matrix manipulations.

THE PROBLEM

The minimum impulsive maneuver must be found which, if applied at a specified time, will permit "safe" passage of the maneuverable satellite past those other satellites for which a "dangerous" passage had been predicted, *and* which, at the same time, avoids moving into "dangerous" passages with any other known satellite orbits. "Safe" passage is defined as existing when the predicted separation distance at the point of closest approach is greater than a specified "stand-off" radius, and a "dangerous" passage is one where the separation distance is less than the "stand-off" radius. The impulsive maneuver will be specified in terms of the magnitude, ΔV, the out-of-plane angle, β, and the pitch angle from the vertical, γ.

One can visualize a "velocity" space at the time of a proposed maneuver. This "velocity" space is shown in Fig. 1. The Cartesian coordinate system is aligned with unit vectors $\underline{U}$ radial and $\underline{W}$ normal to the orbit plane, with $\underline{V}$ forming an orthogonal set and pointed downrange. The origin is the velocity of the maneuverable satellite at the maneuver time without any maneuver. Hence, a point in this space represents a maneuver in terms of the three components of the velocity impulse.

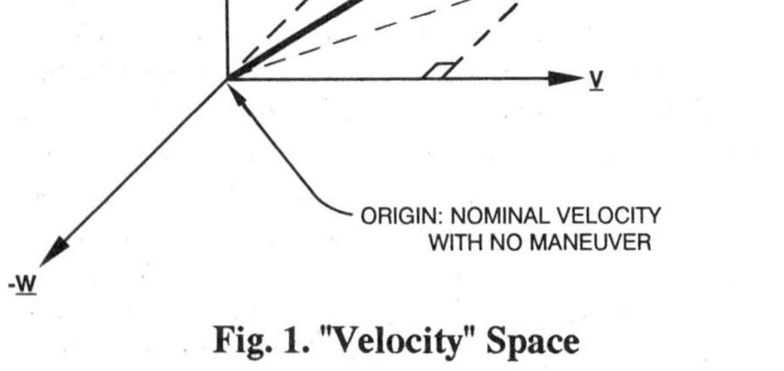

Fig. 1. "Velocity" Space

For each predicted close approach with another satellite at a later time, one can also visualize a surface enclosing a volume in this "velocity" space as shown in Fig. 2. *Inside* this volume, the maneuver velocity is such that the maneuverable satellite would have a "dangerous" passage with the other satellite at the potential conflict time. *Outside* the volume, the maneuver would be such as to make the passage "safe." Any point on the surface of the volume represents a maneuver which will achieve a marginally "safe" passage. The one such point on the surface that is closest to the origin is the minimum ΔV required to just achieve a "safe" passage. Of course, if the volume encloses the *origin*, then a maneuver is required for safe passage. If the volume does *not* enclose the origin, however, then it is only necessary to avoid the volume while trying to determine a maneuver to avoid another close encounter.

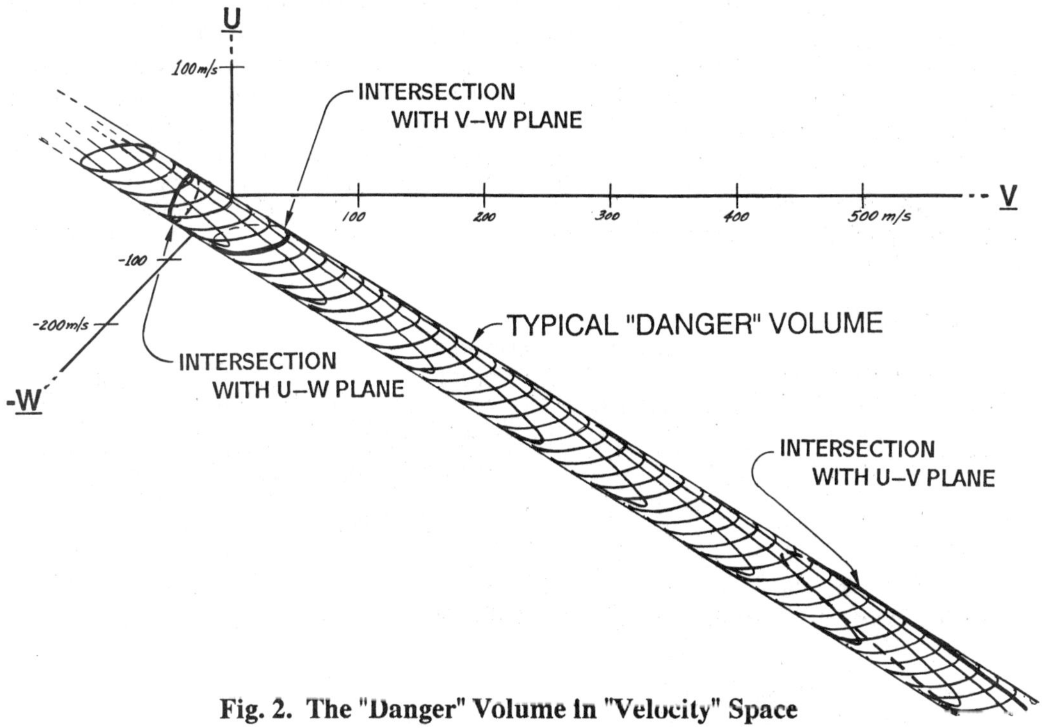

Fig. 2. The "Danger" Volume in "Velocity" Space

To a first approximation, the surface is a generally oriented cylinder of elliptical cross-section. The center-line of the cylinder is the locus of all maneuvers which would achieve intercept. If the surface represents an encounter very long after the maneuver, then the width of the cylinder in the V direction is very small. This is expected because a very small in-track impulse will grow quickly to a significant position departure. Likewise, if an encounter is diametrically opposite the maneuver, then the width in the W direction is very large. This, too, is expected because an out-of-plane maneuver will have no effect with this geometry. For a typical case where the encounter was 211 minutes after the maneuver time, and the stand-off radius was 12 km, the elliptical cross-section was highly eccentric. The smaller axis was about 3 meters per second (m/s) while the larger was about 110 m/s. As the time from maneuver to encounter becomes less, the associated volume grows, especially in the V direction. For another case where the encounter was only 11 minutes after the maneuver, the axis in the V direction grew to about 60 m/s (case shown in Fig. 2).

Of course, in general, there are *many* such volumes with many shapes and sizes in this "velocity" space, each representing a "danger" zone within which a different collision might occur. Those volumes that enclose the origin represent predicted "dangerous" passages. That is, without a maneuver, the passage of the two objects will be "dangerously" close. Those that don't enclose the origin represent potential conflicts if a maneuver *is* performed, but which, in the absence of a maneuver, are "safe." The sizes of the volumes are also directly dependent on the stand-off radius, which is chosen subjectively. Obviously, it is easier to avoid many small volumes than many large, overlapping ones.

In a typical analysis, a COMBO capability was used to establish a baseline of all reasonable encounters over a 12 hour period following the time of a possible maneuver. Points of closest approach were determined for all objects in the catalog which came within six times the stand-off radius (which was chosen to be 12 km). (A multiplier of six was used to reasonably ensure that potential encounters that might be created by a maneuver were included in the baseline.) The process produced four encounters that were considered dangerous in the absence of a maneuver (ie., the point of closest approach was less than the stand-off radius). In addition, 101 other encounters were produced which needed to be avoided after a proposed maneuver. Thus, the "velocity" space for this case contained 105 volumes of various sizes and shapes, four of which enclosed the origin.

THE SOLUTION

Two approaches to a solution were considered: the first, an elegant analytical approach using the characteristics of the volumes in the "velocity" space; and the second, a "brute-force" approach that simply explores the "velocity" space until a solution is found. The first was abandoned reluctantly because, although it had a high intellectual appeal, the potentially large number of "danger" volumes and the discontinuous nature of the problem, made the approach inconsistent with the availability of resources to pursue the problem. The second approach, while inherently reliable and free of insidious convergence problems, was also made to be efficient in terms of computer resources by using a linearized representation and recursive formulas to reduce the computation required at each point examined in the "velocity" space.

Linearizing the Problem

The heart of the solution is the use of partial derivatives that relate position and velocity at the points of closest approach to the proposed maneuver velocity impulse. This permits the linearization of the problem and obviates the need for repeatedly having to redetermine points of closest approach using a COMBO process.

Fig. 3 shows the paths of two satellites in the vicinity of their point of closest approach. We will assume that this region is small enough that we can consider their paths to be straight lines with acceptable accuracy.

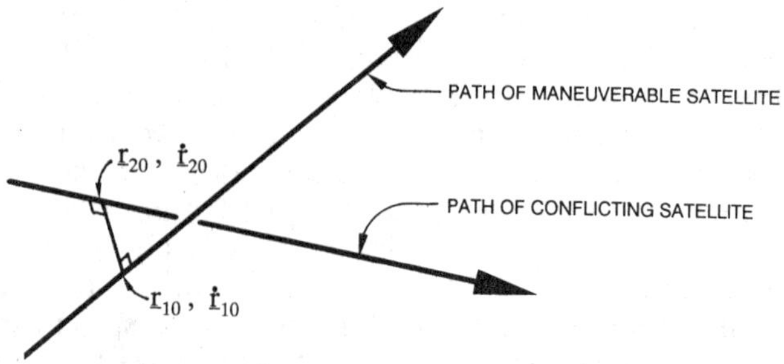

Fig. 3. Geometry at Close Approach

If we know the position and velocity of each of the two objects at a point near the point of closest approach, we can calculate the distance at the point of closest approach using a linear Taylor's series in time as an approximation to the motion.

Assume:

$$\underline{r}_1 = \underline{r}_{10} + \dot{\underline{r}}_{10}(t-t_0) \quad \ldots \quad (1)$$

and

$$\underline{r}_2 = \underline{r}_{20} + \dot{\underline{r}}_{20}(t-t_0) \quad \ldots \quad (2)$$

Then

$$\Delta r^2 = (\underline{r}_2 - \underline{r}_1)\bullet(\underline{r}_2 - \underline{r}_1) = (\underline{r}_{20}-\underline{r}_{10})\bullet(\underline{r}_{20}-\underline{r}_{10}) +$$

$$2\,(\underline{r}_{20}-\underline{r}_{10})\bullet(\dot{\underline{r}}_{20}-\dot{\underline{r}}_{10})\,(t-t_0) + (\dot{\underline{r}}_{20}-\dot{\underline{r}}_{10})\bullet(\dot{\underline{r}}_{20}-\dot{\underline{r}}_{10})\,(t-t_0)^2 \quad \ldots \quad (3)$$

The point of closest approach is where

$$\partial \Delta r^2/\partial t = 0 \quad \ldots \quad (4)$$

The separation distance at the time of closest approach, S, can be determined to be

$$S^2 = (\underline{r}_{20}-\underline{r}_{10})\bullet(\underline{r}_{20}-\underline{r}_{10}) - [(\underline{r}_{20}-\underline{r}_{10})\bullet(\dot{\underline{r}}_{20}-\dot{\underline{r}}_{10})]^2 / [(\dot{\underline{r}}_{20}-\dot{\underline{r}}_{10})\bullet(\dot{\underline{r}}_{20}-\dot{\underline{r}}_{10})] \quad \ldots \quad (5)$$

We now have a general expression for the minimum separation distance as a function of position and velocity *near* the point of closest approach. In order to predict any potential collisions in the absence of a maneuver, we initially perform a COMBO process. If this process is performed with an expanded stand-off radius, then it will also determine nearby potential collision opportunities which should be considered in determining an avoidance maneuver. If we save and use the positions and velocities returned from the COMBO process for each of the potential collisions as the nominal values, we can express the position and velocity as a function of those nominal values plus small increments attributable to an assumed maneuver at some earlier time. These increments can be estimated again with a linear Taylor's series (this time in $\Delta\underline{V}$) using the partial derivative matrices relating position and velocity near the point of closest approach to velocity at the maneuver time. Specifically,

$$\underline{r}_{10} = \underline{r}_{100} + [P_P]\bullet\Delta\underline{V}, \text{ and} \quad \ldots \quad (6)$$

$$\dot{\underline{r}}_{10} = \dot{\underline{r}}_{100} + [P_V]\bullet\Delta\underline{V}, \quad \ldots \quad (7)$$

where the subscript "100" denotes the nominal values for the maneuverable satellite at a potential collision time, $[P_P]$ is the three by three matrix of partials of position at the point of closest approach with respect to the velocity at the maneuver time, $[P_V]$ is a similar matrix relating velocity to velocity, and $\Delta\underline{V}$ is the assumed maneuver velocity. The position and velocity of the conflicting satellite won't change because it is assumed that, for any specific case, only one satellite is maneuverable.

The matrices of partials must be computed from those that are normally used to support differential correction. Differential corrections are always performed on epoch elements and, of course, the maneuver is generally not at epoch. Therefore, the state transition matrix, $[\Gamma_{M,C}]$, relating the state vector in UVW coordinates at the time of conflict to that at maneuver time is computed as

$$[\Gamma_{M,C}] = [\Gamma_{E,C}] \times [\Gamma_{E,M}]^{-1}, \quad \ldots \quad (8)$$

where the subscripts E, M, and C denote epoch, maneuver, and conflict times, respectively. The matrices $[P_P]$ and $[P_V]$ are then simply the appropriate 3 × 3 partitions of $[\Gamma_{M,C}]$.

To summarize, the COMBO process, when an expanded stand-off radius is used, provides a baseline of position and velocity vectors near potential conflict points. When combined with the matrices of partial derivatives (which need to be computed only once for the baseline), a complete linearized description of the danger volumes in "velocity" space is created from which all miss distances can be efficiently computed for any candidate maneuver.

Searching the "Velocity" Space

The simplest logic for exploring the "velocity" space in search of a solution is that of establishing a grid of evenly spaced values of ΔV, β, and γ, and searching for the minimum ΔV that produces no conflicts. The Cartesian components of the maneuver velocity in UVW space are simple trigonometric functions of β and γ. The use of constant increments in the angles, however, permits the use of recursive relationships for the sine and cosine instead of repeated calls to the time-consuming trigonometric functions. Also, since most of the danger volumes are much narrower in the in-track direction, it is prudent to explore that direction first, in both positive and negative senses. One can then explore other values of the angles with the minimum velocity produced in-track to see if an improvement is possible. One might consider more elaborate search schemes, but care must be exercised not to use too large a step size, because small "clear" spaces may exist between adjacent danger volumes. Such schemes would appear to be fraught with logic problems related to special cases that would require special attention.

Fig. 4 contains the "Program Design Language" (PDL) representation of a relatively simple, but computationally efficient, scheme for exploring the "velocity" space for the minimum required maneuver.

TEST RESULTS

Many test cases were run on an experimental version programmed on the IBM 3083JX computer. The case described here is the one previously mentioned wherein it was desired to avoid all conflicts within 12 hours after a candidate maneuver time. The COMBO process produced four encounters that were considered dangerous in the absence of a maneuver (ie., the point of closest approach was less than the stand-off radius of 12 km). In addition, 101 other encounters were produced which needed to be avoided after the proposed maneuver. Thus, the "velocity" space for this case contained 105 volumes of various sizes and shapes, four of which enclosed the origin.

The exploration algorithm used was that described in the PDL of Fig. 4 with a grid size of 0.2 m/s for ΔV, and 15° for both β and γ. The minimum maneuver required to avoid all potential conflicts was determined to be 2.2 m/s in the orbital plane ($\beta = 0$) and at a pitch angle, γ, of 75°. As a check, the COMBO process was rerun with the maneuver assumed to have been applied, and no passages defined as "dangerous" were found. The "velocity" space exploration portion of the process required approximately one minute on the computer. This is considerably less than the time required to run one COMBO process against the entire catalog.

```
Run COMBO against catalog with 6X stand-off radius.
     (Get times, position, and velocity for potential conflicts.)
Build array of partial derivatives for each potential conflict.
β = 0, γ = 0 (Get best positive in-track ΔV)
Loop(1) for ΔV from zero to maximum for this satellite:
   Loop(2) for all potential conflicts:
      Evaluate new separation distance.
           -- (Equations (5), (6), and (7).)
      If Δr too small, exit into loop 1.
           -- (No point in looking at the other potential conflicts.)
           -- (Try next ΔV.)
   End loop(2).
   Save ΔV as smallest (and best) maneuver so far. Exit from loop 1.
End loop(1)
β = π, γ = 0 (Get best negative in-track ΔV)
Loop(3) for ΔV from zero to best for this satellite so far:
   Loop(4) for all potential conflicts:
      Evaluate new separation distance.
           -- (Equations (5), (6), and (7).)
      If Δr too small, exit into loop 3.
           -- (No point in looking at the other potential conflicts.)
           -- (Try next ΔV.)
   End loop(4).
   Save ΔV as smallest (and best) maneuver so far. Exit from loop 3.
End loop(3).
Loop(5) for ΔV from zero to best for this satellite so far:
   Loop(6) for β from 0 to 2π:
      Loop(7) for γ from -π/2 to +π/2:
         Loop(8) for all potential conflicts:
            Evaluate new separation distance.
                 -- (Equations (5), (6), and (7).)
            If Δr too small, exit into loop 7.
                 -- (No point in looking at the other potential conflicts.)
                 -- (Try next γ.)
         End loop(8).
         Save ΔV, β, and γ as smallest (and best) maneuver.
         Exit from loop 5.
              -- (This is the smallest ΔV)
      End loop(7).
   End loop(6).
End loop(5).
Return "best" maneuver for maneuverable satellite.
```

Fig. 4. PDL for Exploring "Velocity" Space

Occasionally, for some cases, the post-maneuver COMBO process will produce some new conflicts after the proposed maneuver that violate the stand-off radius. These may be caused by one of two things: either the linearity assumptions were found to be invalid, or the new conflict was not included in the original COMBO produced baseline of encounters.

The condition where the linearity assumptions are violated has become known as "close encounters of the second kind." These occur only rarely because of a 20% "fudge factor" that is added to the stand-off radius in the "velocity" space search algorithm. This 20% accommodates the non-linearities and doesn't appreciably degrade the process.

The condition where the initial baseline didn't include all the necessary possible encounters has become known as "close encounters of the third kind." This can be a more difficult problem to solve. The problem is reduced if the stand-off multiplier is increased for the initial COMBO process. The penalties for doing this, however, are increases in both storage space for the baseline, and execution time. Experiments were conducted with stand-off radius multipliers of 4, 6, 8, and 12. The value of 4 produced an excessive number of "close encounters of the third kind", while the higher values produced none. The values of 8 and 12, however, produced an excessive number of candidate encounters in the initial baseline with corresponding penalties in storage and throughput requirements. The current value of the multiplier was, therefore, chosen to be 6. That value was found to be effective in that "close encounters of the third kind" occur only occasionally. When they do, they are generally a long time after the maneuver, and an "expiration time" is output, prior to which time, the process should be run again.

Incidentally, "close encounters of the first kind" are those that are both properly included in the initial baseline and properly resolved with the proposed maneuver.

CONCLUSIONS

While not especially elegant, the algorthim described in the paper makes efficient use of mathematical capabilities that are typically available in general surveillance systems to solve the problem of avoiding collisions by maneuvering. The tests of the implemented algorithm indicate that the validity of the results and the use of computer resources are reasonable.

ACKNOWLEDGEMENTS

The development described in this paper was performed by the Ford Aerospace Corporation and was supported by a contract with the United States Air Force.

REFERENCES

1. F.R. Hoots, L.L. Crawford, and R.L. Roehrich, "An Analytical Method to Determine Future Close Approaches Between Satellites," AAS Paper 83–333, presented at the Astrodynamics Specialist Conference at Lake Placid, New York in August, 1983

2. C.G. Hilton, et al, "Mathematical Foundation for SCC Astrodynamic Theory," NORAD Technical Publication TP SCC 008, 4 April 1983, pp. 161–194

3. P.E. Koskela, et al, "Astrodynamic Analysis for the Advanced Orbit/Ephemeris Subsystem," Aeronutronic Publication U–4180, 1 September 1967, pp. 144–178

4. F.R. Hoots and R.L. Roehrich, "Models for Propagation of NORAD Element Sets," SPACETRACK Report No. 3, December 1980

5. F.R. Hoots and R.G. France, "Performance of an Analytic Satellite Theory in a Real World Environment," AAS Paper 83–395, presented at the Astrodynamics Specialist Conference at Lake Placid, New York in August, 1983

AAS 89-400

QUANTIFYING THE INEVITABLE: THE GROWING PERIL OF ORBITAL COLLISIONS

John D. Vedder* and Jill L. Tabor†

An unconventional but general method is described for estimating the probability of collision between an earth orbiting spacecraft and orbital debris. This method uses a Monte Carlo simulation of the orbital motion of the target spacecraft and each discrete debris object to generate an empirical set of distances, each distance representing the separation between the spacecraft and the nearest debris object at random times. Using concepts from the asymptotic theory of extreme order statistics, an analytical density function is fitted to this set of minimum distances. From this function it is possible to generate realistic collision estimates for the spacecraft.

INTRODUCTION

During the past decade increasing attention has been focused on the problem of quantifying the hazard posed by the growing amount of orbital debris. As spacecraft become larger and missions longer in duration, the hazard of potential collisions becomes an ever more serious concern.[1] Plans for permanently orbiting space stations of large dimension in low altitude orbits particularly exacerbate this concern, as does the increasing crowding in such regions as the geosynchronous ring. It is therefore important to be able to realistically estimate the probability of collision for an orbiting object during an arbitrary interval, and such estimates should be based as much as possible on the actual environment which characterizes the object's specific orbit.

Considerable effort has already been expended on solving this problem. Several different methods have been developed which can be used to quantify the hazard from space debris. Although these methods differ in their details, they are generally based on a common underlying approach, i.e., representing the debris population by a spatial debris density characterized by the Poisson distribution and then computing a collision probability based on this debris model. In this paper we discuss a method which implements a substantially different approach.

---

* Staff Engineer, McDonnell Douglas Space Systems Company, Engineering Services, 16055 Space Center Blvd., Houston, Texas 77062-6208.

† Engineer, McDonnell Douglas Space Systems Company, Engineering Services, 16055 Space Center Blvd., Houston, Texas 77062-6208.

APPROACHES TO PROBLEM

An excellent survey and review of the principal methods which have been used by previous investigators to quantify the debris hazard is given in a recent book by Johnson and McKnight.[2] These methods generally convert the aggregate of discrete data describing the orbital debris population into some sort of continuous debris density to enable the calculation of collision probabilities. Unfortunately, in such a transformation of the discrete and dynamic debris environment into a static continuum, a large amount of kinematic information about the debris population is inevitably lost at the outset. This results in an attendant loss of both realism and detail when the probability of collision is estimated for an object in a specific orbit. Furthermore, it is difficult to accurately model the debris population, whose spatial density is both inhomogeneous and time-varying, by a static spatial Poisson distribution, even when the model is multi-zoned. Although multi-zoned models may, at some cost in complexity, model the inhomogeneous nature of the debris population, its time-varying aspect remains unmodeled.

In view of the shortcomings associated with static spatial debris distributions, it is worth considering possible alternatives which would model the debris population differently. It would be advantageous to have a method of determining the collision probability directly from an arbitrary discrete debris population having specified orbits, while also being able to specify the target's orbit arbitrarily. Here the object designated as the target would represent the Space Shuttle, Space Station, or other object of concern, while the debris population would represent a real or hypothetical set of objects, each in its own orbit.

In this paper we develop such a method by taking an unconventional approach to the problem of determining orbital collision probability. Our approach bears little resemblance to previous methods and appears to offer some significant advantages, especially for the lower altitudes of particular concern to the Space Shuttle and Space Station programs. Our method entails applying the asymptotic theory of extreme order statistics[3] to the distribution of distances from the target object to the nearest debris object at random points in time. All of the debris objects are assumed to be moving in specified Keplerian orbits as initially defined either by the U. S. Space Command (USSPACECOM) database of tracked orbital objects or by some hypothetical database. Thus our model characterizes the debris population by a large set of Keplerian orbits, thereby retaining the kinematic information associated with each orbit; in addition, it retains the inhomogeneity of the spatial distribution of the debris. Since our target also has an arbitrarily specifiable Keplerian orbit, we believe that our overall model embodies a substantial degree of realism.

DISCUSSION

The underlying idea of our approach is quite simple, and can be easily understood by considering the actual situation for a target orbiting the earth along with a large population of orbiting debris objects. In such a situation it is clear that at a randomly chosen time there is always some debris object which is the closest one to the target at that particular moment. It follows that by recording the distance from the target to the closest debris at a large number of different times we could estimate the distribution of these distances. Such a distribution would therefore represent the relative fraction of time that the debris object nearest to the target is a given distance away. This distribution would taper off at small values of distance, and would vanish at a distance of zero. Sufficiently small values of distance would represent situations in which the nearest debris object is so close to the target as to collide with it. Hence the problem of estimating the probability of a collision between the target and the debris population is largely equivalent to estimating the initial characteristics of the distribution function of the distance between the target and the nearest debris. Although this estimation may appear to require extrapolation (with its well known dangers) of known data, in this application the extension of the distribution beyond the data region is strongly constrained by both physical and statistical considerations, as will be explained below. Thus the hazards ordinarily associated with extrapolation are avoided.

The basic concept of our approach is therefore to observe the frequency of debris approach distances to a target on a larger, observable scale, and then use statistical methods to extrapolate to frequencies of the virtually unobservable collision proximity distances in which we are interested. To obtain our observable data, we created a Monte Carlo simulation of the orbital dynamics of all USSPACECOM catalogued debris objects and an arbitrarily chosen target object. Using this simulation, we computed the distance between the target and the closest debris object at a random epoch. By repeating this process a large number of times, we obtained a set of minimum distances, each of which represents the closest approach to the target object at some time.

Theoretically, this set of minimum distances should be characterized by a particular distribution function (df), at least asymptotically.[3] Guided by this theory, we have fitted the data to the generalized extreme value distribution with very good results. Once the parameters of the distribution have been determined from the fit to the minimum distance data, the distribution is completely defined. Quantitative estimates of the fraction of an extended time interval during which the target is within a specified distance of a debris object can be directly determined from this distribution. To estimate the number of separate encounters during this interval, it is necessary to define the maximum proximity distance which entails a collision and to derive an expected time per collision, $E(t_c)$; here t_c is a function of several known parameters, and it enables estimating the number of collisions in a given time period for a specified target. Each of these steps is elaborated in greater detail in the sections which follow.

USSPACECOM Tracked Debris Ephemeris Data

To define the orbital debris environment, we obtained ephemeris data generated by the USSPACECOM which represent the orbiting space objects tracked by this agency. We obtained these data for over a dozen different epochs from 1986-1988, thus representing the more recent populations of orbiting space objects. Obviously, these data do not include the countless smaller orbiting objects which cannot currently be tracked but which also pose a collision hazard. Nevertheless, the current population of over 7000 objects constitutes an excellent test environment and includes virtually all larger objects. Based on this known population, investigators have extrapolated to define the total environment, and have also assumed growth rates in the population to represent future debris environments. For our initial purposes we have used only the known population of currently tracked objects. However, we have determined how to scale our results for an n-fold increase or decrease in the debris population, assuming that the additional objects are in similar orbits to the known population.

Monte Carlo Simulation

Implementing our method on a computer may be more time intensive than some of the other methods which have been employed, but it offers a very flexible and realistic representation of the debris environment since each debris orbit is defined by the orbital parameters catalogued in the USSPACECOM database. The orbit of the target object is chosen by the user and can approximate that of any object of concern such as the Space Shuttle or Space Station. The Monte Carlo simulation then assigns a random mean anomaly to the target object and each debris object, in this way specifying the location of each object on its orbit. The resulting sets of Kepler elements are then transformed to Cartesian coordinates, and the distance between the target and each debris object is calculated. The minimum of these distances is then stored along with the corresponding relative velocity. This simulation process is repeated at least several hundred times, with the mean anomaly of each debris object and the target reselected after the generation of each minimum distance. These mean anomalies are always selected from a uniform distribution of random numbers, thereby simulating the relative motion in time of the objects along their orbits. The end result of this entire process is a set of distances, each of which represents the closest approach distance between the target and the debris population at a random point in time.

The transformation of the Kepler elements to Cartesian coordinates necessitates solving Kepler's equation very frequently, and this increases the computation time involved. But we believe that it is essential for realism in the generated data, since many debris orbits have appreciable eccentricities.

It may seem that the optimum method for generating a set of minimum distances would entail starting with a set of state vectors at a given epoch for both the debris objects and the target and actually propagating each of the existing 7000 or so orbits through time. Direct determination of the nearest object to the target at regular time intervals during the propagation would generate the desired minimum distance data set. In practice this concept is both impractical and unnecessary. Huge quantities of computer time would be needed to propagate the thousands of different orbits, and the resulting data would differ insignificantly from that generated by the much faster Monte Carlo process already described. For whenever deterministic orbit propagation is carried out for an extended interval, perturbing forces such as atmospheric drag (never known with high accuracy) inevitably cause a rapid buildup of mostly along-track error, especially at low altitudes. Thus the position along the orbit becomes highly uncertain, and soon is no better than a random guess.

As an indirect test of our assumption that orbit propagation can be replaced with the Monte Carlo simulation, we have applied our method to precisely those debris objects which are common to different epochs. Specifically, we took Kepler data from the USSPACECOM database for two different epochs and eliminated all objects which were not common to both epochs. We then generated minimum distance sets for both epochs, fit the generalized extreme value distribution to the empirical data sets, and calculated quantitative collision frequencies for both epochs. Only minor changes in the collision hazard (for a low altitude target) were observed for the different epochs. From this we conclude that the change in the orbital configuration of the debris population as the orbits evolve in time is relatively unimportant, at least for the epoch separations tested and for low altitude targets. This suggests that little, if anything, would be gained by propagating orbits to generate the minimum distance data, and that the Monte Carlo routine employed is a reasonable simulation of the relative motion of the debris objects and target.

Statistical Characterization of the Data

Once the minimum distance data has been generated by the Monte Carlo program, an analytical distribution is fitted to the empirical data. As already mentioned above, the asymptotic theory of extreme order statistics indicates that the generalized extreme value distribution (GEVD) would be an appropriate choice for a limiting distribution of our data. A detailed exposition of this asymptotic theory is available in a recent monograph.[3]

There are only three possible specific forms of the GEVD: Type I (Gumbel), Type II, and Type III (Weibull) distributions (this particular numbering scheme is not universally used). The fitting method of Hosking[4] et al., which discriminates among the three, has consistently indicated that the Weibull distribution yields the best fit to the data. This was expected on both statistical as well as physical grounds, largely because of the nature of the distribution of distance between any two orbiting objects.

Three parameters completely define the Weibull distribution: the location parameter, α; the scale parameter, β; and the shape parameter, τ. The Weibull df, F(x), for minimum distances takes the following form

$$(1) \quad F(x) = 1 - \exp\{-[(x-\alpha)/\beta]^{\tau}\}$$

where x is distance. For small values of the argument, i.e., whenever $[(x-\alpha)/\beta] << 1$, F(x) can be approximated as

$$(2) \quad F(x) \approx [(x-\alpha)/\beta]^{\tau}$$

Fig. 1 shows the Weibull probability density function (pdf) for three different values of its shape parameter; the range from two to four for this parameter spans the range empirically obtained for our data. It is worth noting that the skewness of the Weibull distribution vanishes for $\tau \approx 3.6$; hence symmetry is maximized for this value of the shape parameter and the distribution closely resembles the normal distribution.

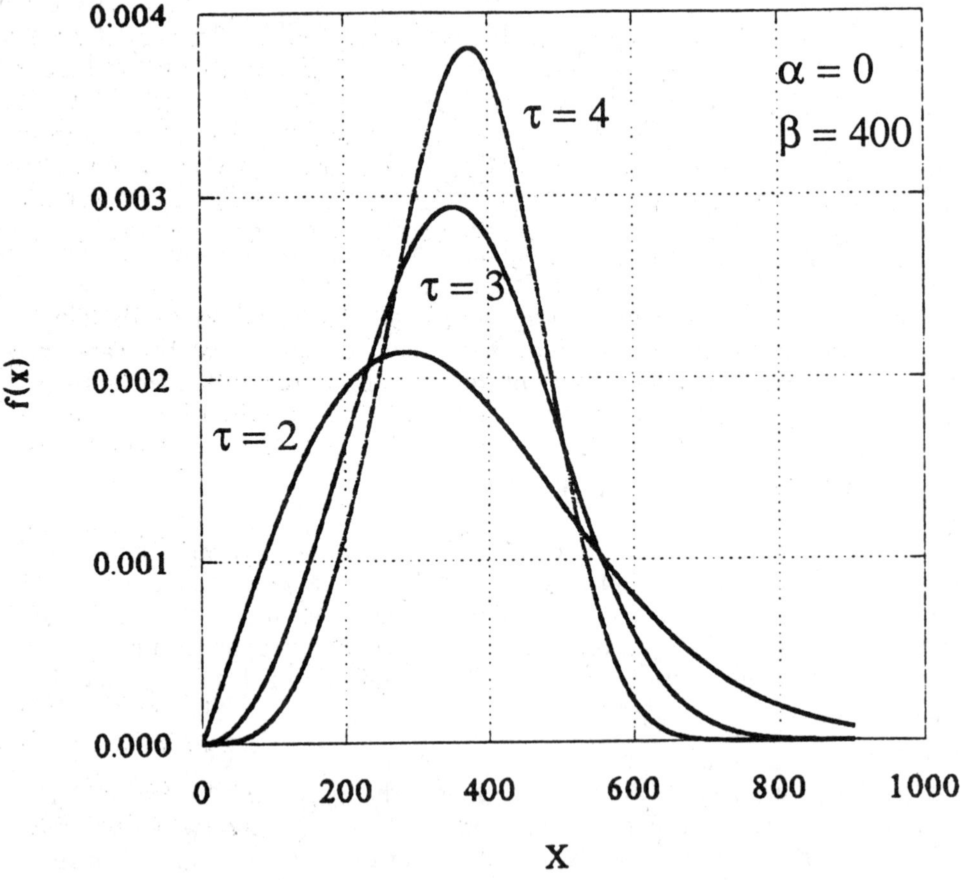

Figure 1 Weibull Probability Density Functions

For many applications the location parameter α vanishes, giving rise to the two-parameter Weibull distribution. A location parameter of zero forces the lower bound of the distribution to be zero, or in other words, it restricts the distribution of the random variable to non-negative values only. It would naturally seem as though, in our application of the Weibull, we could simply use the two-parameter form, since it makes little sense to have a negative minimum distance. As an added test of our method, however, we have chosen to fit not only the two-parameter but also the three-parameter form of the Weibull. When applied to our data, the location parameter should be nearly zero. If, in fact, the value of the location parameter is not close to zero, it would raise questions about our method. Thus far in our investigation, we have indeed observed nearly vanishing values of the location parameters, thereby adding support to the applicability of our method to this type of data.

To convert our data to an analytical form, we have used several fitting techniques: the method of probability weighted moments,[4] the maximum likelihood method, and linear regression. Linear regression fits a straight line to the data when the latter is plotted, in effect, on Weibull probability paper (on which a Weibull distribution appears as a straight line). The first method provides a three parameter fit, while the remaining methods provide two-parameter fits. Two linear regression fits were determined, one based on least squares (LS) and the other on least absolute deviations (LAD). The LAD criterion is much more robust with respect to outliers than is least squares. In essence, these fitting routines simply estimate the parameters of the theoretical distribution, in this case a Weibull, which yield the best fit to the empirical data distribution. The four fitting methods all worked well. Among the four, the two in closest agreement were the maximum likelihood method and linear regression using LAD.

We have also used several goodness of fit statistics (such as the Anderson-Darling, Watson, and Cramer-von-Mises statistics[5-7]) to get an analytical measure of how well the Weibull distribution fits our empirical data. These goodness of fit statistics have generally indicated that the Weibull distribution does in fact yield a good fit to the data in almost all cases, especially when the number of data points is sufficiently large.

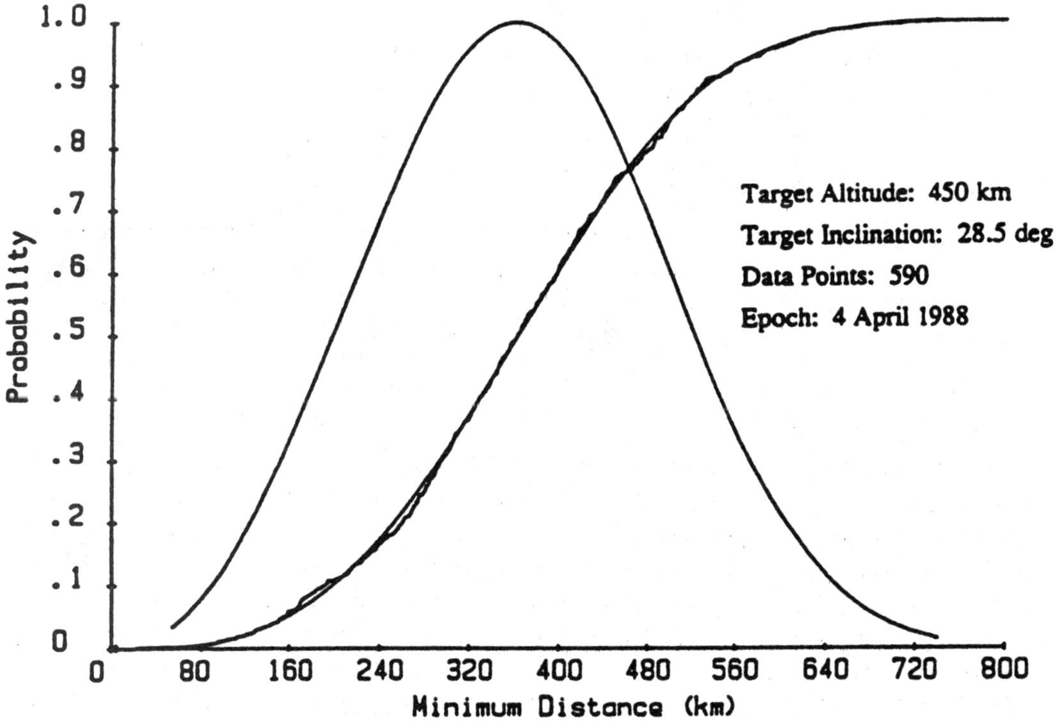

Figure 2 Representative Empirical and Analytical Weibull Distribution Functions

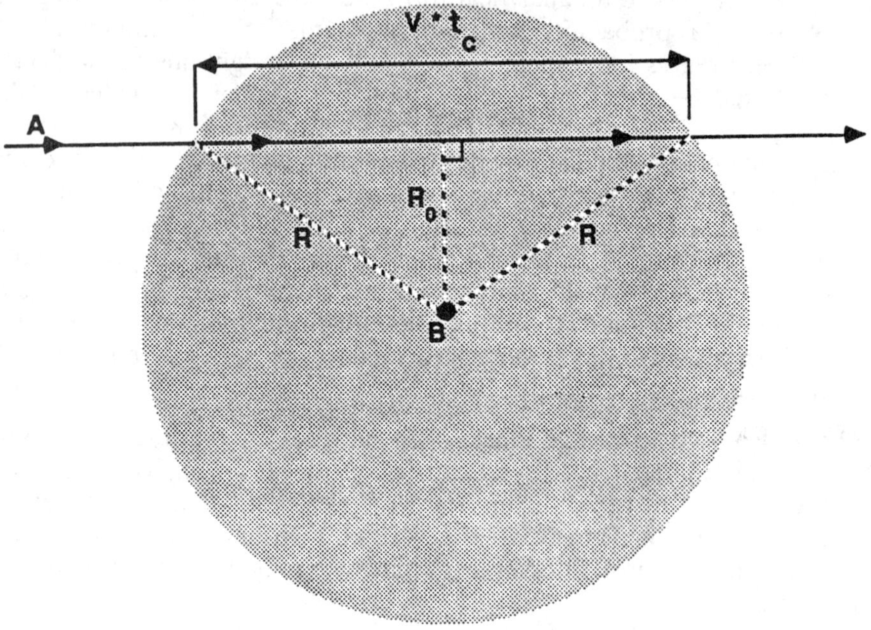

Figure 3   Time Per Target-Debris Encounter

Characteristic Encounter Time

Once the analytical distribution of minimum distances is defined, it can be used to directly estimate the fraction of a given time interval during which the target is within a specified distance of the nearest debris object. This may be more clearly illustrated by utilizing a plot of the distributions: Fig. 2 is a plot of the analytical df and pdf, and also the empirical df, all for a target at an altitude of 450 km and an inclination of 28.5^{o}. The analytical distributions are the smooth curves and the empirical distribution is the jagged curve. As is evident from Fig. 2, the fit of the analytical df to the empirical df is very good. The left hand scale corresponds only to the dfs; the pdf has been scaled to peak at unity only for illustrative clarity. For an estimate, therefore, one would choose a minimum distance on the horizontal axis, move up to the df, and read the corresponding value from the vertical scale. This number represents the fraction of any extended time interval during which the distance between the closest debris object and the target is less than or equal to the minimum distance initially chosen.

As described thus far, our method yields an estimate of the fraction of an extended time interval during which the target is within a specified distance of another orbiting object. To estimate the number of separate encounters during this interval, it is necessary to determine a time per encounter, or characteristic encounter time, t_c. As illustrated in Fig. 3, t_c is defined as the total time that a point

A, moving with a constant relative velocity, V, past another point B, is within a distance R of B. Thus t_c is the total time that point A is within a sphere of radius R centered at B. Fig. 3 illustrates that during the time t_c the shortest distance between points A and B is R_o. Given these assumptions and the previously derived result that R has a Weibull distribution with parameters β and τ, it can then be shown that the pdf, $g(t_c)$, of t_c is given by

$$(3) \quad g(t_c) = (\tau VW/2R)\,[1 - W^2]^{(\tau/2-1)}$$

where $W = (Vt_c/2R)$. It follows that the expected value of t_c, namely $E(t_c)$, is given by

$$(4) \quad E(t_c) = (\tau R/V)\,B(\tau/2,3/2) = (\tau R\pi^{1/2}/2V)\,\Gamma(\tau/2)/\Gamma([\tau+3]/2)$$

where B() is the beta function and Γ() is the gamma function. Furthermore, $G(t_c)$, the df corresponding to (3) is given by

$$(5) \quad G(t_c) = 1 - (1 - W^2)^{\tau/2}$$

The derivations of both (3) and (4) are based on the assumption that the distance R is small compared β; this assumption is realistic since encounter distances of interest are always much less than β. For extremely large encounter distances, a more general (and more complex) form of (3) and (4) could be derived. Thus the characteristic encounter time and its expected value are functions of: the relative velocity during the encounter; the parameters of the extreme value distribution describing the minimum distance data; and the maximum distance characterizing an encounter which entails a collision (or characteristic collision distance, $R = r_c$). Since all of these parameters are known, it is possible to quantify the time per encounter and thus the number of encounters as well. Although the derivation of (3) and (4) is based on point objects, the spherical radius R may of course be assigned a value appropriate to the sizes of the actual target and debris objects.

Collision Probability and Time Between Collisions

Given a characteristic collision distance, r_c and a characteristic collision time, t_c, it now follows that N, the expected number of collisions per unit time is given simply by

$$(6) \quad N = F(r_c)/E(t_c)$$

where F() and E() were given earlier in equations (1), (2) and (4). Substituting (2) and (4) into (6) yields

$$(7) \quad N = (1+1/\tau)(V/\beta\pi^{1/2})\{\Gamma[(\tau+1)/2]/\Gamma(\tau/2)\}(r_c/\beta)^{\tau-1}$$

The expected time between collisions, T_c, is simply the reciprocal of N, and is therefore given by

$$(8) \quad T_c = (1+1/\tau)^{-1}(\beta\pi^{1/2}/V)\{\Gamma(\tau/2)/\Gamma[(\tau+1)/2]\}(r_o/\beta)^{1-\tau}$$

With equations (7) and (8) we have therefore derived explicit expressions for quantifying the collision hazard posed by orbiting debris. These equations also manifest the functional dependence of N and T_c on the relevant parameters β, τ, V, and r_c; obviously the most complex dependency is on τ, the shape parameter of the Weibull distribution. It is worth noting that if the debris were homogeneously distributed in space, then τ would have a value of three. For $\tau=3$, equation (8) becomes simply

$$T_c = (3\pi\beta/8V)(\beta/r_c)^2 \quad (9)$$

As discussed in the ensuing section, the actual values determined for τ tend to cluster about three: the interval from two to four includes all of the values thus far determined.

RESULTS

Although our emphasis in this paper is to describe our method for quantifying the orbital debris collision hazard, we have also used our method to generate some representative numerical results. We have mainly concentrated on generating results of interest to the Space Station and Shuttle programs. In addition to these results, we have generated some results which are useful for comparison with those previously reported in the literature; the latter, however, are not easy to find in any substantial quantity.

The results summarized in Table 1 have been generated based on the USSPACECOM database of tracked orbiting objects at the epoch of 88095 (April 4, 1988). The first two columns of Table 1 specify the altitude and inclination of the target's circular, geocentric orbit. The third column gives the average relative velocity between the target and the nearest debris object; as mentioned earlier, for every minimum distance determined, the corresponding relative velocity is also computed. The fourth column specifies the number of minimum distance data points generated by the Monte Carlo simulation of the relative motion. The next two columns indicate the values of the Weibull distribution parameters determined for the analytical fits of the minimum distance data sets. Finally, the last column gives the estimated time between collisions that the target would experience from the particular debris population assumed. The value of the characteristic collision distance used was 50 meters, corresponding roughly to Space Station dimensions. The target orbit altitudes of 400-500 km and inclination of 28.5° also correspond with those intended for Space Station.

The estimated times between collisions given in Table 1 do not appear to differ radically from those given by other methods, although strictly comparable results are scarce. In agreement with others, our results also show a sharply increasing hazard as a function of increasing altitude, at least up to the well-known peak concentration of larger debris at 800-1000 km altitude. It is interesting to observe that the values of the Weibull shape parameter, τ, cluster on both sides of $\tau=3$. Since, as noted earlier, a homogeneous distribution of debris would imply that $\tau=3$, the actual deviations from this value can be interpreted as a measure of the nonhomogeneity of the debris environment for a particular target orbit. Values above three generally occur at very low altitude, while values below three occur at higher altitudes. At an altitude of 500 km this parameter appears to nearly equal three.

Table 1
REPRESENTATIVE TIMES BETWEEN COLLISIONS

Target Altitude (km)	Target Inclination (deg)	Relative Velocity (km/s)	Number of Points	Shape Parameter tau	Scale Parameter beta	Time Between Collisions (yrs)
300	28.5	8.13	1300	3.658	505.6	95800
400	0	8.52	1200	3.201	452.9	994
400	28.5	8.14	900	3.277	447.4	1980
400	60	8.16	812	3.367	409.2	3290
400	90	10.7	935	3.442	422.7	5490
400	120	11.9	1200	3.372	414.6	2477
400	150	12.3	1215	3.478	437.5	7390
400	180	12.2	895	3.330	451.9	2210
450	28.5	8.06	590	3.102	408.3	313
500	0	8.63	1145	2.986	407.0	103
500	28.5	8.23	1578	3.008	395.5	121
500	60	8.50	1205	3.061	359.5	139
500	90	10.8	720	2.980	367.2	57.3
500	120	11.9	736	3.116	374.0	183
500	150	12.3	1080	3.032	387.0	93.9
500	180	12.1	812	3.046	400.8	120
600	28.5	8.61	973	3.071	356.6	147
700	28.5	8.95	800	2.878	334.6	21.6
800	28.5	8.94	2013	2.795	321.1	9.38
975	28.5	8.93	1000	2.731	331.5	5.89

The values of the Weibull scale parameter, β, tabulated in Table 1 generally range between 300-500 km. Since the mode, m, of the Weibull distribution is given by

$$m = \beta(1-1/\tau)^{1/\tau} \quad (10)$$

it follows that the modes corresponding to the parameter values in Table 1 would tend to be slightly less than β. Since the mode is the most probable value of the distribution, we can conclude that at any given time the nearest tracked debris object is probably less than 300-500 km from the target.

The β parameter is also important for scaling results for a larger or smaller debris population. If a given debris population is assumed to change in size by a factor of n, while retaining essentially the same distribution of orbits, then it can be shown that the only change to equations (7)-(9) is to divide β (wherever it occurs) by a factor of $n^{1/\tau}$. This result enables the estimation of the time between collisions for other than USSPACECOM tracked debris populations, provided the assumption of similarly distributed orbits is considered realistic. Thus the effects of future growth in the debris population can be estimated, as can the effect of the much larger population of smaller, untracked debris.

CONCLUSION AND FUTURE DIRECTIONS

The method we have described for quantifying the orbiting debris collision hazard has a number of distinct advantages, most of which we have already pointed out in the foregoing sections. For a known or hypothetical debris population and a known target orbit, it enables a rather direct, intuitive, and flexible approach to estimating the probability of collisions. This approach can cope easily with an arbitrary target orbit and an inhomogeneous debris distribution. Our method lends itself well to scaling for a debris population which may either change in size in the future, or which may be underestimated because of current observability limitations. It can be implemented readily on a small computer, just as all of our computational work was done on personal computers of modest capability.

As mentioned previously, we have chosen to emphasize our methodology in this paper rather than specific numerical results. If specific results are needed for any particular application, it might be well to implement our method on a faster computer which would allow the generation of larger, and hence more statistically definitive, data sets.

In a possible future effort it would be useful to develop the means of deriving the distribution of T_c, the time between collisions, rather than just a point estimate as given in equation (8). This would necessitate first determining the distribution of V, the relative velocity, since T_c is functionally dependent on V. It would also involve using the distribution of the characteristic encounter time given by equation (3), rather than only its expected value.

Finally, we would note that although our approach to the problem of quantifying the orbital collision hazard differs significantly from previous methods, it is related to certain earlier investigations. The initial application of the asymptotic theory of extreme order statistics to the orbital collision problem appears to be that of McCormick[8] and Vedder, who applied the theory to the problem of predicting collision probability in the geosynchronous ring. Subsequently the theory was applied to assess the collision hazard of colocated geostationary satellites by Liebold[9] and by Härting[10] et al. These earlier investigators all used orbit propagators, rather than Monte Carlo simulations, to generate their minimum distance data. Furthermore, their results were limited to the geosynchronous region, and did not extend to the lower orbital altitudes of particular concern in our present investigation.

ACKNOWLEDGMENTS

We would like to thank Robert H. Brown, NASA Johnson Space Center, for his support of our investigation. We would also like to thank Bernell R. McCormick, David E. DeBusk, and P. L. "Pete" Fardelos, all of McDonnell Douglas Space Systems Company, for reviewing this paper and offering many valuable comments and suggestions. In addition, we would like to thank Donald J. Kessler, NASA Johnson Space Center, and Richard H. Rast, Lockheed Corporation, for supplying us with the USSPACECOM tracked orbiting debris ephemeris data.

REFERENCES

1. *Report on Orbital Debris*, Interagency Group (Space) for National Security Council, Washington, DC, February 1989.

2. N.L. Johnson and D.S. McKnight, *Artificial Space Debris*, Orbit Book Company, Malabar, Florida, 1987.

3. J. Galambos, *The Asymptotic Theory of Extreme Order Statistics*, second ed., Robert E. Krieger Publishing Company, Malabar, Florida, 1987.

4. J.R.M. Hosking, J.R. Wallis and E.F. Wood, "Estimation of the Generalized Extreme-Value Distribution by the Method of Probability Weighted Moments," *Technometrics*, Vol. 27, No. 3, Aug. 1985, pp. 251-261.

5. M.A. Stephens, *The Anderson Darling Statistic*, Technical Report No. 39, Dept. of Statistics, Stanford University, Oct. 31, 1979.

6. M.A. Stephens, *Goodness-of-Fit for the Extreme Value Distribution*, Technical Report No. 247, Dept. of Statistics, Stanford University, July 1977.

7. M. Chandra, N.D. Singpurwalla and M.A. Stephens, "Kolmogorov Statistics for Tests of Fit for the Extreme Value and Weibull Distributions," *JASA*, Vol. 76, No. 375, Sept. 1981, pp. 729-31.

8. B. McCormick, "A New Technique for Predicting Geosynchronous Satellite Collision Probability," *Astrodynamics 1985*; Vol. 58-II, *Advances in the Astronautical Sciences*, (Proceedings of the Conference, AAS 85-374, Vail, CO, Aug. 12-15, 1985), AAS Publications Office, San Diego, CA, 1986, pp. 821-830

9. Liebold, A.F., Type I - Extreme Value Distribution and Weibull Distribution for Modelling Shortest Distance Between Two Colocated Satellites, DFVLR-GSOC, TN 87-2, 1987.

10. Härting, A., Rajasingh, C.K., Eckstein, M.C., Leibold, A.F., and Srinivasa Murthy, K.N., "On the Collision Hazard of Colocated Geostationary Satellites," Astrodynamics 1988; Proceedings of the Conference, AIAA, 88-4239 CP, 1988, pp. 204-212.

AAS 89-375

MASS DISTRIBUTIONS FROM RECENT HYPERVELOCITY IMPACT TESTS*

Christoper B. Brechin† and Darren S. McKnight‡

Mass distributions from recent hypervelocity impact tests are presented. Possible sources of error in these tests and previous tests are discussed. These errors include the miscollection of debris pieces, the creation of secondary debris, and the collection of extraneous debris materials. These uncertainities affect the validity of our present power law model and other modeling efforts based on this type of data. Preliminary analysis has shown that a hyperbolic model fits this new data with more confidence than previous models.

* This paper appears in full in the microfiche supplement to this volume, **AAS Microfiche Series**, Volume 59.

† Captain, U.S. Air Force; Department of Physics, U.S. Air Force Academy, Colorado Springs, Colorado 80840-5151.

‡ Captain, U.S. Air Force; Department of Physics, U.S. Air Force Academy, Colorado Springs, Colorado 80840-5151.

AAS 89-377

BALLISTIC COEFFICIENT ESTIMATION FOR SATELLITE BREAKUP DEBRIS: DEBRIS CLOUD ANALYSIS*

R. A. Glover†, F. R. Hoots‡, and R. G. France**

A method of estimating the ballistic coefficient of orbiting objects using element set data has been developed, calibrated, and applied to the debris from five satellite breakup events. The distributions of ballistic coefficients differ depending on the collisional or explosive nature of the events. The orbital lifetimes of the debris clouds are assessed.

---

* Only an abstract for this paper was available for publication.

† Scientist, General Research Corporation, Colorado Springs, Colorado.

‡ Director, Colorado Springs Operations, General Research Corporation, Colorado Springs, Colorado.

** Senior Engineer, General Research Corporation, Colorado Springs, Colorado.

EARTH ORBITER MISSION ANALYSIS

SESSION 6

Chairman:

Richard Holdaway
Rutherford Appleton Laboratory
Oxfordshire, England

AAS 89-380

DESIGNING GOOD GEOSYNCHRONOUS CONSTELLATIONS*

John M. Hanson† and William Burley Higgins‡

This paper addresses the issue of optimal design of geosynchronous (24-hour) satellite constellations for desired coverage properties. Constellation sizes range from 2 to 10 satellites. Desired coverage properties are single or multiple coverage of six earth areas ranging from global coverage to Northern Hemisphere coverage to coverage of certain regions. Methods of constellation design studied are geostationary, Walker, elliptical, and composite (a combination of elliptical and geostationary). Results show that Walker and elliptical constellations do about equally well for global coverage and for coverage of all the globe except the extreme south. Elliptical constellations are easily the best for Northern Hemisphere coverage and for coverage of the U.S., North Atlantic, and Western Europe. Elliptical and composite constellations have the best potential coverage of Soviet ICBM and SLBM launches, and for coverage of the Northern Hemisphere with extra coverage of the continental United States and Western Europe.

INTRODUCTION

Geostationary satellite constellations are heavily used in the commercial, civil, and military space sectors. Examples include Intelsat, Westar, TDRSS, DSCS, and DSP. Interest in providing redundant stereo coverage of large areas and including coverage of polar regions leads one to consider not only geostationary orbits but the larger set of geosynchronous (24-hour) orbits. Geosynchronous orbits, while not necessarily in a stationary position over the earth, maintain the advantage of repeating ground tracks. They are particularly desirable when one wishes to maintain coverage over certain longitudes of the earth. Draim and Cooper[1] derived some geosynchronous satellite constellations with improved coverage properties for certain areas, relative to geostationary constellations.

The purpose of the present study is to derive geosynchronous constellations that have good coverage properties for a variety of needs, thereby illuminating the design methods that yield the best results. The study considers constellations that range from 2 to 10 satellites in size, providing single or multiple coverage. Each satellite is assumed to be able to see any ground point that is not beyond the earth's

---

* This document reports research sponsored by ANSER's Independent Research and Development Program. The views, opinions, and findings contained in this paper are those of the authors and should not be construed as those of ANSER.

† Senior Aerospace Engineer, Space Systems Division, ANSER, a not-for- profit company, Suite 800, 1215 Jefferson Davis Highway, Arlington, Virginia 22202.

‡ 2nd Lt., United States Air Force, formerly ANSER Research Staff.

limb. Other coverage criteria would lead to different results, but the same methods of analysis are appropriate. The analysis considers six regions of required coverage:

1. Global coverage.
2. Coverage of the globe above -60° latitude.
3. Coverage of the Northern Hemisphere.
4. Coverage of the Northern Hemisphere with extra coverage of the continental United States and Western Europe.
5. Coverage of all potential Soviet ICBM/SLBM launch areas that could threaten the continental United States.
6. Coverage of the United States (including Alaska and Hawaii), the North Atlantic, and Western Europe.

The measure of merit used to choose the most effective constellation is the percentage, averaged over time, of the total area in question that is covered at least once. If the entire area is covered 100 percent of the time, the measure of merit becomes the percentage, averaged over time, of the total area that is covered at least twice, and so on, for multiple levels of coverage.

This analysis contains four approaches to geosynchronous constellation design. First, for a point of comparison, are geostationary constellations. Second are Walker designs,[2] which are often used for designing multiple-satellite constellations. Walker constellations consider circular orbits only, and consist of a number of orbit planes spaced evenly around the equator, with the same number of satellites in each plane, spaced evenly in the plane. The orbital inclination is the same for all satellites, and the relative phasing between satellites in adjacent planes is the same for all planes.

The third approach is to consider elliptical orbits with certain Walker symmetries. All satellites are in 24-hour orbits of the same eccentricity and inclination. Planes are spaced evenly around the equator, and satellites are spaced evenly in mean anomaly within each plane. Relative phasing between satellites in adjacent planes is the same for all planes. This method has some immediate coverage advantages. Placing argument of perigee in the south will increase Northern Hemisphere coverage, for example. Ref. 1 shows a three-satellite, elliptical, geosynchronous constellation that gives continuous Northern Hemisphere coverage, something unobtainable with circular orbits. Draim shows that four satellites in suitable elliptical, 26.5-hour orbits, give complete global coverage, something also not obtainable with circular orbits.[3] He also develops a constellation that gives complete, global, double coverage with six satellites in high-altitude elliptical orbits.[4] Global double coverage with circular constellations requires a minimum of seven satellites.

The final approach is to choose constellations of composite design, which may include some elliptical and some circular orbits, with all orbit parameters besides semimajor axes allowed to vary. Ref. 1 developed one constellation of this type.

TESTING A CONSTELLATION'S COVERAGE

An earth coverage computer model was used to ascertain the level of coverage received from a given constellation. This model breaks the earth up into 5° blocks of longitude and latitude. It propagates the satellites in their orbits and, for each position in time, determines how many satellites can see each ground point in the grid. These data are saved, so that a measure of effectiveness can be determined at the end of the simulation. The grid points are weighted by the cosine of the latitude to account for differences in area. Time points were spaced 19 degrees in mean anomaly for finding the best constellation when iterations were required. Each final constellation was tested with time points spaced 4 degrees apart.

To calculate the measure of effectiveness, the program considers only those grid points that encompass the desired region of coverage. Hence, for coverage criterion 2, the measure of effectiveness includes all longitude points for all latitudes from -60 to $+90^o$, inclusive. Coverage criterion 3 considers latitudes from 0 to $+90^o$, inclusive. Fig. 1 shows the grid points considered for coverage criteria 4 through 6. In the case of coverage criterion 4, the darker shading highlights the areas requiring extra coverage. For coverage criterion 5, the arbitrary choice was made that the Soviet Union and any ocean area within about 5,000 nmi (about 9,000 km) of the continental United States should be included.

CHOOSING THE BEST CONSTELLATION

Geostationary Constellations

For geostationary constellations, satellites were placed at longitudes where they gave the best coverage. This is a relatively simple procedure since the covered region from each satellite has the same size and shape on a map and does not vary with time. The geostationary satellites were required to be 10° or more apart in longitude. This restriction, although somewhat arbitrary, was included since satellites very close to one another might as well be one satellite with extra sensors or redundancy. Satellites would also never provide stereo coverage if they were stationed too close.

Walker Constellations

In a Walker design with fixed altitude and number of satellites, one may vary three parameters in a search for the best constellation: inclination, number of orbit planes, and relative phasing between satellites in adjacent planes. This analysis used a computer program that iterates over these parameters to look for the best coverage, with the best inclination determined to within 5°. Inclinations from 0 to 95° were allowed. The program also rotates each candidate constellation's coverage plot about the polar axis to determine the optimal orientation of ascending nodes relative to the earth, for those coverage criteria that vary as a function of longitude (criteria 4, 5, and 6). The optimal orientation was determined to within 5°.

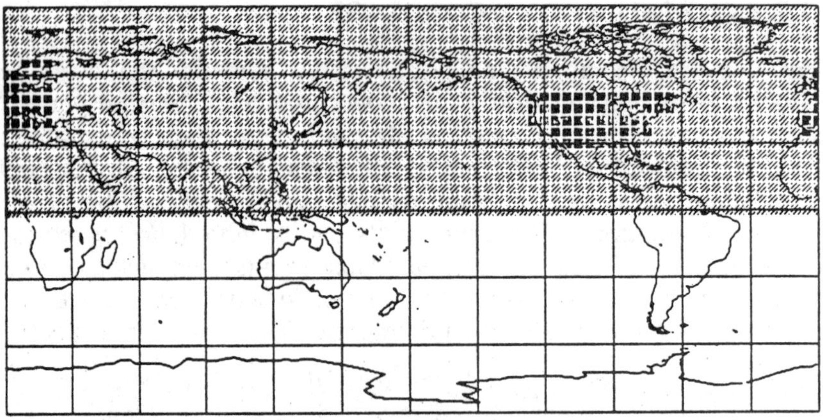

Northern Hemisphere with Extra Coverage of
Continental United States and Western Europe

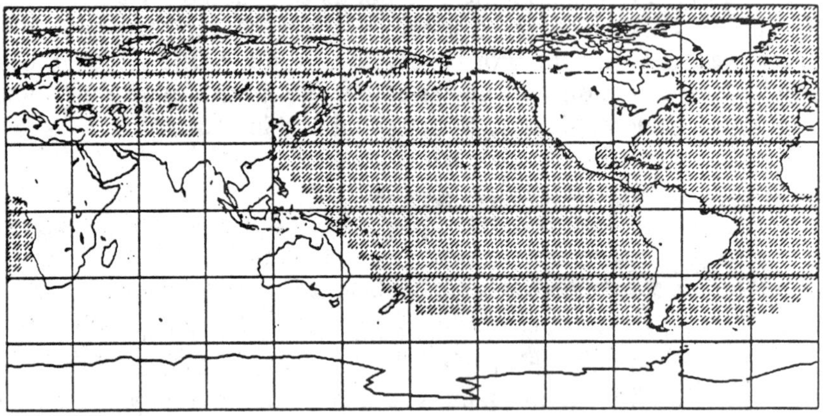

Coverage of Soviet ICBM/SLBM Launches

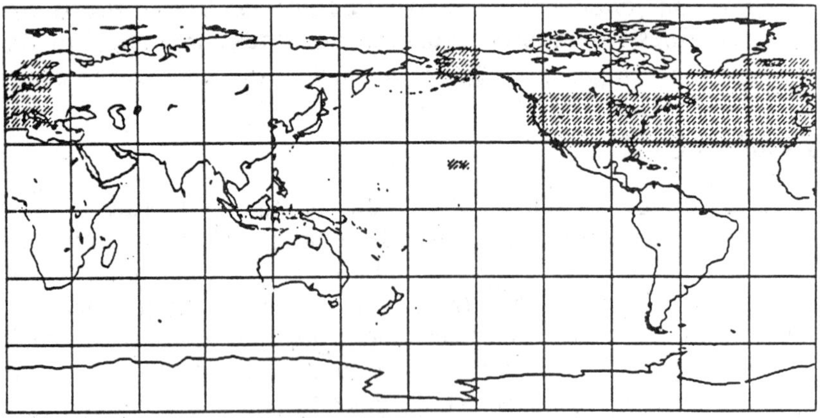

Coverage of the United States, North Atlantic, and Western Europe

Figure 1 Coverage Criteria

Elliptical Constellations

In the elliptical, modified Walker case, this analysis used a program that iterates over eccentricity and over combinations of arguments of perigee for the various satellites in addition to the parameters of iteration for the circular Walker case. Eccentricities were varied from 0.1 to 0.7 in increments of 0.1. Choice of arguments of perigee was more complicated. In cases where all desired coverage is in the Northern Hemisphere (coverage criteria 3, 4, and 6), this analysis assumed all arguments of perigee to be -90° in order to maximize northern coverage. For coverage criteria 1, 2, and 5, arguments of perigee for each satellite were allowed to be either -90° or $+90^o$, with the caveat that there were at least as many satellites with an argument of perigee of -90° as there were with one of $+90^o$. This makes sense for coverage criteria 2 and 5, since they emphasize the north more than the south. In the global coverage case, constellations with more perigees in the north will give the same measure of effectiveness as a mirror-image constellation with more perigees in the south. Arguments of perigee other than $\pm 90^o$ were not considered because of computer run-time limitations.

There are many argument of perigee combinations to be tried when testing for coverage criteria 1, 2, and 5. For each one, the program must iterate over inclination, eccentricity, relative phasing, and number of orbit planes. In addition, the program iterates over time and examines each data point in the 37x72 coverage matrix. The problem quickly becomes unmanageable from a computer run-time standpoint as larger constellations are contemplated. For 10-satellite constellations, there are 638 possible argument of perigee combinations. It turns out, however, that many of the combinations are duplicative. When duplications are taken into account, the case of 10 planes, 1 satellite per plane, has only 67 unique argument of perigee combinations. While this is still a very large number, it is only about one-tenth of the original value.

To illustrate the method of determining unique arguments of perigee, consider the case of six satellites. Fig. 2 shows the unique combinations for six orbit planes and one satellite per plane. As an example, there is only one unique case of one northern perigee and five southern perigees, since all the others are rotations of the first. Recall that the program rotates the coverage matrix about the polar axis to determine the best orientation.

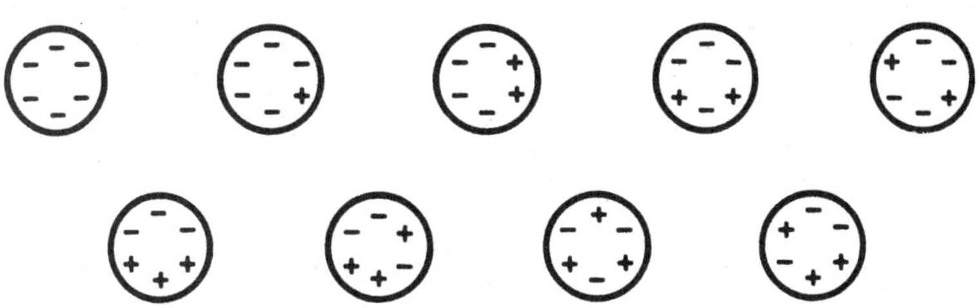

Figure 2 Unique Argument of Perigee Combinations for 6 Satellites, 6 Planes: View is from North Pole, + means $+90^o$, - means -90° (Restriction: No more $+90^o$ than -90°).

After analyzing the one-satellite-per-plane case, next comes two planes, three satellites per plane. The possibilities for orientations within one plane are shown in Table 1. Since there are two planes, these combinations may appear in all the ways illustrated in Fig. 3. Most of the possible combinations are duplications. For example, 5-6 is invalid because there are more northern than southern perigees. 2-1 duplicates 1-2 because of rotation about the polar axis. 1-3 duplicates 1-2 due to rotations within an orbit plane. That is, after satellites have moved ahead in their orbits by 120° in mean anomaly, 1-2 becomes 1-3, so that the satellite with the northern perigee has moved from a mean anomaly of 120° to a mean anomaly of 240°. For the same reason, 4-2 and 3-4 duplicate 2-3, while 3-6 and 4-7 duplicate 2-5. 2-4 duplicates 2-3, since 4-2 duplicates 2-3 and 2-4 is a rotation of 4-2.

Table 1
ARGUMENT OF PERIGEE CHOICES IN A PLANE
3 Satellites, + is +90°, - is -90°

Case	Argument of Perigee Combination
1	- - -
2	- - +
3	- + -
4	+ - -
5	+ + -
6	+ - +
7	- + +
8	+ + +

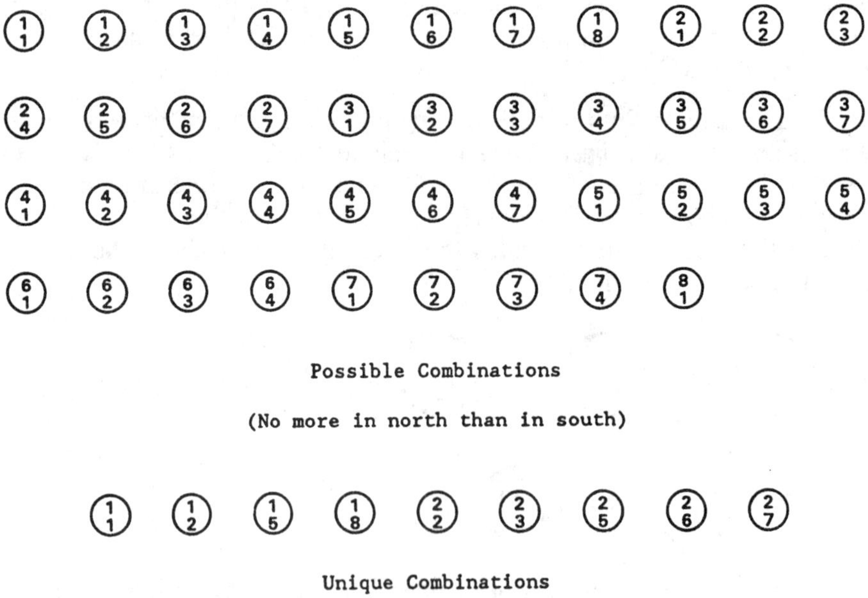

Figure 3 Argument of Perigee Combinations for 6 Satellites, 2 Planes: View from North Pole, Number is Case Number from Table 1.

Composite Constellations

Composite constellations were selected from a study of orbits and their coverage properties, a procedure that does not appear to lend itself to mathematical equations or to numerous iterations on a computer. These may not be the best possible constellations, but were the best found after considerable study of the coverage properties.

One procedure that yielded good coverage of ICBM/SLBM launch points was to choose elliptical satellites in the same ground trace, iterating on eccentricity and on inclination. After choosing one of these constellations to cover the Soviet Union and polar areas, geostationary satellites were carefully placed to cover the Western Hemisphere areas.

A useful procedure for coverage criterion 4 was to start with a constellation that gave a certain level of Northern Hemisphere coverage and to add satellites for extra coverage of Western Europe and the continental United States. In a few cases, satellites were added to the constellation resulting from the elliptical design. The study did not consider composite constellations for coverage criteria 1, 2, 3, and 6, since the authors felt that the elliptical constellations could not be easily improved upon.

RESULTS

Table 2 summarizes the results. Each entry gives the highest level of coverage that is 100 percent covered, plus the coverage percentage for the next higher level. For example, 2-45 means complete, continuous, double coverage plus 45 percent triple or greater coverage. The 45 percent value includes all areas covered by more than two satellites. For coverage criterion 4, the first entry is for the Northern Hemisphere and the second (separated from the first by a comma) is for the continental United States and Western Europe. When there are two rows of data, the second gives results from a different constellation. Results for composite designs are given for coverage criteria 4 and 5 only, since these were the only cases where composite constellations appeared to offer possible improvement over elliptical constellations.

When covering the entire globe or the globe above -60° latitude, circular Walker constellations and elliptical constellations yield approximately equal results. Small variations in results may occur as a result of the 5° grid size. It is possible that some of the cases with nearly a certain level of coverage would give that coverage if the constellations were modified slightly. Modifications of less than 5° in inclination or 0.1 in eccentricity were not modeled. The 10-satellite elliptical constellations that do not quite give triple coverage for the first two coverage criteria give about 80 percent quadruple coverage.

Elliptical constellations typically are the best for coverage of the Northern Hemisphere and for coverage of the United States, North Atlantic, and Western Europe. The differences are quite substantial. The needs of ICBM/SLBM coverage and coverage of the Northern Hemisphere with extra Table 2 coverage of the continental United States and Western Europe are met approximately equally well by the elliptical constellations and by the composite constellations. Each method may be better, depending on the case.

Table 2

COVERAGE RESULTS

COVERAGE CRITERIA	NUMBER OF SATELLITES	METHOD OF CONSTELLATION DESIGN*			
		GEOSTATIONARY	WALKER	ELLIPTICAL WALKER	COMPOSITE
ENTIRE GLOBE	2	0-85.4	0-85.3	0-84.6	N/A
	3	0-98.2	0-98.2	0-98.2	N/A
	4	0-98.7	0-99.9	0-99.995	N/A
	5	0-99.0	1-91.2	1-89.9	N/A
	6	0-99.2	1-98.8	1-98.3	N/A
	7	0-99.3	2-80.8	2-77.2	N/A
	8	0-99.3	2-94.6	2-93.6	N/A
	9	0-99.3	2-99.97	2-99.4	N/A
	10	0-99.3	3-88.1	2-99.998	N/A
GLOBE ABOVE -60° LATITUDE	2	0-86.7	0-86.8	0-86.5	N/A
	3	0-99.0	0-99.0	0-99.3	N/A
	4	0-99.3	0-99.9	0-99.998	N/A
	5	0-99.5	1-93.0	1-94.0	N/A
	6	0-99.5	1-98.7	1-99.7	N/A
	7	0-99.6	2-80.1	2-81.8	N/A
	8	0-99.6	2-94.4	2-97.2	N/A
	9	0-99.6	2-99.97	2-99.9	N/A
	10	0-99.6	3-87.6	2-99.998	N/A
NORTHERN HEMISPHERE	2	0-85.4	0-85.6	0-91.5	N/A
	3	0-98.2	0-98.2	1-46.1	N/A
	4	0-98.7	0-99.9	1-92.2	N/A
	5	0-99.0	1-91.6	1-99.7	N/A
	6	0-99.2	1-98.9	2-86.9	N/A
	7	0-99.3	2-80.1	2-98.4	N/A
	8	0-99.3	2-94.5	3-85.5	N/A
	9	0-99.3	2-99.98	3-96.3	N/A
	10	0-99.3	3-88.0	4-77.6	N/A
NORTHERN HEMISPHERE plus CONUS, WESTERN EUROPE EXTRA	2	0-85.6,1-0.0	0-85.6,1-0.0	0-87.2,1-72.6	N/A
		0-45.5,1-95.0			N/A
	3	0-92.1,2-0.0	0-98.2,1-51.9	1-46.1,1-48.4	1-40.9,1-41.1
		0-48.2,2-84.4		0-48.0,1-98.2	0-89.4,2-57.3
	4	0-51.0,2-95.0	0-99.9,1-71.1	1-92.2,2-85.8	1-87.5,2-98.4
		0-98.8,2-0.0			
	5	0-53.8,3-84.4	1-86.7,1-91.5	1-95.5,3-77.6	1-98.3,3-77.5
		0-92.2,3-0.0		1-99.4,2-68.2	
	6	0-56.5,3-95.0	1-97.4,2-73.0	2-88.0,3-99.95	1-99.5,4-71.8
		0-99.0,3-0.0			2-81.3,3-99.6
	7	0-59.3,4-84.4	1-99.7,2-81.1	2-95.0,4-91.6	2-93.5,4-99.95
			0-99.3,3-4.0	1-99.2,4-99.8	1-91.2,5-27.1
	8	0-62.0,4-95.0	2-82.6,3-59.7	2-98.6,5-99.9	2-97.3,5-91.6
		0-99.3,4-5.7		3-85.5,4-98.0	3-63.2,3-54.3
	9	0-64.8,5-84.4	2-97.1,3-66.8	3-94.9,6-94.7	3-89.2,5-89.7
					2-99.3,6-97.8
	10	0-67.6,5-95.0	2-99.8,4-44.4	3-96.7,7-95.9	3-97.4,7-94.7
		0-99.3,5-78.2		4-78.9,4-95.1	4-75.4,4-73.9
ICBM/ SLBM	2	0-93.7	0-91.2	0-92.2	N/A
	3	0-98.3	0-98.7	0-99.6	0-98.8
	4	0-98.9	0-99.9	1-75.4	1-69.0
	5	0-99.2	1-93.1	1-94.9	1-94.7
	6	0-99.3	1-98.9	2-64.0	2-72.8
	7	0-99.3	2-81.1	2-85.5	2-90.8
	8	0-99.3	2-94.3	2-99.4	2-99.97
	9	0-99.3	2-99.97	2-99.99	3-87.5
	10	0-99.3	3-87.8	3-97.03	3-97.0
UNITED STATES, NORTH ATLANTIC, WESTERN EUROPE	2	1-59.3	0-98.4	1-75.2	N/A
	3	1-93.1	1-34.5	2-63.4	N/A
	4	2-59.3	1-96.2	3-51.5	N/A
	5	2-93.1	2-15.6	4-39.1	N/A
	6	3-59.3	2-59.9	4-99.7	N/A
	7	3-93.1	2-98.0	5-99.0	N/A
	8	4-59.3	3-38.2	6-96.2	N/A
	9	4-93.1	3-83.0	7-90.0	N/A
	10	5-59.3	3-99.96	8-77.4	N/A

* See Tables 3-5 for constellation parameters

Table 3 shows parameters for the best Walker constellations in the cases of coverage of the entire globe and coverage of the globe above -60° latitude. Except for the seven-satellite case, the constellations are the same in both cases. These constellations are very similar, in most cases, to the Walker constellations described by Lang.[5] Differences are probably due to two factors. First, all the constellations in this analysis are restricted to be geosynchronous. Ref. 5, on the other hand, gives minimal constellations for global (single or multiple) coverage without restricting altitude. Second, the measure of merit used in this analysis is different. Rather than finding the smallest number of satellites for double coverage, say, this analysis asks: for a fixed number of satellites giving double coverage, what constellation maximizes the percentage of triple coverage?

Table 3

PARAMETERS FOR SELECTED WALKER CONSTELLATION

NUMBER OF SATELLITES	NUMBER OF PLANES	INCLINATION (DEG)	PHASING (DEG)
ENTIRE GLOBE			
2	2	5	0
3	3	5	120
4	4	75	180
5	5	20	216
6	6	55	240
7	7	60	257.143
8	8	70	270
9	9	55	120
10	10	60	288
GLOBE ABOVE -60° LATITUDE			
2	2	5	0
3	3	5	120
4	4	75	180
5	5	20	216
6	6	55	240
7	7	35	102.857
8	8	70	270
9	9	55	120
10	10	60	288

Table 4 lists parameters for the elliptical, modified Walker constellations. In the argument of perigee column, + means +90° and - means -90°. When there are multiple satellites in each plane, arguments of perigee for all satellites in one plane are listed before those for the next plane. All cases yielded either all negative arguments of perigee or arguments of perigee that alternate, without any other variations. In cases where the longitude of ascending node has an effect on the final results, the values for longitude of ascending node are listed in the table. The constellations for coverage of the Northern Hemisphere plus extra continental United States and Western Europe coverage appear in the same order as their results appear in Table 2.

Some of the constellations contain quite eccentric and inclined orbits. This may not be practical for some applications. However, the same analysis could easily be done with constraints placed on the eccentricity and inclination. Note also that even highly inclined orbits need not require huge launch vehicles. Use of a lunar gravity assist allows one to enter any of these orbits with less fuel expenditure than the usual two-burn transfer to geostationary orbit.

The constellations for coverage of the United States, North Atlantic, and Western Europe generally consist of elliptical paths all in one ground trace. This implies that good coverage will usually result when the desired coverage region is small enough to be covered by elliptical satellites in one ground trace.

One may compare three of these constellations with previously derived constellations. Ref. 1 contains a three-satellite constellation for Northern Hemisphere coverage very similar to the one derived here. The constellation in this study gives a measure of effectiveness result of 1-46.1, slightly better than the 1-40.9 obtained from the constellation in Ref. 1. Ref. 3 gives the Draim constellation, a four-satellite constellation providing complete global coverage with an orbit period of at least 26.5 hours. The constellation derived using the methods in this paper is nearly the same as that constellation (slightly differing eccentricity and inclination with other parameters the same) and gives nearly identical results with 24-hour orbits (0-99.995 versus 0-99.996). Finally, Draim's six-satellite, global, double constellation,[4] if brought down to a geosynchronous semimajor axis, yields 1-97.7, while the best constellation derived here, which is a circular Walker constellation, yields 1-98.8.

Table 4
ELLIPTICAL CONSTELLATION PARAMETERS

NUMBER OF SATELLITES	NUMBER OF PLANES	LONGITUDES OF ASCENDING NODE (DEG)	INCLINATION (DEG)	PHASING (DEG)	ECCENTRICITY	ARGUMENT OF PERIGEE COMBINATION
			ENTIRE GLOBE			
2	2	NO EFFECT	5	180	0.1	- +
3	3		0	0	0.1	- - -
4	4		35	270	0.3	- + - +
5	5		20	72	0.1	- - - - -
6	6		35	300	0.4	- + - + - +
7	7		40	102.857	0.1	- - - - - - -
8	8		70	90	0.1	- + - + - + - +
9	9		45	120	0.1	- - - - - - - - -
10	10		55	36	0.4	- + - + - + - + - +
			GLOBE ABOVE -60° LATITUDE			
2	2	NO EFFECT	5	180	0.1	- -
3	3		10	120	0.1	- - -
4	4		10	90	0.6	- - - -
5	5		15	72	0.1	- - - - -
6	6		15	60	0.2	- - - - - -
7	7		15	51.429	0.2	- - - - - - -
8	8		10	135	0.6	- - - - - - - -
9	9		10	40	0.4	- - - - - - - - -
10	10		55	36	0.4	+ - + - + - + - + -
			NORTHERN HEMISPHERE			
2	2	NO EFFECT	20	0	0.7	ALL NEGATIVE
3	3		45	120	0.3	
4	2		80	90	0.7	
5	5		45	216	0.3	
6	2		65	60	0.7	
7	7		40	257.143	0.4	
8	8		55	135	0.7	
9	9		40	80	0.4	
10	2		50	36	0.6	

Table 4
ELLIPTICAL CONSTELLATION PARAMETERS (Continued)

NORTHERN HEMISPHERE PLUS CONUS/EUROPE EXTRA
(Same order as in TABLE 2)

2	2	343,343	75	180	0.7	ALL NEGATIVE
3	3	154,34,274	45	120	0.3	
3	3	321,321,321	0	240	0.1	
4	2	343,163,253,73	80	90	0.7	
5	5	158,86,14,302, 230	90	216	0.7	
5	5	130,58,346,274, 202	50	216	0.4	
6	2	348,108,228,228, 348,108	70	60	0.7	
7	7	173,70,327,224, 122,19,276	80	205.714	0.7	
7	7	333,127,282,76, 230,25,179	95	102.857	0.7	
8	2	343,73,163,253, 208,298,28,118	80	45	0.7	
8	8	228,48,228,48, 228,48,228,48	55	135	0.7	
9	9	338,98,218,338, 98,218,338,98,218	75	80	0.7	
10	2	343,55,127,199, 271,199,271,343, 55,127	80	36	0.7	
10	2	269,341,53,125, 197,125,197,269, 341,53	55	36	0.6	

ICBM/SLBM

2	2	323,323	30	180	0.7	--
3	3	253,133,13	20	120	0.2	---
4	4	148,58,328,238	20	180	0.2	----
5	5	114,42,330,258, 186	10	216	0.3	-----
6	6	169,289,49,169, 289,49	20	60	0.3	------
7	7	204,307,50,152, 255,358,101	20	51.429	0.3	-------
8	8	149,201,149,201, 149,201,149,201	35	315	0.6	-+-+-+-+
9	3	328,88,208,128, 248,8,288,48,168	10	40	0.7	---------
10	5	329,149,149,329, 329,149,149,329, 329,149	10	108	0.6	----------

USA, NORTH ATLANTIC, EUROPE

2	2	13,13	95	180	0.7	ALL NEGATIVE
3	3	13,13,13	95	240	0.7	
4	4	13,13,13,13	95	270	0.7	
5	5	3,3,3,3,3	95	288	0.7	
6	6	348,348,348,348, 348,348	95	300	0.7	
7	7	8,8,8,8,8,8,8	95	308.571	0.5	
8	8	13,13,13,13,13, 13,13,13	95	315	0.7	
9	9	13,13,13,13,13, 13,13,13,13	95	320	0.7	
10	10	358,358,358,358, 358,358,358,358, 358,358	95	324	0.7	

Table 5
CONSTELLATION PARAMETERS - COMPOSITE CONSTELLATIONS

NUMBER OF SATELLITES	GEOSTATIONARY LONGITUDES (DEG)	ECCENTRICITY	INCLINATION (DEG)	LONGITUDE OF ASCENDING NODE (DEG)	INITIAL MEAN ANOMALIES (DEG)
		NORTHERN HEMISPHERE PLUS CONUS/EUROPE EXTRA			
		(Same order as in TABLE 2)			
3		0.3	30	24,144,264	0,240,120
3		0.7	70	335	0,120,240
4	311.1	0.7	70	203	0,120,240
5	270,311.1	0.7	70	203	0,120,240
6	270,311.1,321.1	0.7	70	203	0,120,240
6	270,311.1,350	0.7	70	203	0,120,240
7	311.1	0.7	70	348,108,228, 228,348,108	0,120,240 60,180,300
7	125,311.1	0.7	70	18	EVEN
8	311.1	0.7	80	173,70,327,224 122,19,276	0,205.714,51.429, 257.143,102.857, 308.571,154.286
8		0.29	33	EVENLY SPACED	0,90,180,270 0,90,180,270
9	80,280,305,340	0.7	70	208	EVEN
9	311.1	0.7	55	228,48,228,48, 228,48,228,48	0,135,270,45, 180,315,90,225
10	311.1	0.7	75	338,98,218,338, 98,218,338,98,218	0,80,160,240,320 40,120,200,280
10	215,255,290,325	0.7	30	98,218,163	EVEN
		ICBM/SLBM			
3	210	0.3	45	64	0,180
4	175	0.6	25	42	0,120,240
5	245,285	0.45	45	140	0,120,240
6	205,255,315	0.7	20	148	0,120,240
7	180,215,265,335	0.6	25	78	0,120,240
8	180,240,280,330	0.73	20	150	0,90,180,270
9	180,210,240,285,330	0.73	20	150	0,90,180,270
10	45,180,210,240,285, 330	0.73	20	150	0,90,180,270

Table 5 shows results for the composite constellations. All arguments of perigee for the elliptical orbits were -90°, although other perigee angles were tested in certain cases and discarded as not yielding better coverage. The constellations for coverage of the Northern Hemisphere plus extra continental United States and Western Europe coverage appear in the same order in which their results appear in Table 2.

CONCLUSIONS

This paper derives geosynchronous constellations that maximize coverage of six different geographic areas. The results illuminate techniques that yield superior constellations. Specifically, for global or near-global coverage, Walker constellations and elliptical designs do about equally well. For coverage of the Northern Hemisphere and for coverage of the United States, North Atlantic, and Western Europe, the elliptical constellations give the best coverage. The implication is that the elliptical constellations are best when one desires coverage either of a hemisphere with no longitude dependence or of a portion of one hemisphere small enough that satellites following one elliptical ground track can cover it. When covering regions that contain all longitudes but vary substantially as a function of longitude, the elliptical constellations and constellations that are composites of geostationary satellites and satellites in elliptical orbits vie to be the best, depending on the individual case.

The authors do not claim that the constellations derived here are optimal, but they were the best found in this study. For other desired coverage regions and for various constraints on the orbit parameters, other constellations will give the best results. A useful follow-on analysis would determine the best constellation when limits are placed on orbit eccentricity.

REFERENCES

1. J. E. Draim and J. K. Cooper, "Satellite Constellations for Optimal Northern Hemisphere Coverage," presented at the Ninth DARPA Strategic Space Symposium, Monterey, CA, 1983.
2. J. G. Walker, "Circular Orbit Patterns Providing Whole Earth Coverage," Royal Aircraft Establishment Technical Report No. 70211, November 1970.
3. J. E. Draim, "A Common Period Four-Satellite Continuous Global Coverage Constellation," *Journal of Guidance, Control, and Dynamics*, Vol 10, No. 5, Sep-Oct 1987.
4. J. E. Draim, "A Six-Satellite Continuous Global Double Coverage Constellation," AAS 87-497, presented at the 1987 AAS/AIAA Astrodynamics Conference, Kalispell, MT, Vol. 55, *AAS Microfiche Series*, American Astronautical Society Publications Office, San Diego, CA, 1988.
5. T. J. Lang, "Symmetric Circular Orbit Satellite Constellations for Continuous Global Coverage," AAS 87-499, presented at the 1987 AAS/AIAA Astrodynamics Conference, Kalispell, MT, Vol. 65-II, *Advances in the Astronautical Sciences*, American Astronautical Society Publications Office, San Diego, CA, 1988.

AAS 89-381

CONTINUOUS GLOBAL N-TUPLE COVERAGE WITH (2N + 2) SATELLITES

John E. Draim*

N-tuple continuous, global coverage of a spherical planet is shown to be possible, using $(2n + 2)$ inclined elliptic orbit satellites. A simple proof is presented, showing that n-tuple coverage is impossible, either instantaneously or continuously, with fewer than $(2n + 2)$ satellites. Examples of single and redundant continuous global coverage constellations developed by the author incorporate rotating, irregular (warping) tetrahedra or prismoids. The constellations are characterized by two groups of satellites, each having approximately circular, overlapping ground tracks on a pseudo-planet having the same period as the satellites. One group of satellites has perigees in the southern hemisphere and their ground tracks on the pseudo-planet appear to rotate counterclockwise. The other group has its perigees in the northern hemisphere and appears to rotate clockwise. A generalized table of orbital parameters for n-tuple coverage constellations is presented.

Nomenclature

A	=	semimajor axis of satellite orbit, in units of the Earth's equatorial radius, *r*
a	=	semimajor axis, nautical miles
e	=	eccentricity
h	=	satellite altitude, nautical miles
i	=	inclination, degrees
M	=	mean anomaly, degrees
P	=	semiparameter, $= a\ (1 - e^2)$
p	=	perpendicular distance from Earth center to satellite plane, nautical miles
r	=	satellite radius, nautical miles
R_e	=	mean radius of Earth = 3442 nautical miles
S_1, S_2, S_3, S_4	=	satellite designator
S'_1, S'_2, S'_3, S'_4	=	satellite surborbital points
T	=	constellation period, *hours*
σ	=	satellite visibility angle ("look angle") above horizon, degrees
ω	=	argument of perigee, degrees
$\dot{\omega}$	=	rotation of perigee, degree/year
Ω	=	right ascension of ascending node, degrees
$\dot{\Omega}$	=	regression of line of nodes, degree/year
μ	=	Earth's gravitational constant = 1.4077×10^{16} ft^3/sec^2

* Director for Advanced Concepts, Science and Technology Associates, Inc., 1700 N. Moore St., Suite 1920, Arlington, Virginia 22209.

1.0 Background

The minimum number of satellites required for continuous global single, double, triple, and quadruple coverage, respectively, using circular orbit Walker-Type constellations, are 5, 7, 9, and 11[1]. The first two four-satellite continuous global single-coverage arrays were developed by the author in 1984 and 1985. The first six-satellite continuous global double-coverage array was developed in 1986. Continuous triple- and quadruple-coverage arrays using eight and ten satellites, respectively, were developed in 1988.

2.0 Introduction

The level of dependence on multi-satellite ordered constellations, or arrays, for a variety of military and civil requirements is increasing every year. Some of these requirements, or missions, include surveillance, communications, data relay, navigation, meteorology and air traffic control. It is not uncommon to find individual satellites in such a system costing in excess of one-hundred million dollars, with the booster necessary to place it in orbit having a like cost. Obviously, any means possible for minimizing the number of satellites (and boosters) to provide single or redundant continuous earth coverage could bring about major cost savings. The use of elliptic orbit constellations can provide more efficient coverage with fewer satellites and without gaps, in many instances, than is possible using circular-orbit, Walker-type constellations.

Several theorems and corollaries have previously been presented which are of assistance in developing and understanding the satellite coverage principles involved in elliptic orbit constellations[2,3]. In dealing with continuous redundant coverage systems, rather than continuous single-coverage systems, it was found necessary to develop additional theorems and corollaries. These theorems and corollaries are grounded in solid geometry, as were the previous ones. Although they are simple to understand and apply, they are only the first step towards the full development and optimization of these optimal-satellite constellations. Optimization would have been impossible to achieve without the availability of the modern, high-speed microcomputer. The majority of the work performed in the development of the triple and quadruple continuous coverage arrays was carried out at STA using an Apple Macintosh II PC, in conjunction with an Apple Laserwriter II printer.

Since continuous coverage implies complete independence from an observer's geographic location, it has been found useful to employ the concept of a "pseudo-planet" whose period is the same as the constellation's period.

3.0 Theorems and Corollaries

The following theorems and corollaries supplement those presented in references 2 and 3. They are useful in addressing the problem of redundant continuous coverage.

Theorem III: If two satellites in a multi-satellite constellation are diametrically opposite one another, a minimum of $(2n + 3)$ satellites is required to obtain instantaneous global n-tuple coverage. (Satellites are assumed to be at a finite altitude.)

Proof: A great circle may be passed through the two satellites which are diametrically opposite and any third satellite. At least $2n$ additional satellites are then needed in order to obtain n-tuple coverage of the two poles normal to the great circle plane.

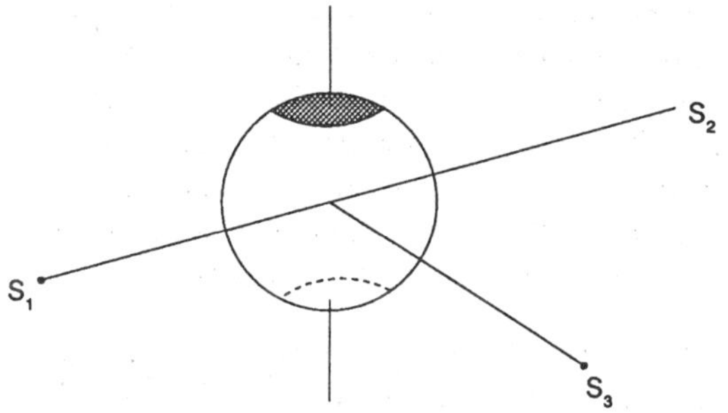

Figure 1. Theorem III

Corollary III: If m satellites are coplanar with a planet's center, then a minimum of $(m + 2n)$ satellites is required to obtain instantaneous n-tuple coverage.

Proof: A great circle may be passed through the m satellites. The two poles to the great circle require an additional $2n$ satellites, at a minimum, to obtain n-tuple coverage.

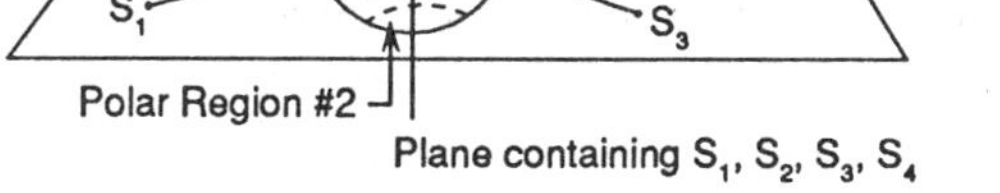

Figure 2. Corollary III

Theorem IV: For a system of synchronous elliptic orbit satellites to have identical ground tracks, the satellites must have the same inclination, eccentricity, and argument of perigee, and the sum of the right ascension of the ascending node and the satellite mean anomaly must be a constant for each satellite in the system for any time instant.

Proof: Since all the satellites' parameters except right ascension of the ascending nodes and mean anomalies are identical, the shape of all the orbital ground tracks are identical, save for a possible longitudinal displacement. Equatorial crossings will occur at the same longitude, resulting in overlying ground tracks, provided any angular increase in the right ascension of a given satellite's line of nodes is offset by an equal but opposite angular decrease of the mean anomaly for that same satellite. Subtracting the mean anomaly for a synchronous system effectively subtracts out the planetary rotation rate and converts the inertial reference frame to a planetary reference frame. Constraining the sum of the right ascension of the ascending node plus the mean anomaly to a unique constant value ensures that any increase in right ascension of the ascending node will occasion a corresponding decrease in mean anomaly. Then, at earlier or later points in time where the other satellites in the system cross the equator at their respective ascending nodes, their mean anomalies will be the same as that of the original reference satellite. The crossing longitudes will thus be the same, since the shape of every satellite's ground track is identical (with corresponding mean anomalies matching with the latitude crossings).

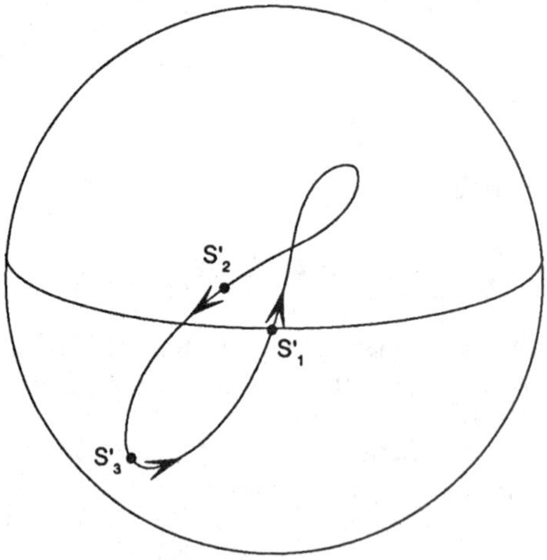

Figure 3. Theorem IV

Theorem V: If three satellites form a plane which intersects a spherical planet, then a minimum of $(2n + 3)$ satellites is required to obtain instantaneous global n-tuple coverage.

Proof: Erect a perpendicular to the triplet plane through the planet's center. We will define the "poles" as the intersections of this perpendicular diameter with the surface of the planet. It may be easily seen that at each pole there is no visibility by any of the triplet satellites (assuming these satellites are at finite altitudes). Thus, n additional satellites are needed to provide instantaneous n-tuple coverage of each of these poles. The minimum total number of satellites required is therefore $(2n + 3)$, representing the three satellites in the intersecting plane and $2n$ more for the two poles.

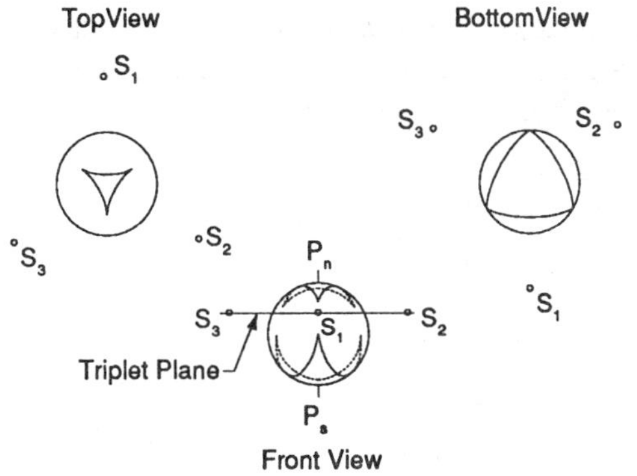

Figure 4. Theorem V

4.0 Common Period Continuous Single Coverage Array (U.S. Patent Pending)

The minimum-satellite array providing continuous single coverage consists of four satellites arranged in a tetrahedral geometry (Figure 5). The orbital parameters for each satellite in the array are listed in Table 1. Satellites in the array are organized into two pairs where both satellites in a pair follow an identical ground track. The two pairs begin an orbital period on opposite hemispheres of the earth and are skewed in relation to each other; a line joining together the first pair would run north-south, while a line joining the second pair's satellites would run east-west.

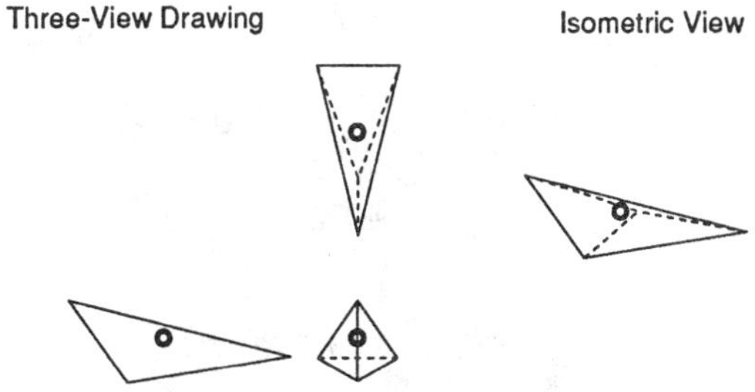

Figure 5. Four-Satellite Tetrahedral Constellation for Continuous Single Coverage (U.S. Patent Pending)

Table 1. Starting Positions, Single-Coverage Constellation; n = 1 (U.S. Patent Pending)

Satellite No.	T (h)	i (deg)	e	ω (deg)	Ω (deg)	M (deg)
1	=> 26.49	31.3	0.263	-90	0	0
2	=> 26.49	↓	↓	+90	-90	+90
3	=> 26.49	↓	↓	-90	-180	+180
4	=> 26.49	↓	↓	+90	-270	+270

Figure 6 illustrates a plot of the loci of satellite positions over an orbital period as seen by an observer rotating in the pseudo-planetary reference frame (the pseudo-planet shown has radius of Earth and is drawn to scale). The frontal view shows clearly the front hemisphere satellite pair loci (perigees in southern hemisphere) and the back hemisphere satellite pair loci (perigees in northern hemisphere). Each pair appears as a single oval shape. It is interesting to note that the side view shows a nearly linear path followed by each satellite pair.

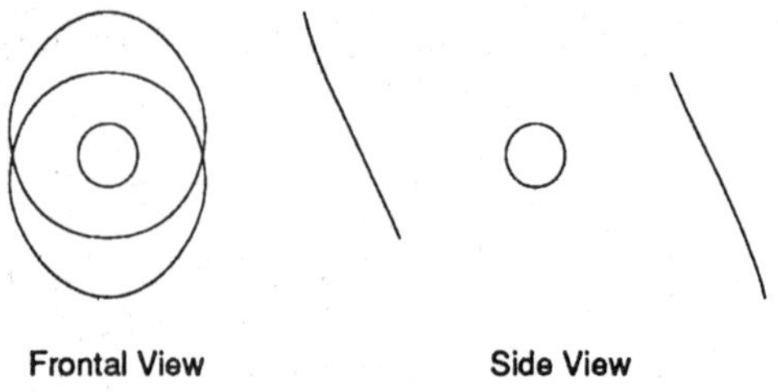

Figure 6. Four-Satellite Tetrahedral Array (Frontal and side views in pseudo-planet reference frame)

Proof of global coverage was accomplished by rigorous examination of instantaneous visibility limit plots for the four-satellite array at discrete intervals spanning an orbital period. Figure 7 shows a rotated visibility limit plot for the four-satellite array at the start of its common orbital period. Figure 8 follows the changing limit curves over an orbital period with a plot taken at every 30-degree interval of mean anomaly. Lowering the array to a synchronous altitude would result in a slightly degraded coverage pattern in which small gaps of non-coverage begin to appear, but most of the globe would still be continuously visible to the array. The minimum period sufficient for global single coverage is 26.5 hours.

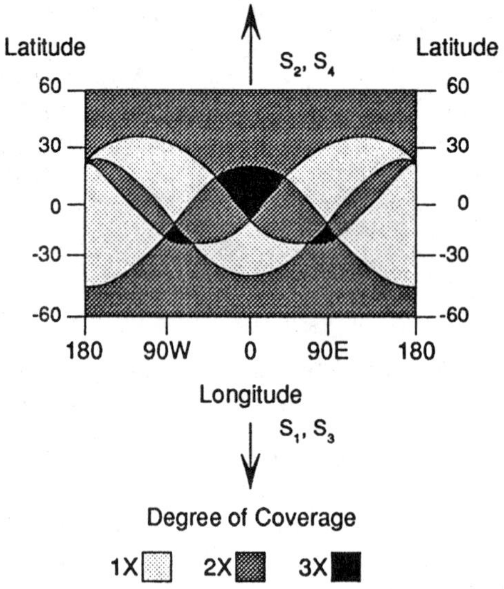

Figure 7. Four-Satellite Visibility Limits, Starting Position of Constellation

The structure of the array insures that with sufficient altitude, no plane constructed from any three of the four satellites in the array will intersect the earth's surface. Such a tetrahedral geometry is unique to the single-coverage model (at higher levels of coverage, minimum-satellite constellation geometries are prismoidal), although the structures of minimum-satellite arrays with higher levels of coverage have the same property of non-intersection.

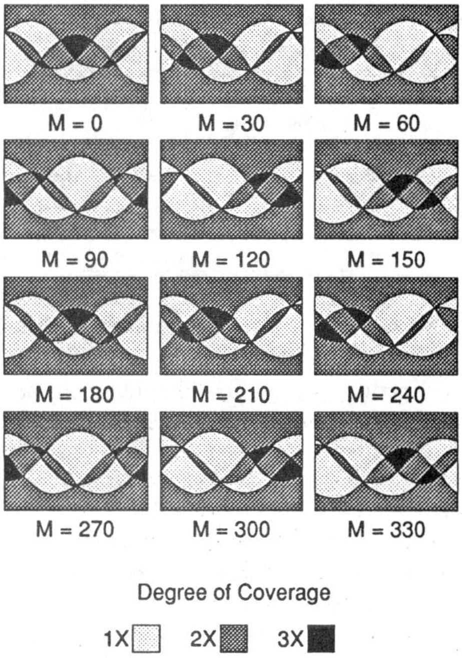

Figure 8. Four-Satellite Visibility Limit Plots at Thirty-Degree Intervals of Mean Anomaly

5.0 Common Period Continuous Double-Coverage Array

The minimum-satellite array providing continuous double coverage[4] consists of six satellites arranged in a triangular prismoidal geometry (Figure 9). This array is the simplest prismoidal array of the family being described. The orbital parameters for each satellite in the array are listed in Table 2. Satellites are organized into two sets of three satellites each, with each set composing one end of the triangular prismoid. The ends of the prismoid are congruent, unlike the four-satellite model, which has skewed pairs. Each set of satellites shares a ground track, so that two ground tracks represent the entire array of satellites.

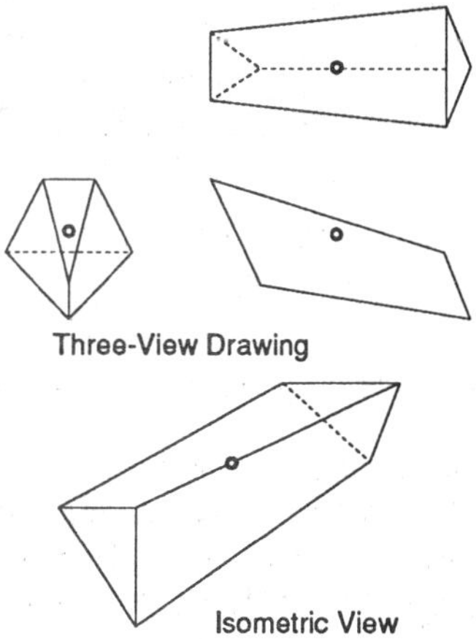

Figure 9. Six-Satellite Visibility Limits, Starting Position of Constellation

Table 2. Starting Positions, Double-Coverage Constellation; n = 2

Satellite No.	T (h)	i (deg)	e	ω(deg)	Ω (deg)	M (deg)
1	= > 102	27.5	0.233	-90	0	0
2	= > 102	↓	↓	+90	-60	+60
3	= > 102	↓	↓	-90	-120	+120
4	= > 102	↓	↓	+90	-180	+180
5	= > 102	↓	↓	-90	-240	+240
6	= > 102	↓	↓	+90	-300	+300

The side and frontal views of the double-coverage array are shown in Figure 10. Proof of coverage was accomplished in the same manner as the single-coverage case, by plotting the satellite visibility limits over an orbital period (Figure 11). The minimum period providing a sufficient constellation altitude for global double coverage is 102 hours.

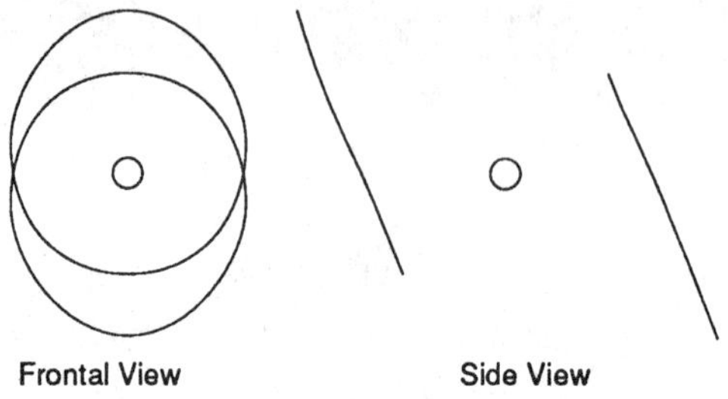

Figure 10. Six-Satellite Triangular Prismoidal Array (Frontal and side views in pseudo-planet reference frame)

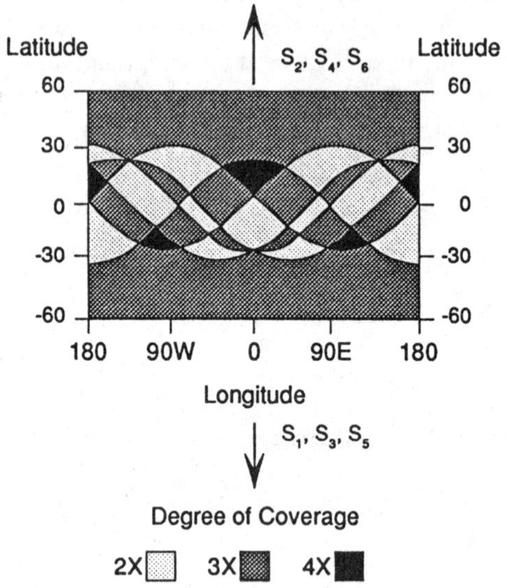

Figure 11. Six-Satellite Visibility Limits, Starting Position of Constellation

6.0 Observations on Single- and Double-Coverage Arrays

The independent discoveries of the four-satellite common period tetrahedral array for continuous global single coverage and the six-satellite prismoidal array for continuous double coverage had only one obvious common link. This was the existence of two counter-rotating sets of roughly circular ground tracks, one set on either side of the pseudo-planet. The phasing, however, was quite different. The four-satellite array assumed a cruciform shape when viewed axially. That is, the two satellites on the far side "split the difference" in the spacing on the near side. By contrast, in the six satellite double-coverage array, the spacings were in parallel, or in alignment. That is, the triangles, representing the two bases of the prismoid, were lined up base to base and vertex to vertex at their starting positions. This fact led, introspectively, to the formulation of Theorem III, when it was observed that "splitting the difference" in spacing with two triangular bases would never achieve the desired double coverage. At the starting position, one key satellite would be at apogee and one satellite (on the other side) at perigee. They would, at that instant, be on opposite ends of a spherical diameter, making the desired double coverage with six satellites impossible, in accordance with Theorem III. (This "skewed" prismoid with opposed triangular bases and 90-degree satellite spacings, is actually an octahedron.) Yet, a regular triangular prism with parallel sides does have the capability of providing instantaneous double coverage, as described in reference (4).

7.0 The Development of N-Tuple Coverage Arrays

In order to avoid the condition of finding satellites lying at opposite ends of a diameter, the author developed the strategy of employing either "skew" or "congruent" prismoidal arrays, depending on whether *n* were even or odd, respectively. The increase of *n* to three (triple coverage) or four (quadruple coverage) would then appear tractable using a "skew" prismoid or a "congruent" prismoid, respectively. This was, in fact, found to be a valid observation, and there is little reason to doubt its generality for all values of *n*.

8.0 Common Period Continuous Triple Coverage Array

Global triple coverage is achieved by arranging eight satellites in a quadrangular prismoidal geometry (Figure 12). The orbital parameters are outlined in Table 3. These satellites are arranged into two sets of four, with each set sharing a single ground track. Each set of four satellites represents one end of the quadrangular prismoid. The two ends are skewed, as in the four-satellite model. This constellation requires a minimum period of 272 hours.

Table 3. Starting Positions, Triple-Coverage Constellation; n = 3

Satellite No.	T (h)	i (deg)	e	ω (deg)	Ω (deg)	M (deg)
1	=> 272	25	0.218	-90	0	0
2	=> 272	↓	↓	+90	-45	+45
3	=> 272	↓	↓	-90	-90	+90
4	=> 272	↓	↓	+90	-135	+135
5	=> 272	↓	↓	-90	-180	+180
6	=> 272	↓	↓	+90	-225	+225
7	=> 272	↓	↓	-90	-270	+270
8	=> 272	↓	↓	+90	-315	+315

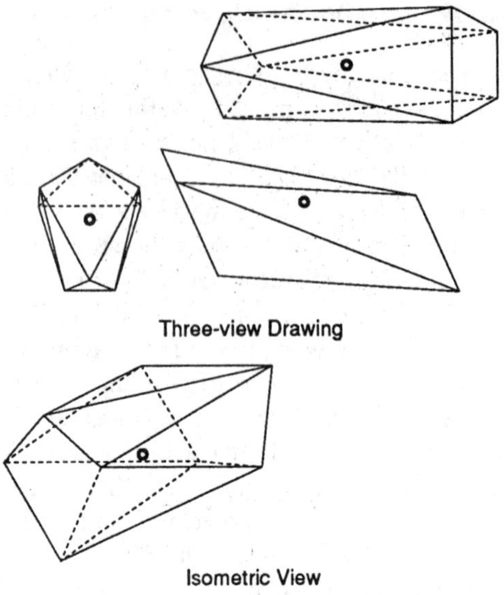

Figure 12. Eight-Satellite Quadrangular Prismoid for Continuous Triple Coverage

The eight-satellite array also produces only two ground tracks. The frontal and side views for the eight-satellite array (Figure 13) show how far out the satellites must be to provide triple coverage (the pseudo-planet is still drawn to scale).

The visibility limit plot for the eight-satellite array's starting position is illustrated in Figure 14.

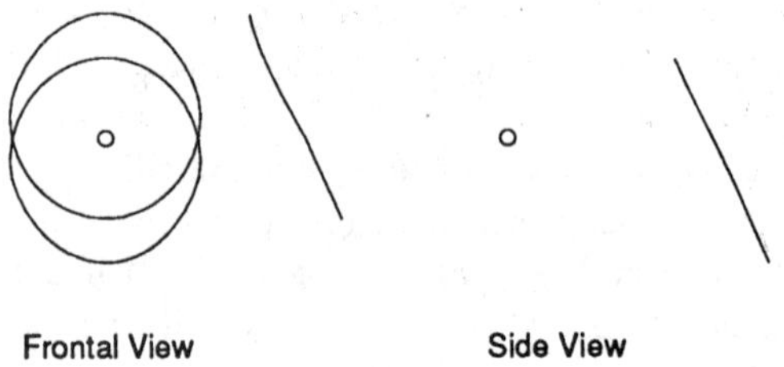

Figure 13. Eight-Satellite Quadrangular Prismoidal Array (Frontal and side views in pseudo-planet reference frame)

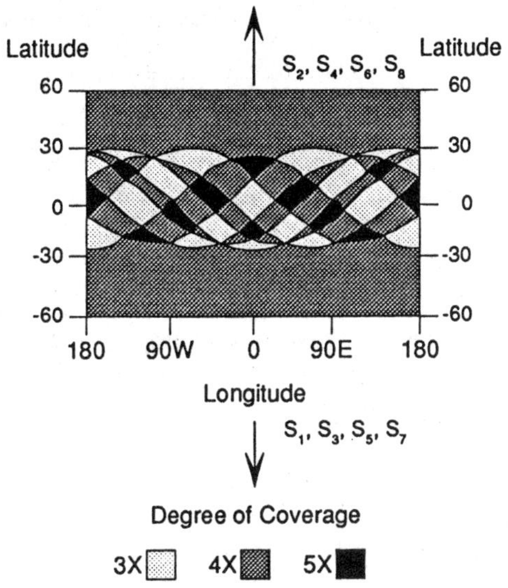

Figure 14. Eight-Satellite Visibility Limits, Starting Position of Constellation

9.0 Common Period Continuous Quadruple-Coverage Array

Global quadruple coverage is achieved by arranging ten satellites in a pentagonal prismoidal geometry (Figure 15). The orbital parameters are outlined in Table 4. These satellites are arranged into two sets of five, with each set sharing a single ground track. Each set of five constitutes one end of the pentagonal prismoid, and the ends of this model are congruent.

The frontal and side views for the ten-satellite array are shown in Figure 16. To provide quadruple coverage, this constellation requires a minimum period of 568 hours, or almost 24 days. As is evident from the visibility limit plot in Figure 17, this array is the most complex of the four that have been described.

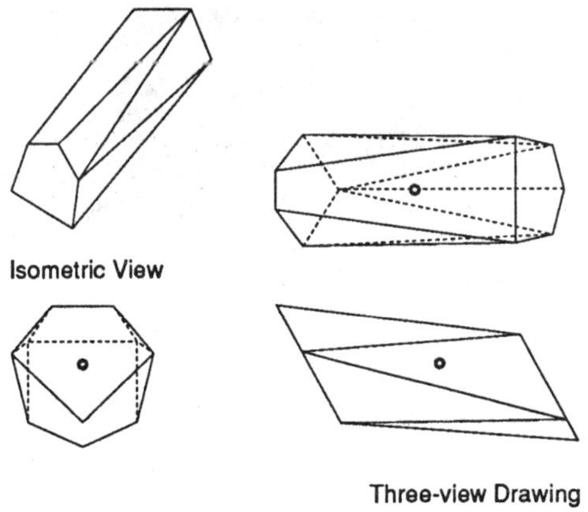

Figure 15. Ten-Satellite Pentagonal Prismoid for Continuous Quadruple Coverage

Table 4. Starting Positions, Quadruple-Coverage Constellation; n = 4

Satellite No.	T (h)	i (deg)	e	ω (deg)	Ω (deg)	M (deg)
1	= > 568	24	0.205	-90	0	0
2	= > 568	↓	↓	+90	-36	+36
3	= > 568	↓	↓	-90	-72	+72
4	= > 568	↓	↓	+90	-108	+108
5	= > 568	↓	↓	-90	-144	+144
6	= > 568	↓	↓	+90	-180	+180
7	= > 568	↓	↓	-90	-216	+216
8	= > 568	↓	↓	+90	-252	+252
9	= > 568	↓	↓	-90	-288	+288
10	= > 568	↓	↓	+90	-324	+324

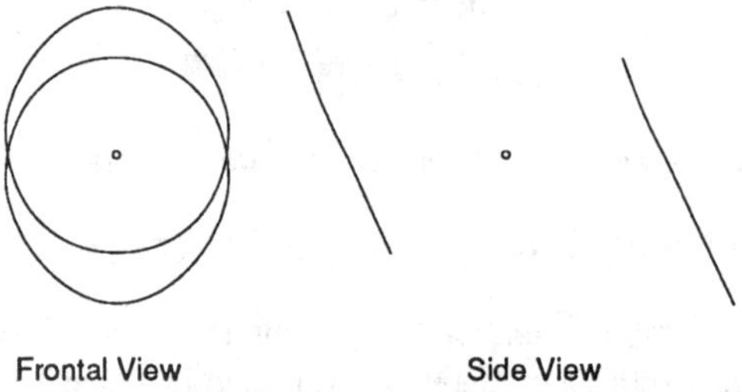

Figure 16. Ten-Satellite Pentagonal Prismoidal Array (Frontal and side views in pseudo-planet reference frame)

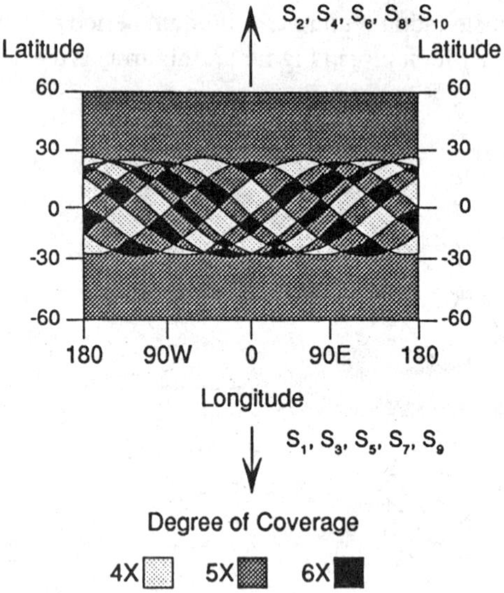

Figure 17. Ten-Satellite Visibility Limits, Starting Position of Constellation

10.0 Satellite Periods, Eccentricities, Apogees, and Perigees

The quadruple-coverage ten-satellite array expresses the limit of our detailed investigation into the geometrical family of minimum-satellite arrays described in this paper. The minimum periods required to meet the redundant coverage levels above $n = 2$ mandate very high orbital altitudes where lunar and solar perturbations would threaten constellation integrity. Table 5 illustrates the great orbital radii involved for higher levels of coverage when compared to the single- and double-coverage models.

Table 5. Satellite Periods, Eccentricities, Apogees and Perigees (n = 1...4)

n =	$T_{\Omega n}$ => (h)	e =	a = (nm)	r_a =(nm)	r_p =(nm)	h_a =(nm)	h_p =(nm)
<1(synch)	**24**	**0.263**	**22767**	**28755**	**16779**	**25313**	**13337**
1	26.49	0.263	24316	30711	17921	27269	14479
2	102	0.233	59735	73653	45817	70211	42375
3	272	0.218	114871	139913	89829	136471	86387
4	568	0.205	187671	226144	149198	222702	145756

11.0 Generalized Table of Orbital Parameters

Development of a family of n-tuple coverage constellations led to the definition of a generalized table of orbital parameters (Table 6) which provides the proper phasing relationships. The table accounts for the skew/congruent relationship, meeting the requirements of Theorem IV. The indexed, generalized table of orbital parameters describes coverage arrays for all values of n. The only parameter values not derivable explicitly are the array's inclination, i, and eccentricity, e. These two parameters appear to be loose functions of n, both of which decrease as the value of n increases. Using computer models, the actual optimal values of i and e (which yield the minimum possible period for the constellation) may be determined.

Table 6. Starting Positions, Generalized Table of Orbital Parameters For N-Tuple Continuous-Coverage Constellations

Satellite No.	i (deg)	e	ω	Ω	M
1	f (n)	g (n)	-90°	0°	0°
⋮	↓	↓	⋮	⋮	⋮
k			$(-1)^k(90°)$	$-\left(\frac{k-1}{n+1}\right)180°$	$+\left(\frac{k-1}{n+1}\right)180°$
⋮	↓	↓	⋮	⋮	⋮
k = n + 2			$(-1)^{n+2}(90°)$	-180°	+180°
⋮	↓	↓	⋮	⋮	⋮
k = 2n + 2	↓	↓	+90°	$-\left(\frac{2n+1}{n+1}\right)180°$	$+\left(\frac{2n+1}{n+1}\right)180°$

12.0 Benefits Derived From Tetrahedral and Prismoidal Arrays

The minimum-satellite, continuous-coverage arrays display a number of important benefits. The first and most obvious is that continuous earth coverage at any level of redundancy can be achieved using fewer satellites than any other system (including Walker-type arrays). Fewer satellites implies fewer boosters and less ground control and support, compounding the cost savings. The use of elliptic orbits in these prismoidal arrays frees more of the orbital parameters for manipulation than are available for circular orbits, so that a high degree of bias control can be exerted to emphasize performance in selected areas of the world. Continuous coverage is critical for surveillance systems, particularly in the polar regions. Cross-link connectivity is also supported, since satellite-to-satellite LOS is perpetually maintained. Lastly, the satellites of these arrays do not reside in the equatorial regions and so avoid the synchronous equatorial belt, which is presently vastly overcrowded, and their large crossing velocities insure minimum interference intervals in crossing situations with other satellites.

13.0 Significance of Results

A new family of satellite arrays has been created that reduces the current minimum-satellite configuration for continuous global coverage by one, resulting in reduced cost for space systems based on such arrays. Higher levels of coverage lead to arrays at very high altitudes which are unattractive due to the effects of lunar perturbations. However, the single-coverage model and possibly the double-coverage model can be placed at altitudes that are directly applicable to current and future missions.There is also evidence that the level of coverage obtained from an array is generally higher than the minimum level required. For example, there is much more double coverage than single coverage obtained from the single-coverage tetrahedral array (see Figure 8).

14.0 Conclusions

The overall conclusions to be drawn from the study on elliptical-orbit constellations are the following:

1) All the constellations are at altitudes greater than earth-synchronous. As the degree of redundancy, n, increases, so does the altitude.

2) From a practical standpoint, only the single- and double-coverage arrays appear useful. The reason for this is that for higher levels of redundancy ($n \geq 3$), the constellation altitudes are so high as to expose the satellite orbits to unacceptable levels of lunar and solar perturbational forces.

15.0 Acknowledgements

The triple and quadruple continuous-coverage models and the generalized theory of prismoidal arrays were developed under the auspices of the Small Business Innovation Research (SBIR) Program, under Contract Number F04701-88-C-0080. The Study Sponsor was the U.S. Air Force Space Division, Los Angeles, California. The views expressed in this paper are those of the author and do not necessarily represent U.S. Air Force official position or policy.

The assistance of Henry Bowers, in computer programming and data reduction, and Kim Westlake, graphics and layout, for this paper, is gratefully acknowledged.

References

1) Walker, J.G., "Continuous Whole Earth Coverage by Circular-Orbit Satellite Patterns," Royal Aircraft Establishment, Tech. Rept. 77044, March 1977.

2) Draim, J.E., "Three- and Four-Satellite Continuous Coverage Constellations," *Journal of Guidance, Control, and Dynamics*, Vol. 6, Nov.-Dec. 1985, pp. 725-730.

3) Draim, J.E., "A Common-Period Four-Satellite Continuous Global Coverage Constellation," *Journal of Guidance, Control, and Dynamics*, Vol. 10, No. 5, Sep.-Oct. 1987, pp. 492-499.

4) Draim, J.E., "A Six-Satellite Continuous Global Double Coverage Constellation," AAS Paper 87-497 AAS/AIAA Astrodynamics Specialist Conference, Aug. 10-13, 1987.

AAS 89-382

SBS-5 MISSION ANALYSIS AND LAUNCH RESULTS*

R. L. Johnson† and K. R. Hartman‡

This paper reviews the prelaunch mission analyses and postflight results of the Satellite Business Systems (SBS)-5 satellite. This communications satellite was launched on September 8, 1988, by an Ariane 3 launch vehicle. SBS-5 is an HS-376 spin-stabilized spacecraft built by Hughes Aircraft Company.

The prelaunch analysis and support included nominal sequence generation, contingency analyses, and mission simulations. Nominal mission planning encompassed launch window analysis, apogee motor size optimization, apogee motor firing (AMF) attitude targeting, and drift orbit analysis. The contingency analysis was used to develop responses to potential mission problems. The simulations familiarized support personnel with the mission sequence of events.

The launch vehicle injected the spacecraft into a low-inclination transfer orbit. AMF attitude was achieved by trim maneuvers. An apogee kick motor placed the spacecraft into its drift orbit. Postburn, reorientation (reor), and spindown maneuvers were performed during the next 25 hours. The spacecraft was on-station 15 days later at 123° W longitude.

---

* This paper appears in full in the microfiche supplement to this volume, **AAS Microfiche Series**, Volume 59.

† Mission Analyst, MCI Telecommunications Corporation, 8283 Greensboro Dr., McLean, Virginia 22102.

‡ Mission Analyst, MCI Telecommunications Corporation, 2201 Comsat Dr., Clarksburg, Maryland 20871.

AAS 89-383

ANALYSIS OF EARTH LIMB VIEWING FROM THE UPPER ATMOSPHERIC RESEARCH SATELLITE*

Donald L. Hitzl†, Jack B. Kumer‡ and Frank Zele**

The Upper Atmospheric Research Satellite (UARS) is a multi-instrumented satellite which will examine the Earth's upper atmosphere. One of the instruments onboard UARS is a Cryogenic Limb Array Etalon Spectrometer (CLAES) developed at the Lockheed Research Laboratory to measure atmospheric infrared emissions. This paper addresses the Sun-Orbit-CLAES viewing geometery during the two year mission. Orbit averages and monthly averages are computed. In particular, the problem of viewing through a partially illuminated atmospheric shell is solved. For this transition situation, part of the look vector is in sunlight while the remainder is in the Earth's shadow.

---

* This paper appears in full in the microfiche supplement to this volume, **AAS Microfiche Series**, Volume 59.

† Staff Scientist, Advanced Systems Studies, Lockheed Research Laboratory, 3251 Hanover Street, Palo Alto, California 94304-1191.

‡ Senior Staff Scientist, Infrared Science Programs, Lockheed Research Laboratory, 3251 Hanover Street, Palo Alto, California 94304-1191.

** Staff Engineer, Electro-Optical Systems Laboratory, Lockheed Research Laboratory, 3251 Hanover Street, Palo Alto, California 94304-1191.

AAS 89-384

SCIENCE MISSION DATA REDUCTION FOR THE STANFORD RELATIVITY GYRO EXPERIMENT*

Xinhua Qun†, John V. Breakwell‡ and Bradford W. Parkinson**

The Stanford Relativity Gyro Experiment is designed to test the General Theory of Relativity. It is to detect two untested relativity effects, *geodetic* and *frame-dragging*, by measuring the spin axis directions of four superconducting gyroscopes carried by a spacecraft. A remarkable accuracy of 1 *milliarc-sec/yr* is desired. The spacecraft is made to roll about the line of sight to a guide star. This reduces the body-dependent errors and modulates the inertial signal to spectrally separate it from low frequency measurement noise. This paper discusses the data reduction simulation which includes spacecraft roll dynamics. These are preliminary results for the science mission. A two step filtering algorithm is developed for this experiment. Studies show that the data reduction can be successfully performed, under a realistic roll disturbance, with only one roll angle measurement per roll period.

---

* This paper appears in full in the microfiche supplement to this volume, **AAS Microfiche Series**, Volume 59.

† Ph.D. Student, Department of Aeronautics and Astronautics, Stanford University, Stanford, California 94305.

‡ Professor, Department of Aeronautics and Astronautics, Stanford University, Stanford, California 94305.

** Research Professor, Department of Aeronautics and Astronautics and High Energy Physics Laboratory, Stanford University, Stanford, California 94305. Program Manager, Relativity Gyro Experiment.

TRAJECTORY OPTIMIZATION APPLICATIONS

SESSION 7

Chairman: Johnny H. Kwok
Jet Propulsion Laboratory

OPTIMAL FINITE-THRUST SPACECRAFT TRAJECTORIES USING COLLOCATION AND NONLINEAR PROGRAMMING

Paul J. Enright* and Bruce A. Conway†

A new method is described for the determination of optimal spacecraft trajectories in an inverse-square field using finite, fixed thrust. The method employs a recently developed optimization technique which uses a piecewise polynomial representation for the state and controls, and collocation, thus converting the optimal control problem into a nonlinear programming problem, which is solved numerically. This technique has been modified to provide efficient handling of those portions of the trajectory which can be determined analytically, i.e., the coast arcs. Among the problems that have been solved using this method are optimal rendezvous and transfer (including multi-revolution cases) and optimal multi-burn orbit insertion from hyperbolic approach.

INTRODUCTION

Historically, research into the optimization of finite-thrust spacecraft trajectories has fallen into two categories. Indirect methods require the derivation of the necessary conditions for optimality using calculus of variations techniques, and result in a two-point boundary value problem (TPBVP), which typically needs to be solved numerically. Direct methods, on the other hand, approximate the continuous problem with a parameter optimization problem, usually requiring mathematical programming for solution. This reduction can involve parameterization of the control variables only, in which case the state trajectory is explicitly integrated (direct shooting), or the states and controls can both be parameterized, requiring some sort of implicit integration scheme to satisfy the equations of motion.

The application of the standard variational optimal control techniques (the indirect method) to the finite-thrust spacecraft trajectory problem results in the "primer vector" formulation of Lawden[1]. The resulting TPBVP can be difficult to solve because of sensitivity problems. Kern and Greenwood[2], Hazelrigg and Lion[3], and more recently, Redding and Breakwell[4] have solved transfer problems with this approach, using shooting methods to solve the TPBVP, and employing estimates of the initial conditions obtained by extrapolation from known impulsive solutions. A different approach to solving the indirect formulation TPBVP was taken by Dickmanns and Well[5], who used a collocation method employing piecewise Hermite

* Graduate Research Assistant, University of Illinois at Urbana-Champaign, 104 S. Mathews Ave., Urbana, Illinois 61801.

† Associate Professor, Department of Aeronautical and Astronautical Engineering, University of Illinois at Urbana-Champaign, 104 S. Mathews Ave., Urbana, Illinois 61801.

cubics. This method avoids the inherent sensitivity of shooting methods, and demands less accurate initial guesses for successful solution.

Many different direct optimization techniques have been proposed and attempted. Direct shooting uses a finite-parameter representation of the control history, and explicitly integrates the state trajectory. In general, mathematical programming is required to optimize the control parameters to minimize the cost and simultaneously satisfy terminal boundary conditions. Sensitivity problems can arise, and many trajectory integrations may be required during the optimization process, which can be costly. A unique approach was taken by Zondervan et. al.[6], where a hybrid (direct/indirect) formulation[7] using nonlinear programming for solution was used to solve some transfer problems. A purely direct approach was described by Johnson[8], and later by Hargraves et. al.[9], who represented the control and state histories using Chebyshev polynomials, and used integral penalty functions to enforce the equations of motion, thus converting the optimal control problem to a problem in unconstrained minimization, which is solved numerically. A different direct approach for solving optimal control problems uses Dickmann's collocation method with Hermite cubics[5] to convert the optimal control problem into a nonlinear programming problem. This is the method used in the current study, and is referred to as Direct Collocation with Nonlinear Programming (DCNLP). Hargraves and Paris[10] used this method to solve several atmospheric trajectory optimization problems, and installed it in their trajectory optimization routine OTIS. This approach was also described by Kraft[11], but the handling of the controls was different than that of Hargraves and Paris[10], and no computational experience was related.

DIRECT COLLOCATION WITH NONLINEAR PROGRAMMING

In the DCNLP approach, the trajectory is approximated by piecewise polynomials, which are represented by state and control values at a number of discrete points ("nodes".) For a given state variable, the state trajectory over a given "segment" between two nodes is taken to be the unique cubic which goes through the endpoints of the segment with the appropriate derivatives that are dictated by the evaluation of the differential equations of motion at the endpoints. (This is the "Hermite cubic" since it is determined by the states and their first derivatives.) Following Hargraves and Paris[10], the control history is taken to be linear across the segment. A collocation point is taken at the center of the segment, where the derivative obtained from the Hermite cubic is compared to the derivative obtained from the evaluation of the equations of motion. The difference is termed the "defect"[5], and is a measure of non-satisfaction of the equations of motion over the given segment, i.e., if the defect is zero, then the differential equation is satisfied at the center collocation point as well as at the end points.

Let the system equations of motion be given by $\underline{x}' = \underline{f}(\underline{x},\underline{u})$, $\underline{x}$ being the state vector and $\underline{u}$ being the control vector, (') denoting differentiation with respect to time. Let the time over a given segment be T. Then the Hermite-interpolated state vector at the center point is [10]

$$\underline{x}_c = (1/2)(\underline{x}_1 = \underline{x}_r) + (T/8)[\underline{f}(\underline{x}_1) - \underline{f}(\underline{x}_r)], \qquad (1)$$

where $\underline{x}_1$ and $\underline{x}_r$ are the state vectors at the left and right nodes (Figure 1). The linearly-interpolated control vector is simply

$$\underline{u}_c = (1/2)(\underline{u}_1 + \underline{u}_r). \qquad (2)$$

The derivative of the interpolating Hermite cubic at the center point is

$$\underline{x}_c' = -[3/(2T)](\underline{x}_1 - \underline{x}_r) - (1/4)[\underline{f}(\underline{x}_1) + \underline{f}(\underline{x}_r)]. \qquad (3)$$

The defect vector is then calculated as

$$\underline{d} = \underline{f}(\underline{x}_c, \underline{u}_c) - \underline{x}_c'. \qquad (4)$$

If $\underline{x}_1$, $\underline{u}_1$, $\underline{x}_r$, and $\underline{u}_r$ are chosen such that the elements of $\underline{d}$ are sufficiently small, the interpolating polynomials become an accurate approximation to the solution of the differential equations of motion (implicit integration.)

Now that the satisfaction of the differential equations of motion has been parameterized, we can proceed to convert the optimal control problem into a nonlinear programming problem. From the states and controls at the nodes, we can calculate the objective function, the defects, and any path or terminal constraints. If the final time is allowed to vary, then the segment times are calculated by some fixed relationship to the final time. The nonlinear programming problem is to minimize the objective subject to the constraints that all the defects are acceptably small, and all other problem constraints are met.

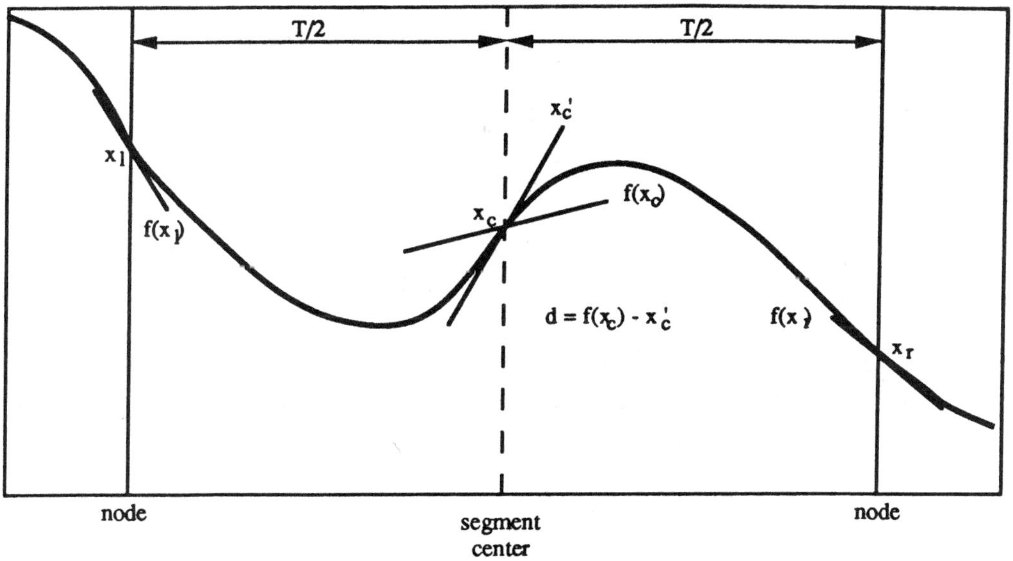

Figure 1 Collocation Using Hermite Polynomials

APPLICATION TO FINITE-THRUST SPACECRAFT TRAJECTORIES

For the current study of finite-thrust spacecraft trajectories, the equations of motion, in two-dimensional polar form (Figure 2), are,

$$
\begin{aligned}
x_1' &= x_3 \\
x_2' &= x_4/x_1 \\
x_3' &= x_4^2/x_1 - 1/x_1^2 + x_5 \sin u_1 \\
x_4' &= -x_3 x_4/x_1 + x_5 \cos u_1 \\
x_5' &= x_5^2/c
\end{aligned}
\qquad (5)
$$

where,

x_1 = radius from center of attraction

x_2 = angle from some reference direction

x_3 = velocity component along the radial direction

x_4 = velocity component along the horizontal direction

x_5 = thrust-acceleration

u_1 = thrust vector angle

c = effective exhaust velocity

and units are chosen such that the gravitational parameter (mu) is 1.

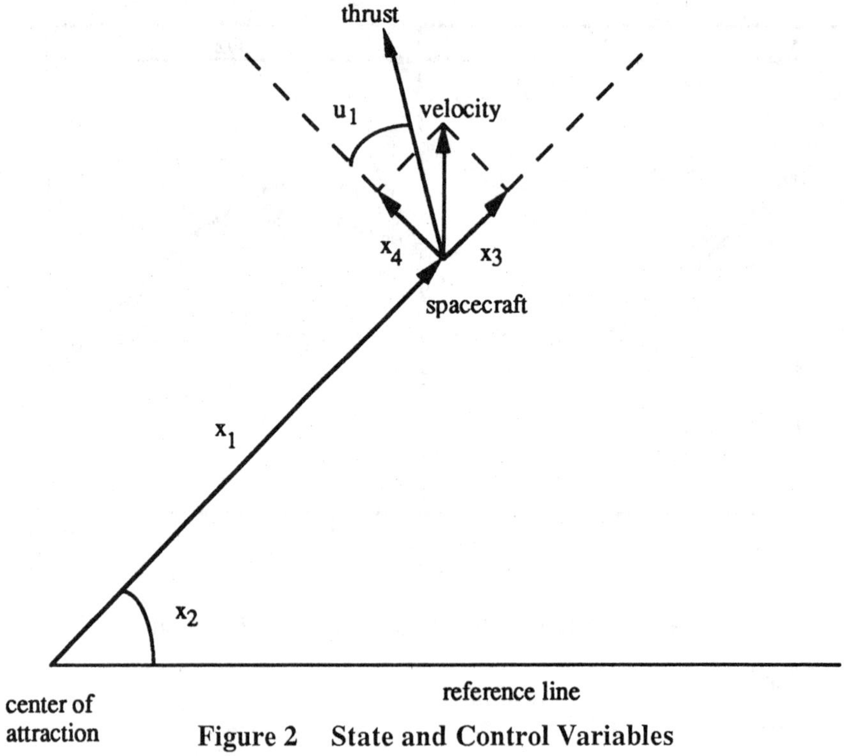

Figure 2 State and Control Variables

The trajectory is first divided into a sequence of thrust arcs (constant thrust) and coast arcs (no thrust.) The thrust arcs are further subdivided (evenly) into segments (Figure 3). At each segment center, collocation defects are calculated from the states and controls at the nodes and the duration of the thrust arc. The coast arcs are not divided into segments because the solution over a coast arc can be determined analytically. Instead, integrals of the motion (e.g. orbital elements) are calculated from the state vectors at each end of the coast arc. If the endpoints are compatible, these integrals are unchanged. Thus a "generalized defect" is taken to be the difference between the integrals at the left and right nodes of the coast arc:

$$q_1 = Q_1(\underline{x}_1) - Q_1(\underline{x}_r),$$

where,

q_1 = i^{th} generalized defect over the given coast arc

Q_1 = i^{th} integral of the motion

$\underline{x}_1$ = state vector at the left node of the coast arc

$\underline{x}_r$ = state vector at the right node of the coast arc

If $\underline{x}_1$ and $\underline{x}_r$ are chosen such that the q_i are acceptably small, then they are endpoints of the same Keplerian coast arc.

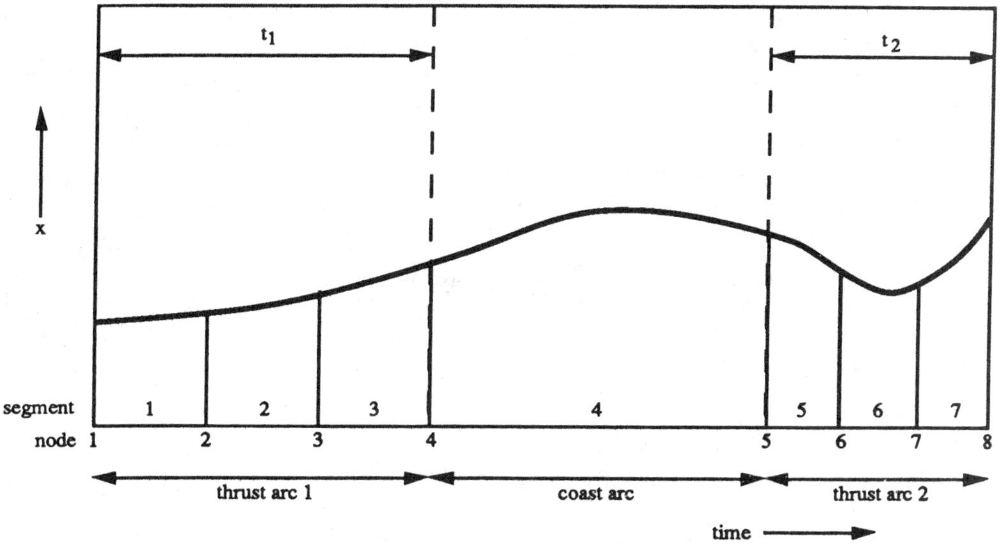

Figure 3 Problem Structure

For the two-dimensional thrust-free orbit equations (eqs. (5) with $x_5 = 0$), there are three such constants of the motion. Angular momentum, energy, and longitude of periapse were usually chosen. A fourth integral is provided by the simple requirement that the thrust-acceleration at the end of the thrust arc which precedes the coast arc equals the thrust-acceleration at the start of the thrust arc which follows. (No mass is expelled during a coast arc.) Using this scheme, the four integrals become:

$$Q_1(\underline{x}) = x_1 x_4$$

$$Q_2(\underline{x}) = (x_3^2 + x_4^2)/2 - 1/x_1 \tag{6}$$

$$Q_3(\underline{x}) = x_2 + \tan^{-1}[-x_3 x_4/(x_4^2 - 1/x_1)]$$

$$Q_4(\underline{x}) = x_5$$

The introduction of the generalized defects may greatly reduce the number of segments required, thus greatly reducing the dimensions of the nonlinear programming problem.

Nonlinear programming routines typically require the evaluation of the derivatives of the objective function and the constraints with respect to the variables. Hargraves and Paris[10] used finite-difference approximations to these derivatives; however, analytical derivatives are easily obtained for the current problem, and are used exclusively. Derivatives of the collocation defects involve chain-rule differentiation of the interpolation and equation of motion relationships, eqs. (1-5). For example, the derivative of a defect vector with respect to the state vector at the node on the segment's left is given by:

$$F(\underline{x}_c,\underline{u}_c)\ [I_5/2 + (T/8)F(\underline{x}_1,\underline{u}_1)] + [3/(2T)]I_5 + (1/4)F(\underline{x}_1,\underline{u}_1) \tag{7}$$

where $F(\underline{x},\underline{u})$ denotes the 5x5 matrix which results from differentiation of the right hand side of the equations of motion, $\underline{f}(\underline{x},\underline{u})$, with respect to the state vector, $\underline{x}$, I_5 is the 5x5 identity matrix. Defect derivatives with respect to the right state vector, the left and right control vectors, and the thrust arc duration are similarly calculated. Derivatives of the generalized defects are obtained by simply differentiating the integrals of the thrust-free motion, eqs. (6), with respect to the state vector. The objective function is usually taken to be the sum of the thrust arc durations, so its derivative is trivial. Finally, any other problem constraints (except constraints expressed as simple bounds) must be differentiated. These are usually constraints on initial or terminal orbital elements (path constraints can also be handled), and their derivatives are found in a manner similar to that of the derivatives of the generalized defects.

STRUCTURE AND SOLUTION OF THE NONLINEAR PROGRAMMING PROBLEM

The variables for the nonlinear programming (NLP) problem are the collected state vectors and control vectors at the nodes, and the thrust arc durations. These quantities are assembled into the NLP state vector $\underline{X}$:

$$\underline{X}^T = [\underline{x}_1^T,\underline{u}_1^T,\underline{x}_2^T,\underline{u}_2^T,\ldots,\underline{x}_n^T,\underline{u}_n^T,t_1,t_2,\ldots,t_k] \quad 8)$$

where n is the total number of nodes, and k is the number of thrust arcs on the trajectory. The defects, generalized defects, and any other problem constraints are collected into the NLP constraint vector $\underline{C}$:

$$\underline{C}^T = [g_1^T,\underline{g}_2^T,\ldots,\underline{g}_{n-1}^T,\underline{w}^T] \tag{9}$$

where g_i is either the defect vector for the corresponding thrust arc segment, or the vector of generalized defects for the corresponding coast arc. $\underline{w}$ is a vector of additional problem constraints.

To solve the NLP problem, a sequential quadratic programming (SQP) method was employed[12]. This is a Lagrangian method which involves the solution of a sequence of quadratic programming problems which approximate the full nonlinear problem. The particular routine used is part of the Numerical Algorithms Group (NAG) library[13], and is essentially identical to the NPSOL routine used by Hargraves and Paris[10]. The NAG routine is a FORTRAN subroutine, which must be called by the user's program. The routine must be provided with subroutines that evaluate the objective, the gradient of the objective with respect to the states, the constraints, and the jacobian of the constraints with respect to the states. In addition, a user's main program must set up algorithm parameters, problem dimensions, any linear constraints or simple bounds, and initial conditions.

These routines were coded in Cray FORTRAN 77 on the Cray X-MP/48 at the National Center For Supercomputing Applications (NCSA) at the University of Illinois. Typical sizes for the user-supplied code were 1000-1500 heavily commented lines.

SOLVED PROBLEMS

Among the problems solved using this method were optimal two-burn transfer, optimal three-burn transfer (multi-revolution), and optimal two-burn insertion (multi-revolution), all using finite, fixed thrust.

The two-burn transfer problem is the finite-thrust analogy to the impulsive Hohmann transfer. Figure 4 depicts the optimal trajectory for this circle-to-circle time-open transfer. Figure 5 shows the optimal thrust pointing angle over each of the burns. The assumed arc sequence was thrust-coast-thrust. Each thrust arc was divided into 12 segments, resulting in a total of 26 nodes. The NLP state vector had dimension 26 nodes x (5 states + 1 control) + 2 thrust arc durations = 158. The only constraints were the defects (5x24) and the generalized defects (4x1), resulting in a constraint vector of dimension 124. The circular terminal orbit conditions can be specified as simple bounds on the state variables, and are not considered constraints, per se, by the NLP routine.

The three-burn transfer problem is similar to the two-burn problem, except that two burns take place at periapse, with a complete-revolution coast arc between them. Figure 6 depicts the optimal trajectory for this solution. The assumed arc sequence was burn-coast-burn-coast-burn. Each thrust arc was represented by 10 segments, resulting in a NLP problem size of 201 states and 158 constraints. This solution is less costly than the two-burn solution as expected, however, both are local time-open optima.

The two-burn insertion problem calls for optimal insertion from a hyperbolic approach trajectory with specified excess velocity, into a circular orbit, using one burn for initial capture, and a subsequent burn for circularization. Figure 7 depicts the optimal solution. The assumed arc sequence was burn-coast-burn. Each thrust arc was represented by 10 segments, resulting in a NLP problem size of 134 states and 106 constraints. Although the circular orbit terminal constraints are treated as simple bounds on the final state vector, the initial conditions are more subtle. The initial hyperbolic trajectory energy was constrained to match the required excess velocity, but the initial angular momentum (and thus the aim point) was free. Thus, the approach targeting, the elements of the intermediate coast ellipse, the thrust arc durations, and the thrust direction over the burns were all simultaneously optimized.

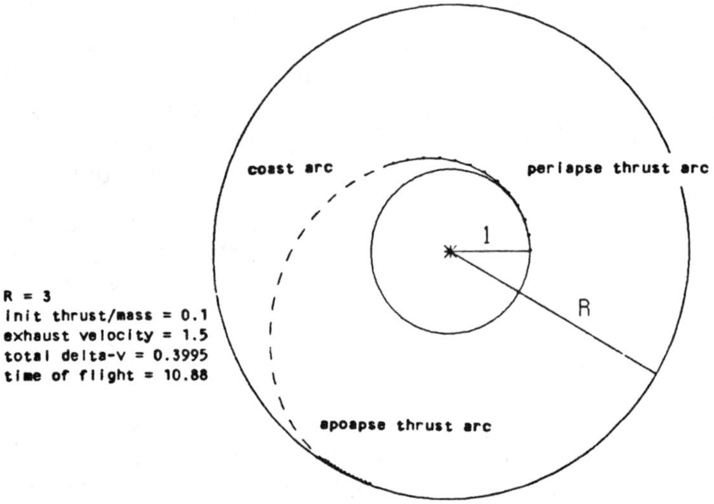

Figure 4 State Trajectory for Optimal Two-Burn Transfer

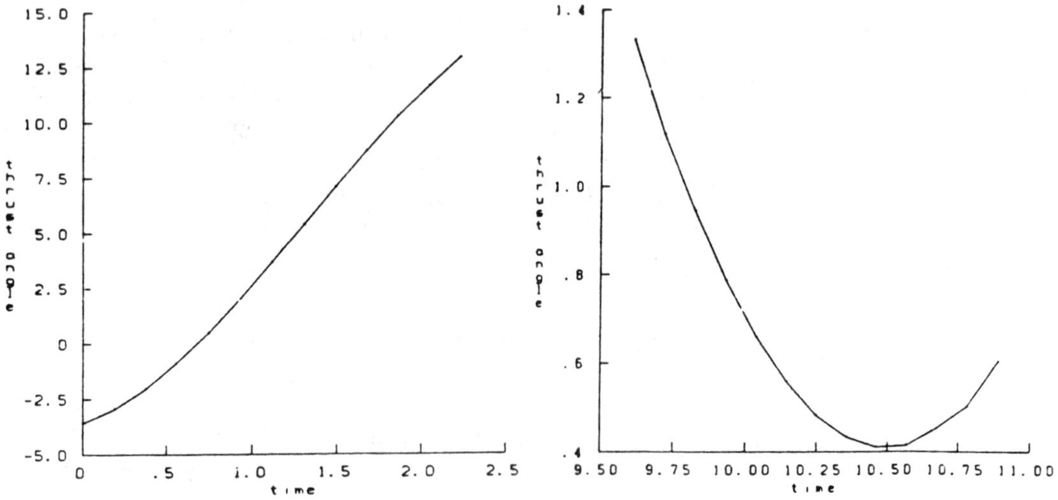

periapse apoapse

Figure 5 Thrust Direction (angle u_1 of Figure 2 in Degrees) for Optimal Two-Burn Transfer

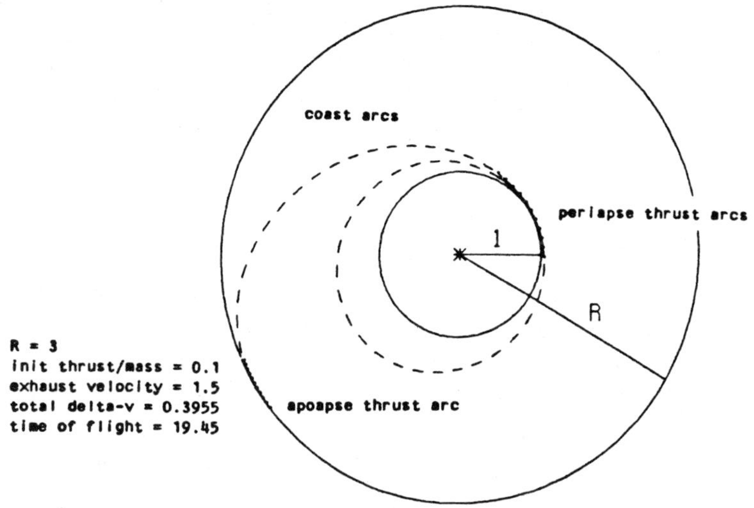

Figure 6 State Trajectory for Optimal Three-Burn Transfer

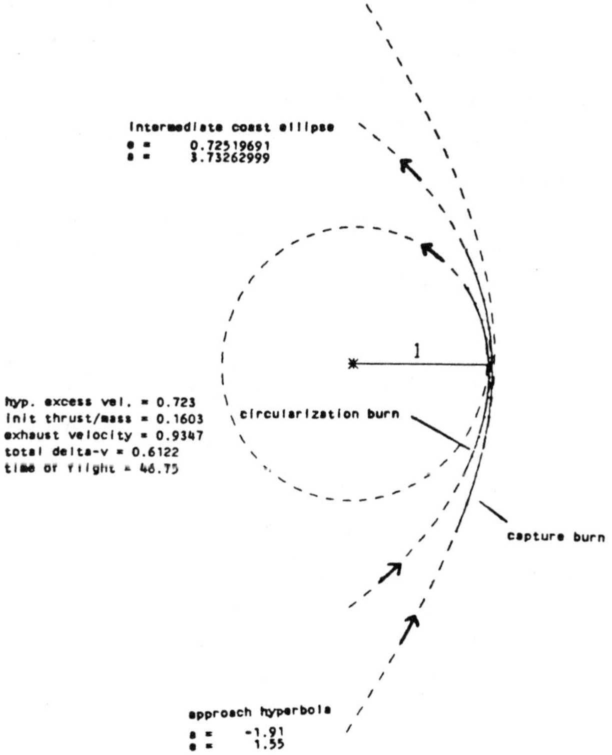

Figure 7 State Trajectory for Optimal Two-Burn Insertion

The particular numbers for this problem were meant to be representative of the injection of the Mars Observer spacecraft. In fact, the problem was inspired by a trial-and-error attempt to optimize the Mars Observer injection described by Parvez and Holman[14].

Many other problems have been solved including circle-to-circle transfers with more than three burns, circle-to-circle rendezvous problems, fixed acceleration problems, and variants of the Mars Observer problem with different constraints on the intermediate coast ellipse. An interesting feature of the rendezvous problems is that the coast arc times must be calculated from the forward Kepler equation and appropriately constrained.

PERFORMANCE

During developments, the method was tested against known solutions where possible. Finite-thrust circle-to-circle transfer results were compared to results from a program[15] based on the Redding-Breakwell[4] approach. For the circle-to-circle cases presented above, the optimal costs (delta-v) matched to 6 decimal places.

Another indication of solution accuracy was obtained from the values of the NLP problem lagrange multipliers at the optimal solution. The multipliers corresponding to the collocation defect constraints were discovered to be discrete approximations to the continuous multipliers of the indirect formulation. The defect multipliers are perfectly analogous to the multipliers of the indirect formulation, which adjoin the difference between the state vector derivative and the equations of motion to the objective function. Using this property, a discrete representation of the primer vector[1] can be assembled over a thrust arc. On an optimal trajectory, the thrust vector is aligned with the primer vector. This agreement was always obtained.

The DCNLP method is remarkably robust to initial guesses, a feature pointed out by Hargraves and Paris[10]. (Dickmanns[5] also reported such robustness for his collocation method applied to the indirect TPBVP.) We have used very poor and sometimes physically unreasonable initial guesses of the solution, yet the optimizer converges to the optimal trajectory.

Perhaps the method's strongest feature is its adaptability. Constraints of almost any type can be added in a very straightforward fashion. An example was the addition of a constraint on the semi-major axis of the intermediate coast ellipse for the Mars Observer problem. This was accomplished, literally, by the addition of a few lines of code.

Execution times on the Cray X-MP/48 were very brief: 2.5 sec for the two-burn transfer, 10.7 sec for the three-burn transfer, and 3.6 sec for the Mars Observer two-burn insertion.

CURRENT AND FUTURE WORK

Current research involves application of the DCNLP method to problems with lower thrust levels. These problems involve many revolutions, and/or very long burn arcs, requiring many segments, and resulting in fairly large NLP problems (an 80-segment 487-variable low thrust escape problem required 225 sec of Cray cpu time.) We would like to improve the efficiency of the solution of these larger problems. An area of particular interest is the banded structure of the constraint jacobian, which is due to the local nature of the collocation method. (Variations of the states at a given node affect only the defects of the adjacent segments.) Hargraves and Paris[10] recognized this sparsity, and avoided the (unnecessary) calculation of the null entries by their finite-difference derivative routine. However, the aspect more critical to us is the possible exploitation of this structure by the NLP routine. (Kraft[11] also mentions the need for such a routine.) SQP routines that are tailored to handle sparse problems are being investigated[16]. Other NLP routines which are designed to handle large, sparse problems[17] may also be considered.

Other possibilities for future investigation are an extension to three-dimensional trajectories, investigation of the utilization of hybrid (high/low thrust) motors, and applications to more complicated systems such as low-thrust problems in earth-moon space.

CONCLUSIONS

A new direct optimization method has been adapted for the optimization of finite-thrust spacecraft trajectories. The method has been improved by the introduction of "generalized defects" which allow known solutions for pieces of the optimal trajectory, e.g., coast arcs, to be used directly rather than solved for numerically, significantly reducing the size of some problems. The method was used to solve several problems, including multiple-finite-burn transfer, rendezvous, and insertion. The method appears to be considerably more powerful and robust than other optimization methods.

ACKNOWLEDGEMENTS

This work was supported by NASA grant NAG 3-805, administered by NASA Lewis Research Center, and utilized the Cray X-MP/48 system at the National Center for Supercomputing Applications, University of Illinois at Urbana-Champaign.

REFERENCES

1. Lawden, D. F., *Optimal Trajectories for Space Navigation,* Butterworths, London, 1963.
2. Kern, E. A., and D. T. Greenwood, "Minimum-Fuel Thrust-Limited Transfer Trajectories Between Coplanar Elliptic Orbits," AIAA Paper 69-914, AIAA/AAS Astrodynamics Conference, Aug. 20-22, 1969, Princeton, N.J.
3. Hazelrigg, G. A., and P. M. Lion, "Analytical Determination of the Adjoint Vector For Optimum Space Trajectories," AIAA Paper 69-916, AIAA/AAS Astrodynamics Conference, Aug. 20-22, 1969, Princeton, N.J.
4. Redding, D. C., and J. V. Breakwell, "Optimal Low-Thrust Transfers to Synchronous Orbit," *AIAA Journal of Guidance and Control,* Vol. 7, March-April 1984, pp.148-155.
5. Dickmanns, E. D., and H. Well, "Approximate Solution of Optimal Control Problems Using Third-Order Hermite Polynomial Functions," *Proceedings of the 6th Technical Conference on Optimization Techniques,* Springer Verlag, New York, IFIP-TC7, 1975.

6. Zondervan, K. P., L. J. Wood, and T. K. Caughey, "Optimal Low-Thrust, Three-Burn Orbit Transfers with Large Plane Changes," *Journal of the Astronautical Sciences,* Vol. 32, July-Sept., 1984, pp.407-427.

7. Hersome, S. E., Dixon, L. C. W., Bartholomew-Biggs, M. C., and J. J. Pocha, "The Optimisation of Spacecraft Trajectories," HSD Report TP 7633, Hawker Siddeley Dynamics Limited, Hatfield, England, 1977.

8. Johnson, F. T., "Approximate Finite-Thrust Trajectory Optimization," *AIAA Journal,* Vol. 7, June 1969, pp.993-997.

9. Hargraves, C. R., and S. W. Paris, and I. Rettie, "Numerical Computation of Optimal Atmospheric Trajectories," *AIAA Journal of Guidance and Control,* Vol. 4, July-Aug 1981, pp.406-414.

10. Hargraves, C. R., and S. W. Paris, "Direct Trajectory Optimization Using Nonlinear Programming and Collocation," *AIAA Journal of Guidance and Control,* Vol. 10, July-Aug. 1987, pp.338-342.

11. Kraft, D., "On Converting Optimal Control Problems Into Nonlinear Programming Problems," NATO ASI Series, Vol. F15, Computational Mathematical Programming, Springer-Verlag, 1985.

12. Fletcher, R., *Practical Methods of Optimization,* John Wiley & Sons, New York, 1987.

13. Numerical Algorithms Group, EO4VCF - NAG FORTRAN Routine Library Document.

14. Parvez, S. A., and J. M. L. Holman, "Low-Thrust Insertion into Orbit Around Mars," *AIAA Journal of Guidance and Control and Dynamics,* Vol. 11, Sept-Oct. 1988, pp.475-477.

15. Marinos, G. M., and L. H. Sentman, "Low Thrust, Optimal, Time-Free Transfer Between Coaxial, Aligned, Noncoplanar Elliptical Orbits," AIAA Guidance and Control Conference 1985, Snowmass, CO.

16. Nickel, R. H., and J. W. Tolle, "A Sequential Quadratic Programming Algorithm for Solving Large, Sparse Nonlinear Programs," Professional Paper, Center For Naval Analyses, May 1984.

17. Murtaugh, B. A., and M. A. Sanders, "A Projected Lagrangian Algorithm and Its Implementation for Sparse Nonlinear Constraints," *Mathematical Programming Study* 16, March 1982, pp.84-117.

AAS 89-351

MINIMUM FUEL TRAJECTORIES FOR A LOW-THRUST POWER-LIMITED MISSION TO THE MOON AND TO LAGRANGE POINTS L_4 AND L_5

Oded M. Golan* and John V. Breakwell†

Minimum fuel trajectories from a low Earth parking orbit to Lagrange points L_4 or L_5 and to the Moon are obtained for a low-thrust limited-power spacecraft, with thrust acceleration levels of the order of $10^{-3}G$. The procedure to find a trajectory to the libration point starts from an analytical description of a slightly elliptical spiral, given by Breakwell and Rauch [1]. The Earth Moon trajectory is found by matching an Earth spiral to a Moon spiral on the sphere of influence. Earth oblateness effect is considered.

1. INTRODUCTION

Electric Low thrust engines have long been considered as an alternative to high thrust chemical rockets for space missions. Previous studies of optimal low thrust missions were primarily focused on Earth orbit transfers, in which a central gravity field can be assumed. Some studies [1],[2] were also conducted on optimal low thrust interplanetary missions, where a simplifying assumption is made by considering the gravity field of one body at a time along different parts of the trajectory. This assumption is not valid in the case of low thrust lunar trajectories, where for part of the trajectory the thrust acceleration is smaller than the gravity pull by both the earth and the moon. Therefore, the two bodies gravity field must be taken into account during at least part of the trajectory.

In this paper the study of two problems is presented: Minimum fuel transfer to Lagrange points L_4 or L_5 from a low circular Earth orbit, and minimum fuel Earth-Moon trajectories. In both cases low-thrust limited-power engines are used with a typical level of $10^{-3}G$ of thrust acceleration.

* Graduate Student, Department of Aeronautics and Astronautics, Stanford University, Stanford, California 94305.

† Professor, Department of Aeronautics and Astronautics, Stanford University, Stanford, California 94305.

Only the planar case, where the motion is in the plane of the lunar orbit, is studied. The first problem is easier since large distance from the moon is kept during the entire transfer. An approximate solution is obtained using the earth gravity field alone, and then a correction for the moon effect is easily found. In the Earth-Moon problem the approach is to consider two separate trajectories; one spiraling away from Earth and the other spiraling into the moon, and to match them at some intermediate point. The trajectory to the Lagrange point serves as a first guess for the earth spiral in the iterative procedure.

In chapter 2 the electric engines equations are introduced and the optimality criterion is formulated. The optimal transfer to the Lagrange points is discussed in chapter 3, and results for L_5 are shown. Chapter 4 describes the solution of the Earth-Moon problem. Some results are given. Finally, chapter 5 shows how to circumvent the earth oblateness effect.

2. THE OPTIMALITY CRITERION

Electric engines generate thrust by accelerating charged particles through an electromagnetic field. It is possible, in principle, to modify both the ejection speed c and the mass flow rate $\dot{m}$, and therefore to control independently the thrust and power, which are known to be:

$$T = |\dot{m}|c \tag{1}$$

$$P = \frac{1}{2}|\dot{m}|c^2 \tag{2}$$

These relations lead to:

$$\frac{1}{m_f} - \frac{1}{m_o} = \int_o^{t_f} \frac{a^2}{2P}\,dt \tag{3}$$

where m is mass and $a = \frac{T}{m}$ is the thrust acceleration.

To minimize the fuel (same as to maximize final mass m_f) it is therefore required to operate with maximum power during the entire transfer, and to minimize the quadratic performance index:

$$J = \frac{1}{2}\int_o^{t_f} a^2\,dt \tag{4}$$

for fixed t_f and specified initial and final orbits.

The variable acceleration level will depend on the allotted time t_f as well as on the initial and final orbits. Not surprisingly, J is a decreasing function of the specified t_f as are the initial and subsequent accelerations. In our numerical examples we shall choose a particular initial acceleration $10^{-3}G$, typical of some engines, and investigate the transfer trajectory and its duration.

3. TRANSFER TO LAGRANGE POINTS L_4/L_5

The Hamiltonian of the optimization problem is

$$H = -\frac{1}{2}a^2 + \lambda_{\vec{r}_1} \cdot \vec{v}_1 + \lambda_{\vec{v}_1} \cdot [\nabla U(\vec{r}_1) + \vec{a}] \tag{5}$$

where $\vec{r}_1$ is a radius vector measured from Earth to the satellite, $\vec{v}_1$ is velocity relative to Earth, $\vec{\lambda} = (\lambda_{\vec{r}_1}, \lambda_{\vec{v}_1})$ is the adjoint vector, which is also the sensitivity of the cost-to-go J to a change in the state $(\vec{r}_1, \vec{v}_1)$. The gravity acceleration due to Earth and Moon is obtained by

$$\nabla U(\vec{r}_1) = -\frac{\mu_1 \vec{r}_1}{r_1^3} - \mu_2 \left(\frac{\vec{r}_2}{r_2^3} - \frac{\vec{r}_{21}}{r_{21}^3} \right) \tag{6}$$

$\vec{r}_2$ is a radius vector measured from the moon to the satellite and $\vec{r}_{21} = \vec{r}_2 - \vec{r}_1$, a radius vector from Moon to Earth, is considered as a fixed length vector. Its direction is time dependent and therefore the Hamiltonian is not constant.

From the necessary conditions for optimality, the sensitivity to velocity (sometimes called "the primer vector") is:

$$\lambda_{\vec{v}_1} = \vec{a} \tag{7}$$

The adjoint equations are:

$$\dot{\lambda}_{\vec{r}_1} = -\frac{\partial H}{\partial \vec{r}_1} \tag{8}$$

$$\dot{\vec{a}} = -\frac{\partial H}{\partial \vec{v}_1} \tag{9}$$

We consider the planar problem (i.e., motion in the plane of the moon orbit around earth), and a departure from a low earth circular parking orbit.

The initial radius and velocity vector are known. The initial acceleration magnitude has been chosen. Another important quantity is the vector sensitivity to rotation defined by

$$\vec{\lambda}_\theta = \vec{r} \times \lambda_{\vec{r}} + \vec{v} \times \lambda_{\vec{v}} \tag{10}$$

For the planar problem, the optimal solution must be stationary with respect to the initial angular position. Therefore, the initial sensitivity to rotation around the earth must vanish. We are thus left with 3 unknown initial conditions, which can be taken as a_{ro}, λ_{ro} and the departure angle θ_{1o}. The final state at the Lagrange points L_4 or L_5 is known. (The velocity is the circular velocity at the moon orbit around Earth). Thus, a solution to the problem can be found by a search over 3 unknown initial values, for a trajectory which satisfies the end conditions.

This procedure can be simplified considerably, if we first find an approximate solution, by neglecting the effect of the moon's gravity field. In this simplified problem the gravity field is spherically symmetrical around Earth, and as a result, the sensitivity λ_θ to rotation vanishes everywhere. The departure angle doesn't enter into the computation, and we need to find only the remaining 2 unknown initial conditions.

The search procedure can be started from an approximate analytical description of a spiral trajectory in a central gravity field for a low thrust spacecraft, which is obtained in powers of r with correction terms which allow the spiral to be slightly elliptical.

This solution was first introduced by Breakwell and Rauch [1]. It takes the form:

$$v_\theta = \sqrt{\frac{\mu_1}{r_1}}\left(1 + \frac{5}{2}\rho^4 + \rho^2 \Delta v_\theta\right) \tag{11}$$

$$v_r = 2\sqrt{\frac{\mu_1}{r_1}}\rho^2\left(1 - 20\rho^4 + \Delta v_r\right) \tag{12}$$

$$a_\theta = a_o\left(1 + \frac{5}{2}\rho^4 + \rho^2 \Delta a_\theta\right) \tag{13}$$

$$a_r = a_o\left(1 - 30\rho^4 + \Delta a_r\right) \tag{14}$$

$\rho = r_1\sqrt{\frac{a_o}{\mu_1}}$, a normalized radius, is unity at a distance where the gravity acceleration equals a_o. The perturbation terms are:

$$\Delta v_\theta = \sum_{i=1}^{2} B_i \rho^{-m_i} \sin(\theta + \beta_i) \tag{15}$$

$$\Delta v_r = \sum_{i=1}^{2} B_i \rho^{-m_i} \cos(\theta + \beta_i) \tag{16}$$

$$\Delta a_\theta = \frac{1}{5}\sum_{i=1}^{2}(42 - 16m_i) B_i \rho^{-m_i} \sin(\theta + \beta_i) \tag{17}$$

$$\Delta a_r = -\frac{1}{5}\sum_{i=1}^{2}(26 - 8m_i) B_i \rho^{-m_i} \cos(\theta + \beta_i) \tag{18}$$

where $m_1 = \frac{3}{2}, m_2 = 2$ and the angular position θ satisfies $\dot{\theta} = \frac{v_\theta}{r}$.

In the case of transfer between 2 circular orbits, the 4 unknown parameters B_i, β_i $(i = 1, 2)$, are obtained from the known initial and final velocity vectors.

This solution gives a good approximation for $\rho < 0.1$. In our case, for $a_o = 10^{-3}G$ and initial parking orbit at $200km$ above sea level, $\rho_o = .0329$ and $\rho_f = 1.9254$.

A starting solution for the search procedure is obtained from this approximation by choosing $\rho_f = 0.1$. a_{ro} is given by eq. 14. From eqs. 9, 7 and 5, we get:

$$\dot{a}_r = -\lambda_r + a_\theta \frac{v_\theta}{r} \qquad (19)$$

which determines λ_{ro}, since the initial rate of a_r in eq. 14 is easily found.

Having found this initial trajectory, we can start a search procedure in which the radius of the final orbit is gradually increased until the required orbit is achieved. At each step in this procedure, we start from an initial guess of the missing initial conditions taken from the previous transfer. A two parameter search based on linear extrapolations is carried out until convergence is achieved, to a specified degree of accuracy.

When the moon's gravity is included, a 3 parameter search, using the same technique of linear extrapolations, can be carried out, starting with the result of the moon-free problem as an initial guess. The search procedure will converge only if the starting trajectory is close enough to the optimal one. For the given level of acceleration this proves to be the case, since the distance from the moon is large during the entire transfer.

The results for L_4 and L_5 are quite close. The cost difference between the two trajectories amount to less than 1%. For this reason only one trajectory, to L_5, is shown.

The trajectory is shown in figure 1 (the unit is Earth radius R_E). Upon arrival to the libration point, the satellite has completed over 36 revolutions around Earth and the transfer time is 10 days. For $P_{max} = 1MWatt$ and specific impulse of 5000sec, the amount of fuel spent during the transfer is 16% of the gross weight.

Figure 2 shows the evolution in time of the distance from Earth, and of the semi-major axis of the osculating orbit along the transfer. It is interesting to note that in the final revolution there is an overshoot in the semi-major axis. This is due to the excess of radial velocity built up in order to bring the satellite from the neighborhood of Earth to the final orbit. This feature would disappear in the case of a lower final orbit, or a lower level of acceleration.

The elliptical character of the spiral is demonstrated by the periodic behaviour of the velocity and acceleration components in figures 3 and 4. As expected, the radial velocity diminishes at the end, when the libration point is reached. The magnitude of the acceleration vector varies between $.1a_o$ and $1.75a_o$.

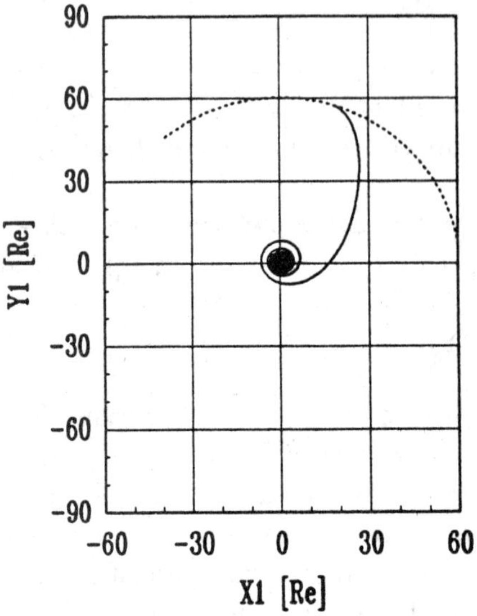

Fig.1 Earth-L_5 trajectory.

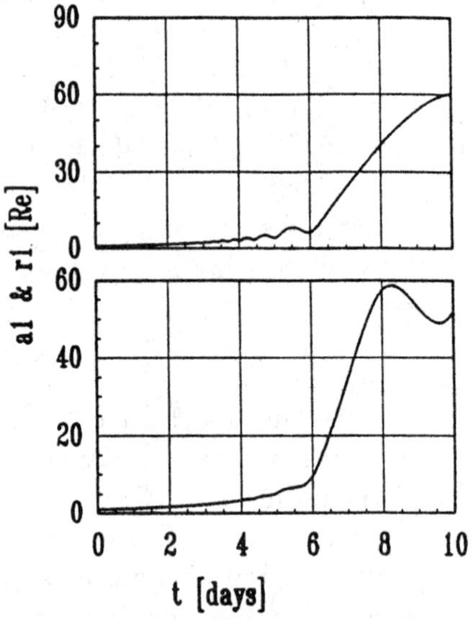

Fig.2 Radius and the semi-major axis along Earth-L_5 trajectory.

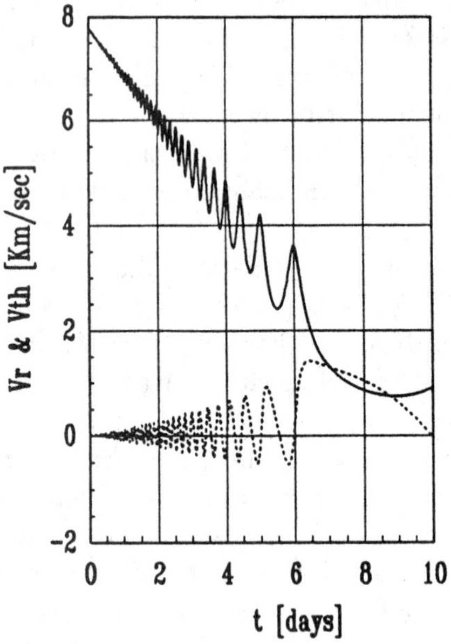

Fig.3 Velocity relative to Earth along Earth-L_5 transfer.

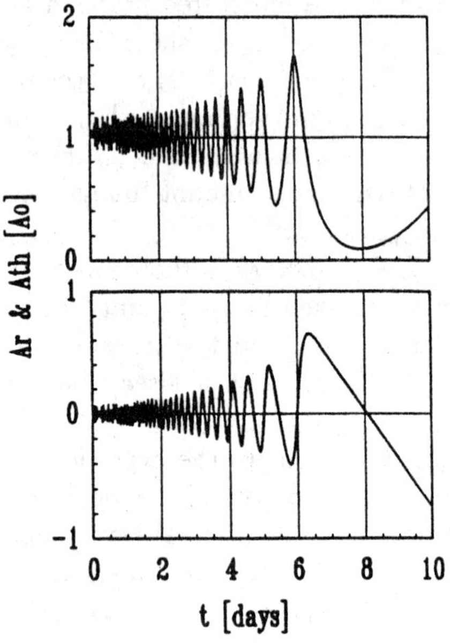

Fig.4 Acceleration along Earth-L_5 transfer.

4. TRANSFER TO THE MOON

In this problem it is more convenient to look at the equations in the rotating frame of the Earth-Moon system. The Hamiltonian is:

$$H = -\frac{1}{2}a^2 + \lambda_{\vec{r}} \cdot \vec{v} + \lambda_{\vec{v}} \cdot [\nabla(U_1 + U_2 + U_c) - 2\vec{n} \times \vec{v} + \vec{a}] \qquad (20)$$

$\vec{v}$ is measured relative to the rotating frame, $\vec{n}$ is the angular rate of the Earth-Moon system, U_1 and U_2 are the gravity potentials of the earth and the moon, and U_c is a centrifugal "potential" $U_c = \frac{1}{2}n^2r^2$; r is measured from the Barycenter.

The Hamiltonian in this problem does not depend explicitly on time and therefore is a constant. As before, the necessary conditions lead to $\lambda_{\vec{v}} = \vec{a}$. The intial and final conditions are circular orbits with specified radius around Earth and Moon. The sensitivities to rotation around Earth and Moon at the end points vanish in the optimal case.

Using these conditions, it is straightforward to show that at the initial point the Hamiltonian becomes:

$$H = \frac{1}{2}a^2 + \vec{a} \cdot \left(\nabla U_2 - \frac{\mu_2 \vec{r}_{21}}{r_{21}^3}\right) \qquad (21)$$

At the final point the indices 1 and 2 will interchange.

Due to the constancy of the Hamiltonian and for the chosen value of $10^{-3}G$ for the initial acceleration magnitude, the final acceleration magnitude is determined; it is almost the same (differs by less than 1%). There remain 6 unknown end conditions which can be chosen as: a_r, λ_r, θ at both ends of the trajectory, using polar coordinates relative to the nearer body.

The optimal initial conditions could be found using a shooting method. However, as could be expected, the problem is extremely sensitive; small changes at the departure conditions close to Earth will cause large changes at the vicinity of the moon.

A more promising approach is to integrate from both ends of the trajectory and match the two portions of the solution at some intermediate point, which will not be as sensitive to the initial parameters. A convenient choice is Laplace's sphere of influence, which is the surface on which the relative perturbations of the moon and earth gravity fields are of the same magnitude.

The surface is approximately a sphere with radius:

$$r_s = r_{12}\left(\frac{m_2}{m_1}\right)^{\frac{2}{5}} \tag{22}$$

In the planar case, there are 7 quantities that need to be matched. These are the position on the sphere of influence, the velocity vector, the acceleration and $\lambda_{\bar{r}}$ which is related to the rate of acceleration. Due to the constancy of the Hamiltonian, it is only necessary to match 6 of the 7 quantities.

The matching is carried out in two steps, starting with a crude matching, by examining Earth and Moon spirals in central gravity fields, i.e., neglecting the gravity effect of the other body. Since the sensitivity to rotation is zero in a spherically symmetric gravity field, the position and velocity direction on the sphere of influence can be matched using geometrical considerations, yielding the initial and final angular positions. It is reasonable to expect that the earth spiral would be close to the one obtained in the libration point problem. That solution can be therefore used as a starting guess. The velocity magnitudes on the sphere of influence of both spirals are close. Adjustment of the parameters λ_r, a_r at the moon can improve the matching of $\vec{a}$. With the full gravity fields included, very precise matching of the 6 quantities has not yet been completed, but figure 5 shows an approximate trajectory in the Earth-Moon frame. The transfer time between a 200km Eastward circular Earth orbit to a 100km Westward circular Moon orbit is 10.6 days. Both initial and final orbits are assumed to lie in the plane of the lunar orbit. The final part of the trajectory almost completes 6 revolutions while spiraling into the moon. For $P_{max} = 1MWatt$ and specific impulse of 5000sec, the amount of fuel spent during the transfer is about 19% of the gross weight.

Another possible solution, terminating on an Eastward Moon orbit, has been considered. However, the result of the approximate match gives a trajectory with a higher cost by a few percent, and therefore this solution seems to be inferior.

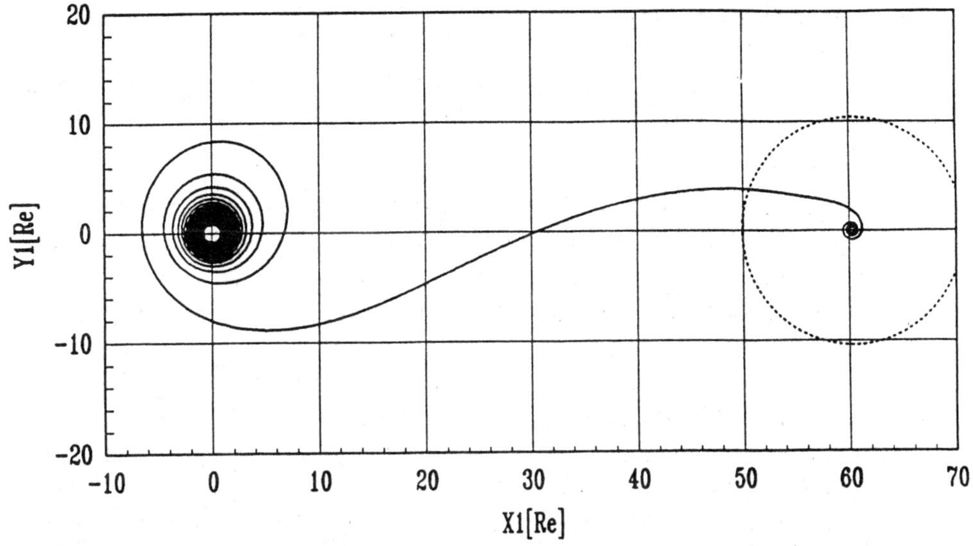

Fig.5 Earth-Moon trajectory, and the sphere of influence.

5. THE EARTH OBLATENESS EFFECT

The only significant effect of the earth oblateness is the out-of-plane motion, described by the rate of change of the right ascension Ω of the ascending node:

$$\dot{\Omega} = -\frac{3}{2}\sqrt{\frac{\mu_1}{r_1^3}}J_2\left(\frac{R_E}{r_1}\right)^2 \cos I \tag{23}$$

J_2 is the oblateness coefficient, R_E is the Earth radius and I is the inclination with respect to the equator. Since this effect dies out rapidly with the distance from Earth, the approximate radial velocity for the near earth spiral (eq. 12) can be used in order to get:

$$\Delta\Omega \approx -\frac{3}{16}\rho_E^{-2}J_2\left(\frac{R_E}{r_{1o}}\right)^4 \cos I \tag{24}$$

where $\rho_E = R_E\sqrt{\frac{a_o}{\mu_1}}$. It is possible to circumvent this effect by starting out with a lead in Ω so that eventually the satellite will reach the lunar plane. This lead angle assumes a value between 8.8° and 9.5°, depending on the inclination of the moon's orbit above the ecliptic at the time of departure.

The in-plane effect of the Earth oblateness, the rotation of the argument of perigee, is negligeable for the near circular spiral.

5. SUMMARY

Minimum fuel trajectories to the Earth-Moon libration points L_4 and L_5 and to the moon, for a low-thrust limited-power spacecraft, were presented. The level of thrust acceleration considered is of the order of $10^{-3}G$. In the second problem a solution is obtained by matching two trajectories, one coming from the earth and the other from the moon, on the sphere of influence. The trajectory to the libration point is used as a starting guess in a search procedure for the earth-moon problem.

The trajectories considered are planar, i.e., all motions are in the lunar's orbit plane. The present approach can be easily extended to handle 3 dimensional lunar trajectories. Such trajectories are required, for example, to achieve a polar Moon's orbit.

The level of thrust acceleration is a sensitive parameter. For higher thrust levels (1-2 orders of magnitude) the approach taken in [1] for interplanetary missions, could be applied. For lower levels of thrust acceleration, the satellite may come close to the moon before the final revolution, to the extent that the method used here, of first obtaining an approximate solution without the moon, may not be adequate. These problems were not examined so far.

Acknowledgement:
This research is sponsored by NASA grant NAG 3-286.

References:

1. J.V.Breakwell, H.E.Rauch, "Asymptotic Matching in Power-Limited Interplanetary Transfers", Space Flight Mechanics Specialist Symposium, AAS Science and Technology Series,Vol.11,1967,pp.281-300.

2. W.G.Melbourne, C.G.Sauer,Jr. "On Performance Computations with Pieced Solutions of Planetocentric and Heliocentric trajectories for Low Thrust Missions", Space Program Summary 37-36 Vol IV, 1965, JPL Pasadena

AAS 89-352

MINIMUM FUEL TRAJECTORY FOR THE AEROSPACE-PLANE

Anne Sauvageot*, Oded Golan* and John V. Breakwell†

The aerospace-plane will take off as a conventional aircraft, fly in the atmosphere up to hypersonic speeds for long durations, subsequently reaching a circular near-earth orbit. The purpose of this study is to minimize the fuel consumption and therefore to determine how to use the four different engines in the most efficient way: turbojet, ramjet, scramjet and rocket.
The usual assumption of a constant mass can no longer be used, because about 50% of the mass is fuel. The general problem is simplified by assuming that the flight-path angle γ will remain small along the trajectory. Subsequently the optimal atmospheric flight is obtained by maximizing the rate of change of the energy with respect to the mass ratio.
The flight in vacuum to orbit consists of a finite burn phase to get to a desired elliptical orbit, a coasting arc along the ellipse and then a short additional thrusting phase to circularize the orbit. The transfer ellipse has been shown to have its apogee on the final orbit.
An overall trajectory is presented for a reference configuration with an initial weight of 234 tons.

INTRODUCTION

The aerospace-plane is a single-stage-to-orbit vehicle which uses a combination of four different engines: turbojet, ramjet, scramjet and rocket engine. The use of airbreathing engines results in a substantial decrease in propellant consumption compared to an all-rocket propelled vehicle but is heavily penalized by the added weights of the ramjet and scramjet engines and structural reinforcements to resist higher aerothermodynamic loads.

The aerospace-plane takes off like a conventional aircraft; however the usual assumption of a constant mass can no longer be used: about 50% of the take-off weight is fuel. The plane is subject to state constraints, such as temperature limit, dynamic pressure limit and acceleration peak. The purpose of this study is to minimize the fuel con-

* Graduate Student, Department of Aeronautics and Astronautics, Stanford University, Stanford, California 94305.

† Professor, Department of Aeronautics and Astronautics, Stanford University, Stanford, California 94305.

sumption and therefore to determine how to use the different engines in the most efficient way. The trajectory is divided into two parts:
- the atmospheric flight deals with the part of the trajectory in the thicker atmosphere, where aerodynamic forces (lift and drag) are significant and airbreather engines may be used.
- the flight to orbit,essentially in vacuum where the airbreather engines are no longer efficient.

ATMOSPHERIC FLIGHT

The flight is assumed to take place in the equatorial plane.The aerospaceplane is subject to gravity, thrust, lift and drag. The dynamic equations in a relative velocity coordinate system are :

$$\dot{v} = \frac{T\cos\alpha - D}{m} - g\sin\gamma \tag{1}$$

$$v\dot{\gamma} = \frac{T\sin\alpha + L}{m}\cos\sigma - g\cos\gamma + \frac{v^2}{R+h}\cos\gamma \tag{2}$$

$$\dot{h} = v\sin\gamma \tag{3}$$

where:
v is the velocity relative to earth
R is the radius of the earth
h is the altitude above sea level
g is the gravity acceleration approximately equal to the gravity acceleration at sea level $g_0 = 9.8\ m/sec^2$
γ is the flight path angle
α is the angle of attack
σ is the bank angle
L is the lift
D is the drag
T is the thrust.

The thrust is assumed to act along the vehicle body axis.

A scalar energy function can be defined as:

$$E = \frac{v^2}{2} + gh \tag{4}$$

Taking the derivative of equation 4 and using equations 1 and 3 gives:

$$\dot{E} = v \frac{T \cos \alpha - D}{m} \tag{5}$$

The equations can therefore be expressed in terms of E instead of v; equation 1 can be replaced by equation 5 and equation 4 gives v.
As thrust is needed all along the trajectory at least to balance drag, the independent variable can be changed to a function of mass:

$$E' = \frac{dE}{d(ln \frac{m_0}{m})} = m \frac{\dot{E}}{|\dot{m}|} \tag{6}$$

$$h' = \frac{dh}{d(ln \frac{m_0}{m})} = m \frac{\dot{h}}{|\dot{m}|} \tag{7}$$

Using equations 3 and 5:

$$E' = v \frac{T \cos \alpha - D}{|\dot{m}|} \tag{8}$$

$$h' = \frac{mv \sin \gamma}{|\dot{m}|} \tag{9}$$

Trajectory Optimization

Ascent performance optimization requires the solution of an optimal control problem. Necessary conditions for optimality arc rather easy to specify by use of the calculus of variations. However, numerical computation of the optimal trajectory can be a time-consuming task, involving iterative computation of a non-linear, two-point boundary value problem. The purpose of this study is to get a trajectory reasonably close to the optimal solution which can be used as a first guess in the iterative procedure just mentioned.
Airbreathing launch vehicles show a tendency for low ascent flight profiles in order to trade efficient use of airbreathing propulsion performance potential against increasing drag losses; therefore the flight-path angle is assumed to remain small along the trajectory: $\cos \gamma$ is taken to be 1 and $\frac{d\gamma}{dt}$ is taken to be 0 in the previous dynamic equations.

The complete set of equations is now:

$$E' = v\frac{T\cos\alpha - D}{|\dot{m}|} \tag{10}$$

$$h' = \frac{mv\sin\gamma}{|\dot{m}|} \tag{11}$$

$$\frac{T\sin\alpha + L}{m}\cos\sigma - g + \frac{v^2}{R+h} = 0 \tag{12}$$

In this set of equations, E and h are the states of the system, whereas T the level of thrust and γ the flight-path angle are control variables. The angle of attack α is usually a function of the other variables and is determined by equilibrium in the vertical plane. (equation 12 with $\sigma = 0$). However in the ram-scram mode, α may be a control variable; for a ramjet or a scramjet the level of thrust depends on the angle of attack and at some stage a higher angle of attack increases the level of thrust and may increase the efficiency. Therefore roll modulation which allows the use of higher angles of attack is used: α is a control variable and σ is determined by equation 12; the roll angle alternates rapidly between $\pm\sigma$.

Optimization Problem

The objective is to reach a prescribed circular orbit with minimum fuel consumption. It is also equivalent to minimize:

$$\int_{\ln(\frac{m_0}{m})_0}^{\ln(\frac{m_0}{m})_f} d(\ln\frac{m_0}{m}) \tag{13}$$

The objective function may be written as:

$$J = \int_0^{\ln(\frac{m_0}{m_f})} d(\ln(\frac{m_0}{m})) \tag{14}$$

The Hamiltonian is defined by:

$$H = -1 + \lambda_E\frac{v}{|\dot{m}|}(T\cos\alpha - D) + \lambda_h\frac{m}{|\dot{m}|}v\sin\gamma \tag{15}$$

The equations for the Lagrange multipliers are:

$$\lambda'_E = -\frac{\partial H}{\partial E} \tag{16}$$

$$\lambda'_h = -\frac{\partial H}{\partial h} \tag{17}$$

Since the Hamiltonian is linear in $\sin\gamma$, from the necessary conditions either $\sin\gamma = +/-1$ or $\lambda_h = 0$. As γ is expected to be small, the solution will lie on a singular arc on which λ_h remains zero; H may now be written as:

$$H = -1 + \lambda_E \frac{v}{|\dot{m}|}(T\cos\alpha - D) \tag{18}$$

$$\lambda'_E = -\lambda_E \frac{\partial}{\partial E}\{\frac{v}{|\dot{m}|}(T\cos\alpha - D)\} = -\lambda_E \frac{\partial E'}{\partial E} \tag{19}$$

$$\lambda'_h = -\lambda_E \frac{\partial}{\partial h}\{\frac{v}{|\dot{m}|}(T\cos\alpha - D)\} = -\lambda_E \frac{\partial E'}{\partial h} = 0 \tag{20}$$

Necessary conditions on alpha

When α is a control variable, it must be such that:

$$\frac{\partial H}{\partial \alpha} = \lambda_E \frac{\partial E'}{\partial \alpha} = \lambda_E v \frac{\partial}{\partial \alpha}\{\frac{T\cos\alpha - D}{\dot{m}}\} = 0 \tag{21}$$

This equation can be solved for α by searching extremum for the function $\frac{T\cos\alpha - D}{\dot{m}}$ at each step (holding h and E constant).

Determination of $\sin\gamma$

The flight-path angle γ is determined by the singular arc condition $\lambda_h = 0$. Equation 20 implies that h should be such that E′ is maximized; therefore h can be selected by a numerical search holding E constant and $\sin\gamma$ solved from equation 11, after determining h′ numerically.

Aerodynamic Model

The lift and drag are related to the lift coefficient C_L and drag coefficient C_D.

$$L = \frac{1}{2}\rho v^2 S_{ref} C_L \tag{22}$$

$$D = \frac{1}{2}\rho v^2 S_{ref} C_D \tag{23}$$

where S_{ref} is a reference area and ρ is the density.

The dependence of the coefficients on angle of attack α and on Mach number M has been provided by NASA Langley Research Center for a reference vehicle. The speed of sound involved in calculating M is obtained from a continuous atmospheric model derived from U.S. Standard Atmosphere. The lift coefficient $C_L(\alpha, M)$ is calculated from $C_{L_\alpha}\alpha$ where $C_{L_\alpha}(M)$ is represented by a cubic spline function; the drag coefficient $C_D(\alpha, M)$ is obtained from $C_{D_0} + K\alpha^2$ where both $C_{D_0}(M)$ and K(M) are represented by cubic spline functions which approximate the aerodynamic data input of the reference vehicle.

Propulsion Model

Turbojet.

The thrust T(M,h) and the fuel rate $|\dot{m}|(M, h)$ are characterized by a strong, highly nonlinear sensitivity to the flight conditions M and h. The properties of a reference turbojet engine module determined by NASA Lewis Research Center are used to define a turbojet model by means of bicubic spline functions. The number of turbojet modules is chosen according to an initial thrust loading of 0.7.
An afterburner may be switched on, which increases the level of thrust and the level of fuel rate significantly: it is very useful when the aerospace-plane has to go through the sonic barrier where the drag is much higher.

Ramjet-Scramjet

This engine can be used from Mach 3 to Mach 25 and limits due to maximum engine pressure (10 atm) and minimum combustor pressure (4.5 psia) determine a corridor in the (Mach,altitude) plane within which the plane is allowed to fly.

The thrust T and the fuel rate $|\dot{m}|$ are functions of the Mach number M, the altitude h, the angle of attack α and the fuel-air ratio ϕ. For a given combination of Mach number,altitude and angle of attack, the thrust can be approximated by a piecewise linear function of ϕ; $|\dot{m}|$ is linear with ϕ. For a given combination of Mach number, altitude and ϕ, the thrust and the fuel rate can be approximated by linear functions of the angle of attack α. Different bicubic splines in the (M,h) plane were obtained for different reference ϕ's and α's and interpolation between the different splines was then used to evaluate the level of thrust and the fuel rate at a given point.

Initial Conditions

The initial weight at take-off is 234 tons. A mass breakdown at initial T/W = 0.7 is given in Table (1). The initial conditions are v=0 and h=0. The best policy is to integrate the equations forward keeping h=0 until the singular arc is met. Then the plane will take off.

Table 1: Mass estimate of some major subsysytems.

SUBSYSTEM:	MASS (POUNDS):
WEIGHT OF WING GROUP	5932.
WEIGHT OF TAIL GROUP	4009.
WEIGHT OF BODY GROUP	23054.
WEIGHT OF THERMAL PROTECTION SYSTEM	10841.
WEIGHT OF LANDING GEAR	7669.
WEIGHT OF TURBOJETS	30093.
WEIGHT OF RAM-SCRAMJETS	148500.
WEIGHT OF RCS PROPULSION SYSTEM	2286.
WEIGHT OF PRIME POWER SYSTEM	3683.
WEIGHT OF ELECTRICAL SYSTEM	12558.
WEIGHT OF SURFACE CONTROLS	1099.
WEIGHT OF AVIONICS	1606.
WEIGHT OF ECS	912.
WEIGHT OF PERSONAL PROVISIONS	302.
MARGIN	16745.
DRY WEIGHT	269280.
WEIGHT OF PERSONNEL	600.
WEIGHT OF RESIDUAL FLUIDS	1863.
WEIGHT OF PAYLOAD	2500.
OMS & RCS PROPELLANT	1936.
FUEL (LIQ. HYDROGEN)	228195.
OXIDIZER (LIQ. OXYGEN)	11498.
GROSS WEIGHT	515881. POUNDS (234 TONS)

CONFIGURATION:	
FORWARD CONE LENGTH	158 FEET
REFERENCE AREA	2769 $FEET^2$

RESULTS

The values quoted and presented in the figures that follow, can only be regarded as an indicator to relative performance i.e. they are qualitative results.

The trajectory is constrained by a maximum axial acceleration of 3 g's and a dynamic pressure limit was imposed, nominally 2000 lb_f/ft^2. Eastward equatorial launch to low Earth orbit was assumed. (see figure 1, 2, 3 and 4)

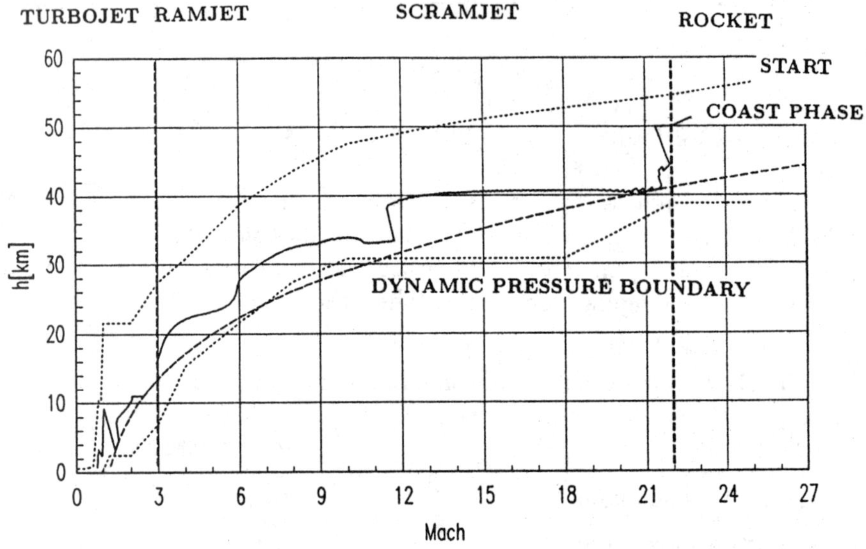

Figure 1.

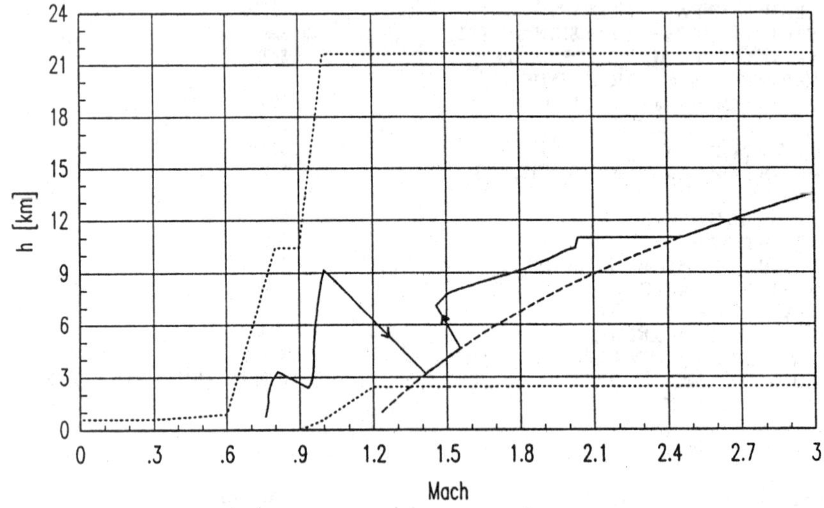

Figure 1(a)

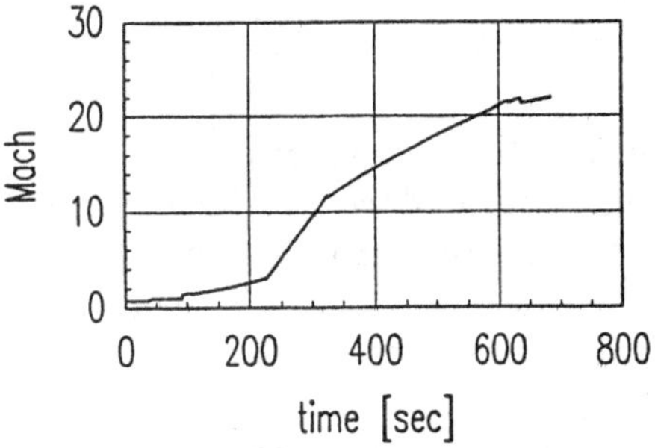

Figure 2(a).

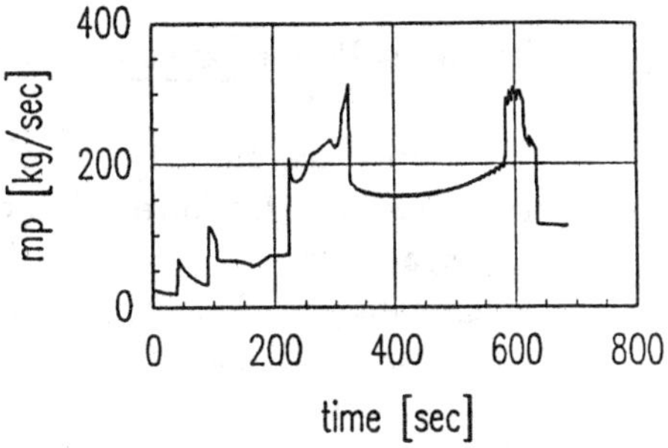

Figure 2(b).

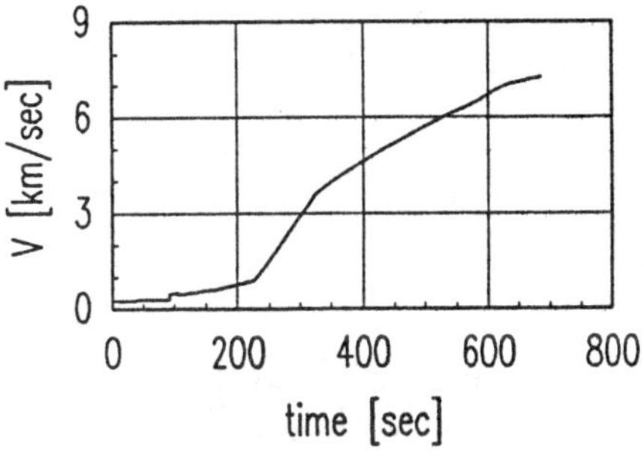

Figure 2(c).

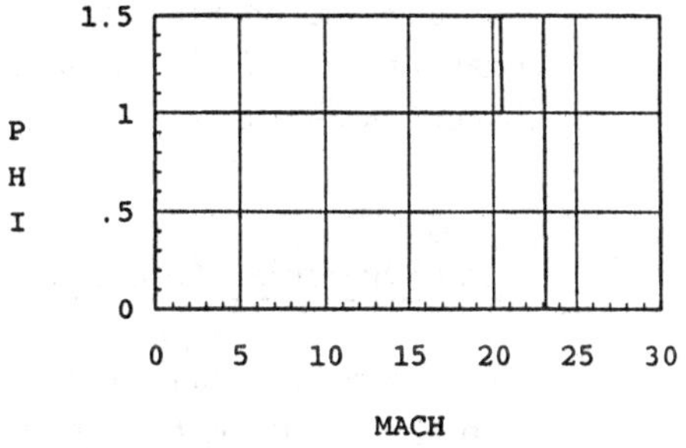

Figure 2(d).

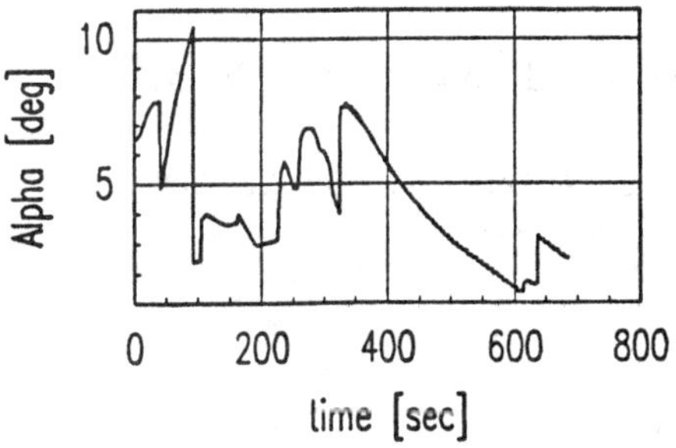

Figure 3(a).

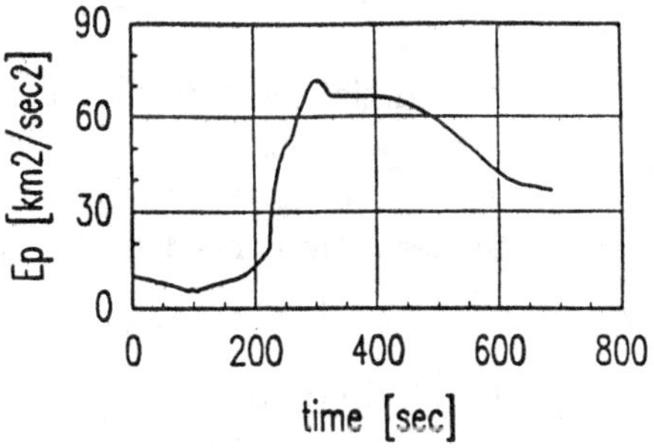

Figure 3(b).

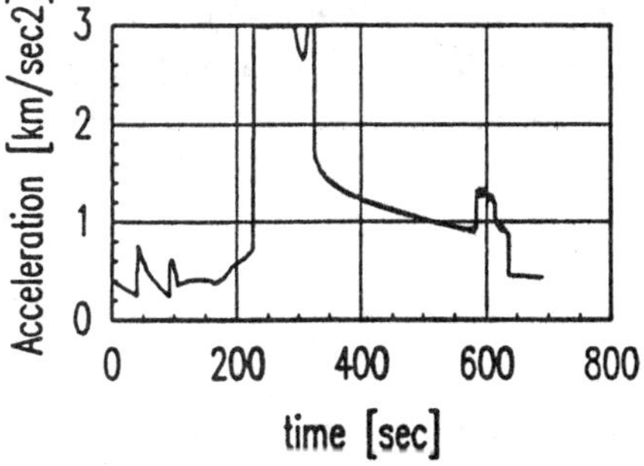

Figure 3(c).

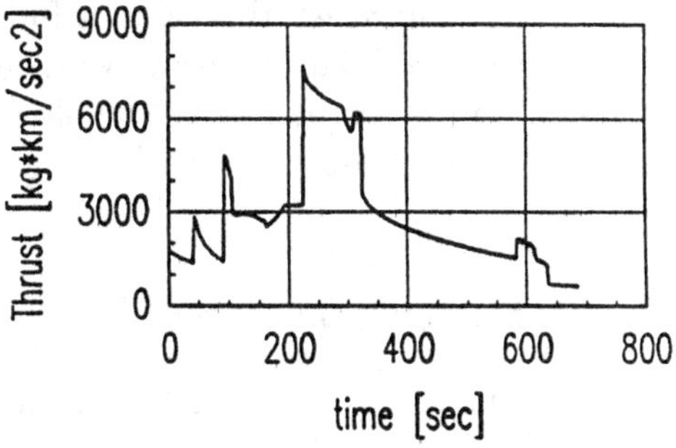

Figure 3(d).

Turbojet Performance.

The aerospace-plane takes off at Mach 0.75 with an angle of attack of 6.5 deg. The afterburner is switched on just before Mach 1 to help the aerospace-plane to go through the sonic barrier; the plane has to dive significantly at Mach 1 due to heavy drag: the thrust level is higher at a lower altitude. Once the aerodynamic loads decrease, the plane climbs up again.
The trajectory flattens at 11 km: this feature is related to the speed of sound which stops decreasing with altitude and levels off; E′ is maximized at the same altitude for different energy levels.
The acceleration remains below 3 g's; however the dynamic pressure constraint is met and the plane will fly along this boundary from Mach 2.3 on.(see figure 1(a))

Ramjet-Scramjet Performance.

The ramjet is turned on at Mach 3 as soon as it can be operated: E′ from the ramjet is barely higher than E′ from the turbojet. The acceleration constraint is active at the beginning for about 100 sec from Mach 3 to 9, whereas the q limit is met later at around Mach 20. The trajectory flattens at 50 km: at that point the speed of sound presents a peak.
The angle of attack remains below 8 deg and during most of the flight is below 6 deg.
The stochiometric fuel-air ratio ($\phi = 1$) is used up to Mach 20, then a higher ratio is needed ($\phi = 1.5$). (see figure 2(d))

Roll Modulation.

α becomes a control variable when roll modulation is used. A trajectory including roll modulation was computed for the same configuration; however the gain in fuel consumption was not very significant: this modulation was advantageous only above Mach 19 and only one ton was saved over the whole trajectory; therefore roll modulation need not be taken into consideration.

FLIGHT TO ORBIT

The rocket engine is ignited at high altitudes where the airbreather engines are no longer efficient; this second flight phase experiences only minor aerodynamic effects due to high altitude and is controlled by

thrust vectoring. There is a finite burn time to reach a desired elliptical transfer orbit, a coasting arc along the ellipse and an impulsive change of velocity to circularize the orbit.
The optimal transfer ellipse can be shown to have its apogee on the final orbit. The optimization problem is a two-point boundary value problem with a final constraint which can be reduced to a two parameter search. The first initial guess is obtained from a near-Hohmann transfer case and then is improved by linear interpolation.
The rocket specific impulse was taken to be 500 seconds and the fuel rate 200 kg/s. The best option is to switch on the rocket engine at Mach 22 for a short time: 28 seconds.

DISCUSSION

Variation of assumptions:

The results presented here are dependent on many specific variables used in the calculations. For example:
(1) Performance for turbojets and scramjets is influenced by the q limit used.
(2) The data used for the scramjet might be optimistic especially in the high Mach number range; in this case, the rocket should be ignited at a lower Mach number, e.g. around Mach 18.
(3) The flight path-angle remains reasonably small except at discontinuities in the computed h as well as in Mach number and α.(see the arrows in figure 1 and 1(a))
A full state treatment, with h and γ included as states, would require a smooth flight-path angle history throughout dive and climb.

CONCLUSION

An overall trajectory was computed for a single-stage-to-orbit vehicle using four different engines: turbojet,ramjet, scramjet and rocket engine. Optimization techniques using the assumption of a small flight-path angle have been applied to optimize ascent performance and to evaluate the effects of constraints. The computerized methods used avoid long iterative procedures: one numerical search gave the optimized trajectory which presents different features:
- (1) Dive around Mach 1 to go through the sonic barrier, and climb around Mach 1.5 and 11.5.
- (2) Level-off at 11 km and 50 km due to the changes in the vertical gradient of the speed of sound.

- (3) Dynamic pressure constraint (2000 $lb_f/feet^2$) active near Mach 2.5 and 20.
- (4) Maximum acceleration between Mach 3 and Mach 9.

Roll modulation was shown to save very little fuel during the trajectory.
Only a short burn phase was necessary to get to a transfer ellipse before circularizing the orbit.
The results suggest that the combined use of airbreathing propulsion and aerodynamic lift make the aerospace-plane a very feasible vehicle but the trajectory is heavily dependent on the ramjet-scramjet propulsion model, which needs to be verified.

ACKNOWLEDGEMENT

This research is sponsored by NASA grant NAG 3-908.

REFERENCES

1. N.X. Vinh, "Optimal Trajectories in Atmospheric Flight", Elsevier Scientific Publishing Company, Chapter 11.
2. F.W. Boltz, "Comparison of Horizontal and Vertical Launch into Orbit", AIAA Paper 87-2569
3. J.R. Olds, "A Conceptual Design for a Single-Stage-to-Orbit Space Station Service Vehicle", IAF-ST-87-07
4. G. Knip and J.L. Allen, "Analysis of Booster Systems with a Recoverable Hypersonic Airplane First Stage", AIAA Paper 64-543
5. G. Knip and J.D. Eisenberg, "Performance Estimates for Space Shuttle Vehicles Using a Hydrogen or a Methane-Fueled Turboramjet-Powered First Stage", NASA TN D-6634
6. L.C. Hartung, J.M. Karkow, W.L. Ordway, D.C. Pickett and A.D. Muras, "Orbit-on-Demand Vehicle Propelled by ATR/R Engines", AIAA Paper 86-1847
7. P.J. Waltrup, "Liquid-Fueled Supersonic Combustion Ramjets: A Research Perspective", AIAA Paper 86-0158
8. P.J. Waltrup, "Supersonic Combustion Ramjet(Scramjet) Engine Development in the United States" NASA TN
9. R.A. Jones and P.W. Huber, "Toward Scramjet Aircraft", Aeronautics and Astronautics February 1978
10. G.E. Dorrington, "Optimum Directions for Air-Breathing Launch Vehicle Design", AIAA Paper 87-1818
11. H.Kuczera and D.E. Koelle, "Propulsion Systems Optimization for a Two- Stage Winged Launcher System", IAF 88-246

AAS 89-354

OPTIMAL ORBITAL RENDEZVOUS USING HIGH AND LOW THRUST

Catherine A. Larson* and John E. Prussing†

Optimal control theory is used to examine a specific class of spacecraft trajectory problems where high- and low-thrust propulsion systems are utilized. These problems assume a spacecraft is in an established orbit about a planet. It is desired to execute an intercept of a pre-determined position in space in a specified amount of time using an optimal high-thrust program. The spacecraft then returns to the original orbit station using the low-thrust propulsion system in an optimal fashion. A minimum fuel solution is sought using the linearized equations of motion, known as the CW equations, which simplify the necessary computations. Solutions are obtained for problems with a fixed final time. However, for the time-open case, the optimal solution is for the final time to be infinite. With a weighted function of the final time in the performance index, a limited range of optimal single impulse solutions for the time -open case can also be found.

INTRODUCTION

Since the early 1950's much attention has been focused on computing optimal space trajectories for a variety of systems. Early work, represented by Lawden[1], Edelbaum[2], and Prussing[3,4], concentrated on high-thrust or impulsive trajectories. Low-thrust or continuous thrust trajectories were examined as early as 1959 by Breakwell[5] and by Irving[6] and also by Lawden[7], Melbourne and Sauer[8], Billik[9], Hinz[10], and recently by Carter[12] and Widhalm[13]. Fewer studies exist for trajectories which utilize both high- and low-thrust. Two papers, one by Edelbaum[14] and the other by Vahlberg and Edelbaum[15] compare the performance of high-thrust, low-thrust, and dual-thrust programs by evaluating the payload

* Graduate Student, Department of Aeronautical and Astronautical Engineering, University of Illinois at Urbana-Champaign, 104 S. Mathews Ave., Urbana, Illinois 61801. Member AIAA.

† Professor of Aeronautical and Astronautical Engineering, University of Illinois at Urbana-Champaign, 104 S. Mathews Ave., Urbana, Illinois 61801. Senior Member AAS, Associate Fellow AIAA.

capacity for each type. Whereas these two papers examine the case when both high- and low-thrust systems are in use at the same time, the research presented here assumes a combination of the two types with one following the other.

In general spacecraft trajectories are classified into the two types previously mentioned: high-thrust and low-thrust. For high-thrust, the ratio of thrust acceleration to gravitational acceleration is typically greater than one. An example would be any engine/motor that can be used as a booster. Up to now, most space missions have had high-thrust programs. High-thrust trajectories have relatively short thrust times when compared to the time between thrusting. As a result, these high-thrust programs are approximated by impulsive thrusts. For low-thrust, the ratio of thrust acceleration to gravitational acceleration is on the order of less than one-tenth. Such low-thrust trajectories also tend to require longer thrusting times. Here, an optimal spacecraft trajectory is one which consumes the least amount of fuel. Therefore, it is necessary to determine first, the type of engine which is to be used for a particular trajectory, and second, how to measure the amount of fuel used by such an engine.

Spacecraft engines can be categorized as: constant specific impulse (CSI) and variable specific impulse (VSI) engines. Constant specific impulse (or equivalently constant exhaust velocity) engines are modulated by the mass-flow rate they produce. In general, they can be either high- or low-thrust where the high-thrust engines are predominantly impulsive. CSI engines that are low-thrust are sometimes referred to as "thrust-limited" because their mass flow rate has a maximum value which limits the amount of thrust the engine is capable of producing. VSI engines are primarily low-thrust and are controlled by varying the exhaust velocity of the spacecraft's engine. They are constrained by the amount of kinetic power contained in the exhaust propellant, therefore, these are referred to as "power-limited."

Typically, high thrust systems are used to boost a spacecraft into a low altitude orbit and then a low-thrust system is used for the remainder of the mission. However, for this research, it is assumed that the spacecraft is already in an established orbit, and that a high-thrust system is used to execute an intercept of a pre-determined position in space in a specified amount of time. The low-thrust system is then used to return the spacecraft to its original orbit station. This type of mission can help prolong the life of a satellite when used in collision avoidance, can be used to perform platform surveillance and scientific observation for an orbiting space station, and can assist in orbit correction maneuvers.

PROBLEM STATEMENT AND ANALYSIS

The objective of this research is to solve for the minimum fuel trajectory for an impulsive intercept followed by a low-thrust return on-station. Initially, the spacecraft executes the necessary impulse required to have the spacecraft intercept a pre-determined position in space in a specified amount of time. This single impulse is followed by a low-thrust, power-limited (variable specific impulse) rendezvous with the original orbit station. Using the necessary conditions for an optimal trajectory developed by Lawden[7], the optimal solution is obtained[16]. The problem is treated as having interior path constraints, where the pre-determined position, $\underline{r}_1$, at some intermediate time, t_1, is the constraint. Similar studies was conducted by Clifton[17], Murphy[18] and Widhalm[13]. The first two used an entirely impulsive thrust program and the third used all continuous thrust.

The computations necessary to obtain an optimal solution for this trajectory problem can be greatly simplified if the spacecraft's equations of motion are linear. This can be accomplished by linearizing a planet's inverse-square gravitational force about a local vertical coordinate frame. This approximation is valid as long as the distance of the spacecraft from the original orbit station is small compared with the orbital radius. The coordinate frame has its origin in a circular reference orbit which moves at a constant speed (Fig. 1). Independently developed by Wheelon19 and Clohessy and Wiltshire[20] the resulting equations are commonly referred to as the "CW equations." They are also referred to as "Hill's equations" because they belong to a class of

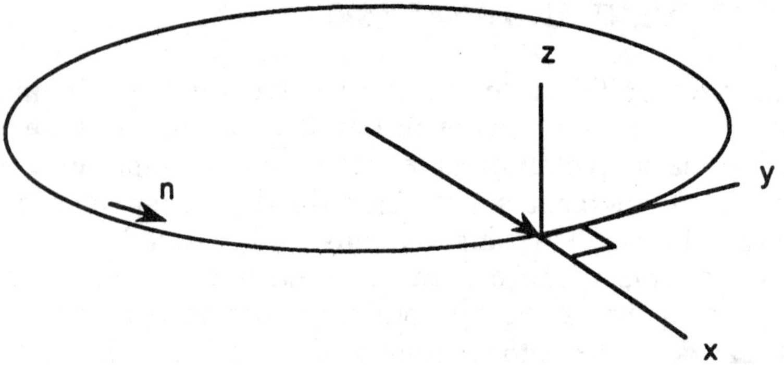

FIG. 1 Circular Reference Orbit

equations originally studied by Hill in the 19th century. The equations are

$$\ddot{x}(t) = 3n^2x(t) + 2n\dot{y}(t) + \Gamma_x$$
$$\ddot{y}(t) = -2n\dot{x}(t) + \Gamma_y$$
$$\ddot{z}(t) = -n^2z(t) + \Gamma_z \qquad (1)$$

where n is the mean motion of the reference frame, the variables x, y, and z, represent radial, circumferential and out-of-plane relative displacements, with the components of the thrust acceleration, Γ, in the direction of the displacements also included.

Let the mean motion, n, be equal to one, thereby changing the independent variable to nondimensional time and giving the reference frame orbit a period of 2π. The spacecraft's equations of motion can now be written as

$$\begin{pmatrix} \dot{\underline{r}} \\ \dot{\underline{v}} \end{pmatrix} = \begin{pmatrix} 0_3 & I_3 \\ A & B \end{pmatrix} \begin{pmatrix} \underline{r} \\ \underline{v} \end{pmatrix} + \begin{pmatrix} \underline{0} \\ \underline{\Gamma} \end{pmatrix} \qquad (2)$$

where 0_3 and I_3 are three-by-three zero and identity matrices, respectively, and A and B are given by

$$A = \begin{pmatrix} 3 & 0 & 0 \\ 0 & 0 & 0 \\ 0 & 0 & -1 \end{pmatrix} \qquad B = \begin{pmatrix} 0 & 2 & 0 \\ -2 & 0 & 0 \\ 0 & 0 & 0 \end{pmatrix} \tag{3}$$

from Eq. (1).

The performance index for this minimum fuel trajectory is the sum of the cost (fuel expended) for the impulsive segment and the cost for the low-thrust segment of the trajectory. Because both the impulsive and low-thrust parts measure performance independent of each other, the problem can be approached in segments. Each segment can then be related to the other using the interior path constraint.

The Impulsive Segment

The impulsive segment spans the time frame from zero to t_1 and from the point of origin to the intermediate position, $\underline{r}_1$. If the original orbit station is defined to be the origin of the local vertical coordinate frame, then the boundary conditions for this segment are:

$$\begin{aligned} &\underline{r}(0) = \underline{0} \qquad \underline{v}(0) = \underline{0} \\ &\underline{r}(t_1) = \underline{r}_1 \\ &t_1 \text{ specified} \end{aligned} \tag{4}$$

For an impulsive trajectory, the thrust acceleration is considered instantaneous, therefore, its contribution to the position and velocity state equations is also instantaneous. Thus, the solution to the unthrusted equations of motion, represents the spacecraft's motion immediately following an impulse and can be written as

$$\underline{X}(t) = \begin{pmatrix} \underline{r}(t) \\ \underline{v}(t) \end{pmatrix} = \Phi(t)\, \underline{X}_o^+ = \begin{pmatrix} M(t) & N(t) \\ S(t) & T(t) \end{pmatrix} \begin{pmatrix} \underline{r}_o^+ \\ \underline{v}_o^+ \end{pmatrix} \tag{5}$$

where $\underline{X}_o^+$, $\underline{r}_o^+$ and $\underline{v}_o^+$ are evaluated after the impulse at the initial time, $t_o = 0$, and $\Phi(t)$, is the transition matrix. Now, the performance index,

$$J = \int_{t_o}^{t_f} \Gamma \, dt \tag{6}$$

coupled with the above equations of motion results in

$$H = \Gamma + \underline{\lambda}_r^T \underline{v} + \underline{\lambda}_v^T [A\underline{r} + B\underline{v} + \underline{\Gamma}] \tag{7}$$

for the impulsive segment Hamiltonian. The adjoint equations are

$$\dot{\underline{\lambda}}_r^T = - \frac{\partial H}{\partial \underline{r}} = - \underline{\lambda}_v^T A \tag{8}$$

$$\dot{\underline{\lambda}}_v^T = - \frac{\partial H}{\partial \underline{v}} = - \underline{\lambda}_r^T - \underline{\lambda}_v^T B \tag{9}$$

where, if the primer vector is defined as in Lawden[7], and noting that $A^T = A$ and $B^T = - B$, the equation of the primer vector can be written

$$\ddot{\underline{p}}(t) = A \underline{p} + B \dot{\underline{p}} \tag{10}$$

which is the same form as the equations of motion, Eq. (2). The primer vector equation has a similar solution as given for $\underline{X}(t)$ in Eq. (5)

$$\begin{pmatrix} \underline{p}(t) \\ \dot{\underline{p}}(t) \end{pmatrix} = \Phi(t) \begin{pmatrix} \underline{p}_o \\ \dot{\underline{p}}_o \end{pmatrix} = \begin{pmatrix} M(t) & N(t) \\ S(t) & T(t) \end{pmatrix} \begin{pmatrix} \underline{r}_o \\ \dot{\underline{p}}_o \end{pmatrix} \tag{11}$$

and, to first order, is independent of the state. The primer history, therefore, is determined by the initial primer and primer-rate vectors. The initial primer vector is defined to be the unit vector in the direction of the initial impulse. The initial primer-rate is determined using the terminal boundary conditions in the low-thrust segment. This formulation differs from that of Jezewski[21].

It is given that an impulse occurs at time zero. Immediately following the impulse, the position is unchanged. However, the velocity just after the impulse becomes $\underline{v}_o^+$ and is found using Eq. (5) to be

$$\underline{r}(t_1) = \underline{r}_1 = M(t_1)\underline{r}_o + N(t_1)\underline{v}_o^+ \quad \rightarrow\rightarrow \quad \underline{v}_o^+ = N^{-1}(t_1)\,\underline{r}_1 \qquad (12)$$

Requiring there to be only one impulse occurring at time t_o, the state equations can be written as

$$\underline{X}(t) = \begin{pmatrix} \underline{r}(t) \\ \underline{v}(t) \end{pmatrix} = \Phi(t) \begin{pmatrix} \underline{0} \\ \underline{v}_o^+ \end{pmatrix} \qquad (13)$$

so that the position and velocity can be determined at any time between zero and t_1.

The necessary impulse required to move the spacecraft from $\underline{r}_o$ to $\underline{r}_1$ in the given time t_1 is measured as the change in velocity, $\Delta\underline{v}$. For the initial impulse, the change in velocity is

$$\Delta\underline{v}_o = \underline{v}_o^+ - \underline{v}_o^- = N^{-1}(t_1)\,\underline{r}_1 - \underline{0} = N^{-1}(t_1)\,\underline{r}_1 \qquad (14)$$

where the initial condition on the velocity (just prior to the impulse), given in Eq. (4) is $\underline{v}_o^-$. Then, the initial primer vector as defined earlier can be written

$$\underline{p}_o = \frac{\Delta\underline{v}_o}{\Delta v_o} \qquad (15)$$

and can be used to determine the primer vector history as described in Eq. (11).

The interior constraint which relates the impulsive segment to the low-thrust segment at time $t = t_1$ is

$$\underline{P} = \begin{pmatrix} \underline{r}(t_1) - \underline{r}_1 \\ t_1 - \underline{Q} \end{pmatrix} = \begin{pmatrix} \underline{0} \\ 0 \end{pmatrix} \qquad (16)$$

where Q is the pre-determined value for the time t_1. Similar to the initial velocity, t_1^- signifies the time just before t_1 and t_1^+ signifies the time immediately after. The necessary conditions (Bryson and Ho[22]) required for interior constraints are

$$\underline{\lambda}_r^T(t_1^-) = \underline{\lambda}_r^T(t_1^+) + \underline{\pi}_r^T \frac{\partial \underline{P}}{\partial \underline{r}(t_1)} \tag{17a}$$

$$\underline{\lambda}_v^T(t_1^-) = \underline{\lambda}_v^T(t_1^+) + \underline{\pi}_v^T \frac{\partial \underline{P}}{\partial \underline{v}(t_1)} \tag{17b}$$

$$H(t_1^-) = H(t_1^+) - \underline{\pi}_h^T \frac{\partial \underline{P}}{\partial t_1} \tag{17c}$$

where $\underline{\pi}_i$ are a necessary additional set of multipliers. Now, substituting from Eq. (16) for $\underline{P}$ into the above equations gives

$$\underline{\lambda}_r^T(t_1^-) = \underline{\lambda}_r^T(t_1^+) + \underline{\pi}_r^T \tag{18a}$$

$$\underline{\lambda}_v^T(t_1^-) = \underline{\lambda}_v^T(t_1^+) \tag{18b}$$

$$H(t_1^-) = H(t_1^+) - \pi_h \tag{18c}$$

Note, that both the adjoint vector for the position, $\underline{\lambda}_r$, and the Hamiltonian, H, are discontinuous at t_1. The adjoint for the velocity, however, is not. These equations, along with the terminal boundary conditions on the low-thrust segment, will enable a solution for the initial primer-rate to be found.

<u>The Low-Thrust Segment</u>

The low-thrust segment of the trajectory spans the time frame from the intermediate time, t_1, to the final time, t_f, and from the position $\underline{r}_1$, to the rendezvous with the original orbit station. The boundary conditions for this segment are

$$\underline{r}(t_1) = \underline{r}_1$$
$$t_1 \text{ specified}$$
$$\underline{r}(t_f) = \underline{0} \qquad \underline{v}(t_f) = \underline{0} \tag{19}$$

As in the impulsive segment, the equations of motion are given by Eqs. (2) and (3) where $\underline{\Gamma}$, instead of representing the impulsive thrust acceleration, is the thrust acceleration due to a power-limited (VSI) engine. The solution to the equations of motion for the low-thrust segment must include the contribution due to the thrust acceleration because the optimal solution has the engine on continuously unless the primer magnitude is equal to zero (Larson, 1989). The solution can be written as

$$\begin{pmatrix} \underline{r}(t) \\ \underline{v}(t) \end{pmatrix} = \Phi(t - t_1) \begin{pmatrix} \underline{r}_1 \\ \underline{v}_1 \end{pmatrix} + D(t - t_1)\, \underline{q} \tag{20}$$

where $D(t - t_1)\, \underline{q}$ is the contribution due to the thrust acceleration.

The Hamiltonian for this segment is

$$H = \frac{1}{2}\Gamma^2 + \underline{\lambda}_r^T \underline{v} + \underline{\lambda}_v^T [A\underline{r} + B\underline{v} + \underline{\Gamma}] \tag{21}$$

with adjoint equations identical to that in the impulsive segment Eqs. (8) and (9) and an identical equation for the primer vector Eq. (10). The Hamiltonian, in terms of the primer vector and its derivative is

$$H = \frac{1}{2}\Gamma^2 + \dot{\underline{p}}^T \underline{v} - \underline{p}^T [A\underline{r} + \Gamma\underline{u}] \tag{22}$$

As can be found in Prussing (1988), the solution for an optimal thrust acceleration is that the profile should be identical to the primer vector history on this segment of the trajectory. The primer vector history is given by

$$\begin{pmatrix} \underline{p}(t) \\ \dot{\underline{p}}(t) \end{pmatrix} = \Phi(t-t_1) \begin{pmatrix} \underline{p}_1 \\ \dot{\underline{p}}_1 \end{pmatrix} = \begin{pmatrix} M(t-t_1) & N(t-t_1) \\ S(t-t_1) & T(t-t_1) \end{pmatrix} \begin{pmatrix} \underline{p}_1 \\ \dot{\underline{p}}_1 \end{pmatrix} \tag{23}$$

where $\underline{p}_1$ and $\dot{\underline{p}}_1$ are found using the terminal conditions on the state Eq. (19) and also from Eq. (20),

$$\underline{q} = - D^{-1}(t_f - t_1)\ \Phi(t_f - t_1)\ \underline{X}_1 \tag{24}$$

where the vector $\underline{q}$ has elements corresponding to the primer and primer-rate at t_1. For a fixed terminal time, t_f, the solution for the vector $\underline{q}$, and hence, the remainder of the problem, is analytically simple. However, for an unspecified terminal time, it is necessary to determine, using the transversality condition[22], the optimal terminal time for the given boundary conditions. The transversality condition for this problem is

$$\Omega = \frac{\partial \Phi[\underline{x}(t_f), t_f]}{\partial tf} + H(t_f) = 0 = - \frac{1}{2}\Gamma^2(t_f) \tag{25}$$

which indicates that the optimal value for the final thrust acceleration, as well as the final primer vector, is zero. Note that for non-optimal values of the final time, the transversality condition will always have a value less than zero. This indicates that a finite optimal value for the final time does not exist, and that a graph of cost versus time will approach zero in the limit as the final time tends to infinity. The cost is related to the transversality condition by

$$\Omega = \frac{\partial J}{\partial t_f} \tag{26}$$

or, in other words, omega is the slope of the cost with respect to the final time when all the other necessary conditions are satisfied. A graph of omega versus the final time, t_f, shows indirectly how J will approach zero as the final time tends to infinity (Fig. 2). To be able to make a time versus fuel cost trade-off in an optimal manner, the performance index for this segment is modified to include a penalty on the final time.

Omega vs. Final Time

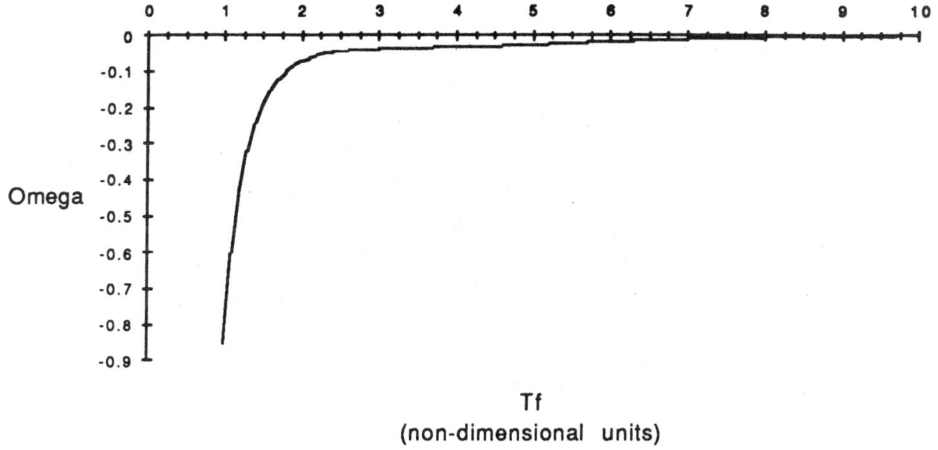

FIGURE 2

The new performance index for the low-thrust segment is

$$J = \frac{1}{2} \int_{t_o}^{t_f} \Gamma^2 \, dt + \frac{1}{2} K^2 t_f \tag{27}$$

where K is a weighting factor which has units of acceleration. The only necessary condition which is effected by this change is the transversality condition

$$\Omega = \frac{\partial \Phi[\underline{x}(t_f), t_f]}{\partial t_f} + H(t_f) = 0 = \frac{1}{2} K^2 - \frac{1}{2} \Gamma^2(t_f) \tag{28}$$

where the optimal value of the final thrust acceleration is now K. The value of K is chosen based on the importance given to the value of the final time in relation to the fuel cost. With a value of K chosen, there are now a sufficient number of equations to obtain the optimal solution for this problem.

In summary, the optimal final time can be found from

$$\underline{p}(t_f) = [M(t_f - t_1) \; N(t_f - t_1)] \begin{pmatrix} \underline{p}_1 \\ \dot{\underline{p}}_1 \end{pmatrix} \tag{29}$$

$$\underline{p}^T(t_f)\,\underline{p}(t_f) = K^2 \tag{30}$$

$$\underline{q} = -\,D^{-1}(t_f - t_1)\,\Phi(t_f - t_1)\,\underline{X}_1 \tag{31}$$

where the state $\underline{X}_1$ is determined from the impulse segment and t_f, $\underline{p}_1$, and $\dot{\underline{p}}_1$ are the seven unknowns in the above seven equations. The final time cannot be solved for analytically. However, using any one of a number of successive approximation schemes, the solution may be found. In this study, a Newton iteration was used with an analytically determined derivative. The primer history for the entire trajectory can, therefore, be written as

$$\begin{pmatrix} \underline{p}(t) \\ \dot{\underline{p}}(t) \end{pmatrix} = \Phi(t) \begin{pmatrix} \underline{p}_o \\ \dot{\underline{p}}_o \end{pmatrix} \qquad 0 \le t \le t_1 \tag{32a}$$

$$\begin{pmatrix} \underline{p}(t) \\ \dot{\underline{p}}(t) \end{pmatrix} = \Phi(t-t_1) \begin{pmatrix} \underline{p}_1 \\ \dot{\underline{p}}_1 \end{pmatrix} \qquad t_1 \le t \le t_f \tag{32b}$$

where $\underline{p}_1$ and $\dot{\underline{p}}_1$ are found from the vector $\underline{q}$, Eq. (24). The vector $\dot{\underline{p}}_o$ is found using the interior constraint condition that the primer is continuous at t_1, Eq. (18b) along with Eq. (32a) evaluated at t_1

$$\dot{\underline{p}}_o = N(t_1)^{-1}\,[\underline{p}_1 - M(t_1)\underline{p}_o] \tag{33}$$

recalling that $\underline{p}_o$ is given in Eq. (15). The state can be obtained from

$$\underline{X}(t) = \begin{pmatrix} \underline{r}(t) \\ \underline{v}(t) \end{pmatrix} = \Phi(t) \begin{pmatrix} \underline{0} \\ \underline{v}_o^+ \end{pmatrix} \qquad 0 \le t \le t_1 \tag{34a}$$

$$\underline{X}(t) = \begin{pmatrix} \underline{r}(t) \\ \underline{v}(t) \end{pmatrix} = \Phi(t - t_1)\,\underline{X}_1 + D(t - t_1)\,\underline{q} \qquad t_1 \le t \le t_f \tag{34b}$$

where $\underline{v}_o^+$ is given in Eq. (12).. Using the above equations, all the elements of the trajectory can now be found.

RESULTS

A computer program was developed to evaluate the optimal solution for an impulsive intercept followed by a low-thrust return on-station. For a given set of initial input; $\underline{r}_1$, t_1 and the weighting factor, K, the program computes the optimal state vector, primer vector, primer-rate vector and Hamiltonian and evaluates them at a chosen number of increments of the total time.

Recall from the problem statement that initially it was assumed a single impulse would be used to execute the flyby of $\underline{r}_1$. It was found, however, that for a given $\underline{r}_1$ and t_1, a critical value of the weighting factor, K^*, exists where, for values greater than K^*, the primer vector magnitude exceeds one on the impulsive segment. Figure 3 shows the primer magnitude versus time for several values of K. Note that K^* is defined as the value of K where the primer magnitude becomes exactly one at t_1. The weighting factor, K, is due to the transversality condition which requires the magnitude of the final primer vector Eq. (28) to be positive rather than zero. Physically, a larger value of K forces a smaller final time, t_f, due to the increase in the cost. When the final time becomes sufficiently small, the single-impulse coast trajectory to the intermediate point apparently does not leave enough time for an optimal low-thrust return. In these situations, the violation of the primer magnitude condition seems to indicate that an alternative mission profile is optimal. It is possible that either a second impulse is needed to reach $\underline{r}_1$ in the time t_1, or that the low-thrust engine should be turned on sooner than t_1.

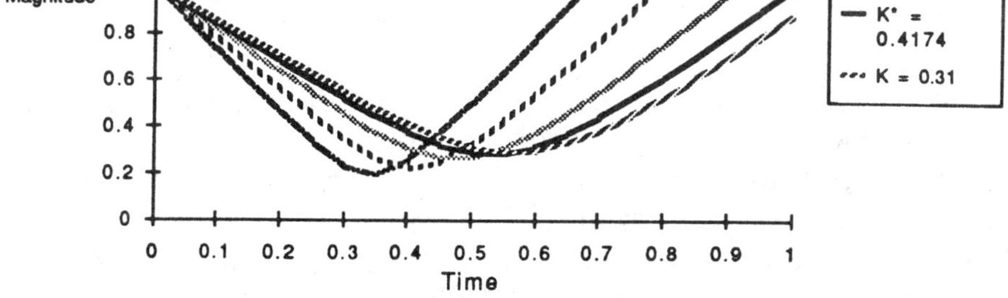

FIGURE 3 Primer Magnitude History for Impulsive Segment

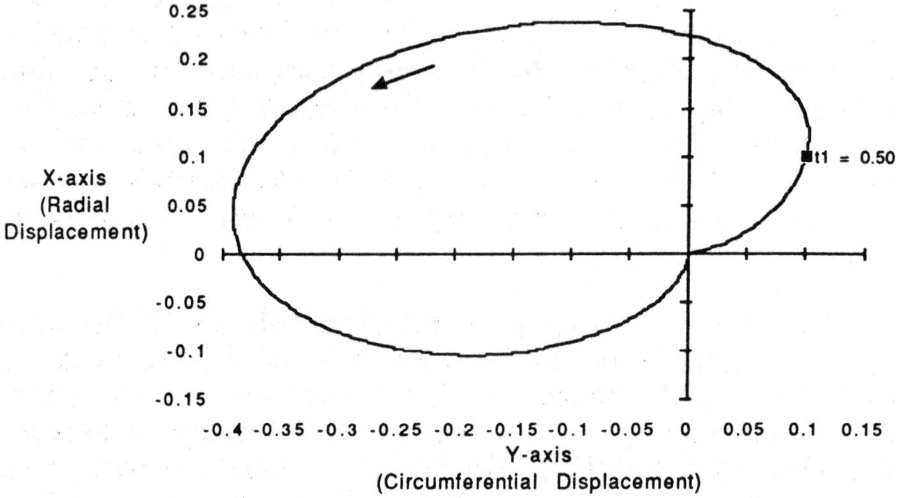

FIGURE 4 X-Y Plane Projection for an Equidistant R1

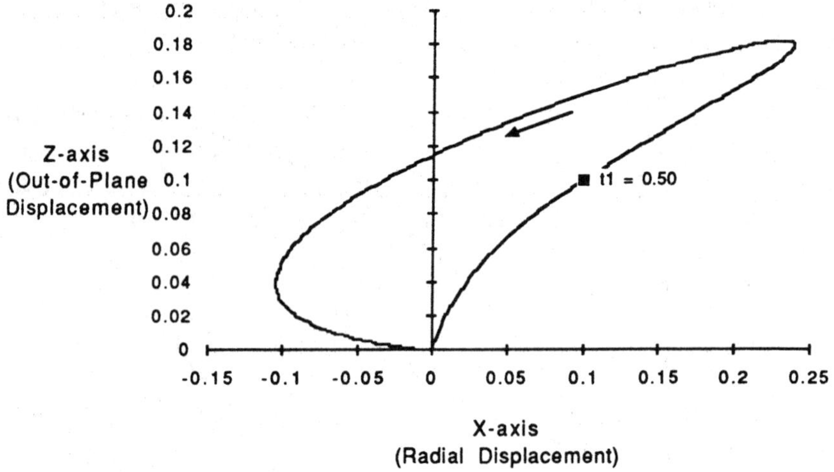

FIGURE 5 X-Z Plane Projection for an Equidistant R1

For each value of $K \leq K^*$ an optimal solution can be found. A three-dimensional trajectory is projected in Figs. 4 and 5. In this case, the intermediate position is equidistant from all axes and the intermediate time is equal to 0.50 non-dimensional time units. The weighting factor, K, is 0.3464 , the optimal final time, t_f, is 4.4711 time units, and the initial impulse (measured as the change in velocity) is $\Delta \underline{v} =$ [0.0664, 0.2783, 0.2086]. Figure 4 shows the x-y plane projection of the position as it changes throughout the total time of flight. Note that the earth is located in the negative x direction. The spacecraft moves in a counter-clockwise direction as indicated by the arrow, and the intermediate position, $\underline{r}_1$, is marked. As the spacecraft moves radially away from the earth (positive x direction), it begins to fall behind the original orbit station, i.e., the y coordinate decreases. This results from the fact that higher altitude orbits have longer orbit periods. Figure 5 is the x-z plane projection of the same trajectory. Note, the low-thrust segment of the trajectory is longer than the impulsive segment.

The following figures show the control history for this trajectory. Figure 6 is a graph of the primer vector, and hence the thrust acceleration vector, in the x and y directions. The discontinuity in the third quadrant occurs at the time t_1, and is due to the necessary condition at the intermediate point, Eq. (18a), which results in a discontinuous primer rate. Comparing this to the x-y position graph (Fig. 4), it can be seen that at the intermediate position, the thrust direction is negative in both the x and y directions. The initial thrust required to arrive at the desired position in the allotted time carries the spacecraft beyond its destination. The reverse thrust is needed to prevent the spacecraft from traveling too great a distance radially out from the intermediate position.

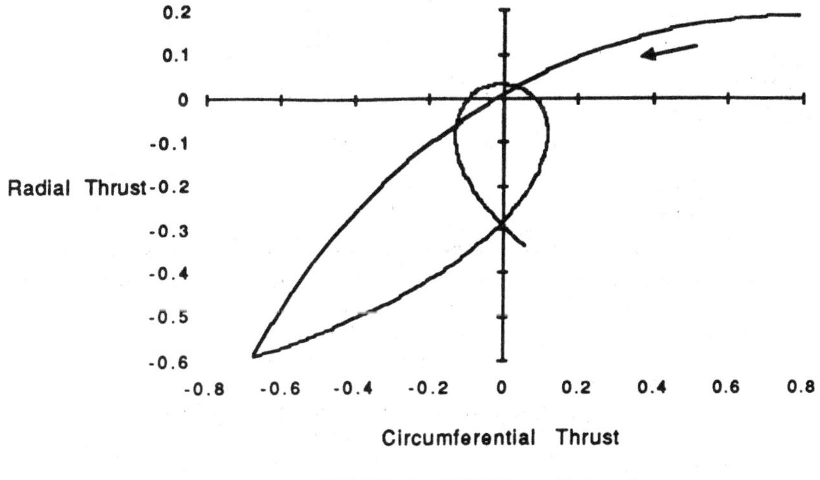

FIGURE 6 X-Y Primer History for an Equidistant R1

During the last segment of the trajectory, the thrust direction changes rapidly in order to allow the spacecraft to return to the original station with zero displacement and zero velocity. Figure 7 is the thrust profile corresponding to the x-z displacement (Fig. 5). Note that, as in the x-y plane, the thrust direction becomes negative before reaching the intermediate position. After the low-thrust engines are turned on, the magnitude of the thrust in the $\pm z$ direction does not vary greatly. The largest thrust along the z-axis occurs at the initial time when an impulse is required to move the spacecraft a specified distance in a constrained amount of time. A plot of the primer magnitude history for this trajectory is also included (Fig. 8). The weighting factor, K, for this particular case was chosen such that the primer magnitude on the impulsive segment is always less than one after the initial impulse.

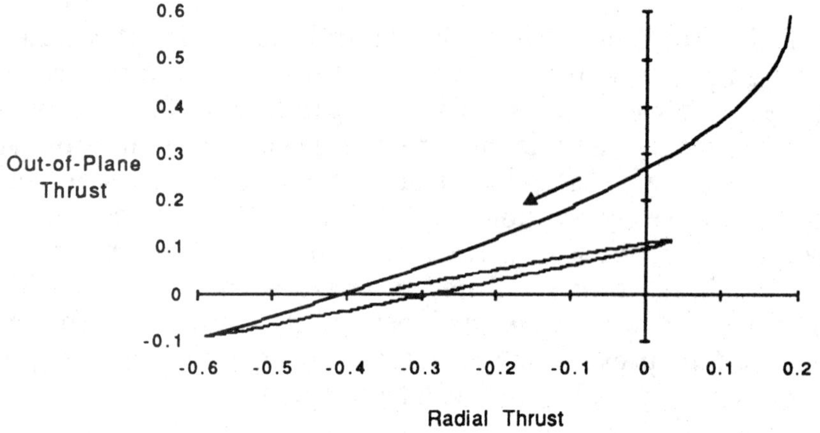

FIGURE 7 X-Z Primer History for an Equidistant R1

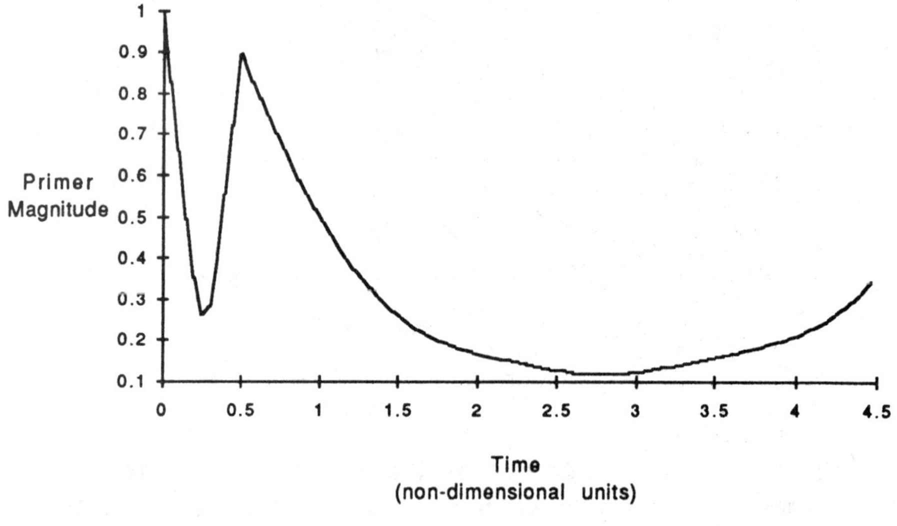

FIGURE 8 Primer Magnitude History

The next trajectory (Figs. 9-12) is distinctly different than the one previously presented in that the total time of flight is 23.8837 time units. This is roughly 3.8 reference orbit periods which means that the spacecraft is away from the original orbit station for nearly four of its revolutions. The value of the weighting constant ($K = 0.0001$) was chosen to be small so that the total flight time would be much less significant than the overall fuel consumption. Figures 9 and 10 show the orbit path in the x-y and x-z planes. As with the previous orbit projections, it is in terms of the rotating reference frame, however, it is much more evident in this case. Figures 11 and 12 diagram the corresponding thrust profiles. Note, the relatively small thrust magnitude on the low-thrust segment of the trajectory.

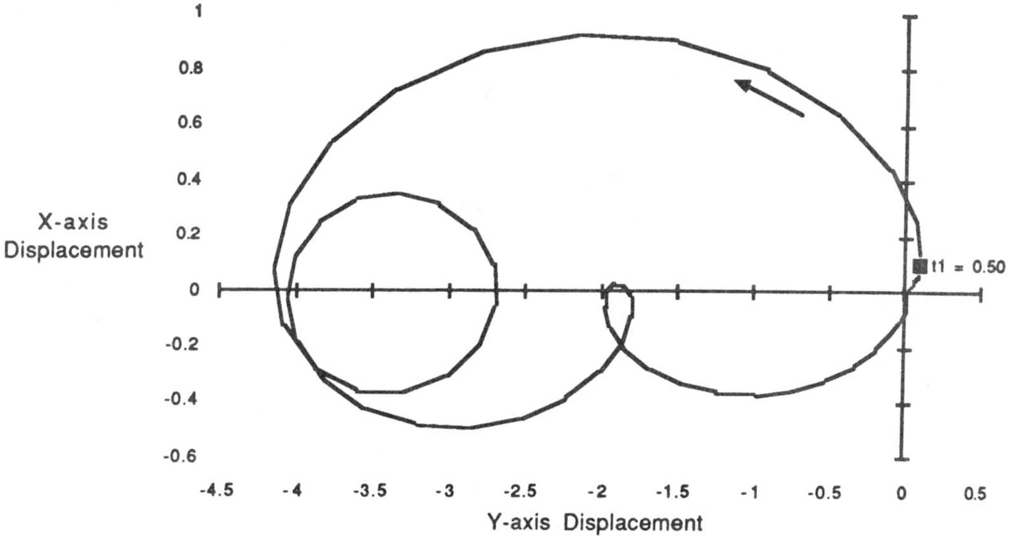

FIGURE 9 X-Y Plane Projection for Multi-Revolution Trajectory

CONCLUDING REMARKS

From the results presented, several conclusions can be made: 1) the problem of an optimal impulsive intercept maneuver followed by a low-thrust return can be solved for a fixed final time; 2) for the time-open case, the optimal solution is for the final time to be infinite; and 3) with a weighted function of the final time in the performance index, a limited range of optimal single impulse solutions for the time-open case can be found. The range can be identified by values of the weighting factor, K, which satisfy the condition $K \leq K^*$, where K^* is a critical value of the weighting factor.

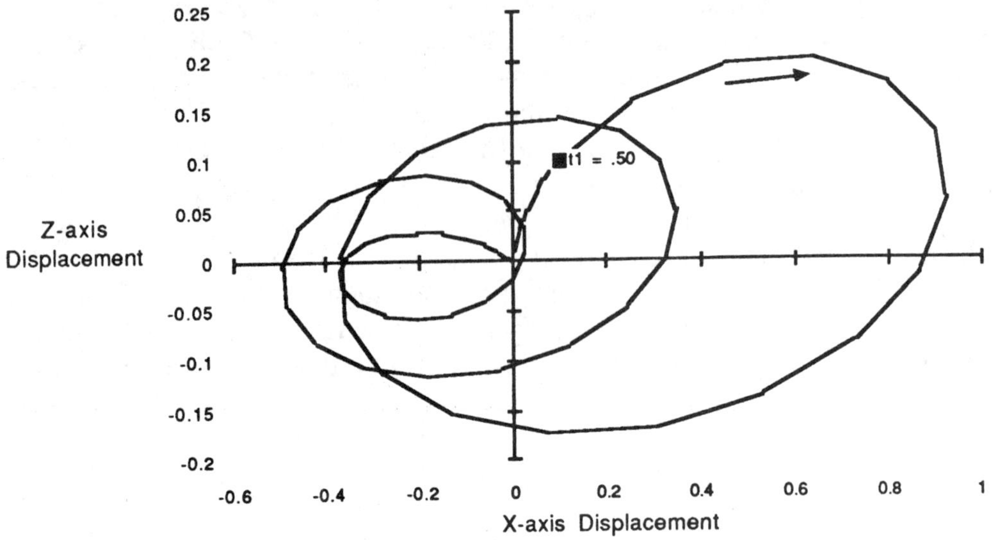

FIGURE 10 X-Z Plane Projection for Multi-Revolution Trajectory

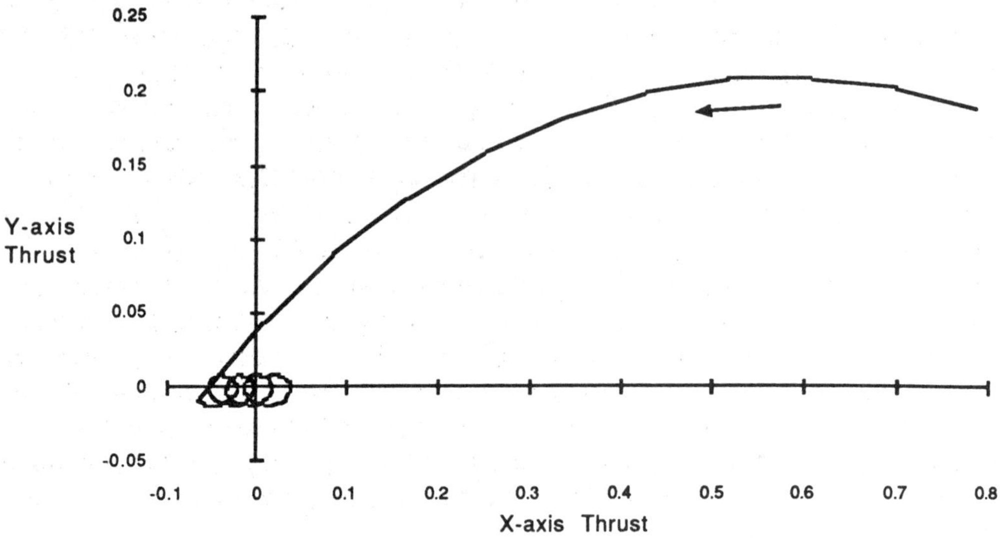

FIGURE 11 X-Y Primer History for Multi-Revolution Trajectory

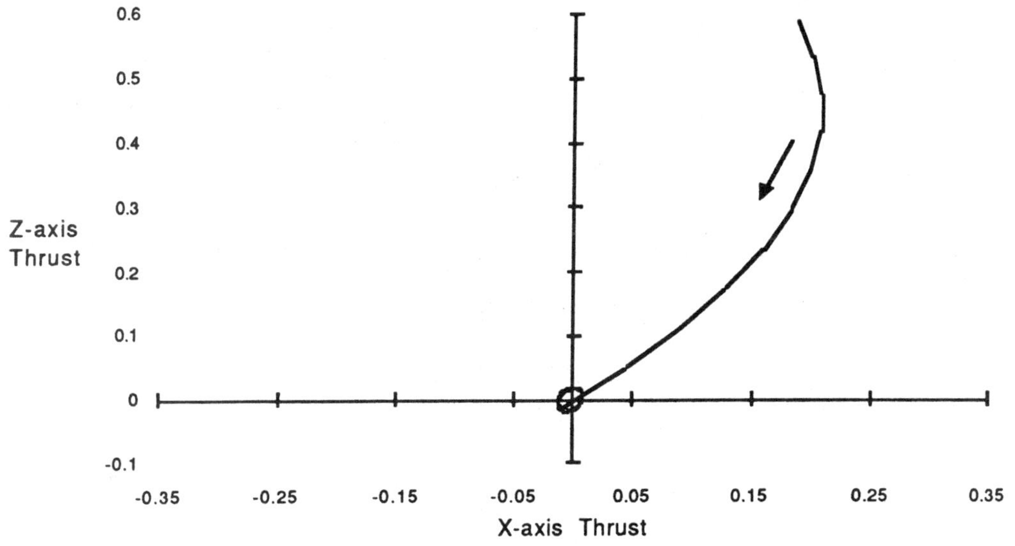

FIGURE 12 X-Z Primer History for Multi-Revolution Trajectory

The research presented here can be further enhanced or modified by examining the following areas: multiple impulse intercepts, dual high and low-thrust programs, and thrust-limited (as opposed to power-limited) low-thrust engines. Allowing multiple impulses during the impulsive segment would change the effect that the weighting factor has on the primer history, and perhaps eliminate the existence of a critical weighting factor, K^*, altogether. Dual-thrust programs have been studied and have been shown to be more effective than individual high- and low-thrust programs when their individual performances are the same. It may be that for this problem, using a dual-thrust program when the necessary conditions on the primer vector (during the impulsive segment) are violated, lowers the overall cost for some trajectories. The use of a thrust-limited low-thrust engine in place of the power-limited one used for this study, will change the solution to the problem, and may, as suggested for the dual-thrust case, effect the overall cost of the trajectories.

ACKNOWLEDGMENT

This research was supported by the NASA Lewis Research Center through NASA Grant NAG 3-805.

REFERENCES

1. Lawden, D. F., "General Motion of a Rocket in a Gravitational Field", *Journal of the British Interplanetary Society,* Vol. 6, Dec. 1947, pp.187-191.

2. Edelbaum, T. N., "Minimum-Impulse Transfers in the Near Vicinity of a Circular Orbit", *Journal of the Astronautical Sciences,* Vol. 14, April 1967, pp.66-73.

3. Prussing, J. E., "Optimal Four-Impulse Fixed-Time Rendezvous in the Vicinity of a Circular Orbit", *AIAA Journal,* Vol. 7, No. 5, May 1969, pp.928-935.

4. Prussing, J. E., "Optimal Two- and Three-Impulse Fixed-Time Rendezvous in the Vicinity of a Circular Orbit", *AIAA Journal,* Vol. 8, No. 7, July 1970, pp.1221-1228.

5. Breakwell, J. V., "The Optimization of Trajectories", *SIAM Journal,* 1959.

6. Irving, T. H., "Low-Thrust Flight: Variable Exhaust Velocity in Gravitational Fields", *Space Technology,* Wiley and Sons, New York, 1959.

7. Lawden, D. F., *Optimal Trajectories for Space Navigation,* Butterworths, London, 1963.

8. Melbourne, W. G., and Sauer, C. G., Jr., "Optimum Thrust Programs for Power-Limited Propulsion Systems", *Astronautica Acta,* Vol. 8, 1961, pp.205-227.

9. Billik, B. H., "Some Low-Acceleration Rendezvous Maneuvers", *AIAA Journal,* Vol. 2, No. 3, March 1964, pp.510-516.

10. Hinz, H. K., "Optimal Low-Thrust Near-Circular Orbit Transfer", *AIAA Journal,* Vol. 1, No. 6, June 1963, PP.1367-1371.

11. Carter, T. E., "Fuel-Optimal Maneuvers of a Spacecraft Relative to a Point in Circular Orbit", *Journal of Guidance and Control,* Vol. 7, No. 5, Nov. 1984, pp.710-716.

12. Carter, T. E., "Effects of Propellant Mass Loss on Fuel-Optimal Rendezvous near Keplerian Orbit", *Journal of Guidance, Control and Dynamics,* Vol. 13, No. 1, Jan.-Feb. 1989, pp.19-26.

13. Widhalm, J. W. and Eide, S. A., "Optimal Continuous Thrust In-Plane Orbital Evasive Maneuvers", *AIAA 26th Aerospace Sciences Meeting,* Reno NV, Jan. 1988, Paper AIAA 88-0374.

14. Edelbaum, T. N., "The Use of High and Low-Thrust Propulsion in Combination for Space Missions", *Journal of the Astronautical Sciences,* Vol. 9, No. 2, 1961, pp.49-60.

15. Vahlberg, J. C., Jr. and Edelbaum, T. N., "Transfers Between Coplanar Coaxial Ellipses Using the Combination of High and Low-Thrust Propulsion", *AAS Microfiche Series,* Vol. 20, Paper Number AAS 71-368, 1972.

16. Larson, C. A. "Optimization of an Impulsive Intercept Maneuver Followed by a Low-Thrust Return", M.S. Thesis, University of Illinois a Urbana-Champaign, 1989.

17. Clifton, R. S., "Optimal Multiple-Impulse ASAT Avoidance Maneuvers", M.S. Thesis, University of Illinois at Urbana-Champaign, 1987.

18. Murphy, G. O., "Optimal Multiple-Impulse ASAT Interception", M.S. Thesis, University of Illinois at Urbana-Champaign, 1988.

19. Wheelon, A. D., "Midcourse and Terminal Guidance", *Space Technology,* Wiley and Sons, New York, 1959.

20. Clohessy, W. H., and Wiltshire, R. S., "Terminal Guidance Systems for Satellite Rendezvous", *Journal of the Astronautical Sciences,* Sept. 1960, pp.653-658, 674.

21. Jezewski, D., "Primer Vector Theory Applied to the Linear Relative Motion Equation", *Optimal Control Applications and Methods,* Vol. 1, 1980, pp.387-401.

22. Bryson, A. E., Jr. and Ho, Y. C., *Applied Optimal Control,* Hemisphere Publishing, Washington D.C., 1975.

AAS 89-355

A HOMOTOPY METHOD FOR SPACE FLIGHT RENDEZVOUS PROBLEMS

Gopal Vasudevan*, Frederick H. Lutze† and Layne T. Watson‡

The fuel-optimal rendezvous problem which assumes a finite number of Keplarian coasting arcs, each one separated from a previous one by a velocity impulse, can be approached from the point of view of a nonlinear programming problem with the components of the velocity impulse and the change of true anomaly along the coasting arcs as the parameters. A homotopy approach to the solution of this problem is presented which includes a summary of the supporting theory and a novel method for dealing with inequality constraints. As a result, two homotopy maps are proposed. The first map has the property that all solutions along the homotopy path are optimal solutions to some rendezvous problem, the final point being the solution of interest, and the second map provides a means to bridge the point where the inequality constraint set changes. Some discussion of implementing the method is given including the selection of a convenient set of variables. Selected results are presented.

INTRODUCTION

A problem which has received considerable attention in recent years is that associated with intercept and rendezvous in space. In particular one would like to perform the maneuver in a timely manner and at the same time keep fuel expenditures at a minimum. Additional complications enter when constraints in the form of equalities and inequalities are introduced. Of the many techniques used to solve this problem one version assumes a finite number of Keplarian coasting arcs, each one separated from the previous one by a velocity impulse (ΔV). With this formulation the fuel-optimal rendezvous problem can be approached from the point-of-view of a parameter optimization or nonlinear programming problem. Under these conditions the components of each ΔV and the lengths of each coasting arc for both the interceptor and the target become the optimization parameters to be selected[1,2]. Consequently many of the tools developed for the solution of nonlinear programming problems can be applied to the space flight rendezvous problem.

* Graduate Student, Department of Aerospace and Ocean Engineering, Virginia Polytechnic Institute and State University, Blacksburg, Virginia 24061-0203.

† Professor, Aerospace and Ocean Engineering, Virginia Polytechnic Institute and State University, Blacksburg, Virginia 24061-0203.

‡ Professor, Computer Science, Virginia Polytechnic Institute and State University, Blacksburg, Virginia 24061-0203.

Typically a solution procedure consists of making an initial guess of the optimization parameters, determining the cost function to be optimized and then applying some algorithm to update the guess in such a manner that the cost is reduced. The procedure is repeated until a proper solution is obtained. The success of such a procedure often depends upon the initial guess, a poor guess leading to slow or non-convergence. An often used heurestic approach is to use the solution to a previously solved problem which has similar characteristics to the problem of interest as the initial guess. The hope is that since the characteristics are essentially unchanged, the solution to the problem of interest is in the neighborhood of the initial guess. This idea, called continuation, can in principle be repeated until a path of solutions is generated from some simple known problem to a more general problem of interest. Unfortunately, as many will attest, the procedure often fails to provide a final solution. The likelihood of success is further diminished when inequality constraints are present in the formulation and the active constraint set changes as the solution moves along the continuation path.

However a procedure which on the surface appears similar to a continuation method but has a firmer mathematical basis, called a homotopy procedure, can alleviate many of the difficulties discussed above. The first part of this paper will present the essence of the supporting theory for homotopy procedures and summarize the important ingredients of a desirable homotopy map. Particular attention is given to the problems associated with inequality constraints and to some extent the selection of an initial guess. The second part of the paper discusses the implementation of the homotopy procedure for the space–flight–rendezvous problem. Particular attention is given to the use of a new set of variables to describe the coasting arcs in space and the associated ΔV between them.

FIRST ORDER NECESSARY CONDITIONS FOR OPTIMALITY

In general terms the constrained optimization problem takes the form:

$$C(\overline{x}) = \underset{x \epsilon X}{Min}\, C(x);$$

$$\overline{x} \epsilon X = \{x | x \epsilon R^n,\ g(x) \geq 0,\ h(x) = 0\} \tag{1}$$

where

$$C : R^n \rightarrow R, \quad g : R^n \rightarrow R^{ni} \text{ and } h : R^n \rightarrow R^{ne}$$

are twice continuously differentiable functions.

The first order necessary conditions for optimality are given by the generalized Kuhn–Tucker Conditions or equivalently the Fritz-John Conditions(3):

If $\overline{x}$ is a solution to the above problem then there exists a scalar $\overline{\mu}_0$ and, vectors $\overline{\lambda} \epsilon R^{ne}$ and $\overline{\mu} \epsilon R^{ni} = \{\overline{\mu}_1, \ldots, \overline{\mu}_{ni}\}$ such that:

$$\overline{\mu}_0 \nabla C(\overline{x}) - \overline{\lambda} \nabla h(\overline{x}) - \overline{\mu} \nabla g(\overline{x}) = 0, \tag{2}$$

$$h(\overline{x}) = 0, \tag{3}$$

$$\overline{\mu} g(\overline{x}) = 0, \tag{4}$$

$$g(\overline{x}) \geq 0, \quad \overline{\mu} \geq 0, \tag{5}$$

$$\overline{\mu}_0 \geq 0 \;\; [\overline{\lambda}, \overline{\mu}, \overline{\mu}_0] \neq 0. \tag{6}$$

We normally assume $\overline{\mu}_0 = 1$ (the Kuhn–Tucker conditions), hoping that the Kuhn–Tucker constraint qualification[3] will be satisfied at $\overline{x}$. Here $C(x)$ is the cost function and $h(x)$ and $g(x)$ are the equality and inequality constraints, respectively.

HOMOTOPY METHODS

The object of the homotopy procedure is to track the solution of eqs. (2-6) as a set of system parameters change. These system parameters can appear in the cost function, equality and inequality constraints. It is convenient to scale these parameters in such a way that only one parameter, the homotopy parameter σ, needs to be monitored. Typically one could use

$$p = \sigma p_1 + (1 - \sigma)p_0 \tag{7}$$

where p_0 is the system parameter vector for the known or starting problem and p_1 the vector for the final or desired values. Then if one could track the solution to eqs. (2-6) as σ is varied from 0 to 1, not only would the desired problem be solved but so would all the problems associated with the system parameter $p(\sigma)$, $0 \leq \sigma \leq 1$. Unfortunately using only this map, chances of complete success are slim as documented in the literature[4].

The Chow–Yorke algorithm[5] based on a parameterized version of Sard's theorem[6,7] in conjunction with the Implicit function theorem[8] provides the basis for constructing a homotopy map which yields a smooth, non–bifurcating, and non–stopping homotopy path. A key ingredient is that the Jacobian matrix of the system, eqs. (2-6), cannot become singular along the path. Another key ingredient is the inclusion of a constant parameter vector to be selected somewhat randomly.

In most cases when the active constraint set changes, the Jacobian matrix of the system will indeed become singular. Hence it is desirable to change the inequality to equality constraints in some fashion. The requirement that either the constraint or its multiplier must be zero and if not zero must be greater than zero, eqs. (4), (5), is called a complementarity condition. An equivalent condition given in terms of equalities only is given in Mangasarian's complemetarity theorem[9], and is used here.

With these introductory remarks two homotopy maps will be presented, the first which is useful when the active constraint set does not change, and the second which makes full use of the freedom associated with the Chow–Yorke algorithm to enable it to deal with constraint set changes and arbitrary initial conditions.

Map 1.

An equivalent set of necessary conditions can be written as

$$\nabla C(x, \sigma) - \lambda \nabla h(x, \sigma) - \mu \nabla g(x, \sigma) = 0, \tag{2a}$$

$$h(x, \sigma) = 0, \tag{3a}$$

$$\theta(|g_i(x, \sigma) - \mu_i|) - \theta(g_i(x, \sigma)) - \theta(\mu_i) = 0, \quad i = 1, \ldots, ni \tag{8a}$$

where $\theta : R \to R$ is Mangasarian's complementarity function and is any strictly increasing function which is at least C^2 and $\theta(0) = 0$.[9] Here $\theta(x) = x^3$ was used exclusively, being the simplest C^2 function satisfying all requirements.

<u>Map 2</u>.

The second map is equivalent to the necessary conditions when $\sigma = 1$ and is given by

$$\sigma\left[\nabla C(x,\sigma) - \lambda\nabla h(x,\sigma) - \mu\nabla g(x,\sigma)\right] + (1-\sigma)(x-\bar{b}) = 0, \tag{2b}$$

$$\sigma[h(x,\sigma)] + (1-\sigma)(\lambda - \bar{c}) = 0, \tag{3b}$$

$$-\sigma[\theta(|g(x,\sigma) - \mu|) - \theta(g(x,\sigma)) - \theta(\mu)] + (1-\sigma)(\mu - \bar{a}) = 0, \tag{8b}$$

where $\bar{a}\varepsilon R^{n_i}$, $\bar{b}\varepsilon R^{n}$, $\bar{c}\varepsilon R^{n_e}$.

These maps exhibit the following properties:

1. Map 1 satisfies the necessary conditions for all values of σ. Hence all solutions are optimal solutions to some problem.

2. Map 2 satisfies the necessary conditions only for values of $\sigma = 1$.

3. Map 2 can be satisfied at $\sigma = 0$ by selecting the proper values for $\bar{a}$, $\bar{b}$ and $\bar{c}$ which are the arbitrary vectors identified in the Chow–Yorke algorithm.

Additional details of these maps are given in Refs. 10 and 11.

APPLICATION TO SPACE–FLIGHT RENDEZVOUS

The problem is to find a minimum fuel rendezvous trajectory between two bodies, one of which is non-maneuvering (target) and the interceptor. The interceptor trajectory consists of Keplarian coasting arcs separated by impulsive thrusting, characterized by a change in velocity (magnitude and direction). The maneuver must be completed within some prespecified time and the trajectories must avoid passing through the Earth, i.e., they must not violate a minimum radius constraint. The fuel-optimal problem translates to minimizing the total change in the velocity (Characteristic Velocity).

The variables are — the coasting angles on each arc including a possible initial coast, components of the velocity change vector, and the coasting angle of the Target.

A certain elegance in the equations governing the motion of a satellite about the Earth is obtained if one introduces a particular regularizing transformation. The formulation used is similar to the Burdet oscillator equations[12] modified recently by Schumacher[13] and Jezewski[14].

Define a unit vector triad $\hat{r}$, $\hat{r}'$, and $\hat{h}$ where $\hat{r}$ is a unit vector in the radial direction, $\hat{h}$ is a unit vector normal to the plane of motion in direction of angular momentum vector, and $\hat{r}'$ is a unit vector in the plane of motion perpendicular to the radius vector. Note that $()'$ means the derivative with respect to the change in true anomaly. If u is the reciprocal of the radius vector and h the magnitude of the angular momentum, the position and velocity of the satellite is given by

$$\bar{r}(\cdot) = \hat{r}(\cdot)/u(\cdot), \tag{9}$$

$$\overline{V}(\cdot) = h(\cdot)\left[u(\cdot)\hat{r}'(\cdot) - u'(\cdot)\hat{r}(\cdot)\right]. \tag{10}$$

The equations of motion along a Keplarian arc are given by

$$u'' + u = \mu/h^2, \tag{11}$$

$$\hat{r}'' + \hat{r} = 0. \tag{12}$$

The solutions for the various terms in eqs. (9) and (10) are in the form of simple harmonic oscillators

$$u(\eta) = \mu/h^2 + \left[u(0) - \mu/h^2\right]\cos(\eta) + u'(0)\sin(\eta), \tag{13}$$

$$u'(\eta) = -\left[u(0) - \mu/h^2\right]\sin(\eta) + u_0'(0)\cos(\eta), \tag{14}$$

$$\hat{r}(\eta) = \hat{r}(0)\cos(\eta) + \hat{r}'(0)\sin(\eta), \tag{15}$$

$$\hat{r}'(\eta) = -\hat{r}(0)\sin(\eta) + \hat{r}'(0)\cos(\eta), \tag{16}$$

$$h(\eta) = h(0). \tag{17}$$

The time of flight along any Keplarian arc is obtained by integrating

$$T(\eta) = \int_o^\eta \frac{1}{hu^2(\theta)} d\theta, \tag{18}$$

where η is the change in true anomaly along the arc.

The initial conditions for these equations can be determined from the initial position, $\bar{r}_0$, and velocity, $\overline{V}_0$, vectors along each coasting arc as follows:

$$\begin{array}{ll} i. & u(0) = 1/\|\bar{r}_o\| \\ ii. & \hat{r}(0) = u(0)\bar{r}_o \\ iii. & \bar{h}(0) = \bar{r}_o \times \overline{V}_o \\ iv. & h(0) = \|\bar{h}(0)\| \\ v. & \hat{h}(0) = \bar{h}(0)/h(0) \\ vi. & \hat{r}'(0) = \hat{h}(0) \times \hat{r}(0) \\ vii. & u'(0) = -(\bar{r}_o \cdot \overline{V}_o)u(0)/h(0) \end{array} \tag{19}$$

At a velocity impulse the variables u and $\hat{r}$ remain unchanged and the impulse is characterized by the change in magnitudes of u' and h and by the change in directions of $\hat{r}'$ and $\hat{h}$. Since $\hat{r}$ is fixed, the only change, if any, in $\hat{r}'$ and $\hat{h}$ is the rotation ϕ about $\hat{r}$. Thus an impulse vector $[\Delta V_x, \Delta V_y, \Delta V_z]$ is characterized by the three variables $[\Delta u', \Delta h, \phi]$, where Δ indicates the change in the variable across the impulse.

Mathematically the space flight rendezvous problem can be described as choosing a sequence of variables $[\eta, \Delta u', \Delta h, \phi]$ so that the characteristic velocity (total velocity change), which provides a measure of fuel consumed, is minimized. More formally we can write the problem with a cost function, equality and inequality constraints.

Cost Function.

The cost function can be expressed as

$$\min_{x} \Delta V(x), \tag{20}$$

where $\{x\} = \{(\eta, \Delta u', \Delta h, \phi)_j, \eta_t\}$, $j = 1, nim$,
where nim is a prespecified number of impulses, and the subscript t refers to the target variable. Characteristic velocity ΔV can be expressed in terms of the regularized variables as (see eq. (10))

$$\Delta V = \sum_{j=1}^{nim} \left\{ u_j^2 \left[h_{j+1}^2 - 2h_{j+1}h_j \cos(\phi_j) + h_j^2 \right] + \left[h_{j+1}u'_{j+1} - h_j u_j \right]^2 \right\}^{1/2}, \tag{21}$$

where the subscript j indicates the values at the end of the jth subarc before the impulse, and $j+1$ the initial values of the variables on the next subarc after the impulse. Across the impulse $\Delta u'_j = u'_{j+1} - u'_j$ and $\Delta h_j = h_{j+1} - h_j$. In addition the following constraints must be satisfied:

Equality Constraints.

For rendezvous we must satisfy a final position match constraint,

$$h_1(x) = \bar{r}_f - \bar{r}_t(\eta_t) = 0, \tag{22}$$

a final velocity match constraint,

$$h_2(x) = \overline{V}_f - \overline{V}_t(\eta_t) = 0, \tag{23}$$

and a matching time-of-flight constraint,

$$h_3(x) = T_i - T_t = 0, \tag{24}$$

where the subscript f refers to the conditions on the interceptor trajectory after the final impulses and i refers to the interceptor.

Inequality Constraints.

Certain inequality constraints must be included, some due to physical restrictions such as requiring non-earth-penetrating transfer orbits and others to overcome ambiguities in the model such as time or true anomaly changes becoming negative. Hence we have the following set of inequality constraints:

Non-negativity of the coasting arcs of the interceptor,

$$g_i(x) \equiv \eta_i \geq 0. \quad i = 1, nim \tag{25}$$

Non-negativity of the coasting arc of the target,

$$g_{nim+1}(x) \equiv \eta_t \geq 0. \tag{26}$$

Time-of-flight constraint,

$$g_{nim+2}(x) \equiv T_{max} - T_i \geq 0. \tag{27}$$

Minimum radius constraint, for each coasting arc except the initial coast arc of the interceptor trajectory,

$$g_j(x) \equiv u_o - u_{max} \geq 0, \ j = nim + 3, 2nim + 1.$$

Non-negativity of the radius,

$$g_j(x) \equiv u_{min} \geq 0, \ j = 2nim + 2, 3nim. \tag{28}$$

The final constraint is included to prevent the mathematically possible negative radius that may occur if the true anomaly exceeds the asymptotic value in a hyperbolic orbit.

The minimum radius constraint requires some discussion. It is required that all transfer arcs must lie outside a circle of radius $r_{min} = 1/u_{max}$. This is essentially a semi-infinite constraint, but from the nature of the transfer arcs, i.e., conic sections, the minimum radius on any subarc is given by:

1. Perigee radius in the case of perigee passage

2. $Min(r_{initial}, r_{final})$

Unfortunately the minimum radius constraint as given above is not C^2 and consequently for the moment we have chosen the stiffer constraint that requires the perigee radius of any transfer arc to be greater than the minimum allowable radius. Such a restriction precludes any "pop up" type orbits which generally are not globally optimal anyhow[1].

Under these restrictions the value of the minimum radius or equivalently u_{max} can be obtained from eq. (13) and is found to be,

$$u_{max} = \mu/h^2 + \sqrt{(u(0) - \mu/h^2) + u'(0)^2}. \tag{29}$$

For circular orbits

$$u = u_{max} = u(0) = \mu/h^2. \tag{30}$$

It would be desirable to use the expression in eq. (29) for the minimum radius constraint. However on circular orbits its derivatives are undefined and the Jacobian matrix of the system becomes singular, an occurrence to be avoided. The objective is to generate a function which is C^2 over the whole domain of circular and non-circular orbits and at the same time gives a good approximation to the original constraints for both circular and non-circular orbits. After considerable searching one function which provides these properties best (to date) is given by

$$Tanh^2(\gamma e^2)/u_{max} + h^2/\mu\left[1 - Tanh^2(\gamma e^2)\right] - 1/U_{max} \geq 0 \tag{31}$$

where U_{max} is the specified minimum radius value, u_{max} is given by eq. (29), and e is the orbit eccentricity

$$e = h^2/\mu\sqrt{[u(0) - \mu/h^2]^2 + u'(0)^2}. \tag{32}$$

Although this expression underestimates the minimum radius constraint the true constraint can be approximated as closely as required by making γ large. In particular it is convenient to introduce another homotopy in the expression such as $\gamma = \sigma\gamma_1 + (1-\sigma)\gamma_0$ to insure the desired tightness in the constraint at the desired solution.

HOMOTOPY MAPS FOR THE SPACE FLIGHT RENDEZVOUS PROBLEM

Typically for a two impulse problem one might start with the known solution such as a simple Hohmann transfer and then slowly vary the time–of–flight, until the solution to a desired problem is obtained. If Map 1 is used every solution along the zero curve is a solution to a rendezvous problem with specified time–of–flight. Typically as the maximum time–of–flight is reduced, the minimum radius constraint becomes active and the Jacobian matrix associated with Map 1 becomes singular. At this point any previous solution along the zero curve may be used as an initial solution for Map 2. The result of using Map 2 will be a solution with the minimum radius constraint active. Once the solution has been obtained, it can serve as an initial solution for the original homotopy, Map 1, as the maximum time for maneuver is further decreased. A similar procedure can be used for variations in initial phase angle, target orbit inclination or some other parameter. The results will give two–impulse trajectories which satisfy the Kuhn–Tucker necessary conditions.

If a three (or more) impulse trajectory is required, the initial problem must be made to look like a three (or more) impulse problem. This requirement is best achieved by splitting the single coast arc of the Hohmann (or any other solution) into two (or more) equal parts separated by a zero impulse. This procedure works better than splitting an impulse into two impulses separated by an arc length of zero. These changes may alter the character of the problem, for example, the active constraint set may be different from that of the initial problem. It is advisable to start with Map 2 for this reason.

The homotopy Maps 1 and 2 show the homotopy parameter σ imbedded in cost and constraint functions. To see how this is done we note the following: The initial conditions of the interceptor at the start of the maneuver arc described by $u(0), u'(0), h(0), \hat{r}(0), \hat{r}'(0)$ and $\hat{h}(0)$ as determined from eq. (19). If the changes in true anomaly for each subarc are known along with the changes $\Delta u', \Delta h$, and ϕ at the end of each coasting arc, then the entire trajectory can be determined using eqs. (13)-(18) along with (9) and (10). Consequently the cost, eq. (21) and the constraints, eqs. (22)-(28) can easily be determined. If it were of interest to investigate different initial conditions for a problem, a natural way to imbed the homotopy σ would be in a linear map for the scalar initial conditions.

$$u(0) = \sigma u_1(0) + (1-\sigma)u_0(0) \tag{33}$$

$$u'(0) = \sigma u_1'(0) + (1-\sigma)u_0'(0) \tag{34}$$

$$h(0) = \sigma h_1(0) + (1-\sigma)h_0(0) \tag{35}$$

where the subscripts 0 and 1 designate initial conditions for the known and desired problem, respectively.

An additional homotopy map is needed for the unit vector triad $(\hat{r}, \hat{r}', \hat{h})$. A linear map is not advocated since the unit vector properties would not be preserved for the values of σ not equal to 1 or 0. We can generate the desired map by using a sequence of Euler angle rotations to represent the rotation from $(\hat{r}(0), \hat{r}'(0), \hat{h}(0))_0$ to $(\hat{r}(0), \hat{r}'(0), \hat{h}(0))_1$. The map is given by

$$\begin{bmatrix} \hat{r}(0) \\ \hat{r}'(0) \\ \hat{h}(0) \end{bmatrix} = \begin{bmatrix} I & 0 & 0 \\ 0 & \cos(\sigma\chi)I & \sin(\sigma\chi)I \\ 0 & -\sin(\sigma\chi)I & \cos(\sigma\chi)I \end{bmatrix}$$

$$\begin{bmatrix} \cos(\sigma\zeta)I & 0 & -\sin(\sigma\zeta)I \\ 0 & I & 0 \\ \sin(\sigma\zeta)I & 0 & \cos(\sigma\zeta)I \end{bmatrix} \cdot \begin{bmatrix} \cos(\sigma\psi)I & \sin(\sigma\psi)I & 0 \\ -\sin(\sigma\psi)I & \cos(\sigma\psi)I & 0 \\ 0 & 0 & I \end{bmatrix} \cdot \begin{bmatrix} \hat{r}_0(0) \\ \hat{r}'(0) \\ \hat{h}(0) \end{bmatrix}, \tag{36}$$

where I is a 3×3 identity matrix and ψ is the rotation about the $\hat{h}_0(0)$ axis, ζ the rotation about the $\hat{r}'$ axis, and χ the rotation about $\hat{r}$ axis to obtain $[\hat{r}_1(0), \hat{r}_1'(0), \hat{h}_1(0)]$ for the desired problem. The three angles can be determined from

$$\tan(\psi) = [(\hat{r}_1(0) \cdot \hat{r}_0'(0))/(\hat{r}_1(0) \cdot \hat{r}_0(0)], \tag{37}$$

$$\sin(\zeta) = -\hat{h}_0(0) \cdot \hat{r}_1(0), \tag{38}$$

with

$$\cos(\zeta) = \hat{r}_0(0) \cdot \hat{r}_1(0)/\cos(\psi), \tag{38a}$$

or

$$\sin(\zeta) = \hat{r}_0'(0) \cdot \hat{r}_1'(0)/\sin(\psi), \tag{38b}$$

depending on the value of ψ. Also

$$\tan\chi = (\hat{h}_0(0) \cdot \hat{r}_1'(0)/(\hat{h}_0(0) \cdot \hat{h}_1(0)). \tag{39}$$

In a similar manner the homotopy parameter σ can be introduced into the initial conditions of the target.

The maximum time-of-flight can be mapped using a linear homotopy as

$$\sigma T_{max1} + (1-\sigma)T_{max0} - T_I \geq 0, \tag{40}$$

where

T_I is total time to rendezvous for the interceptor

T_{max} is prescribed maximum time to rendezvous limit.

The minimum radius constraint can be treated similarly by replacing the U_{max} in eq. (31) with

$$U_{max} = \sigma U_{max1} + (1-\sigma)U_{max0} \tag{41}$$

Using the above imbedding procedures one can use Map 1 to obtain solutions corresponding to variations in one or more of the above mentioned system constants. If a singular Jacobian matrix is encountered then one switches to Map 2. There is no guarantee that the solution to Map 2 will be a continuation of the solutions found by Map 1 prior to the singular condition.

The homotopy equations to be solved, either Map 1 or Map 2 require the gradients of the cost function, eq. (21) and the constraints, eqs. (22)-(28). These gradients and their derivatives with respect to the variables

$$\{[\eta, \Delta u', \Delta h, \phi]_i, \eta_T\} \quad i = 1, \ldots, nim$$

are obtained using the chain rule. The form of the solutions given by eqs. (13)-(18) make the derivatives of the initial conditions on any subarc reasonably straight forward. Derivatives of the cost function and constraints with respect to these initial conditions are also easily calculated. The derivatives of the time–of–flight equation (18) are not so straight forward but are tractable in analytic form[11].

EXAMPLES

In order to use the homotopy method, it is necessary to track the solutions to the equations in Map 1 and Map 2 as the homotopy parameter changes from 0 to 1. This curve tracking was performed using the normal flow algorithm FIXPNF from a set of algorithms given in HOMPACK due to Watson, Billups, and Morgan[15]. The algorithm requires the computation of the Jacobian matrix of the homotopy map to be tracked.

Several examples were considered using this technique including coplanar circle–to–circle rendezvous, non–coplanar circle–to–circle rendezvous and non–circular non–coplanar rendezvous. All solutions were obtained in a systematic manner starting from known solutions made available from an earlier effort. In many cases these earlier solutions were those of a reference Hohmann transfer. Initial and final conditions were then changed to meet the requirements of the desired problem by using the appropriate maps in eqs. (33)-(41).

Circle–to–Circle Coplanar Rendezvous.

In general fuel–optimal rendezvous trajectories are composed of near tangential velocity impulses. Therefore the interceptor trajectories are characterized by the requirement to "speed–up" or "slow–down" so that the last two impulses resemble a Hohmann transfer. A "speed–up" or type A orbit requires an initial impulse which brings the interceptor inside the inner orbit until a proper phase angle is obtained for the Hohmann–like rendezvous. These types of orbits encounter the minimum radius constraint as maximum time–of–flight is reduced. "Slow down" or type B orbits tend to loft the interceptor outside the outer orbit leading to trajectories which resemble bi–elliptic transfers.

Results presented here pertain to two coplanar circular orbits with the interceptor initially in a circular orbit of 1.2 Earth radii (DU) and the target in a circular orbit of 1.45 DU. In Figure (1) the change in characteristic velocity as the maximum time–of–flight is changed is presented for the case of an initial phase angle (β_T) of 0°. The solid lines represent type A trajectories and the dashed lines type B trajectories. Figure (2) presents a similar graph for the case where the phase angle (β_T) is 180°. Here the asterisk, *, indicates where the minimum radius constraint (100) is encountered as time-of-flight is reduced. If further reduction in time-of-flight causes a second arc to meet the minimum radius constraint, a second asterisk is shown. Figure (3) shows the 4 impulse orbit where the second coasting arc encounters the minimum radius constraint. The total flight time of 11.2 TU falls between the two asterisks in the corresponding curve of Figure (2).

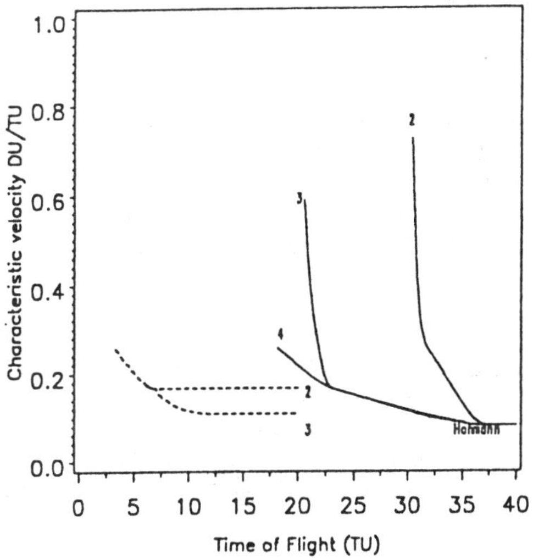

Figure 1. Characteristic Velocity for $\beta_T = 0^o, i_T = 0^o$

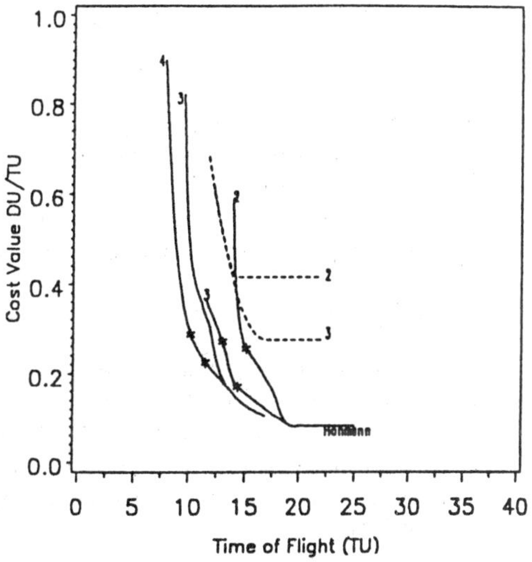

Figure 2. Characteristic Velocity for $\beta_T = 180^o, i_T = 0^o$

Circle-to-Circle Non-Coplanar Rendezvous.

Here the two circular orbits are the same size as in the previous section. However the target orbit is at inclinations of 0.5 and 1.0 radian with respect to the inner interceptor orbit. Figures (4) and (5) for an orbit inclination of 0.5 rad correspond to Figures 1 and 2 for the coplanar problem and show the effect of maximum time-of-flight on characteristic velocity. Similar information is presented in Figures (6) and (7) for the case where the target is inclined 1.0 radians with respect to the interceptor orbit. The interceptor initiates the maneuver from the target ascending node of the target-interceptor orbit plane intersection and the target phase angle β_T is measured from this point to the target location in the plane of the target orbit.

An interesting 3-impulse orbit which represents the global minimum for the times shown for the case of a 1.0 radian target inclination angle is shown in Figure (8).

General Orbital Rendezvous Examples .

The number of parameters associated with a general non-circular non-coplanar rendezvous problem can be quite large. Here we will assign the interceptor to an

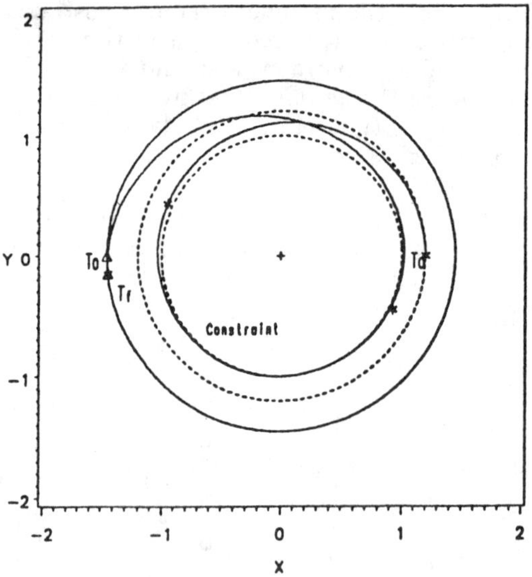

Figure 3. 4-Impulse Trajectory,
$\beta_T = 180^o, i_T = 0.0$ rad.

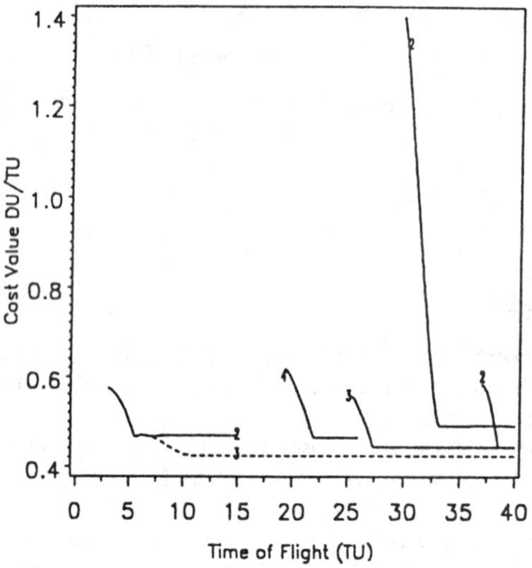

Figure 4. Characteristic Velocity for
$\beta_T = 0^o, i_T = 0.5$ rad.

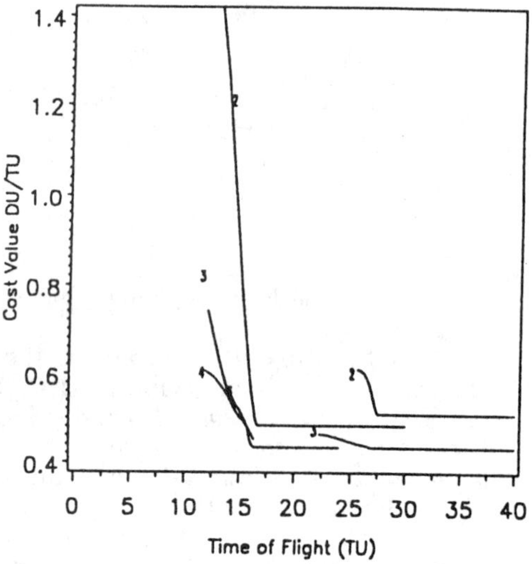

Figure 5. Characteristic Velocity for
$\beta_T = 180^o, i_T = 0.5$ rad.

elliptic orbit with the classical orbit parameters given by $a = 1.4$ DU, $e = 0.2$, and $i = 0$. The ascending node Ω is defined to be zero and the argument of perigee $\omega = 0$. The interceptor starts at perigee. The target orbital elements are given by $a_t = 2.0$

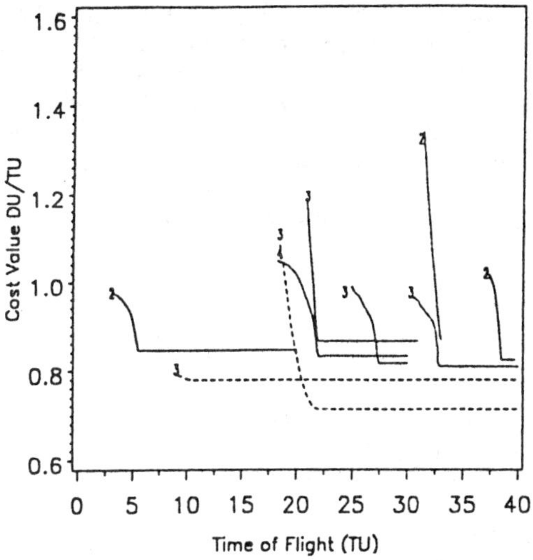

Figure 6. Characteristic Velocity for $\beta_T = 0^o, i_T = 1.0$ rad.

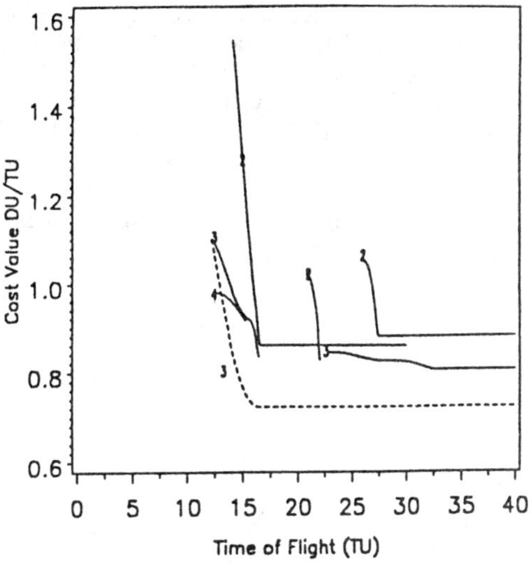

Figure 7. Characteristic Velocity for $\beta_T = 180^o, i_T = 1.0$ rad.

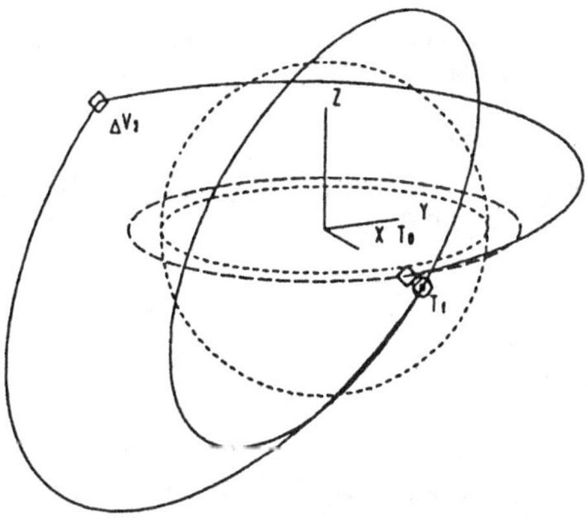

Figure 8. 3-Impulse Trajectory, $\beta_T = 0^o, i_T = 1.0$ rad.

DU, $e_t = 0.1, \Omega_t = 0^o$ and $\omega_t = 45^o$. Of interest is the characteristic velocity for rendezvous with a target at different inclinations. Again β_T is measured from the target ascending node.

Selected results are given for target–orbit inclinations of 0.0, 0.5 and 1.0 radians in Figures (9), (10) and (11), respectively. These were generated by using the same two homotopy maps as before. Here however we started with circle–to–circle results obtained previously, but increasing the inner and outer orbit radii until they were at 1.4 DU and 2.0 DU, respectively. These results thus served as the initial known problem for the elliptic case where e was mapped linearly to the desired value.

A 3-impulse trajectory for the case of relative orbit inclination of 1.0 radians is shown in Figure (12).

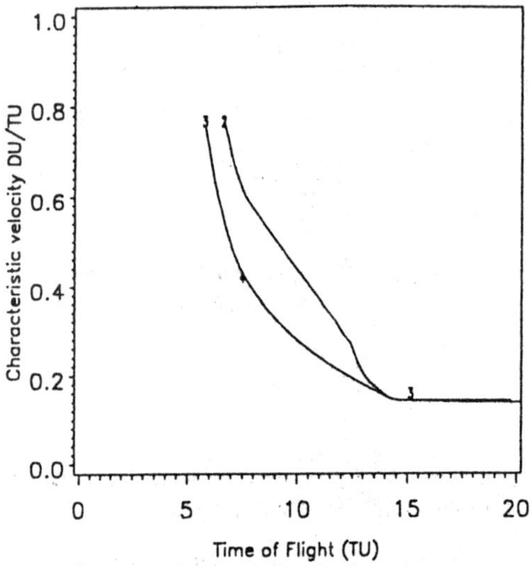

Figure 9. Characteristic Velocity for $\beta_T = 120^o, i_T = 0.0$ rad.

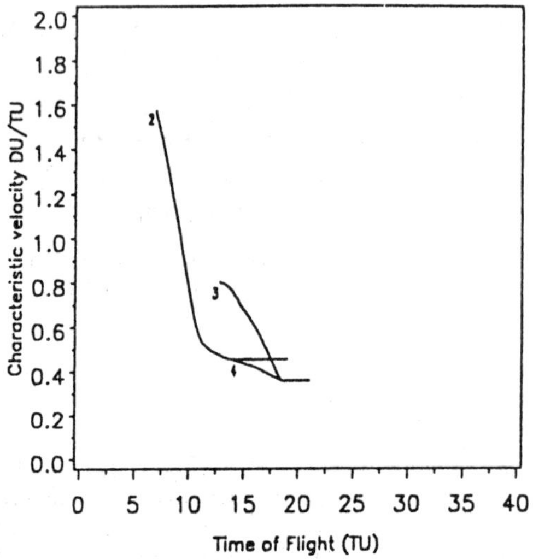

Figure 10. Characteristic Velocity for $\beta_T = 120^o, i_T = 0.5$ rad.

CONCLUDING REMARKS

Homotopy methods have been applied to the space–flight rendezvous problem. Two maps were provided, one yielding the solution to some optimal rendezvous problem for all intermediate steps leading to the solution of the desired problem, and the other providing solutions when singularities occur in the first map due to a change in the active constraint set.

The equations of motion of the satellites are expressed in terms of Burdet oscillator type variables facilitating the evaluation of the first and second partial derivatives required for tracking the homotopy zero curve. A homotopy map specific to the space–flight rendezvous problem was constructed and applied to numerous problems. The method provides a systematic procedure to obtain the variation of some system characteristic such as cost function with respect to some system parameter such as maximum time–of–flight using Map 1, with Map 2 used to bridge the singular situation. Map 2 can be used by itself to seek the solution to some problem directly with an arbitrary initial guess. However since Map 2 does not satisfy all the requirements of the Chow–Yorke algorithm, it is not always as robust as would be desired.

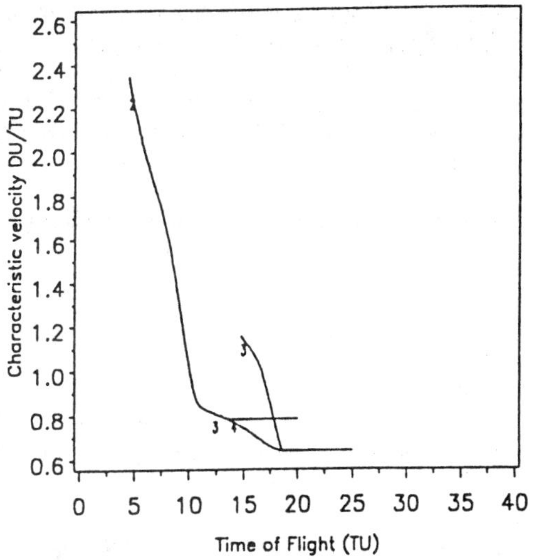

Figure 11. Characteristic Velocity for $\beta_T = 120^o, i_T = 1.0$ rad.

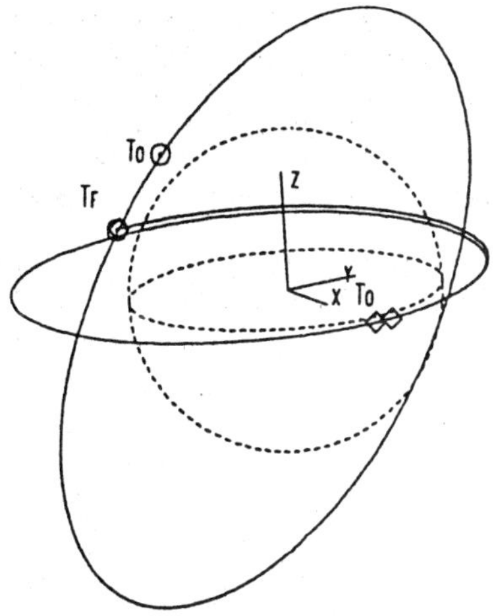

Figure 12. 3-Impulse Trajectory, $\beta_T = 120^o, i_T = 1.0$ rad.

REFERENCES

1. G. Vasudevan, "Fuel Optimal Rendezvous Including a Radial Constraint," M.S. Thesis, Dept. of Aerospace and Ocean Engineering, VPI&SU, 1986

2. G. Vasudevan and F.H. Lutze, "On Verification of Fuel-Optimal Space Trajectories," *Opt. Cont. Appl. Meth.*, Vol. 9, 1988, pp 285–302

3. O.L. Mangasarian and S. Fromovitz, "The Fritz–John Necessary Optimality Conditions in the Presence of Equality and Inequality Constraints," *J. Math. Anal. and Appl.*, Vol. 17, 1967, pp 37–47

4. H. Gfrerer, J. Guddat and H.J. Wacker, "A Globally Convergent Algorithm Based on Imbedding the Parametric Optimization," *Computing*, Vol. 30, 1983, pp 225–252

5. S.N. Chow, J. Mallet–Paret and J.A. Yorke, "Finding Zeros of Maps: Homotopy Methods that are Constructive with Probability One," *Mathematics of Computation*, Vol. 32, 1978, pp 887–899

6. L.T. Watson, "A Globally Convergent Algorithm for Computing Fixed Points of C^2 Maps," *Appl. Math. and Comput.*, Vol. 5, 1979, pp 297–311

7. L.T. Watson, "Engineering Applications of the Chow–Yorke Algorithm," *Appl. Math. and Comput.*, Vol. 9, 1981, pp 111-133

8. J.M. Ortega and W.C. Rheinboldt, *Iterative Solution of Nonlinear Equations in Several Variables*, Academic Press, Orlando, FL, 1970

9. O.L. Mangasarian, "Equivalence of the Complementarity Problem to a System of Non-Linear Equations," *SIAM J. Appl. Math.*, Vol. 31, 1976, pp 89-92

10. G. Vasudevan, L.T. Watson and F.H. Lutze, "A Homotopy Approach for Solving Constrained Optimization Problems," Proceedings American Control Conference, Pittsburgh, PA, June 1989, pp 780–785

11. G. Vasudevan, "A Homotopy Approach to the Solutions of Minimum-Fuel Space-Flight Rendezvous Problems," Ph.D. Dissertation, Dept. of Aerospace and Ocean Engr., VPI&SU, 1989

12. C.A. Burdet, "Regularization of the Two Body Problem," *Zeitshrift fur Angewandt Mathematik und Physik*, Vol. 18, 1967, pp 434–438

13. P.W. Schumacher, "Results of True-Anomaly Regularization in Orbital Mechanics," Ph.D. Dissertation, Dept. of Aerospace and Ocean Engr., VPI&SU, 1987

14. D.J. Jezewski, "An Ideal Reference Frame for Perturbed Orbital Motion," Proceedings AIAA/AAS Astrodynamics Conference, Minneapolis, MN, 1988, pp 62–68

15. L.T. Watson, S.C. Billups and A.P. Morgan, "HOMPACK: A Suite of Codes for Globally Convergent Homotopy Algorithms," GMRL Research Publication GMR-5344, General Motors Research Labs, Warren, MI, June 1986

AAS 89-353

AUTONOMOUS, EXOATMOSPHERIC, TRAJECTORY OPTIMIZATION*

D. J. Jezewski†

A general formulation for the optimal n-burn, exoatmospheric trajectory problem is developed for the two-body problem with three degrees of freedom in an inverse-square force field. The principal goal is to develop an efficient and autonomous method for obtaining optimal solutions to mission planning trajectory transfer problems. This is accomplished by combining (1) an optimal n-impulse preprocessor, (2) a thrust moment-generating method, and (3) the perturbation matrices of the solution to develop the unknown parameters of the problem that satisfy the boundary conditions (BC) and transversality conditions (TC) for an optimal n-burn solution. The state and costate vectors, and the variational matrix are propagated numerically on the burn arcs with a unique Runge-Kutta Nystrom integrator and analytically on the coast arcs. A combined gradient/Newton-Raphson interator is used in the convergence process of the optimal n-burn solution; it combines the sureness of the gradient method and the quadratic characteristics of the Newton-Raphson method.

* This paper appears in full in the microfiche supplement to this volume, **AAS Microfiche Series**, Volume 59.

† McDonnell Douglas Space Systems Company, 16055 Space Center Blvd., Houston, Texas 77062.

PLANETARY SPACECRAFT ORBIT DETERMINATION

SESSION 10

Chairman: Daryl G. Boden
Department of Astronautics
U. S. Air Force Academy

AAS 89-403

A STATISTICAL STUDY OF THE DIFFERENTIAL VERY LONG BASELINE INTERFEROMETRY DELAY RATE ERROR BUDGET

Charles D. Edwards*

The Delta Differential One-way Doppler (ΔDOD) observable provides target-relative tracking information for the navigation of planetary orbiters such as the Magellan mission to Venus. To better understand the ΔDOD error budget, twelve years of dual-frequency Deep Space Network (DSN) Very Long Baseline Interferometry (VLBI) data have been analyzed. Simulated ΔDOD spacecraft-quasar observables are formed by differencing residual delay rates for pairs of quasar observations. The resulting distribution of ΔDOD residuals for 71,098 quasar pairs has an RMS width of 3.16×10^{-13} sec/sec.

The dependence of the RMS residuals on a variety of observation parameters is examined to reveal the signatures of dominant ΔDOD error sources. The ΔDOD residuals are observed to depend only very weakly on the angular separation between quasars and the temporal separation between observations. However, a strong signature is observed in the dependence of ΔDOD errors on elevation angle and total atmospheric pathlength. These results are consistent with the hypothesis that the ΔDOD errors are dominated by short-term temporal troposphere fluctuations.

Ionosphere and solar plasma errors are studied by examining the dependence of ΔDOD residuals on ionospheric longitude and on sun-earth-quasar angle. The relative importance of having dual-band ionospheric calibrations is addressed by quantifying the degradation in the ΔDOD RMS residual when only single-band data are used.

* Jet Propulsion Laboratory, California Institute of Technology, Mail Stop 238-700, 4800 Oak Grove Drive, Pasadena, California 91109.

INTRODUCTION

The Magellan mission to Venus is the first NASA mission to use differential VLBI delay rate measurements as a primary tracking data type. The Delta Differential One-way Doppler (ΔDOD) observable, formed by differencing delay rate measurements for a spacecraft and a nearby quasar, contains information about the angular spacecraft rate on the plane of the sky. The signature of a spacecraft's orbital motion about a planet on its apparent VLBI angular rate provides a strong constraint on the target-relative orbit [1]. To satisfy its navigational orbit determination requirements, Magellan requires a one-sigma ΔDOD accuracy of 2.4×10^{-13} sec/sec on a time scale of 600 sec [2]. Error analysis indicates that this goal can be met, but with little or no margin [3]. The dominant error source is expected to be due to wet troposphere fluctuations, except at low sun-earth-probe angles, where solar plasma errors should dominate.

To better understand the impact of tropospheric and charged particle error sources on the ΔDOD observable, this paper examines delay rate residuals for a large data base of DSN VLBI quasar observations. Simulated ΔDOD observables are formed from pairs of quasar observations, and the resulting ΔDOD residuals are studied to reveal the signature of limiting error sources.

The ΔDOD Observable

In a typical Delta Differential One-way Doppler (ΔDOD) observation, VLBI data are recorded simultaneously at two DSN antennas, first of the received spacecraft signal, and then of an angularly close quasar. Fig. 1 depicts the ΔDOD observation geometry.

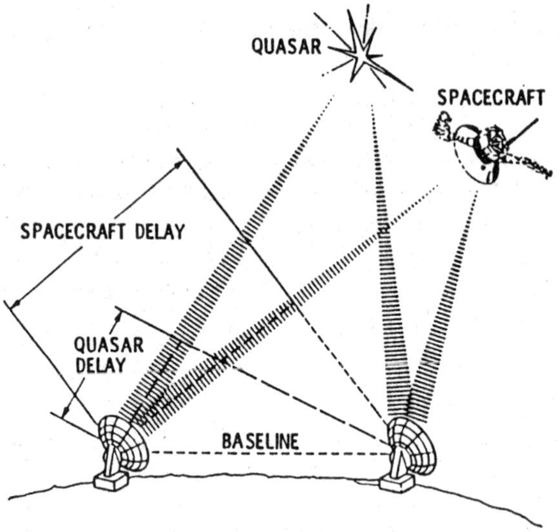

Fig. 1. Observation geometry for ΔDOD tracking

For a single radio source, the differenced one-way delay in the direction $\hat{s}$, observed at two stations separated by a baseline $\vec{B}$, is equal to:

$$\tau = \frac{1}{c}\vec{B} \cdot \hat{s} \tag{1}$$

where c is the velocity of light [4]. The differential one-way Doppler, or VLBI delay rate, in dimensionless units, can therefore be expressed as:

$$y \equiv \dot{\tau} = \frac{1}{c}(\dot{\vec{B}} \cdot \hat{s} + \vec{B} \cdot \dot{\hat{s}}) \qquad (2).$$

For observations at an RF frequency ν, the actual observed differential frequency shift will be $y\nu$.

The measurement of y will be corrupted by a number of errors, dominated by uncertainties in the clock rates at the two stations, as well as time variations in the line-of-sight propagation media delays. The clock rate offset error (which is typically a few parts in 10^{-13} sec/sec for DSN hydrogen masers), as well as a portion of the propagation media errors, cancel when we difference two delay rate observations to form the ΔDOD observable:

$$\Delta y \equiv y_{S/C} - y_Q \qquad (3)$$

where the subscripts indicate the spacecraft and quasar observations.

Based on our *a priori* knowledge of the baseline B and its time-evolution $\dot{B}$ due to earth rotation, as well as the *a priori* directions $\hat{s}_{S/C}$ and $\hat{s}_Q$ and their derivatives, we can calculate expected model values for $y_{S/C}$ and y_Q, thereby forming a ΔDOD observation residual

$$\Delta\tilde{y} \equiv (y_{S/C}^{obs} - y_{S/C}^{calc}) - (y_Q^{obs} - y_Q^{calc}). \qquad (4)$$

Throughout this paper we will be dealing with the statistics of $\Delta\tilde{y}$, the dimensionless ΔDOD residual.

In the limit of perfect *a priori* knowledge of B, $\dot{B}$, $\hat{s}_Q$, and $\dot{\hat{s}}_Q$, the ΔDOD residual will be simply

$$\Delta\tilde{y} = \frac{1}{c}\vec{B} \cdot \delta\dot{\hat{s}}_{S/C} + \frac{1}{c}\dot{\vec{B}} \cdot \delta\hat{s}_{S/C} \qquad (5)$$

where $\delta\hat{s}_{S/C}$ and $\delta\dot{\hat{s}}_{S/C}$ represent the difference between the true and *a priori* values of the spacecraft direction and its derivative. Prior orbit determination analysis [1] has demonstrated how the angular and angular rate information content in the ΔDOD observable can be used to constrain the orbital elements for a planetary orbiter such as Magellan, and in particular, to determine the node of the orbit in the plane of the sky, to which Doppler data are completely insensitive.

The Magellan Error Budget

A previous error analysis for Magellan delay rate observations indicates that tropospheric errors will dominate the ΔDOD error budget [3]. Table 1 is derived from [3] and shows the contribution of each error source. Troposphere errors are indicated for several different values of θ_{min}, the minimum elevation angle for the four relevant ΔDOD lines of sight (spacecraft and quasar at each of two stations.) The total RSS ΔDOD error for $\theta_{min} = 10$ deg is 2.4×10^{-13}, with the troposphere alone contributing nearly 2.3×10^{-13}. This error budget assumes that ΔDOD data are simultaneously obtained at 2.3 and 8.4 GHz, allowing calibration of charged particle errors. At sun-earth-probe (SEP) angles less than 5 deg, it becomes impossible to reliably track the 2.3 GHz spacecraft phase due to large phase scintillations in the solar plasma, and so dual-frequency charged particle calibration is no longer possible.

In this paper we will be attempting to verify some of these features of the Magellan error budget, in particular the role of troposphere fluctuations and charged particle errors, by simulating spacecraft - quasar ΔDOD observations using quasar pairs. For these quasar observations, the angular positions of both radio sources are known to a high accuracy of 10-20 nrad, and are constant. Thus the size of the quasar-quasar delay rate residuals will reflect the magnitude of limiting ΔDOD errors.

Tropospheric Effects

Before continuing, it will be helpful to briefly review the nature of tropospheric delay and delay rate errors. The errors associated with the troposphere can be separated into two classes. The first error arises in the context of modeling the troposphere as a static distribution of dry and wet atmospheric constituents whose spatial density is only a function of altitude. According to this model, the propagation delay at an arbitrary elevation angle θ_{el} is related to the zenith propagation delay by a mapping function $f(\theta_{el})$, which to first order can be approximated as $1/\sin\theta_{el}$. A more precise treatment, accounting for the spherical nature of the Earth's atmospheric shell, can be found in [5]. From a value of one at zenith, f increases to nearly 10 at 6 deg elevation angle, which corresponds to the minimum elevation angle typically allowed in the VLBI observations. Similarly, $\dot{f}$ can be approximated as $-\dot{\theta}\cos\theta_{el}/\sin^2\theta_{el}$. This has a value of zero at zenith, while at 6 deg elevation it lies between $\pm 6.6 \times 10^{-3}$, depending on whether the source is rising, setting, or transiting.

According to this model, then, the troposphere error on a ΔDOD observation is simply a function of the uncertainties in the zenith troposphere delay at each station and the derivatives of the mapping function along the four ΔDOD lines of sight:

$$\Delta\tilde{y}_{trop} = (\dot{f}_{1A} - \dot{f}_{1B})\delta\hat{\tau}_{trop}(1) - (\dot{f}_{2A} - \dot{f}_{2B})\delta\hat{\tau}_{trop}(2) \quad (6)$$

where the subscripts 1,2 designate stations, the subscripts A, B designate radio sources, and $\delta\hat{\tau}_{trop}(i)$ is the zenith tropospheric delay error at station i. On intercontinental baselines, the zenith troposphere uncertainties at the two stations can be assumed to be completely uncorrelated. Surface meteorological data typically provide *a priori* zenith troposphere delays with uncertainties of about 3-4 cm for DSN sites [6].

This static troposphere model neglects both spatial and temporal variations in the atmospheric refractivity. The wet component of the troposphere is particularly affected by such inhomogeneities since water vapor is not a well-mixed atmospheric constituent. Tatarski [7] and Ishimaru [8] have shown that over a broad range of scales, the structure function of atmospheric refractivity follows a power law based on Kolmogorov turbulence. Treuhaft and Lanyi [9] have numerically integrated this refractivity structure function for pairs of VLBI raypaths to obtain spatial and temporal statistics of the effect of water vapor fluctuations on VLBI delay and rate observables.

Some of the major conclusions of this latter analysis include: (1) the power-law behavior of the temporal fluctuations statistics have a breakpoint at around 120 sec, corresponding to the 1 km scale height of the wet troposphere divided by the typical 8 m/s tropospheric windspeed at DSN sites; (2) on time scales much less than 120 sec, the single site delay rate RMS scales as $T^{-1/6}f^{1/2}$ where T is the scan length, and f is the atmospheric pathlength; (3) on time scales much greater than 120 sec, the single site delay rate RMS scales as $T^{-2/3}f$; (4) at the time scale of 200 sec, corresponding to the typical scan lengths of the

Table 1
THE MAGELLAN BLOCK I ΔDOD ERROR BUDGET

Error Source	**ΔDOD Error (sec/sec)**
System Noise - Quasar (70m/70m, 0.2 Jy, 600 sec)	6.7×10^{-15}
System Noise - S/C (34m, superior conjunction)	$< 3 \times 10^{-15}$
Quasar Position (30 nrad/component)	4.0×10^{-14}
Station Oscillator ($\sigma_y(90\,\text{min}) = 10^{-14}$)	4.7×10^{-14}
Instrumental Dispersion ($\Delta\phi = 0.1$ cyc)	3.0×10^{-14}
Baseline Error (50 cm/component, $\Delta\theta = 20$ deg)	4.3×10^{-14}
Earth Orientation (30 cm/component, $\Delta\theta = 20$ deg)	2.7×10^{-14}
Static Troposphere (1 cm zenith *a priori* , $\Delta\theta_{el} = 10$ deg)	5.0×10^{-14} for $\theta_{min} = 10$ deg (1.0×10^{-14} for $\theta_{min} = 20$ deg) ($< 1.0 \times 10^{-14}$ for $\theta_{min} = 30$ deg)
Fluctuating Wet Troposphere (Treuhaft & Lanyi model [9])	2.3×10^{-13} for $\theta_{min} = 10$ deg (1.3×10^{-13} for $\theta_{min} = 20$ deg) (1.0×10^{-13} for $\theta_{min} = 30$ deg)
Ionosphere (S/X calibration)	$< 3 \times 10^{-15}$
Solar Plasma (S/X calibration)	$< 3 \times 10^{-15}$
RSS ΔDOD Error	2.4×10^{-13} for $\theta_{min} = 10$ deg (1.6×10^{-13} for $\theta_{min} = 20$ deg) (1.3×10^{-13} for $\theta_{min} = 30$ deg)

(Adapted from Ref. [3]. $\Delta\theta$ is the separation angle of the quasar and spacecraft; $\Delta\theta_{el}$ is the difference in elevation angles for the spacecraft and quasar, at the station for which the difference is maximum; θ_{min} is the minimum elevation angle for the spacecraft and quasar lines of sight at the two stations.)

Block 0 data used in this analysis, we are in the transition region between the short and long time behavior; the local power law dependence on elevation at this time scale is roughly $f^{0.6}$; (5) the delay rate observable is particularly affected by temporal fluctuations since it is a differential phase measurement over short time scales; (6) for two non-simultaneous observations at two lines of sight, the resulting fluctuation-induced rate errors are nearly uncorrelated, even if the lines of sight are angularly close [10].

This last point implies that differencing rate observations for two sources will achieve little if any cancellation of the stochastic troposphere errors, even if the observations are close together both spatially and temporally. This will be tested by examining the dependence of $\Delta\tilde{y}_{ij}$ RMS residuals on angular source separation and observation time separation.

THE JPL BLOCK 0 VLBI DATASET

The data used in this analysis come from twelve years of VLBI quasar observations collected on the California - Spain and California - Australia intercontinental DSN baselines, collected between 1978 and 1989. All of the data are from dual-band experiments with observations at S-band (2.3 GHz) and X-band (8.4 GHz). The Block 0 VLBI data acquisition system was used to sample a 2 MHz received bandwidth at the Nyquist rate of 4 MHz and record the single-bit-quantized digital data on videotape. After shipment of the recorded tapes to JPL, the data were subsequently processed at the Caltech/JPL Block 0 correlator. For these Block 0 experiments, the 2-MHz observation bandwidth was time-multiplexed between three S-band and three X-band channels, with 80% of the time spent at X-band. (The multiple channels at each band were used to obtain accurate VLBI delay observables using the technique of bandwidth synthesis [11].)

These data were acquired under the aegis of the JPL Catalog Maintenance and Enhancement program, with the goals of establishing a high-accuracy inertial reference frame defined by source positions of several hundred distant quasars, and monitoring DSN station locations in that frame. Typical observing sessions were 24 hours long, with several hundred quasar observations, the vast majority of which were about 200 sec in length. Observations were schedule over a wide range of azimuths and elevations to break correlations between station locations, troposphere delays, and station clocks. Differential quasar-quasar observables were not a goal of these experiments; however, many pairs of quasar observations can be found which are close in terms of both angular separation on the sky and difference in observing epoch. Further descriptions of the dataset can be found in [12].

QUASAR-QUASAR ANALYSIS

Formation of Quasar-Quasar Observables

The basic philosophy of this analysis was to use the best *a priori* models to form the ΔDOD residuals, without fitting for any model parameters. This is in the spirit of typical navigation measurements, in which a minimal set of observations is made, precluding the possibility of solving for session-specific parameters such as zenith tropospheres, station locations, UTPM, etc.

In the first step of the analysis, the observed S- and X-band delay rates for each scan were linearly combined to form an observable free from charged particle errors:

$$y_{S/X} = \frac{\nu_X^2}{\nu_X^2 - \nu_S^2} y_X - \frac{\nu_S^2}{\nu_X^2 - \nu_S^2} y_S . \tag{7}$$

Next, a model delay rate was calculated for and subtracted from the observed S/X-calibrated delay rate for each of the 8730 scans in the 1978-1989 VLBI dataset. The *a priori* model troposphere delay rates were based on the Chao troposphere model of average monthly dry and wet troposphere delays at DSN sites [6]. These Chao monthly averages have an expected RMS error of 3-4 cm. The earth orientation *a priori* model was based on measurements from the IRIS VLBI network [13], which conducts periodic measurements of UT1-UTC and polar motion.

This preliminary processing provides a residual delay rate $\tilde{y}_i$ for each scan i. The next step in the analysis is to form all quasar-quasar pairs (i, j) that satisfy two simple criteria:

$$\#1: \quad S_{ij} < 60 \text{ deg},$$

where S_{ij} is the angular separation (arclength on the celestial sphere) between the two radio sources, and

$$\#2: \quad \Delta T_{ij} < 90 \text{ minutes},$$

where ΔT_{ij} is the time difference between the scan midpoints for the two observations. For the Magellan mission, the quasar and spacecraft observations will typically be separated by about 90 minutes, which is half the orbital period of the spacecraft about Venus.

From the 8730 quasar observations, 71,768 pairs of scans satisfy these two criteria. For each pair (i, j), the ΔDOD variable $\Delta\tilde{y}_{ij} \equiv \tilde{y}_i - \tilde{y}_j$ is formed by differencing the delay rate residuals for the individual quasar scans. The resulting distribution of $\Delta\tilde{y}_{ij}$ then characterizes the ΔDOD errors. As no data editing has been performed so far, the $\Delta\tilde{y}_{ij}$ distribution at this point has long non-Gaussian tails.

The resulting variance of the distribution is dominated by the relatively small fraction of outliers in the tails. As a result it was decided at this stage to place a hard cut on $\Delta\tilde{y}_{ij}$, introducing a third quasar-quasar criterion:

$$\#3: \quad |\Delta\tilde{y}_{ij}| < 2.0 \times 10^{-12}.$$

This final requirement rejected 670 of the 71,768 quasar-quasar pairs, or less than 1%.

RMS ΔDOD Residual

The $\Delta\tilde{y}_{ij}$ distribution is shown in Fig. 2a for the remaining 71,098 pairs, for which the RMS standard deviation is 3.16×10^{-13}. The cuts at $\pm 2.0 \times 10^{-12}$ are clearly well-separated from the central peak of the distribution.

Fig. 2b shows an exceedence plot for all of the data satisfying the first two quasar-quasar selection criteria, along with the theoretical exceedence plot for a Gaussian with $\sigma = 3.16 \times 10^{-13}$. The exceedence plot shows the probability that the absolute value of a ΔDOD observation residual will exceed a given threshold. We see evidence for a significant non-Gaussian tail in the last 5-10% of the $\Delta\tilde{y}_{ij}$ distribution.

(While it is somewhat unsatisfying to introduce this third *ad hoc* criterion, it is the simplest way of eliminating "bad" data points, corresponding to either experimental problems during the observation, or errors in correlation and fringe-fitting. It is a "soft" cut in that it rejects only a small fraction of the data, and to some extent it represents the way real navigation data is processed: points with large residuals are subjected to greater scrutiny and are usually rejected if, after further observations, their post-fit residuals remain more than 2-3 times the RMS postfit scatter. And although one worries that, along with

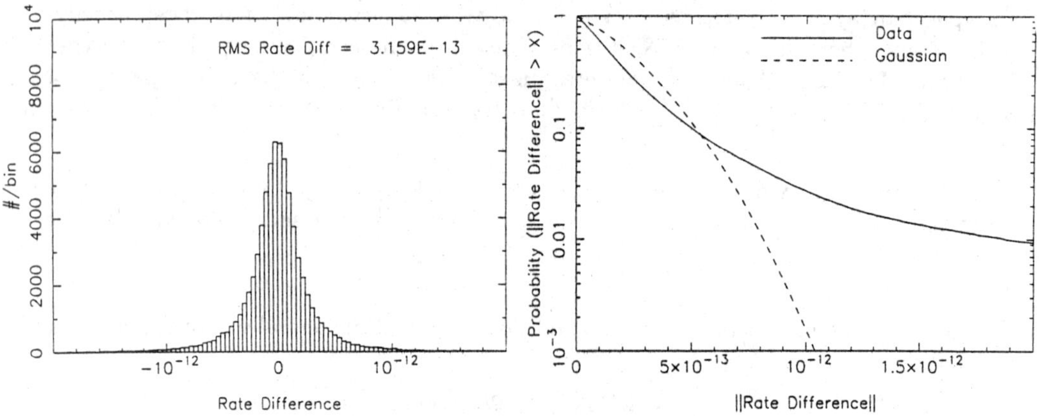

Fig. 2a. Distribution of $\Delta\tilde{y}_{ij}$ for S/X-calibrated quasar pair data. The RMS width is 3.16×10^{-13}. **Fig. 2b.** Exceedence plot for S/X-calibrated quasar pair data, along with theoretical curve (dashed line) for a Gaussian of width 3.16×10^{-13}.

"bad" data, one may be also removing the tails of the true distribution describing the limiting error sources, because of the simplicity of this cut, one can easily calculate its effect on the standard deviation for any given *a priori* distribution. For example, for a zero-mean Gaussian with a true width of 5.0×10^{-13}, rejecting data outside the range $\pm 2.0 \times 10^{-12}$ would reduce the calculated standard deviation by only 0.11%.)

In addition to the S/X delay rate residuals, the S- and X-band data were processed separately to evaluate the degradation in ΔDOD quality when dual frequency charged particle calibration is not possible. The X-band data actually had a slightly smaller RMS width of 3.12×10^{-13}. Two factors explain this. First, in forming the S/X-calibrated linear combination, the effect of system noise is increased by nearly 10%. More important, however, is that at low SEP angles or for very large ionospheric fluctuations, S-band data begin to suffer cycle slip errors before X-band, as the charged particle effects scale as $1/\nu$ in phase. Supporting this latter hypothesis, the distribution of S-band ΔDOD residuals had a much larger RMS width. As a result, for the S-band data the third quasar-quasar selection criterion was modified to

$$\#3(S): \quad |\Delta\tilde{y}_{ij}| < 1.0 \times 10^{-11}.$$

With this larger limit, the percentage of S-band scans removed by this cut remained at just under 1%. The RMS width of the resulting distribution was 1.22×10^{-12}, nearly a factor of four larger than for the X and S/X data.

For the S/X-calibrated data, statistics were also determined separately for the California - Spain (C-S) and California - Australia (C-A) DSN baselines. The RMS ΔDOD residual for the C-S baseline was 2.79×10^{-13}, while for the C-A baseline, the RMS residual rose to 3.59×10^{-13}. This result is most likely due to the longer C-A baseline length, rather than any inherent problems with the Australia data. The longer baseline forces a higher percentage of the C-A observations to be conducted at very low elevation angles, where tro-

Table 2
RMS ΔDOD RESIDUALS

Data Type	# Q-Q Pairs	Cutoff	% Exceeding Cutoff	RMS Residual
S/X-Calibrated	71,098	2.0×10^{-12}	0.93%	3.16×10^{-13}
S/X Calif-Aus	30,636	2.0×10^{-12}	1.43%	3.59×10^{-13}
S/X Calif-Spain	40,462	2.0×10^{-12}	0.55%	2.79×10^{-13}
S-band	71,152	1.0×10^{-11}	0.86%	1.22×10^{-12}
X-band	71,272	2.0×10^{-12}	0.69%	3.12×10^{-13}

pospheric and ionospheric effects are largest. Table 2 summarizes the RMS ΔDOD residual results for the various cases considered.

These results portray the overall RMS delay rate error for the 1978-1989 VLBI dataset. The remainder of the analysis presented here consists of binning the quasar-quasar observations according to various observation parameters, calculating the RMS ΔDOD residual for the observations in each data bin, and looking for systematic dependencies of the ΔDOD errors on these observation parameters.

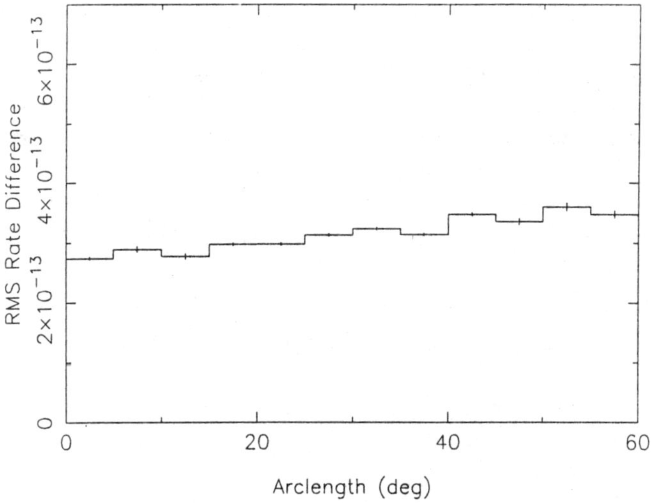

Fig. 3. Dependence of RMS $\Delta\tilde{y}_{ij}$ on arclength separation of radio source pair.

Angular Source Separation

Geometric errors due to earth orientation and baseline uncertainties, as well as static propagation media errors, should cancel in a differential observation as the separation angle between the radio sources is reduced. If these are dominant errors, we should see a signature in the dependence of the RMS ΔDOD residual on arclength separation. On the other hand, if tropospheric fluctuations dominate the rate errors, we should not see a strong separation dependence, since the rate errors decorrelate even for nearby raypaths if the observations are not simultaneous. In Fig. 3, the quasar pair data have been binned according to the arclength separation of the two radio sources, and an RMS width calculated of the $\Delta\tilde{y}_{ij}$ distribution for each bin. Only a very weak dependence on arclength is observed, with the RMS ΔDOD residual increasing by only about 25% as the arclength varies from 0 to 60 deg. This is substantially different than the case for VLBI differential delay observations, where, as reported in a previous quasar pair analysis [14], the RMS ΔDOR error exhibited a strong, roughly linear dependence on the source separation angle. As discussed above, this is due to the differing nature of static troposphere errors, which dominate the delay error budget, and fluctuating troposphere errors, which dominate the rate error budget.

Time Separation Between Observations

Due to the planned sequence of Venus surface radar mapping and subsequent data telemetry to Earth, Magellan ΔDOD tracking will consist of a quasar scan and a spacecraft scan separated by approximately 90 minutes. Over this long time interval, various stochastic parameters can change, including station frequency standards, station instrumentation, and propagation media errors. To assess the importance of these errors, the quasar pair data were binned according to the time separation between the midpoints of the two quasar scans. The resulting dependence of the RMS $\Delta\tilde{y}_{ij}$ residual on this time separation is shown in Fig. 4a. Essentially no dependence on time separation is observed, except in the very first bin, which includes scans with time separations of less than 200 sec. This weak dependence of the rate residuals on temporal separation is in accord with expectations from the troposphere fluctuation model of [9]. Fig. 4b shows this short time scale region with the data binned at higher resolution. Due to the low slew rates of large DSN antennas, short scan separation times correspond to close angular proximity; in fact, every pair with time separation less than 200 sec had an angular separation less than 20 deg. Hence these scans sample tropospheric fluctuations on short temporal and spatial scales. Only for the closest spatial and temporal separations do we observe any appreciable differential cancellation of the fluctuating troposphere errors.

Elevation Angle and Atmospheric Pathlength Dependence

Another way to isolate tropospheric effects is to bin the data according to elevation angle. Each ΔDOD observation involves four lines of sight, so it is difficult to characterize an observation with a single parameter. In Fig. 5a, the RMS $\Delta\tilde{y}_{ij}$ residual is plotted against θ_{min}, the minimum elevation angle for the four lines of sight. Here we see a very strong dependence, with the rate error varying by a factor of four between the lowest and highest values of θ_{min}. A different parametrization is used in Fig. 5b, where the RMS residual is plotted as a function of the total mapping function, *i.e.*, the sum Σf of the mapping functions for the four lines of sight. Again we observe a strong dependence. A fit of the RMS residual σ to a power-law dependence $(\Sigma f)^\alpha$ gave an exponent of $\alpha = 0.50 \pm 0.05$, close to the value of about 0.6 expected according to the Treuhaft-Lanyi tropopshere fluctuation model [9] for scan lengths of 200 sec.

If static troposphere errors were dominant, we would expect to see a functional dependence based on the expression in Eq. 6, which relates the ΔDOD error to uncertainties in the zenith troposphere at each site. For an uncertainty of σ_{trop} in the zenith troposphere

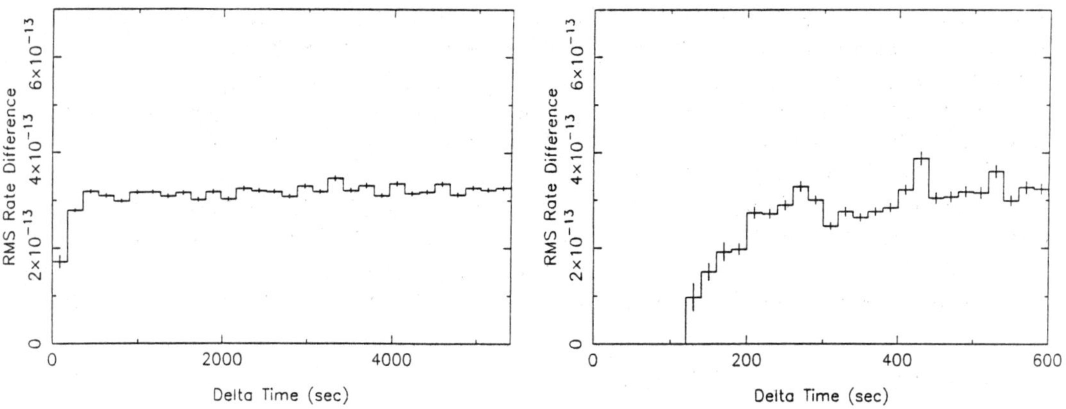

Fig. 4a. Dependence of RMS $\Delta\tilde{y}_{ij}$ on time separation between scan midpoints of radio source pair. **Fig. 4b.** The range 0-600 sec from Fig. 4a, viewed with finer bin resolution (20 sec bins).

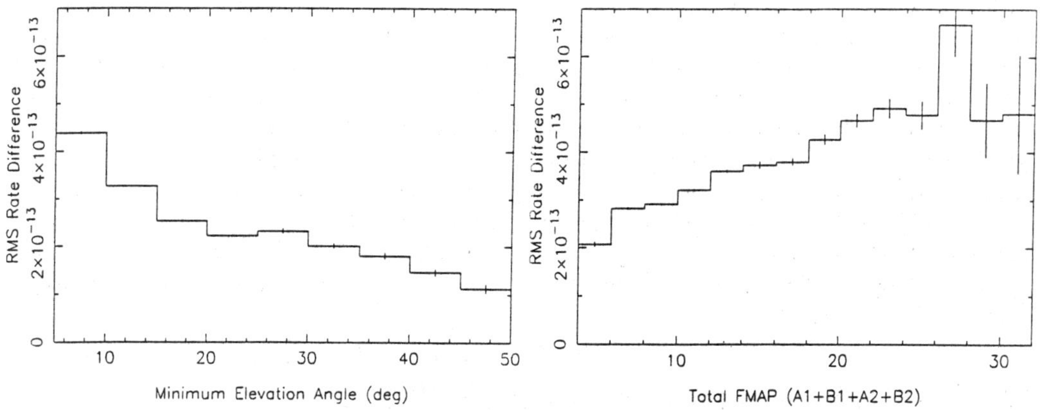

Fig. 5a. Dependence of RMS $\Delta\tilde{y}_{ij}$ on the minimum elevation angle θ_{min} among the four ΔDOD lines of sight. **Fig. 5b.** The RMS $\Delta\tilde{y}_{ij}$ residual as a function of the total atmospheric pathlength Σf.

at each site, and no correlation between the two sites, the resulting ΔDOD uncertainty is

$$\begin{aligned} \sigma_{\Delta \mathrm{DOD}} &= \sigma_{trop}\sqrt{(\dot{f}_{1A} - \dot{f}_{1B})^2 + (\dot{f}_{2A} - \dot{f}_{2B})^2} \\ &\equiv g_{static}\sigma_{trop} \end{aligned} \tag{10}$$

where we have defined g_{static} to be the combination of mapping function derivatives that is proportional to the static troposphere ΔDOD error. Figure 6 shows the $\Delta\tilde{y}_{ij}$ RMS residual dependence on g_{static}, along with the predicted dependence for $\sigma_{trop} = 3$ cm $= 100$ psec, assuming that static tropophere errors are the dominant error source.

Several interesting observations can be made from Fig. 6. First of all, even for observing geometries for which g_{static} is near zero, where the static troposphere error should be very small, the RMS $\Delta\tilde{y}_{ij}$ residual is still large. In fact, for the first bin of Fig. 6, where $g_{static} < 2\times10^{-4}$, (corresponding to a ΔDOD error of 2×10^{-14},) the RMS $\Delta\tilde{y}_{ij}$ residual is 2.41×10^{-13}, over 75% of the RMS value for the entire dataset. The bulk of the simulated ΔDOD observations correspond to a small value of g_{static}: nearly 3/4 of the data have $g_{static} < 10^{-3}$. Thus static errors are not dominant for the majority of observations. On the other hand, as g_{static} increases, the RMS residual does indeed rise, and at a slope roughly consistent with the expectation based on a 3 cm *a priori* zenith troposphere uncertainty. For the small subset of observations where the observing geometry leads to large values of g_{static}, it appears that the static troposphere errors do contribute significantly.

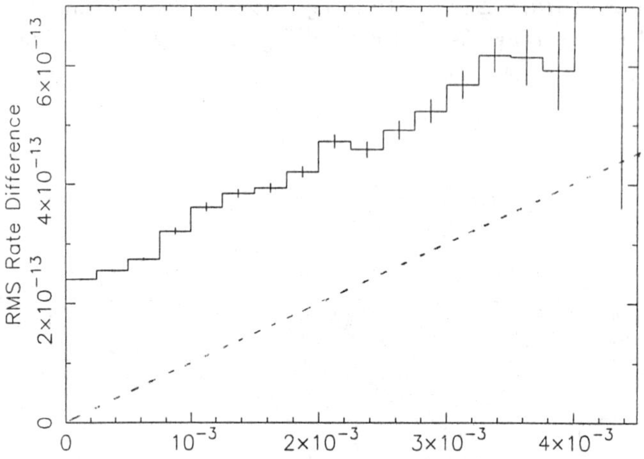

Fig. 6. Dependence of RMS $\Delta\tilde{y}_{ij}$ on the function g_{static} of pathlength derivatives. The dotted line indicates the expected static troposphere error for an uncorrelated 3 cm zenith troposphere path delay error at each station.

Solar Plasma Effects

Magellan will spend much of its mission in close angular proximity to the Sun, as viewed from Earth. The plasma of charged particles in the solar wind introduces phase scintillations and path delay errors which affect VLBI observables [15], and these effects increase as the sun-earth-probe angle decreases. Because the raypaths from the two stations pass through different regions of the solar plasma, spatial and temporal fluctuations in the charged particle distribution will lead to errors on the spacecraft differential Doppler observable. In addition, rapid temporal fluctuations can lead to difficulties in phase tracking the received spacecraft signal, resulting in possible cycle slips in the tracked phase. For instance, a one-cycle phase jump in the middle of a scan of length T at observing frequency ν will cause a delay rate error of approximately $2/\nu T$. For $T = 200$ sec, this corresponds to delay rate errors of 4.4×10^{-12} and 1.2×10^{-12} at S-band and X-band respectively.

Because charged particle phase errors scale as $1/\nu$, (and delay and dimensionless delay rate errors as $1/\nu^2$,) phase tracking problems are most serious at low frequencies. Thus, as the sun-earth-probe angle is decreased, S-band data will be affected before X-band data are. This is clearly evident in Fig. 7, where the RMS $\Delta\tilde{y}_{ij}$ residual is plotted for S, X, and S/X quasar pair data as a function of the minimum sun-earth-quasar (SEQ) angle for the two quasars in each simulated ΔDOD observation. For SEQ angles of less than about 20 deg, the S-band residuals rise very steeply. In addition, even for SEQ$>$ 20 deg, there is still a visible, albeit much weaker, correlation with SEQ angle. This latter effect may be due not to solar plasma errors but rather to a correlation between SEQ angle and ionospheric errors, since ionospheric activity is higher during the daytime. The strong behavior inside of SEQ = 20 deg, however, does seem to be related to the solar plasma.

The X-band and S/X data show a much less pronounced residual increase over the range 5-20 deg and essentially no SEQ dependence for SEQ $>$ 20 deg. In the first bin, 0 $<$ SEQ $<$ 5 deg, the X and S/X data do show a marked increase in residuals, although the RMS value is still much less than for the S-band data. The S/X data do not appear to be of better quality than the X-band only data; in fact, for SEQ $<$ 5, the S/X-calibrated data have a higher RMS residual than the X-band only, presumably due to the deleterious effect of the highly corrupted S-band rates.

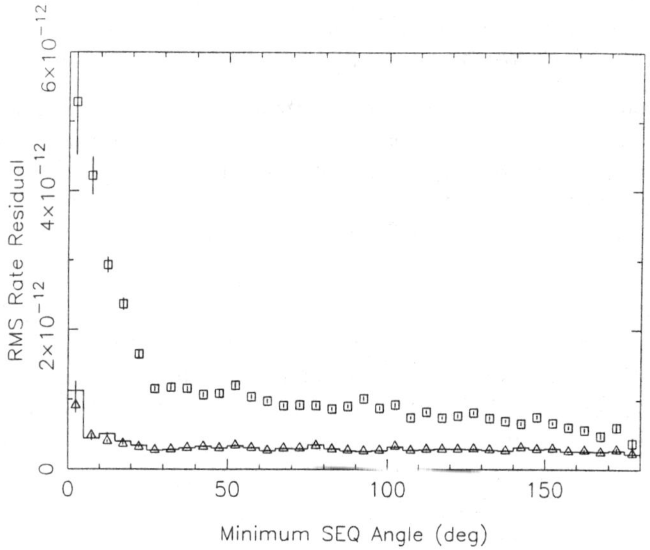

Fig. 7. Dependence of RMS $\Delta\tilde{y}_{ij}$ residual on the minimum sun-earth-quasar (SEQ) angle for S-band (squares), X-band (triangles), and S-X calibrated (solid line) quasar pair data.

Ionospheric Effects

In addition to the solar plasma, charged particles in the Earth's ionosphere also contribute errors to the ΔDOD observable [16]. The total columnar electron content in the ionosphere varies diurnally by factors of 3-8 with a maximum in the early afternoon. In addition, the daytime maximum varies by a factor of about 2 depending on season, and also varies by factors of 2-3 over the 11-year solar cycle. Like the troposphere, one can distinguish two types of ionospheric errors: a static error due to uncertainty in the zenith ionospheric delay, and a stochastic error due to spatial and temporal variations in the line-of-sight electron content. However, a simple mapping function does not characterize the elevation dependence nearly as well for the ionosphere as for the troposphere.

The *a priori* rate model used in forming the delay rate residuals for this analysis included no ionosphere modeling; *i.e.*, the *a priori* electron content was assumed to be zero. Some reduction of the static component of the ionospheric errors for the single-band data could probably be achieved by including an *a priori* model value for the zenith ionospheric content at each station as a function of local time, season, and latitude. However, such models are of only very limited accuracy, due to the high variability of the ionosphere. The S/X-calibrated data should be essentially free from ionospheric effects.

To isolate the ionospheric contribution to the ΔDOD errors in the quasar-quasar dataset, the RMS $\Delta\tilde{y}_{ij}$ residual was plotted against the ionospheric longitude. The ionospheric longitude is defined as the geocentric longitude in a sun-fixed reference frame of the point where an observation line-of-sight intersects the ionospheric shell (assumed to be at an altitude of 350 km). The sun-fixed spherical coordinate system on the ionospheric shell has its pole aligned with the Earth's pole, but has its origin of longitude fixed in the direction towards the sun. This is a useful coordinate system for studying ionospheric effects, as the ionospheric total electon content depends systematically on the ionospheric longitude. The total electron content reaches a maximum near 0-60 deg east ionospheric longitude, corresponding to early afternoon in local time.

For each quasar-quasar pair, one can calculate a minimum ionospheric longitude, *i.e.*, the ionospheric longitude whose absolute value is closest to zero among the four ΔDOD lines of sight. In Fig. 8 the RMS $\Delta\tilde{y}_{ij}$ residual is plotted against this minimum ionospheric longitude. To more clearly separate the ionospheric effects from those due to solar plasma, Fig. 8 only includes quasar-quasar pairs for which the minimum SEQ angle is greater than 20 deg.

Again, we see a clear signature in the S-band data showing a correlation of the RMS residual with the daytime maximum in total electron content. The X-band residuals are well below the S-band data, and show no dependence on ionospheric longitude, indicating that they are essentially unaffected by ionospheric effects. Not suprisingly, then, the S/X-calibration does not lead to any improvement over the results using only X-band.

DISCUSSION AND CONCLUSIONS

Comparing the results of this ΔDOD analysis to an earlier analysis of ΔDOR errors points out marked differences in the relative importance of various VLBI errors sources on the two types of observables. Whereas ΔDOR observations appear to be dominated currently by uncertainties in clock offsets, zenith troposphere delays, and earth orientation, this analysis demonstrates that ΔDOD observations are dominated by short-term stochastic temporal troposphere fluctuations.

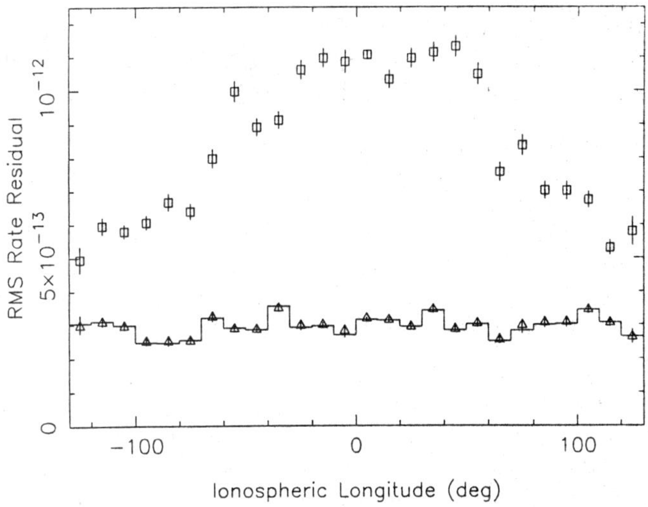

Fig. 8. Dependence of RMS $\Delta\tilde{y}_{ij}$ residual on the minimum ionospheric longitude for S-band (squares), X-band (triangles), and S-X calibrated (solid line) quasar pair data.

Observing Strategies

Recent studies [17] have shown how expanding the spacecraft-quasar observation to include many spatially separated quasars, thereby forming a "local reference frame," allows one to solve for and remove the systematic errors that dominate the ΔDOR delay errors. This technique can only work, however, when the limiting systematic errors can be well-modeled by a small number of parameters. The stochastic nature of the ΔDOD rate errors implies that such an approach will not be as applicable in ΔDOD observations. We did observe that for extreme geometries, large errors could be attributed to static troposphere errors, but these represented a minority of the quasar-quasar data. For this subset of observations, it may be possible to reduce the static troposphere errors by using a local reference frame observing technique to solve for corrections to the zenith troposphere delays at each station. However, for the bulk of the observations, the fluctuating troposphere dominates the ΔDOD error; this stochastic error is not reduced by the local reference frame technique.

The lack of dependence on scan time separation simply reinforces the known high stability of the H-maser frequency references and local oscillators used at the DSN stations. Even at time scales of 90 minutes, where environmental effects could start to important, and where H-masers begin to rise from their flicker-frequency noise floor, there is still no sign of instrumental stability errors competing with tropospheric instabiity in their effects on ΔDOD observables.

The very weak dependence of the ΔDOD error on angular source separation, which can be understood in the context of the fluctuating troposphere model, means that quasars with larger angular separations from a spacecraft can be used as navigation references with very little increase in ΔDOD error. By increasing the allowable separation angle $\Delta\theta$, the area on the sky from which a quasar can be selected is increased by roughly $(\Delta\theta)^2$, enabling far more quasars to be considered as potential navigation reference sources. This should allow brighter quasars to be used for tracking; for Magellan this means that observations could be shifted from the 70-m antenna to the 34-m antenna, using a brighter quasar to make up for the reduced antenna gain. In addition, the freedom to select a quasar from a larger region of the sky should make it possible to avoid observing geometries for which the static troposphere error is large, and to more easily find sources angularly far from the sun.

(It is important to reemphasize that these arguments apply only to the ΔDOD observable because the rate errors are dominated by troposphere fluctuations. The ΔDOR differential delay observable, when only a single reference quasar is used, is still dominated by geometric errors due to earth orientation and static propagation media, whose error contributions scale linearly with angular source separation. As a result, small angular separations are still crucial in reducing ΔDOR tracking errors, and this requires a dense catalog of radio sources, particularly along the ecliptic.)

The high quality of the single-frequency X-band data suggests that good ΔDOD tracking can be obtained for spacecraft with an X-band-only downlink capbility. In fact, dual-frequency S/X calibration of the quasar data actually led to a slight increase in the observed ΔDOD residuals, presumably due to occasional cycle slips in the S-band data for observations with large charged particle errors. Using only S-band data, however, led to a drastic increase in the observed RMS ΔDOD residuals, particularly near the sun and at daytime ionospheric maximum. Of course, other mission requirements, including ΔDOR tracking or precise ranging for radio science, might provide independent justification for a dual-band spacecraft transponder.

It is worth mentioning here that, simply due to the huge volume of data in the quasar VLBI dataset, the observations used in this analysis have not received the same scrutiny that typical Block I navigation measurements receive. Closer examination of the phase residuals within each quasar scan might have allowed detection and correction of cycle slips. (Neither the Block 0 nor the Block I system currently attempts to use the X-band phase residuals to aid phase tracking of the S-band data when charged particles begin to corrupt the lower frequency. This is one way in which the reliability of the S-band data could be improved.)

Extrapolation of Block 0 Results to the Block I

There are several important differences between the Block 0 VLBI data acquisition system, which was used to collect the data presented in this paper, and the Block I system, which is used operationally for spacecraft tracking. The most important difference involves the recorded bandwidths of the two systems. The Block 0 samples and records a 2 MHz band, while the Block I only records a 250 KHz band. (This reduced bandwidth allows the Block I to store the observed signals on disk and transfer them in near-realtime to a correlation facility at JPL, enabling fast tracking turn-around for operational spacecraft navigation applications.)

Because the signal to noise ratio of a VLBI observation varies as the square root of the product of recorded bandwidth and observing time, the Block I system typically uses significantly longer scan lengths than the Block 0. Most of the Block 0 scans in this analysis were about 200 sec in duration. Typical Block I scans, however, are 600 seconds long. The difference in scan lengths will cause stochastic errors to affect the two systems differently, depending on the underlying spectral characteristics of the error source.

Except at very low SEQ angles, the data presented here strongly support the hypothesis that the ΔDOD errors are dominanted by wet troposphere fluctuations. The model of

[9] predicts how the RMS delay rate error varies with scan length. Fig. 9 shows the results of numerical integrations based on the model of [9] for the single station RMS delay rate at a variety of elevation angles. For a ΔDOD observation, one would add in quadrature the errors on the four lines of sight, at least to the extent that the fluctuation errors were uncorrelated (as they appear to be in this dataset, except for the smallest temporal and spatial separations.)

At high elevations, the delay rate error decreases quickly with longer scan times, dropping 45% at zenith and 42% at 45 deg elevation as the scan length is increased from 200 sec up to 600 sec. However, at lower elevations, where troposphere errors are largest, the temporal dependence is weaker. At 10 deg elevation, the rate error drops 14%, and at 6 deg elevation, the decrease is only 8%. Thus the explicit scan length dependence will be a function of the observing geometry. If we take the approximate median elevation angle of 20 deg for the scans in this dataset as a guide, we expect the ΔDOD errors to be reduced by 28% for the longer Block I scans. This reduces the overall ΔDOD RMS error from 3.16×10^{-13} down to 2.28×10^{-13}, just within the Magellan project requirement of 2.4×10^{-13}. However, it is evident that there is very little margin, particularly since lower elevation data may be of poorer quality. Hopefully the added flexibility due to relaxing the angular separation constraint can help to select observing geometries with smaller total pathlength, thereby reducing the ΔDOD troposphere error.

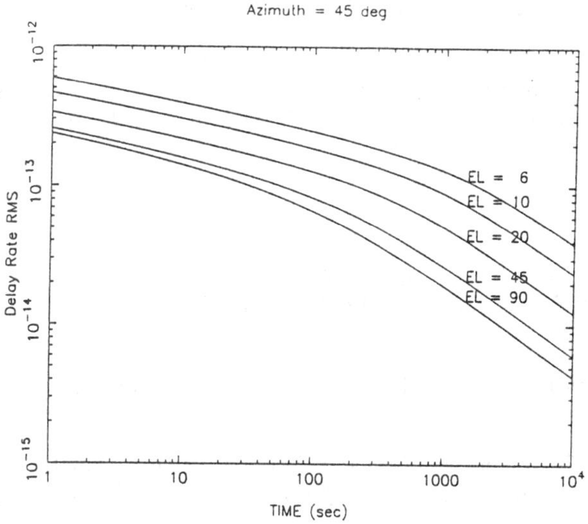

Fig. 9. Predicted single-station delay rate error as a function of scan length, based on the model of [9], for several different elevation angles.

Another significant difference between Block 0 and Block I observations relates to time multiplexed observations and its effect on cycle slips. Block 0 data typically have a six-second multiplexing period: each of the six frequency channels are observed every six seconds, for either 0.4 sec (S-band) or 1.6 sec (X-band). For Block I ΔDOD observations, on the other hand, only a single passband is typically observed for S-band and at X-band, and these can be power combined so that no multiplexing is required, allowing continuous phase tracking. By providing better phase tracking difficult during high phase dynamics, this strategy can reduce cycle slips for Block I ΔDOD observations.

Troposphere Calibrations

Before closing, some mention should be made of the potential role of improved tropospheric calibrations. Water vapor radiometer (WVR) calibrations have demonstrated the potential to obtain cm-level zenith troposphere accuracy, which would represent a significant improvement over the 3-4 cm uncertainties in the *a priori* zenith delays which result when the troposphere model is based solely on surface meteorological data. However, for the majority of ΔDOD observations, the static errors due to zenith delay uncertainty are dominated by the fluctuating component of the troposphere.

For WVR measurements to provide useful ΔDOD calibrations, it is crucial that they be optimized for line-of-sight calibrations of the fluctuating delay over the individual spacecraft and quasar observations. This requires WVRs with excellent short-term gain stability. In addition, narrow WVR beamwidths will be required to reduce the spatial filtering of small scale spatial and temporal fluctuations, and to enable line-of-sight operation at the low elevation angles used in intercontinental VLBI observations. An ideal system might couple the WVR optics to the primary antenna to provide true line-of-sight calibrations coincident with the tracking antenna beamwidth [18].

Acknowledgement

I would like to thank J. Border and R. Treuhaft for several stimulating discussions and suggestions.

The research described in this paper was carried out by the Jet Propulsion Laboratory, California Institute of Technology, under a contract with the National Aeronautics and Space Administration.

References

[1] S. N. Mohan and P. B. Esposito, "Venus Radar Mapper Orbit Accuracy Analysis," Paper AIAA-84-1986 presented at the AIAA/AAS Astrodynamics Conference, Aug 20-22, 1984.

[2] Magellan Navigation Plan 630-51 Rev. B, JPL D-1480 Rev. B, (internal document), in press.

[3] J. S. Border, "An Error Analysis for Magellan Differential Delay Rate Measurements," JPL Engineering Memorandum 335-96, (internal document), February 23, 1987.

[4] A.R. Thompson, J.M Moran, and G.W. Swenson, Jr., *Interferometry and Synthesis in Radio Astronomy,* New York, John Wiley and Sons, 1986.

[5] G.E. Lanyi, "Tropospheric Delay Effects in Radio Interferometry," *TDA Progress Report 42-78,*pp. 152-159, Jet Propulsion Laboratory, Pasadena, CA, April-June, 1984.

[6] C. C. Chao, "The Tropospheric Calibration Model for Mariner Mars 1971," JPL Technical Report 32-1587.

[7] V. I. Tatarski, *Wave Propagation in a Turbulent Medium,* New York, Dover, 1961.

[8] A. Ishimaru, *Wave Propagation and Scattering in Random Media,* New York, Academic Press, 1978.

[9] R. N. Treuhaft and G. E. Lanyi, "The Effect of the Dynamic Wet Troposphere on Radio Interferometric Measurements," *Radio Sci,* 22, 251-265, 1987.

[10] C. D. Edwards, "The Effect of Spatial and Temporal Wet-Troposphere Fluctuations on Connected Element Interferometry," *TDA Progress Report 42-97,* Jet Propuslion Laboratory, Pasadena, CA, pp. 47-57, May 15, 1989.

[11] A.E.E. Rogers, "Very Long Baseline Interferometry with Large Effective Bandwidth for Phase-Delay Measurements," *Radio Sci,* 5, 1239-1247, 1970.

[12] O. J. Sovers, C. D. Edwards, C. S. Jacobs, G. E. Lanyi, K. M. Liewer, R. N. Treuhaft, "Astrometric Results of 1978-85 Deep Space Network Radio Interferometry: The JPL 1987-1 Extragalactic Source Catalog," *Astron. J.,* 95, 1647-1658, 1988.

[13] W. E. Carter and D. S. Robertson, in *International Symposium on Space Techniques for Geodynamics, Proceedings 1,* Research Institute of the Hungarian Academy of Sciences, Sopron, Hungary, pp. 214-222, 1984.

[14] B. K. Trinkle and S. M. Lichten, "Differential Very Long Baseline Interferometry for 50 Nanoradian Deep Space Navigation: Results from Quasar Pair Experiments," presented at the AAS/AIAA Astrodynamics Specialist Conference, Vail, CO, August 12-15, 1985.

[15] R. D. Kahn and J. S. Border, "Precise Interferometric Tracking of Spacecraft at Low Sun-Earth-Probe Angles," Paper 88-0572, AIAA Aerospace Sciences Meeting, Reno, NV, January 11-14, 1988.

[16] P. S. Callahan, "Ionospheric Variations Affecting Altimeter Measurements: A Brief Synopsis," *Marine Geodesy,* 8, 249-263, 1984.

[17] R. N. Treuhaft, "Deep Space Tracking in Local Reference Frames," *TDA Progress Report 42-94,* Jet Propuslion Laboratory, Pasadena, CA, pp. 1-15, August 15, 1988.

[18] G.M. Resch, D.E. Hogg, and P.J. Napier, "Radiometric Correction of Atmospheric Path Length Fluctuations in Interferometric Experiments," *Radio Sci,* 19, 411-422, 1984.

AAS 89-404

AN APPROACH FOR TARGETING LANDERS AND PENETRATORS USING ORBITAL OPTICAL NAVIGATION*

Tseng-Chan Wang†, Thomas C. Duxbury‡, Stephen P. Synnott** and Kathleen Edwards††

A new approach for targeting landers, rovers and penetrators to specific points on the surfaces of planetary bodies using on–board, orbital optical navigation data is presented. Earth-based radiometric tracking of the orbiting spacecraft provides a very precise determination of the spacecraft orbit about the primary planetary body. On-board optical navigation images are then used to relate the spacecraft orbit to the surface of the target body. This paper concentrates on the problem associated with using the on-board optical navigation data which will have to estimate, in a near-real time orbit determination process, a large set of parameters including the spacecraft orbit and primary body gravity field, the rotational properties of the planetary body, the coordinates of surface features, and the camera pointing and orientation of each picture. A batch-sequential formulation of the standard least squares problem is used with the addition of backward smoothing. A square-root information filter is also used for numerical precision and stability and to minimize computer memory. To illustrate this approach, over 100 images of Phobos were processed to estimate about 2000 parameters including the camera pointing angles, the rotational properties of Phobos and the locations of surface features.

INTRODUCTION

Current and future planetary missions will involve the delivery of landers, rovers and penetrators to specific points on the surfaces of planetary bodies. These missions include the Soviet Phobos 88 mission which was planned to place a short-term hopper and a long-term lander on Phobos, the Comet Rendezvous and Asteroid Flyby[1] (CRAF) mission which will place penetrators into the surface of a comet, and the Soviet Mars 94 and U.S. Mars Rover Sample Return[2] (MRSR) missions which will place landers and rovers on Mars.

The safety and survivability of these probes to the surface will depend on first selecting safe landing sites, and then precisely targeting the probes to specific points on the surface and performing terminal guidance to land at these points.[1,3] Both CRAF and MRSR have

* This work was carried out at the Jet Propulsion Laboratory, California Institute of Technology, Pasadena, California, under contract to the National Aeronautics and Space Administration.

† Member of Technical Staff, Jet Propulsion Laboratory, California Institute of Technology, 4800 Oak Grove Drive, Pasadena, California 91109.

‡ Senior Scientist, Jet Propulsion Laboratory, California Institute of Technology, 4800 Oak Grove Drive, Pasadena, California 91109.

** Optical Systems Analysis Group Supervisor, Jet Propulsion Laboratory, California Institute of Technology, 4800 Oak Grove Drive, Pasadena, California 91109.

†† Cartographer, U. S. Geological Survey, 601 E. Cedar Avenue, Flagstaff, Arizona 86001.

requirements to land relative to known features to accuracies at the meter level.[1,4] This level of accuracy ensures missing known hazards (e.g., boulders, chasms) and places the probes in areas of favorable local ground slopes.

Earth-based radiometric tracking (e.g., 2 and 3-way doppler, ranging, VLBI) of the orbiting spacecraft carrying the landers/rovers/penetrators provide a very precise determination of the spacecraft orbit about the primary planetary body (e.g., Mars, a comet). On-board optical navigation images are then used to relate the spacecraft orbit to the surface of the target body. Also, in the case of Phobos 88, it was planned to use the optical data to tie the orbit of the spacecraft about Mars to both the orbit of Phobos and the surface of Phobos.

This paper concentrates on the problem associated with using the on-board optical navigation data. The optical data will be used to produce cartographic map products which include topography, boulder sizes, and the distribution and the location of other hazards to assess the relative safety and traversability of various landing sites. Relative topographic information is more important than absolute information in this application. Once a landing site is selected, the absolute, inertial coordinates of the spacecraft orbit and desired landing point are used to target the probes to the surface. Once deployed, MRSR may use a terminal guidance scheme which uses a pre-computed map of the desired landing site to control the landing to within a meter.

PROBLEM STATEMENT

The on-board, orbital optical navigation data will have to estimate the following parameters in a near-real time orbit determination process:

1. The spacecraft orbit and primary body gravity field. The earth-based radiometric tracking will be the primary data source for determining these parameters with the optical data providing additional refinement;
2. The orbit of a secondary body such as Phobos if the landing is to be on that body rather than the primary body;
3. The rotational properties of the planetary body including spin axis direction, rotation rate, precession, nutation, libration, pole wondering, wobble, etc. to relate the position and orientation of the surface to inertial space over time;
4. The body-fixed coordinates (e.g., latitude, longitude, radius) of surface control points (e.g., craters, boulders, albedo marks) used to produce maps of the landing sites; and
5. The camera pointing and orientation of each picture.

The widely varying stochastic nature of these parameters and the total number of parameters to be estimated restrict the application of current techniques to this problem. A new approach is presented which takes advantage of the known stochastic characteristics of the problem to obtain a feasible solution; however, this solution will require a computer with the capability of today's supercomputer class to produce solutions within the near-real time mission operational time constraints.

PROPOSED SOLUTION

The spacecraft orbit and primary body gravity field coefficients represent tens of parameters to be estimated. The gravity field coefficients are constant for all time and have no stochastic variation unless sub-sets of the total gravity field model are used. The spacecraft orbital parameters can be grouped in batches separated by orbit trim maneuvers causing orbit discontinuities or shorter arcs if a reduced gravity field model is used. Non-gravitational forces acting on the spacecraft due to solar pressure or gas leaks have varying stochastic behaviors and may also place limits on batch sizes.

The rotational properties of planetary bodies have been modelled with constant parameters for past missions. However, when dealing with sub-meter observational accuracies of any planetary body or comet having a chaotic rotation,[5] rotational parameters (5 to 10 variables) may also have to be treated as stochastic processes.

The largest set of parameters to be estimated in the orbit determination process is the set of coordinates of surface features. Three positional coordinates for each feature are needed and with thousands to hundreds of thousands of features in the control network, one can easily fill the memory (non-virtual) of any contemporary computer when attempting to manipulate the covariance associated with this problem. Therefore, a much smaller subset of features will be used to establish a global control network of the body and determine the rotational properties. Additional smaller, local control networks will be established at each potential landing site and processed separately, after the orbits and rotational parameters have been determined from the global control network.

Finally, three camera pointing angles are estimated for each picture. Using thousands of pictures for navigation makes this a large set of parameters to be estimated also. However, the pointing errors are completely random from picture to picture allowing a backward smoothed solution for these parameters to be produced after the main orbit determination process has been completed. This method provides a fully optimal solution of the camera pointing angles while not requiring these parameters to be in computer memory with the orbit determination process.

A batch-sequential formulation of the standard least squares problem will be used with the addition of backward smoothing.[6] A square-root information filter,[7] using Bierman's Estimation Subroutine Library,[8] will be used for numerical precision and stability and to minimize computer memory.

The time span for filter estimation is selected from t_0 to t_N, where t_0 is the epoch time and t_N is the final time*. This interval is subdivided into batches of varying time increments depending on the known stochastic characteristics of the problem and can be specified by the user. In the measurement updates, the *a priori* information about the uncertainties in the parameters being estimated is included as additional data to the information accumulation process. The Householder orthogonal transformations are applied to obtain an upper triangular square-root information matrix.[6,7] The data equation for batch j in $t_{j-1} \leq t < t_j$, where $0 < j \leq N$, is given by

$$\begin{pmatrix} \mathbf{R}_X & \mathbf{R}_{XP} & \mathbf{R}_{XY} \\ \mathbf{0} & \mathbf{R}_P & \mathbf{R}_{PY} \\ \mathbf{0} & \mathbf{0} & \mathbf{R}_Y \end{pmatrix}_j \begin{pmatrix} \mathbf{X}_j \\ \mathbf{P}_j \\ \mathbf{Y} \end{pmatrix} = \begin{pmatrix} \mathbf{z}_X \\ \mathbf{z}_P \\ \mathbf{z}_Y \end{pmatrix}_j , \tag{1}$$

* The definitions of all the mathematical symbols used in this paper will be summarized later in a table of mathematical notations.

where $\mathbf{R}$'s are the associated elements of the upper triangular square-root information matrix with respect to parameters $\mathbf{X}$, $\mathbf{P}$ and $\mathbf{Y}$. $\mathbf{X}_j$ is the spacecraft "pseudo-epoch" state corrections for batch j, $\mathbf{P}_j$ is the stochastic parameters that are piecewise constant having discontinuities at t_j, $\mathbf{Y}$ is the bias parameters which have constant values over the entire time span, and the $\mathbf{z}$'s are the adjusted residual vectors.

The system dynamics in terms of the pseudo-epoch state formulation from batch j to batch $(j+1)$ are given as follows:[6]

$$\begin{pmatrix} \mathbf{X}_{j+1} \\ \mathbf{P}_{j+1} \\ \mathbf{Y} \end{pmatrix} = \begin{pmatrix} \mathbf{I}_X & \mathbf{V}_P(j) & \mathbf{0} \\ \mathbf{0} & \mathbf{M}_j & \mathbf{0} \\ \mathbf{0} & \mathbf{0} & \mathbf{I}_Y \end{pmatrix} \begin{pmatrix} \mathbf{X}_j \\ \mathbf{P}_j \\ \mathbf{Y} \end{pmatrix} + \begin{pmatrix} \mathbf{0} \\ \mathbf{w}_j \\ \mathbf{0} \end{pmatrix}, \tag{2}$$

where $\mathbf{w}_j$ is the filter process noise that derives the colored noise, and $\mathbf{M}_j$ is a diagonal transition matrix computed for each batch as follows:

$$\mathbf{M}_j = \text{Diag}\left[\exp(-\Delta t_j/\tau_1), \ldots, \exp(-\Delta t_j/\tau_{np})\right],$$

where $\Delta t_j = t_{j+1} - t_j$, τ_k is the user input correlation time for the k-th stochastic parameter, and np is the number of stochastic parameters. The matrix $\mathbf{V}_P(j)$ (i.e., $(\partial \mathbf{X}_{j+1}/\partial \mathbf{P}_j)_{P_j=0}$) is the partials of the spacecraft state with respect to the parameters $\mathbf{P}$ from batch j to batch $(j+1)$. Eq. (2) is used in the following two filter time updating processes in the order of deterministic update and then colored noise update.

$$\text{Deterministic Update :} \qquad \mathbf{X}_{j+1} = \mathbf{X}_j + \mathbf{V}_P(j)\,\mathbf{P}_j \tag{3}$$

$$\text{Colored Noise Update :} \qquad \mathbf{P}_{j+1} = \mathbf{M}_j\,\mathbf{P}_j + \mathbf{w}_j \tag{4}$$

Let p_j be a stochastic parameter in $\mathbf{P}_j$. The results of the colored noise component update one at a time[7] for p_j are given as follows:

$$\sigma_p^*\, p_j + \begin{pmatrix} R_{pX}^* & r_{pp}^* & R_{pY}^* \end{pmatrix} \begin{pmatrix} \mathbf{X}_{j+1} \\ p_{j+1} \\ \mathbf{Y} \end{pmatrix} = z_w^*, \tag{5}$$

and

$$\begin{pmatrix} \mathbf{R}_X & R_{Xp} & \mathbf{R}_{XY} \\ \mathbf{0} & r_p & R_{pY} \\ \mathbf{0} & 0 & \mathbf{R}_Y \end{pmatrix}_{j+1} \begin{pmatrix} \mathbf{X}_{j+1} \\ p_{j+1} \\ \mathbf{Y} \end{pmatrix} = \begin{pmatrix} \mathbf{z}_X \\ z_p \\ \mathbf{z}_Y \end{pmatrix}_{j+1} \tag{6}$$

The coefficients with an "*" in Eq. (5) are saved for backward smoothing.

The algorithm for one colored noise component update at a time is applied to the succeeding stochastic parameters in $\mathbf{P}_j$ until the filter time updating process is completed. The measurement updates are then applied at the completion of the time updates for batch $(j+1)$. These two filtering procedures, measurement updates and time updates, are applied iteratively until the last requested batch N is processed.

Similar to the filter, the smoother operates in a backwards recursive manner from the final batch N to the epoch of the first batch. The smoothed stochastic parameters $\widetilde{\mathbf{P}}_j$ at batch j are determined by computing each smoothed stochastic parameter, $\widetilde{p}_j$, using the smoothing coefficients (with an "*") saved in Eq. (5) and the smoothed estimate at batch $(j+1)$ as follows:[6]

$$\tilde{p}_j = \begin{pmatrix} z_w^* \\ \sigma_p^* \end{pmatrix}_j - \frac{1}{(\sigma_p^*)_j} (R_{pX}^* \;\; r_{pp}^* \;\; R_{pY}^*)_j \begin{pmatrix} \tilde{\mathbf{X}}_{j+1} \\ \tilde{p}_{j+1} \\ \mathbf{Y} \end{pmatrix}. \tag{7}$$

The resulting smoothed stochastic parameters $\tilde{\mathbf{P}}_j$ and the smoothed spacecraft state $\tilde{\mathbf{X}}_{j+1}$ at batch $(j+1)$ are used to compute the smoothed state at batch j by reversing the deterministic update from Eq. (3) as follows:

$$\tilde{\mathbf{X}}_j = \tilde{\mathbf{X}}_{j+1} - \mathbf{V}_P(j)\, \tilde{\mathbf{P}}_j. \tag{8}$$

This completes the smoothed estimate step for batch j. Note that the bias parameters in $\mathbf{Y}$ are not changed in the smoothing process. The detailed recursion for the smoothed covariance reversing colored noise update and then deterministic update is given in Ref. (6).

The solution to the system of equations represented by Eqs. (1) and (2) will require the use of advanced radiometric tracking of the spacecraft from earth (doppler, ranging, VLBI, ΔDOR, ΔVLBI, etc.) as well as on-board optical, SAR and possibly altimetry data. The on-board optical data used in this process would include limb/terminator observations to infer the target body center location and possibly a select few surface landmarks whose positions are well known.[9]

This paper pursues a method of taking the results of this process as *a priori* to processing hundreds to thousands of control points to produce high precision maps of potential landing sites which are accurately tied to inertial space and to the spacecraft trajectory. The separation of these two processes is proposed to keep the orbit determination task manageable and numerically stable. The processing of the control points separately represents a sub-optimal approach, but it is felt that little accuracy is lost while substantial gains are made in numerical stability and ease of filter implementation for a problem involving thousands to tens of thousands of estimated parameters.

By separating the two processes, the second process can only concern itself with estimating small corrections to the spacecraft trajectory produced by the first process and does not include the difficult dynamic stochastic parameters (e.g., non-gravitational forces) which were handled by the first process. Therefore, the second process, which described in this paper, only estimates parameters which have constant values (the state transition matrix is an identity matrix) over the entire data set. Parameters like corrections to the spacecraft trajectory and camera pointing have fixed batch sizes and are handled as backward smoothed parameters to minimize the number of parameters being carried in the filter equations.

To effectively perform the second process, the data equation for the on-board optical observables is formulated as follows:

$$\begin{pmatrix} \mathbf{R}_C & \mathbf{R}_{C\hat{X}} & \mathbf{R}_{CY} \\ 0 & \mathbf{R}_{\hat{X}} & \mathbf{R}_{\hat{X}Y} \\ 0 & 0 & \mathbf{R}_Y \end{pmatrix} \begin{pmatrix} \mathbf{C} \\ \hat{\mathbf{X}} \\ \mathbf{Y} \end{pmatrix} = \begin{pmatrix} \mathbf{z}_C \\ \mathbf{z}_{\hat{X}} \\ \mathbf{z}_Y \end{pmatrix}. \tag{9}$$

The $\mathbf{R}$'s are the associated elements of the upper triangular square-root information matrix with respect to parameters $\mathbf{C}$, $\hat{\mathbf{X}}$ and $\mathbf{Y}$. $\mathbf{C}$ is the corrections to the camera pointing angles,

$\hat{\mathbf{X}}$ is the corrections to the current spacecraft state, $\mathbf{Y}$ is the bias parameters which have constant values over the entire data set, such as pole and prime meridian parameters and coordinates of surface features, and the $\mathbf{z}$'s are the adjusted residual vectors.

It is assumed that the corrections to camera pointing angles are independent from picture to picture. Therefore, the information related to camera pointing angles in Eq. (9) is saved at the end of processing each picture as follows: (see Fig. 1)

$$(\mathbf{R}_C\ \mathbf{R}_{C\hat{X}}\ \mathbf{R}_{CY}) \begin{pmatrix} \mathbf{C} \\ \hat{\mathbf{X}} \\ \mathbf{Y} \end{pmatrix} = \mathbf{z}_C. \tag{10}$$

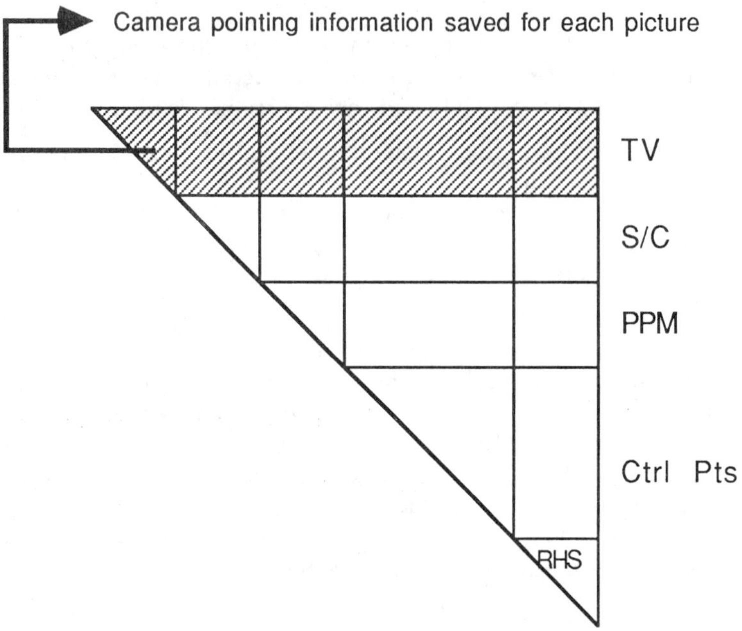

Fig. 1 Information related to camera pointing corrections saved for each picture.

Eq. (10) will be used in the backward smoothing process to obtain the camera pointing corrections for the corresponding pictures. After saving the information array in Eq. (10), the camera pointing information in Eq. (9) is re-initialized to continue data processing for the next picture as follows:

$$\begin{pmatrix} \mathbf{R}_{C0} & 0 & 0 \\ 0 & \mathbf{R}_{\hat{X}} & \mathbf{R}_{\hat{X}Y} \\ 0 & 0 & \mathbf{R}_Y \end{pmatrix} \begin{pmatrix} \mathbf{C} \\ \hat{\mathbf{X}} \\ \mathbf{Y} \end{pmatrix} = \begin{pmatrix} 0 \\ \mathbf{z}_{\hat{X}} \\ \mathbf{z}_Y \end{pmatrix}, \tag{11}$$

where $\mathbf{R}_{C0}$ is the *a priori* information matrix for camera pointing angles containing only diagonal elements.

It is also assumed that the corrections to the current spacecraft state are independent from orbit to orbit. Therefore, the information related to spacecraft state in Eq. (9) is saved at the end of processing each spacecraft orbit as follows: (see Fig. 2)

$$(\mathbf{R}_{\widehat{X}}\ \mathbf{R}_{\widehat{X}Y}) \begin{pmatrix} \widehat{\mathrm{X}} \\ \mathrm{Y} \end{pmatrix} = \mathbf{z}_{\widehat{X}}. \tag{12}$$

Spacecraft state correction information saved for each orbit

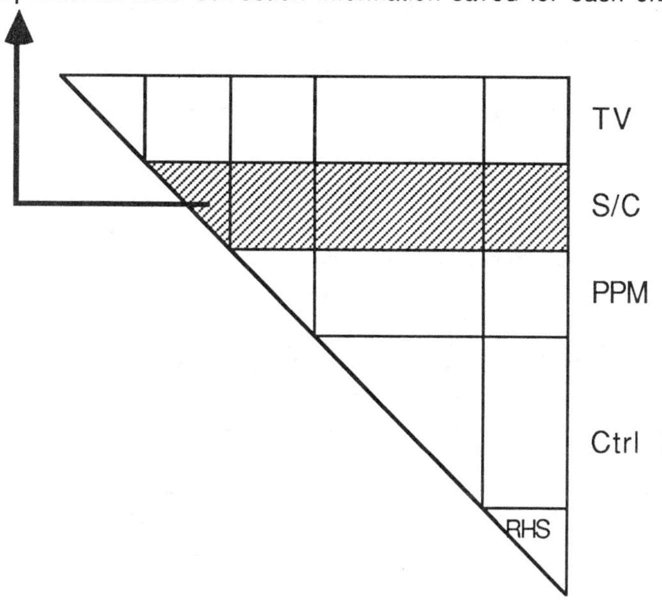

Fig. 2 Information related to spacecraft state corrections saved for each orbit.

Eq. (12) will be used in the backward smoothing process to obtain the corrections to the spacecraft state for the corresponding orbits. After saving the information array in Eq. (12), the information related to spacecraft state in Eq. (9) is re-initialized to continue data processing for the next orbit as follows:

$$\begin{pmatrix} \mathbf{R}_{C0} & 0 & 0 \\ 0 & \mathbf{R}_{\widehat{X}0} & 0 \\ 0 & 0 & \mathbf{R}_Y \end{pmatrix} \begin{pmatrix} \mathrm{C} \\ \widehat{\mathrm{X}} \\ \mathrm{Y} \end{pmatrix} = \begin{pmatrix} 0 \\ 0 \\ \mathbf{z}_Y \end{pmatrix}, \tag{13}$$

where $\mathbf{R}_{\widehat{X}0}$ is the *a priori* information matrix containing only diagonal elements for the spacecraft state. Note that Eq. (13) shows that the end of data processing for each spacecraft orbit coincides with the end of each picture processing.

After processing the entire data sets, the solutions for the estimated bias parameters Y are computed by

$$\mathrm{Y} = \mathbf{R}_Y^{-1}\ \mathbf{z}_Y. \tag{14}$$

and the covariance matrix (Γ) for Y is computed by

$$\Gamma = \mathbf{R}_Y^{-1}\ \mathbf{R}_Y^{-\mathrm{T}}. \tag{15}$$

By applying the backward smoothing process to the information saved in Eq. (12) and using the solution of $\mathbf{Y}$ in Eq. (14), the corrections to the spacecraft state for each orbit can be obtained as follows:

$$\widehat{\mathbf{X}} = \mathbf{R}_{\widehat{X}}^{-1}(\mathbf{z}_{\widehat{X}} - \mathbf{R}_{\widehat{X}Y}\ \mathbf{Y}). \tag{16}$$

Similarly, applying the backward smoothing process to the information saved in Eq. (10) and using the solution of $\widehat{\mathbf{X}}$ (where the picture was taken) and $\mathbf{Y}$ in Eqs. (16) and (14), respectively, the camera pointing corrections for each picture can be obtained as follows:

$$\mathbf{C} = \mathbf{R}_{C}^{-1}(\mathbf{z}_{C} - \mathbf{R}_{C\widehat{X}}\ \widehat{\mathbf{X}} - \mathbf{R}_{CY}\ \mathbf{Y}). \tag{17}$$

The technique of batching various parameters together and the use of the backward smoothing algorithm allow fully optimal solutions for most parameters while significantly reducing the amount of computer memory (virtual and non-virtual). Future problems will need to solve for hundreds of thousands of parameters; howerer, only a few thousands of parameters will be resident in the computer's memory at a time.

EXAMPLE

The following example was produced for Phobos, the inner moon of Mars, processing Viking Orbiter imaging data. Over 100 images of Phobos were processed to estimate about 2000 parameters including camera pointing, the rotational properties of Phobos, and the locations of surface features. These estimated results were used to generate cartographic map products useful for landing site selection.

Coordinate Systems

The following coordinate systems are established for processing surface feature images in the Phobos pictures taken by Viking Orbiter.

Body–Fixed Reference Coordinates. A body–fixed $\bar{\mathbf{x}}\bar{\mathbf{y}}\bar{\mathbf{z}}$ coordinate system (Fig. 3), used to define the control points, is chosen to align with the three principal moments-of-inertia of Phobos.[10,11] In this system, $\bar{\mathbf{z}}$ is along the spin axis, $\bar{\mathbf{x}}$ is normal to $\bar{\mathbf{z}}$ and in the direction of the longest axis and defines the prime meridian, and $\bar{\mathbf{y}}$ completes the orthogonal, right-handed system.

Let $\bar{\mathbf{u}}$ (in *km*) be a Phobos–centered position vector to the surface of a control point (e.g., crater, boulder, albedo mark) in $\bar{\mathbf{x}}\bar{\mathbf{y}}\bar{\mathbf{z}}$ and it is expressed (see Fig. 3) in terms of cartographic latitude (ϕ, in *deg*) and longitude (λ, in *deg*) as follows:

$$\bar{\mathbf{u}} = \begin{pmatrix} u_{\bar{x}} \\ u_{\bar{y}} \\ u_{\bar{z}} \end{pmatrix} = u \begin{pmatrix} \cos\phi\ \cos\lambda \\ \cos\phi\ \sin\lambda \\ \sin\phi \end{pmatrix} = u\ \hat{\mathbf{u}}, \tag{18}$$

where $\hat{\mathbf{u}}$ is the unit vector of $\bar{\mathbf{u}}$, $(u_{\bar{x}}\ u_{\bar{y}}\ u_{\bar{z}})^{\mathrm{T}}$ are the components of $\bar{\mathbf{u}}$ along the axes of $\bar{\mathbf{x}}\bar{\mathbf{y}}\bar{\mathbf{z}}$, and u is the radius to the point on the surface.

Inertial Reference Coordinates. The IAU standard Earth Mean Equator and Equinox of Besselian 1950.0 inertial reference system (referred to as B1950) is used to define the inertial orientation of $\bar{\mathbf{x}}\bar{\mathbf{y}}\bar{\mathbf{z}}$ as a function of time.[12] The relationship between the body–fixed coordinates and the inertial coordinates is defined by the transformation, $\mathbf{T}^I_{BF}$, from B1950 ($\overline{\mathbf{XYZ}}$) to $\bar{\mathbf{x}}\bar{\mathbf{y}}\bar{\mathbf{z}}$ as a function of 3 rotation angles representing the right ascension (α) and declination (δ) of the spin axis of Phobos, and also the hour angle (W) of the prime meridian. The transformation, $\mathbf{T}^I_{BF}$, from B1950 ($\overline{\mathbf{XYZ}}$) to $\bar{\mathbf{x}}\bar{\mathbf{y}}\bar{\mathbf{z}}$ is given by[12]

$$\mathbf{T}^I_{BF} = [W]_3\ [90 - \delta]_1\ [\alpha + 90]_3, \tag{19}$$

where $\mathbf{T}^I_{BF}$ is a 3×3 orthogonal rotation matrix and the form $[\theta]_i$ represents a 3×3 rotation matrix of angle θ about ith axis ($i = 1, 2,$ or 3) as expressed by

$$[\theta]_1 = \begin{pmatrix} 1 & 0 & 0 \\ 0 & \cos\theta & \sin\theta \\ 0 & -\sin\theta & \cos\theta \end{pmatrix}, \qquad [\theta]_2 = \begin{pmatrix} \cos\theta & 0 & -\sin\theta \\ 0 & 1 & 0 \\ \sin\theta & 0 & \cos\theta \end{pmatrix},$$

and

$$[\theta]_3 = \begin{pmatrix} \cos\theta & \sin\theta & 0 \\ -\sin\theta & \cos\theta & 0 \\ 0 & 0 & 1 \end{pmatrix}.$$

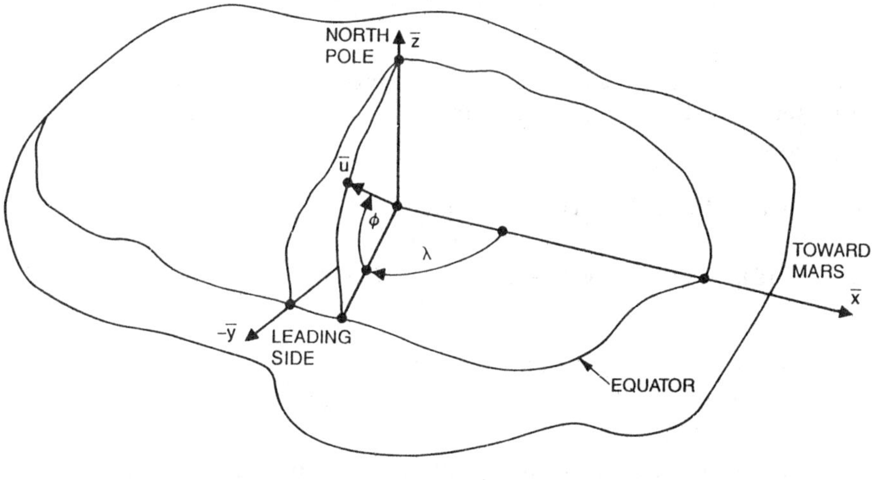

Fig. 3 Cartographic coordintates of a point on the surface of Phobos in body-fixed coordinate system.

<u>TV Camera Coordinates.</u> A camera–fixed **xyz** coordinate system is defined with **x** on the face of the vidicon and in the direction of increasing sample number, **y** on the face of the vidicon and in the direction of increasing line number, and **z** completing the orthogonal, right–hand system and along the optical axis. The origin of the camera–fixed **xyz** is on the face of the vidicon and located at the optical principal point.

To define the camera pointing and orientation in B1950, the following transformation, $\mathbf{T}^I_{TV}$, from B1950 ($\overline{\mathbf{XYZ}}$) to **xyz** as a function of 3 rotation angles representing the right ascension (α_c) and declination (δ_c) of the optical axis (camera **z** axis), and the orientation angle (κ) of camera axes **x**-**y** about **z**, is given by

$$\mathbf{T}^I_{TV} = [\kappa]_3\ [90 - \delta_c]_1\ [\alpha_c + 90]_3. \tag{20}$$

Image (Picture) Coordinates. Each of the Viking Orbiter twin TV cameras had a 1.4×1.7 *deg* field-of-view and each camera was scanned using 1056 lines with 1204 pixels per line. Given a vector **d**, where

$$\mathbf{d} = (d_x\ d_y\ d_z)^{\mathrm{T}}, \tag{21}$$

in TV **xyz** coordinates, the vidicon (x, y) coordinates associated with **d**, assuming no optical distortions, are defined by[9]

$$\begin{pmatrix} x \\ y \end{pmatrix} = \frac{f}{d_z} \begin{pmatrix} d_x \\ d_y \end{pmatrix}, \tag{22}$$

where f is the camera focal length.

The sample and line coordinates (s, l) associated with the vidicon (x, y) coordinates, computed for an object viewed through the optics or for a reseau whose vidicon coordinates have been measured, are given by

$$\begin{pmatrix} s \\ l \end{pmatrix} = \mathsf{K}(x, y) + \begin{pmatrix} s_0 \\ l_0 \end{pmatrix}, \tag{23}$$

where s_0 and l_0 are the image coordinates of the central reseau which is assumed to be the optical principal point and $\mathsf{K}(x, y)$ is the transformation from vidicon coordinates to sample and line coordinates.[13]

Transforming Points on Reference Surface to Images in the Picture

From Eq. (19), transforming a vector $\overline{\mathbf{b}}$ in B1950 to $\bar{\mathbf{u}}$ in $\bar{\mathbf{x}}\bar{\mathbf{y}}\bar{\mathbf{z}}$ is accomplished by

$$\bar{\mathbf{u}} = \mathsf{T}^{I}_{BF}\, \overline{\mathbf{b}} = \mathsf{T}^{I}_{BF} \begin{pmatrix} b_{\bar{X}} \\ b_{\bar{Y}} \\ b_{\bar{Z}} \end{pmatrix}, \tag{24}$$

where $b_{\bar{X}}$, $b_{\bar{Y}}$, and $b_{\bar{Z}}$ are the positional components of $\overline{\mathbf{b}}$ along the axes of B1950. Therefore, transforming $\bar{\mathbf{u}}$ to $\overline{\mathbf{b}}$ is

$$\overline{\mathbf{b}} = \mathsf{T}^{BF}_{I}\, \bar{\mathbf{u}}, \tag{25}$$

where T^{BF}_{I}, also a 3×3 orthogonal rotation matrix, is the transpose (inverse) of T^{I}_{BF}. Note that $\overline{\mathbf{b}}$ is defined as a surface feature position vector with respect to the center of Phobos in the B1950 system as follows:

$$\overline{\mathbf{b}} = \bar{\mathbf{a}} - \bar{\mathbf{t}}, \tag{26}$$

where $\bar{\mathbf{a}}$ is a position vector from the spacecraft to the surface feature in B1950 and $\bar{\mathbf{t}}$ is the Phobos position vector with respect to the spacecraft in B1950. Here, $\bar{\mathbf{t}}$ can be derived from the spacecraft and Phobos ephemerides as follows:

$$\bar{\mathbf{t}} = \bar{\mathbf{s}} - \bar{\mathbf{x}}, \tag{27}$$

where $\bar{\mathbf{s}}$ is the Phobos position vector with respect to Mars in B1950 and $\bar{\mathbf{x}}$ is the spacecraft position vector with respect to Mars in B1950.

From Eqs. (20) and (26), transforming the surface feature position vector (with respect to spacecraft in B1950), $\bar{\mathbf{a}}$, to $\mathbf{d}$ in TV $\mathbf{xyz}$ coordinates is given by

$$\mathbf{d} \;=\; \mathsf{T}_{TV}^{I}(\bar{\mathbf{b}}+\bar{\mathbf{s}}-\bar{\mathbf{x}}). \tag{28}$$

From Eqs. (25) and (28), transforming $\bar{\mathbf{u}}$ in Phobos body–fixed $\bar{\mathbf{x}}\bar{\mathbf{y}}\bar{\mathbf{z}}$ coordinates to $\mathbf{d}$ in TV $\mathbf{xyz}$ coordinates is given by

$$\mathbf{d} \;=\; \mathsf{T}_{TV}^{I}\;\mathsf{T}_{I}^{BF}\;\bar{\mathbf{u}}+\mathsf{T}_{TV}^{I}(\bar{\mathbf{s}}-\bar{\mathbf{x}}). \tag{29}$$

Since $\mathbf{d}$ can be transformed to the vidicon (x, y) coordinates in the millimeter (mm) space as shown in Eqs. (21) and (22) and then transformed to the image (s, l) coordinates in the sample and line space as shown in Eq. (23). Therefore, from Eqs. (21)–(23) and (29), a vector $\bar{\mathbf{u}}$ in Phobos body–fixed $\bar{\mathbf{x}}\bar{\mathbf{y}}\bar{\mathbf{z}}$ coordinates can be transformed to a point (s, l) in the image plane. This is the foundation for processing the surface features using the control network.

Parameter Estimation

The Phobos control network was used to establish the cartographic coordinates of surface features on Phobos.[10] The picture coordinates of the control point images were measured interactively using a display of the digital data on a frame buffer.[10] Over 100 Phobos pictures with a total of 542 control points were measured. Each control point was observed at least twice, from different directions. An example of identifying Phobos control points on Viking Orbiter images is given in Fig. (4).

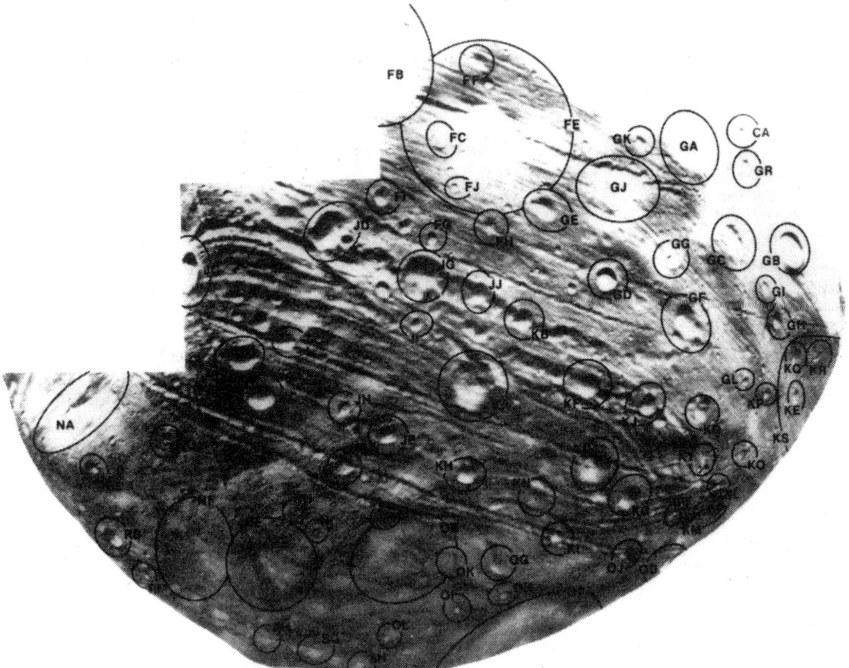

Fig. 4 Identification of Phobos control points in Viking Orbiter images.

In this example, it is assumed that there is no ephemeris error and that the spacecraft orbit, Phobos orbit, and Mars gravity field were not estimated in the orbit determination process. Instead, the parameters to be estimated were the cartographic coordinates (ϕ, λ, u) for each control point, the camera pointing and orientation angles (α_c, δ_c, κ) for each picture, and modelled corrections to the pole and prime meridian angles (α, δ, W). An ephemeris of Phobos was determined using data from over 100 years of earth-based observations, as well as Mariner 9 and Viking imaging data.[10] This process also included corrections to the spacecraft positions as well.

The initial values* of ϕ, λ and u were taken as the average values of the control point latitude, longitude and radius from the image location measurement process. The initial values of α, δ and W were taken from Ref. (10).

A priori camera pointing angles for each picture were obtained from the tracking data and also from star images when available. All camera parameters were determined using hundreds of star images prior to the image location measurement process.

Using the above *a priori* values, the predicted image locations of each control point were computed in order to generate residuals and linear partial derivatives for the parameter estimation process. The total number of parameters to be estimated was about 2,000. However, because of the stochastic nature of camera pointing angles, the backward smoothing algorithm was used to reduce the problem to about 1,650 parameters. This feature will be important in future missions where thousands of pictures will have to be processed. Therefore, the coordinates of surface features (control points) were the largest set of parameters to be estimated.

This problem had n ($< 1,650$) unknowns with m ($> 4n$) data equations (observables) and was solved using a combination of a square-root information filter (SRIF) and a backward smoothing algorithm.† SRIF was used to form the upper triangular square-root information matrix from the normal equations for minimizing the computer memory space and also to obtain the solutions for numerical precision and stability.[6,7,8] With the use of backward smoothing algorithm and the SRIF solutions, more than 300 additional camera pointing angles were obtained.

A batch sequential data processing algorithm was used for parameter estimation. The control points identified in each picture were grouped as a data batch. Because the camera pointing angles were assumed to be random from picture to picture, the information related to camera pointing angles in the data equations was saved at the end of each batch for later use in the backward smoothing process. The series of steps performed in the parameter estimation process is summarized as follows:

1. The initial values and the appropriate level of uncertainties are selected for the estimated parameters. These uncertainties are used to initialize the SRIF array.

---

* All the data sets used in this example were defined relative to the B1950 system. A futrue application of this concept should use the data sets which are defined in the new IAU standard Earth Mean Equator and Equinox of J2000 inertial reference system.

† The conjugate gradient iterative method has also been used by other researchers[14,15,16] to solve large control network problems. In general, the control network was divided into several overlapping blocks. The normal equations in each block were solved to adjust points in the overlapping regions. The block adjustment processes were performed iteratively until all the control point coordinates in the overlapping regions converged.[14,15]

2. For each batch (one batch = one picture), the following steps are performed to accumulate data into the SRIF array.

   a. Predicted image locations of each control point are computed as follows:

      - Compute the body-fixed position vector $\bar{\mathbf{u}}$ using Eq. (18).
      - Transform $\bar{\mathbf{u}}$ to $\overline{\mathbf{b}}$ in the B1950 system using Eq. (25).
      - Input the spacecraft and Phobos positions with respect to Mars in B1950.
      - Transform $(\overline{\mathbf{b}} + \bar{\mathbf{s}} - \bar{\mathbf{x}})$ to $\mathbf{d}$ in TV camera coordinates using Eq. (28).
      - Compute image locations (s, l) using Eqs. (22) and (23).

   b. Linear partial derivatives of sample and line image coordinates with respect to the estimated parameters are computed as follows:

      - Compute camera pointing partials by

$$\frac{\partial(s,l)}{\partial(\alpha_c,\delta_c,\kappa)} = \frac{\partial(s,l)}{\partial(x,y)} \frac{\partial(x,y)}{\partial(d_x,d_y,d_z)} \frac{\partial(d_x,d_y,d_z)}{\partial(\alpha_c,\delta_c,\kappa)}$$

      - Compute pole and prime meridian partials by

$$\frac{\partial(s,l)}{\partial(\alpha,\delta,W)} = \frac{\partial(s,l)}{\partial(x,y)} \frac{\partial(x,y)}{\partial(d_x,d_y,d_z)} \cdot \frac{\partial(d_x,d_y,d_z)}{\partial(b_{\bar{X}},b_{\bar{Y}},b_{\bar{Z}})} \frac{\partial(b_{\bar{X}},b_{\bar{Y}},b_{\bar{Z}})}{\partial(\alpha,\delta,W)}$$

      - Compute cartographic coordinates partials by

$$\frac{\partial(s,l)}{\partial(\phi,\lambda,u)} = \frac{\partial(s,l)}{\partial(x,y)} \frac{\partial(x,y)}{\partial(d_x,d_y,d_z)} \frac{\partial(d_x,d_y,d_z)}{\partial(b_{\bar{X}},b_{\bar{Y}},b_{\bar{Z}})} \cdot \frac{\partial(b_{\bar{X}},b_{\bar{Y}},b_{\bar{Z}})}{\partial(u_{\bar{x}},u_{\bar{y}},u_{\bar{z}})} \frac{\partial(u_{\bar{x}},u_{\bar{y}},u_{\bar{z}})}{\partial(\phi,\lambda,u)}$$

   c. Residuals are computed by differencing the measured and predicted image locations.

   d. The data reduction (or data fitting process) is carried out using a square-root information filter to accumulate the linear partial derivatives and residuals into an upper triangular square-root information matrix.

3. Information that is related to the camera pointing angles for each picture is saved at the end of each batch process as shown in Fig. (1). The data array saved will be used in the backward smoothing process to obtain the corrections to camera pointing angles for each corresponding picture. After saving the camera pointing information, these values are re-initialized to zero except that the diagonal elements are assigned values equal to the reciprocals of the *a priori* sigmas of camera pointing angles (see Eq. 11). The resulting SRIF array is used to process the next picture.

4. After picture data processing is completed, the corrections to the pole and prime meridian parameters and the cartographic coordinates for all the control points processed are obtained by inverting the augmented SRIF array. The square-root of the error covariance ($\Gamma^{1/2}$) for the estimated parameters is saved for producing the cartographic contour map of Phobos.

5. With the solutions obtained in step (4) and the camera pointing information saved in step (3), the backward smoothing algorithm is applied to obtain the corrections to camera pointing angles for all the pictures.

6. The resulting solutions for the estimated parameters as shown in steps (4) and (5) may then be used to re-initialize step (1), and the series of sub-functions (steps 1 through 6) may be carried out iteratively until the solution adjustment converges.

This problem required about 4 thousand double precision words (16 MBytes) of virtual memory on a VAX 11/780 computer and over 24 hours of CPU for each iteration performed. It is anticipated that advances in computer technology will enable this same problem to be solved in minutes by the mid-1990s. It is noted that existing supercomputers could do the processing in about an hour.

Cartographic Accuracy of Phobos

The square-root of the error covariance ($\Gamma^{1/2}$) saved for the Phobos control network was used to compute the mapped absolute error covariance for each control point as follows:

$$\Psi_i = \left(\frac{\partial \bar{\mathbf{u}}_i}{\partial(\lambda,\phi,u)_i}\right)^{\mathrm{T}} \Lambda_i \, \frac{\partial \bar{\mathbf{u}}_i}{\partial(\lambda,\phi,u)_i}, \qquad i = 1,\ldots,nc, \tag{30}$$

where nc is the number of control points, Ψ_i is the mapped error covariance for the i-th control point, Λ_i is the error covariance derived from the original error covariance Γ for the i-th control point, and $\partial\bar{\mathbf{u}}_i/\partial(\lambda,\phi,u)_i$ is the transformation matrix from cartographic coordinates to body-fixed position vector for the i-th control point.

The uncertainty associated with the i-th control point, σ_i, was computed by

$$\sigma_i = \sqrt{\Psi_i^2(1,1) + \Psi_i^2(2,2) + \Psi_i^2(3,3)}. \tag{31}$$

These uncertainties along with cartographic coordinates of each control point were used to generate a contour map of the entire Phobos surface. Note that the control points are not evenly distributed. In order to produce a contour map of the entire surface, a set of evenly-spaced grid points on latitude and longitude coordinates had to be generated for our contour plotters. It is assumed that the contribution of the control point uncertainties to the uncertainty of a given grid point is inversely proportional to the 4th power of the angular separation between the grid point and each control point. The grid point uncertainty was computed by interpolating each control point uncertainty as follows:

$$\sigma_k = \frac{\sum_{i=1}^{nc} \sigma_i/\vartheta_{ik}^4}{\sum_{i=1}^{nc} 1/\vartheta_{ik}^4}, \qquad k = 1,\ldots,ng \tag{32}$$

where ng is the number of grid points, σ_k is the level of uncertainty for the k-th grid point, σ_i is the level of uncertainty for the i-th control point as defined in Eq. (31), and ϑ_{ik} is the angle between the control point unit position vector ($\hat{\mathbf{u}}_i$) and the grid point unit position vector ($\hat{\mathbf{u}}_k$) and is computed by

$$\vartheta_{ik} = \cos^{-1}(\hat{\mathbf{u}}_i \cdot \hat{\mathbf{u}}_k). \tag{33}$$

Fig. (5) shows the surface features (534 control points) selected in the Phobos control network and the accuracy of the resultant processing displayed on a contour map. The "x" marks shown in Fig. (5) are the locations of the Phobos control points. Using a sinusoidal map projection,[18,19] the results shown in the contour map of Fig. (5) are displayed in Fig. (6). These maps are needed to determine the level of cartographic accuracy for selecting landing sites.

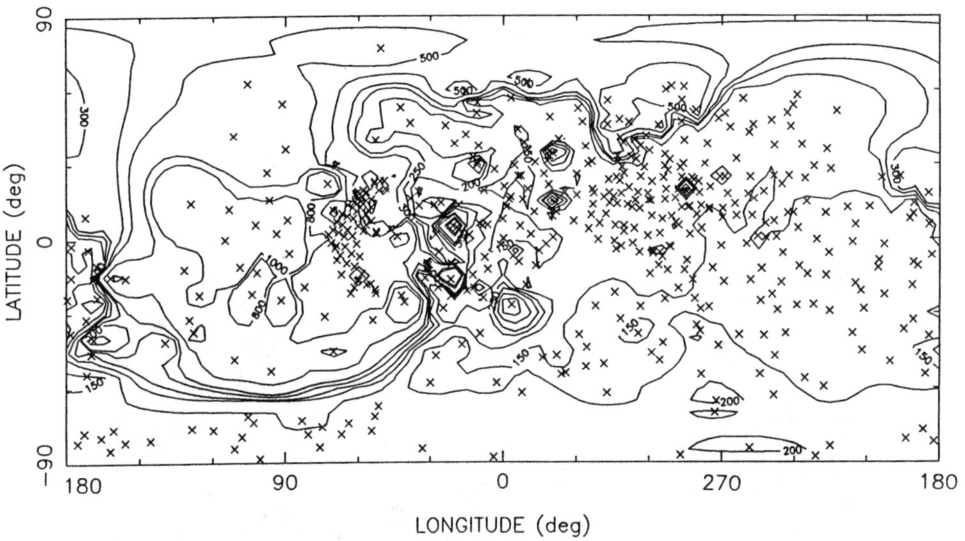

Fig. 5 Cartographic accuracy (in meters) of Phobos in a Mercator projection.

In addition to these maps, topographic maps of Phobos can also be generated using spherical harmonic expansion[11,17] of the radius to the Phobos surface in terms of cartographic coordinates. A Digital Terrain Model[18] (DTM), which is a computer file of radius or measurement uncertainty data at the discrete latitude and longitude grid points, was used to generate cartographic and topographic maps of Phobos. The DTM maps of Phobos in a sinusoidal projection are given in Fig. (7). The DTM topographic map on the top as shown in Fig. (7) was generated using a topographic model for the surface of Phobos which includes an 8th degree and order harmonic expansion model, and Stickney and crater models.[17] The Stickney and crater models are used to better fit Stickney and other large craters. The results of using Stickney and crater models at the control points are shown in the middle of Fig. (7). The corresponding DTM cartographic accuracy is shown in the bottom of Fig. (7). Using the combination of topographic maps and cartographic accuracy maps, a favorable landing site can be selected for targeting landers and penetrators.

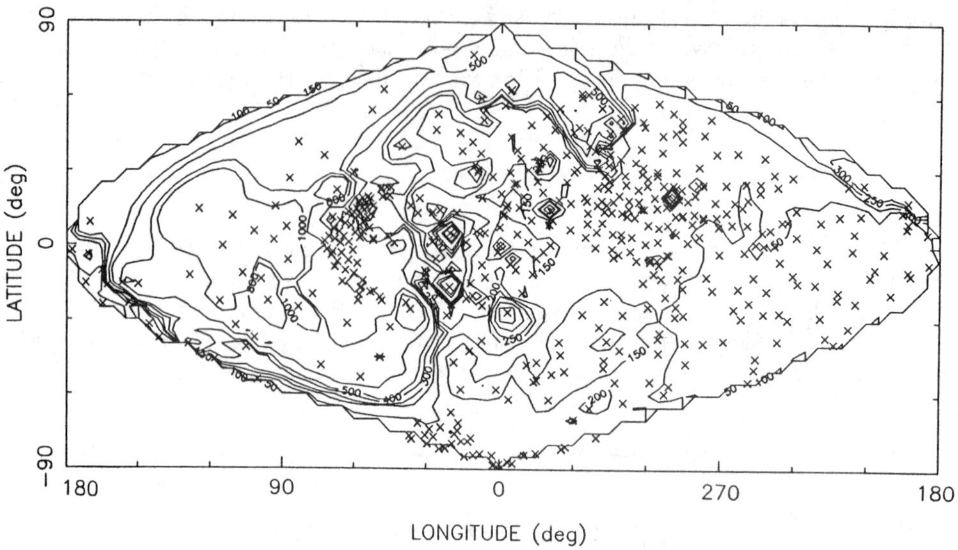

Fig. 6 Cartographic accuracy (in meters) of Phobos in a sinusoidal projection.

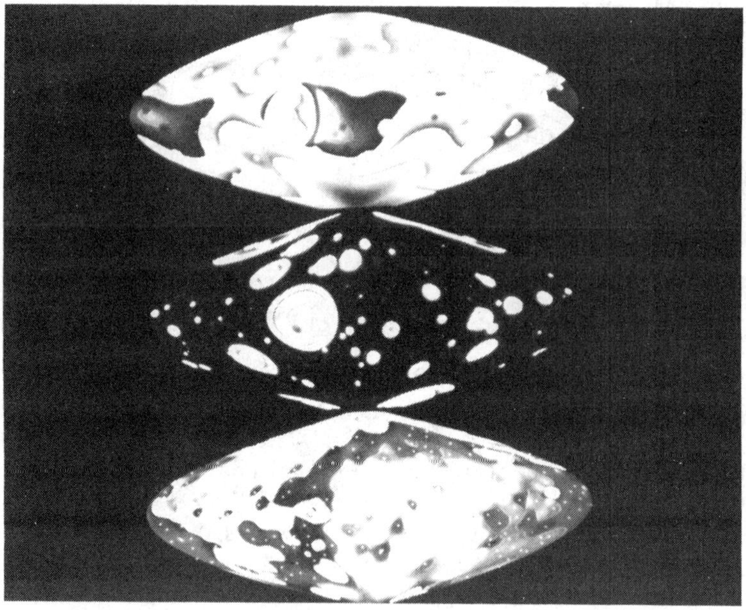

Fig. 7 Maps of Phobos using a Digital Terrain Model.

CONCLUSIONS

On-board orbital optical navigation data can be used to generate topographic maps for selecting a landing site. The accuracy for these maps depends on the success of a near-real time orbit determination process which solves for a large set of parameters including cartographic coordinates of surface features and camera pointing and orientation angles of each picture. This new approach for targeting landers, rovers, and penetrators using orbital optical navigation in a near-real time operation deserves further study. Future work should include studies using supercomputers to solve large orbit determination problems in near-real time operational environments.

Since the relative topographic information is more important than absolute information for selecting a landing site, future work should also include generating maps of the relative cartographic accuracy in the areas where the potential landing sites are to be selected.

ACKNOWLEDGEMENT

The authors would like to thank Chuck Acton, Manager of the JPL Navigation Ancillary Information Facility (NAIF) for providing the NAIF Toolkit, SPICELIB. The data and image analysis was performed at NAIF. The authors would also like to thank Peter Breckheimer, Eric Cannell and Ian Underwood of JPL for reviewing this paper and for valuable discussions. This work was carried out at the Jet Propulsion Laboratory, California Institute of Technology, Pasadena California under contract to the National Aeronautics and Space Administration.

NOTATION

Symbol	Description
$\bar{a}$	Surface feature position vector *wrt* spacecraft in B1950;
$\bar{b}$	Surface feature position vector *wrt* Phobos center in B1950;
$b_{\bar{X}}, b_{\bar{Y}}, b_{\bar{Z}}$	Components of $\bar{\mathbf{b}}$ along the axes of the B1950 system;
$\mathbf{C}$	Camera pointing parameters;
$\mathbf{d}$	Surface feature position vector in camera-fixed coordinate system;
d_x, d_y, d_z	Components of $\mathbf{d}$ along the axes of camera-fixed coordinates;
f	camera focal length;
$\mathbf{I}_X$, $\mathbf{I}_Y$	Identity matrices corresponding to sizes of $\mathbf{X}$ and $\mathbf{Y}$, respectively;
j	Batch number;
$(\ )_j$	The given quantity is associated with batch j;
K	Transformation matrix from vidicon coordinates to sample and line coordinates;
M	Diagonal transition matrix;
N	Total number of batches;
nc	Number of control points processed;
ng	Number of grid points used for generating a contour map;
np	Number of stochastic parameters;
$\mathbf{P}$	Stochastic parameters;
p	A given stochastic parameter in $\mathbf{P}$;
$\widetilde{\mathbf{P}}$	Smoothed stochastic parameters;
$\tilde{p}$	A given smoothed stochastic parameter in $\widetilde{\mathbf{P}}$;

NOTATION (continued)

Symbol	Description
$\mathbf{R}_X$, $\mathbf{R}_P$, $\mathbf{R}_Y$	Elements of the upper triangular square-root information matrix corresponding to **X**, **P**, and **Y**, respectively;
$\mathbf{R}_{XP}$, $\mathbf{R}_{XY}$, $\mathbf{R}_{PY}$	Cross elements of the upper triangular square-root information matrix associated with **X**, **P**, and **Y**;
$\mathbf{R}_C$, $\mathbf{R}_{\widehat{X}}$, $\mathbf{R}_Y$	Elements of the upper triangular square-root information matrix corresponding to **C**, $\widehat{\mathbf{X}}$, and **Y**, respectively;
$\mathbf{R}_{C\widehat{X}}$, $\mathbf{R}_{CY}$, $\mathbf{R}_{\widehat{X}Y}$	Cross elements of the upper triangular square-root information matrix associated with **C**, $\widehat{\mathbf{X}}$, and **Y**;
$\mathbf{R}_{C0}$	A priori information matrix for camera pointing parameters;
$\mathbf{R}_{\widehat{X}0}$	A priori information matrix for spacecraft state parameters;
$(R^*_{pX} \quad r^*_{pp} \quad R^*_{pY})_j$	Smoothing coefficients for $\widetilde{p}$ in batch j associated with $(\widetilde{\mathbf{X}} \quad \widetilde{p} \quad \mathbf{Y})^{\mathrm{T}}$ in batch $(j+1)$;
$\overline{\mathbf{s}}$	Phobos position vector *wrt* Mars in B1950;
s, l	Sample and line (image) coordinates;
s_0, l_0	Image coordinates of the central reseau;
$(\ )^{\mathrm{T}}$	Transpose of a matrix;
$\overline{\mathbf{t}}$	Phobos position vector *wrt* spacecraft in B1950;
t_0	Epoch time;
t_N	Final time;
Δt_j	Delta time from t_j to t_{j+1};
$\mathbf{T}_I^{BF}$	Transformation matrix from body-fixed coordinates to B1950;
$\mathbf{T}_{BF}^I$	Transformation matrix from B1950 to body-fixed coordinates;
$\mathbf{T}_{TV}^I$	Transformation matrix from B1950 to camera-fixed coordinates;
$\overline{\mathbf{u}}$	Phobos-center position vector to the surface of a control point in body-fixed $\bar{\mathbf{x}}\bar{\mathbf{y}}\bar{\mathbf{z}}$ coordinate system;
$u_{\bar{x}}, u_{\bar{y}}, u_{\bar{z}}$	Components of $\overline{\mathbf{u}}$ along the axes of body-fixed coordinates;
$\hat{\mathbf{u}}$	Unit vector of $\overline{\mathbf{u}}$;
u	Radius to the control point on the surface;
$\mathbf{V}_P(j)$	Matrix of partials of the spacecraft state with respect to parameters **P** from batch j to batch $(j+1)$;
$\mathbf{w}$	Filter process noise that derives the colored noise;
$\mathbf{X}$	Spacecraft "pseudo–epoch" state corrections;
$\widetilde{\mathbf{X}}$	Smoothed spacecraft "pseudo–epoch" state corrections;
$\widehat{\mathbf{X}}$	Spacecraft current epoch state corrections;
$\overline{\mathbf{XYZ}}$	The IAU standard Earth Mean Equator and Equinox of Besselian 1950.0 inertial reference system (referred to as B1950);
$\overline{\mathbf{x}}$	Spacecraft position vector *wrt* Mars in B1950;
$\mathbf{xyz}$	Camera-fixed coordinate system;
$\bar{\mathbf{x}}\bar{\mathbf{y}}\bar{\mathbf{z}}$	Body-fixed coordinate system;
x, y	Vidicon coordinates in *mm* space;
$\mathbf{Y}$	Bias parameters;
$\mathbf{z}_C$, $\mathbf{z}_{\widehat{X}}$, $\mathbf{z}_Y$	Adjusted residual vectors corresponding to **C**, $\widehat{\mathbf{X}}$ and **Y**;
$\mathbf{z}_X$, $\mathbf{z}_P$, $\mathbf{z}_Y$	Adjusted residual vectors corresponding to **X**, **P** and **Y**;

NOTATION (continued)

Symbol	Description
z_w^*	Smoothing coefficient for $\widetilde{p}$ associated with the normalized process noise estimate;
Γ	Error covariance of the estimated bias parameters;
Λ_i	Error covariance derived from Γ for the ith control point;
Ψ_i	Mapped error covariance for the ith control point;
α, δ, W	Pole and prime meridian parameters;
$\alpha_c, \delta_c, \kappa$	Camera pointing and orientation angles;
τ_k	Correlation time for the k-th stochastic parameter;
σ_i	Uncertainty of the ith control point;
σ_k	Uncertainty computed for the kth grid point;
σ_p^*	Smoothing coefficient for $\widetilde{p}$;
ϑ_{ik}	Angle between the ith control point unit position vector and the kth grid point unit position vector;
ϕ, λ	Cartographic latitude and longitude of a control point;

REFERENCES

1. R. F. Draper, "The Mariner Mark II Program", *AIAA Paper No.* 88–0067, AIAA 26th Aerospace Sciences Meeting, Reno, Nevada, January, 1988
2. R. D. Bourke, J. H. Kwok and A. Friedlander, "Mars Rover Sample Return Mission", *AIAA Paper No.* 89–0417, AIAA 27th Aerospace Sciences Meeting, Reno, Nevada, January, 1989
3. P. H. Carter and R. S. Smith, "Mars Rover Sample Return Lander Performance", *AIAA Paper No.* 89–0633, AIAA 27th Aerospace Sciences Meeting, Reno, Nevada, January, 1989
4. D. S. Pivirotto, T. J. Penn and W. C. Dias, "Mars Rover 1988 Concepts", *AIAA Paper No.* 89–0419, AIAA 27th Aerospace Sciences Meeting, Reno, Nevada, January, 1989
5. J. K. Miller, C. J. Weeks and L. J. Wood, "Orbit Determination of the Comet Rendezvous/Asteroid Flyby Mission: Post-Rendezvous", *AIAA Paper No.* 89–0348, AIAA 27th Aerospace Sciences Meeting, Reno, Nevada, January, 1989
6. T. C. Wang, J. B. Collier, J. E. Ekelund and P. J. Breckheimer, "Applications of Square–Root Information Filtering and Smoothing in Spacecraft Orbit Determination", *Proceedings* of the 27th IEEE Conference on Decision and Control, Austin, Texas, December 1988
7. G. J. Bierman, *Factorization Methods for Discrete Sequential Estimation*, Academic Press, New York, New York, 1977
8. G. J. Bierman and K. H. Bierman, "Estimation Subroutine Library User's Guide (Preliminary)", *Factorized Estimation Applications, Inc.*, Report No. 81584, Studio City, California, August 1984
9. S. P. Synnott, A. J. Donegan, J. E. Riedel and J. A. Stuve, "Interplanetary Optical Navigation: Voyager Uranus Encounter", *AIAA Paper No.* 86–2113-CP, 1986 AIAA/AAS Astrodynamics Conference, Williamsburg, Virginia, August 1986

REFERENCES (continued)

10. T. C. Duxbury and J. D. Callahan, "Phobos and Deimos Control Networks", *Icarus*, Vol. 77, pp. 275–268, 1989

11. T. C. Duxbury, "The Figure of Phobos", *Icarus*, Vol. 78, pp. 169–180, 1989

12. M. E. Davies,V. K. Abalakin, M. Bursa, T. Lederle, J. H. Lieske, R. H. Rapp, P. K. Seidelmann, A. T. Sinclair, V. G. Teifel, and Y. S.Tjuflin, "Report of the IAU/IAG/ COSPAR Working Group on Cartographic Coordinates and Rotational Elements of Planets and Satellites: 1985", *Celest. Mech.*, Vol. 39, pp. 103–113, 1986

13. R. W. Gaskell, "Digital Identification of Cartographic Control Points", *Photogramm. Eng. Remote Sens.*, Vol. 56, pp. 723–727, 1988

14. M. E. Davies and D. W. G. Arthur, "Martian Surface Coordinates", *J. Geophys. Res.*, Vol. 78, pp. 4355–4394, 1973

15. M. E. Davies, T. A. Hauge, F. Y. Katayama and J. A. Roth, "Control Networks for the Galilean Satellites: November 1979", *R–2532/JPL/NASA*, The RAND Corporation, November 1979

16. M. E. Davies, T. R. Colvin, F. Y. Katayama and P. C. Thomas, "The Control Networks of the Satellites of Uranus", *Icarus*, Vol. 71, pp. 137–147, 1987

17. T. C. Duxbury, "An Analytic Model for the Phobos Surface", *in preparation*, 1989

18. R. M. Batson, "Digital Cartography of the Planets: New Methods, Its Status and Its Future", *Photogramm. Eng. Remote Sens.*, Vol. 53, pp. 1211–1218, 1987

19. K. Edwards, "Geometric Processing of Digital Images of the Planets", *Photogramm. Eng. Remote Sens.*, Vol. 53, pp. 1219–1222, 1987

AAS 89-405

ATMOSPHERIC DRAG PERTURBATIONS ON THE MARS OBSERVER ORBITER

P. B. Esposito and S. Demcak*

The nominal launch date of the Mars Observer spacecraft is 09-16-92, with Mars encounter on 08-19-93 and the mapping orbit established on 12-06-93. Because this orbit is nearly circular at low altitude (~361 km), both gravity and atmospheric drag will dominate the orbital evolution and are also the major error sources. This paper provides: a) a summary of atmospheric density and errors for the Mars Observer orbit and time frame, b) an assessment of spacecraft-position errors using current density estimates and c) an estimate of atmospheric density modeling improvement directly from the doppler data acquired during flight operations. With an improved gravity field model, reconstructed spacecraft-position errors are at the level of 0.5 km (one sigma). However when predicting the spacecraft's position over 14 days, the density error (at 90% confidence level) dominates yielding a down-track error of almost 70 km. By modeling the mean atmospheric density, this error can be reduced by at least a factor of two.

INTRODUCTION

The Mars Observer launch period is 09-16-92 to 10-06-92 with a corresponding insertion into an elliptical orbit about Mars during 08-19-93 to 09-05-93. During approximately the next 120 days, propulsive maneuvers will nearly circularize the orbit at low altitude (~361 km) and sun-synchronism will be established. On 12-16-93 the mapping phase (or science data acquisition) will begin with the orbit possessing the elements given in Table 1. In this orbit, the gravity field will strongly perturb the spacecraft's motion. Existing gravity field models, determined from Mariner 9 and Viking 1 and 2 radiometric data, provide a starting point from which a more extensive model will be necessary. A gravity calibration orbit (GCO) and period have been defined for this purpose and reported upon in Ref. 1. With the development of this gravity model, reconstructed spacecraft-position errors, at the level of 0.5 km one-sigma, will be adequate for the analysis of science data. However for predicting spacecraft-location errors longer than 2-3 days, the

* Jet Propulsion Laboratory, California Institute of Technology, 4800 Oak Grove Drive, Pasadena, California 91109.

atmospheric drag perturbation becomes the dominant error source and leads to substantial errors. This paper provides: a) a summary of atmospheric density and errors for the Mars Observer orbit and time frame, b) an assessment of spacecraft-position errors using current density models and c) an estimate of atmospheric density modeling improvement directly from the doppler data acquired during flight operations, and the resultant reduction in predicted position error.

Table 1
BASELINE MAPPING ORBIT ELEMENTS

Element	Mean Value per Orbit	Osculating Value
a (km)	3758.317	3749.288
e	0.00705	0.00382
Node (deg)	-28.216+0.524T*	-28.216+0.524T*
Inclination (deg)	92.818	92.821
Arg. of Periapsis (deg)	-90.0	-90.0
Mean Anomaly (deg)	Zero	Zero

* T = Number of earth days past 1 August 1993 00:00:0.0 (JD = 244 9200.5). Angles referenced to Mars' equator and equinox of date.

MARS ATMOSPHERIC DENSITY MODEL

Profiles of the mean atmospheric density over a three year period and an orbital period are given in Figs. 1 and 2 respectively. These have provided the a priori information for investigating the drag perturbation acting on the Mars Observer spacecraft.

Atmospheric density measurements, using accelerometer and upper atmosphere, mass spectrometer data, were made during the Viking 1 and 2 descents to the martian surface. However, they only cover an altitude range of 0-200 km (Ref. 2). Density estimates at higher altitudes (up to 600 km) which includes the Mars Observer satellite altitude are given in Ref. 3. This paper utilizes a broad source of input information, such as atmospheric constituents, atmospheric dynamics including dust storm activity, the 10.7 cm solar radiation intensity, direct density data mentioned above as well as a variety of indirect data related to the evaluation of an atmospheric density model. Implicit in this model are the following cycles: the 11 year solar cycle as well as the solar rotation (26.35 days) and the Mars annual, seasonal and diurnal cycles. Refinements to the model given in Ref. 3 have continued (Ref. 4) with the current densities summarized in Figs. 1 and 2. The error sources in this model are summarized as follows: a) the 10.7 cm solar intensity, b) the temperature at the exosphere, c) upper atmospheric constituents especially atomic oxygen, d) determination of the turbopause altitude and e) the influence of dust storms in the atmosphere (Ref. 5).

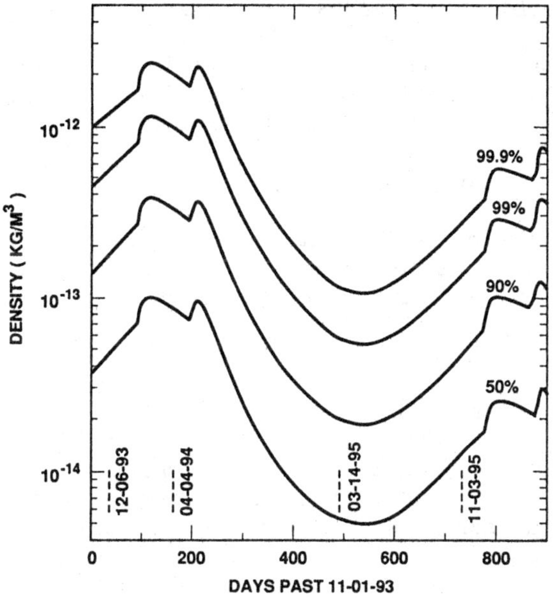

Figure 1. Mars' atmospheric density throughout mapping. Values for the nominal density and three confidence levels (90%, 99% and 99.9%) were generated using mean elements for the spacecraft's orbit and a flattened figure of Mars. The broad peaks occur during Mars' perihelion (4-4-94).

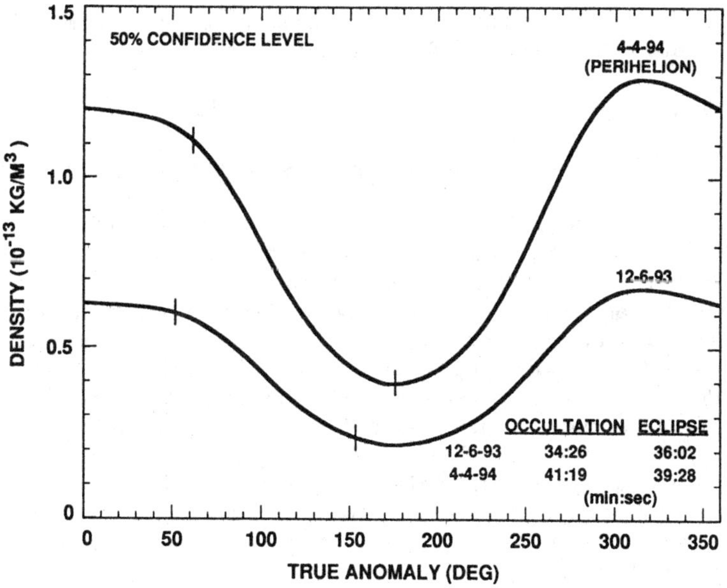

Figure 2. Mars' atmospheric density -- orbital variations. This variation is at the nominal or 50% confidence level and applies to the mapping orbit at 12-6-93 and 4-4-94.

ORBITAL PERTURBATIONS DUE TO ATMOSPHERIC DRAG

Based upon the previous density profiles, we have determined a reference density (ρ_o) applicable at a reference altitude (h_o) and epoch (t_o). The density variation with altitude is:

$$\rho = \rho_o \exp\left[- (h-h_o)/H\right] \qquad (1)$$

where h is the altitude above a flattened figure of Mars (f = 0.005216) and H is the density scale height. The acceleration acting on the spacecraft as a result of atmospheric drag is:

$$\ddot{\bar{r}} = \frac{1}{2} \rho v_r^2 \; C_D A/m \; \hat{v}_r \qquad (2)$$

where v_r is the spacecraft's velocity relative to the atmosphere and the mean area-to-mass ratio has a value of 0.016 m^2/kg with the coefficient of drag assumed to be 2.0. While this area-to-mass ratio is accurate for this study, the spacecraft's configuration is being revised and minor changes can be expected.

Monitoring the decay of the semi-major axis (Δa) is important for controlling the orbit. This perturbation is given by (Refs. 6,7)

$$\Delta a(\text{meters/orbit}) = -2\pi a^2 \frac{C_D A}{m} (1-d)^2 \;\; \rho_P \exp(-c) \; [B_o I_o(c)+B_1 I_1(c)]. \qquad (3)$$

In this equation, ρ_P is the density at periapsis, $c=ae/H$, B_i and I_i are functions given in Ref. 6 ($B_o I_o + B_1 I_1 \doteq 1.25$) and d is a term accounting for atmospheric rotation which we shall ignore.

The altitude above a flattened figure of Mars is

$$h = r-R = a(1-e \cos E) - R_o \; (1-f \sin^2 i \sin^2 \beta) \qquad (4)$$

where R_o is the equatorial radius (3397.2 km), i the orbit's inclination and $\beta = w+\Theta$ that is, the sum of the argument of periapsis and the true anomaly. Values for the mean atmospheric density at several epochs are given in Table 2 and the corresponding semi-major axis decay in Table 3. Thus on 12-06-93, the spacecraft orbital decay ranges from -0.15 to -3.75 meters per orbit depending upon the true value of the mean atmospheric density. Note that Eq. 3 requires the density at periapsis, ρ_P. One calculates the altitude at periapsis (hp) from Eq. 4 using the osculating elements given in Table 1. This value is substituted in Eq. 1 along with the reference values given in Table 2 to determine ρ_P. The semi-major axis perturbation (Δa) can be converted to an orbital period perturbation by $\Delta P/P = 1.5 \;\; \Delta a/a$. Thus a decay of -3.75 meters per orbit converts to $\Delta P = -0.0105$ seconds/orbit.

For one case (12-06-93, $\rho(3\sigma) = 1.20 \times 10^{-12}$ kg/m^3), we numerically integrated the equations of motion (central body and drag only) and determined a semi-major axis decay of -3.80 meters per orbit which compares favorably with the value of -3.75 given in Table 3.

Table 2

REFERENCE DENSITIES AT VARIOUS CONFIDENCE LEVELS (X 10**-12 kg/m**3).

Epoch	50%	90.0% (1.28σ)	99.0% (2.33σ)	99.9% (3.0σ)	H(km)	h_o(km)
12-06-93	0.047	0.19	0.58	1.20	46.9	364.2
01-07-94	0.060	0.23	0.72	1.47	43.2	364.2
04-04-94	0.091	0.35	1.04	2.09	44.4	364.2
05-16-94	0.094	0.35	1.06	2.13	43.2	364.2
01-04-95	0.0074	0.029	0.090	0.18	47.8	364.2

Table 3

DRAG PERTURBATION OR DECAY OF THE SEMI-MAJOR AXIS (METERS/ORBIT) AS A FUNCTION OF DENSITY SHOWN IN TABLE 2 (OSCULATING ELEMENTS).

Epoch	50%	90.0% (1.28σ)	99.0% (2.33σ)	99.9% (3.0σ)
12-06-93	-0.15	-0.59	-1.83	-3.75
01-07-94	-0.19	-0.74	-2.28	-4.65
04-04-94	-0.29	-1.09	-3.28	-6.58
05-16-94	-0.30	-1.12	-3.35	-6.73
01-04-95	-0.023	-0.091	-0.28	-0.58

Finally we completed one more numerical integration accounting for central body motion as well as a 6X6 global gravity field and atmospheric drag and determined Δa=-3.078 meters per orbit. The 6X6 gravity field provides a strong perturbation to the orbit (even over one orbit) whereas Eq. 3 implicitly assumes that the orbit elements are constant over one orbit. For example, the short-period perturbation of J_2 to the semi-major axis is (Ref. 8).

$$\Delta a(\mathrm{km}) = \frac{J_2 R^2}{a} \left\{ (1-1.5 \sin^2 i)\left[\left(\frac{a}{r}\right)^3 - (1-e^2)^{-3/2}\right] + 1.5\left(\frac{a}{r}\right)^3 \sin^2 i \cos 2\beta \right\} . \quad (5)$$

Thus, Δa has an amplitude of 9.27 km and a period of one-half the orbital period. Based upon such variations in the orbit elements, we re-evaluated Table 3 using mean orbit elements (along with the corresponding value of ρ_P) with the results given in Table 4.

Table 4
DRAG PERTURBATION OR DECAY OF THE SEMI-MAJOR AXIS (METERS/ORBIT) AS A FUNCTION OF DENSITY SHOWN IN TABLE 2 (MEAN ELEMENTS).

Epoch	50%	90.0% (1.28σ)	99.0% (2.33σ)	99.9% (3.0σ)
12-06-93	-0.13	-0.51	-1.60	-3.28
01-07-94	-0.16	-0.64	-1.97	-4.03
04-04-94	-0.25	-0.95	-2.85	-5.72
05-16-94	-0.26	-0.97	-2.91	-5.84
01-04-95	-0.02	-0.08	-0.25	-0.51

SPACECRAFT - POSITION ERROR

Over a fourteen day interval, the position error of the spacecraft, due to drag, can become substantial. In this section, we give current position errors which are a function of current errors associated with Mars' atmospheric density (Fig. 1). This is followed by an analysis to determine how those errors can be improved (made smaller) by estimating daily mean-density profiles directly from the radiometric data acquired during flight operations.

This was accomplished as follows. On 12-06-93, we simulated one tracking pass (8 hours) of doppler data during the mapping orbit and estimated the accuracy of the state and a small subset of gravity field parameters. Using current errors in the gravity field and atmospheric density along with the state errors just determined, we propagated these errors throughout a fourteen day interval. This was done a) using a 90% confidence level (1.28σ) density and b) under the assumption that the density was known perfectly. This simulation was repeated but utilizing the gravity field results (i.e. reduced errors) from the GCO analysis (Refs. 1 and 9). Finally once the gravity field update (resulting from the GCO) is available for flight operations, mean-density profiles can be determined directly from the doppler acquired over a single tracking pass. This results in improved density modeling which reduces spacecraft-position error throughout most of the fourteen day prediction interval. Since the drag perturbation produces a secular decrease in the semi-major axis, longer data arcs should, in principle, yield more accurate estimates of the mean density. Thus we simulated one case in which data from two consecutive passes (16 hours) were analyzed in order to improve density modeling. The results of these six cases are shown in Fig. 3 in which we plot the downtrack position error of the spacecraft over a fourteen day interval. These results are also summarized in Table 5.

Solutions 1A and 1B in Fig. 3 provide an envelope of downtrack errors assuming current gravity field errors coupled with a 90% confidence level (1.28σ) density error and no density error. For the former case, the error growth is substantial; for the latter, the one-sigma downtrack error at the end of fourteen days is 10.5 km. In solutions 2A and 2B, we utilized the gravity field errors as a result of the GCO and analysis instead of the current errors used in solns 1A and 1B. Solution 2A shows an initial smaller error, due to the improved gravity field knowledge, but after several days of error propagation the drag error dominates and reaches the same level as soln 1A. Solution 2B assumes no drag error as well as an improved gravity model. The downtrack error is initially small (0.23 km, 1σ) and exhibits a slight error growth after fourteen days (0.96 km, 1σ). While this predicted error is unattainable, it vividly illustrates the dominance of the drag error. Solution 3 utilizes the GCO gravity field error model and an estimate of the reference density (ρ_o; Eq.1) directly from one pass of doppler data. The improved density is directly responsible for the improved downtrack error although at the end of fourteen days the error is still large (36.8 km, 1σ).

Finally solution 4 shows the downtrack error improvement if we utilize two consecutive tracking passes (16 hrs.) to estimate the state and the reference density. At the end of fourteen days, the error, perhaps optimistically, is 12.1 km (one sigma). Although this analysis requires additional work, it does appear possible to approach this level of error.

Figs. 4 and 5 show the corresponding crosstrack and radial position errors.

Note that the downtrack position error can be approximated as follows: a) from the density error, determine Δa and ΔP, b) the accumulated timing error after n orbits is $\Delta t = 0.5\, n(n+1)\, \Delta P$ and c) the downtrack error is $\epsilon(DT) = \bar{v}\, \Delta t$ where the mean spacecraft speed ($\bar{v}$) is 3.37 km/s. For example if $\rho = 1.20X10^{-12} kg/m^3$ then $\Delta a = -3.078$ m/orbit and $\Delta P = -0.00858$ sec/orbit. Thus, the predicted downtrack (DT) errors are:

After ΔT (days)/n orbits	DT error (km) for $\rho = 1.20x10^{-12} kg/m^3$	DT error (km) scaled to $\rho = 0.188x10^{-12} kg/m^3$
7 / 86	108.	17
14 / 172	431.	67.5
21 / 259	975.	152.8
28 / 345	1729.	270.8

Table 5

MARS OBSERVER RECONSTRUCTED AND PREDICTED POSITION ERROR (KM)

Position * Error (1σ)	Downtrack					Crosstrack					Radial				
Propagation Interval (days past 12-06-93)	0	1	3	7	14	0	1	3	7	14	0	1	3	7	14
1A	1.43	1.31	3.36	16.9	68.5	1.47	1.47	1.48	1.80	2.56	.018	.025	.079	.56	1.73
1B	1.43	1.29	2.03	4.83	10.5	1.47	1.47	1.48	1.80	2.56	.018	.024	.079	.56	1.73
2A	0.23	0.35	2.68	16.2	67.8	0.24	0.24	0.25	0.28	0.34	.0045	.0057	.020	.11	.36
2B	0.23	0.26	0.37	0.63	0.96	0.24	0.24	0.25	0.28	0.34	.0045	.0045	.017	.11	.36
3	0.23	0.34	1.62	8.99	36.8	0.24	0.24	0.25	0.28	0.34	.0045	.0048	.016	.11	.35
4	0.17	0.19	0.39	2.75	12.1	0.19	0.19	0.19	0.21	0.25	.0044	.0045	.015	.11	.35

Solution	Comments	Parameters Estimated
1A	Current gravity field errors, 90% density (0.188×10^{-12} kg/m^3)	State, gravity anomalies (8)
1B	Current gravity field errors, no density error	State, gravity anomalies (8)
2A	GCO gravity field errors, 90% density (0.188×10^{-12} kg/m^3)	State only
2B	GCO gravity field errors, no density error	State only
3	GCO gravity field errors, a priori density error = 0.188×10^{-12} kg/m^3 one tracking pass (N=93); $\sigma\rho_o$(consider) = 0.101×10^{-12} kg/m^3	State, ρ_o
4	GCO gravity field errors, a priori density error = 1.20×10^{-12} kg/m^3 two tracking passes (N=196); $\sigma\rho_o$(consider) = 0.035×10^{-12} kg/m^3	State, ρ_o

*These are one-sigma errors except when the 90% density error is used (solns 1A and 2A - see text).

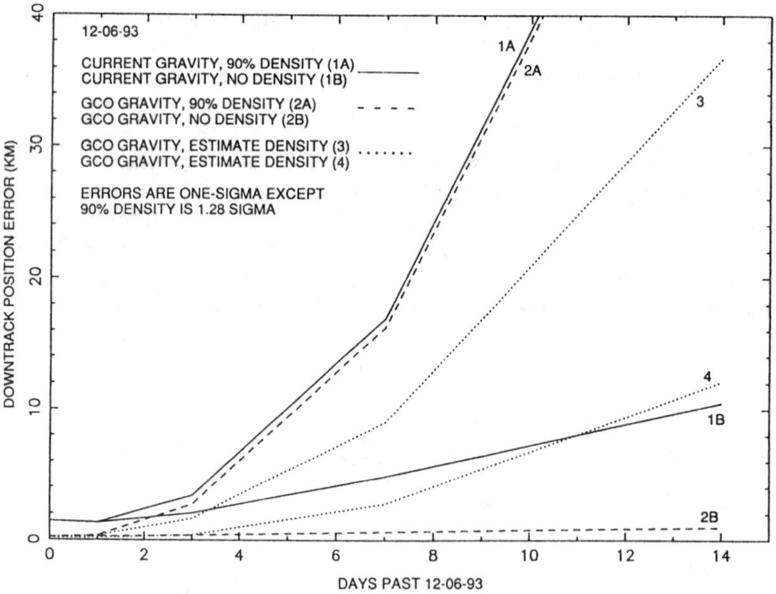

Figure 3. Error (one sigma; see text for cases 1A and 2A) in the downtrack component of the spacecraft's position. These are given initially (that is, as a result of analyzing doppler from a tracking pass and are called reconstructed errors) and are then propagated over a 14 day interval.

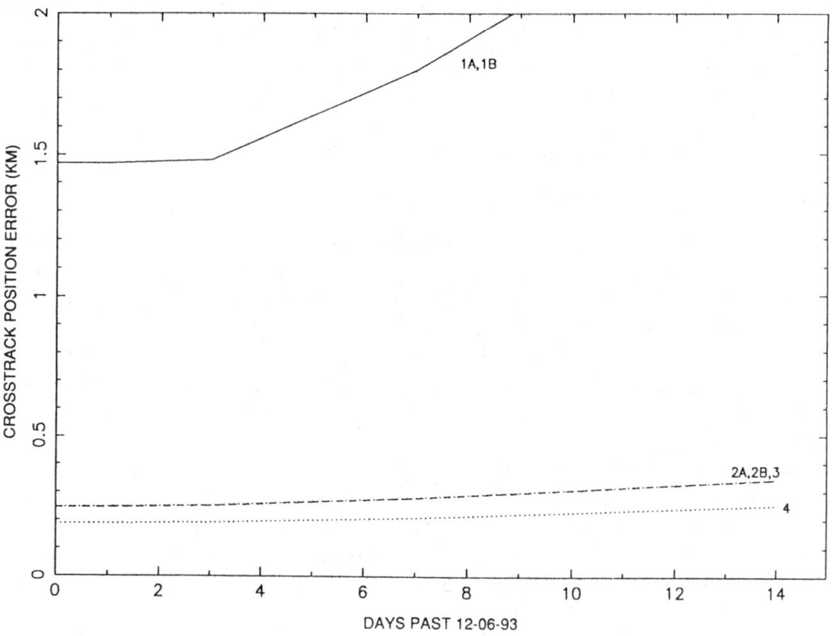

Figure 4. Error (one sigma; see text for cases 1A and 2A) in the crosstrack component of the spacecraft's position.

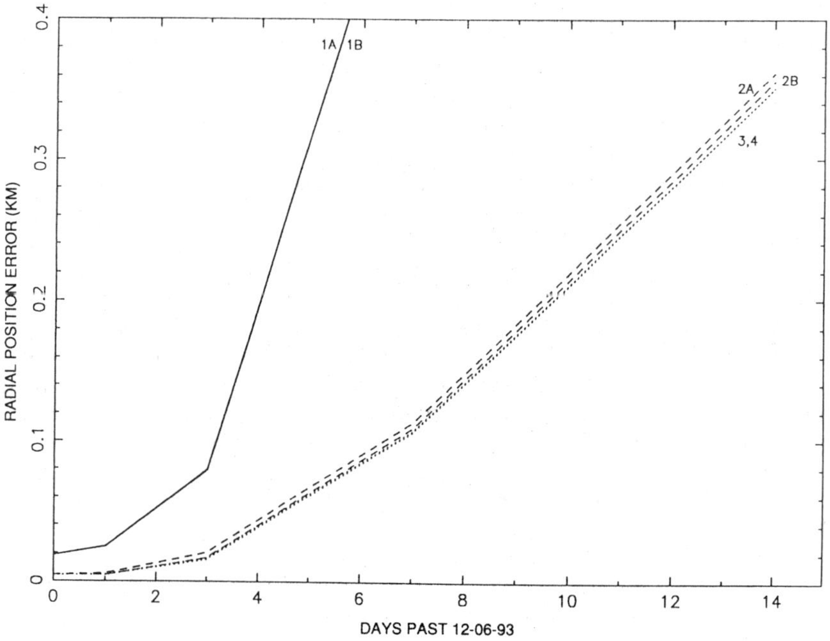

Figure 5. Error (one sigma; see text for cases 1A and 2A) in the radial component of the spacecraft's position.

After 7 and 14 days, the downtrack error is 17.0 km and 67.5 km respectively for a density error of $0.188 \times 10^{-12} kg/m^3$. This compares favorably with the corresponding errors given in Table 5, soln 1A; that is, 16.9 km and 68.5 km.

CONCLUSIONS AND DISCUSSION

It should be clear that estimating the atmospheric density directly from the radiometric data acquired during flight operations can reduce the predicted down track error considerably. We note that the errors given in Table 5 result from a consider analysis; the consider parameters or effects are: a) a 6x6 global gravity field, b) seven additional gravity anomalies (for solns 1A and 1B) for a total set of fifteen anomalies, c) solar radiation pressure and a constant level of non-gravitational accelerations, and d) tracking station locations. Planetary ephemeris errors were investigated and determined to be negligible. We also note that from a flight operational point of view, the implicit assumption has been made that multiple and consecutive orbits (of data) can be analyzed in a single-batch analysis. If the number of orbits is small, this assumption appears reasonable. As the number of orbits becomes larger, the rate of convergence of the solution will have to be investigated.

At this time (i.e. 12-06-93 or immediately after entering the mapping orbit), the Navigation commitment for reconstructed and predicted spacecraft-position error is given by case 1A. This is so because the GCO-derived gravity field will not be available until 01-17-94 and without this field daily density profiles cannot be accurately determined.

When considering the atmospheric density error, we usually use a 90% confidence level of error (which corresponds to a 1.28 σ high value). As expected almost all of this error propagates into a downtrack error. The 99% density error (2.33σ) is three times larger than the 90% density error (see Table 2). Thus whenever the downtrack error is driven primarily by the density error, the downtrack error corresponding to the 99% density is three times larger than the value corresponding to the 90% density.

Under present Mars Observer planning guidelines, a propulsive maneuver may be executed every week during the mapping phase. Most of these are period adjusts (under the Longitude Grid Control strategy) and the maneuver duration is of the order of seconds and thus near-instantaneous. An estimate of the downtrack (DT) error due to the velocity-correction error (or maneuver error) in the DT direction is determined by: a) relate the velocity-correction error to the semi-major axis and period perturbations by $2v\ \Delta v = \mu \Delta a/a^2$ and $\Delta P/P = 1.5\ \Delta a/a$, b) the accumulated timing error after n orbits is $\Delta t = n\ \Delta P$ and c) the DT error after n orbits is estimated as $\epsilon(DT) = \vec{v}\ \Delta t \doteq 3nP\Delta v$. Typical DT propagated errors (km) as a function of time and velocity errors are:

ΔT(days)/n orbits	Δv maneuver error (mm/sec) 10	16.67	50
1 / 12	2.5	4.2	12.6
3 / 36	7.6	12.6	37.8
5 / 60	12.6	21.0	63.0
7 / 86	18.1	30.1	90.3
14 /172	36.1	60.2	180.6

At present, the 16.67 mm/sec error represents the one-sigma maneuver execution error. Thus in order to minimize the influence of maneuver errors on DT prediction errors, one must a) determine the corrected orbit (as a result of a maneuver) immediately after the maneuver, b) execute the maneuver less frequently than once per week and c) reduce the magnitude of the maneuver execution errors.

At present, there is some analysis underway to change the baseline mapping orbit parameters to those similar to the gravity calibration orbit. This would increase the orbital semi-major axis by about 18 km with a corresponding reduction in atmospheric density and thus drag. In Table 6, we compare some orbit elements, atmospheric density, semi-major axis perturbation and downtrack position errors after 7 and 14 days for both orbits. As shown, the reduction in density results in a 24% reduction in downtrack error after 7 days of error propagation. This reduction in DT error would prevail over the interval in which we cannot estimate daily mean-density profiles (basically 12-06-93 to 01-17-94). Thereafter the density will be estimated (because of the availability of the GCO-derived gravity field) and the accuracy of that estimate is nearly the same for both orbits.

Table 6
THE BASELINE MAPPING AND GRAVITY CALIBRATION ORBITS

ELEMENT	MAPPING ORBIT	GCO
a(km)	3758.317	3776.709
e	0.00705	0.0072

ATMOSPHERIC DENSITY (10^{-12} kg/m^3)	90%*	99%	99.9%	ORBIT
12-06-93	0.19	0.58	1.20	MAPPING
	0.13	0.42	0.88	GC
Δa(m/orbit)	-0.51	-1.60	-3.28	MAPPING
	-0.37	-1.16	-2.42	GC

PROPAGATED* DOWNTRACK POSITION ERROR (km)	7 DAYS	14 DAYS	
	17.0	67.5	MAPPING
	12.9	51.2	GC
	(24% REDUCTION)		

REFERENCES

1. P.B. Esposito and S. Demcak, Mars Observer Orbital Accuracy Analysis:II, Paper AAS 87-504 AAS/AIAA Astrodynamics Specialist Conference, August 10-13, 1987.

2. A. Seiff and D. Kirk, Structure of the Atmosphere of Mars in Summer at Mid-Latitudes, JGR, 82, 4364-4378, 1977.

3. R. Culp and A. Stewart, Time-Dependent Model of the Martian Atmosphere for Use in Orbital Lifetime and Sustenance Studies, J. Astronautical Sciences, 32, 329-341, 1984.

4. A. Stewart, Revised Time-Dependent Model of the Martian Atmosphere for Use in Orbital Lifetime and Sustenance Studies, Final Report to JPL dated 3-26-87.

5. M. Dwortzan, Mars Atmospheric Density, JPL internal memo, 08-20-87.

6. T. Sterne, An Introduction To Celestial Mechanics, Interscience Publishers, Inc., New York, 1960.

7. F. Kalil, Effects of An Oblate Rotating Atmosphere On The Eccentricity, Semi-Major Axis and Period Of A Close Earth Satellite, AIAA Journal, 1, 1963.

8. A.E. Roy, Orbital Motion, A. Hilger Ltd., 1982.

9. P.B. Esposito, Mars Observer Navigation Plan-Update for the 1992 Launch, JPL Internal Report D-3820, 12-16-88.

ACKNOWLEDGMENT

The research described in this paper was carried out by the Jet Propulsion Laboratory, California Institute of Technology, under a contract with the National Aeronautics and Space Administration.

AAS 89-406

GALILEO ASTEROID ENCOUNTER NAVIGATION

D. W. Murrow, P. W. Chodas and P. H. Kallemeyn*

The Galileo spacecraft will be targeted to encounter one or more asteroids during its cruise to Jupiter. Accurate navigation will maximize science return from these asteroid flyby opportunities. Navigation errors for these encounters are dominated by uncertainties in the asteroid ephemeris, which is obtained from fits to ground-based observations. As the spacecraft approaches, on-board optical navigation dramatically improves knowledge of the spacecraft-relative asteroid position normal to the line of sight, while correlations in the asteroid ephemeris provide moderate improvement along the approach direction. The remaining uncertainty in encounter time can be further reduced only by improving the ground-based asteroid ephemeris. Uncertainties perpendicular to the line of sight can be reduced by improving the timing of optical navigation images and their placement with respect to the star background. At the closest approach to the asteroid Gaspra, the one-sigma errors in knowledge of the spacecraft position are less than 10 km in position and 25 seconds in encounter time.

INTRODUCTION

The Galileo spacecraft will be targeted to encounter one or more main belt asteroids during its cruise to Jupiter, becoming the first spacecraft to perform closeup scientific observations of these small bodies. Asteroids are of great scientific interest because of the clues they may provide concerning the origin and evolution of the solar system. Having undergone very few of the modifications which have affected the larger planets and moons, they may preserve the chemical and physical characteristics of the original planetesimals from which the inner solar system was formed.

Under the primary mission plan, Galileo will make close flybys of two asteroids — the asteroid Gaspra in October 1991 and the asteroid Ida in August 1993. These objects are estimated to be about 16 and 33 km in diameter, respectively. The observation of these asteroids is a unique scientific opportunity and provides science results early in the Galileo mission. If Galileo is launched late in its launch period, it will no longer be possible to encounter Gaspra or Ida. In this case, Galileo will be targeted for an encounter with the asteroid 1938 SD1 in March 1992. This alternate target, which has not yet received an official name, is estimated to be about 18 km in diameter. As the most likely target for Galileo, the Gaspra encounter will be focused on in this paper. The Ida flyby is similar to the Gaspra flyby,

* Members of the Technical Staff, Navigation Systems Section, Jet Propulsion Laboratory, California Institute of Technology, 4800 Oak Grove Drive, Pasadena, California 91109.

and certain aspects translate directly to a 1938 SD1 encounter. The differences which make the 1938 SD1 encounter unique will also be examined.

The navigation of a spacecraft to an encounter with a nearly massless body presents significant challenges. Classical planetary encounter navigation techniques, with their reliance upon the gravitational signature of the target body, are ineffective. Optical navigation imaging is the only source of target-relative information. Earth-based radiometric data determines the heliocentric spacecraft trajectory, and combining the two data types improves the accuracies of both the spacecraft trajectory and the asteroid ephemeris.

The a priori knowledge of the asteroid ephemeris, determined from ground-based astronomical observations, is also of great importance. While the optical navigation images provide two-dimensional spacecraft-relative position information perpendicular to the approach direction, the third dimension of position is determined by the asteroid ephemeris. As a result, the accuracy with which the time of encounter can be predicted depends strongly on the accuracy of the asteroid ephemeris.

Accurate navigation to the desired encounter position is required to maximize the science return during the encounter. The primary consideration is the asteroid position uncertainty relative to the camera field of view, which determines how many pictures must be scheduled to ensure coverage of the asteroid. This position uncertainty is determined by the timing of the last optical navigation image, presently planned for one day before closest approach.

GALILEO MISSION OVERVIEW

The Galileo spacecraft is scheduled to be launched by the space shuttle Atlantis in the fall of 1989, on a mission to study Jupiter's planetary system. The launch period begins on October 12, and ends on November 21. The spacecraft will be deployed from the shuttle, and injected into its interplanetary trajectory using the Boeing Inertial Upper Stage (IUS) booster.

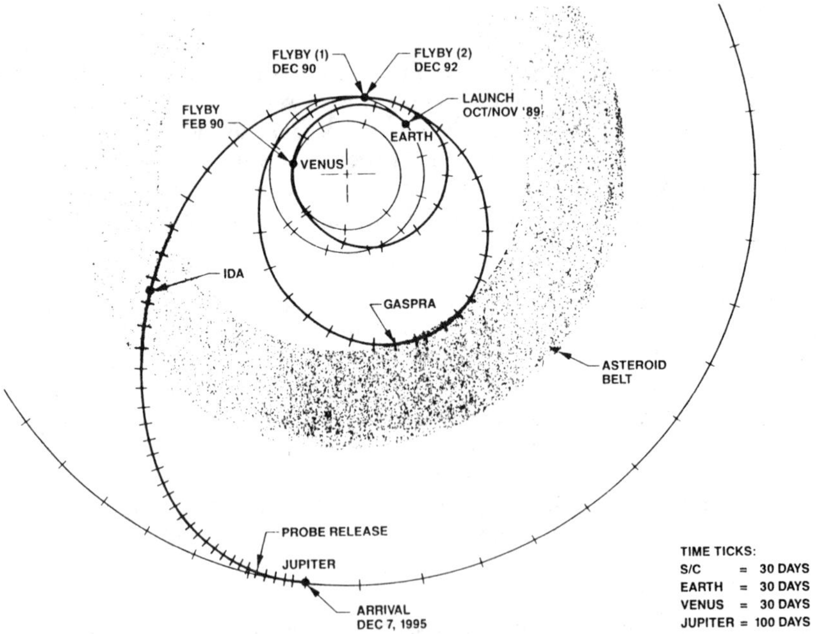

Figure 1: Galileo VEEGA Trajectory

To acquire the energy required to get to Jupiter, the trajectory makes use of three gravity assists, one at Venus and two at the Earth. The trajectory, called Venus-Earth-Earth Gravity Assist (VEEGA), takes Galileo through the asteroid belt twice, and reaches Jupiter in December, 1995. Fig. 1 shows the nominal VEEGA trajectory and the locations of the Gaspra and Ida encounters [1].

The scientific instrument package on-board Galileo will cover the electromagnetic spectrum from ultraviolet to near infrared. For the asteroid encounters, the wavelengths covered by the Solid State Imaging (SSI) subsystem and the Near Infrared Mapping Spectrometer (NIMS) are particularly useful. The SSI is also used to provide the optical navigation (OPNAV) images.

Upon arrival at Jupiter, Galileo will deliver a probe into the Jovian atmosphere and begin a twenty-two month tour of the Galilean satellite system. The probe will relay data on the Jovian atmospheric environment to the orbiter for retransmission to Earth. The orbiter will be targeted to encounter the four Galilean satellites, to take measurements of their composition and surface characteristics. Following the satellite tour, Galileo will be placed into a "tail petal" orbit, an elliptical orbit with the apojove in Jupiter's magnetotail, to provide information on the Jovian magnetic environment.

ASTEROID EPHEMERIS DEVELOPMENT

The ephemerides of asteroids are computed using sets of astronomical observations gathered over many years from dozens of observatories around the world. The three Galileo asteroid candidates have long observational histories: Gaspra, Ida, and 1938 SD1 were discovered in 1916, 1884, and 1938, respectively, although only observations as far back as 1905 were used in the orbit solutions. The observations themselves are referred to as *astrometric* observations. Each consists of a measurement of the right ascension and declination of the asteroid at a particular time. Typically, a photographic plate is taken of the region around the asteroid, and a measuring machine is used to determine the positions of the asteroid image with respect to several star images on the plate. Given the known coordinates of the stars, the coordinates of the asteroid can be computed. Measurements typically have a one-sigma error of 1 arc second. The dominant error source is usually the star catalog itself (i.e. errors in the coordinates of the stars due to measurement errors, uncertainties in proper motion, etc.).

Using data sets of 54, 105, and 48 astrometric observations, respectively, precise orbits of Gaspra, Ida, and 1938 SD1 were computed using JPL's Comet and Minor Planet Orbit Determination program. Perturbations due to all 9 planets were accounted for in the solutions. The post-fit root-mean-square residual was about 1 arc-sec for all three solutions, indicating good fits to the data. Once the orbits were computed, accurate ephemerides were produced for use in targeting the Galileo spacecraft. Table 1 lists some of the orbital characteristics of these asteroids.

Table 1: Some Orbital Parameters for the Galileo Asteroids

	Gaspra	Ida	1938 SD1
Perihelion distance, AU	1.826	2.741	1.856
Eccentricity	.1737	.0422	.1737
Inclination to ecliptic, deg	4.100	1.135	3.518
Period, yr	3.286	4.842	3.366

The expected position uncertainties for each asteroid at the time of encounter were also computed. These uncertainties depend on the number, timing, and measurement errors of the astrometric observations. Each analysis accounted for not only the existing observations, but also the future observations which are expected to be made between now and the time of the encounter. For Gaspra, 6 future observations were simulated in each of its two remaining oppositions before encounter (early 1990 and mid-1991). All observations were assumed to have one-sigma errors of 1.5 arcsec, a value larger than the solution RMS value to account for unmodelled error sources. Figure 2 shows the evolution of the one-sigma position uncertainties for Gaspra, expressed in the Radial-Transverse-Normal (RTN) coordinate frame. This frame is defined by the radial unit vector **R** from the Sun to the asteroid, the unit vector **N** normal to the asteroid's orbit plane, and the transverse unit vector **T** that completes the right-handed orthogonal system, **T** = **N**×**R**. The stepping patterns in the plot are caused by the incorporation of observations into the solution. The corresponding plots for Ida and 1938 SD1 are similar.

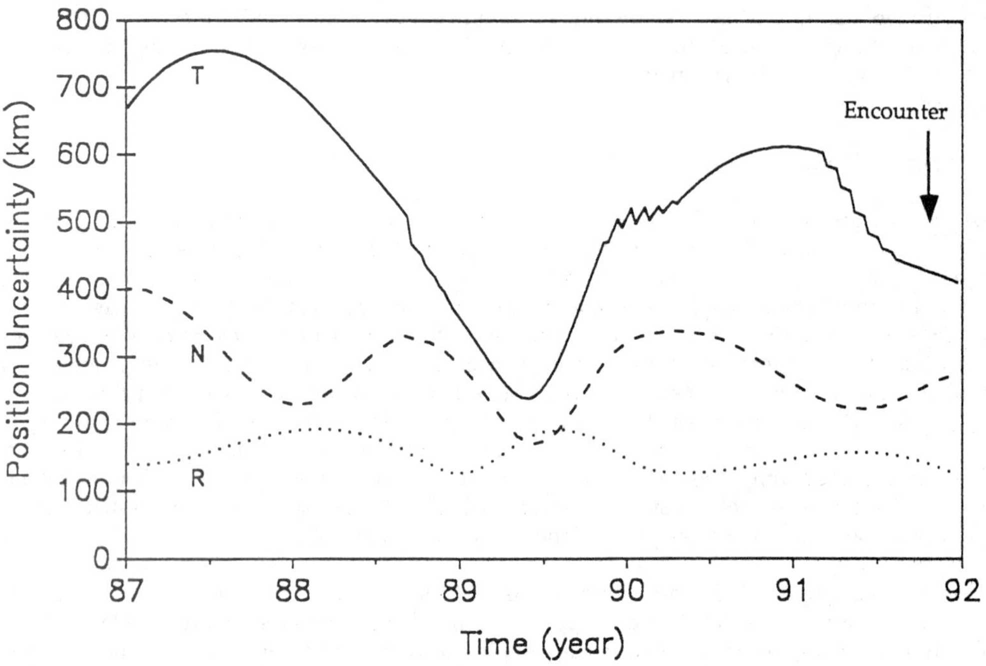

Figure 2: Evolution of One-Sigma Position Uncertainties for Gaspra

The three components of uncertainty define an error ellipsoid, which near Gaspra encounter is quite elongated in the transverse direction (approximately in the direction of the asteroid's motion). This orientation for an asteroid's error ellipsoid is quite common: errors in the along-track velocity tend to spread the position ellipsoid along the direction of motion. Table 2 lists the position uncertainties for the three Galileo asteroids at their respective encounter times. Just 3 observations are assumed in each of the 4 future oppositions of Ida and 2 future oppositions of 1938 SD1.

These uncertainties, based on astrometric observations alone, are not the final encounter uncertainties: they become the a priori uncertainties for the spacecraft observation phase of the encounter navigation. As will be discussed later, optical navigation during the approach phase will be used to greatly reduce the two components of these uncertainties perpendicular to the line of sight. The components of the uncertainties along the line of sight will retain their a priori values.

Table 2: One-Sigma Asteroid Position Uncertainties at Galileo Encounter (km)

	Radial	Transverse	Normal
Gaspra	139	425	263
Ida	156	447	322
1938 SD1	285	304	250

A second analysis was performed for Gaspra to examine a "worst case" in which no future observations were made. The resulting error ellipsoid was about 20 percent larger, indicating that the astrometric observations close to encounter are especially important in reducing uncertainties at encounter. Conversely, increasing the number of observations assumed for coming oppositions of Gaspra should reduce the size of the error ellipsoid. Analyses supporting this conclusion will be discussed later in this paper, along with other strategies for improving upon the baseline results listed in Table 2.

NAVIGATION OBJECTIVES

The asteroid approach trajectories are designed to satisfy two conditions. The spacecraft is targeted to allow viewing of the sunlit side of the asteroid, and the flyby distance is chosen to be 1000 km. For navigation, the consequence of the close flyby is that the optical navigation images are along nearly the same viewing direction until some hours before closest approach. The actual flyby distance, for which the results in this paper are valid, will be chosen closer to the encounter time.

The encounter targeting parameters are defined using the so-called B-plane, which is the plane perpendicular to the approach asymptote and containing the center of mass of the target body (Fig. 3). **B** is the vector from the center of mass of the target body to the point of intersection of the arrival asymptote. The **S**-axis is along the direction of the approach asymptote, the **T**-axis is perpendicular to **S** and parallel to the ecliptic, and the **R**-axis completes the right-hand triad such that **R** = **S**×**T**.

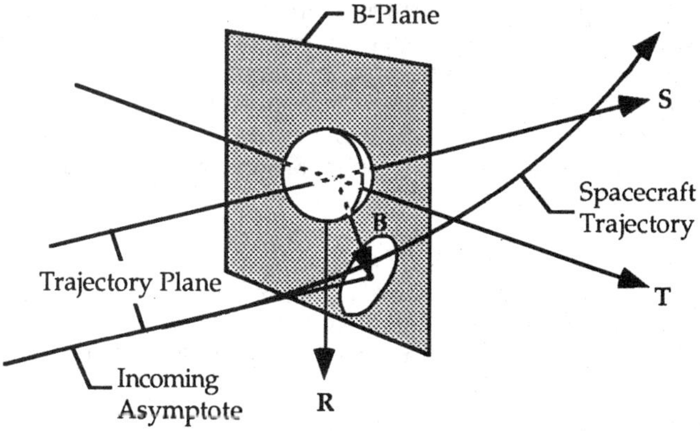

Figure 3: B-Plane Defined at Target Body

To ensure coverage, the science picture sequences must cover all of the possible positions of the asteroid, even though many empty frames will result. Accurate navigation is needed to reduce the number of wasted science images. The science sequence will be designed to point the camera at the first possible position of the asteroid. The camera is then held fixed while images are taken, and the asteroid drifts past, resulting in a "strip" of images. An accurate prediction of the encounter time, determined by the navigation time of flight error, will reduce the number of frames which must be scheduled in these strips. Increases in out-of-plane, or B•R, errors will require more strips to be laid side by side to ensure coverage of the entire asteroid. B•T errors correspond to errors in the flyby distance, but do not result in wasted science images.

The final targeting maneuvers are scheduled at 20 and 5 days before encounter (E). By this time, the asteroid position will have been measured in the optical navigation images, so these maneuvers are designed based on an improved asteroid ephemeris. Navigation errors remaining at E – 5 days will not be corrected, but their effect can be compensated for with camera pointing. This is accomplished with a science pointing update, based on images taken up to E – 24 hours. The delivery errors are required to be less than 50 km, so that the magnitude of this update will be no more than 3 degrees (one-sigma). The errors remaining after the science pointing update represent the final navigation accuracy for the asteroid science opportunity.

DATA AND ERROR SOURCE ASSUMPTIONS

Two passes per week of coherent Doppler and range data are assumed to be available, increasing to continuous coverage (3 passes per day) at E – 30 days. These data have assumed noise levels of 1 mm/sec and 1 km respectively. The nominal optical image sequence contains 28 images, beginning at 52 days before closest approach, with the final picture at one day before closest approach. Other OPNAV sequences have been studied, and will be discussed later in this paper. The images are scheduled at approximate two day intervals up to the last four frames, which are taken at one day intervals. One–pixel asteroid image weighting is used to account for possible albedo effects and irregularity in shape. This somewhat conservative assumption is due also to the fact that the asteroid approach will be the first operational use of Galileo's optical navigation system.

Estimated parameters account for errors in asteroid ephemeris, the Sun line component of solar pressure, and maneuver execution errors. The asteroid ephemeris is the dominant error source when it enters at its a priori level, and is improved through the use of optical navigation frames. The effects of maneuver execution errors and the Sun line component of the solar pressure are solved for, since those perturbations on the trajectory are large enough to be resolved with the data. Additionally, stochastic and bias accelerations along the Earth-spacecraft direction are estimated to account for the signature of spacecraft pointing corrections present in the data. The effects of all of these error sources are contained in the computed statistics. Other model errors, including tracking station location, atmospheric media, and the cross components of the solar pressure acceleration, are treated as consider effects, i.e., errors which will corrupt the solution but cannot be improved by estimation.

The simulated data and error models are combined in JPL's Orbit Determination Program, which implements a least squares filter with the square root information array formulation following Bierman [2]. The results presented here are obtained with covariance analysis techniques, and represent the expected accuracies which will be obtained while tracking the Galileo spacecraft.

RESULTS

As discussed earlier, the asteroid ephemeris errors dominate the orbit determination accuracy until the first OPNAV images are acquired at E – 52 days. At this point, the optical data begins to provide information on the spacecraft-relative Gaspra position. This information together with the determination of the heliocentric spacecraft state from the radiometric data is used to improve the heliocentric asteroid state. The one-sigma uncertainties at the time of the final maneuver are 31 km in B•R, 26 km in B•T, and 21 seconds in time of flight. At the science pointing update, one day before closest approach, the one-sigma B-plane uncertainties are 6.3 km in B•R, 5.8 km in B•T, and 20 seconds in time of flight. The uncertainty in encounter time corresponds to a one-sigma linear uncertainty of 160 km along the flight path. For comparison, the SSI frame size is 9 km at 1000 km.

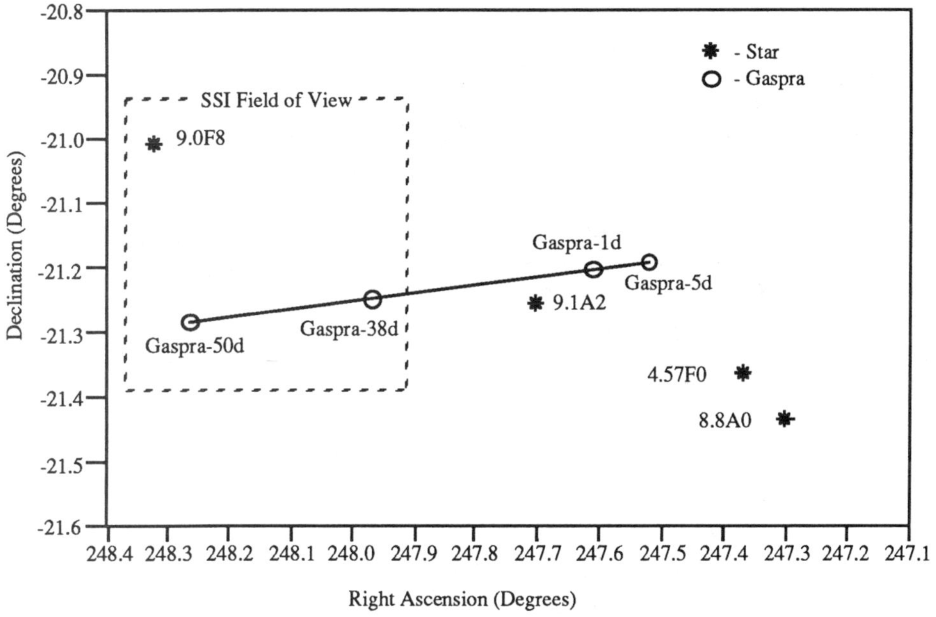

Figure 4: Motion of Gaspra, Travelling from Left to Right and Reversing at E – 5 days, as seen from Galileo During Approach

The improvement of the asteroid ephemeris comes from pixel-level position information in a plane perpendicular to the spacecraft viewing direction. Each pixel subtends 10 μrad, which is scaled by the spacecraft-asteroid distance into a B-plane error. The information contained in each optical image is in essentially the same plane, becoming more accurate as the spacecraft-asteroid range decreases. This is due to the small motion of Gaspra across the background star field during the Galileo approach, illustrated in Fig. 4. As seen from the spacecraft, the asteroid moves from left to right from Gaspra E – 52 days to E – 5 days, then reverses its path. The dashed-line box represents the SSI field of view, and the numbers next to the stars indicate their magnitude and type. Due to the relative scarcity of usable stars, the first 18 pictures contain only one star. A one-star picture provides information on the spacecraft position in only one direction in the B-plane. Since the asteroid-star line in the first several pictures (9.0F8) is nearly orthogonal to the line in the next few pictures (9.1A2), however, the measurement information can be resolved into two independent directions.

Figure 5 shows one-sigma errors in the three B-plane components over time, and illustrates the information contained in the one-star pictures. At Gaspra E – 45 days, the sequence switches from star 9.0F8 to 9.1A2. The separation between the star and asteroid at this time is primarily in right ascension, as seen in Fig. 4. The in-plane information provides continuing improvement in B•T, whereas the B•R uncertainty is nearly constant. This continues until the asteroid passes above 9.1A2, when the B•R uncertainty again begins to decrease. Upon acquisition of two-star pictures, both components are improved.

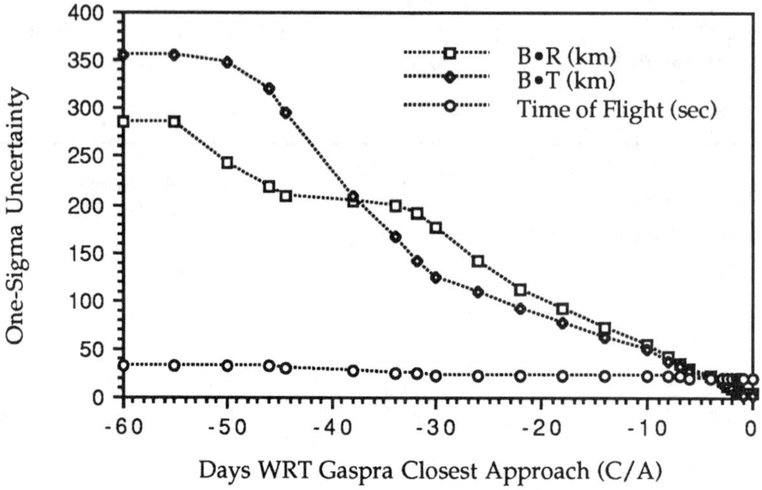

Figure 5: Improvement of B-Plane Accuracy

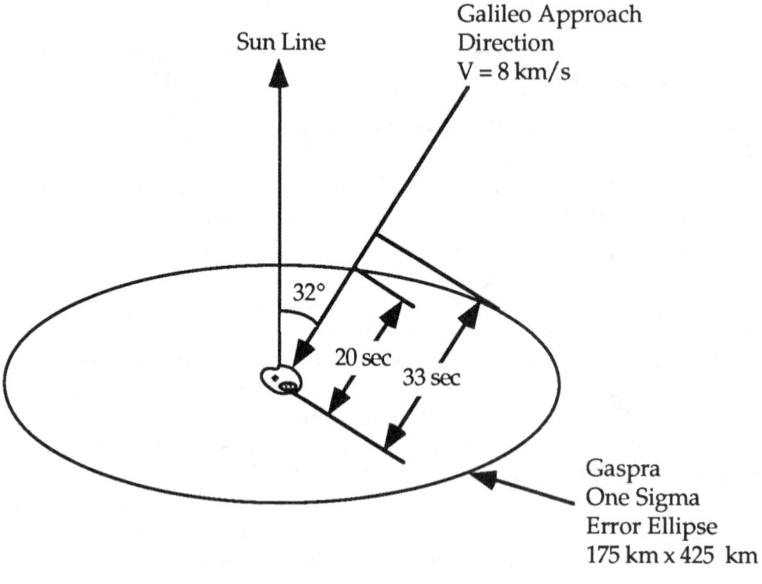

Figure 6: One-Sigma Gaspra Orbit Plane Error Ellipse

The trajectory plane projection of the one-sigma a priori Gaspra error ellipse is shown in Fig. 6. The projection of the ellipse along the Galileo approach direction represents the down track encounter error, which determines the time of flight error. Since the direction of the approach is at an angle of 32 degrees to the minor axis of the error ellipse, there is a strong correlation between the B•T errors and the time of flight errors. This correlation results in a reduction in time of flight error as optical data is acquired, contrary to the usual situation where the time of flight errors are not improved by optical data. An initial reduction in the time of flight error is seen from its a priori level of 33 seconds, but as the correlation is removed a lower limit is reached at approximately 20 seconds.

ALTERNATE OPTICAL IMAGE SEQUENCE STRATEGIES

The scheduling and pointing of the OPNAV images is significant in the time evolution of the B-plane uncertainties. In this section, three strategies involving different picture sequences and optical error modeling are discussed. The goal is to develop a method for acquiring data that maximizes the determination of the B-plane parameters, while accounting for possible errors in the optical system. Table 3 summarizes the B-plane errors at the science pointing update for each method.

Table 3: B-plane Errors (One-Sigma) for Gaspra at Science Pointing Update

	B•R	B•T	Time of Flight
Nominal Picture Sequence	6.3 km	5.8 km	20 seconds
2nd Picture Sequence	3.7 km	3.2 km	20 seconds
2nd Picture Sequence + Optical Error Model	5.6 km	5.0 km	20 seconds
3rd Picture Sequence + Optical Error Model	4.0 km	3.6 km	20 seconds

By designing a sequence of pictures timed closer together than the nominal case during the last few days, the random component of error can be reduced. This was shown in the Gaspra approach study by replacing the nominal picture sequence with a second sequence in which pictures are taken more often at closer proximity to Gaspra. This new sequence contains 26 images, of which the last 8 occur between one and two days out from Gaspra. The effect this sequence has on the B-plane uncertainties is shown on line 2 of Table 3. The later pictures contain more information due to the decreased distance from the asteroid, and thus improve knowledge of the B-plane components. The time of flight uncertainty remains fixed due to the inability of the pictures to provide information along the line of sight.

This case is optimistic, however, since it assumes that the camera pointing will be completely solved for using the stars in the navigation image. This can be made more realistic by modeling random camera pointing, and solving for the pointing angles using star images in the optical data. Figure 7 depicts the camera-centered coordinate system used for this study [3]. In this system, the **L**-axis is in the viewing direction of the camera, the **N**-axis points from the camera toward the center of the spacecraft, and the **M**-axis completes the right-hand system such that

M=N x L. "Clock" and "cone" angles are produced via rotation about the **M** and **N** axes, while "twist" is rotation about **L.**

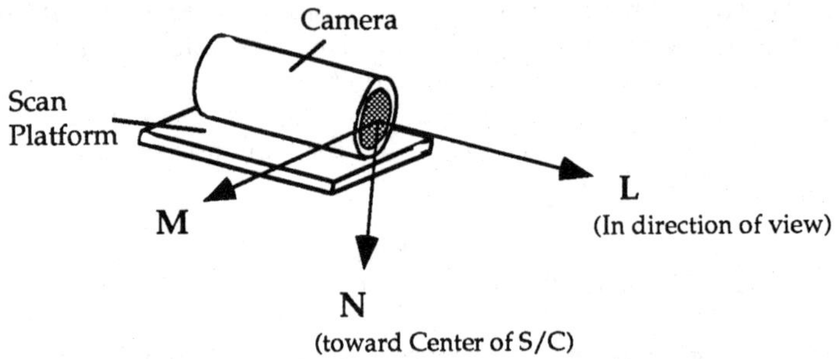

Figure 7: Galileo Camera-Centered Coordinate System

A priori uncertainties were set at 1 mrad for clock and cone angles and 11 mrad for twist. Star positions in each image were weighted at 0.5 pixels. Center-finding ability and line/pixel bias for both Gaspra and star positions were included as consider effects. All other error sources remained the same as in the nominal case.

Line 4 of Table 3 shows the results from these optical error assumptions. Again, time of flight uncertainty remained unchanged since the optical data give no information along the line-of-sight. The increase in B•R and B•T uncertainties is driven in part by error in solving for camera twist angle during the last 22 days before encounter. Images taken during this time contain two stars, separated by less than 1.7 mrad. As a result, camera twist can only be solved for to about 0.05 mrad.

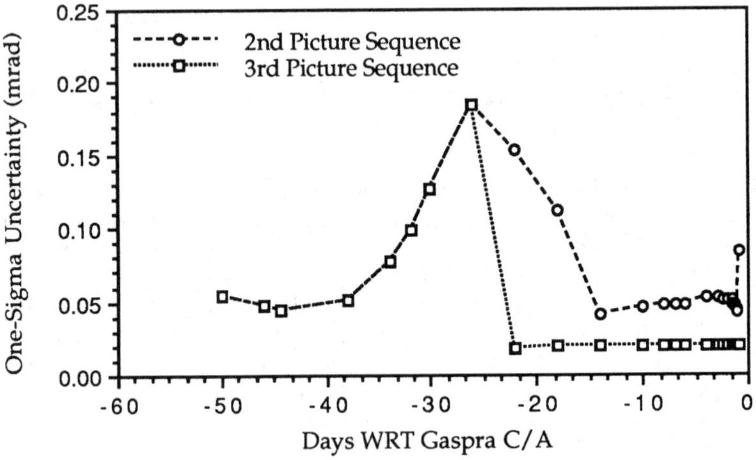

Figure 8: Evolution of Camera Twist Error

To improve the solution for camera twist, a third picture sequence was generated utilizing maximum star separation in each picture. As seen in Fig. 4, the 4.57F0-8.8A0 star pair, used in the first and second picture sequences, is quite close together. When the final 22 days of images were instead pointed to the 9.1A2-4.57F0 pair, the solution for camera twist was improved. Figure 8 illustrates the difference in camera twist error obtained from the second and third picture sequences. At E – 22 days, the two-star images reduce the twist uncertainty to 0.025 mrad. This result follows since the center of rotation will be approximately halfway between the two stars, which constrain the rotation. As the center of rotation moves away from the star, the one-half pixel uncertainty on the star position implies a smaller angle of rotation. This improved twist solution leads to better knowledge of the B-plane parameters, since the contribution of twist error to asteroid position error is reduced. The final B-plane uncertainties at the science pointing update for this third sequence are given on line 4 of Table 3.

IMPROVING THE ASTEROID EPHEMERIS

Because the a priori ephemeris uncertainty is the largest contributor to the time of flight uncertainty, and because this latter uncertainty has a strong bearing on the amount of science data which can be acquired at the encounter, it is most important to reduce the a priori uncertainty as much as possible, especially the radial component. To this end, several error analyses were performed to test strategies which could reduce the a priori asteroid position uncertainties.

The first strategy studied was to increase the number of astrometric observations assumed for future oppositions, and the time span over which they are assumed to be made. Instead of 6 observations in each of the remaining two Gaspra oppositions, 35 and 77 observations were assumed, respectively. These rather optimistic numbers of observations can be realized only if a concerted ground-based observing campaign is organized.

The second strategy for improving the asteroid ephemeris is to attempt to reduce measurement errors. As was mentioned earlier, the dominant error source in astrometric observations is due to inaccuracies in the star catalog. This error could be reduced by producing a special high-accuracy catalog for just those star fields through which the asteroid will pass in the coming oppositions. This involves taking special photographic plates of these regions of the sky, and carefully measuring the positions of the thousands of stars near the asteroid's expected track. The resulting special star catalog would be more accurate than current catalogs because the epoch of the catalog will be close to the time of the asteroid observations, so that uncertainties in proper motion will be insignificant. Also, modern measuring machines are superior to those used for older catalogs, resulting in more accurate position measurements. Once produced, this special catalog would have to be distributed to active astrometric observers, who would have to be encouraged to use it in their data reductions.

The effect of applying these two strategies separately and in combination is shown in Fig. 9, which depicts cross-sections of the error ellipsoids in the asteroid's orbit plane. Ellipse 1 is the nominal asteroid error ellipse, 2 results from using the additional observations, 3 results from using the improved star catalog, and 4 represents applying both strategies. In cases 2 and 4, it is assumed that the improved star catalog is used for all future observations, and that it reduces measurement errors by a third. As was discussed earlier, the time of flight error is represented graphically as the radius of the error ellipse along the approach line. The reduction in the time of flight error when both strategies are applied is approximately 30%, from an error of 20 seconds to an error of 14 seconds. The B-plane errors remain unchanged due to their dependence on OPNAV images.

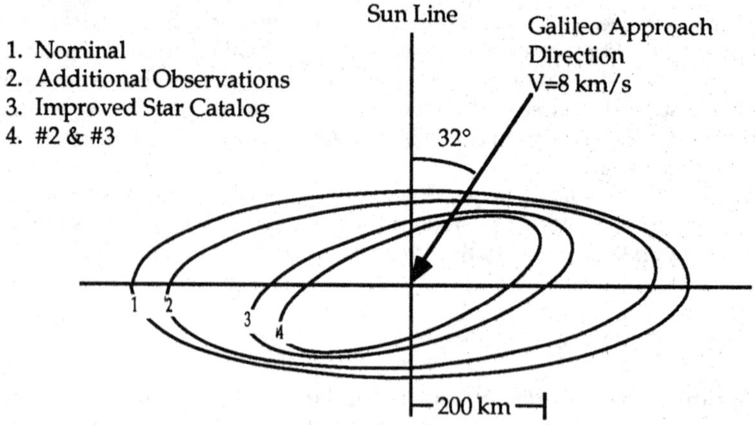

Figure 9: Comparison of One-Sigma Gaspra Orbit Plane Error Ellipses Under Different Observing Strategies

1938 SD1 AND IDA FLYBY OPPORTUNITIES

For dates late in the launch period, the Galileo spacecraft will be targeted for an encounter with the asteroid 1938 SD1 in March, 1992. The flyby strategy is the same as for the Gaspra encounter. The pointing requirements, which are determined by the out-of-plane errors, are also the same as for the Gaspra encounter. The downtrack or time of flight error determines the length of the science picture sequences at closest approach.

The 1938 SD1 flyby differs from the Gaspra and Ida flybys in two respects. First, the approach to 1938 SD1 is at a lower relative velocity than the other asteroid approach trajectories. The flyby velocity is 4.8 km/s, compared to 8.0 km/s for the Gaspra encounter and 12.4 km/s for the Ida flyby. This low relative velocity scales any encounter position error into a larger time of flight error. Conversely, the 24 hour data cutoff implies that the last OPNAV images are acquired closer to the asteroid, reducing the B-plane error which results from the angular error in the image. This scaling by the smaller spacecraft-asteroid distance at the data cutoff will reduce the encounter statistics by the ratio of the encounter velocities.

Also, the asteroid ephemeris error projected into the trajectory plane is roughly circular at the encounter time (Fig. 10). This results from the time variation of the asteroid orbit covariance. At the encounter time, the radial error is increasing while the transverse error is decreasing, making the two components' errors nearly equal at the encounter time. This near equality implies an absence of correlation between position components in that plane, and prevents the cross line of sight information contained in the optical data from improving the time of flight determination. This is illustrated in Fig. 10, where the trajectory plane projection of the 1938 SD1 error ellipse is shown. As shown, the projection of the error ellipse onto the approach direction is 300 km. The 4.8 km/s approach velocity translates this position error into a 62.5 second encounter time error.

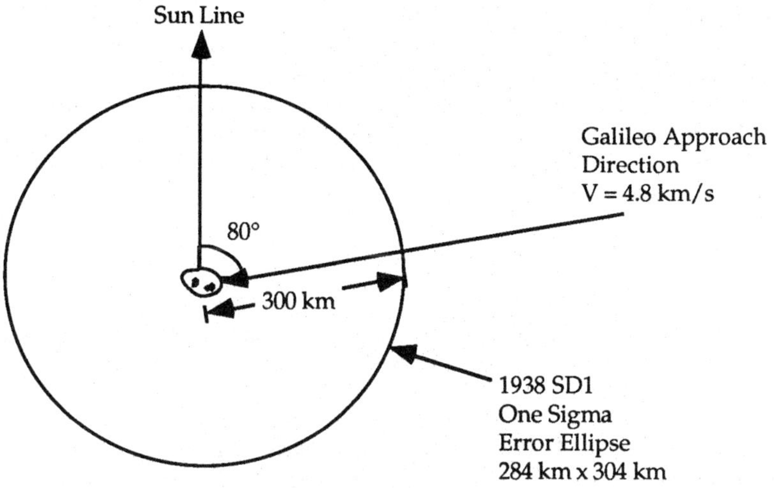

Figure 10: One-Sigma 1938 SD1 Orbit Plane Error Ellipse

In the baseline Galileo trajectory, the spacecraft is targeted to a second asteroid flyby, of Ida, in August 1993. The characteristics of the Ida flyby are similar to those of the Gaspra flyby, with two exceptions. Galileo approaches the asteroid more nearly along the minor axis of the asteroid error ellipse, providing a reduction in the time of flight uncertainty. The encounter is also at a larger relative velocity, which has an effect opposite that for the 1938 SD1 flyby. The larger flyby velocity reduces the time of flight error by scaling the downtrack position error, but the larger spacecraft-asteroid distance at the last picture causes an increase in the B-plane position errors. The nominal one-sigma B-plane errors at the science pointing update for the three Galileo asteroid opportunities are shown in Table 4.

Table 4: One-Sigma B-Plane Errors for Galileo Asteroid Encounters
SSI Frame Size at closest approach is 9 km

	B•R	B•T	Time of Flight
Gaspra	6.3 km	5.8 km	20 seconds
Ida	7.5 km	6.7 km	14 seconds
1938 SD1	3 km	3 km	63 seconds

CONCLUSIONS

Accurate navigation is required to maximize science return from the Galileo asteroid flyby opportunities. This requires optical navigation as the only source of target-relative spacecraft position information. An accurate a priori asteroid ephemeris is required to provide line of sight targeting information. Further reductions in the time of flight uncertainty at encounter can be realized only through improvements in the ground-based asteroid ephemeris. Reductions in the B-plane uncertainties can be achieved by improving the timing and pointing of the optical navigation images.

ACKNOWLEDGEMENTS

The authors of this paper wish to acknowledge the support and help of F. T. Nicholson, D. K. Yeomans, W. M. Owen and many others in preparing and understanding the results presented in this paper. The research described in this paper was carried out by the Jet Propulsion Laboratory, California Institute of Technology, under a contract with the National Aeronautics and Space Administration.

REFERENCES

1) Johannesen, J. R., "Reference Integrated Interplanetary, Probe and Orbiter Trajectories for 1989 VEEGA with Gaspra and Ida Asteroid Flybys", JPL Internal Document, Interoffice Memorandum 312/88.4-1461, 25 Apr 1988.

2) Bierman, G. J., *Factorization Methods for Discrete Sequential Estimation,* Academic Press, New York, 1977.

3) Owen, W. M., Jr., "Galileo Attitude and Camera Models", JPL Internal Document, Engineering Memorandum 314-323, 14 Nov 1983.

AAS 89-402

AN ANALYTIC DEVELOPMENT OF ORBIT DETERMINATION FOR A DISTANT PLANETARY ORBITER*

R. K. Russell and S. W. Thurman†

With the advent of the Mariner '71 Mission, we have been sending spacecraft to orbit various distant bodies within the solar system. At present, we still have no adequate theory describing the inherent state estimation accuracy, based on two-way, coherent range-rate data. The purpose of this development is to lay the groundwork for a general elliptic theory, and in addition to provide an approximate analytic solution for the special case of circular orbits.

An important aspect of this orbital estimation problem is to pose it in a proper coordinate system. This shows that the main contributor to position error is just the error in the longitude of ascending node relative to the plane-of-the-sky. Since there will be some parallactic motion between the observer and the spacecraft, it is possible to obtain a "reasonable" solution for this parameter.

From this point, the paper proceeds to develop a model of the observable, obtain the partial derivatives of that observable with respect to the spacecraft state, accumulate the information array, and in the circular orbit case to invert a 4x4 sub-portion of that array. This step leads to some surprising conclusions of which orbit determination analysts should be aware. The general elliptic problem has been treated by performing numerical inversions of the information array, yielding some interesting estimation behavior as a function of the orbit orientation.

---

* This paper appears in full in the microfiche supplement to this volume, **AAS Microfiche Series**, Volume 59.

† Jet Propulsion Laboratory, California Institute of Technology, 4800 Oak Grove Drive, Pasadena, California 91109.

SATELLITE ORBIT CONTROL

SESSION 11

Chairman: Christopher J. Scolese
NASA Goddard Space Flight Center

AAS 89-407

ELIMINATING GPS STATIONKEEPING MANEUVERS BY CHANGING THE ORBIT ALTITUDE

C. C. Chao* and D. L. Schmitt†

This analysis provides an in-depth investigation of the Global Positioning System (GPS) longitude drift at the nominal GPS altitude, and how a ±2-deg longitude stationkeeping tolerance can result in a worst-case 8-deg change in relative phasing among the Block II GPS satellites. Simulation results show that the navigation performance of the optimized 21-satellite constellation is sensitive to the longitude drift between stationkeeping maneuvers.

The long-term orbital motion and longitude variations at altitudes slightly different from the nominal GPS altitude with repeating ground tracks are studied. Both analytical and numerical findings show that a slight altitude change can significantly reduce the resonant tesseral harmonics effects. By changing the altitude slightly, the stationkeeping maneuvers can be eliminated and the resulting longitude deviations will remain within ±1 deg. Thus, the navigation performance of the GPS constellation can be maintained at its desirable value and the ground control load will be reduced significantly without having to perform orbit maintenance maneuvers. Impacts on autonomous momentum management system and coverage for Yuma test are discussed.

INTRODUCTION

Over the history of the GPS program, various satellite constellations have been considered as the baseline. Early in the program, a uniform arrangement of 24 satellites in 3 orbit planes was the baseline constellation (it was referred to as "The Walker 24/3/2"). Other considerations included a uniform arrangement of 18 satellites in 6 orbit planes with 3 active spares placed in every other plane. This

* Manager, Orbit Dynamics Section, Astrodynamics Department, The Aerospace Corporation, Mail Station M4/946, P.O. Box 92957, Los Angeles, California 90009-2957. Member AIAA.

† Project Engineer, SSTS Program Office, The Aerospace Corporation, P.O. Box 92957, Los Angeles, California 90009-2957.

was referred to as "The Walker 18/6/2 plus three" (Table 1). The current baseline was developed by optimizing the positions of the 18/6/2 plus three to give nearly complete worldwide coverage. This is referred to as "The Optimized 21" satellite constellation (Ref. 1) and is shown in Table 2.

Table 1
THE WALKER 18/6/2 + 3 SV GPS CONSTELLATION
(classical elements in km and deg)

Epoch = 1989, 4, 1., 0., 0., 0.,

	a	e	i	Ω	ω	M
A1	26560	0.	55.0	335.0	0.	0
A2	26560	0.	55.0	335.0	0.	120
A3	26560	0.	55.0	335.0	0.	240
B1	26560	0.	55.0	35.0	0.	40
B2	26560	0.	55.0	35.0	0.	160
B3	26560	0.	55.0	35.0	0.	280
C1	26560	0.	55.0	95.0	0.	80
C2	26560	0.	55.0	95.0	0.	200
C3	26560	0.	55.0	95.0	0.	320
D1	26560	0.	55.0	155.0	0.	120
D2	26560	0.	55.0	155.0	0.	240
D3	26560	0.	55.0	155.0	0.	0
E1	26560	0.	55.0	215.0	0.	160
E2	26560	0.	55.0	215.0	0.	280
E3	26560	0.	55.0	215.0	0.	40
F1	26560	0.	55.0	275.0	0.	200
F2	26560	0.	55.0	275.0	0.	320
F3	26560	0.	55.0	275.0	0.	80
A4	26560	0.	55.0	335.0	0.	30
E4	26560	0.	55.0	215.0	0.	310
C4	26560	0.	55.0	95.0	0.	170

Table 2
THE OPTIMAL 21 SV GPS CONSTELLATION
(classical elements in km and deg)

Epoch = 1989, 4, 1., 0., 0., 0.,

	a	e	i	Ω	ω	M
A1	26560	0.	55.0	335.0	0.	55.5
A2	26560	0.	55.0	335.0	0.	194.5
A3	26560	0.	55.0	335.0	0.	311.75
B1	26560	0.	55.0	35.0	0.	125.5
B2	26560	0.	55.0	35.0	0.	223.5
B3	26560	0.	55.0	35.0	0.	354.5
C1	26560	0.	55.0	95.0	0.	154.5
C2	26560	0.	55.0	95.0	0.	293.5
C3	26560	0.	55.0	95.0	0.	37.25
D1	26560	0.	55.0	155.0	0.	192.75
D2	26560	0.	55.0	155.0	0.	311.25
D3	26560	0.	55.0	155.0	0.	81.5
E1	26560	0.	55.0	215.0	0.	236.75
E2	26560	0.	55.0	215.0	0.	333.5
E3	26560	0.	55.0	215.0	0.	112.25
F1	26560	0.	55.0	275.0	0.	267.5
F2	26560	0.	55.0	275.0	0.	37.75
F3	26560	0.	55.0	275.0	0.	156.25
A4	26560	0.	55.0	335.0	0.	89.0
E4	26560	0.	55.0	215.0	0.	15.5
C4	26560	0.	55.0	95.0	0.	260.0

When the optimized 21-satellite constellation was adopted as the baseline by the GPS Program Office, questions were raised concerning the effects of stationkeeping and long-term orbit perturbations on the performance of the improved constellation. Results of an early analysis by Chao and Bowen (Ref. 2) indicated that the long-term perturbations caused by the coupling J_2 and luni-solar effects on inclination and node were significant to the Walker 18/6/2 constellation. These effects can be largely removed by a proper biasing of the initial values of inclination. The stationkeeping effects on the optimized 21 constellation were found to be significant from two independent studies recently conducted by the authors of this paper and Massatt. Massatt's findings were based on pure numerical simulations. His study also revealed that changing the orbit altitude slightly can greatly reduce the resonant effects due to earth gravity tesseral harmonics (Ref. 1).

The purposes of this analysis are: (1) to perform an in-depth investigation of the stationkeeping and perturbation effects on the performance of the optimized constellation based on both analytical and numerical examinations, and (2) to provide a clear understanding of the constellation stability at various altitudes, which will aid in selecting the most favorable altitude at which the longitude stationkeeping maneuvers can be eliminated.

One of the measurements used to evaluate constellation performance in this study is called constellation value (CV). The CV is the fraction of the earth, averaged over 24 hr, in which the GPS satellites provide an acceptable navigational solution. For the purposes of this study, an acceptable navigational solution occurs when the PDOP value is less than or equal to six. [PDOP is the Positional Dilution of Precision which, when multiplied by the User-Range Error (URE), gives the three-dimensional rms error.] The CV calculation for this study also assumed a minimum elevation angle of 5 deg.

STATIONKEEPING AND PERTURBATIONS EFFECTS

Longitude Drift and Stationkeeping Effects

The nominal GPS orbit has a period of nearly 12 hr with repeating ground tracks. The longitude of the ascending node of a given GPS satellite will repeat on every other revolution. Due to the resonance effects of the Earth gravity harmonics (the tesseral harmonics,* J_{32}, J_{44}, J_{52}, etc.), the longitude of node gradually drifts away from its initial value. The drift rate depends on the initial longitude and semimajor axis. The theoretical explanation of the resonant tesseral harmonics effects on semimajor axis and longitude of node can be understood through Kaula's representation (Ref. 3). Resonance occurs when the sum of the angular rates of a particular term in the harmonics is vanishingly small, i.e.

$$(\ell - 2p)\,\dot{\omega} + (\ell - 2p + q)\,\dot{M} + m\,(\dot{\Omega} - \dot{\alpha}_G) \approx 0 \qquad (1)$$

where $\dot{\omega}$, $\dot{M}$, $\dot{\Omega}$ and $\dot{\alpha}_G$ are the time rate of argument of perigee, mean anomaly, right ascension of ascending node, and Greenwich hour angle, respectively. The integers ℓ, p, m, q, are the index of a given term. For a GPS orbit with repeating ground tracks, the dominant terms which cause resonant effects on semimajor axis can be combined as follows:

---

*The J_{22} effects are not important due to the small eccentricity of the GPS orbit.

$$da/dt = 6.17 \sin 2(\lambda_\Omega - \lambda_{32})$$

$$-1.52 \sin 4(\lambda_\Omega - \lambda_{44})$$

$$+0.19 \sin 2(\lambda_\Omega - \lambda_{52}) \text{ m/day} \qquad (2)$$

where

$$\lambda_\Omega = \frac{M + \omega}{2} + \Omega - \alpha_G$$

and

λ_Ω = longitude of ascending node

λ_{32} = 27.37°E

λ_{44} = 30.31°E

λ_{52} = 31.93°E

The above equation can be converted to the following equation for longitude of ascending node.

$$\ddot{\lambda}_\Omega = \frac{1}{2}\frac{dn}{dt} = -\frac{3}{4}\,\frac{n}{a}\,\frac{da}{dt}$$

$$= -0.000126 \text{ (deg/day}^2\text{) } [\sin 2(\lambda_\Omega - \lambda_{32})$$

$$-0.246 \sin 4(\lambda_\Omega - \lambda_{44}) + 0.031 \sin 2(\lambda_\Omega - \lambda_{52})] \qquad (3)$$

The above nonlinear differential equation is a pendulum-type equation with longitude λ_Ω slowly librating about the stable point near 27.4 deg and 27.4 deg + 180 deg. The amplitude and period of the long-period variations are a function of initial λ_Ω. The two solid curves in Fig. 1 are the longitude histories of two GPS orbits with initial longitude of ascending node at 70 deg E and 38 deg E. These long-period librations indicate that periodic station-keeping maneuvers are required to maintain the longitude variations within a desired tolerance.

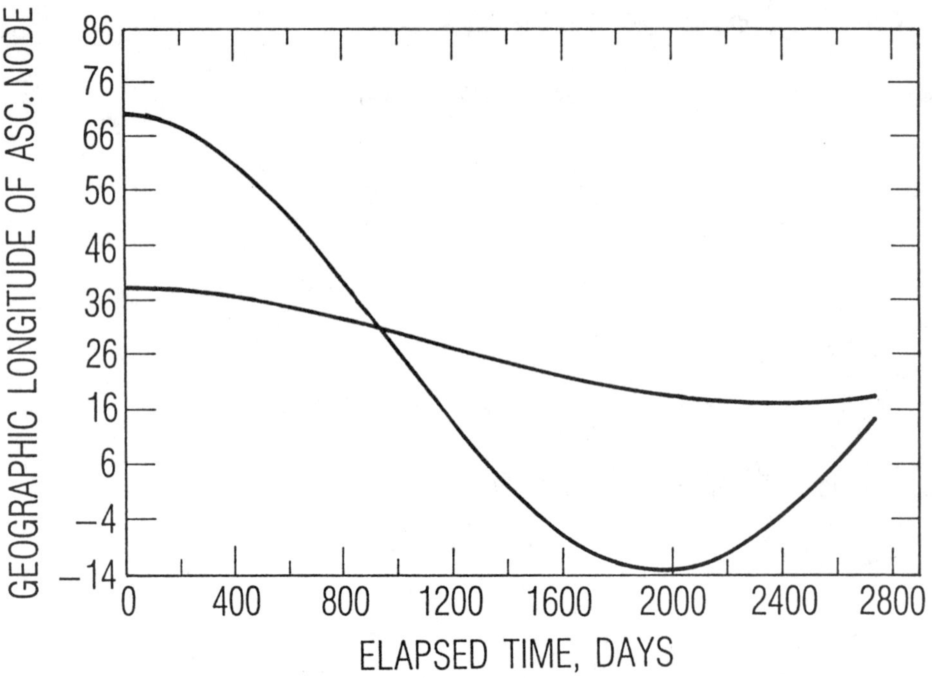

Fig. 1 Time History of Longitude of Ascending Node of a GPS Orbit Without Stationkeeping

The current stationkeeping technique is to apply small ΔV maneuvers at the times when the longitude of node reaches the boundary of the $\pm$2-deg tolerance as shown in Fig. 2. The magnitude and the sign of the ΔV are predetermined so that the scallop-shaped variation is always within the tolerance and the time between maneuvers is maximized. When the longitude variation is confined within the specified small tolerance of $\pm$2 deg, the above equation [Eq. (3)] can be integrated twice to a quadratic solution by $\ddot{\lambda}_\Omega$ being constant. With the proper boundary conditions satisfied, a simple equation for predicting the time interval between two adjacent stationkeeping maneuvers can be obtained as follows

$$T = 2[2(\lambda_u - \lambda_\ell)/|\ddot{\lambda}_\Omega|]^{1/2} \quad (4)$$

where λ_u and λ_ℓ are the upper and lower limits of longitude tolerance of stationkeeping and $\ddot{\lambda}_\Omega$ can be computed from Eq. (3) for a given longitude of ascending node. The maximum time interval is a function of stationkeeping tolerance and the nominal longitude as shown by the two types of curves in Fig. 3. The solid curves are for a $\pm$1.8-deg tolerance, and the broken curves are for a $\pm$0.9-deg tolerance. The time between stationkeeping maneuvers ranges from 330 days to 1000 days or longer, depending on tolerance and longitude. The curves in Fig. 3 and the four equilibrium points, 27.4 deg E, 117.4 deg E,

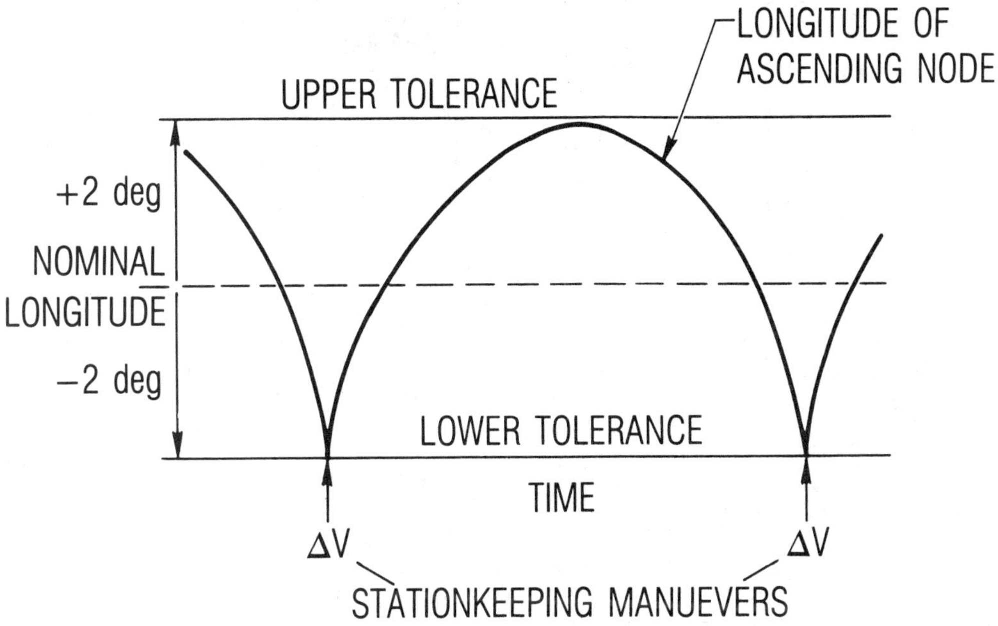

Fig. 2 Stationkeeping Algorithm

207.4 deg E, and 297.4 deg E, were derived based on a simplified equation only considering the J_{32} term. Through boundary conditions of longitude variations between maneuvers and vis viva integral, one can derive the following useful relations for computing ΔV:

$$\Delta V = \frac{a}{3}|\ddot{\lambda}_{\Omega}|T \tag{5}$$

or

$$\Delta V = \frac{16}{3} \frac{a(\lambda_u - \lambda_\ell)}{T} \tag{6}$$

where a is semimajor axis of the GPS orbit.

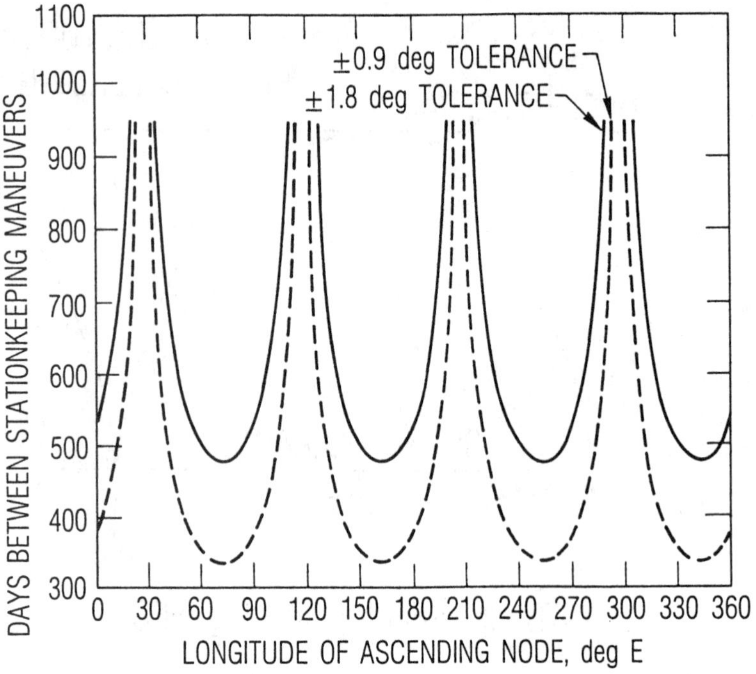

Fig. 3 Estimated Durations between Stationkeeping Maneuvers for GPS Orbits at Nominal Altitude

With the above understanding of the longitude drift and stationkeeping, the effects of stationkeeping on constellation performance can now be examined. The GPS orbit motion is twice as fast as the Earth rotation, so a ±2-deg longitude tolerance translates into a ±4-deg tolerance in in-plane phasing. Among the 21 satellites, nearly half of the satellites will have a longitude acceleration with a different sign than the remaining satellites. Thus, there are times when some satellites will be near the opposite longitude boundaries resulting in a significant 8-deg change in relative phasing from the nominal. How the 8-deg change in relative phasing will affect the constellation performance can be answered by simulation.

Long-term Orbit Perturbation Effects

The nodal drift, induced by the Earth oblateness (J_2), is a function of the semimajor axis and inclination. For GPS orbits at 55-deg inclination, the nodal drift is -14.665 deg per year. There will be no net effect on the constellation because all of the orbits regress at the same rate. However, the luni-solar perturbations gradually change the orbit inclination, with the variation depending on the right ascension of the ascending node. Consequently, the six orbit planes will regress at slightly different nodal rates, and the accumulated unequal nodal separation of the six orbit planes will alter the symmetry of the constellation. According to the analysis in

Reference 2, the largest deviations in inclination and nodal separation for a 10-year mission were 2.1 and 7.8 deg, respectively. As a result, the navigation degradation for the Walker 18/6/2 constellation decreased the CV from 0.995 to 0.975 (Ref. 2).

The above effects can be largely eliminated by properly biasing the initial inclination and by stationkeeping the argument of latitude of the satellites. These two methods are explained in detail in Ref. 2.

Results of Simulation

In order to understand the effect on performance that long-term stationkeeping has on the GPS constellation, the stationkept constellations in Tables 3 through 6 were examined in terms of CV.

Table 7 shows the CVs for both the Walker 18/6/2 plus three and the optimized constellations in a variety of cases. The cases included in the table show data for both constellations after 2.5 and 7.5 years of stationkeeping, with 1- and 2-deg longitude tolerances, and both a "worst" case and an "average" case of satellite positions due to stationkeeping. The satellites' positions in each of these cases are listed in Tables 3 through 6.

Several trends appear in Table 7. The first trend is that the performance for a given constellation does not change much from 2.5 to 7.5 years. Another trend is that the average cases clearly perform better than their worst-case counterparts (as expected). For the optimized constellation there is more degradation in performance if the longitude tolerance is increased from 1 to 2 deg when compared to the baseline. This indicates a greater sensitivity for the optimized constellation to changes in satellite mean anomalies. However, the average case optimized constellation always has a higher CV than the corresponding average case for the Walker 18/6/2 plus three constellation.

It is important to note here that in the cases examined the PDOPs are no longer bounded by a reasonable value. In no case is there a maximum PDOP below 45 and, in most cases, the maximum PDOP is several hundred. The constellations stationkept with a 1-deg tolerance typically had a smaller maximum PDOP than constellations stationkept with a 2-deg tolerance.

Though the CV is a good indicator of overall constellation performance, it is also helpful to have an understanding of the relation between CV and the percentage of the earth that experiences an outage.

For example, the nominal CV of the optimized constellation (0.9999) seems to indicate almost no degraded regions. In fact, 3.3% of the Earth has degraded performance for up to 22 min over a 24-hr period. A CV of 0.9995 is four times worse. Table 8 lists some selected CV values from Table 7, along with the corresponding time and area of degraded coverage. All the data in Table 8 are based on the optimized

Table 3
ICs FOR THE OPTIMIZED CONSTELLATION AT AN INSTANT 2.5 YEARS AFTER THE EPOCH OF 1 APRIL 1989 -"AN AVERAGE CASE"

	Inclination	RAAN	Argument of Latitude: 2-deg Longitude Tolerance	Argument of Latitude: 1-deg Longitude Tolerance
A1	55.168	298.849	132.0	130.5
A2	55.265	298.427	93.5	95.0
A3	55.202	298.682	355.0	353.5
A4	55.526	298.811	232.0	233.5
B1	55.449	358.159	163.0	164.5
B2	55.635	358.612	38.0	36.5
B3	55.631	358.820	267.0	265.5
C1	55.093	58.197	74.5	76.0
C2	55.043	57.873	331.0	332.5
C3	55.039	57.687	298.5	300.0
C1	55.068	57.830	198.0	196.5
D1	54.478	118.329	125.0	123.5
D2	54.692	118.744	348.5	350.0
D3	54.746	118.756	230.0	231.5
E1	54.906	178.893	155.5	154.0
E2	55.031	179.006	58.0	57.0
E3	54.654	178.594	11.5	13.0
E4	54.786	178.856	274.0	275.5
F1	54.758	238.395	81.0	79.5
F2	54.955	238.624	311.0	309.5
F3	54.777	238.390	193.5	195.0

ICs AT AN INSTANT 7.5 YEARS AFTER THE EPOCH

	Inclination	RAAN	Argument of Latitude
A1	54.155	231.329	
A2	54.257	231.075	
A3	54.191	231.221	
A4	54.518	231.886	
B1	54.788	291.350	
B2	54.987	292.113	
B3	54.987	292.320	
C1	55.879	351.993	
C2	55.822	351.584	
C3	55.815	351.389	Same as the 2.5-year values
C4	55.846	351.898	
D1	55.452	51.445	
D2	55.656	52.214	
D3	55.708	52.317	
E1	55.110	112.926	
E2	55.231	113.252	
E3	54.867	112.194	
E4	54.993	112.683	
F1	54.440	171.591	
F2	54.636	172.152	
F3	54.458	171.618	

Table 4
ICs FOR THE OPTIMIZED 21 CONSTELLATION AT AN INSTANT 2.5 YEARS AFTER THE EPOCH OF 1 APRIL 1989 - "THE WORST CASE"

	Inclination	RAAN	Argument of Latitude: 2-deg Longitude Tolerance	Argument of Latitude: 1-deg Longitude Tolerance
A1	55.168	298.849	132.5	131.0
A2	55.265	298.427	93.0	94.5
A3	55.202	298.682	356.0	354.0
A4	55.526	298.811	231.0	233.0
B1	55.449	358.159	162.0	164.0
B2	55.635	358.612	39.0	37.0
B3	55.631	358.820	268.0	266.0
C1	55.093	58.197	73.5	75.5
C2	55.043	57.873	331.0	332.0
C3	55.039	57.687	297.5	299.5
C1	55.068	57.830	199.0	197.0
D1	54.478	118.329	126.0	124.0
D2	54.692	118.744	347.5	349.5
D3	54.746	118.756	229.0	231.0
E1	54.906	178.893	156.5	154.5
E2	55.031	179.006	59.0	57.5
E3	54.654	178.594	11.0	12.5
E4	54.786	178.856	273.0	275.0
F1	54.758	238.395	82.0	80.0
F2	54.955	238.624	312.0	310.0
F3	54.777	238.390	192.5	194.5

ICs AT AN INSTANT 7.5 YEARS AFTER THE EPOCH

A1	54.155	231.329	
A2	54.257	231.075	
A3	54.191	231.221	
A4	54.518	231.886	
B1	54.788	291.350	
B2	54.987	292.113	
B3	54.987	292.320	
C1	55.879	351.993	
C2	55.822	351.584	
C3	55.815	351.389	Same as the 2.5-year values
C4	55.846	351.898	
D1	55.452	51.445	
D2	55.656	52.214	
D3	55.708	52.317	
E1	55.110	112.926	
E2	55.231	113.252	
E3	54.867	112.194	
E4	54.993	112.683	
F1	54.440	171.591	
F2	54.636	172.152	
F3	54.458	171.618	

Table 5

ICs FOR THE WALKER 18/6/2 PLUS THREE CONSTELLATION AT AN INSTANT 2.5 YEARS AFTER THE EPOCH OF 1 APRIL 1989 - "AN AVERAGE CASE"

			Argument of Latitude	
	Inclination	RAAN	2-deg Longitude Tolerance	1-deg Longitude Tolerance
A1	55.168	298.849	14.5	16.0
A2	55.265	298.427	133.5	135.5
A3	55.202	298.682	260.0	258.5
A4	55.526	298.811	49.0	48.0
B1	55.449	358.159	60.0	58.5
B2	55.635	358.612	174.0	175.5
B3	55.631	358.820	300.0	298.5
C1	55.093	58.197	94.0	95.5
C2	55.043	57.873	331.0	332.5
C3	55.039	57.687	334.0	335.5
C1	55.068	57.830	184.5	185.5
D1	54.478	118.329	140.0	138.5
D2	54.692	118.744	254.5	255.5
D3	54.746	118.756	14.0	15.5
E1	54.906	178.893	180.0	178.5
E2	55.031	179.006	294.5	296.0
E3	54.654	178.594	54.0	55.0
E4	54.786	178.856	329.5	328.0
F1	54.758	238.395	214.0	215.5
F2	54.955	238.624	340.0	338.5
F3	54.777	238.390	100.0	98.5

ICs AT AN INSTANT 7.5 YEARS AFTER THE EPOCH

A1	54.155	231.329	
A2	54.257	231.075	
A3	54.191	231.221	
A4	54.518	231.886	
B1	54.788	291.350	
B2	54.987	292.113	
B3	54.987	292.320	
C1	55.879	351.993	
C2	55.822	351.584	
C3	55.815	351.389	Same as the 2.5-year values
C4	55.846	351.898	
D1	55.452	51.445	
D2	55.656	52.214	
D3	55.708	52.317	
E1	55.110	112.926	
E2	55.231	113.252	
E3	54.867	112.194	
E4	54.993	112.683	
F1	54.440	171.591	
F2	54.636	172.152	
F3	54.458	171.618	

Table 6

ICs FOR THE WALKER 18/6/2 PLUS THREE CONSTELLATION AT AN INSTANT 2.5 YEARS AFTER THE EPOCH OF 1 APRIL 1989 - "THE WORST CASE"

	Inclination	RAAN	Argument of Latitude	
			2-deg Longitude Tolerance	1-deg Longitude Tolerance
A1	55.168	298.849	14.0	15.5
A2	55.265	298.427	133.0	135.5
A3	55.202	298.682	261.0	259.0
A4	55.526	298.811	50.0	48.5
B1	55.449	358.159	61.0	59.0
B2	55.635	358.612	173.0	175.0
B3	55.631	358.820	301.0	299.0
C1	55.093	58.197	94.0	95.5
C2	55.043	57.873	221.0	219.0
C3	55.039	57.687	333.0	335.0
C1	55.068	57.830	70.0	68.5
D1	54.478	118.329	141.0	139.0
D2	54.692	118.744	253.0	255.5
D3	54.746	118.756	13.0	15.5
E1	54.906	178.893	181.0	179.0
E2	55.031	179.006	293.0	295.0
E3	54.654	178.594	54.0	55.5
E4	54.786	178.856	90.0	88.5
F1	54.758	238.395	213.0	215.0
F2	54.955	238.624	341.0	339.0
F3	54.777	238.390	101.0	99.0

ICs AT AN INSTANT 7.5 YEARS AFTER THE EPOCH

A1	54.155	231.329	
A2	54.257	231.075	
A3	54.191	231.221	
A4	54.518	231.886	
B1	54.788	291.350	
B2	54.987	292.113	
B3	54.987	292.320	
C1	55.879	351.993	
C2	55.822	351.584	
C3	55.815	351.389	Same as the 2.5-year values
C4	55.846	351.898	
D1	55.452	51.445	
D2	55.656	52.214	
D3	55.708	52.317	
E1	55.110	112.926	
E2	55.231	113.252	
E3	54.867	112.194	
E4	54.993	112.683	
F1	54.440	171.591	
F2	54.636	172.152	
F3	54.458	171.618	

Table 7
CONSTELLATION VALUES* OF OPTIMIZED 21 AND THE WALKER 18 PLUS 3 CONSTELLATIONS AFTER 2.5 AND 7.5 YEARS OF STATIONKEEPING

Longitude Tolerance	Satellite Positions Due to Stationkeeping	Time from Initial Satellite Placement			
		2.5 Years		7.5 Years	
		Optimized 21	Walker 18 + 3	Optimized 21	Walker 18 + 3
1°	"Worst" case	0.9996	0.9987	0.9996	0.9987
	"Average" case	0.9997	0.9986	0.9997	0.9985
2°	"Worst" case	0.9984	0.9988	0.9989	0.9986
	"Average" case	0.9992	0.9987	0.9994	0.9986

*CV based on PDOP $\leq$ 6 and a 5-deg minimum elevation angle.

Table 8
A SAMPLING OF CONSTELLATION VALUES* AND THE CORRESPONDING PERCENT OF THE EARTH EXPERIENCING DEGRADED PERFORMANCE

CV	Percent of the Earth Experiencing Degraded Performance over 24 hr
0.9999	3.3% for up to 22 min
0.9997	10.4% for up to 29 min
0.9993	15.4% for up to 28 min
0.9989	20.1% for up to 27 min
0.9985	23.5% for up to 37 min

*Based on a 5-deg mask angle and a PDOP $\leq$ 6.

constellation. It is important to note that although two constellations may have identical CVs, their distributions and maximum duration of degraded regions may be significantly different.

Results of computer simulation show that the optimized 21-satellite constellation is sensitive to the $\pm$2-deg longitude variations between stationkeeping maneuvers. The constellation value of the optimized constellation degraded to 0.9992 from 0.9999 for an average case of $\pm$2-deg longitude tolerance and to 0.9984 for a worst-case simulation with the same tolerance (Table 7). When the longitude tolerance was tightened to $\pm$1 deg, the worst-case CV was significantly improved to 0.9996. However, this will increase the frequency of stationkeeping maneuvers and thus increase the total satellite downtime over the entire mission, which is highly undesirable. The downtime for each vehicle is 3 to 5 days per maneuver. Downtime is the time that a satellite is out of service.

Results of this study also show that the Walker 18 plus three was not sensitive to the $\pm$2-deg stationkeeping tolerance, and the long-term orbit perturbation effects were found to be insignificant once the initial inclinations of the six planes were properly biased.

LONGITUDE STABILITY AT DIFFERENT ALTITUDES

The above results show that the optimized 21-satellite constellation is sensitive to the nominal longitude stationkeeping tolerance of $\pm$2 deg. This effect cannot be desensitized by simply reducing the tolerance, because this would increase the frequency of stationkeeping maneuvers and thus increase the downtime of each spacecraft, which is highly undesirable. Alternative methods have been considered, and the concept of changing the constellation altitude by a small amount, about 50 km, to reduce or eliminate the stationkeeping effects has been found to be promising (Ref. 1).

The purpose of the second portion of this analysis is to study the long-term orbital motion and longitude variations at altitudes slightly different from the current altitude with repeating ground tracks. The results of this study should provide a clear understanding of the constellation stability at various altitudes, and help select the most favorable altitude.

Background

As stated previously, the long-term longitude variation of the GPS orbits is caused by the resonant effects due to the Earth gravity tesseral harmonics. At GPS orbit altitude, the dominant term of the tesseral harmonics is J_{32} (C_{32} and S_{32}). This term will induce a long-period motion in the semimajor axis through the following relation from Eq. (2):

$$da/dt = 6.17 \sin[\omega + M + 2(\Omega - \alpha_G) - 2\lambda_{32}] \text{ m/day} \quad (7)$$

with the resonant condition

$$\dot{\omega} + \dot{M} + 2(\dot{\Omega} - \dot{\alpha}_G) = 0 \qquad (8)$$

where ω, M, Ω and α_G are argument of perigee, mean anomaly, right ascension of ascending node, and right ascension of Greenwich, respectively. The constant angle λ_{32} with a value of 27.37 deg E was determined from the coefficients C_{32} and S_{32}. Equation (7) and the resonant condition (8) suggest that the semimajor axis, a, has a long-period variation induced by the J_{32} harmonic. As a result, the orbit period and the longitude of ascending node have a long-term variation having the same period (Fig. 1).

Equation (8) indicates that the resonant condition can be destroyed by altering the value of mean motion $\dot{M}$ through changing the orbit altitude (or semimajor axis a). Once the sum of the four angular rates does not approach zero, Eq. (7) can be integrated and the variation of the semimajor axis δa, and longitude of ascending node $\delta\lambda_L$, can be predicted by the following equations:

$$\delta a = - \frac{6.17}{[\dot{\omega} + \dot{M} + 2(\dot{\Omega} - \dot{\alpha}_G)]} \cos[\omega + M + 2(\Omega - \alpha_G) - 2\lambda_{32}] \qquad (9)$$

and

$$\delta\lambda_\Omega = \frac{3}{4}\frac{n}{a} \frac{6.17}{[\dot{\omega} + \dot{M} + 2(\dot{\Omega} - \dot{\alpha}_G)]^2} \sin[\omega + M + 2(\Omega - \alpha_G) - 2\lambda_{32}] \qquad (10)$$

where n = mean motion of orbit

After substituting the values of those angular rates in the above equations, one can estimate the amplitudes and period of variations in semimajor axis and longitude of ascending node, $\delta\lambda_\Omega$. Table 9 shows the corresponding values at selected values of altitude change. Figure 4 shows the amplitude of the longitude deviation as a function of altitude change using the above equation.

Simulation Analysis

The above analytical findings suggest that the orbit altitude should be changed by at least 25 km to avoid the significant effect due to the J_{32} tesseral harmonics. The values of altitude change used in this simulation study were $\pm$25, $\pm$40, $\pm$50, and $\pm$60 km. For each case, the orbit was integrated with two different initial nodes, 220 and 320 deg (best and worst case), and two values of inclination, 55 and 63.4 deg. Program MEANELT, which is a high-precision long-term orbit propagator with an 8 by 8 WGS 84 gravity model and luni-solar effects, was used for the 7.5-year integration.

Table 9
ESTIMATED LONGITUDE VARIATIONS AT VARIOUS ALTITUDES

Altitude Change from GPS Nominal (km)*	Amplitude of Semimajor Axis (m) δa	Amplitude of Longitude (deg) $\delta\lambda_{\Omega}$	Period of Variation (days)
10	1000	4.3	1000
15	650	1.34	638
25	375	0.45	369
50	182	0.11	179
75	120	0.05	118

*Results are the same if the altitude change from nominal is negative.

A Summary of Simulations Results

Maximum deviations in semimajor axis and in-plane motion over a period of 7.5 years are summarized in Table 10. The findings of the simulation analysis can be listed as follows:

1. Table 10 shows that over the 7.5-year mission duration the largest in-plane deviation, which is a combination of the long bulge-type variation and the sinusoidal variation, is found to be between 4.1 to 4.6 deg for the 55-deg inclination orbits depending on the altitude change. These maximum deviations in in-plane motion correspond to a 2 to 2.3-deg maximum deviation in longitude of ascending node.

2. At a higher inclination (63.4 deg), Table 10 shows that the largest combined variation for the 50 km altitude change decreases to 3.4 deg in in-plane motion or to 1.7 deg in longitude of node.

3. The long bulge-type variations (Fig. 5) are caused by the long-term inclination changes due to sun-moon gravitational perturbations and are not a function of altitude change.

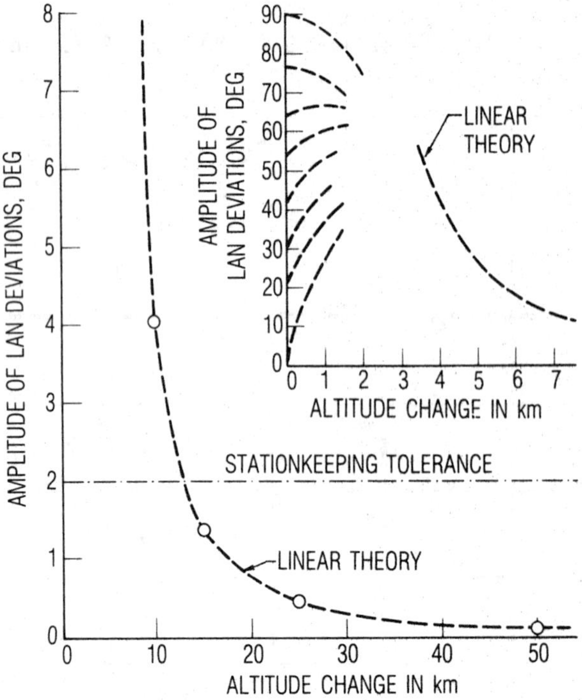

Fig. 4 Amplitude of Longitude of Ascending Node Deviations at Various Orbit Altitudes

4. The sinusoidal variations in Fig. 5 are caused by the Earth gravity tesseral harmonic (J_{32}). The amplitude and period of these variations decrease as the orbit altitude change increases. When the altitude change is 50 km or greater, Table 9 shows that the amplitude of the variation becomes insignificant (0.1 deg or less) compared to the long bulge-type variations.

Reduction in Stationkeeping Effects

The above simulation results show that the worst-case peak-to-peak variation in longitude of the node is 2 deg over the 7.5-year mission life when the orbit altitude is altered by 50 km. Therefore, by biasing the initial longitude of node, the deviations from a reference longitude can, in principle, remain within ± 1 deg without having to perform any stationkeeping maneuvers. However, it was estimated that errors in initial orbit injection would require, at most, one stationkeeping maneuver for each satellite during the entire mission.

Table 10
MAXIMUM DEVIATIONS IN SEMIMAJOR AXIS AND
IN-PLANE MOTION OVER 7.5 YEARS AT VARIOUS ALTITUDES

$\Delta a = a - a_0 \quad \Delta\theta = \theta(t) - n(t - t_0)$

Altitude Change from GPS Nominal* (km)	Inclination = 55 deg		Inclination = 63.4 deg	
	Ω_0 = 220 deg	Ω_0 = 320 deg	Ω_0 = 220 deg	Ω_0 = 320 deg
25	0.36** 0.7**	0.36 4.6	0.33 0.6	0.33 3.6
40	0.22 0.5	0.22 4.1	0.20 0.4	0.11 3.2
50	0.22 0.5	0.20 4.1	0.17 0.4	0.10 3.4
60	0.22 0.6	0.18 4.1	0.10 0.3	0.10 3.15

*The results are the same if the altitude change from nominal is negative.
**The upper values are Δa(km), and the lower values are $\Delta\theta$(deg).

The current stationkeeping tolerance is ± 2 deg, and the average number of stationkeeping maneuvers required for each satellite during the mission is about five. Assuming a 3-day downtime for each maneuver, there will be 300 days during the 7.5 years on which the constellation experiences one satellite failure. By changing the altitude 50 km, not only can the stationkeeping tolerance be tightened to ± 1 deg but also the number of stationkeeping maneuvers can be eliminated. It should be pointed out that the sensitivity of the optimized 21-satellite constellation being stationkept with a ± 2-deg stationkeeping tolerance did not include the effects due to maneuver downtime.

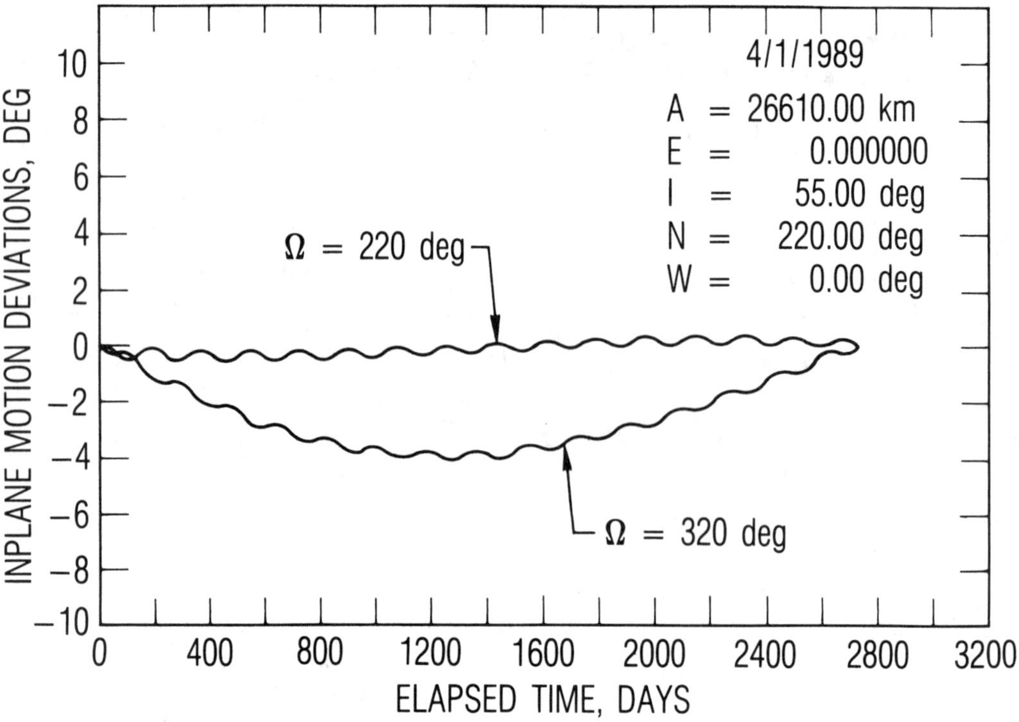

Fig. 5 Time History of In-plane Motion Deviations from a Reference Orbit with 50-km Altitude Increase from GPS Nominal (i=55 deg)

Impact on Autonomous Momentum Management and Yuma Test

The Autonomous Momentum Management System (AMMS) onboard the Block II GPS spacecraft was designed for the nominal orbit with repeating ground tracks. The orbit rate or orbit period was hardwired in the AMMS. Changing the orbit altitude will change the orbit rate. The accumulated error will affect the function of AMMS. It is a costly procedure to change the orbit rate in the hardware design of the Block II spacecraft. Current effort is being made to determine the magnitude of the accumulated orbit rate error as a function of altitude increase and number of days of desired autonomy. The desired interval of autonomous momentum management is 180 days. It is hoped that a new value of altitude increase (< 50 km) could be determined, which can minimize both the impact on AMMS and effects due to stationkeeping without changing the hardwired orbit rate.

The Yuma test requires that a favorable GDOP geometry will occur at Yuma during a certain time of day for several months. Changing the altitude by 50 km, the GDOP geometry over Yuma will change over time. However, a simulation study shows that the effects on a Yuma test due to altitude change are not significant. The impact can, in fact, be minimized by delaying the altitude change until the Yuma test is completed.

CONCLUSIONS

A $\pm$2-deg longitude stationkeeping tolerance can result in a worst-case 8-deg change in relative phasing among the Block II GPS satellites. Simulation results show that the navigation performance of the optimized 21-satellite constellation is sensitive to the relative phasing variations between stationkeeping maneuvers.

The slow variations in relative phasing among satellites decrease rapidly as the orbit altitude is increased or decreased from the nominal GPS altitude. Both analytical and numerical findings show that the desired altitude change which can significantly reduce the resonant tesseral harmonics effects is about $\pm$50 km. By changing the altitude 50 km, the stationkeeping maneuvers can be eliminated and the resulting longitude deviations will remain within $\pm$1 deg. Thus, the navigation performance of the GPS constellation can be maintained at its desirable value, and the ground control load will be reduced significantly.

REFERENCES

1. Green, G.B., Massatt, P.D., and Rhodus, N.W., "The GPS 21 Primary Satellite Constellation," presented at the ION Conference in Colorado Springs, CO, September 1988

2. Chao, C.C. and Bowen, A.F., "Effects of Long-Term Orbit Perturbations and Injection Errors on GPS Constellation Values," AIAA Paper 86-2173-CP, paper presented at the 1986 AIAA/AAS Astrodynamics Conference held at Williamsburg, VA, August 1986

3. Kaula, W. M., Theory of Satellite Geodesy, Blaisdell Publishing Co, London, 1966.

AAS 89-408

TOPEX/POSEIDON ORBIT MAINTENANCE MANEUVER DESIGN*

R. S. Bhat†, R. B. Frauenholz† and Patrick E. Cannell‡

In June 1992 the TOPEX/POSEIDON mission will begin a three-year global study of the earth's oceans from a near-circular orbit at an altitude of ~1335 km and an inclination of ~64.6 deg. Science objectives require the satellite 10-day repeat ground track be maintained within ±1 km on 95% of the planned verification site overflights during the first six months; the remainder of the mission requires control within a fixed 2-km equatorial bandwidth. Supporting maneuvers are constrained to occur between 10-day precision orbit determination solution arcs, and no more frequently than once every 30 days.

An analytic orbit propagation model includes fifth-order zonal harmonics, luni-solar gravity, and atmospheric drag to meet the stringent ground track control requirements. Maneuver design early in the mission is influenced primarily by drag modelling uncertainties arising from expected solar flux prediction errors. Late in the mission as solar cycle 22 minimum approaches, orbit determination and maneuver execution accuracies become the dominant factors.

INTRODUCTION

The Ocean Topography Experiment (TOPEX/POSEIDON) will be launched by an Ariane 42P from French Guiana in June 1992 and begin a three-year global study of the Earth's ocean circulation and its variability. These space-borne observations are important to improving global climate prediction and long-range weather forecasting. To accomplish these science objectives requires a precise measure of ocean surface height to an accuracy of 14 cm (1σ) using satellite on-board altimetry and precision orbit determination[1] (POD) from ground-based laser tracking data.

The TOPEX/POSEIDON Project, hereafter called TOPEX, is jointly sponsored by the National Aeronautics and Space Administration (NASA) and the Centre National d'Etudes Spatiales (CNES) of France. NASA provides the satellite and all on-board instrumentation, including a duel-frequency radar altimeter; CNES provides the Ariane launch vehicle and a solid-state radar altimeter.

The Jet Propulsion Laboratory (JPL) manages the TOPEX Project for NASA and will conduct mission operations, including all orbit maintenance and satellite control functions. The NASA Goddard Space Flight Center (GSFC) provides operational orbit determination support to JPL for mission operations using radio metric tracking data acquired via the NASA Tracking and Data Relay Satellite System (TDRSS).

---

* The research described in this paper was carried out by the Jet Propulsion Laboratory, California Institute of Technology, under a contract with the National Aeronautics and Space Administration.

† Member of Technical Staff, Jet Propulsion Laboratory, California Institute of Technology, 4800 Oak Grove Drive, Pasadena, California 91109. Member AIAA.

‡ Member of the Professional Staff, Sterling Federal Systems, Inc., Pasadena, California. Member AIAA.

The Ariane launch vehicle will place TOPEX in a biased orbit lower than the operational orbit to avoid possible later collision with the third stage, and to facilitate timely achievement of the desired operational orbit. Maneuver strategies during the first 20 days will retarget TOPEX from the biased orbit to the operational orbit. This paper describes only the maintenance of the resulting operational orbit.

Kechichian[2] first developed TOPEX orbit maintenance maneuver strategies for a 1986 launch near expected low solar activity when atmospheric drag influences would be minimal. This work predicts maneuver spacing times for ground track maintenance maneuvers in terms of orbit determination and maneuver execution error sources. More recently, Shapiro[3,4] has described the maneuver design used to maintain an exact ground track repeat for GEOSAT, indicating the limiting factors to be orbit determination errors and atmospheric density prediction uncertainties.

The scheduled TOPEX launch in June 1992 follows the predicted peak of solar cycle 22 (early 1990) when atmospheric drag levels are expected to significantly influence orbit maintenance maneuver design. This paper summarizes the orbit control requirements and the allocation of navigation system errors between orbit determination and prediction sources. Analytic ground track propagation and prediction error models include the effects of geopotential perturbations, luni-solar gravity, and atmospheric drag. A maneuver design philosophy consistent with the ground track control requirements, major perturbations, and modelling errors is then described with particular attention to the relative influences of solar activity and luni-solar gravity.

ORBIT REQUIREMENTS

Early mission and orbit design investigations by Frautnick and Cutting[5] identified the need for accurate control of an exactly repeating satellite ground track in order to meet TOPEX science objectives. Farless[6] later defined a detailed orbit design space from which the reference operational orbit will be selected. The current baseline orbit provides an exact repeat of the satellite ground track every 10 sidereal days and 127 orbits, during which it will overfly one NASA and one CNES ground verification site. The verification site currently planned for use by NASA is the Harvest oil platform located offshore from Pt. Conception, California; the candidate CNES site lies between Lampedusa and Lampione islands near Sicily in the Mediterranean Sea. Both sites are within one degree of a 35-deg northern latitude.

In addition, the use of a frozen orbit has been proposed to restrict the variation in the orbit eccentricity e and the argument of periapse ω, with the benefit of limiting the variability in satellite altitude for enhanced on-board altimetry performance, while also reducing the need for dedicated maneuvers to control these parameters. The techniques required to design frozen orbits have been previously applied to SEASAT[7], GEOSAT[8], and LANDSAT[9]. Recently, Smith[10] and Tang[11] have investigated the possible use of a frozen orbit for TOPEX.

Reference Orbit

Since a final orbit has not yet been approved, a representative sample orbit has been chosen to illustrate typical orbit maintenance maneuver strategies. The mean elements presented in Table 1 provide a near-circular orbit at an mean altitude of ~1335 km and an inclination of ~64.6 deg. The orbit semi-major axis a_o and inclination i_o provide a 10-day ground track repeat cycle and the desired verification site overflights in the presence of fifth-order zonal harmonics. These elements also provide a frozen orbit from the near-cancellation of higher-order perturbations on ω by secular variations, while the lower-order perturbations on e vanish when ω_o = 90 or 270 deg. When higher-order harmonics are considered at the design inclination of 64.6 deg, an ω_o value of 270 deg is required to obtain an acceptably small eccentricity.

Table 1. Sample Mean Orbital Elements

Parameter	Value
Reference Epoch, t_o	21 June 1992, 0^h GMT
Semi-major Axis, a_o	7713.386,9 km
Eccentricity, e_o	0.000,982,5
Inclination, i_o	64.606 deg
Right Ascension of Ascending Node, Ω_o	139.552 deg
Argument of Periapse, ω_o	270 deg
Mean Anomaly, M_o	0.0 deg

Orbit Control Requirements

During the first six months following launch, an Initial Verification Phase will be conducted for satellite altimetry calibrations via the NASA and CNES ground verification sites, and for earth gravity model improvement by POD (see Marsh, et al[12]). To satisfy these initial objectives requires precise control of the satellite ground track and minimum interference to POD from orbit maintenance maneuvers. Accurate altimetry requires that the orbit eccentricity remain $\leq$0.001. Maneuver strategies must maintain the satellite ground track within $\pm$1 km on 95% of the planned verification site overflights. These maneuvers are nominally to be placed between adjacent 10-day repeat cycles to avoid possible interference with POD processing, while assuring with 95% probability that maneuver spacing is at least 30 days.

The Observational Phase follows the Initial Verification Phase and ends three years after launch. During this second phase, 95% of the satellite ground tracks must be maintained within a 2-km bandwidth in equatorial longitude, while the orbit eccentricity remains $\leq$0.001. Attendant orbit maintenance maneuvers are to be scheduled "as infrequent as practical" to avoid possible interference with continuing science activities.

MANEUVER DESIGN PHILOSOPHY

An accurate prediction of the satellite ground track is essential to meet the control and maneuver spacing requirements. For an exact repeat of the ground track, the equatorial longitude spacing between successive ground tracks D_R* would remain constant over the entire repeat cycle. The value of D_R depends on N, the number of orbits in a single ground track repeat cycle, and on d, the length of the repeat cycle in sidereal days.

$$D_R = (360 \times d)/N \tag{1a}$$

so that $\simeq 28.347$ deg/orbit, since $N = 127$ and $d = 10$,

and $$D = \tau_n(\omega_e - \dot{\Omega}) \tag{1b}$$

The actual spacing between successive ground tracks D varies due to perturbations which change the nodal period τ_n and the nodal precession rate $\dot{\Omega}$. When $D > D_R$, the actual ground track drifts west of the reference; when $D < D_R$, the drift is eastward.

---

* *the notations section on p. 22 defines all mathematical symbols not defined in text*

The primary systematic contributors to changes in D are atmospheric drag and luni-solar gravitational perturbations. Luni-solar perturbations induce both secular and long periodic variations in the nodal period and the precession rate, whereas atmosphere drag continuously reduces the nodal period. The luni-solar perturbations depend on the relative positions of the sun, moon, and the geocentric orientation of the satellite orbit. However, precise modelling of drag acceleration during the prediction period presents the greatest challenge because of the strong dependence on difficult-to-predict solar activity. To convey the maneuver design concept we shall temporarily ignore contributions due to luni-solar perturbations and examine only the influence of atmospheric drag.

Atmospheric drag is a function of the satellite orbital position, atmospheric density ρ, satellite velocity V, and the drag reference area A. Escobal[13] defines the decay rate of the semi-major axis of a near-circular orbit under constant drag to be

$$da/dt = -C_D A \rho V^2 / M_s n \tag{2}$$

which causes the satellite ground track to drift eastward. Assuming a constant decay rate, the accumulated change in satellite equatorial longitude after time t is

$$\Delta\lambda_d = \frac{3}{4}\omega_e \frac{C_D A}{M_s} \rho V t^2 \tag{3}$$

Periodic maneuvers become necessary to compensate for drag to maintain the ground track within required limits. The accumulated change in equatorial longitude after time t due to the combined effects of drag and a compensating maneuver ΔV is

$$\Delta\lambda = -3\omega_e \frac{\Delta V}{V} t + \frac{3}{4}\omega_e \frac{C_D A}{M_s} \rho V t^2 \tag{4}$$

Fig.(1) shows the variation in $\Delta\lambda$ with t and ΔV when $\rho \simeq 2.3 \times 10^{-6}$ kg/km^3. This density is the orbit-average value at ~1335 km mean altitude when the 81-day mean solar flux is 225, the expected 95-percentile value at TOPEX launch. Since the maneuver must induce westward drift in the ground track to offset drag, all maneuvers begin at the east boundary of the 2-km control bandwidth to provide maximum maneuver spacing. The resulting maneuver spacing increases with maneuver magnitude. However, when the magnitude is too large the ground track crosses the western boundary, as shown by examples in Fig.(1).

The maximum maneuver magnitude ΔV_{max} provides maximum maneuver spacing T_{max}, while maintaining the ground track just inside the control bandwidth $\Delta\lambda_{max} = 2$ km. This balance of drag by the maneuver occurs at $T_{max}/2$ when $d(\Delta\lambda)/dt = 0$ (from Eq. 4). In the Fig.(1) example, this balance is achieved by $\Delta V_{max} \simeq 4.40$ mm/s, which results in $T_{max} \simeq 108$ days.

$$\Delta V_{max} = \sqrt{\frac{C_D A}{M_s} \frac{\rho V}{3} \left(\frac{\Delta\lambda_{max}}{\omega_e} \right)} \tag{5}$$

$$\text{and } T_{max} = \frac{4 M_s \Delta V_{max}}{C_D A \rho V} \tag{6}$$

Longitude Targeting

The above strategy provides *longitude targeting* because it uses the full 2-km bandwidth to maximize maneuver spacing. The primary reference for this strategy is a_o, the mean semi-major axis providing the nodal period τ_n necessary for an exact repeat of the satellite ground track (see Table 1). When ΔV_{max} is applied, the semi-major axis increases by Δa to $(a_o + \Delta a)$; the resulting higher nodal period causes the ground track to drift westward from the eastern bandwidth boundary. Drag continuously reduces the nodal period at a constant rate until the ground track becomes tangent to the western boundary at $T_{max}/2$; at this time the nodal period and semi-major axis are again nominal (τ_n and a_o). Under the continued influence of constant drag, the ground track then reverses eastward, returning to the east boundary at T_{max} with a semi-major axis of $(a_o - \Delta a)$.

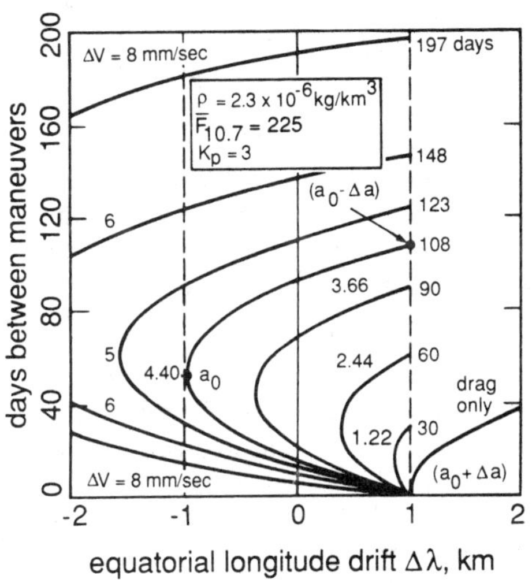

Fig. 1.
Longitude vs Time Targeting

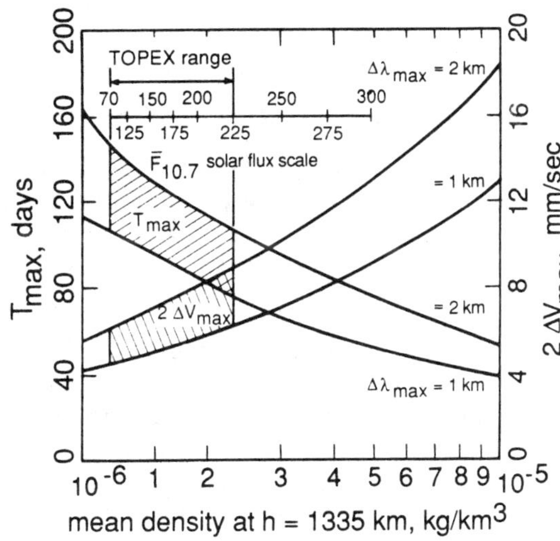

Fig. 2.
Longitude Targeting vs Solar Flux

The semi-major axis must now be increased by $2\Delta a$ to achieve $(a_o + \Delta a)$, the value required to repeat the previous ground track traversal in T_{max} in the presence of the same constant drag level. The magnitude of all future maneuvers remains $2\Delta V_{max}$ with a maneuver spacing T_{max} as long as the drag level remains constant.

Fig.(2) shows the variation in *longitude targeting* within the nominal 2-km bandwidth for other constant drag levels. Density values for mean solar flux levels between 225 and 70 indicate the expected levels at the beginning and end of the mission, respectively. The corresponding maneuver requirements for $2\Delta V_{max}$ vary between ~6 and 9 mm/s, whereas T_{max} varies between ~108 and 146 days.

Time Targeting

Another possible maneuver strategy provides a specified maneuver spacing using *time targeting*. The selected maneuver spacing must be less than T_{max} to maintain the ground track inside the 2-km bandwidth, so ΔV is also always less than ΔV_{max}. Use of this strategy requires maneuvers be placed between adjacent 10-day precision orbit determination data arcs. Accordingly, candidate *time targeting* maneuver spacings are multiples of 10 days beginning with 30 days, but not exceeding T_{max}.

Fig.(1) compares the maneuver magnitudes for sample maneuver spacings of 30, 60, and 90 days with the maximum value required for the *longitude targeting* strategy. The required ΔV magnitudes are 1.22, 2.44, and 3.66 mm/s, respectively, compared to $\Delta V_{max} \simeq 4.40$ mm/s and $T_{max} \simeq 108$ days for *longitude targeting.*

These results suggest the 30-day minimum maneuver spacing requirement is easily satisfied, but this has been demonstrated under the unrealistic assumption of constant drag acceleration. In fact, the maneuver strategies must not only account for variable drag acceleration, but also for the presence of secular and long periodic luni-solar gravitational perturbations, orbit determination errors, and maneuver execution errors. These perturbations combine to restrict the useful ground track bandwidth, with an attendant reduction in the maximum maneuver spacing. This trend is evident in Fig.(2) where the maneuver requirements constrained by a 1-km bandwidth are compared with the 2-km baseline.

GROUND TRACK PREDICTION MODEL

To optimize the maneuver design requires a prediction model that considers all perturbations that cause significant variations in the satellite ground track. To effectively account for these variations, an analytical model has been developed for use as a rapid and efficient *preliminary* orbit propagation tool to condition maneuver requirements for subsequent precision evaluation by numerical integration techniques.

This analytic algorithm accepts initial orbital elements that provide an exact repeat ground track in the presence of fifth-order zonal harmonics (see Table 1). The dynamic model includes the effects of these zonal harmonics, atmospheric drag, and luni-solar gravitational perturbations. The effects of solar radiation pressure have been omitted in this preliminary design because the resulting ground track variations have been determined to be significantly smaller than the other perturbations; their omission does not alter the character of the maneuver design.

Ground track maintenance begins when the operational orbit has been achieved and the actual and reference ground tracks become nearly aligned. These conditions are scheduled to be achieved on 21 June 1992, 20 days after launch. At this epoch, designated as t_o for these studies, the mean Keplerian elements are a_o, e_o, i_o, Ω_o, ω_o, and M_o, with values listed in Table 1. The actual and reference values of the equatorial longitude crossing at t_o are λ_o and λ_{ro}; their actual and reference values for the ith orbit are λ_i and λ_{ri}.

The accumulated drift in the equatorial longitude of the actual satellite ground track from the reference value, $\Delta\lambda$, is determined by summing the individual orbit deviations over the total number of orbits during the period t_o to the ith orbit:

$$\Delta\lambda = \Delta\lambda_o + \sum_{i=1}^{i^{th}\ orbit} \Delta\lambda_i \tag{7}$$

$$\begin{aligned} \text{where} \quad \Delta\lambda_o &= \lambda_o - \lambda_{ro} \\ \Delta\lambda_i &= \lambda_i - \lambda_{ri} \\ \text{and} \quad \lambda_i &= \lambda_{i-1} - D_{i-1} \\ \lambda_{ri} &= \lambda_{r_{i-1}} - D_R \end{aligned}$$

Recall from Eqs.(1a,b) that the reference ground track moves a fixed equatorial longitude distance D_R every orbit, and the actual ground track moves a variable distance D_i each ith orbit. The value of D_i, which depends on the current nodal period τ_{ni} and nodal precession rate $\dot{\Omega}_i$, is computed at the mid-point of each orbit to ensure it represents the mean value of all orbital variations. Therefore, at epoch t_o the mean elements of the reference orbit are advanced to the mid-point of the first orbit, and then advanced by one nodal period for each subsequent orbit.

$$
\begin{aligned}
a_i &= a_{i-1} + (da/dt)_{i-1}\tau_{n_{i-1}} \\
e_i &= e_{i-1} \\
i_i &= i_o + \Delta i_i \\
\Omega_i &= \Omega_{i-1} + \dot{\Omega}_{i-1}\tau_{n_{i-1}} \\
&= \Omega_{i-1} + (\dot{\Omega}_g + \dot{\Omega}_{ls})_{i-1}\tau_{n_{i-1}} \\
\omega_i &= \omega_{i-1} + \dot{\omega}_{i-1}\tau_{n_{i-1}} \\
&= \omega_{i-1} + (\dot{\omega}_g + \dot{\omega}_{ls})_{i-1}\tau_{n_{i-1}} \\
M_i &= M_{i-1} + n_{i-1}\tau_{n_{i-1}} \\
&= M_{i-1} + (n_g + n_{ls})_{i-1}\tau_{n_{i-1}} \\
\tau_{n_{i-1}} &= 2\pi/(n_{i-1} + \dot{\omega}_{i-1})
\end{aligned}
\tag{8}
$$

Orbital elements a_i, e_i, Ω_i, and ω_i are obtained using the mean elements and their rates of change from the $(i-1)^{th}$ orbit. The rates of change are evaluated using a perturbation model described later.

Table 2 shows that some perturbations affect only a few elements. For example, atmospheric drag has an appreciable effect on semi-major axis for a near-circular orbit like TOPEX, whereas it has a negligible effect on the other elements. The orbital inclination is affected only by luni-solar perturbations.

Table 2. Effects of Modelled Perturbations on the Mean Orbital Elements

Parameter	Geopotential Perturbations	Luni-Solar Perturbations	Atmospheric Drag
Semi-major Axis, a	–	–	√
Eccentricity, e	√*	–	–
Inclination, i	–	√	–
RA of Ascending Node, Ω	√	√	–
Argument of Periapse, ω	√*	√	–
Mean Anomaly, M	√	√	–

* *these effects small for frozen orbit*

The updated mean elements are then used to compute the current nodal period τ_{ni} and D_i, and the actual equatorial longitude spacing between the ith and the $(i+1)^{th}$ orbits. While computing the semi-major axis decay rate da/dt at the mid-point of each orbit, the drag reference area and atmospheric density values are also updated to reflect the time-dependence of these parameters; these models are described later.

Eqs.(7,8) update the expression for $\Delta\lambda$ presented previously by Eq.(4) to now account for time-dependent variations in the satellite ground track due to the combined effects of geopotential perturbations, luni-solar gravity, and drag. The modelling assumptions used to define these individual perturbations follow.

Geopotential Perturbations

Since our sample reference orbit is frozen, geopotential perturbations do not change the argument of periapse, so $\dot{\omega}_g = 0$. However, evaluation of D_i depends on the geopotential contributions to the mean motion n_{gi} and the nodal precession rate $\dot{\Omega}_{gi}$. Expressions for n_{gi} and $\dot{\Omega}_{gi}$, listed in the Appendix as Eqs.(A) and (B), have been derived using fifth-order zonal harmonics (Escobal[13] and Kaula[14]).

Luni-Solar Gravity Perturbations

At TOPEX altitude, the luni-solar gravity effects become significant because of the stringent ground track control requirements. Luni-solar gravity induces periodic perturbations in all orbital elements except the semi-major axis, and secular variations in Ω, ω, and M. The magnitude of the disturbing force and its periodic effect on each orbital element depends on the relative positions of the sun, moon, and the satellite orbit orientation with respect to orbital planes of the sun and moon about earth.

A simple analytic planetary ephemeris, based on ecliptic mean elements developed by Escobal[15] and by Kwok,[16,17] was used to predict the sun and moon positions and verified by comparison with a precision planetary ephemeris (DE118). The disturbing function U due to luni-solar gravitational perturbations was developed by Kaula[18] in terms of Keplerian elements (see Eq. C in Appendix).

Taking appropriate partials of U and substituting into the Lagrange planetary equations (e.g., Kaula[14]) provides the rate of change of each orbital element. The explicit expressions for the associated inclination and eccentricity functions derived by Kaula[14] and by Hughes[19] were used, as only lower-order perturbations are required. The rate of change of inclination was integrated from epoch time t_o to time t assuming the angle Y (see Eq. D in Appendix) varies linearly (suggested by Born, et al[20]). As expected, the variation in eccentricity was found to be negligible. Expressions for Δi, $\dot{\Omega}_{ls}$, n_{ls}, and $\dot{\omega}_{ls}$ used in the model are summarized by Eqs.(D–G) in the Appendix.

Figs.(3) and (4) show the nodal rate and inclination variations for the three-year period beginning 21 June 1992. The total nodal rate $\dot{\Omega}$ is $(\dot{\Omega}_g + \dot{\Omega}_{ls})$, the sum of the contributions due to earth gravity and to luni-solar gravity. The inclination variation is determined by adding the contribution due to luni-solar gravity to the mean inclination at epoch: $i(t) = i(t_o) + \Delta i(t)$. The primary variations in inclination are the sum of three effects having amplitudes of 2.1 mdeg, 1.2 mdeg, and 0.54 mdeg over periods of ~164, 57, and 12 days. The major contributor is lunar gravity, including a secular variation of 0.16 mdeg/day in $\dot{\Omega}$, in addition to periodic variations dominated by a 12-day component with an amplitude of 0.13 mdeg/day.

To maintain the ground track modelling accuracy within a few tens of meters over a long duration (60 to 150 days) it was necessary to include the small contributions of luni-solar gravity to the mean motion n_{ls}, and to the rate of change in the argument of periapse $\dot{\omega}_{ls}$. Mean motion has a secular component of ~0.19 mdeg/day in addition to periodic variations dominated by the 12-day component with an amplitude of 0.88 mdeg/day. The secular variation in $\dot{\omega}$ is 0.015 mdeg/day; the only significant periodic components are 0.5 mdeg/day and 0.3 mdeg/day, with periods of approximately 12 and 164 days, respectively.

Fig.(5) shows the three-year variation in the satellite ground track due to luni-solar perturbations. The periodic behavior of the westward ground track drift may be easily correlated with the variations in $\dot{\Omega}$ and i shown in Figs.(3) and (4).

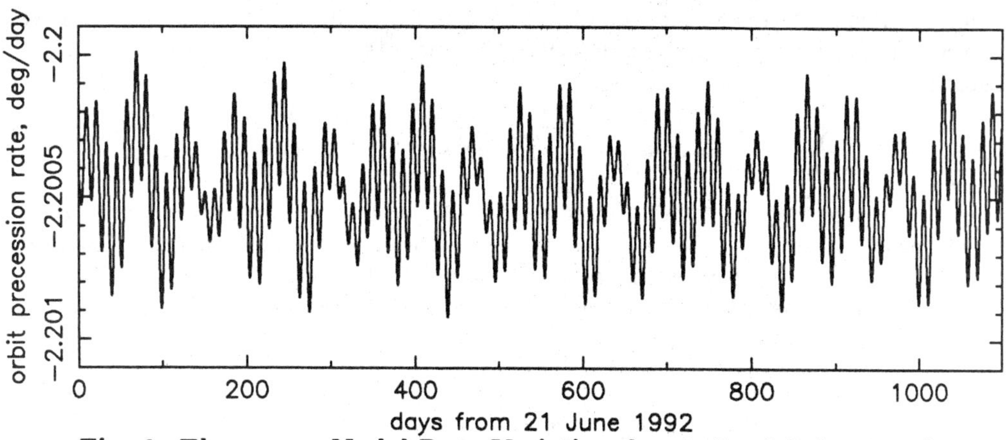

Fig. 3. Three-year Nodal Rate Variation due to Luni-Solar Gravity

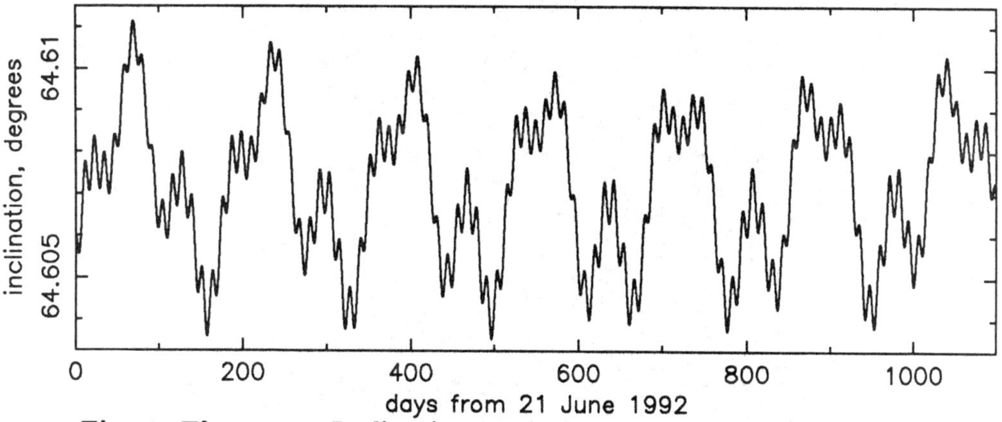

Fig. 4. Three-year Inclination Variation due to Luni-Solar Gravity

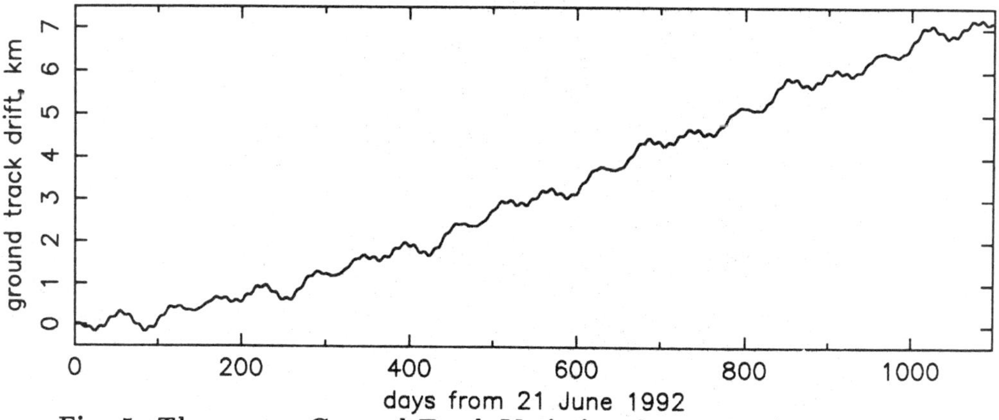

Fig. 5. Three-year Ground Track Variation due to Luni-Solar Gravity

Atmospheric Drag Perturbations

In the context of this ground track maintenance maneuver design, the greatest influence of atmospheric drag is the effect on the orbit semi-major axis, which in turn directly affects the nodal period. The rate of change of semi-major axis da/dt, including earth-relative velocity and daily values of atmospheric density, is computed using the expression derived by Escobal[15].

$$(da/dt)_i = -[C_D A_i \rho_d V_i^2 / M_s n_i]\left[1 - (\omega_e/n_i)\cos i_i\right]^2 \qquad (9)$$

When constant average values of drag reference area A and atmospheric density ρ are assumed, the orbit decay rate varies between the constant extremes of 0.6 cm/day when the solar flux is 70, and 40 cm/day when the solar flux is 225. However, both area and density have time-dependent variations that must be modelled to accurately predict the satellite ground track.

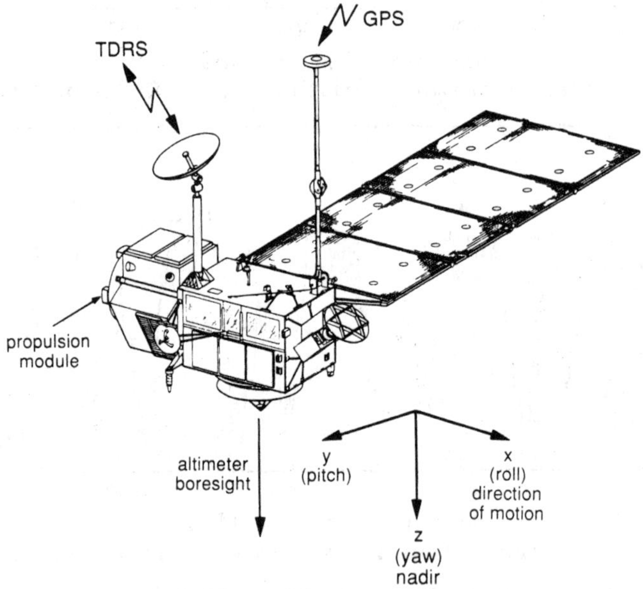

Fig. 6. TOPEX Satellite

Atmospheric Drag Reference Area Modelling. Although TOPEX is a three-axis stabilized satellite, near-continuous yaw steering about the local nadir and solar panel pitching are required to maintain the dominant 28 m^2 solar panel area pointed near the sun for power optimization (see Fig. 6). This attitude articulation strategy causes the satellite drag area to continuously vary—as much as 4:1 during a single orbit period. Fig.(7) shows the envelope of this "Continuously Variable Area" (CVA) evaluated over each orbit for the 90-day period beginning 21 June 1992. The insert in Fig.(7) shows the CVA variation as a rapid function of the periodic orbit angle ξ, and a slowly varying function of β', the angle between the orbit plane and earth-sun direction. A single β' cycle repeats in ~112 days during each orbit nodal precession period; the amplitude varies seasonally depending on sun position, with a maximum value of ~88 deg. Because of the symmetrical behavior of β', the variation in satellite drag area repeats during half the β' cycle, or ~56 days, as is evident in Fig.(7).

The maximum CVA amplitude during each β' cycle occurs when the satellite orbit plane passes through the earth-sun line and β' is zero. When $-10 \leq \beta' \leq 10$ deg a fixed yaw attitude is used to avoid excessive yaw rates (see intervals in Fig. 7). The minimum CVA amplitude occurs when β' reaches a local maximum. For the 90-day period shown in Fig.(7), the local maximum β' angle occurs when the earth-sun direction is ~45 deg *below* the orbit plane.

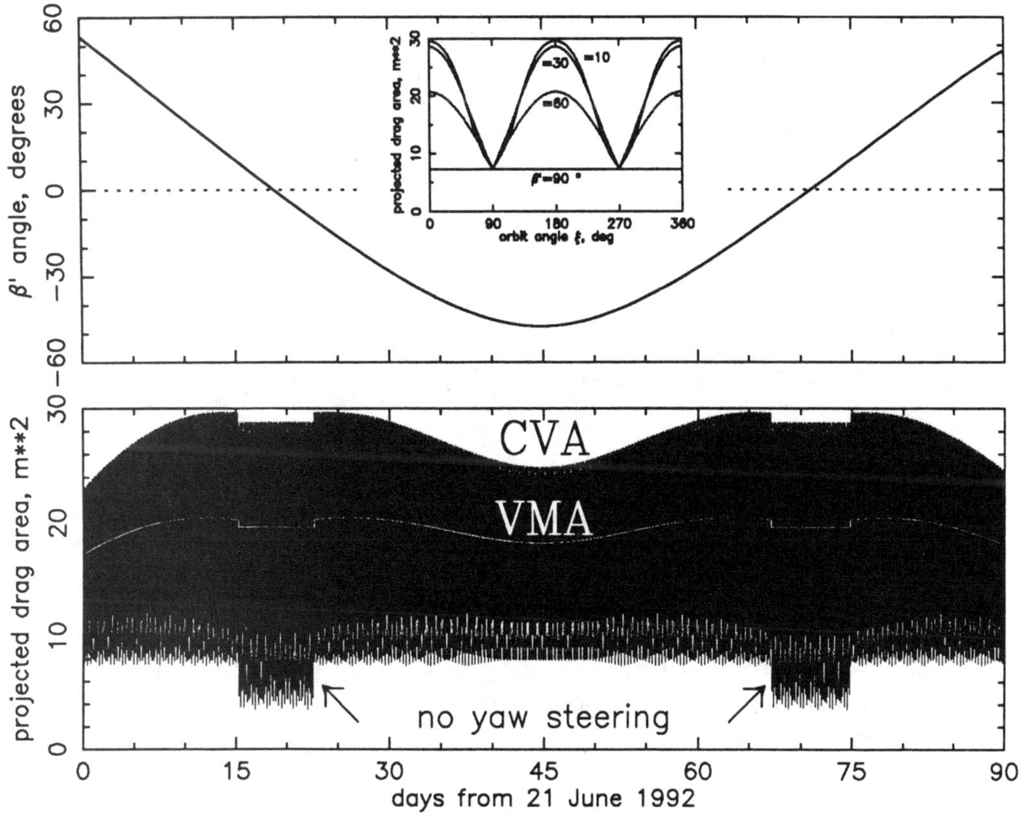

Fig. 7. Satellite Drag Reference Area vs Time and Angle β'

Studies indicate variations in the drag reference area must be modelled to meet the orbit prediction accuracy requirements. However, use of the CVA model is computationally intense, requiring a dynamic algorithm which continuously updates the satellite area to match the ever-changing attitude. The CVA model has been approximated by an efficient and sufficiently accurate "Variable Mean Area" (VMA) model which defines the mean area per orbit as a tabular function of the angle β'. This simple area representation, defined as A_i in Eq.(9) of our orbit propagation model, is compared with the CVA model in Fig.(7).

Atmospheric Density Modelling. Atmospheric density is modelled as a function of solar activity and several geometric parameters including satellite altitude, latitude, and diurnal variations. Since the TOPEX orbit is near-circular, only the orbital average density at the mean altitude of 1335 km need be considered here. When treated in this manner the orbit-average atmospheric density becomes a function only of solar activity parameters.

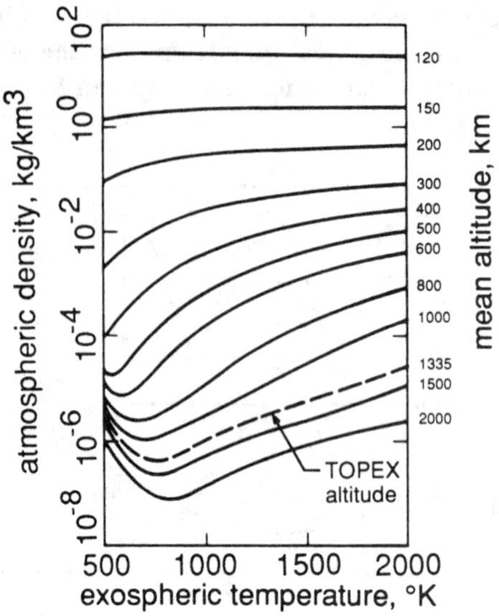

Fig. 8. Jacchia '71 Density vs Exospheric Temperature and Altitude

Density models used for orbit propagation are usually derived empirically from actual flight data in terms of orbital geometric parameters and daily values of the $F_{10.7}$ solar flux, the geomagnetic index K_p, and the 81-day mean solar flux $\overline{F}_{10.7}$. Unfortunately, candidate density models do not reflect flight data at TOPEX altitude, as none are presently available. One candidate, the Jacchia '71 model,[21] graphically represents atmospheric density as a function of mean altitude and the exospheric temperature T_∞. In Fig.(8) this information has been used to estimate the density-temperature variation at 1335 km altitude for representation by the following fifth-order polynomial:

$$\rho = \left[a_o + a_1 T_\infty + a_2 T_\infty^2 + \ldots + a_5 T_\infty^5 \right] \times 10^{-6} \text{ kg/km}^3 \qquad (10a)$$

where

$$a_o = 76.368$$
$$a_1 = -0.358$$
$$a_2 = +6.589 \times 10^{-4}$$
$$a_3 = -5.971 \times 10^{-7}$$
$$a_4 = +2.679 \times 10^{-10}$$
$$a_5 = -4.710 \times 10^{-14}$$

and

$$T_\infty = T_c + \Delta T_\infty \quad \text{in °K} \qquad (10b)$$
$$T_c = 379 + 3.24\overline{F}_{10.7} + 1.3(F_{10.7} - \overline{F}_{10.7})$$
$$\Delta T_\infty = 28K_p + 0.03e^{K_p}$$

The 81-day mean solar flux $\overline{F}_{10.7}$ exhibits a long-term variation between extremes of ~70 and ~280 × 10^{-22} watts/m^2/cycle/s over cycles that average 11 years. Fig.(9) shows this variation for solar cycle 19, beginning in 1955, through June 1989 of current cycle 22, which began in September 1986. The National Oceanic and Atmospheric Administration

(NOAA) estimates the peak of cycle 22 will occur in early 1990 when $\overline{F}_{10.7}$ is predicted to be ~240.[22] By TOPEX launch in June 1992, the current predicted $\overline{F}_{10.7}$ level is expected to drop below 200. The end-of-mission in mid-1995 coincides with the predicted end of cycle 22 when the flux level is expected to be at a near-minimum value of ~70.

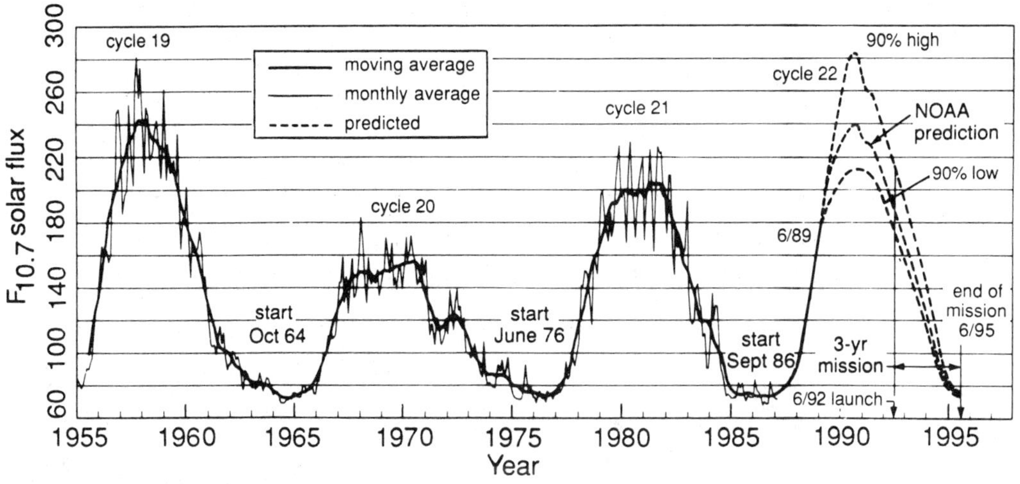

Fig. 9. Recent Solar Cycles vs Current Cycle 22

Summary of Perturbations

Table 2 identified the perturbations affecting each mean orbital element. The effects of these perturbations on the ground track are now examined for a 100-day period using high and low combinations of drag and luni-solar gravity to establish extremes in ground track variability. The drag extremes are easily characterized using constant minimum and maximum values of $\overline{F}_{10.7}$ expected during the mission (70 and 225).

The ground track variability due to luni-solar gravity is much more complex than drag, dependent primarily on relative sun, moon, and satellite geocentric orbit geometries. Two extremes in luni-solar phasing are compared in Fig.(10) to illustrate the effect of these geometries on the ground track. Each example uses the reference orbit (Table 1) for initial conditions and includes only luni solar perturbations. The first case begins on 21 June 1992; the second on 1 August 1992. When the prediction begins on 21 June, the ground track drift oscillates about zero with a maximum excursion of ~270 m over 100 days. When the prediction period begins on 1 August, the ground track drifts steadily westward, accumulating nearly 2 km after 100 days.

Extremes in Ground Variability. Fig.(11) compares the extremes in ground track variability over 100 days for combinations of constant drag at $\overline{F}_{10.7}$ values of 70 and 225 and luni-solar gravity phasing beginning on 21 June and on 1 August 1992. Each ground track is untargeted, beginning with the same reference orbital elements (Table 1). These elements have been tuned for geopotential perturbations, so the presence of earth gravity perturbations has a negligible effect on the ground track (i.e., $D \simeq D_R$).

When luni-solar gravity is added to geopotential perturbations, the ground track drifts slowly, the direction and rate dependent on the relative sun, moon, and satellite orbit geometries. Curves 1 and 3 in Fig.(11) isolate the predicted luni-solar effects for initial epochs of 1 August 1992 and 21 June 1992, respectively, repeating the behavior already described in Fig.(10).

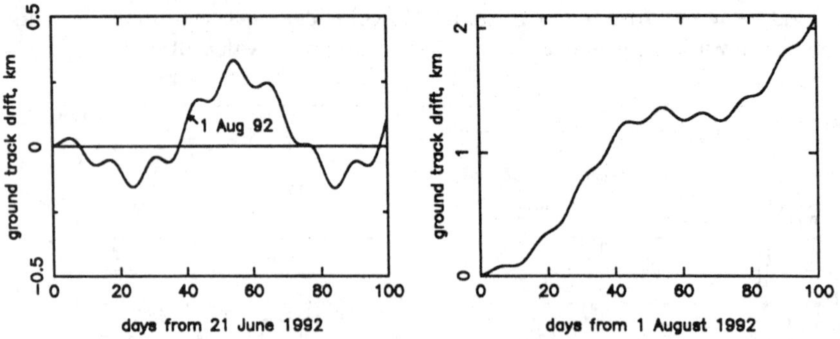

Fig. 10. Ground Track Variation with Luni-Solar Gravity for 100 days

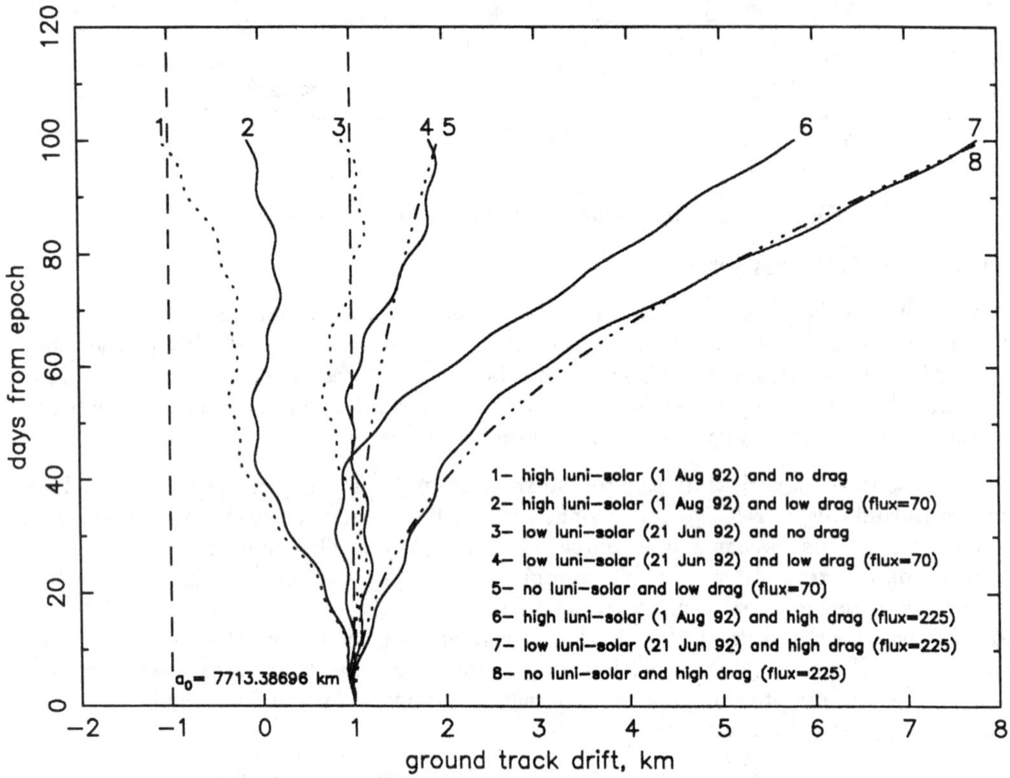

Fig. 11. Ground Track Variations for High and Low Combinations of Drag and Luni-Solar Gravity

If atmospheric drag is added to the geopotential perturbations instead of luni-solar gravity, the ground track drifts continuously eastward at a rate that increases with the solar flux level. Curves 5 and 8 in Fig.(11) compare the extremes in ground track drift over 100 days for constant solar flux levels of 70 and 225.

Curves 2, 4, 6, and 7 in Fig.(11) show the predicted effects of high and low combinations of drag and luni-solar gravity. The two extremes are curves 2 and 7 which combine high luni-solar gravity with low drag, and low luni-solar gravity with high drag. In each case the larger perturbation controls the ground track drift rate. Comparisons of curves 4 and 7 reveal that both high and low drag levels dominate the ground track drift when the luni-solar influence is low. Comparison of curves 2 and 6 show the effect of drag in the presence of high luni-solar perturbations. In curve 6, high drag and high luni-solar gravity are nearly balanced for the first 40 days; afterwards drag becomes the dominant perturbation.

TARGETING WITH ALL MODELLED PERTURBATIONS

The *longitude targeting* strategy described earlier is now applied in the presence of the combined geopotential, drag, and luni-solar perturbations. Figs.(12a,b) compare these results for high and low luni-solar gravity when $\overline{F}_{10.7} = 70$; Figs.(12c,d) when $\overline{F}_{10.7} = 225$. Low luni-solar gravity assumes targeting occurs on 21 June 1992; targeting occurs on 1 August for the high luni-solar cases.

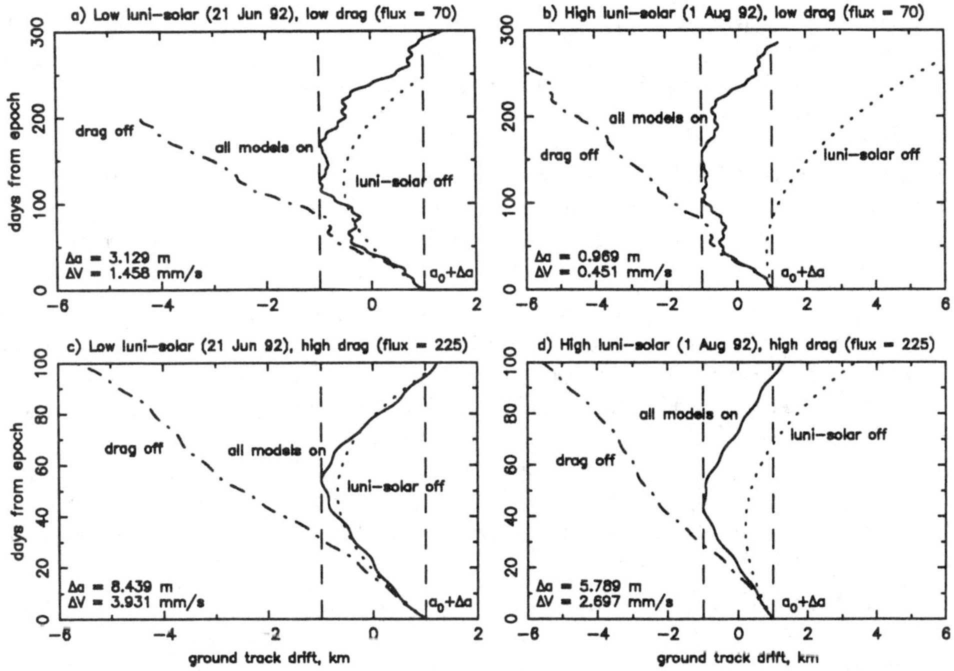

Fig. 12. Longitude Targeting for High and Low Combinations of Drag and Luni-Solar Gravity

Table 3 presents a summary of the longitude targeting requirements, indicating the achieved maneuver spacing depends primarily on drag, while maneuver magnitude depends on both drag and luni-solar levels. These results show that high luni-solar cases require magnitudes more than ~1 mm/s *less* than the low luni-solar cases. This result is expected because the higher westward drift rate of high luni-solar gravity reduces the maneuver necessary to offset the eastward drift induced by drag.

The combination of low drag and low luni-solar perturbations requires the minimum maneuver magnitude. Additional reduction in the magnitude is anticipated when orbit prediction errors are considered in the targeting process; the resulting magnitudes may approach the minimum achievable magnitude on the order of ~0.4 mm/s. When this occurs *longitude targeting* strategies may not be satisfactory and other alternatives must be considered. A better appreciation of possible alternatives comes from an understanding of how orbit prediction errors affect the maneuver design.

Table 3. Summary of Maneuver Requirements for Longitude Targeting

Targeting Epoch	$\overline{F}_{10.7} = 70$ ΔV(mm/s)	T_{max}(days)	$\overline{F}_{10.7} = 225$ ΔV(mm/s)	T_{max}(days)
21 June 1992	1.458	~290	3.931	~94
1 August 1992	0.451	~280	2.697	~94

GROUND TRACK PREDICTION ERRORS

The ground track control and maneuver spacing requirements must be satisfied in the presence of expected prediction errors arising from orbit determination, maneuver execution, and orbit prediction modelling error sources. The maneuver design must account for 95-percentile errors to satisfy these requirements. Here we are concerned with 95-percentile error envelopes about the nominal ground track, assuming a Gaussian distribution. Future studies will establish the actual error distribution using Monte Carlo sampling techniques.

Error Budget

Separate prediction error budgets have been established for orbit determination and orbit prediction since these functions are performed separately by GSFC and JPL, respectively. This error management policy assures interface compatibility between these highly interactive functions.

The total allowable one-sigma error in predicted equatorial longitude after 30 days is 250 m.[23] This prediction error includes 75 m due to all orbit determination error sources when $\overline{F}_{10.7} \leq 225$ during the definitive data arc, a requirement intended to cover all solar flux conditions expected during the mission (see Fig. 9). The root-sum-square balance of 238 m is comprised of all other modelling error sources during the 30-day prediction interval, including especially, maneuver execution and atmospheric drag modelling errors. This requirement also applies only when $\overline{F}_{10.7} \leq 225$. When these errors are assumed Gaussian, 95-percentile errors correspond to 1.645σ, or 411 m in equatorial longitude after 30 days.

While both orbit determination and prediction accuracies are also sensitive to geopotential modelling uncertainties, these effects have been neglected here since extensive pre-flight modelling[12] and planned on-orbit gravity tuning during the first six months are expected to reduce this contribution to a small level compared to other error sources.

Since ground track maintenance maneuvers are primarily concerned with control of nodal period, errors in semi-major axis are particularly important. The semi-major axis has an initial uncertainty at the targeting epoch t_o due to orbit determination and maneuver execution errors. After the maneuver, semi-major axis errors arise from changes in the expected orbit decay rate due to atmospheric density prediction errors.

Initial Semi-major Axis Errors

The semi-major axis error at the targeting epoch includes contributions due to orbit determination and maneuver execution error sources. These error sources also contribute to errors in the mean motion n_o and nodal period τ_{n_o}. Updates to orbital elements needed to propagate these errors remain consistent with Eq.(8), and ground track drift behavior remains consistent with Eq.(7). The total one-sigma error in semi-major axis at t_o due to independent orbit determination and maneuver execution error sources is

$$\sigma_{a_{t_o}} = \sqrt{\sigma^2_{a_{OD}} + \sigma^2_{a_{ex}}} \tag{11}$$

Orbit Determination Errors. Errors in the determination of semi-major axis, defined by $\sigma_{a_{OD}}$ in Eq.(11), map directly into errors in the maneuver design. For example, a one meter error in the determination of semi-major axis results in a maneuver magnitude error of ~0.46 mm/s, an error which results in an equatorial longitude error in the ground track of ~230 m after only 30 days. The maneuver magnitude must be adjusted to compensate for this semi-major axis error.

If the one-sigma 30-day prediction error budget of 75 m for orbit determination were totally allocated to semi-major axis error, then the allowable error would be $\sigma_{a_{OD}} = 0.333$ m. This value has been adopted for nominal maneuver design. However, the effects of larger semi-major axis errors will also be examined and discussed.

Maneuver Execution Errors. As noted in Fig.(6), the propulsion module is mounted on the aft end of the satellite. This module includes four one-newton thrusters that provide velocity changes along the positive roll axis for ground track maintenance maneuvers. These thrusters can be fired simultaneously, or in either of two fixed pairs. Maneuver execution errors result from thruster proportional and fixed velocity magnitude errors, and satellite pointing uncertainties.

The one-sigma pointing control requirement for maneuvers is 1.5 deg, including contributions due to satellite pointing control, thruster misalignments, and uncertainties in satellite center of mass location. The effect of this pointing error on maneuver execution accuracy is proportional to maneuver magnitude. Since orbit maintenance maneuvers are executed in the orbit plane and are small, expected pointing errors introduce negligible orbit errors and have been ignored in this analysis.

The satellite is required to provide velocity changes with a one-sigma fixed magnitude error $\sigma_{\delta V_f} = 0.133$ mm/s, and a proportional magnitude error $\sigma_{\Delta V_p} = 0.5\%$.[23] For these error levels, a maneuver magnitude of ~27 mm/s results in equal fixed and proportional error components. Since nearly all planned ground track maintenance maneuvers are well below 27 mm/s (Fig. 2), fixed magnitude errors are expected to be the dominant maneuver execution error source. A fixed velocity error of 0.133 mm/s results in a semi-major axis error of ~0.3 m, about the same magnitude assumed for orbit determination.

The standard deviation in semi-major axis due to these independent maneuver execution errors is $\sigma_{a_{ex}}$

$$\begin{aligned} \sigma_{a_{ex}} &= \frac{da_o}{dV}\sqrt{\sigma^2_{\delta V_f} + (\Delta V \sigma_{\Delta V_p})^2} \\ &= \frac{2a_o}{V}\sqrt{\sigma^2_{\delta V_f} + (\Delta V \sigma_{\Delta V_p})^2} \end{aligned} \tag{12}$$

The maneuver ΔV magnitude is determined iteratively by the targeting process using ΔV_{max} as an initial estimate based on the value of $\overline{F}_{10.7}$ at t_o (see Eqs. 5,10).

Atmospheric Density Modelling Errors. Errors in density arise from physical mis-modelling, and from uncertainties in the values of $F_{10.7}$, $\overline{F}_{10.7}$, and K_p used in the density model. Density errors change the decay rate of the semi-major axis, which in turn changes the nodal period and the ground track drift rate. The decay rate from Eq.(9), rewritten to now include the standard deviation in the *daily* density error σ_{ρ_d} is

$$(da/dt)_i = -\left(C_D A_i \rho_d [1 \pm \sigma_{\rho_d}/\rho_d] V_i^2 / M_s n_i\right) \left[1 - (\omega_e/n_i)\cos i_i\right]^2 \tag{13}$$

The decay rate is updated every i^{th} orbit using daily updates of ρ_d and σ_{ρ_d} based on current values of solar activity parameters. Positive values of σ_{ρ_d} increase the decay rate and the eastward ground track drift rate; negative values reduce both rates.

The density error $\Delta\rho_d$ is easily derived from Eqs.(10a,b) in terms of errors in each solar activity parameter.

$$\Delta\rho_d = \frac{\partial\rho}{\partial T_\infty}\left[\frac{\partial T_\infty}{\partial \overline{F}_{10.7}}\Delta\overline{F}_{10.7} + \frac{\partial T_\infty}{\partial(F_{10.7} - \overline{F}_{10.7})}\Delta(F_{10.7} - \overline{F}_{10.7}) + \frac{\partial T_\infty}{\partial K_p}\Delta K_p\right] \tag{14}$$

$$= \frac{\partial\rho}{\partial T_\infty}\left[3.24\Delta\overline{F}_{10.7} + 1.3\Delta(F_{10.7} - \overline{F}_{10.7}) + (28 + 0.03K_p e^{K_p})\Delta K_p\right]$$

The expected value of $(\Delta\rho_d)^2$ is the variance $\sigma^2_{\rho_d}$. For brevity, temporarily let $\overline{F}_{10.7} = \alpha$ and $(F_{10.7} - \overline{F}_{10.7}) = \eta$, then the standard deviation in density error σ_{ρ_d} becomes

$$\sigma_{\rho_d} = \left|\frac{\partial\rho}{\partial T_\infty}\right| \sqrt{10.5\sigma_\alpha^2 + 8.42\rho_{\alpha\eta}\sigma_\alpha\sigma_\eta + 1.69\sigma_\eta^2 + [28 + 0.03K_p e^{K_p}]^2 \sigma_{K_p}^2} \tag{15}$$

The values of $\partial\rho/\partial T_\infty$ and K_p in Eq.(15) are evaluated using nominal values of the solar activity parameters. Variations in K_p are assumed independent of variations in $\overline{F}_{10.7}$ and $(F - \overline{F}_{10.7})$, and therefore uncorrelated. This assumption is substantiated by observed daily values of $F_{10.7}$ and K_p from solar cycles 21 and 22 (e.g., Ref. 22). However, variations in $\overline{F}_{10.7}$ are fully correlated with variations in $(F - \overline{F}_{10.7})$, so $\rho_{\alpha\eta} = 1$.

Errors in $\overline{F}_{10.7}$ have the greatest influence on density, whereas K_p errors have the least influence. These errors vary during the prediction interval. Density errors early in the prediction interval have the greatest effect on orbit prediction accuracy because they act for the longest time. Fortunately, $\overline{F}_{10.7}$ is known best at the beginning of the prediction interval because it is based entirely on observations. However, the uncertainty in $\overline{F}_{10.7}$ increases during the prediction interval as it becomes increasingly dependent on predicted values of $F_{10.7}$. After 81 days, $\overline{F}_{10.7}$ is based entirely on predicted values.

Our model assumes $\sigma_{\overline{F}_{10.7}}$ is negligible at the beginning of the prediction period and grows linearly with time to a 3σ level of 20% of $\overline{F}_{10.7}$ after 81 days; beyond this time the error remains constant at 20% of $\overline{F}_{10.7}$. Based on these variations, the standard deviation in $\overline{F}_{10.7}$ then becomes

$$\begin{aligned} \sigma_{\overline{F}_{10.7}} &= \tfrac{1}{3} \times 2.5 \times 10^{-3}\overline{F}_{10.7}(t-1) \quad \text{when } 1 \le t \le 81 \text{ days} \\ &= \tfrac{1}{3} \times 0.20 \times \overline{F}_{10.7} \quad \text{when } t > 81 \text{ days} \end{aligned} \tag{16}$$

Errors in the daily value $(F_{10.7} - \overline{F}_{10.7})$ are based entirely on predictions. Recent flux history like that shown in Fig.(9) suggests the expected 3σ variability in $(F_{10.7} - \overline{F}_{10.7})$ is on the order of $\pm 25\%$ of $\overline{F}_{10.7}$ at high flux levels, and slightly less at lower flux levels. The standard deviation in $(F_{10.7} - \overline{F}_{10.7})$ then becomes

$$\begin{aligned} \sigma_{(F_{10.7} - \overline{F}_{10.7})} &= \tfrac{1}{3} \times 0.25 \times \overline{F}_{10.7} \quad \text{when } \overline{F}_{10.7} = 225 \\ &= \tfrac{1}{3} \times 0.20 \times \overline{F}_{10.7} \quad \text{when } \overline{F}_{10.7} = 70 \end{aligned} \tag{17}$$

Observed values of K_p from solar cycles 21 and 22 suggest a 3σ variability of unity which is independent of solar flux. Accordingly, a standard deviation $\sigma_{K_p} = 0.333$ has been adopted for use throughout this study.

When $\overline{F}_{10.7} = 225$, solar activity parameter errors result in 95-percentile density errors of ~13% early in the prediction period, up to a maximum of ~40% after 81 days. When $\overline{F}_{10.7} = 70$, the 95-percentile density error range is ~8 to 28%.

TARGETING WITH PREDICTION ERRORS

The *longitude targeting* strategy is now examined for two extreme cases using all modelled perturbations and prediction errors. The first case corresponds to the nominal mission conditions with a targeting epoch of 21 June 1992 (low luni-solar gravity) and high drag ($\overline{F}_{10.7} = 225$). The second case examines high luni-solar gravity (1 August targeting epoch) and low drag ($\overline{F}_{10.7} = 70$).

Figs.(12b,c) and Table 3 summarized the longitude targeting characteristics for these cases using all modelled perturbations, but without prediction errors. We now wish to target these cases using an initial semi-major axis error due to orbit determination and maneuver execution error sources, and semi-major axis decay rate variations induced by atmospheric density errors.

Since targeting is constrained by a 95-percentile probability level, $\pm 1.645\sigma$ errors about the nominal ground track establish envelopes of the expected prediction errors. Prediction errors west of nominal arise from error sources that increase the initial semi-major axis and/or decrease drag; eastward ground track errors result from the opposite error combinations.

Longitude targeting requires a maneuver ΔV magnitude resulting in the maximum maneuver spacing in the presence of western ground track errors. With this constraint, the nominal and eastern error ground tracks become a direct consequence of this targeting process.

Recall from Table 3 that longitude targeting without prediction errors requires 3.931 mm/s, resulting in a maximum maneuver spacing of ~94 days. Fig.(13) shows that including prediction errors reduces the maneuver ΔV magnitude to 3.18 mm/s and the nominal maneuver spacing to ~79 days. The resulting 95-percentile high and low maneuver spacing times are ~123 and 65 days, respectively. These results satisfy the 30-day minimum maneuver spacing requirement for 95-percentile ground track prediction errors.

The ground track prediction error contribution due to atmospheric density errors was isolated by removing this effect while retaining all other error sources. The resultant ground track errors, shown as the dotted contours in Fig.(13), are due solely to initial semi-major axis errors caused by orbit determination and maneuver execution error sources. Since these contours do not deviate significantly from the nominal ground track, density errors are clearly the dominant source of ground track prediction errors.

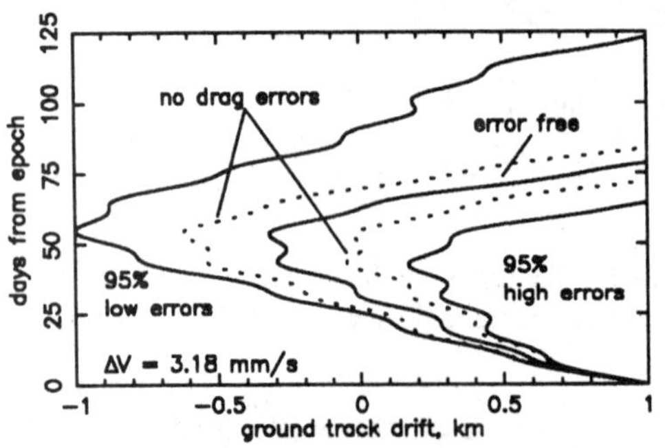

Fig. 13.
Longitude Targeting with Prediction Errors (21 June 92/$\overline{F}_{10.7} = 225$)

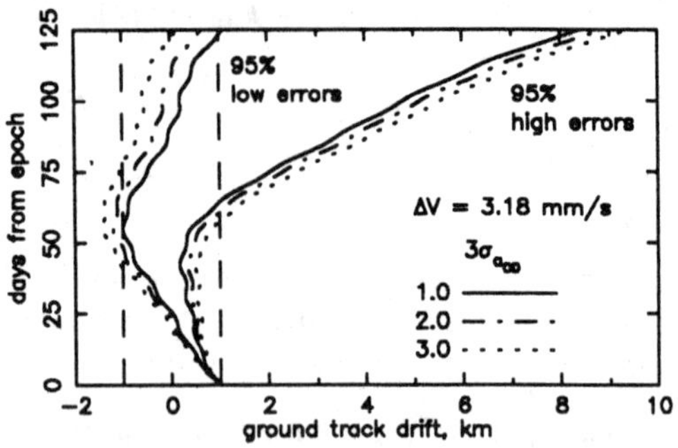

Fig. 14.
Effect of Semi-major Axis Errors due to Orbit Determination

The effect of increases in initial semi-major axis error due to orbit determination was assessed using larger 3σ values of 2 and 3 m, while retaining all other errors at their nominal values. Fig.(14) shows that these increases have little effect on the ground track, indicating maneuver design at high solar flux is relatively insensitive to larger orbit determination errors assumed here. This conclusion is less valid at low solar flux.

Time Targeting with Low Drag

Selection of a targeting epoch of 1 August 1992 and $\overline{F}_{10.7} = 70$ requires maneuver magnitudes that approach the fixed magnitude execution error of 0.4 mm/s assumed in this analysis (Table 3). Even smaller magnitudes become necessary when expected ground track prediction errors are also considered. A minimum acceptable maneuver magnitude must be established that provides needed orbit control in the presence of expected execution errors. A practical limitation would restrict magnitudes to be outside the range $-1 < \Delta V < 1$ mm/s. Positive ΔV magnitudes increase the orbit semi-major axis, whereas negative values (retro to satellite velocity) reduce the semi-major axis.

The subject targeting example was examined with these maneuver magnitude constraints in mind. Retro maneuvers of at least 1 mm/s used to perform *longitude targeting* from the east boundary are much larger than required to offset the influence of luni-solar gravity. In fact, westward ground track drift does not even occur, so retro maneuvers are not possible.

Time targeting to a specified maneuver spacing time which is a 10-day multiple greater than 30 days is possible. A maneuver spacing time of 40 days requires 1.16 mm/s when targeting 95% low prediction errors, as shown in Fig.(15). Here, the prediction error is due ONLY to the combined effects of orbit determination and maneuver execution errors. West boundary arrival times for the nominal and 95% high prediction error cases are ~50 and 85 days, respectively.

The prediction errors for this example are nearly equally divided between fixed magnitude execution errors and orbit determination errors. When the 3σ orbit determination errors increases from 1 m to 2 and 3 m, as shown in Fig.(16), the west boundary crossing times are not significantly different. *Time targeting* is relatively insensitive to the orbit determination errors assumed here.

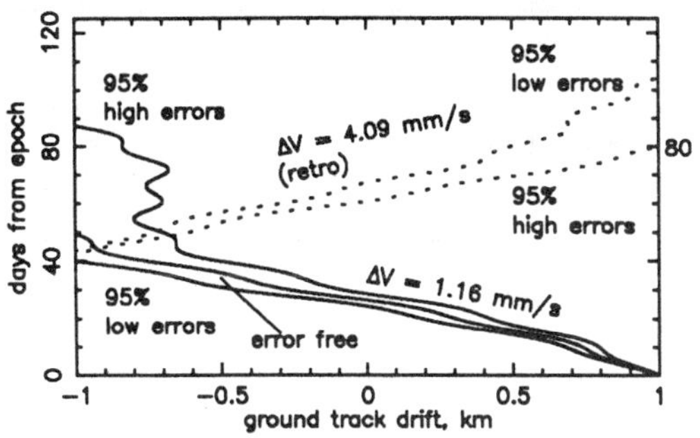

Fig. 15.
Time Targeting with Prediction Errors (1 Aug 92/$\overline{F}_{10.7}$ = 70)

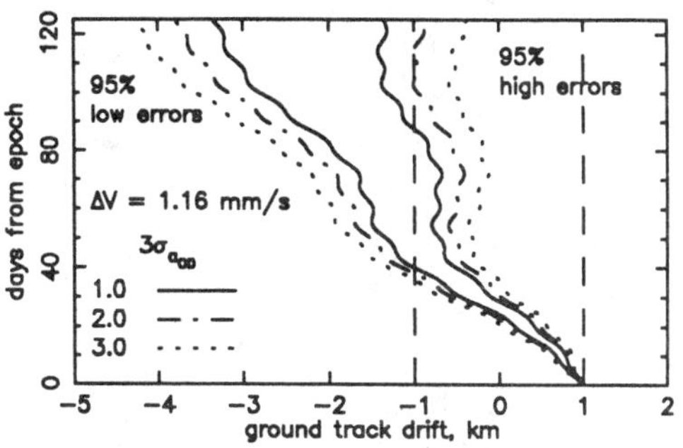

Fig. 16.
Effect of Semi-major Axis Errors due to Orbit Determination

A Second *Time Targeting* Maneuver. A second maneuver applied at the west boundary shows the 'ping-pong' character of a possible maneuver sequence during periods of low drag and high ground track drift rate due to luni-solar gravity. Here, a retro maneuver is required to lower the orbit semi-major axis to begin an *eastward* ground track drift under the control of drag. A magnitude of at least 4 mm/s is necessary to reverse the ground track eastward; magnitudes greater than 5 mm/s cause the ground track to reach the east boundary before 30 days. The dotted contours in Fig.(15) show targeting to day 80 with 95% high prediction errors; the retro maneuver magnitude is ~4.09 mm/s. The 95% low prediction errors increase the predicted maneuver spacing time to ~105 days.

CONCLUSIONS

This maneuver design study indicates atmospheric drag modelling errors are the dominant ground track prediction error source early in the mission during expected high solar flux. The maximum maneuver spacing time achieved with *longitude targeting* is primarily a function of the drag level. Luni-solar perturbations reduce the maneuver magnitude required to offset eastward ground track drift due to drag, the amount dependent on current luni-solar phasing geometry. The high drag level causes the maneuver design to be relatively insensitive to expected orbit determination errors, optimistically assumed here to be only in semi-major axis.

Low solar flux levels expected late in the mission require smaller maneuver magnitudes, causing fixed execution errors to become the dominant source of ground track prediction errors. Instead of longitude targeting, a *minimum time targeting* strategy, constrained to limit the minimum maneuver magnitude, has been designed to satisfy the 30-day maneuver spacing requirement. At low drag, maneuver design becomes more sensitive to semi-major axis determination errors, although without significant variability in the achieved maneuver spacing times.

These maneuver design studies reflect conservative prediction errors defined by 95-percentile envelopes. Predicted density variations are also conservative. Planned studies will develop the expected distributions in maneuver magnitude and spacing times using Monte Carlo techniques. The resulting error distributions can then be more realistically compared with the ground track control and maneuver spacing requirements.

NOTATION

Symbol	Definition
a_o	reference orbit semi-major axis
e_o	reference orbit eccentricity
i_o	reference orbit inclination
Ω_o	reference orbit right ascension of ascending node
ω_o	reference orbit argument of periapse
M_o	reference orbit mean anomaly
a_i	semi-major axis of ith orbit
e_i	eccentricity of ith orbit
i_i	inclination of ith orbit
Ω_i	right ascension of ascending node of ith orbit
ω_i	argument of periapse of ith orbit
M_i	mean anomaly of ith orbit
a_d	semi-major axis of sun or moon
e_d	eccentricity of sun or moon
i_d	inclination of sun or moon
Ω_d	right ascension of ascending node of sun or moon
ω_d	argument of periapse of sun or moon
M_d	mean anomaly of sun or moon
A_i	orbit-average satellite reference drag area during i^{th} orbit
C_D	satellite drag coefficient, design value 2.3
d	number of sidereal days of one ground track repeat cycle, design value 10
$(da/dt)_i$	decay rate of satellite semi-major axis during i^{th} orbit
D_i	actual ground track spacing between the ith and the $(i+1)^{th}$ orbits
D_R	reference longitude spacing between two successive ground tracks
$F_{10.7}$	10.7 cm solar flux (10^{-22} watts/m^2/cycle/s)
$\overline{F}_{10.7}$	81-day mean solar flux (10^{-22} watts/m^2/cycle/s)
$F_{lmp}(i)$	inclination function for satellite orbit
$F_{lmh}(i_d)$	satellite inclination function for sun or moon orbit
$G_{lhj}(e_d)$	eccentricity function for sun and moon orbit
$H_{lpq}(e)$	eccentricity function for satellite orbit
J_2 to J_5	earth zonal harmonics
K_p	geomagnetic index
k	operator ($k = 1$ if $m = 0$ and $k = 2$ if $m \neq 0$)
$l, m, p,$	
M_s	satellite mass, design value 2400 kg
n_o	Keplerian mean motion of satellite, $\sqrt{\mu/a_o^3}$
n_i	mean motion of satellite in i^{th} orbit
n_d	mean motion of sun or moon
n_{gi}	mean motion of satellite due to earth gravity during i^{th} orbit
n_{ls}	contribution to satellite mean motion from luni-solar gravity
N	number of orbits in one ground track repeat cycle, design value 127
P	semi-latus rectum of satellite Keplerian orbit
q, h, j	running indices for luni-solar disturbing function
R_e	mean equatorial radius of the earth (6378.14 km)
t_o	epoch of reference orbital elements; also time of maneuver
T_c	contribution to exospheric temperature due to solar flux
T_∞	exospheric temperature
ΔT_∞	correction to exospheric temperature due to geomagnetic heating
V_i	satellite velocity during i^{th} orbit, $\sqrt{\mu/a_i}$

NOTATION (cont)

Symbol	Definition
β'	angle between the satellite orbit plane and the earth-sun direction
Δi_i	change in orbit inclination during i^{th} orbit due to luni-solar gravity
λ_i	actual equatorial longitude of satellite ground track during i^{th} orbit
λ_{ri}	reference equatorial longitude of satellite ground track during i^{th} orbit
μ	gravitational constant of the earth
μ_d	gravitational constant of sun or moon
ρ_d	orbit-average atmospheric density based on daily solar activity
τ_n	reference orbit nodal period, $2\pi/(n+\dot{\omega})$
τ_{ni}	actual satellite nodal period in i^{th} orbit
$\dot{\Omega}_i$	nodal precession rate during i^{th} orbit $(\dot{\Omega}_i = \dot{\Omega}_{gi} + \dot{\Omega}_{ls_i})$
$\dot{\Omega}_{gi}$	geopotential contribution to satellite nodal precession rate during i^{th} orbit
$\dot{\Omega}_{ls_i}$	luni-solar contribution to satellite nodal precession rate during i^{th} orbit
ω_e	earth rotation rate
$\dot{\omega}_i$	satellite apsidal rotation rate during i^{th} orbit $(\dot{\omega}_i = \dot{\omega}_{gi} + \dot{\omega}_{ls})$
$\dot{\omega}_{gi}$	geopotential contribution to satellite apsidal rotation rate during i^{th} orbit
$\dot{\omega}_{ls}$	luni-solar contribution to satellite apsidal rotation rate during i^{th} orbit

REFERENCES

1. Tapley, B. D., "Orbit Determination Requirements for TOPEX," AAS/AIAA Astrodynamics Specialist Conference, August 1987, Kalispell, Montana, AAS–87–429.

2. Kechichian, J. A., "TOPEX Orbit Sustenance Maneuver Design," AIAA 20th Aerospace Sciences Meeting, January 1982, Orlando, Florida, AIAA–82–0202.

3. Shapiro, Bruce, Pino, Antonino, "Maintenance of an Exact Repeat Ground Track – The GEOSAT ERM," AIAA/AAS Astrodynamics Conference, August 1988, Minneapolis, Minnesota, AIAA–88–4301–CP.

4. Shapiro, Bruce, "The GEOSAT Orbit Adjust," *The J. Astronautical Sciences*, Vol. 36, No. 4, pp. 407-424, October-December 1988.

5. Frautnick, J. C., and Cutting, E., "Flight Path Design Issues for the TOPEX Mission," AIAA 21st Aerospace Sciences Meeting, January 1983, Reno, Nevada, AIAA–83–0197.

6. Farless, D. L., "The Application of Periodic Orbits to TOPEX Mission Design," AAS/AIAA Astrodynamics Specialist Conference, August 1985, Vail, Colorado, AAS–85–301.

7. Cutting, E., Born, G. H., and Frautnick, J. C., "Orbit Analysis for SEASAT," *The J. Astronautical Sciences*, Vol. 24, January-March 1976, pp. 55-90.

8. Born, George, "Analysis of GEOSAT Repeat Track Capability," University of Texas at Austin Memorandum, Austin, Texas, September 26, 1983.

9. McIntosh, J. R., and Hassett, P.J., "LANDSAT 4 Orbit Adjust Maneuver Report," Computer Sciences Corporation, CSCTM-82-6232.

10. Smith, J. C., "Analysis and Application of Frozen Orbits for the TOPEX Mission," AIAA/AAS Astrodynamics Conference, August 1986, Williamsburg, Virginia, AIAA–86–2069–CP.

11. Tang, C. C. H., "An Accurate and Efficient Satellite Long-Term Orbit Predictor Employing "Fictitious" Mean Orbital Elements," AIAA/AAS Astrodynamics Conference, August 1988, Minneapolis, Minn., AIAA–88–4243–CP.

12. Marsh, J. G., et al, "A New Gravitational Model for the Earth from Satellite Tracking Data: GEM-T1," *J. Geophysical Research*, Vol. 93, No. B6, pp. 6169–6215, June 10, 1988.

13. Escobal, P. R., *Methods of Orbit Determination*, Robert E. Krieger Publishing Company, Huntington, New York, 1976.

14. Kaula, W. M., *Theory of Satellite Geodesy*, Blaisdell Publishing Company, Massachusetts, 1966.

15. Escobal, P. R., *Methods of Astrodynamics*, Robert E. Krieger Publishing Company, Inc., Huntington, New York, 1979.

16. Kwok J. H., "The Artificial Satellite Analysis Program (ASAP), Version 2.0," JPL EM 312/87-153, 20 April 1987.

17. Kwok, J. H., "The Implementation of Two Satellite Programs in a Microcomputer Environment," AIAA/AAS Astrodynamics Conference, Williamsburg, Virginia, August 1986. AIAA–86–2165–CP.

18. Kaula, W. M., "Development on the Lunar and Solar Disturbing Functions for a Close Satellite," *Astronomical Journal*, Vol. 67, pp. 300-303, June 1962.

19. Hughes, S., "Satellite Orbits Perturbed by Direct Solar Radiation Pressure; General Expansion of the Disturbing Function," *Planetary and Space Science*, Vol. 25, pp. 809-815, Pergamon Press, 1977.

20. Born, G. H., et al, "Orbit Analysis for the GEOSAT – ERM," *The J. Astronautical Sciences*, Vol. 36, No. 4, pp. 425-446, October-December 1988.

21. Jacchia, L. G., "Revised Static Models of the Thermosphere and Exosphere with Empirical Temperature Profiles," Smithsonian Astrophysical Observatory. Special Report 332, 5 May 1971.

22. "Preliminary Report and Forecast of Solar Geophysical Data," U.S. Department of Commerce, National Oceanic and Atmospheric Administration, Boulder, Colorado, SESC PRF 719, 13 June 1989,

23. "TOPEX/POSEIDON Mission and Systems Requirements," JPL Doc. No. 633-103, April 1989.

APPENDIX

Geopotential Perturbations:

$$
\begin{aligned}
n_g = n_o \Big(& 1 + (3/2) J_2 \sqrt{(1-e^2)} (R_e/P)^2 [1 - (3/2)\sin^2 i] \\
& + (3/128) J_2^2 (R_e/P)^4 \sqrt{(1-e^2)} \{16\sqrt{(1-e^2)} + 25(1-e^2) - 15 \\
& + [30 - 96\sqrt{(1-e^2)} - 90(1-e^2)] \cos^2 i + [105 + 144\sqrt{(1-e^2)} + 25(1-e^2)] \cos^4 i\} \\
& - (45/128) J_4 (R_e/P)^4 \sqrt{(1-e^2)} e^2 [3 - 30\cos^2 i + 35\cos^4 i] \Big) \\
& + (3/2) n_o J_3 (R_e/P)^3 (1-e^2)^{3/2} \{\sin i/e\} \{(5/4)\sin^2 i - 1)\} \sin\omega \\
& + (15/16) n_o J_5 (R_e/P)^5 \sqrt{(1-e^2)} (4 + 5e^2)(\sin i/e) \{1 - (7/2)\sin^2 i \\
& + (21/8)\sin^4 i\} \sin\omega \qquad (A)
\end{aligned}
$$

$$\dot{\Omega}_g = -(3/2)J_2(R_e/P)^2 n\cos i\left[1 + (3/2)J_2(R_e/P)^2\right.$$

$$\left.[(3/2) + (e^2/6) - 2\sqrt{(1-e^2)} - \{(5/3) - (5/24)e^2 - 3\sqrt{(1-e^2)}\}\sin^2 i]\right]$$

$$- (35/8)J_4(R_e/P)^4 n_o\{1 + (3/2)e^2\}\{(12 - 21\sin^2 i)/14\}\cos i \qquad (B)$$

Perturbations due to luni-solar gravity:

In this model, the indices *l, m, p, h, q,* and *j* were set to vary as follows: *l=2; m, p, h=0, 2; q= (2p-l), and j=0.* The index *q* is always set to *(2p - l)* to neglect short periodic effects. The disturbing potential of luni-solar gravity is given by:

$$U = (\mu_d a^l)/(a_d^{l+1}) \sum_{m=0}^{l} k\ (l-m)!/(l+m)!$$

$$\sum_{p=0}^{l}\sum_{h=0}^{l}\sum_{q=-\infty}^{\infty}\sum_{j=-\infty}^{\infty} F_{lmp}(i)F_{lmh}(i_d)H_{lpq}(e)G_{lhj}(e_d)\cos Y \qquad (C)$$

where $Y = \{(l-2p)\omega - (l-2h)\omega_d - (l-2h+j)M_d + m(\Omega - \Omega_d)\}$

The expressions for Δi, $\dot{\Omega}_{ls}$, $\dot{\omega}_{ls}$ and n_{ls} are given by:

$$\Delta i = (\mu_d a^{l-2})/\{a_d^{l+1}(1-e^2)n\sin i\} \sum_{m=0}^{l} k\ (l-m)!/(l+m)!$$

$$\sum_{p=0}^{l}\sum_{h=0}^{l}\sum_{q=-\infty}^{\infty}\sum_{j=-\infty}^{\infty} F_{lmp}(i)$$

$$F_{lmh}(i_d)\ H_{lpq}(e)\ G_{lhj}(e_d)\{(l-2p) - \cos i\}\ (\cos Y/\dot{Y})|_{t_o}^{t} \qquad (D)$$

$$\dot{Y} = \{(l-2p)\dot{\omega} - (l-2h)\dot{\omega}_d - (l-2h+j)n_d + m(\dot{\Omega} - \dot{\Omega}_d)\}$$

$$\simeq \{m\dot{\Omega} - (l-2h+j)n_d\}$$

$$\dot{\Omega}_{ls} = (\mu_d a^{l-2})/\{a_d^{l+1}\sqrt{(1-e^2)}n\sin i\} \sum_{m=0}^{l} k\ (l-m)!/(l+m)!$$

$$\sum_{p=0}^{l}\sum_{h=0}^{l}\sum_{q=-\infty}^{\infty}\sum_{j=-\infty}^{\infty} \partial F_{lmp}(i)/\partial i\ F_{lmh}(i_d)\ H_{lpq}(e)\ G_{lhj}(e_d)\ \cos Y \qquad (E)$$

$$n_{ls} = -(\mu_d a^{l-2})/a_d^{l+1} \sum_{m=0}^{l} k\ (l-m)!/(l+m)!$$

$$\sum_{p=0}^{l} \sum_{q=-\infty}^{\infty} F_{lmp}(i)\ \{\partial H_{lpq}(e)/\partial e\ (1-e^2)/e + 2l H_{lpq}(e)\}$$

$$\sum_{h=0}^{l} \sum_{j=-\infty}^{\infty} F_{lmh}(i_d)\ G_{lhj}(e)\ \cos Y \qquad (F)$$

$$\dot{\omega}_{ls} = (\mu_d a^{l-2})/a_d^{l+1} \sum_{m=0}^{l} k\ (l-m)!/(l+m)!$$

$$\sum_{p=0}^{l} \sum_{q=-\infty}^{\infty} \{(\sqrt{1-e^2}/e)\ \partial H_{lpq}(e)/\partial e\ F_{lmp}(i) - \ \partial F_{lmp}(i)/\partial i\ H_{lpq}(e)/(\tan i \sqrt{1-e^2})\}$$

$$\sum_{h=0}^{l} \sum_{j=-\infty}^{\infty} F_{lmh}(i_d)\ G_{lhj}(e_d)\ \cos Y \qquad (G)$$

ORBIT SYNTHESIS FOR TARGET SATELLITES*

Charles K. Wilkinson†

The orbit synthesis process resolves the many conflicts among mission requirements, system design constraints, and test operation realities. This subject is addressed for an inexpensive non-attitude-controlled LEO satellite whose mission is to provide a prototype target for and/or score the performance of a ground-based or direct-ascent-based sensor or weapon system. Specific examples are given for the Instrumented Test Vehicles (ITVs) launched in December 1985. Factors considered in the synthesis process include: launch vehicle, test site/arena, altitude and illumination cycles, test schedules, orbit lifetime, communications, ground trace repeatability, detailed test constraints, and satellite design constraints.

INTRODUCTION

The Instrumented Test Vehicle (ITV) was conceived in 1978 as an inexpensive target for the LEO testing of Anti-Satellite (ASAT) weapon systems. The ASAT could be either a direct-ascent-based Kinetic Energy Weapon (KEW) or a ground-based Directed Energy Weapon (DEW). This paper illustrates the orbit synthesis process considering all significant factors, several of which are unique to target satellites.

The association of a target satellite with a particular sensor and test site or arena introduces some unique mission requirements and constraints. However, it is the interaction of the test program operational realities -- target launch vehicle capabilities, imposed parameter constraints, scheduling of range resources, etc. -- with orbit dynamics which is the primary focus of the orbit synthesis process. Several techniques (mathematical algorithms with graphical output) have been developed to analyze and illustrate the impact of these constraints on the selected orbit. Each mission requirement and constraint category is discussed, and the associated analysis technique is introduced. The ingredients of the methods are discussed, but the algorithms themselves are not described. Finally, the synthesis process is illustrated. Both ground-based and direct-ascent-based weapon systems are considered. The

* This work was performed in conjunction with Contract F04701-78-C-0125 for the United States Air Force, Space Division.

† Principal Engineer, Flight Dynamics Department, Textron Defense Systems, 201 Lowell Street, Wilmington, Massachusetts 01887. Member AIAA.

examples given are appropriate to typical constraints of a test program. Vandenburg AFB (VAFB) is used as the test arena for a direct-ascent-based KEW and the Southwest U.S. as the test location for a ground-based DEW. The numerical values were selected to illustrate trade-offs relative to competing constraints. There is no intended correlation to the operational capabilities of any existing or proposed weapon system.

Two ITVs were launched on a Scout launch vehicle from Wallops Island, Virginia in December 1985. One of the two completed an extensive set of operations in its deployed test configuration beginning in December, 1986. Its orbital demise occurred in August, 1987. The other ITV became a victim of solar cycle 22 when it decayed in May of 1989. It was still in its packaged state, but with batteries depleted. The next section gives a brief description of the ITV vehicle. Its design philosophy is primarily simplicity. For this paper, its capabilities and limitations are treated as mission constraints.

TARGET DESCRIPTION

The overall system weighing 184 kg, comprises two ITV payloads and a dispenser assembly that mates to the Scout Launch vehicle. Each ITV payload weighs 81 kilograms and consists of a canister (1.1 meters long by 0.5 meters in diameter) containing a deployable shell and a core package. When deployed to its full 2 meter diameter, the Kevlar cloth shell can be chemically heated by command to a desired LWIR signature. To conserve fuel (required for active thermal control in its deployed test configuration) and to maximize orbit lifetime, the ITV can remain in its canister (packaged) configuration until shortly before a test. The principal subsystems and their functions are, in brief summary:

Electrical subsystem. Provides power for all on-board functions.

Communication subsystem. Receives uplink commands and provides data on health and status and on test conditions via downlink. Most uplink commands can be stored for execution at a specified time. Downlink communications are real-time. With the exception of diagnostic thermal history data there is no provision for on-board data storage.

Scoring subsystem. Provides specific test data.

Thermal-control subsystem. Maintains all components within the required temperature range in all operational conditions.

Signature-control subsystem. Provides and maintains known, selectable temperature levels for test using a closed-loop, or feedback, scheme.

Hydrazine Storage Subsystem. Holds up to 12.25 kg (27 lb) of hydrazine for purposes of either signature or thermal control in the test configuration.

Fig. 1 shows the ITV in its deployed condition for test. The pertinent subsystems are designed to be omnidirectional so that attitude control is not required or attempted. However, telemetered data from the three-axis magnetometer can be analyzed to reconstruct the attitude history. Obviously, the ITV has no orbit adjust capability.

ORBIT SYNTHESIS PROCESS

The goal of the orbit synthesis process is very simple. It is to define a set of orbital parameters which "optimizes" the mission requirements and constraints. The orbit parameters requiring definition are four in number. They are summarized in Table 1.

Table 1
ORBIT PARAMETERS REQUIRING DEFINITION

Orbit Inclination
Perigee Altitude
Apogee Altitude
Launch Times of Day (Window)

The launch times of day, or launch window, are dependent on the launch site and launch azimuth (if more than one is available). This defines the allowable initial right ascensions of the ascending node. The initial argument of perigee is fixed by the launch site/launch azimuth combination assuming that perigee of the orbit coincides with burnout of the launch vehicle. This is a very appropriate assumption for the Scout launch vehicle. The launch date is not included in Table 1 since it is in general not a constraint for target satellite orbits.

Each mission requirement and constraint must be examined individually over the matrix of possible orbits to determine a measure of performance. The performance of a given orbit relative to a given requirement can vary from unacceptable, marginally acceptable, acceptable, or ideal. In theory, an optimum orbit can be determined mathematically by assigning numerical values to the measures of performance and weighting factors to each competing requirement. In practice this does not work, since each player in the decision has a different set of weighting factors. The conflicts resulting from the competing requirements can only be resolved through sessions for negotiation and compromise among all concerned.

The key mission requirements and constraints for target satellite orbits are summarized in Table 2.

Table 2
KEY REQUIREMENTS AND CONSTRAINTS

Test Arena	Ground Trace Repeatability
Launch Vehicle	General Test Constraints
Satellite Design Constraints	Altitude
Communications	Illumination
Thermal	Schedule
Orbit Lifetime	Detailed Test Constraints

The principal dynamical perturbations to the orbit are[1,2]:

1. Nodal regression due to J_2,
2. Perigee rotation due to J_2,
3. Altitude decay due to atmospheric drag.

The following sections address each of the mission requirements and constraints, though not necessarily in the order given in Table 2. The interplay with the dynamics of the orbit is included where appropriate.

TEST ARENA AND COMMUNICATIONS CONSIDERATIONS

The location and extent of the test site or arena has a major influence on the selection of the orbit inclination. An inclination nearly equal to or slightly larger than the latitude of the test site is test cooperative in that it maximizes the number of daily test opportunities (passes over the site or arena). In the absence of an orbit adjust capability for ground trace control, this allows the maximum flexibility in detailed test planning.

Ground-based DEW

Fig. 2 illustrates the geometry differences between a near polar orbit and one inclined slightly larger than the latitude of hypothetical DEW instrumentation located in the southwestern U.S. The instrumentation is to be tested at low altitude and requires high elevation angles to minimize the path length through the atmosphere. This yields a very small test arena which has a high probability of being "skipped over" by the passes of a polar orbit if the orbit period is not carefully controlled. The "skip-over" can occur for many consecutive days if the ground-trace is nearly repeatable over one, or a few days. Algorithms to compute the average number of opportunities per day for a given arena (constrained by elevation angle, altitude, and time in the region) are easily generated for circular orbits. The intervals of ascending nodes (earth referenced) of orbits passing through the region for both ascending (northerly heading) and descending (southerly heading) passes are determined using spherical trigonometry. The sum of the two intervals is divided by the longtitude shift per rev (which is a function of the orbital period, earth rotation rate, and nodel regression rate) to obtain the average test opportunity frequency per day. The geometry is illustrated in Fig. 3. Fig. 4 illustrates the average opportunity frequency per day as a function of orbit inclination for selected orbit altitudes. The site is the same as that in Fig. 2 and the target is constrained to be above 60° elevation for at least 30 seconds. The opportunities are maximized at an orbit inclination (I for prograde and 180° – I for retrograde orbits) slightly larger than the latitude of the site. The results pertain to circular orbits but are perfectable acceptable for trade-offs associated with preliminary planning for low altitude elliptical orbits.

Real-time communication constraints can impact the mission design. The ITV communication subsystem is compatible with the Air Force Satellite Control Facility (AFSCF) network of Remote Tracking Stations (RTSs). Sites VTS, HTS, NHS, and TTS will be referred to in this paper.

Fig. 5 illustrates that the RTS site closest to the test site of Fig. 2 is VTS. It provides marginal coverage (below 5° elevation is considered marginal for preliminary planning purposes) at low altitudes. This is a signal to 1) carefully

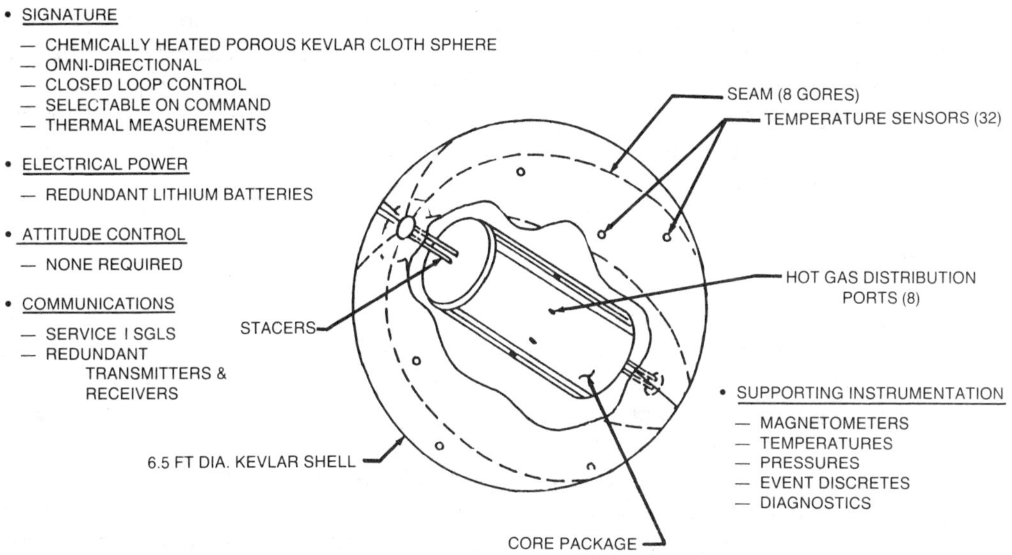

Fig. 1 Design Description of ITV (Shown Deployed)

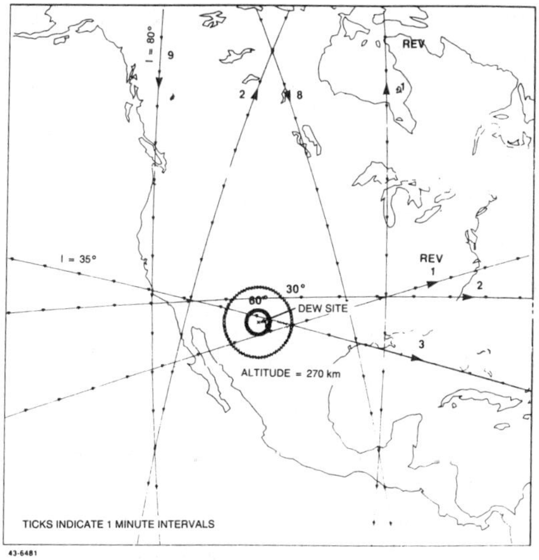

Fig. 2 Test Compatible Compared to Polar Orbits

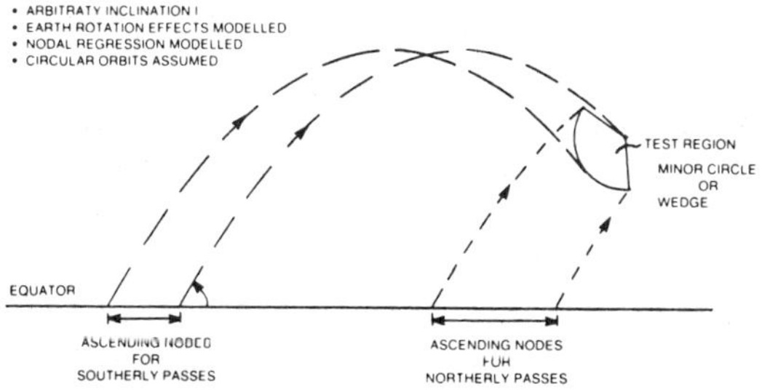

Fig.3 Geometry For Passes Through the Test Region

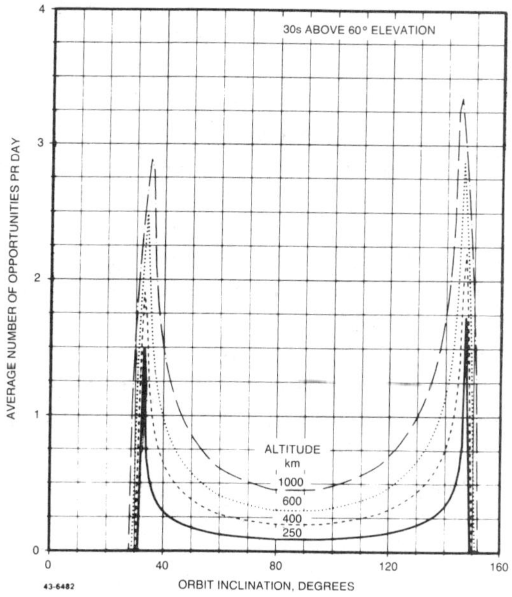

Fig. 4 Test Opportunities are Sensitive to Orbit Inclination

examine the communcations link from 0° to 5° elevation, 2) consider the use of a transportable receiver co-located with the instrumentation for recording of telemetry to supplement VTS, or 3) change the instrumentation site (preliminary planning phase) to improve communications. A prograde orbit is preferable to a retrograde orbit for the illustrated scenario since precious time will be used by the RTS to acquire the downlink signal.

Direct-Ascent-Based KEW

The test scenario for a direct-ascent-based KEW against a LEO target can be quite different from that of a ground-based DEW. The test arena is defined by considerations of range safety, tracking of both vehicles, and telemetry from both vehicles. A typical test arena is illustrated in Fig. 6. It consists of a wedge-shaped area exceeding 30° elevation from VAFB/VTS and an up-coast receiver for either, or both, tracking and telemetry and is offset from the coast for purposes of range safety. A further constraint concerns the desire to minimize the probability of wasting an expensive KEW against a malfunctioning inexpensive target. To accomplish this, the target is commanded to its test state within 10 to 20 minutes of the test and is validated by comparing real-time telemetry with pre-established criteria. A GO/ABORT decision is issued at the latest possible moment prior to the KEW launch. This constraint requires the use of an RTS other than VTS for the GO/ABORT decision. Fig. 6 illustrates the possible GO/ABORT sites:

HTS for prograde orbits;
TTS for near-polar orbits;
NHS for retrograde orbits.

A typical groundrule for satisfactory GO/ABORT decision time is a pass yielding two to three minutes above 5° elevation over the GO/ABORT site.

The algorithms required to compute opportunity frequency for this scenario are similar to, but more complex than, those for the single site constrained by elevation angle. The test is modeled by multiple wedges (for the VAFB arena there is typically one each for the up-coast receiver and VTS and an intermediate one for the range safety coast-line offset) constrained by elevation and azimuth angles. The intervals of ascending node longitudes are computed for each wedge and the GO/ABORT site. The set of acceptable passes corresponds to the intersection, over all areas modelled, of the sets of longtitude intervals. This approach identifies passes which go through each of the areas modelled. It is not necessarily true that a pass has a portion simultaneously within each of the areas representing the test arena. However, careful selection of the wedges can identify those passes which do indeed go through the desired test arena. Fig. 7 summarizes a typical variation in KEW test opportunity frequency as a function of orbit inclination and altitude. HTS is the GO/ABORT site for inclinations between 30° and 75°, TTS for inclinations between 75° and 120°, and NHS for inclinations between 120° and 150°. As with the single site frequency, an orbit inclination slightly larger than the latitude of the center of the test arena maximizes the frequency of test opportunities.

To summarize, the location of the test arena has a major influence on the selection of the orbit inclination. With no orbit adjust capability and inaccurate orbit attainment (refer to the next Section), an orbit inclination approximately equal to the latitude of the center of the test arena is required for some certainty in advance test planning, particularly for low altitude tests. Prograde and retrograde orbits are possible alternatives. The location of the test arena must be correlated with existing RTSs if real-time commanding and/or display of telemetry is required. Alternately, a transportable telemetry data recorder is required since the ITV has no on-board storage or play-back capability.

LAUNCH VEHICLE CONSTRAINTS

During the initial stages of the ITV program, a number of candidate launch vehicles were considered. The Scout launch vehicle was selected. It is nearly ideal for launching inexpensive targets because of payload-capability compatibility, availability, and low cost and risk. Once this decision was made, three important factors had to be addressed: 1) number of payloads per launch, 2) the launch site, and 3) Scout orbit achievement accuracy.

Payloads per launch

From the outset, cost effectiveness and schedule considerations dictated that there be either 2 or 3 ITVs per launch vehicle. Designs for both options were carried until they were sufficiently mature to give reliable payload estimates. It was then evident that the 3-payload design was too limiting in orbit capability to justify the added cost associated with two separate designs.

Launch Site

The Scout launch vehicle has two launch sites within the Continental U.S.: VAFB and Wallops Flight Center (WFC) in Virginia. Table 3 summarizes the orbit capabilities (perigee and apogee altitudes) of the Scout for various orbit inclinations achievable from these two launch sites.

Table 3
SCOUT/ITV ORBIT FLEXIBILITY

Orbit Inclination deg	Launch Site	Payload Mass kg	Hydrazine Loading kg	Achievable Orbits km
35	WFC	184	12.25	To 600 Circular
37.7	WFC	184	12.25	250 x 2400
				400 x 1900
				600 x 1200
				740 Circular
55	WFC	184	12.25	To 360 Circular
60	WFC	184	12.25	To 400 Circular
80	VAFB	184	12.25	To 460 Circular
		174	7.75	To 520 Circular
98	VAFB	174	7.75	To 420 Circular

A due east launch from WFC (I = 37.7°) makes maximum use of the earth's rotation and hence can achieve the highest orbits. Inclinations from 37.7° to 41° and 50° to 53° require no plane change maneuvers. Inclinations below 37.7° (the latitude of the launch site) require dogleg maneuvers to change the orbit plane. A large plane change maneuver significantly reduces achievable orbit capability. Inclinations between 41° and 50° and above 53° require dogleg maneuvers during launch to satisfy range safety constraints. Inclinations between 77° and 145° are achievable with a launch from VAFB with no plane change required for range safety. However, the achievable orbit altitudes are low for near-polar orbit inclinations even with off-loading 4.5 kg (10 lb) of hydrazine. Retrograde orbits are not feasible for the two-payload design. They would be fine for launching a single ITV.

The high orbit altitudes achievable from WFC combined with the high average number of opportunities per day led to the selection of WFC as the primary launch site for ITV.

Orbit Accuracy

The Scout orbit achievement accuracies play a significant role in the orbit synthesis process. The principal error sources are the velocity and flight path angle errors at orbit injection*. The injection altitude error is smaller by comparison, but cannot be ignored. They affect primarily the perigee and apogee altitudes and the initial argument of perigee. Fig. 8[(3)] illustrates typical accuracies, based on total Scout flight experience, in terms of isoprobability contours of apogee-perigee deviations. The achieved Scout orbit has a 75% (95%) probability of being within the 75% (95%) isoprobability contour for the indicated nominal perigee/apogee combination. Single error sources of 40 m/s in injection velocity or 0.9° in flight path angle at injection yield orbits on the 75% isoprobability contour. The former is associated with a +160 km deviation in the altitude opposite to the injection altitude.

The magnitudes of these deviations have a significant impact on orbit selection. A large perigee deviation for a low-altitude near-circular orbit will significantly affect orbit lifetime. It will be seen later that they also have a significant impact on test scheduling when general test parameters, e.g. altitude, are tightly constrained.

Launch Window Duration

The Scout can be launched at a specified time and there is no minimum required launch window duration. However, a launch day slip due to minor weather or hardware delays can be avoided with the selection of a 1 to 2 hour window. It is imperative that the constraint, whatever it is, be clearly pre-defined for all contingencies so that the decisions are clear-cut on launch day if there are delays.

* *It is assumed throughout that nominal injection occurs at perigee.*

ESTABLISHMENT OF OFF-NOMINAL ORBITS

For the ITV program, it was decided to establish a set of off-nominal orbits whose mission performance was to be weighted equally with the nominal orbit within the orbit synthesis process. The off-nominal orbits were constrained, conservatively, to be on the 95% apogee-perigee isoprobability contours. The isoprobability contours used, however, were not the published ones[(3)] which pertained to the total Scout flight experience, but a separate set[(4)] generated to reflect the performance of the 18 Scout flights using the Algol III first stage, the configuration used for the ITV flights. The 95% isoprobability contours for that set are not very different from the 75% isoprobability contours of Reference 3. The five selected off-nominal orbits are described in Table 4.

Table 4
DESCRIPTION OF OFF-NOMINAL ORBITS

Acronym	Description
+SMA	Maximum semi-major axis;
-SMA	Minimum semi-major axis;
+FPA	Maximum shift in argument of perigee opposite to the direction of satellite motion, caused primarily by by a positive flight path angle at orbit injection;
-FPA	Reverse of +FPA;
+HP	Maximum perigee altitude.

The minimum perigee altitude category is generally close to either -SMA or +FPA and would be redundant. The +HP category gave performance similar to +SMA and was not always included in the analysis.

ALGORITHMS FOR OFF-NOMINAL ORBITS

The large number of candidate orbits examined made it necessary to create algorithms which would generate the off-nominal orbit parameters from specification of a nominal orbit. The parameters used were velocity, altitude and flight path angle, i.e., the three key injection error sources. Judgement was necessarily used in the generation of the algorithms since there could be several solutions to a given off-nominal orbit. Generally, it was desirable to have one major injection error source contributing to a given off-nominal. Also it was desirable that contributions from various error sources be consistent with the bivariate 95% apogee-perigee isoprobability contours. For normally distributed, independent error sources, the maximum contribution for a single error source occurs along an axis and represents approximately 2.45σ. Although the Scout injection errors are not independent, the following guideline was used:

> An acceptable algorithm should have the RSS (root sum squared) of the sigmas from the three injection error sources ≤ 2.45.

Standard deviations in the error sources (σ_V, σ_h, σ_γ) and isoprobability contours over a range of apogee and perigee altutudes were supplied in Ref. 5. The algorithms generated using this data are summarized in Table 5.

Table 5
ALGORITHMS FOR OFF-NOMINAL ORBITS

Off-Nominal Orbit	Contributions from each Parameter		
	ΔV	Δh	$\Delta\gamma$
+SMA	+2.20 σ_V	0.0	0.0
-SMA	-1.50 σ_V	-1.50 σ_h	-0.60 σ_γ
+FPA	+0.55 σ_V	-1.00 σ_h	-2.25 σ_γ
-FPA	+0.55 σ_V	-1.00 σ_h	-2.25 σ_γ
+HP	+0.75 σ_V	+1.65 σ_h	0.0

The algorithms gave orbits which matched the desired values on the isoprobability contours to within 2.5% in all cases except for the -SMA off-nominal for nearly circular nominal orbits. In those cases the errors were up to 10% too high. These differences were considered to be perfectly acceptable in light of the relatively small number of flights in the data base and the conservatism placed in the off-nominal orbit definitions. The algorithms were incorporated into the appropriate computer codes allowing the user to select an arbitrary isoprobability level, not just 95%.

THERMAL CONTROL CONSIDERATIONS

The thermal control subsystem on a satellite must keep the satellite components within established temperature bounds. The core package of the ITV is extremely well insulated. Its thermal time constant is approximately nine days. Thus, the long term temporal variation in the eclipse period is the major factor affecting the thermal control design.

Characteristics of Eclipse Periods

The eclipse period of a LEO satellite is a function of its altitude and the angle between the direction to the sun and the normal to the satellite orbit plane. For circular orbits, the equations are straightforward[2]. The percentage of time the satellite is in the shadow varies from a maximum of 40% at an altitude of about 200 km, to a minimum of 0% if the direction to the sun is sufficiently out of the orbit plane. For moderately elliptical LEO orbits, a good estimate of the eclipse period is obtained from the circular orbit equations using for altitude that of the satellite at the position on the orbit making the largest angle to the direction to the sun.

It is instructive to illustrate the temporal variation in the eclipse period for selected orbits. Two techniques are shown. Fig. 9 illustrates an eclipse period history for a moderately elliptical (280 x 1100 km) orbit inclined at 37.7° to the equator. The history was generated using a general-perturbation simulation with a 1-rev step size assuming a point source sun. Typically, low inclination orbits are "cool" with periods of 20 to 30 or more days with maximum shadow (near 40%) separated by short "warm" periods when the sun is at its maximum angle out of the orbit plane (occurs when the right ascension of the sun is 90° from the right ascension of the node of the orbit). The eclipse period is

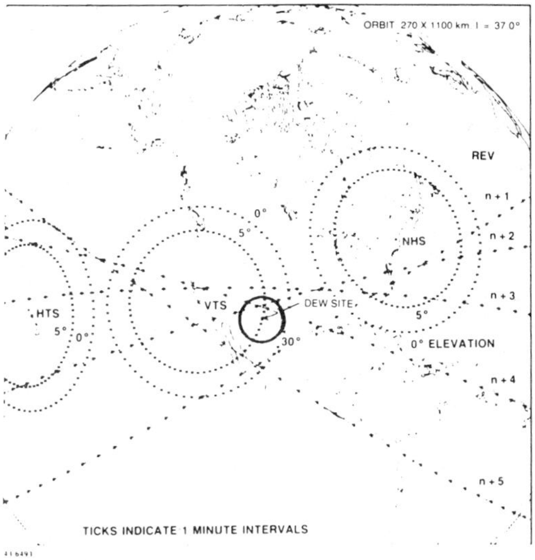

Fig. 5 Direct SGLS Communications are Marginal for Low Altitude DEW Tests

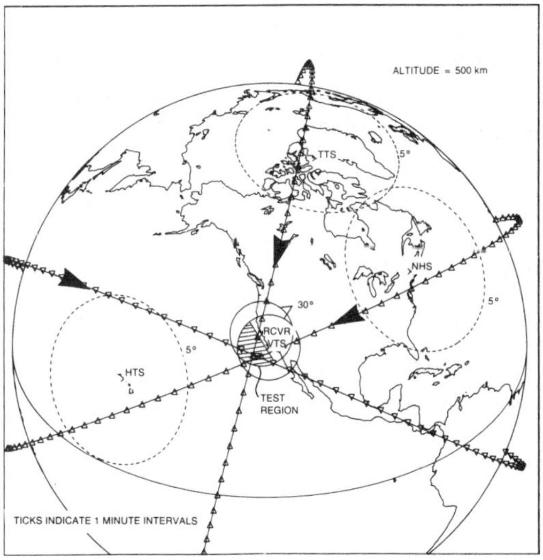

Fig. 6 Go/Abort Site Places Constraints on Orbit Inclination

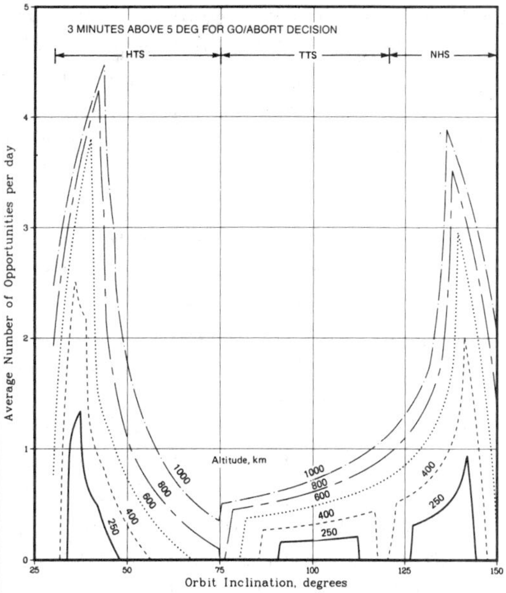

Fig. 7 Test Opportunity Frequency Variation with Go/Abort Site

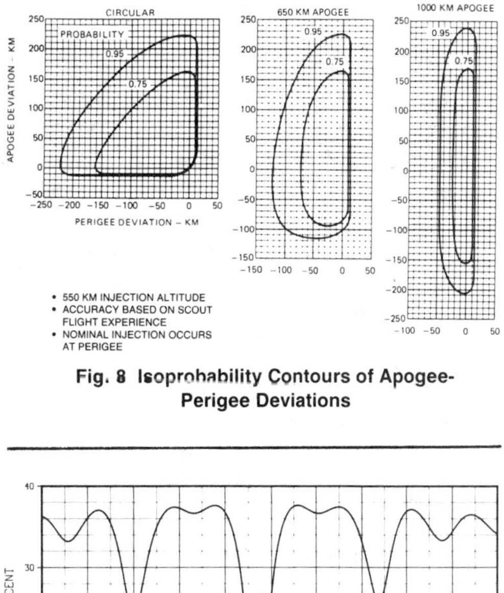

Fig. 8 Isoprobability Contours of Apogee-Perigee Deviations

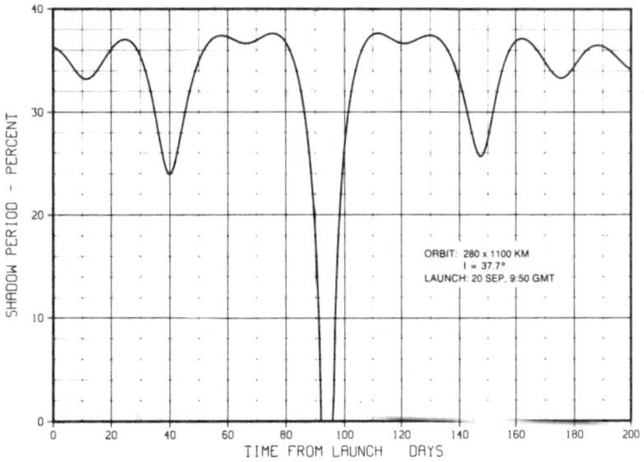

Fig. 9 Shadow History for Hot Transient Orbit

minimized when this geometry occurs at a solstice, and the argument of perigee is either $+90°$ (winter) or $-90°$ (summer) placing apogee in the same hemisphere as the sun. The specific geometry is illustrated in Fig. 10. The launch date and time of Fig. 9 were selected specifically to give this geometry at the winter solstice. It yields approximately 4 days of continuous sun on the orbit.

Fig. 11 illustrates a typical eclipse period variation for a high inclination (non sun-synchronous) orbit. It delineates the latitude excursions of the satellite over which it is in the earth's shadow, distinquishes between ascending (S-N) and descending (N-S) azimuths, and indicates the eclipse length in minutes. This form has proven useful for early planning of observations or tests where solar illumination of the satellite is important. Algorithms assuming a point source sun and spherical earth geometric figure were used to generate the information presented for this 500 km, 77° inclined circular orbit. Generally, high inclination orbits oscillate between cool periods of high percent shadow and warm periods of many days of continuous sun. A sun-synchronous orbit inclination (between 96° and 100° for a LEO satellite with an average altitude below 1000 km) matches the nodal regression to the apparent motion of the sun (due to the earth's orbital revolution) and yields essentially constant (except for seasonal effects) sun geometry. The launch time controls whether the orbit is cool (high percent shadow) or warm (continuous sun).

Fig. 12 gives a summary of the maximum period of continuous sun as a function of orbit inclination and circular orbit altitude. For a specified inclination and altitude, algorithms make an initial estimate for the maximum continuous full sun period and converge to the solution using successive approximations. Key assumptions are a spherical earth figure and circular earth orbit about the sun. The results give good estimates for moderately elliptical orbits where the altitude used is that at apogee.

ITV Thermal Control Design

The ITV has different thermal properties and associated control in its packaged and deployed states. In the deployed state, the Kevlar cloth shell has properties dictated by signature requirements. These properties lead to acceptable ITV core temperature levels in a continuous sun environment, but require heating in a high percent shadow condition. The vehicle is designed to periodically monitor temperature conditions and fire the heater system for a short period if required to maintain a minimum temperature.

Thermal control in the packaged state is accomplished passively through the use of bands of black paint. The design had to accommodate a "cool" orbit of approximately 40% shadow: the Scout injection error characteristics were such that no matter what nominal orbit was selected (including a warm sun-synchronous orbit) there was a reasonable probability the vehicle would experience a cool orbit period within its design lifetime of 6 months. Required conservative margins on analysis calculations indicated a maximum acceptable continuous sun period of about 3 days for the design just meeting the "cool" orbit environment*. With the earlier elimination of consideration of launches from VAFB, this constraint was not serious. However, analysis had to be performed on potential Wallops-launched high-apogee orbits to determine what launch date/launch time combinations gave transient warm conditions.

ORBIT LIFETIME CONSIDERATIONS

The orbit lifetime of target vehicles plays an important role in the selection of apogee altitude. (Perigee altitude is generally dictated by general test requirements.) Fig. 13, generated using standard algorithms[2,5] and atmosphere models[6], shows typical ITV orbit lifetimes for a low solar activity period (mid-1980s). The lifetime of a deployed ITV in a 300 km circular orbit is less than 10 days. The lifetime is increased to 300 days by raising apogee to 1100 km. The corresponding lifetimes are a factor of 3 to 4 lower when perigee is reduced by 50 km. The lifetime in the packaged state is (see scale at right) about 10 times that in the deployed state. For low altitude testing, it is obviously desirable to increase apogee and delay deployment until just prior to the testing period.

Solar activity will be significantly higher in the 1989 to 1992 time period. The atmospheric density at a given altitude will increase (and expected orbit lifetime correspondingly decrease) by a factor of 2 to 5. Orbit lifetime (and correspondingly, degraded ephemeris prediction accuracy) will have an increased impact on the orbit synthesis process.

GROUND TRACE REPEATABILITY

It is generally desirable to avoid orbits with repeating ground traces. This is almost mandatory for high inclination orbits or low altitude tests. Even when there are a number of test opportunities per day satisfying general test constraints, particular longitudes of the ascending node may violate detailed test constraints. Ref. 7 gives a very complete discussion of the subsatellite trace repetition parameter and associated algorithms. Fig. 14 summarizes the average-altitude/inclination combinations giving ground-trace repeats in 1 through 4 days. Indicated on the plot is the selection of an average altitude for a target orbit, at a 37.7° inclination, which is farthest from a daily repeating ground trace. The +SMA and -SMA off-nominals associated with the 95% Scout isoprobability contour are also illustrated. A semi-major axis of 650 km, or slightly higher to allow for semi-major axis decay due to drag effects in the deployed state, is appropriate for a WFC launch.

---

* *ITV on-orbit experience has proved the packaged ITV to be operating 3K to 6K cooler than anticipated. It is still well above lower temperature limits for cool, 38% shadow, conditions. Analysis shows that it will be well below upper temperature limits for a warm, continuous-sun orbit. Thus, the ITV passive thermal control design is essentially optimal, i.e. it works for all low earth orbits.*

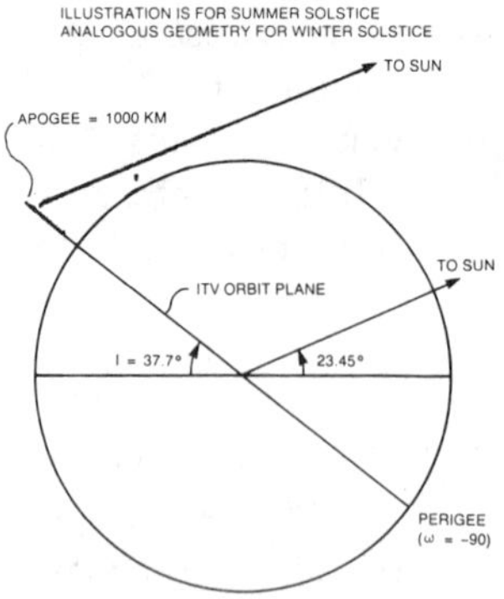

Fig. 10 Geometry for Continuous Sun (No Eclipse)

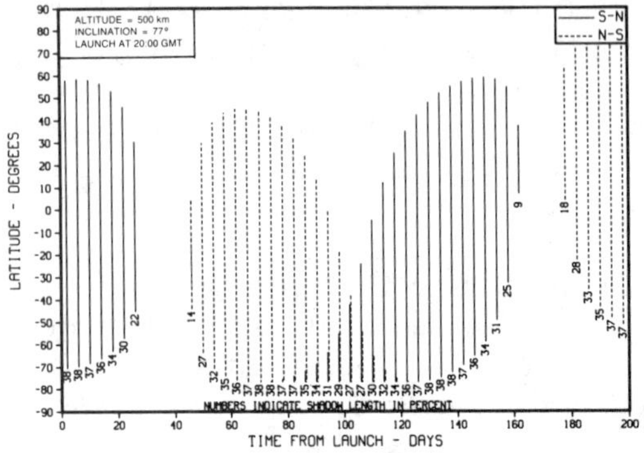

Fig. 11 Typical Eclipse Periods for a High Inclination Orbit

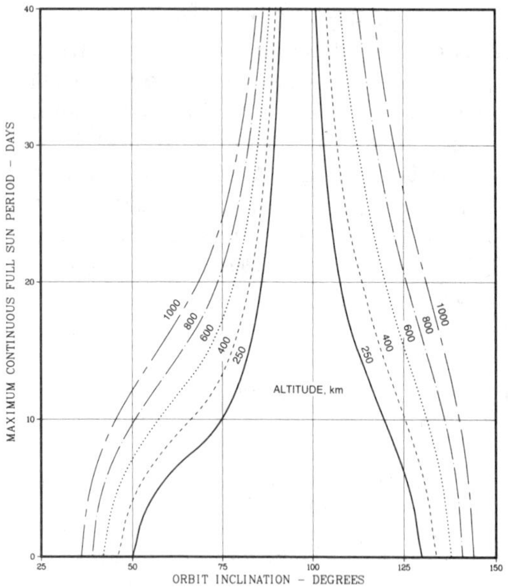

Fig. 12 Variation of Full Sun Period with Orbit Parameters

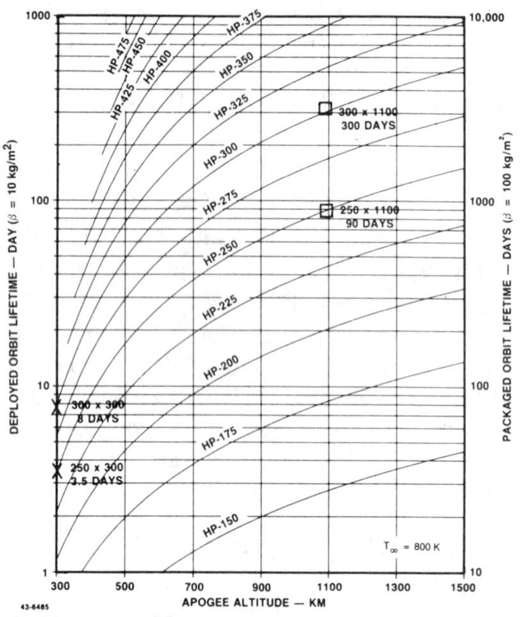

Fig. 13 Typical ITV Orbit Lifetimes, Mid 1980'S

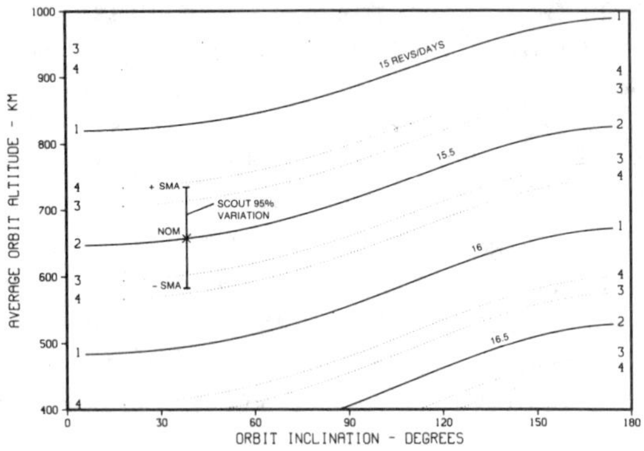

Fig. 14 Repeat Cycle of Ground Traces

ALTITUDE AND ILLUMINATION CYCLES

Low altitude orbits experience altitude and illumination cycles caused primarily by the nonspherical gravitational attraction (J_2) of the earth, i.e., perigee rotation and nodal regression effects mentioned earlier. The earth's revolution about the sun is also a factor in the illumination cycle. The cycles vary primarily with orbit inclination and secondarily with orbit size and shape. The periods of these cycles are illustrated in Fig. 15 for an orbit altitude of 600 km. An orbit inclination, I, may be selected on the basis of constant illumination conditions (sun-synchronous, I = 98°), no apsidal rotation (I = 63.6° or 116.4°), or equal altitude and illumination periods (I = 42°, 76.5°, 110°, or 128.5°). If there are overriding reasons for selecting other inclinations, the altitude and illumination cycles will affect inclination optimization, launch time selection, and test scheduling. For example, a due east launch from WFC (I = 37.7°) gives an altitude cycle of 45 days and an illumination cycle of 52 days.

The initial location of a satellite in the illumination cycle is controlled by the launch time of day. The location in the altitude cycle is controlled by the launch to orbit injection sequence. This is usually not flexible for a direct-ascent expendable launch vehicle, e.g. Scout, where orbit injection coincides with perigee of the orbit. Some control can be exercised if there are two alternative launch azimuths available to achieve the desired orbit inclination. The situation does not occur for the large majority of inclinations achievable with launches from WFC or VAFB.

GENERAL TEST CONSTRAINTS

A general test variable is a non-test-arena parameter satisfying the following two conditions:

a) It has an important influence on the selected orbit parameters (Refer to Table 1); and

b) It is test specific, i.e., it may change from test to test.

It is the nature of planners running a test program to generate an overall test matrix with tightly constrained parameters for each test. A discussion of this process in beyond the scope of this paper. Rather, the intent is to illustrate the effects of selected constraints on the synthesis process. For simplicity, the examples will be restricted to those where the general test constraints are the same for both target vehicles launched together on a single booster. Table 6 lists candidate general target-related test parameters:

Table 6
GENERAL TEST PARAMETERS

Altitude Band	Sensor-Target-Sun Aspect Angle
Target Illumination	Ground-Sensor Illumination
Sunlit	Daytime
Eclipsed	Twilight
Proximity to terminator	Night

Altitude is related to weapon reach capability and performance. The illumination parameters are related to constraints imposed by, and performance sensitivities of the sensor(s) associated with, the weapon system.

Test schedule has not been included as a general test parameter. It is treated as a measure of orbit performance. Test schedules should meet the following general criteria:

a) There should be opportunities for multiple retests to allow for possible aborts or delays;

b) Schedules should be predictable to allow for effective planning and use of range resources.

DETAILED TEST CONSTRAINTS

Detailed test constraints are associated with detailed weapon system test requirements which serve as a final screen on a potential test opportunity. These may be associated with: detailed range safety constraints of a direct-ascent-based KEW; mutual inclination between vehicles; celestial background constraints associated with the moon, planets, or stars; earth limb angle constraints; constraints on visible magnitude of the target; gimbal limits on DEW associated sensors constraining the elevation at culmination; weather; atmospheric turbulence, etc. These constraints, in general, would not influence the orbit synthesis process. However, they could be switched to the general category if warranted. The most likely impact would be constraints on launch date to avoid a) unfavorable celestial backgrounds or b) periods of statistically likely poor weather conditions. These detailed constraints will not be addressed in the orbit synthesis discussions.

METHODOLOGYS FOR ASSESSING ORBIT PERFORMANCE

Two separate and distinct methodologies are used for assessing the orbit performance. The first is a general perturbation simulation technique which determines and displays, for a specified orbit, the characteristics of all test opportunities satisfying the general test constraints. The second is an algorithmic procedure used to generate and display the schedules of opportunities satisfying the general test constraints.

Simulation of Test Opportunities

A general perturbation method with a one rev. step size is used to propagate the ephemeris. Brouwer Theory[8] is used to model the perturbing effects of the zonal harmonic coefficients J_2 through J_5. The altitude decay model uses algorithms from References 2 and 5. Although several atmosphere models are available for use, a Harris-Priester model[8] with the appropriate solar activity level is generally used. A data set containing the characteristics of all opportunities satisfying the test arena constraints is created by the simulation code and passed to a "constraints" code. This code applies the appropriate general test constraints and produces graphical output describing the remaining acceptable test opportunities. Figures 16 and 17 illustrate typical output of the simulation procedure. They pertain to actual Signature Confirmation Measurement (SCM) opportunities for the ITV (#2) which was deployed in December 1986 and tested over the MOTIF observation site on Mt. Haleakela, Maui. Figure 16 includes all opportunities which pass over MOTIF exceeding 30° elevation for at least 2 minutes. The 45 day altitude (perigee rotation) cycle, 52 day illumination (sun-referenced nodal regression) cycle and altitude decay effects are very apparent in this figure. Fig. 17 illustrates the application of 2 constraints appropriate for acquisition of the ITV in the visible waveband: a) the ITV must be sunlit, and b) it must pass over the site between 30 minutes after sunset and 30 minutes before sunrise. Fig. 18 interprets the symbols.

Algorithmic Code For Schedules

Algorithms have been developed and coded in a computational/graphical program called CYCLES to compute intervals (boundaries and length in days) of satellite test (or observation) windows satisfying specified constraints. The constraints currently include four categories:

1. Minimum and maximum altitude;
2. Constraints relative to satellite terminator crossings;
3. Constraints on time of day (relative to sunrise, sunset, or midnight) at the test site; and
4. Constraints on the angle, termed the coarse sun angle, between the satellite's negative velocity vector and the line-of-site to the sun.

The code does not compute the likely number of acceptable passes per day. This is done in a separate algorithmic code described earlier.

The program concept is relatively simple. The test windows satisfying a particular constraint comprise a set of intervals of consecutive days. Window sets are computed separately for each combination of constraint, pass type, and test arena latitude. The pass type may be ascending (south to north heading), or descending (north to south heading). The test arena can be specified in two ways: 1) latitude of the center of the arena, or 2) latitudes of the southern and northern extents of the arena. Each of the up to 16 combinations (4x2x2) of constraints, pass types, and arena latitudes result in test window sets encompassing the time period of interest. The final set of test windows is obtained by taking the appropriate unions and intersections of those sets associated with the various combinations.

The earth figure is assumed to be spherical and non-rotating and the sun a point source. The sun's ephemeris is computed from algorithms[9] which are accurate to about 1 minute of arc. The dynamics of the orbit are modelled. The location of the satellite within the orbit is not. The nodal regression, perigee rotation, and altitude decay perturbations are included in the algorithmic model. The equations are designed to estimate the time of occurrence of a given constraint and converge to a solution using successive approximations. Two or three iterations are generally sufficient to give an accuracy of 0.1

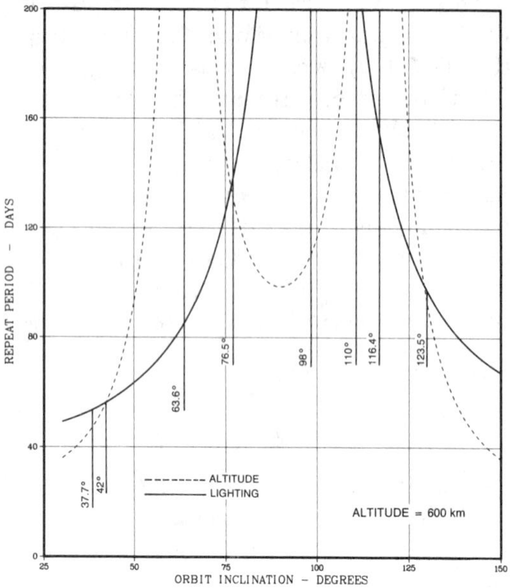

Fig. 15 Altitude and Illumination Repeat Periods

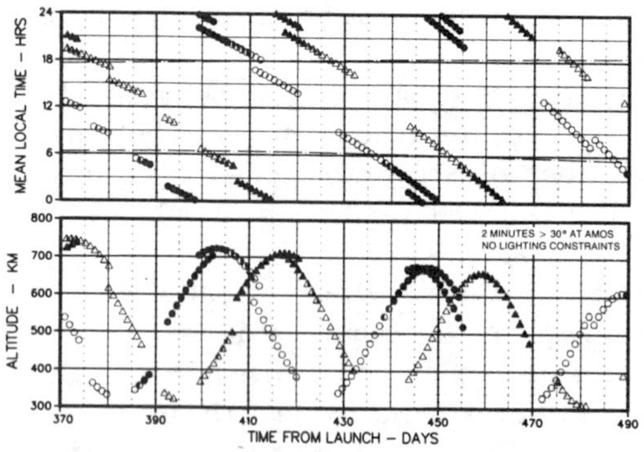

Fig. 16 SCM Opportunities for ITV #2

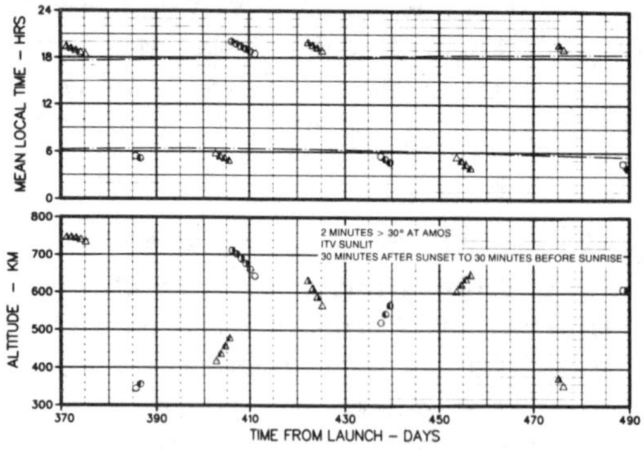

Fig. 17 Twilight SCM Opportunities for ITV #2

SYMBOL	SATELLITE ILLUMINATION CONDITIONS
△ ○	TEST CONTACT ENTIRELY IN SUNLIGHT.
▲ ●	TEST CONTACT ENTIRELY IN DARKNESS.
◭ ◐	TEST CONTACT BEGINNING IN DARKNESS AND TERMINATING IN SUNLIGHT.
◮ ◑	TEST CONTACT BEGINNING IN SUNLIGHT AND TERMINATING IN DARKNESS

TRIANGLES REPRESENT DESCENDING PASSES
CIRCLES REPRESENT ASCENDING PASSES

Fig. 18 Plotting Symbols for Solar Illumination Conditions

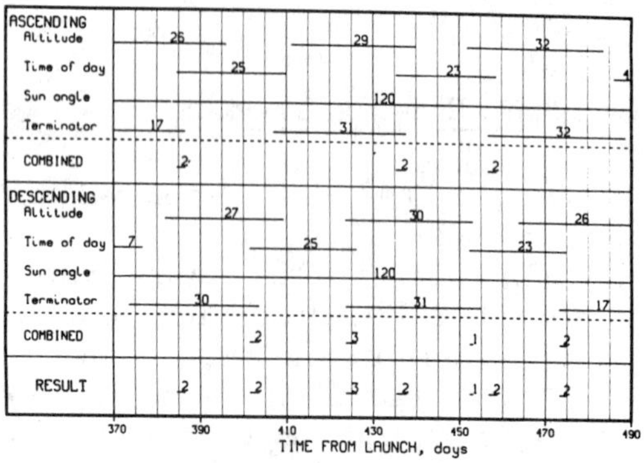

Fig. 19 AMOS Twilight SCM Schedules for ITV #2; Altitude <600 Km; 30 Minutes After Sunrise or Before Sunset

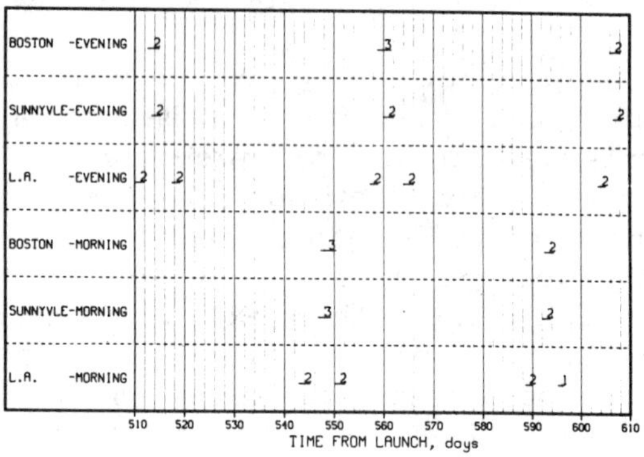

Fig. 20 Periods for Visible Sightings of Object 2867 Satellite Sunlit; 45m Before Sunrise or After Sunset

days. Computationally, CYCLES can process 100 days (2 to 3 altitude and illumination cycles) of a low inclination orbit for 0.1 cpu seconds on an IBM 3033.

Fig. 19 is the explanatory CYCLES version of Fig. 17 with an added 600 km constraint on altitude. Note the excellent agreement except for the ascending passes at day 460 (Fig. 19) where resonance (daily ground-trace repeat) causes the ascending passes of the ITV to skip-over the MOTIF region for approximately 15 consecutive days (see Fig. 16). Test cases comparing CYCLES output to simulation results have shown excellent agreement. Any differences were attributable to simulated opportunities which marginally met (or violated) one or, frequently, more than one of the constraints. Fig. 20 illustrates a more practical application of the CYCLES code. It indicates periods when interested individuals in Los Angeles, Sunnyvale, and Boston might possibly observe the deployed ITV with the naked eye or the aid of binoculars. CYCLES identifies the potential observation periods for planning purposes. Shortly before the period updated ephemerides can be input into a simulation code to determine the observation details (i.e. histories of visible magnitude, azimuth/elevation and right ascension/declination). *

ORBIT SYNTHESIS EXAMPLE

The orbit synthesis process is illustrated for a postulated direct-ascent KEW mission. The constraints discussed earlier are all applicable. The primary test constraints are:

a) Test altitude at 400 ± 40 km;
b) Target is sunlit;
c) KEW is launched between sunrise and sunset (for photography);
d) Coarse sun angle ≧ 30°;
e) A variety of ascending and descending passes are to be included as potential test opportunities; and
f) There should be 3 consecutive test windows, each with minimum/desired lengths of 10/15 days. The two ITV targets are to be tested in separate windows. This allows for a global schedule slip of one test window, and local schedule shifts of 1 to 2 weeks within a test window.

Note that constraint b) is a necessary condition for constraint c).

Test Opportunity Influence on Orbit Inclination

Constraint e), in conjunction with the low altitude test constraint a), has an interesting impact on the orbit inclination. Fig. 21 illustrates the geometry for an orbit inclination of 37.7° (associated with a due east launch from WFC). The descending passes over VTS (VAFB) pass north of the HTS 5° elevation region and hence violate the GO/ABORT constraint. Fig. 22 gives the variation in the total number of ascending and descending opportunities with altitude

* *The deployed ITV was observed, without the aid of binoculars, from both Los Angeles and Sunnyvale in early January, 1987.*

and orbit inclination. Fig. 23 illustrates that the passes being eliminated at the higher inclinations and lower altitudes are indeed the descending ones passing through the central and southern portions of the test arena. Thus, constraint e) suggests that the orbit inclination should be less than approximately 37° to ensure that descending passes are included as potential test opportunities.

Test Window Concept

A test window is defined as a period of consecutive days when the orbit geometry is conducive to yielding test opportunities satisfying the general test constraints. Fig. 24^{+} illustrates the concept for an elliptical nominal orbit with perigee in the middle of the desired test altitude band.

The effects of perigee rotation and nodal regression on altitude and illumination conditions over the test arena are graphed. The solid line indicates when the vehicle is in the sunlight with the coarse sun angle constraint being satisfied. Fig. 24 pertains to ascending passes; the corresponding curve for the descending passes is shifted to the right by about 3 days for the 37.0° orbit inclination. The nominal orbit, and the four 95% isoprobability off-nominal orbits illustrated, each pass through the test altitude band over the test arena for periods of approximately 10 to 12 days spaced approximately 45 days apart. The launch time of 0700 GMT was selected to center the daylight conditions at the second test window (day 80). With this selection, windows 1 and 3 are nearly within the acceptable daylight illumination conditions for most orbits. The biggest violators are the +FPA off-nominal for window 1 and the -FPA off-nominal for window 3. The test schedules, i.e. collection of test windows, for this nominal and associated off-nominal orbits, are summarized in the 2nd subframe (from the top) of Fig. 25.

Selection of Apogee Altitude

Fig. 25 also shows the effect of apogee on the test schedules. The higher apogees give slightly shorter test window lengths on average, but less schedule variation for the off-nominals. The low apogee (450 km) gives very long window lengths for the nominal orbit and +SMA (95%) off-nominals. However, the +FPA (95%) off-nominals give significant variations in schedule and short (8 day at best) window lengths. Fig. 26 illustrates the altitude history for the ascending passes. (The descending passes, not shown, are shifted 3 days to the right in altitude and sunlight/shadow illumination; the sun angles are somewhat different). The perigee/apogee deviations for the +FPA off-nominals result in the altitude over the arena traversing the test altitude band (due to perigee rotation) in 5 days. When the constraints are satisfied, the descending and ascending passes combine to give an 8 day window. A decision on apogee altitude based solely on test schedules is highly subjective. It depends upon the weighting given to window length, schedule consistency, and off-nominal orbits of various

---

$^{+}$ *The altitude history and test schedule plots in this section do not model the altitude decay perturbation since the target deployment time is not fixed.*

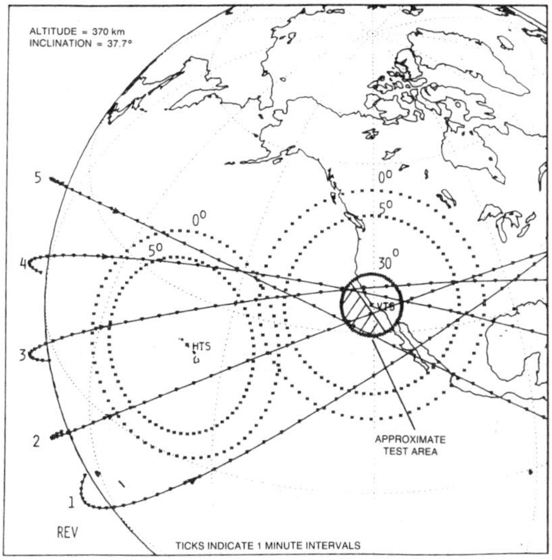

Fig. 21 Descending Opportunities Pass North of HTS 5° Elevation Region

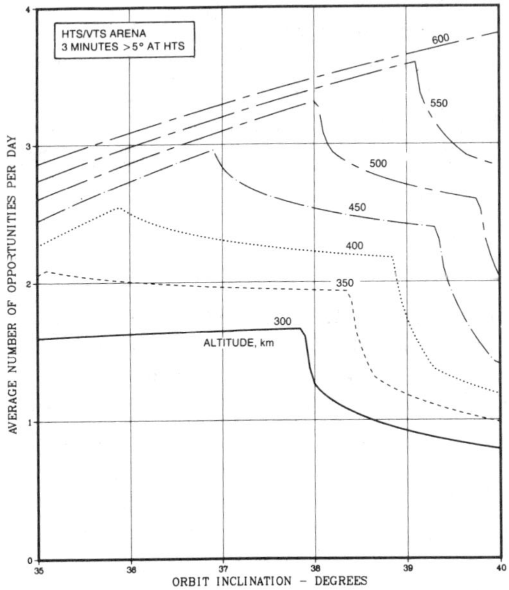

Fig. 22 Frequency of All Ascending and Descending Passes

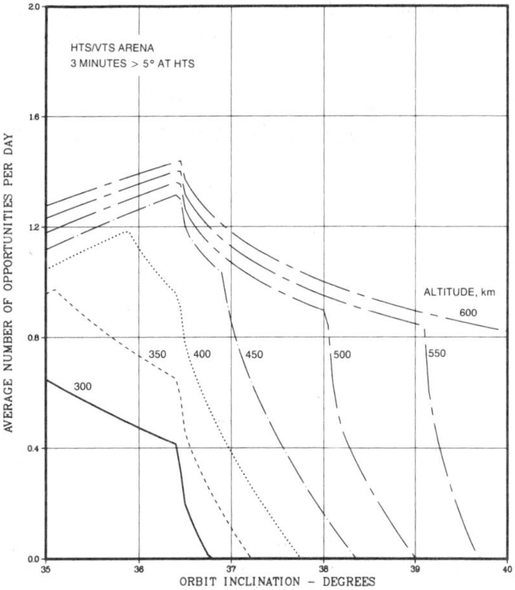

Fig. 23 Frequency of Descending (N/S) Passes South of 36.5° Latitude

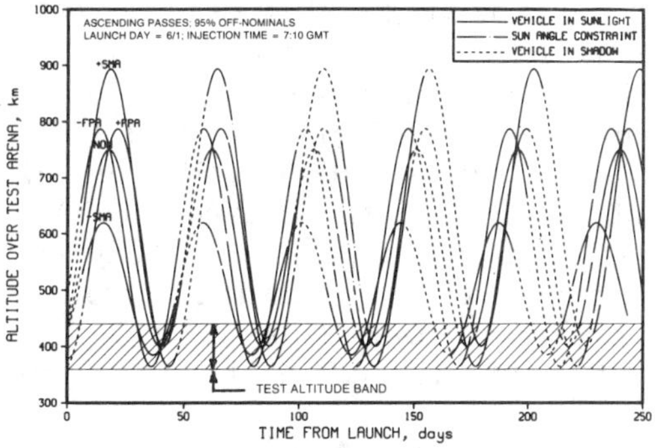

Fig. 24 Altitude History Over Arena for 400 x 750 km, I = 37.0°

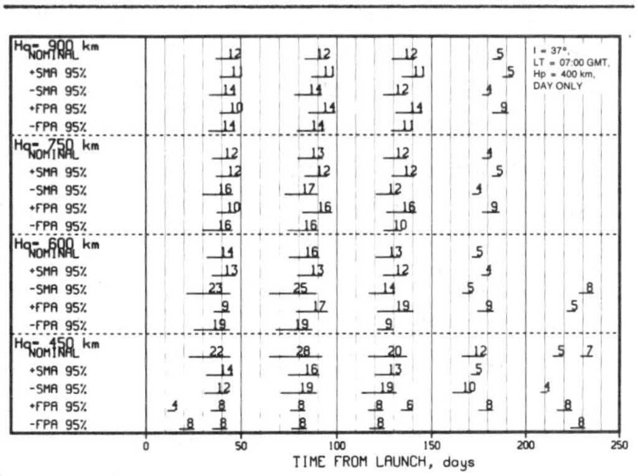

Fig. 25 Effect of Apogee Altitude on Test Schedule

isoprobability levels. However, there are two other factors which could drive the decision: a) orbit lifetime and b) ground-trace repeat.

The deployed orbit lifetimes associated with the nominals and 95% off-nominals for three apogee altitudes are summarized in Table 7. The deployed operating lifetime requirement is generally about two months to allow for potential use over two test windows. Additional orbit lifetime may be required for ephemeris-prediction accuracy requirements, which are inversely related to perigee altitude and orbit lifetime.

Table 7
COMPARISON OF ORBIT LIFETIME

CATEGORY	H_a = 450 km		H_a = 600 km		H_a = 750 km	
	$H_p \times H_a$ km x km	Lifetime days	$H_p \times H_a$ km x km	Lifetime days	$H_p \times H_a$ km x km	Lifetime days
NOM	400 x 450	150	400 x 600	500	400 x 750	900
+SMA	400 x 585	400	400 x 740	900	400 x 895	1000
-SMA	315 x 400	30	380 x 480	150	385 x 620	400
+FPA	320 x 535	75	350 x 650	250	365 x 785	600

T_∞ = 800 K, β_d = 10 kg/m^2 (See Fig. 13)

Even higher apogee altitudes are required to avoid the possibility of a daily repeat of the ground trace. Referring to Fig. 14 for an orbit inclination of 37.7°, an average orbit altitude a) exceeding 500 km is required to keep Q below 16 revs/day and b) equal to 650 km yields Q = 15.5 revs/day. For a perigee altitude of 400 km (ignoring altitude decay effects), criteria a) implies that the apogee altitude exceed 750 km and criteria b) that it equal 900 km. The nominal initial apogee altitude should be higher to account for decay due to atmospheric drag. For the remaining discussions, a nominal apogee altitude of 750 km is used. This choice is, of course, inconsistent with avoidance of a daily repeating ground trace.

Selection of Orbit Inclination

Previous discussion indicated that an orbit inclination of 37° or less was required to ensure that descending passes met the GO/ABORT criteria. Fig. 27 summarizes the test schedules as a function of orbit inclination for the associated near optimum launch time. Compared to $I = 37°$, an inclination of 36° gives slightly shorter test window lengths and loses the 4th window almost completely because of the shorter altitude and illumination cycles and larger mismatch between the two cycle lengths. For the higher inclinations, the smaller mismatch between cycle lengths improves the 4th and even 5th windows. However, the longer cycle lengths are not sufficient to overcome the loss of the descending passes and the window lengths do not increase.

The improvement is significant for higher orbit inclinations with the elimination of the GO/ABORT constraint (Fig. 28). An orbit inclination of 37° is selected for comparison of the other orbit parameters.

Selection of Perigee Altitude

For low test altitudes, the perigee altitude should be no lower than the desired test altitude in order to maximize orbit lifetime and ephemeris prediction accuracy and minimize the effect of orbit injection errors. Fig. 29 illustrates that the test schedules are not terribly sensitive to small changes in the nominal perigee altitude. A decrease in perigee altitude towards the lower end of the test altitude band slightly increases the window lengths for the nominal and ±SMA off-nominals. However, it causes perigee of the ±FPA off-nominals to drop below the altitude band and slightly decrease the window lengths. A perigee altitude of 395 km would keep perigee of the 95% off-nominals within the test altitude constraint band.

Selection of the Launch Window

The launch window is defined as the interval of acceptable launch times. Earlier discussion indicated that there is no minimum requirement on launch window duration. However, it is important to have it firmly defined. It is necessary to define the opening of the window, the primary window length, and criteria for extending the window real-time. Fig. 30 summarizes the variation in the test schedules with launch time. Possible launch windows and criteria are:

a) 0630 to 0730 GMT: Meets minimum criteria;
b) 0600 to 0800 GMT: Meets minimum criteria with exception of one off-nominal;
c) 0700 to 0800 GMT: Same as b) but for third window only.

Each of the above could include a real-time option to extend the window to 0900 or later, thus losing some third window capability, dependent upon weather predictions and schedule limitations.

Effect of Launch Date

Seasonal illumination variations generally have a small effect on the target orbit synthesis process. The optimum launch time may vary by an hour, and the daylight constrained test windows will be longer in summer and shorter in winter. (The launch date used for the examples in this paper was 1 June). It is possible, however for tight constraints on certain illumination parameters, e.g. coarse sun angle, to induce constraints on launch date.

Validation via Simulation

The orbit synthesized by algorithmic procedures is generally analyzed in detail to validate the process. The simulation/constraint codes described above are used to identify the potential test opportunities for selected nominal and off-nominal orbits. Fig. 31 illustrates typical output for a selected test window; the opportunities are consistent with test frequency and schedule results developed above. Each opportunity is then fully screened to determine the viability of the orbit and/or to identify changes to the general test constraints for an additional iteration on the synthesis process.

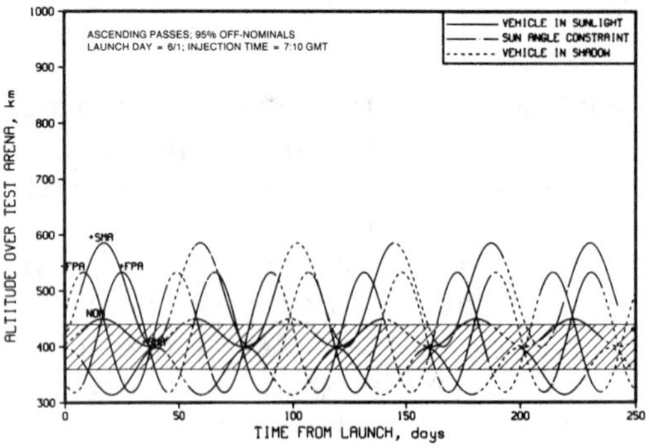

Fig. 26 Altitude History Over Arena for 400 X 450 km, I = 37.0°

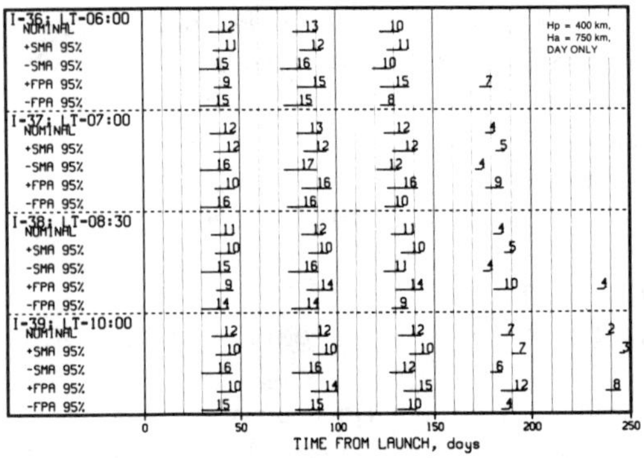

Fig. 27 Effect of Orbit Inclination on Test Schedules, With Go/Abort Constraint

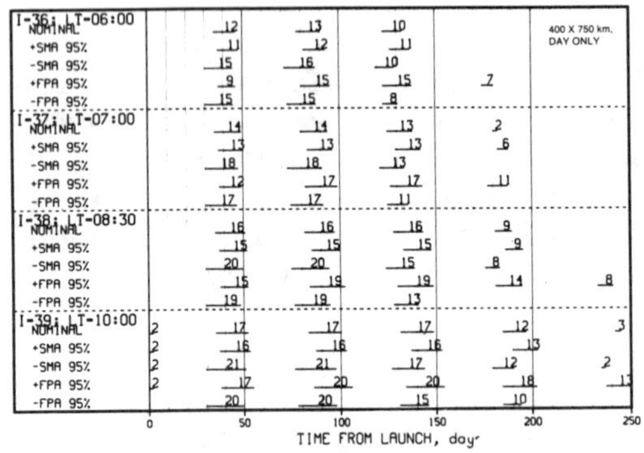

Fig. 28 Effect of Orbit Inclination on Test Schedules, No Go/Abort Constraint

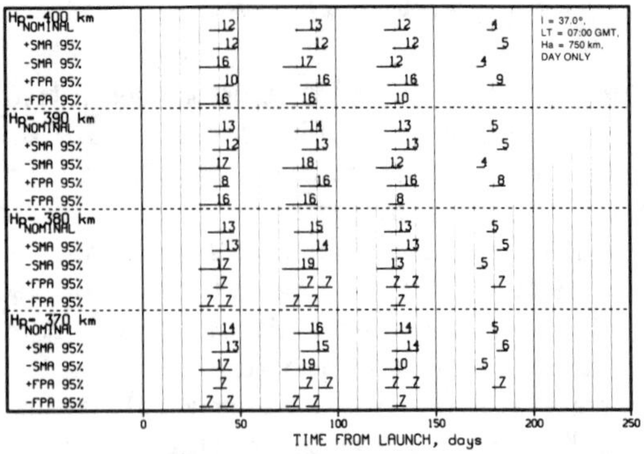

Fig. 29 Effect of Perigee Altitude on Test Schedules

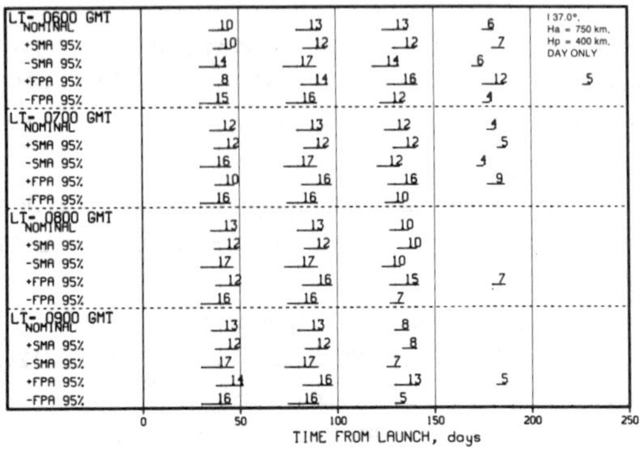

Fig. 30 Effect of Launch Time on Test Schedule

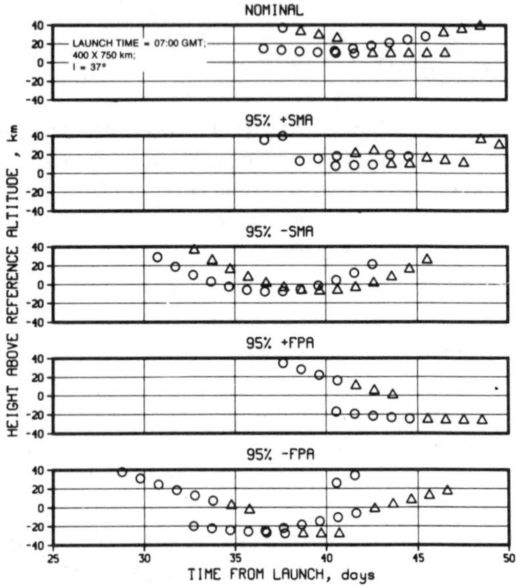

Fig. 31 Opportunities Satisfying General Test Constraints

SUMMARY

The orbit synthesis process has been illustrated for a hypothetical test of a direct-ascent-based KEW against an ITV. Algorithmic techniques were introduced and used to synthesize orbit parameters which met general test constraints, typical of a test program, in consonance with test arena geometry, target design limitations and target launch-vehicle accuracy capability. Competing parameters and associated trade-offs were identified. Test arena, GO/ABORT coverage, test opportunity frequency, and schedule (number of test windows and window length) requirements compete in the selection of orbit inclination. For low test altitudes, considerations of launch vehicle accuracy, orbit lifetime, and ephemeris prediction accuracy demand that the perigee altitude be essentially equal to the desired test altitude. They also influence apogee altitude selection as does ground trace pattern characteristics and test window length. Finally, the launch window (interval of acceptable launch times) is dependent on the test illumination constraints.

ACKNOWLEDGEMENT

The author would like to thank Mr. Stan Kaminski of the Aerospace Corporation for his contributions to the orbit synthesis process during the early phases of the ITV contract, Capt. Jeff Lindsey of the U. S. Air Force Space Division and Mr. William O'Donnell of Textron Defense Systems for their helpful guidance, and to Mr. John Traxler of the Aerospace Corporation for his helpful suggestions on this manuscript and long-standing support.

REFERENCES

1. R. H. Battin, *Astronautical Guidance*, McGraw-Hill, Inc. 1964
2. *Flight Performance Handbook for Orbital Operations*, John Wiley and Sons, Inc., Revised and Corrected Printing, 1963
3. *Scout User's Manual*, Vought Corp., Dallas, Texas, 1 January 1980
4. "Scout Orbital Accuracy Data for ITV", Vought Internal Release 23-DIR-2381, Vought Corp., Dallas, Texas, 29 July 1983
5. D. G. King-Hele, "Methods for Predicting Satellite Orbit Lifetimes", Royal Aircraft Establishment, TR-77111, 28 July 1977
6. *U.S. Standard Atmosphere Supplements*, 1966, ESSA/NASA/USAF, U.S. Government Printing Office
7. H. K. Karrenberg, E. Levin, and R. D. Luders, "Orbit Synthesis", The Journal of the Astronautical Sciences, Vol. XVII, No. 3, Nov.-Dec. 1969, pp. 129-177
8. *Mathematical Theory of the Goddard Trajectory Determination System*, Goddard Space Flight Center, NASA-TM-X-71106, April 1976
9. *Constants and Related Information for Astrodynamic Calculations*, 1968, Jet Propulsion Laboratory, NASA-TR-32-1306, 15 July 1968

AAS 89-411

ORBIT CONTROL STRATEGY FOR A CONSTELLATION OF THREE SATELLITES IN TUNDRA ORBITS

G. Rondinelli*, A. Cramarossa* and F. Graziani†

A constellation composed by three satellites placed in highly inclined geosynchronous eccentric orbits (TUNDRA orbit) suitably phased and constantly spaced can provide different services to cover high-latitude areas. Satellites on such orbits will spend most of the time around apogee.

In order to satisfy the mission requirements and to continuously provide the selected services at high latitude, it is necessary to maintain the constellation geometry.

Due to the perturbation effect the nominal constellation geometry and phasing will degrade; hence, an orbit control strategy has to be implemented in order to maintain the constellation architecture.

This paper deals with a low-cost orbit-control strategy addressed only to control the orbit subtrack (latitude and longitude of the operational loop of the orbit) and to control the apogee passage time-shift, which impacts on the synchronization between the satellites.

This control strategy exclusively based on in-plane maneuvers, consists of adjusting the apogee radius by perigee burns obtaining a semiaxis variation which cancels the total subtrack shift or puts the subtrack on the other side with respect to the nominal position.

Keywords: High inclined orbit, critically inclined orbits, geosynchronous orbits, station-keeping strategies, in-plane maneuvers, TUNDRA orbit.

INTRODUCTION

A constellation composed of three satellites placed in highly inclined geosynchronous eccentric orbits (TUNDRA orbit) suitably phased and constantly spaced can provide continuous communications service over a large area close to the subtrack longitude including high-latitude zones and the North Pole (Fig. 1).

Other services can also be supported by the same constellation: paging, mailbox, fixed or mobile service communications and search and rescue.

* ITALSPAZIO, Via V.E. Orlando, 83 - 00185 Roma, Italy.

† Scuola di Ingegneria Aerospaziale - Universita' "La Sapienza", Via Eudossiana, 18 - 00184 Roma, Italy.

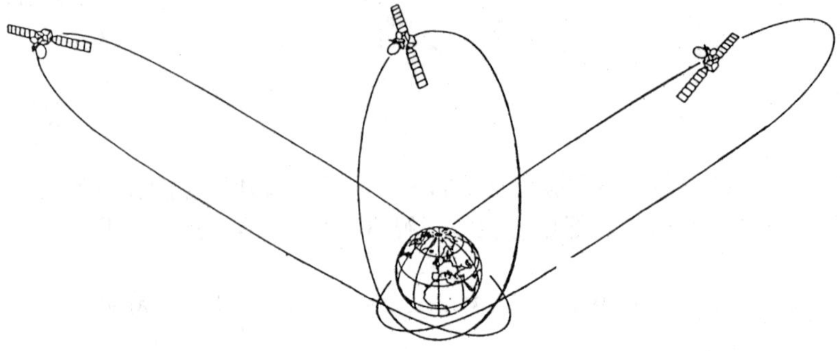

Figure 1 Three Satellite Constellation in TUNDRA Orbits

The time spent by each satellite in the operational loop is about eight hours and every eight hours one satellite is at apogee, so at least one satellite is always visible from selected areas.

When the first satellite is at perigee, the second is entering the operational loop around apogee and the third is leaving the loop.

In order to satisfy the mission requirements and to continuously provide the selected services at high latitude, it is necessary to maintain the constellation geometry and the subtrack position; that is, the phasing between the satellites and the longitude and latitude of subsatellite points in the operational loop.

However, orbit perturbations mainly due to Earth oblateness and Sun and Moon gravity gradient will modify the orbits and the mission requirements (geometry between satellites and with respect to the Earth). Due to the high perigee altitude the effect of atmospheric drag is neglected.

It is required to perform suitable maneuvers to maintain the constellation architecture and to fulfill the mission requirements.

A station-keeping maneuver based on the correction of all the orbital parameters requires a very high delta-V value, which can negatively impact on the feasibility of the mission [Ref. 1].

This paper deals with a low-cost orbit control strategy addressed only to control the orbit subtrack (latitude and longitude of the operational loop of the orbit) and to control the apogee passage time-shift, which impacts on the synchronization between the satellites.

This control strategy, exclusively based on in-plane maneuvers, consists of adjusting the apogee radius by perigee burns thus obtaining a semiaxis variation which cancels the total subtrack shift or puts the subtrack on the other side with respect to the nominal position.

ORBIT PERTURBATIONS

The constellation geometry is defined by the parameters of Table 1.

Table 1
CONSTELLATION ORBIT PARAMETERS

SEMIAXIS	42164 Km
ECCENTRICITY	0.2687
INCLINATION	63.4°
ARGUMENT OF PERIGEE	270°
RIGHT ASCENSION OF ASCENDING NODE	45°/165°/285°
PERIOD	geosynchronous

The orbital elements are selected also to compensate the effect of perturbations.

The critical inclination implies no apsis line secular rotation due to Earth oblateness.

Choosing the satellite orbital period slightly lower than the geosynchronous it is possible to counteract the nodal regression of the node produced by the second harmonic of the Earth potential field (J2), so the subtrack can be frozen at the desired longitude at least in respect to the Earth oblateness effects.

As a consequence, the resulting perturbations on the satellites are mainly due to luni-solar effects.

STATION-KEEPING STRATEGIES

Two different strategies to counteract the orbital perturbation effects are analyzed.

The first could be considered as an **Orbit-Tended Strategy** and consists of small impulses applied as soon as an orbital element exceeds the prefixed range of tolerances.

Each correction must be executed in a proper position of the orbit in order to ensure that the impulse provided to correct one parameter does not produce an undesired perturbation on the other parameters.

The other strategy could be considered as a **Mission-Tended Strategy**. We want first of all to fulfill the mission requirements (phasing between satellites and latitude and longitude of subsatellites, point in the operational loop) even if the orbit parameters are not all strictly controlled in a very small range of tolerances; that is, some of the orbit parameters could be relaxed if the mission requirements are obtained.

The small impulses strategy requires a very high delta-V value which is unfavorable to the feasibility of the mission.

The second strategy is exclusively based on in-plane maneuver and consists of adjusting the apogee radius by perigee burns, which are, as it is well known, more economical than apogee burns. So we control only the semiaxis providing a low-cost maneuver.

The semiaxis variation can cancel the subtrack shift or can put the subtrack on the other side with respect to the nominal position.

SMALL IMPULSES STRATEGY

This method consists of small impulses applied when an orbital parameter exceeds its prefixed range of tolerance.

The delta-V required depends strongly on the tolerances associated with each orbital parameter around its nominal value.

A typical value of total delta-V necessary for this correction is about 150 m/s/year [Ref. 2]. This value might represent a critical design aspect.

IN-PLANE MANEUVERS

Key Parameters Selection

To reduce total delta-V required for orbit-keeping it is better to control only some parameters which are strictly mission-related without correcting a global set of six orbital parameters, thus avoiding corrections out of plane which are very expensive.

The key parameters of the mission are the satellite phasing and the subtrack position in terms of latitude and longitude of the operational handover point.

As far it concerns the satellite phasing, it is required to maintain the synchronization among the three satellites within a desired tolerance.

The synchronization can be defined as the time interval between the entrance of one satellite in the orbital operational loop at the handover and the exit of the previous satellite from the operational loop.

It means that only the mutual relative delay among the three satellites is the key parameter, while the absolute delay of each satellite can reach higher values.

Previous analyses of free orbit evolution [Ref. 1] have shown that the subtrack latitude presents negligible variation from the nominal value. Then the longitude shift and the time shift at the handover point can be chosen as the key parameters for orbit control.

Key Parameters Variations

At each orbit the longitude shift ($d\lambda$) at apogee is:

$$d\lambda \cong d\Omega + \frac{1}{\cos^2\omega + \cos^2 L \ \mathrm{sen}^2\omega} \cos L \ d\omega - \omega_E dT \qquad (1)$$

Where:
- Ω is the R.A.A.N.
- ω is the argument of perigee
- T is the orbit period
- L is the latitude
- ω_E is the earth angular velocity

then if $\omega \cong \frac{3}{2}\pi$ and $L \cong 60^o$

$$d\lambda \cong d\Omega + 2d\omega - \omega_E dT = dL_{RAAN} + dL_{ARG} + dL_a$$

where the first term is due to the orbital plane rotation from its nominal value ($d\Omega$). The longitude shift dl_{RAAN} is equal to the plane rotation ($d\Omega$).

The second term is due to the apsis line rotation ($d\omega$) and represents the longitude shift.

The last term is due to the semiaxis displacement (da) from its nominal value.

A constant semiaxis displacement generates a longitude shift (dl_a) proportional to time; semiaxis reduction yields an eastward longitude shift, while a semiaxis increase yields a westward longitude shift.

The synchronization error is due to the semiaxis (period) variation ($dt_{TOT} = dt_a$). Clearly, for a 24-hour orbit, if the orbital period variation is one second, after ten days the "de-synchronization" will be 10 seconds.

It can be noted that the part (dl_a) of the total longitude shift (d_{TOT}) and the global synchronization error ($dt = dt_a$) have the same origin: the period variation due to the semiaxis variation.

Thus, the above errors relevant to semiaxis variation (dl_a and dt_a) can simultaneously be eliminated, or at least strongly reduced, with the same appropriate semiaxis correction. In any case, with this correction, it is impossible to eliminate also the remaining part ($dl_{RAAN} + dl_{ARG}$) of the longitude shift.

CONTROL STRATEGY

Using only in-plane corrections the position and orientation of orbital plane (ascending node and inclination) are not corrected.

Two approaches can be followed by correcting the semiaxis:

1. Correct the semiaxis in order to eliminate exactly the total synchronization error and the part of the longitude shift dl_a due to the semiaxis.

   In this case after one new orbit revolution the key parameters errors will be:

   $dl_{TOT} = dl_{RAAN} + dl_{ARG} + dl'_a$

   $dt_{TOT} = 0$

   The period variation dT to be applied is $\pm dT = dt_a$. The period variation will be negative if the synchronization error is due to satellite delay and positive if the satellite anticipates its nominal phase at the selected time.

2. Give a different appropriate semiaxis correction in order to cancel completely the total subtrack longitude shift or to put the subtrack on the other side w.r.t. the nominal position. In this case, a certain value of the synchronization error (dt') will be introduced.

   Therefore the key parameters errors will be after one revolution:

   $dl_{TOT} = 0$

   $dt_{TOT} = dt'$

However after a new orbital revolution a new correction should be necessary in order to reestablish the exact nominal period.

But if only a part of total period variation dT is applied (e.g. only 1/n) the correction end after n revolutions, and the nominal period reestablishment is no longer necessary and a strong fuel saving can be achieved.

The period variation to be applied is evaluated integrating the eg. (1); it yields

$$l_f - l_i = (\dot{\Omega} + 2\dot{\omega})n_d - W_e(Dt_i\ n_d + 1/2\ \dot{T}_i\ n_d^2)$$

Where:

l_f is the longitude at the time of the next correction ($ln \pm Dln$)

l_i is the actual longitude ($ln \pm Dln$)

$\dot{\Omega}$ is the actual derivative of the ascending node at the time of the correction.

$\dot{\omega}$ is the actual derivative of the argument of the perigee at the time of the correction.

W_e is the earth angular velocity.

DT_i is the initial displacement of the period, from its nominal value (Ts) after the actual correction.

n_d is the number of days of the time interval between the actual and the next correction.

T_i is the actual derivative of the orbital period at the time of correction.

Hence, in order to obtain the desired subtrack longitude shift ($D_l = l_f - l_i$) it is necessary

1. to determine the orbital evolution at the time of the burn in terms of $\dot{\Omega}$, $\dot{\omega}$, T;
2. to assume the same behavior of the three above parameters in the time interval succeeding the orbit-keeping maneuver:
3. to suitably select the number of days during which the correction $D\lambda$ is desired:
4. by means of the above formula, to evaluate the initial orbital period displacement DT_i.

RESULTS

Two analyses have been carried out in order to confirm the applicability of this strategy: The first analysis deals with the synchronization control and the second with the longitude control.

In the first approach a requirement of ±2 minutes was imposed on the time delay while the longitude shift was left free.

In the second analysis a maximum longitude shift of 5 degrees was imposed while the synchronization was not controlled.

The synchronism control results point out (Fig. 2) a displacement of the subtrack longitude greater then 5 degrees, while the lack of synchronism among the three satellites could be maintained within its prefixed threshold.

Then, the investigations were addressed toward the case of longitude control.

The evolution of all three orbits of the constellation has been analyzed.

The results are presented in Figs. 3 to 5 that show that the subtrack longitude (at 15°) is perfectly maintained with the ±5 degrees interval for the three orbits.

Moreover, the lack of synchronism is an advance for the three cases with respect to the nominal apogee (or handover) passage.

Finally, the third plot of the above-mentioned tables shows the evolution of the semi-major axis and its relevant corrections.

The total delta-V necessary to perform the required sequence of corrections is respectively 17, 25, 10 m/sec for the three satellites that corresponds to 2.4, 3.5, 1.4 m/sec/year.

At this point it is important to highlight that:

1. the proposed in-plane maneuver strategy really allows strict control the subtrack longitude,
2. the lack of synchronism of the three orbits presents a quite similar behavior.

Actually, an advance of the apogee passage is experienced in the three cases. This means that the mutual delay among the satellites can be maintained within the allowed tolerance.

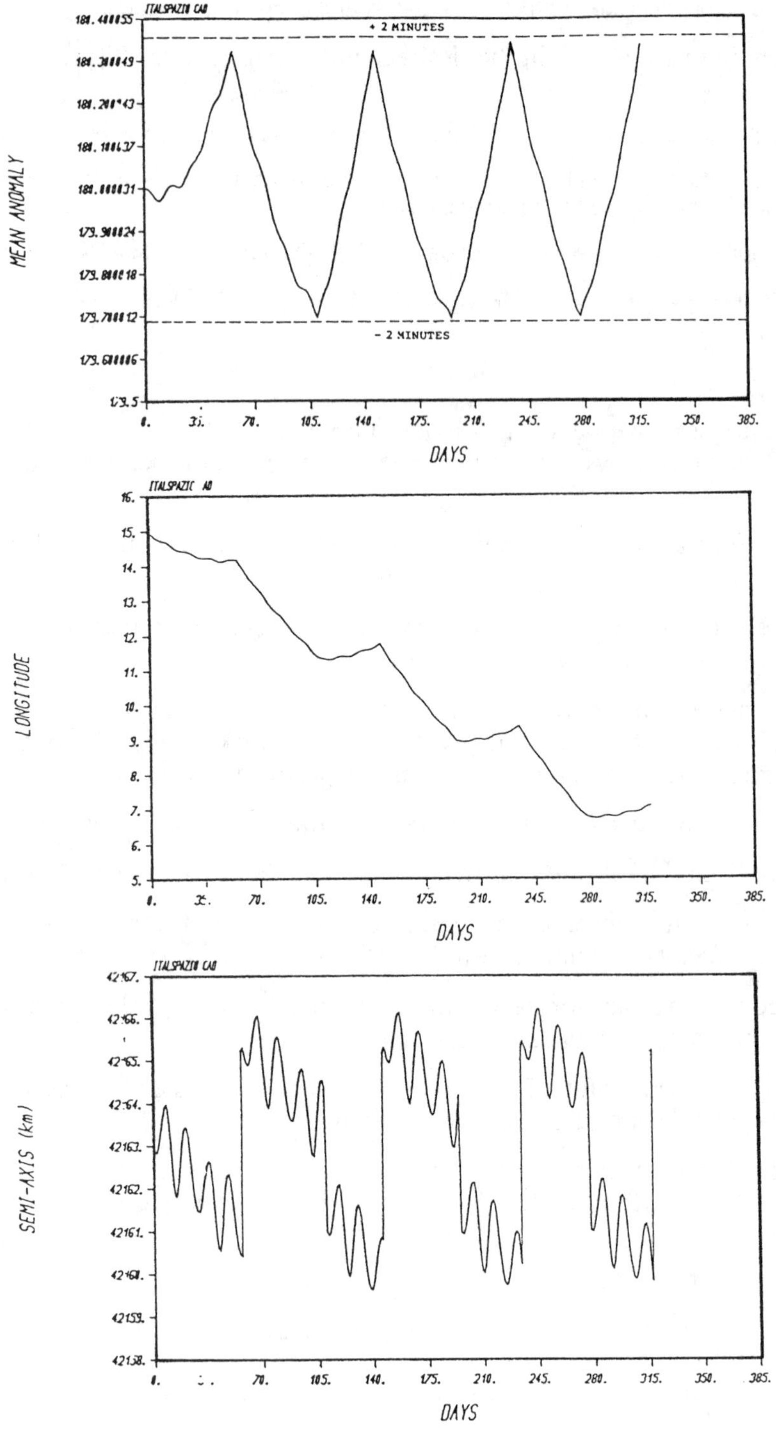

Figure 2 Example of Synchronization Control (±2 Minutes)

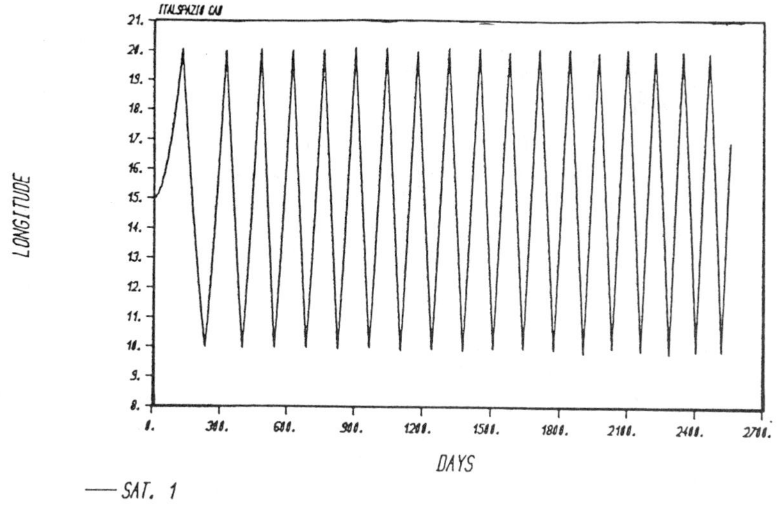

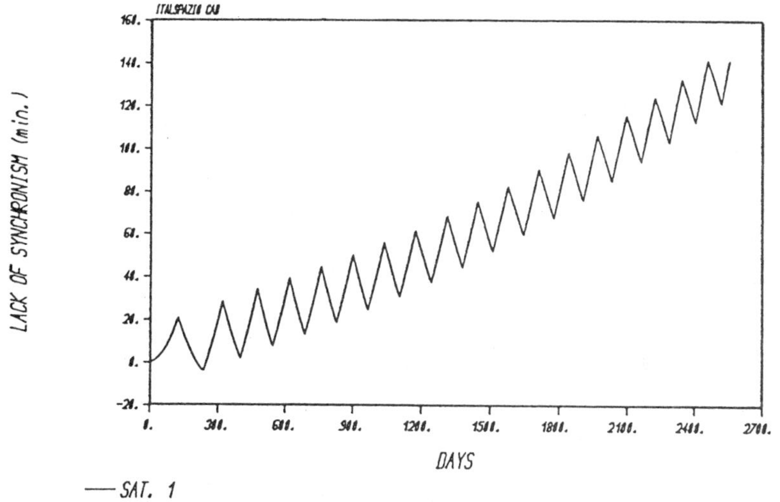

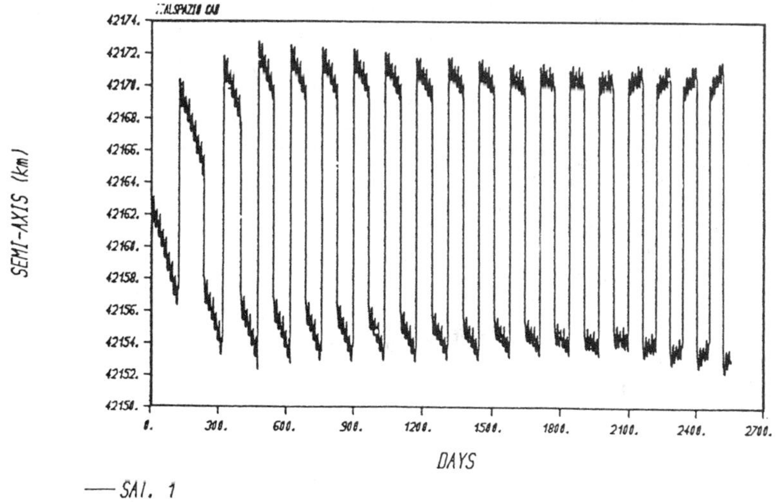

Figure 3 Longitude Control of Satellite 1 (RAAN = 285°)
Key Parameters Evolution

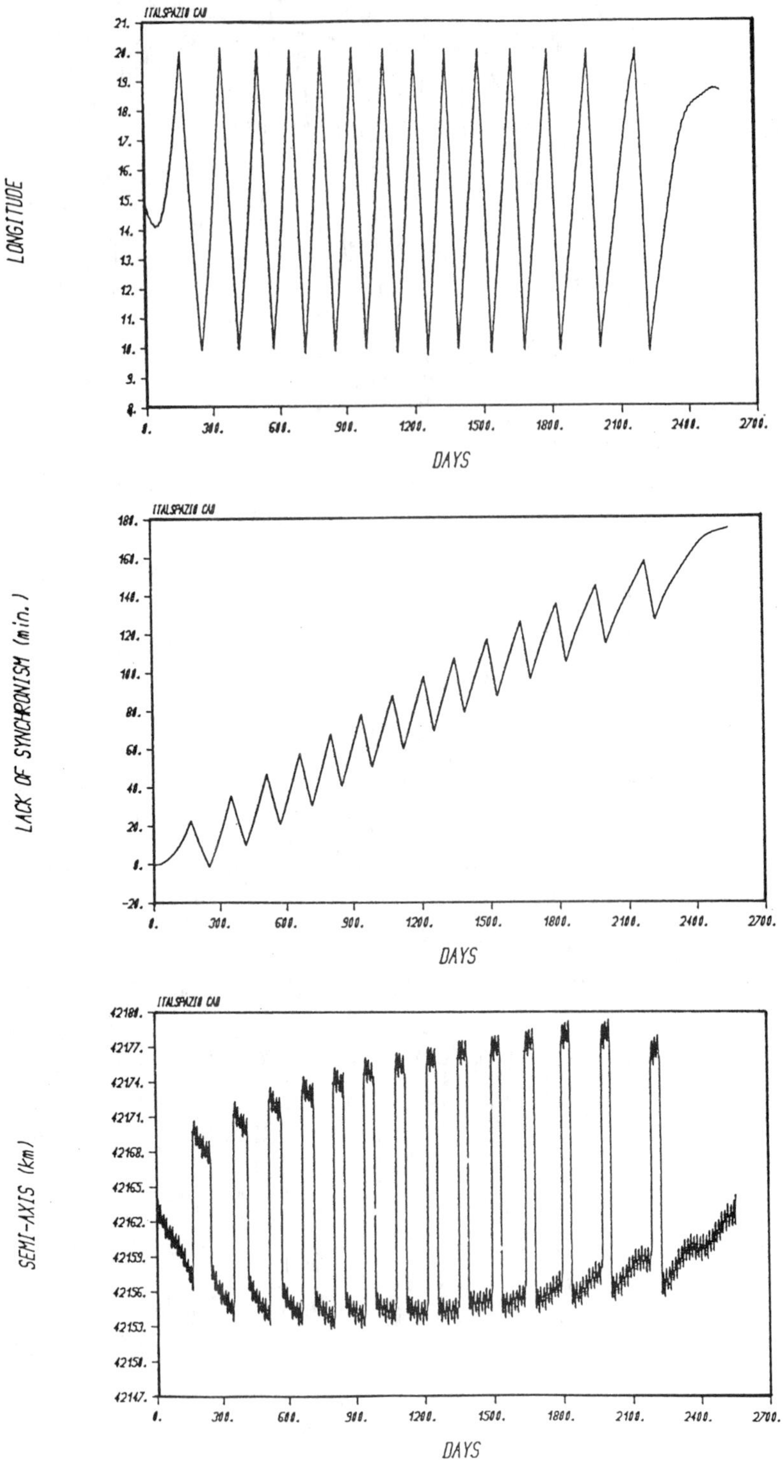

**Figure 4 Longitude Control of Satellite 2 (RAAN = 45^0)
Key Parameters Evolution**

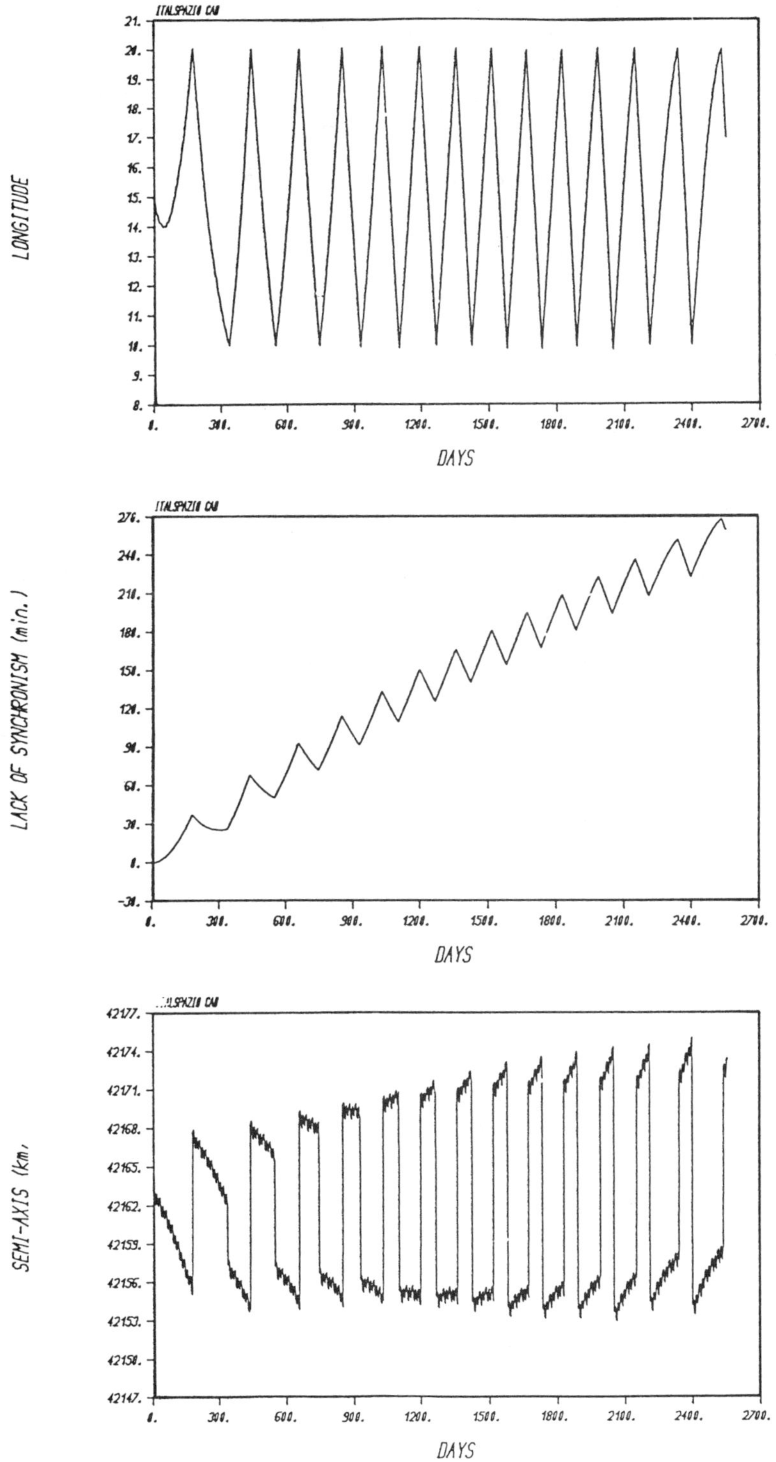

Figure 5 Longitude Control of Satellite 3 (RAAN = 165^0)
Key Parameters Evolution

Only the third satellite (RAAN = 165 deg.) shows a greater advance with respect to the other two satellites. In any case, it has to be considered that these results have been obtained resorting to the same automatic procedure of control for the three satellites. In other words, the greater advance of the third satellite can be reduced if in some circumstance a different control logic were adopted.

The most important result obtained with the above analyses resides in the fact that, for both cases, delta-V values are approximately 3 m/sec/year. This means that the proposed in-plane strategy really provides a drastic reduction of propellant consumption also for orbit-keeping of a constellation of three satellites.

ACKNOWLEDGMENT

The authors wish to acknowledge the contribution and the support of Dr. A. Teofilatto, President of ITALSPAZIO.

REFERENCES

1. Comparative analysis of highly inclined orbits for mobile communications and navigation system, ITS-TR-073.A/87, May 1987.

2. A. Cramarossa, G. Barresi, G. Rondinelli, F. Graziani: Low Cost Station Keeping Maneuvers for a Small-Satellites Constellation in Tundra Orbit, Proceedings of the 2nd Annual AIAA/USU Conference on Small Satellites, September 1988.

AAS 89-409

THE NON-OPTIMALITY OF FORCED KEPLERIAN MOTION IN AN ATMOSPHERE*

I. M. Ross† and R. G. Melton‡

This paper considers the problem of optimal, low-altitude orbital transfer, including the effects of density-and gravity-gradients. The Mayer problem is formulated and singular extremals are considered. The singular feedback control is derived for the special case of collinear thrust and velocity vectors. It is shown that Keplerian motion is not optimal. The problem of minimum fuel orbital maintenance is considered and a numerical treatment suggests that a simple bang-bang scheme is also not optimal. Thrust-vectoring and the possibility of optimal singular subarcs are proposed.

---

* This paper appears in full in the microfiche supplement to this volume, **AAS Microfiche Series**, Volume 59.

† Graduate Research Assistant, Department of Aerospace Engineering, The Pennsylvania State University, University Park, Pennsylvania 16802. Student member AIAA.

‡ Associate Professor, Department of Aerospace Engineering, The Pennsylvania State University, University Park, Pennsylvania 16802. Member of AAS, Senior Member AIAA.

AAS 89-412

COLLOCATION ANALYSIS OF TWIN SATELLITES SHARING A GEOSTATIONARY PARKING SLOT AND A COORDINATED STATION-KEEPING STRATEGY*

B. Srinivasan†

The increasing satellite population in geostationary orbits necessitates a venture of collocation. The collocation mission is comprised of two segments:

1) Collocation Mission Planning in which the feasibility of collocation is analyzed, taking into account the tracking measurement accuracies achievable and the reduced longitude band available. And secondly a collocation strategy is evolved which may be used in post-launch mission operations. In this case, a strategy involving operation of satellites within non-overlapping subslots (of the total slot) has been conceived. The results of a Monte-Carlo simulation process show a 3-sigma error of 0.016 degree in longitude prediction falling well within the half slot of ±0.05 degree, thus supporting the feasibility of collocation of twin satellites in the longitude slot of ±0.01 degree.

2) Mission Operations in which a coordinated station-keeping strategy (already evolved as mentioned above) has to be implemented; the orbit determination and longitude prediction done regularly, separation monitored and corrective action to be taken, if necessary.

---

* Only an abstract of this paper was available for publication.

† Scientist/Engineer, Department of Space, Government of India, Hassan-573201, India.

AAS 89-413

ECCENTRICITY CONTROL OF COLLOCATED GEOSTATIONARY SATELLITES*

B. Srinivasan†

Geostationary satellites must be controlled to follow the nominal orbits in terms of the semi-major axis to control the drift rate in longitude, the eccentricity to control the oscillations in the longitude within a day and the inclination to control the declination varying within a day. The eccentricity control needs to be done less frequently than the other two. The situation however, tightens if collocation of even two satellites is involved.

This paper comes out with an eccentricity control procedure compatible with the coordinated station-keeping strategy for collocation described in an earlier paper by the author on collocation. Here, the sun-pointing strategy is not feasible as the eccentricity has to be maintained within one-third the sun-pointing eccentricity, after allowing for other uncertainities. The dynamics of eccentricity evolution has been studied analytically as well as computer simulated so as to maximize the duration of residence of the eccentricity within a specified limit. For a typical existing satellite, the eccentricity can be kept within 0.0002 for a maximum of 1/6 of a year, i.e., 61 days. The requirement of fuel for the eccentricity-turn maneuvers is that for a delta-V of 0.35 meters/sec per year per satellite for a typical satellite considered above.

Finally, the eccentricity cycle should be so timed that the minimum value occurs at the dead-bands of the longitude for maximum utilization of the slot. The above cycle time may be chosen to be a proper fraction of the drift cycle.

* Only an abstract of this paper was available for publication.

† Scientist/Engineer, Department of Space, Government of India, Hassan-573201, India.